HYPERSPECTRAL REMOTE SENSING
OF VEGETATION
SECOND EDITION
VOLUME IV

Advanced Applications in Remote Sensing of Agricultural Crops and Natural Vegetation

Hyperspectral Remote Sensing of Vegetation
Second Edition

Volume I: Fundamentals, Sensor Systems, Spectral Libraries, and Data Mining for Vegetation

Volume II: Hyperspectral Indices and Image Classifications for Agriculture and Vegetation

Volume III: Biophysical and Biochemical Characterization and Plant Species Studies

Volume IV: Advanced Applications in Remote Sensing of Agricultural Crops and Natural Vegetation

HYPERSPECTRAL REMOTE SENSING OF VEGETATION

SECOND EDITION

VOLUME IV

Advanced Applications in Remote Sensing of Agricultural Crops and Natural Vegetation

Edited by

Prasad S. Thenkabail

John G. Lyon

Alfredo Huete

CRC Press is an imprint of the
Taylor & Francis Group, an **informa** business

CRC Press
Taylor & Francis Group
6000 Broken Sound Parkway NW, Suite 300
Boca Raton, FL 33487-2742

First issued in paperback 2022

ISBN 13: 978-1-03-247587-5 (pbk)
ISBN 13: 978-1-138-36476-9 (hbk)

DOI: 10.1201/9780429431166

Publisher's Note
The publisher has gone to great lengths to ensure the quality of this reprint but points out that some imperfections in the original copies may be apparent.

Visit the Taylor & Francis Web site at
http://www.taylorandfrancis.com

and the CRC Press Web site at
http://www.crcpress.com

Dr. Prasad S. Thenkabail, *Editor-in-Chief of these four volumes would like to dedicate the four volumes to three of his professors at the Ohio State University during his PhD days:*

1. **Late Prof. Andrew D. Ward**, *former professor of The Department of Food, Agricultural, and Biological Engineering (FABE) at The Ohio State University,*

2. **Prof. John G. Lyon**, *former professor of the Department of Civil, Environmental and Geodetic Engineering at the Ohio State University, and*

3. **Late Prof. Carolyn Merry**, *former Professor Emerita and former Chair of the Department of Civil, Environmental and Geodetic Engineering at the Ohio State University.*

Contents

SECTION VII Conclusions

Foreword to the First Edition

The publication of this book, *Hyperspectral Remote Sensing of Vegetation*, marks a milestone in the application of imaging spectrometry to study 70% of the Earth's landmass which is vegetated. This book shows not only the breadth of international involvement in the use of hyperspectral data but also in the breadth of innovative application of mathematical techniques to extract information from the image data.

Imaging spectrometry evolved from the combination of insights from the vast heterogeneity of reflectance signatures from the Earth's surface seen in the ERTS-1 (Landsat-1) 4-band images and the field spectra that were acquired to help fully understand the causes of the signatures. It wasn't until 1979 when the first hybrid area-array detectors, mercury-cadmium-telluride on silicon CCD's, became available that it was possible to build an imaging spectrometer capable of operating at wavelengths beyond 1.0 μm. The AIS (airborne imaging spectrometer), developed at NASA/JPL, had only 32 cross-track pixels but that was enough for the geologists clamoring for this development to see *between* the bushes to determine the mineralogy of the substrate. In those early years, vegetation cover was just a nuisance!

In the early 1980s, spectroscopic analysis was driven by the interest to identify mineralogical composition by exploiting absorptions found in the SWIR region from overtone and combination bands of fundamental vibrations found in the mid-IR region beyond 3 μm and the electronic transitions in transition elements appearing, primarily, short of 1.0 μm. The interests of the geologists had been incorporated in the Landsat TM sensor in the form of the add-on, band 7 in the 2.2 μm region based on field spectroscopic measurements. However, one band, even in combination with the other six, did not meet the needs for mineral identification. A summary of mineralogical analyses is presented by Vaughan et al. in this volume. A summary of the historical development of hyperspectral imaging can be found in Goetz (2009).

At the time of the first major publication of the AIS results (Goetz et al., 1985), very little work on vegetation analysis using imaging spectroscopy had been undertaken. The primary interest was in identifying the relationship of the chlorophyll absorption red-edge to stress and substrate composition that had been seen in airborne profiling and in field spectral reflectance measurements. Most of the published literature concerned analyzing NDVI, which only required two spectral bands.

In the time leading up to the 1985 publication, we had only an inkling of the potential information content in the hundreds of contiguous spectral bands that would be available to us with the advent of AVIRIS (airborne visible and infrared imaging spectrometer). One of the authors, Jerry Solomon, presciently added the term "hyperspectral" to the text of the paper to describe the "...multidimensional character of the spectral data set," or, in other words, the mathematically, over-determined nature of hyperspectral data sets. The term hyperspectral as opposed to multispectral data moved into the remote sensing vernacular and was additionally popularized by the military and intelligence community.

In the early 1990s, as higher quality AVIRIS data became available, and the first analyses of vegetation using statistical techniques borrowed from chemometrics, also known as NIRS analysis used in the food and grain industry, were undertaken by John Aber and Mary Martin of the University of New Hampshire. Here, nitrogen contents of tree canopies were predicted from reflectance spectra by regression techniques using reference measurements from laboratory wet chemical analyses of needle and leaf samples acquired by shooting down branches. At the same time, the remote sensing community began to recognize the value of "too many" spectral bands and the concomitant wealth of spatial information that was amenable to information extraction by statistical techniques. One of them was Eyal Ben-Dor who pioneered soil analyses using hyperspectral imaging and who is one of the contributors to this volume.

As the quality of AVIRIS data grew, manifested in increasing SNR, an ever-increasing amount of information could be extracted from the data. This quality was reflected in the increasing number of nearly noiseless principal components that could be obtained from the data or, in other words, its dimensionality. The explosive advances in desktop computing made possible the application of image processing and statistical analyses that revolutionized the uses of hyperspectral imaging. Joe Boardman and others at the University of Colorado developed what has become the ENVI software package to make possible the routine analysis of hyperspectral image data using "unmixing techniques" to derive the relative abundance of surface materials on a pixel-by-pixel basis.

Many of the analysis techniques discussed in this volume, such as band selection and various indices, are rooted in principal components analysis. The eigenvector loadings or factors indicate which spectral bands are the most heavily weighted allowing others to be discarded to reduce the noise contribution. As sensors become better, more information will be extractable and fewer bands will be discarded. This is the beauty of hyperspectral imaging, allowing the choice of the number of eigenvectors to be used for a particular problem. Computing power has reached such a high level that it is no longer necessary to choose a subset of bands just to minimize the computational time.

As regression techniques such as PLS (partial least squares) become increasingly adopted to relate a particular vegetation parameter to reflectance spectra, it must be remembered that the quality of the calibration model is a function of both the spectra and the reference measurement. With spectral measurements of organic and inorganic compounds under laboratory conditions, we have found that a poor model with a low coefficient of determination (r^2) is most often associated with inaccurate reference measurements, leading to the previously intuitive conclusion that "spectra don't lie."

Up to this point, AVIRIS has provided the bulk of high-quality hyperspectral image data but on an infrequent basis. Although Hyperion has provided some time series data, there is no hyperspectral imager yet in orbit that is capable of providing routine, high-quality images of the whole Earth on a consistent basis. The hope is that in the next decade, HyspIRI will be providing VNIR and SWIR hyperspectral images every 3 weeks and multispectral thermal data every week. This resource will revolutionize the field of vegetation remote sensing since so much of the useful information is bound up in the seasonal growth cycle. The combination of the spectral, spatial, and temporal dimensions will be ripe for the application of statistical techniques and the results will be extraordinary.

Dr. Alexander F. H. Goetz PhD
Former Chairman and Chief Scientist
ASD Inc.
2555 55th St. #100
Boulder, CO 80301, USA
303-444-6522 ext. 108
Fax 303-444-6825
www.asdi.com

REFERENCES

Goetz, A. F. H., 2009, Three decades of hyperspectral imaging of the Earth: A personal view, *Remote Sensing of Environment*, 113, S5–S16.

Goetz, A.F.H., G. Vanc, J. Solomon and B.N. Rock, 1985, Imaging spectrometry for Earth remote sensing, *Science*, 228, 1147–1153.

BIOGRAPHICAL SKETCH

Dr. Goetz is one of the pioneers in hyperspectral remote sensing and certainly needs no introduction. Dr. Goetz started his career working on spectroscopic reflectance and emittance studies of the Moon and Mars. He was a principal investigator of Apollo-8 and Apollo-12 multispectral photography studies. Later, he turned his attention to remote sensing of Planet Earth working in collaboration with Dr. Gene Shoemaker to map geology of Coconino County (Arizona) using Landsat-1 data and went on to be an investigator in further Landsat, Skylab, Shuttle, and EO-1 missions. At NASA/JPL he pioneered field spectral measurements and initiated the development of hyperspectral imaging. He spent 21 years on the faculty of the University of Colorado, Boulder, and retired in 2006 as an Emeritus Professor of Geological Sciences and an Emeritus Director of Center for the Study of Earth from Space. Since then, he has been Chairman and Chief Scientist of ASD Inc. a company that has provided more than 850 research laboratories in over 60 countries with field spectrometers. Dr. Goetz is now retired. His foreword was written for the first edition and I have retained it in consultation with him to get a good perspective on the development of hyperspectral remote sensing.

Foreword to the Second Edition

The publication of the four-volume set, *Hyperspectral Remote Sensing of Vegetation*, second edition, is a landmark effort in providing an important, valuable, and timely contribution that summarizes the state of spectroscopy-based understanding of the Earth's terrestrial and near shore environments. Imaging spectroscopy has had 35 years of development in data processing and analysis methods. Today's researchers are eager to use data produced by hyperspectral imagers and address important scientific issues from agricultural management to global environmental stewardship. The field started with development of the Jet Propulsion Lab's Airborne Imaging Spectrometer in 1983 that measured across the reflected solar infrared spectrum with 128 spectral bands. This technology was quickly followed in 1987 by the more capable Advanced Visible Infrared Imaging Spectrometer (AVIRIS), which has flown continuously since this time (albeit with multiple upgrades). It has 224 spectral bands covering the 400–2500 nm range with 10 nm wavelength bands and represents the "gold standard" of this technology. In the years since then, progress toward a hyperspectral satellite has been disappointingly slow. Nonetheless, important and significant progress in understanding how to analyze and understand spectral data has been achieved, with researchers focused on developing the concepts, analytical methods, and spectroscopic understanding, as described throughout these four volumes. Much of the work up to the present has been based on theoretical analysis or from experimental studies at the leaf level from spectrometer measurements and at the canopy level from airborne hyperspectral imagers.

Although a few hyperspectral satellites have operated over various periods in the 2000s, none have provided systematic continuous coverage required for global mapping and time series analysis. An EnMap document compiled the past and near-term future hyperspectral satellites and those on International Space Station missions (EnMap and GRSS Technical Committee 2017). Of the hyperspectral imagers that have been flown, the European Space Agency's CHRIS (Compact High Resolution Imaging Spectrometer) instrument on the PROBA-1 (Project for On-Board Autonomy) satellite and the Hyperion sensor on the NASA technology demonstrator, Earth Observing-1 platform (terminated in 2017). Each has operated for 17 years and have received the most attention from the science community. Both collect a limited number of images per day, and have low data quality relative to today's capability, but both have open data availability. Other hyperspectral satellites with more limited access and duration include missions from China, Russia, India, and the United States.

We are at a threshold in the availability of hyperspectral imagery. There are many hyperspectral missions planned for launch in the next 5 years from China, Italy, Germany, India, Japan, Israel, and the United States, some with open data access. The analysis of the data volumes from this proliferation of hyperspectral imagers requires a comprehensive reference resource for professionals and students to turn to in order to understand and correctly and efficiently use these data. This four-volume set is unique in compiling in-depth understanding of calibration, visualization, and analysis of data from hyperspectral sensors. The interest in this technology is now widespread, thus, applications of hyperspectral imaging cross many disciplines, which are truly international, as is evident by the list of authors of the chapters in these volumes, and the number of countries planning to operate a hyperspectral satellite. At least some of the hyperspectral satellites announced and expected to be launched in this decade (such as the HyspIRI-like satellite approved for development by NASA with a launch in the 2023 period) will provide high-fidelity narrow-wavelength bands, covering the full reflected solar spectrum, at moderate (30 m pixels) to high spatial resolution. These instruments will have greater radiometric range, better SNR, pointing accuracy, and reflectance calibration than past instruments, and will collect data from many countries and parts of the world that have not previously been available. Together, these satellites will produce an unprecedented flow of information about the physiological functioning (net primary production, evapotranspiration, and even direct measurements related to respiration), biochemical characteristics (from spectral indices and from radiative transfer first principle methods), and direct measurements of the distributions of plant and soil biodiversity of the terrestrial and coastal environments of the Earth.

This four-volume set presents an unprecedented range and scope of information on hyperspectral data analysis and applications written by leading authors in the field. Topics range from sensor characteristics from ground-based platforms to satellites, methods of data analysis to characterize plant functional properties related to exchange of gases CO_2, H_2O, O_2, and biochemistry for pigments, N cycle, and other molecules. How these data are used in applications range from precision agriculture to global change research. Because the hundreds of bands in the full spectrum includes information to drive detection of these properties, the data is useful at scales from field applications to global studies.

Volume I has three sections and starts with an introduction to hyperspectral sensor systems. Section II focuses on sensor characteristics from ground-based platforms to satellites, and how these data are used in global change research, particularly in relation to agricultural crop monitoring and health of natural vegetation. Section III provides five chapters that deal with the concept of spectral libraries to identify crops and spectral traits, and for phenotyping for plant breeding. It addresses the development of spectral libraries, especially for agricultural crops and one for soils.

Volume II expands on the first volume, focusing on use of hyperspectral indices and image classification. The volume begins with an explanation of how narrow-band hyperspectral indices are determined, often from individual spectral absorption bands but also from correlation matrices and from derivative spectra. These are followed by chapters on statistical approaches to image classification and a chapter on methods for dealing with "big data." The last half of this volume provides five chapters focused on use of vegetation indices for quantifying and characterizing photosynthetic pigments, leaf nitrogen concentrations or contents, and foliar water content measurements. These chapters are particularly focused on applications for agriculture, although a chapter addresses more heterogeneous forest conditions and how these patterns relate to monitoring health and production.

The first half of Volume III focuses on biophysical and biochemical characterization of vegetation properties that are derived from hyperspectral data. Topics include ecophysiological functioning and biomass estimates of crops and grasses, indicators of photosynthetic efficiency, and stress detection. The chapter addresses biophysical characteristics across different spatial scales while another chapter examines spectral and spatial methods for retrieving biochemical and biophysical properties of crops. The chapters in the second half of this volume are focused on identification and discrimination of species from hyperspectral data and use of these methods for rapid phenotyping of plant breeding trials. Lastly, two chapters evaluate tree species identification, and another provides examples of mapping invasive species.

Volume IV focuses on six areas of advanced applications in agricultural crops. The first considers detection of plant stressors including nitrogen deficiency and excess heavy metals and crop disease detection in precision farming. The second addresses global patterns of crop water productivity and quantifying litter and invasive species in arid ecosystems. Phenological patterns are examined while others focus on multitemporal data for mapping patterns of phenology. The third area is focused on applications of land cover mapping in different forest, wetland, and urban applications. The fourth topic addresses hyperspectral measurements of wildfires, and the fifth evaluates use of continuity vegetation index data in global change applications. And lastly, the sixth area examines use of hyperspectral data to understand the geologic surfaces of other planets.

Susan L. Ustin
Professor and Vice Chair, Dept. Land, Air and Water Resources
Associate Director, John Muir Institute
University of California
Davis California, USA

REFERENCE

EnMap Ground Segment Team and GSIS GRSS Technical committee, December, 2017. *Spaceborne Imaging Spectroscopy Mission Compilation*. DLR Space Administration and the German Federal Ministry of Economic Affairs and Technology. http://www.enmap.org/sites/default/files/pdf/Hyperspectral_EO_Missions_2017_12_21_FINAL4.pdf

BIOGRAPHICAL SKETCH

Dr. Susan L. Ustin is currently a Distinguished Professor of Environmental and Resource Sciences in the Department of Land, Air, and Water Resources, University of California Davis, Associate Director of the John Muir Institute, and is Head of the Center for Spatial Technologies and Remote Sensing (CSTARS) at the same university. She was trained as a plant physiological ecologist but began working with hyperspectral imagery as a post-doc in 1983 with JPL's AIS program. She became one of the early adopters of hyperspectral remote sensing which has now extended over her entire academic career. She was a pioneer in the development of vegetation analysis using imaging spectrometery, and is an expert on ecological applications of this data. She has served on numerous NASA, NSF, DOE, and the National Research Council committees related to spectroscopy and remote sensing. Among recognitions for her work, she is a Fellow of the American Geophysical Union and received an honorary doctorate from the University of Zurich. She has published more than 200 scientific papers related to ecological remote sensing and has worked with most of the Earth-observing U.S. airborne and spaceborne systems.

Preface

This seminal book on *Hyperspectral Remote Sensing of Vegetation* (Second Edition, 4 Volume Set), published by Taylor and Francis Inc.\CRC Press is an outcome of over 2 years of effort by the editors and authors. In 2011, the first edition of *Hyperspectral Remote Sensing of Vegetation* was published. The book became a standard reference on hyperspectral remote sensing of vegetation amongst the remote sensing community across the world. This need and resulting popularity demanded a second edition with more recent as well as more comprehensive coverage of the subject. Many advances have taken place since the first edition. Further, the first edition was limited in scope in the sense it covered some very important topics and missed equally important topics (e.g., hyperspectral library of agricultural crops, hyperspectral pre-processing steps and algorithms, and many others). As a result, a second edition that brings us up-to-date advances in hyperspectral remote sensing of vegetation was required. Equally important was the need to make the book more comprehensive, covering an array of subjects not covered in the first edition. So, my coeditors and myself did a careful research on what should go into the second edition. Quickly, the scope of the second edition expanded resulting in an increasing number of chapters. All of this led to developing the seminal book: *Hyperspectral Remote Sensing of Vegetation*, Second Edition, 4 Volume Set. The four volumes are:

Volume I: Fundamentals, Sensor Systems, Spectral Libraries, and Data Mining for Vegetation
Volume II: Hyperspectral Indices and Image Classifications for Agriculture and Vegetation
Volume III: Biophysical and Biochemical Characterization and Plant Species Studies
Volume IV: Advanced Applications in Remote Sensing of Agricultural Crops and Natural Vegetation

The goal of the book was to bring in one place collective knowledge of the last 50 years of advances in hyperspectral remote sensing of vegetation with a target audience of wide spectrum of scientific community, students, and professional application practitioners. The book documents knowledge advances made in applying hyperspectral remote sensing technology in the study of terrestrial vegetation that include agricultural crops, forests, rangelands, and wetlands. This is a very practical offering about a complex subject that is rapidly advancing its knowledge-base. In a very practical way, the book demonstrates the experience, utility, methods, and models used in studying terrestrial vegetation using hyperspectral data. The four volumes, with a total of 48 chapters, are divided into distinct themes.

- **Volume I**: There are 14 chapters focusing on hyperspectral instruments, spectral libraries, and methods and approaches of data handling. The chapters extensively address various preprocessing steps and data mining issues such as the Hughes phenomenon and overcoming the "curse of high dimensionality" of hyperspectral data. Developing spectral libraries of crops, vegetation, and soils with data gathered from hyperspectral data from various platforms (ground-based, airborne, spaceborne), study of spectral traits of crops, and proximal sensing at field for phenotyping are extensively discussed. Strengths and limitations of hyperspectral data of agricultural crops and vegetation acquired from different platforms are discussed. It is evident from these chapters that the hyperspectral data provides opportunities for great advances in study of agricultural crops and vegetation. However, it is also clear from these chapters that hyperspectral data should not be treated as panacea to every limitation of multispectral broadband data such as from Landsat or Sentinel series of satellites. The hundreds or thousands of hyperspectral narrowbands (HNBs) as well as carefully selected hyperspectral vegetation indices (HVIs) will help us make significant advances in characterizing, modeling, mapping, and monitoring vegetation biophysical, biochemical, and structural quantities. However, it is also important to properly understand hyperspectral data and eliminate redundant bands that exist for every application and to optimize computing as well as human resources to enable seamless and efficient handling enormous volumes of hyperspectral data. Special emphasis is also put on preprocessing and processing of Earth Observing-1 (EO-1) Hyperion, the first publicly available hyperspectral data from space. These methods, approaches, and algorithms, and protocols set the stage for upcoming satellite hyperspectral sensors such as NASA's HyspIRI and Germany's EnMAP.

- **Volume II**: There are 10 chapters focusing on hyperspectral vegetation indices (HVIs) and image classification methods and techniques. The HVIs are of several types such as: (i) two-band derived, (ii) multi-band-derived, and (iii) derivative indices derived. The strength of the HVIs lies in the fact that specific indices can be derived for specific biophysical, biochemical, and plant structural quantities. For example, you have carotenoid HVI, anthocyanin HVI, moisture or water HVI, lignin HVI, cellulose HVI, biomass or LAI or other biophysical HVIs, red-edge based HVIs, and so on. Further, since these are narrowband indices, they are better targeted and centered at specific sensitive wavelength portions of the spectrum. The strengths and limitations of HVIs in a wide array of applications such as leaf nitrogen content (LNC), vegetation water content, nitrogen content in vegetation, leaf and plant pigments, anthocyanin's, carotenoids, and chlorophyll are thoroughly studied. Image classification using hyperspectral data provides great strengths in deriving more classes (e.g., crop species within a crop as opposed to just crop types) and increasing classification accuracies. In earlier years and decades, hyperspectral data classification and analysis was a challenge due to computing and data handling issues. However, with the availability of machine learning algorithms on cloud computing (e.g., Google Earth Engine) platforms, these challenges have been overcome in the last 2–3 years. Pixel-based supervised machine learning algorithms like the random forest, and support vector machines as well as object-based algorithms like the recursive hierarchical segmentation, and numerous others methods (e.g., unsupervised approaches) are extensively discussed. The ability to process petabyte volume data of the planet takes us to a new level of sophistication and makes use of data such as from hyperspectral sensors feasible over large areas. The cloud computing architecture involved with handling massively large petabyte-scale data volumes are presented and discussed.
- **Volume III**: There are 11 chapters focusing on biophysical and biochemical characterization and plant species studies. A number of chapters in this volume are focused on separating and discriminating agricultural crops and vegetation of various types or species using hyperspectral data. Plant species discrimination and classification to separate them are the focus of study using vegetation such as forests, invasive species in different ecosystems, and agricultural crops. Performance of hyperspectral narrowbands (HNBs) and hyperspectral vegetation indices (HVIs) when compared with multispectral broadbands (MBBs) and multispectral broadband vegetation indices (BVIs) are presented and discussed. The vegetation and agricultural crops are studied at various scales, and their vegetation functional properties diagnosed. The value of digital surface models in study of plant traits as complementary\supplementary to hyperspectral data has been highlighted. Hyperspectral bio-indicators to study photosynthetic efficiency and vegetation stress are presented and discussed. Studies are conducted using hyperspectral data across wavelengths (e.g., visible, near-infrared, shortwave-infrared, mid-infrared, and thermal-infrared).
- **Volume IV**: There are 15 chapters focusing on specific advanced applications of hyperspectral data in study of agricultural crops and natural vegetation. Specific agricultural crop applications include crop management practices, crop stress, crop disease, nitrogen application, and presence of heavy metals in soils and related stress factors. These studies discuss biophysical and biochemical quantities modeled and mapped for precision farming,

hyperspectral narrowbands (HNBs), and hyperspectral vegetation indices (HVIs) involved in assessing nitrogen in plants, and the study of the impact of heavy metals on crop health and stress. Vegetation functional studies using hyperspectral data presented and discussed include crop water use (actual evapotranspiration), net primary productivity (NPP), gross primary productivity (GPP), phenological applications, and light use efficiency (LUE). Specific applications discussed under vegetation functional studies using hyperspectral data include agricultural crop classifications, machine learning, forest management studies, pasture studies, and wetland studies. Applications in fire assessment, modeling, and mapping using hyperspectral data in the optical and thermal portions of the spectrum are presented and discussed. Hyperspectral data in global change studies as well as in outer planet studies have also been discussed. Much of the outer planet remote sensing is conducted using imaging spectrometer and hence the data preprocessing and processing methods of Earth and that of outer planets have much in common and needs further examination.

The chapters are written by leading experts in the global arena with each chapter: (a) focusing on specific applications, (b) reviewing existing "state-of-art" knowledge, (c) highlighting the advances made, and (d) providing guidance for appropriate use of hyperspectral data in study of vegetation and its numerous applications such as crop yield modeling, crop biophysical and biochemical property characterization, and crop moisture assessment.

The four-volume book is specifically targeted on hyperspectral remote sensing as applied to terrestrial vegetation applications. This is a big market area that includes agricultural croplands, study of crop moisture, forests, and numerous applications such as droughts, crop stress, crop productivity, and water productivity. To the knowledge of the editors, there is no comparable book, source, and/or organization that can bring this body of knowledge together in one place, making this a "must buy" for professionals. This is clearly a unique contribution whose time is now. The book highlights include:

1. Best global expertise on hyperspectral remote sensing of vegetation, agricultural crops, crop water use, plant species detection, crop productivity and water productivity mapping, and modeling;
2. Clear articulation of methods to conduct the work. Very practical;
3. Comprehensive review of the existing technology and clear guidance on how best to use hyperspectral data for various applications;
4. Case studies from a variety of continents with their own subtle requirements; and
5. Complete solutions from methods to applications inventory and modeling.

Hyperspectral narrowband spectral data, as discussed in various chapters of this book, are fast emerging as practical most advanced solutions in modeling and mapping vegetation. Recent research has demonstrated the advances and great value made by hyperspectral data, as discussed in various chapters in: (a) quantifying agricultural crops as to their biophysical and harvest yield characteristics, (b) modeling forest canopy biochemical properties, (c) establishing plant and soil moisture conditions, (d) detecting crop stress and disease, (e) mapping leaf chlorophyll content as it influences crop production, (f) identifying plants affected by contaminants such as arsenic, and (g) demonstrating sensitivity to plant nitrogen content, and (h) invasive species mapping. The ability to significantly better quantify, model, and map plant chemical, physical, and water properties is well established and has great utility.

Even though these accomplishments and capabilities have been reported in various places, the need for a collective "knowledge bank" that links these various advances in one place is missing. Further, most scientific papers address specific aspects of research, failing to provide a comprehensive assessment of advances that have been made nor how the professional can bring those advances to their work. For example, deep scientific journals report practical applications of hyperspectral narrowbands yet one has to canvass the literature broadly to obtain the pertinent facts. Since several papers report this, there is a need to synthesize these findings so that the reader gets the correct picture of the best wavebands for their practical applications. Also, studies do differ in exact methods most suited for detecting parameters such as crop moisture variability, chlorophyll content, and stress levels. The professional needs this sort of synthesis and detail to adopt best practices for their own work.

In years and decades past, use of hyperspectral data had its challenges especially in handling large data volumes. That limitation is now overcome through cloud-computing, machine learning, deep learning, artificial intelligence, and advances in knowledge in processing and applying hyperspectral data.

This book can be used by anyone interested in hyperspectral remote sensing that includes advanced research and applications, such as graduate students, undergraduates, professors, practicing professionals, policy makers, governments, and research organizations.

Dr. Prasad S. Thenkabail, PhD
Editor-in-Chief
Hyperspectral Remote Sensing of Vegetation, Second Edition, Four Volume Set

Acknowledgments

This four-volume *Hyperspectral Remote Sensing of Vegetation* book (second edition) was made possible by sterling contributions from leading professionals from around the world in the area of hyperspectral remote sensing of vegetation and agricultural crops. As you will see from list of authors and coauthors, we have an assembly of "**who is who**" in hyperspectral remote sensing of vegetation who have contributed to this book. They wrote insightful chapters, that are an outcome of years of careful research and dedication, to make the book appealing to a broad section of readers dealing with remote sensing. My gratitude goes to (mentioned in no particular order; names of lead authors of the chapters are shown in bold): **Drs. Fred Ortenberg** (Technion–Israel Institute of Technology, Israel), **Jiaguo Qi** (Michigan State University, USA), **Angela Lausch** (Helmholtz Centre for Environmental Research, Leipzig, Germany), **Andries B. Potgieter** (University of Queensland, Australia), **Muhammad Al-Amin Hoque** (University of Queensland, Australia), **Andreas Hueni** (University of Zurich, Switzerland), **Eyal Ben-Dor** (Tel Aviv University, Israel), **Itiya Aneece** (United States Geological Survey, USA), **Sreekala Bajwa** (University of Arkansas, USA), **Antonio Plaza** (University of Extremadura, Spain), **Jessica J. Mitchell** (Appalachian State University, USA), **Dar Roberts** (University of California at Santa Barbara, USA), **Quan Wang** (Shizuoka University, Japan), **Edoardo Pasolli** (University of Trento, Italy), (Nanjing University of Science and Technology, China), **Anatoly Gitelson** (University of Nebraska- Lincoln, USA), **Tao Cheng** (Nanjing Agricultural University, China), **Roberto Colombo** (University of Milan-Bicocca, Italy), **Daniela Stroppiana** (Institute for Electromagnetic Sensing of the Environment, Italy), **Yongqin Zhang** (Delta State University, USA), **Yoshio Inoue** (National Institute for Agro-Environmental Sciences, Japan), Yafit Cohen (Institute of Agricultural Engineering, Israel), **Helge Aasen** (Institute of Agricultural Sciences, ETH Zurich), **Elizabeth M. Middleton** (NASA, USA), **Yongqin Zhang** (University of Toronto, Canada), **Yan Zhu** (Nanjing Agricultural University, China), **Lênio Soares Galvão** (Instituto Nacional de Pesquisas Espaciais [INPE], Brazil), **Matthew L. Clark** (Sonoma State University, USA), **Matheus Pinheiro Ferreira** (University of Paraná, Curitiba, Brazil), **Ruiliang Pu** (University of South Florida, USA), **Scott C. Chapman** (CSIRO, Australia), **Haibo Yao** (Mississippi State University, USA), **Jianlong Li** (Nanjing University, China), **Terry Slonecker** (USGS, USA), **Tobias Landmann** (International Centre of Insect Physiology and Ecology, Kenya), **Michael Marshall** (University of Twente, Netherlands), **Pamela Nagler** (USGS, USA), **Alfredo Huete** (University of Technology Sydney, Australia), **Prem Chandra Pandey** (Banaras Hindu University, India), **Valerie Thomas** (Virginia Tech., USA), **Izaya Numata** (South Dakota State University, USA), **Elijah W. Ramsey III** (USGS, USA), **Sander Veraverbeke** (Vrije Universiteit Amsterdam and University of California, Irvine), **Tomoaki Miura** (University of Hawaii, USA), **R. G. Vaughan** (U.S. Geological Survey, USA), Victor Alchanatis (Agricultural research Organization, Volcani Center, Israel), Dr. Narumon Wiangwang (Royal Thai Government, Thailand), Pedro J. Leitão (Humboldt University of Berlin, Department of Geography, Berlin, Germany), James Watson (University of Queensland, Australia), Barbara George-Jaeggli (ETH Zuerich, Switzerland), Gregory McLean (University of Queensland, Australia), Mark Eldridge (University of Queensland, Australia), Scott C. Chapman (University of Queensland, Australia), Kenneth Laws (University of Queensland, Australia), Jack Christopher (University of Queensland, Australia), Karine Chenu (University of Queensland, Australia), Andrew Borrell (University of Queensland, Australia), Graeme L. Hammer (University of Queensland, Australia), David R. Jordan (University of Queensland, Australia), Stuart Phinn (University of Queensland, Australia), Lola Suarez (University of Melbourne, Australia), Laurie A. Chisholm (University of Wollongong, Australia), Alex Held (CSIRO, Australia), S. Chabrillant (GFZ German Research Center for Geosciences, Germany), José A. M. Demattê (University of São Paulo, Brazil), Yu Zhang (North Dakota State University, USA), Ali Shirzadifar (North Dakota State University, USA), Nancy F. Glenn (Boise State University, USA), Kyla M. Dahlin (Michigan State

University, USA), Nayani Ilangakoon (Boise State University, USA), Hamid Dashti (Boise State University, USA), Megan C. Maloney (Appalachian State University, USA), Subodh Kulkarni (University of Arkansas, USA), Javier Plaza (University of Extremadura, Spain), Gabriel Martin (University of Extremadura, Spain), Segio Sánchez (University of Extremadura, Spain), Wei Wang (Nanjing Agricultural University, China), Xia Yao (Nanjing Agricultural University, China), Busetto Lorenzo (Università Milano-Bicocca), Meroni Michele (Università Milano-Bicocca), Rossini Micol (Università Milano-Bicocca), Panigada Cinzia (Università Milano-Bicocca), F. Fava (Università degli Studi di Sassari, Italy), M. Boschetti (Institute for Electromagnetic Sensing of the Environment, Italy), P. A. Brivio (Institute for Electromagnetic Sensing of the Environment, Italy), K. Fred Huemmrich (University of Maryland, Baltimore County, USA), Yen-Ben Cheng (Earth Resources Technology, Inc., USA), Hank A. Margolis (Centre d'Études de la Forêt, Canada), Yafit Cohen (Agricultural research Organization, Volcani Center, Israel), Kelly Roth (University of California at Santa Barbara, USA), Ryan Perroy (University of Wisconsin-La Crosse, USA), Ms. Wei Wang (Nanjing Agricultural University, China), Dr. Xia Yao (Nanjing Agricultural University, China), Keely L. Roth (University of California, Santa Barbara, USA), Erin B. Wetherley (University of California at Santa Barbara, USA), Susan K. Meerdink (University of California at Santa Barbara, USA), Ryan L. Perroy (University of Wisconsin-La Crosse, USA), B. B. Marithi Sridhar (Bowling Green University, USA), Aaryan Dyami Olsson (Northern Arizona University, USA), Willem Van Leeuwen (University of Arizona, USA), Edward Glenn (University of Arizona, USA), José Carlos Neves Epiphanio (Instituto Nacional de Pesquisas Espaciais [INPE], Brazil), Fábio Marcelo Breunig (Instituto Nacional de Pesquisas Espaciais [INPE], Brazil), Antônio Roberto Formaggio (Instituto Nacional de Pesquisas Espaciais [INPE], Brazil), Amina Rangoonwala (IAP World Services, Lafayette, LA), Cheryl Li (Nanjing University, China), Deghua Zhao (Nanjing University, China), Chengcheng Gang (Nanjing University, China), Lie Tang (Mississippi State University, USA), Lei Tian (Mississippi State University, USA), Robert Brown (Mississippi State University, USA), Deepak Bhatnagar (Mississippi State University, USA), Thomas Cleveland (Mississippi State University, USA), Hiroki Yoshioka (Aichi Prefectural University, Japan), T. N. Titus (U.S. Geological Survey, USA), J. R. Johnson (U.S. Geological Survey, USA), J. J. Hagerty (U.S. Geological Survey, USA), L. Gaddis (U.S. Geological Survey, USA), L. A. Soderblom (U.S. Geological Survey, USA), and P. Geissler (U.S. Geological Survey, USA), Jua Jin (Shizuoka University, Japan), Rei Sonobe (Shizuoka University, Japan), Jin Ming Chen (Shizuoka University, Japan), Saurabh Prasad (University of Houston, USA), Melba M. Crawford (Purdue University, USA), James C. Tilton (NASA Goddard Space Flight Center, USA), Jin Sun (Nanjing University of Science and Technology, China), Yi Zhang (Nanjing University of Science and Technology, China), Alexei Solovchenko (Moscow State University, Moscow), Yan Zhu, (Nanjing Agricultural University, China), Dong Li (Nanjing Agricultural University, China), Kai Zhou (Nanjing Agricultural University, China), Roshanak Darvishzadeh (University of Twente, Enschede, The Netherlands), Andrew Skidmore (University of Twente, Enschede, The Netherlands), Victor Alchanatis (Institute of Agricultural Engineering, The Netherlands), Georg Bareth (University of Cologne, Germany), Qingyuan Zhang (Universities Space Research Association, USA), Petya K. E. Campbell (University of Maryland Baltimore County, USA), and David R. Landis (Global Science & Technology, Inc., USA), José Carlos Neves Epiphanio (Instituto Nacional de Pesquisas Espaciais [INPE], Brazil), Fábio Marcelo Breunig (Universidade Federal de Santa Maria [UFSM], Brazil), and Antônio Roberto Formaggio (Instituto Nacional de Pesquisas Espaciais [INPE], Brazil), Cibele Hummel do Amaral (Federal University of Viçosa, in Brazil), Gaia Vaglio Laurin (Tuscia University, Italy), Raymond Kokaly (U.S. Geological Survey, USA), Carlos Roberto de Souza Filho (University of Ouro Preto, Brazil), Yosio Edemir Shimabukuro (Federal Rural University of Rio de Janeiro, Brazil), Bangyou Zheng (CSIRO, Australia), Wei Guo (The University of Tokyo, Japan), Frederic Baret (INRA, France), Shouyang Liu (INRA, France), Simon Madec (INRA, France), Benoit Solan (ARVALIS, France), Barbara George-Jaeggli (University of Queensland, Australia), Graeme L. Hammer (University of Queensland, Australia), David R. Jordan (University of Queensland, Australia), Yanbo Huang (USDA, USA), Lie Tang (Iowa State

University, USA), Lei Tian (University of Illinois. USA), Deepak Bhatnagar (USDA, USA), Thomas E. Cleveland (USDA, USA), Dehua ZHAO (Nanjing University, USA), Hannes Feilhauer (University of Erlangen-Nuremberg, Germany), Miaogen Shen (Institute of Tibetan Plateau Research, Chinese Academy of Sciences, Beijing, China), Jin Chen (College of Remote Sensing Science and Engineering, Faculty of Geographical Science, Beijing Normal University, Beijing, China), Suresh Raina (International Centre of Insect Physiology and Ecology, Kenya and Pollination services, India), Danny Foley (Northern Arizona University, USA), Cai Xueliang (UNESCO-IHE, Netherlands), Trent Biggs (San Diego State University, USA), Werapong Koedsin (Prince of Songkla University, Thailand), Jin Wu (University of Hong Kong, China), Kiril Manevski (Aarhus University, Denmark), Prashant K. Srivastava (Banaras Hindu University, India), George P. Petropoulos (Technical University of Crete, Greece), Philip Dennison (University of Utah, USA), Ioannis Gitas (University of Thessaloniki, Greece), Glynn Hulley (NASA Jet Propulsion Laboratory, California Institute of Technology, USA), Olga Kalashnikova, (NASA Jet Propulsion Laboratory, California Institute of Technology, USA), Thomas Katagis (University of Thessaloniki, Greece), Le Kuai (University of California, USA), Ran Meng (Brookhaven National Laboratory, USA), Natasha Stavros (California Institute of Technology, USA).

Hiroki Yoshioka (Aichi Prefectural University, Japan), My two coeditors, **Professor John G. Lyon** and **Professor Alfredo Huete**, have made outstanding contribution to this four-volume *Hyperspectral Remote Sensing of Vegetation* book (second edition). Their knowledge of hyperspectral remote sensing is enormous. Vastness and depth of their understanding of remote sensing in general and hyperspectral remote sensing in particular made my job that much easier. I have learnt a lot from them and continue to do so. Both of them edited some or all of the 48 chapters of the book and also helped structure chapters for a flawless reading. They also significantly contributed to the synthesis chapter of each volume. I am indebted to their insights, guidance, support, motivation, and encouragement throughout the book project.

My coeditors and myself are grateful to **Dr. Alexander F. H. Goetz** and **Prof. Susan L. Ustin** for writing the foreword for the book. Please refer to their biographical sketch under the respective foreword written by these two leaders of Hyperspectral Remote Sensing.

Both the forewords are a must read to anyone studying this four-volume *Hyperspectral Remote Sensing of Vegetation* book (second edition). They are written by two giants who have made immense contribution to the subject and I highly recommend that the readers read them.

I am blessed to have had the support and encouragement (professional and personal) of my U.S. Geological Survey and other colleagues. In particular, I would like to mention Mr. Edwin Pfeifer (late), Dr. Susan Benjamin, Dr. Dennis Dye, and Mr. Larry Gaffney. Special thanks to Dr. Terrence Slonecker, Dr. Michael Marshall, Dr. Isabella Mariotto, and Dr. Itiya Aneece who have worked closely with me on hyperspectral research over the years. Special thanks are also due to Dr. Pardhasaradhi Teluguntla, Mr. Adam Oliphant, and Dr. Muralikrishna Gumma who have contributed to my various research efforts and have helped me during this book project directly or indirectly. I am grateful to Prof. Ronald B. Smith, professor at Yale University who was instrumental in supporting my early hyperspectral research at the Yale Center for Earth Observation (YCEO), Yale University. Opportunities and guidance I received in my early years of remote sensing from Prof. Andrew D. Ward, professor at the Ohio State University, Prof. John G. Lyon, former professor at the Ohio State University, and Mr. Thiruvengadachari, former Scientist at the National Remote Sensing Center (NRSC), Indian Space Research Organization, India, is gratefully acknowledged.

My wife (Sharmila Prasad) and daughter (Spandana Thenkabail) are two great pillars of my life. I am always indebted to their patience, support, and love.

Finally, kindly bear with me for sharing a personal story. When I started editing the first edition in the year 2010, I was diagnosed with colon cancer. I was not even sure what the future was and how long I would be here. I edited much of the first edition soon after the colon cancer surgery and during and after the 6 months of chemotherapy—one way of keeping my mind off the negative thoughts. When you are hit by such news, there is nothing one can do, but to be positive, trust your doctors, be thankful to support and love of the family, and have firm belief in the higher spiritual being (whatever your beliefs are). I am so very grateful to some extraordinary people who helped me through this difficult life event: Dr. Parvasthu Ramanujam (surgeon), Dr. Paramjeet K. Bangar (Oncologist), Dr. Harnath Sigh (my primary doctor), Dr. Ram Krishna (Orthopedic Surgeon and family friend), three great nurses (Ms. Irene, Becky, Maryam) at Banner Boswell Hospital (Sun City, Arizona, USA), courage-love-patience-prayers from my wife, daughter, and several family members, friends, and colleagues, and support from numerous others that I have not named here. During this phase, I learnt a lot about cancer and it gave me an enlightened perspective of life. My prayers were answered by the higher power. I learnt a great deal about life—good and bad. I pray for all those with cancer and other patients that diseases one day will become history or, in the least, always curable without suffering and pain. Now, after 8 years, I am fully free of colon cancer and was able to edit the four-volume *Hyperspectral Remote Sensing of Vegetation* book (second edition) without the pain and suffering that I went through when editing the first edition. What a blessing. These blessings help us give back in our own little ways. To realize that it is indeed profound to see the beautiful sunrise every day, the day go by with every little event (each with a story of their own), see the beauty of the sunset, look up to the infinite universe and imagine on its many wonders, and just to breathe fresh air every day and enjoy the breeze. These are all many wonders of life that we need to enjoy, cherish, and contemplate.

Dr. Prasad S. Thenkabail, PhD
Editor-in-Chief
Hyperspectral Remote Sensing of Vegetation

Editors

Prasad S. Thenkabail, Research Geographer-15, U.S. Geological Survey (USGS), is a world-recognized expert in remote sensing science with multiple major contributions in the field sustained over more than 30 years. He obtained his PhD from the Ohio State University in 1992 and has over 140+ peer-reviewed scientific publications, mostly in major international journals.

Dr. Thenkabail has conducted pioneering research in the area of hyperspectral remote sensing of vegetation and in that of global croplands and their water use in the context of food security. In hyperspectral remote sensing he has done cutting-edge research with wide implications in advancing remote sensing science in application to agriculture and vegetation. This body of work led to more than ten peer-reviewed research publications with high impact. For example, a single paper [1] has received 1000+ citations as at the time of writing (October 4, 2018). Numerous other papers, book chapters, and books (as we will learn below) are also related to this work, with two other papers [2,3] having 350+ to 425+ citations each.

In studies of global croplands in the context of food and water security, he has led the release of the world's first Landsat 30-m derived global cropland extent product. This work demonstrates a "paradigm shift" in how remote sensing science is conducted. The product can be viewed in full resolution at the web location www.croplands.org. The data is already widely used worldwide and is downloadable from the NASA\USGS LP DAAC site [4]. There are numerous major publication in this area (e.g. [5,6]).

Dr. Thenkabail's contributions to series of leading edited books on remote sensing science places him as a world leader in remote sensing science advances. He edited three-volume *Remote Sensing Handbook* published by Taylor and Francis, with 82 chapters and more than 2000 pages, widely considered a "magnus opus" standard reference for students, scholars, practitioners, and major experts in remote sensing science. Links to these volumes along with endorsements from leading global remote sensing scientists can be found at the location give in note [7]. He has recently completed editing *Hyperspectral Remote Sensing of Vegetation* published by Taylor and Francis in four volumes with 50 chapters. This is the second edition is a follow-up on the earlier single-volume *Hyperspectral Remote Sensing of Vegetation* [8]. He has also edited a book on *Remote Sensing of Global Croplands for Food Security* (Taylor and Francis) [9]. These books are widely used and widely referenced in institutions worldwide.

Dr. Thenkabail's service to remote sensing community is second to none. He is currently an editor-in-chief of the *Remote Sensing* open access journal published by MDPI; an associate editor of the journal *Photogrammetric Engineering and Remote Sensing* (PERS) of the American Society of Photogrammetry and Remote Sensing (ASPRS); and an editorial advisory board member of the International Society of Photogrammetry and Remote Sensing (ISPRS) *Journal of Photogrammetry and Remote Sensing*. Earlier, he served on the editorial board of *Remote Sensing of Environment* for many years (2007–2017). As an editor-in-chief of the open access *Remote Sensing* MDPI journal from 2013 to date he has been instrumental in providing leadership for an online publication that did not even have a impact factor when he took over but is now one of the five leading remote sensing international journals, with an impact factor of 3.244.

Dr. Thenkabail has led remote sensing programs in three international organizations: International Water Management Institute (IWMI), 2003–2008; International Center for Integrated Mountain Development (ICIMOD), 1995–1997; and International Institute of Tropical Agriculture (IITA),

1992–1995. He has worked in more than 25+ countries on several continents, including East Asia (China), S-E Asia (Cambodia, Indonesia, Myanmar, Thailand, Vietnam), Middle East (Israel, Syria), North America (United States, Canada), South America (Brazil), Central Asia (Uzbekistan), South Asia (Bangladesh, India, Nepal, and Sri Lanka), West Africa (Republic of Benin, Burkina Faso, Cameroon, Central African Republic, Cote d'Ivoire, Gambia, Ghana, Mali, Nigeria, Senegal, and Togo), and Southern Africa (Mozambique, South Africa). During this period he has made major contributions and written seminal papers on remote sensing of agriculture, water resources, inland valley wetlands, global irrigated and rain-fed croplands, characterization of African rainforests and savannas, and drought monitoring systems.

The quality of Dr. Thenkabail's research is evidenced in the many awards, which include, in 2015, the American Society of Photogrammetry and Remote Sensing (ASPRS) ERDAS award for best scientific paper in remote sensing (Marshall and Thenkabail); in 2008, the ASPRS President's Award for practical papers, second place (Thenkabail and coauthors); and in 1994, the ASPRS Autometric Award for outstanding paper (Thenkabail and coauthors). His team was recognized by the Environmental System Research Institute (ESRI) for "special achievement in GIS" (SAG award) for their Indian Ocean tsunami work. The USGS and NASA selected him to be on the Landsat Science Team for a period of five years (2007–2011).

Dr. Thenkabail is regularly invited as keynote speaker or invited speaker at major international conferences and at other important national and international forums every year. He has been principal investigator and/or has had lead roles of many pathfinding projects, including the ~5 million over five years (2014–2018) for the global food security support analysis data in the 30-m (GFSAD) project (https://geography.wr.usgs.gov/science/croplands/) funded by NASA MEaSUREs (Making Earth System Data Records for Use in Research Environments), and projects such as Sustain and Manage America's Resources for Tomorrow (waterSMART) and characterization of Eco-Regions in Africa (CERA).

REFERENCES

1. Thenkabail, P.S., Smith, R.B., and De-Pauw, E. 2000b. Hyperspectral vegetation indices for determining agricultural crop characteristics. *Remote Sensing of Environment*, 71:158–182.
2. Thenkabail, P.S., Enclona, E.A., Ashton, M.S., Legg, C., and Jean De Dieu, M. 2004. Hyperion, IKONOS, ALI, and ETM+ sensors in the study of African rainforests. *Remote Sensing of Environment*, 90:23–43.
3. Thenkabail, P.S., Enclona, E.A., Ashton, M.S., and Van Der Meer, V. 2004. Accuracy assessments of hyperspectral waveband performance for vegetation analysis applications. *Remote Sensing of Environment*, 91(2–3):354–376.
4. https://lpdaac.usgs.gov/about/news_archive/release_gfsad_30_meter_cropland_extent_products
5. Thenkabail, P.S. 2012. Guest Editor for Global Croplands Special Issue. *Photogrammetric Engineering and Remote Sensing*, 78(8).
6. Thenkabail, P.S., Knox, J.W., Ozdogan, M., Gumma, M.K., Congalton, R.G., Wu, Z., Milesi, C., Finkral, A., Marshall, M., Mariotto, I., You, S. Giri, C. and Nagler, P. 2012. Assessing future risks to agricultural productivity, water resources and food security: how can remote sensing help? *Photogrammetric Engineering and Remote Sensing*, August 2012 Special Issue on Global Croplands: Highlight Article. 78(8):773–782. IP-035587.
7. https://www.crcpress.com/Remote-Sensing-Handbook---Three-Volume-Set/Thenkabail/p/book/9781482218015
8. https://www.crcpress.com/Hyperspectral-Remote-Sensing-of-Vegetation/Thenkabail-Lyon/p/book/9781439845370
9. https://www.crcpress.com/Remote-Sensing-of-Global-Croplands-for-Food-Security/Thenkabail-Lyon-Turral-Biradar/p/book/9781138116559

John G. Lyon, educated at Reed College in Portland, OR and the University of Michigan in Ann Arbor, has conducted scientific and engineering research and carried out administrative functions throughout his career. He was formerly the Senior Physical Scientist (ST) in the US Environmental Protection Agency's Office of Research and Development (ORD) and Office of the Science Advisor in Washington, DC, where he co-led work on the Group on Earth Observations and the USGEO subcommittee of the Committee on Environment and Natural Resources and research on geospatial issues in the agency. For approximately eight years, he was director of ORD's Environmental Sciences Division, which conducted research on remote sensing and geographical information system (GIS) technologies as applied to environmental issues including landscape characterization and ecology, as well as analytical chemistry of hazardous wastes, sediments, and ground water. He previously served as professor of civil engineering and natural resources at Ohio State University (1981–1999). Professor Lyon's own research has led to authorship or editorship of a number of books on wetlands, watershed, and environmental applications of GIS, and accuracy assessment of remote sensor technologies.

Alfredo Huete leads the Ecosystem Dynamics Health and Resilience research program within the Climate Change Cluster (C3) at the University of Technology Sydney, Australia. His main research interest is in using remote sensing to study and analyze vegetation processes, health, and functioning, and he uses satellite data to observe land surface responses and interactions with climate, land use activities, and extreme events. He has more than 200 peer-reviewed journal articles, including publication in such prestigious journals as *Science* and *Nature*. He has over 25 years' experience working on NASA and JAXA mission teams, including the NASA-EOS MODIS Science Team, the EO-1 Hyperion Team, the JAXA GCOM-SGLI Science Team, and the NPOESS-VIIRS advisory group. Some of his past research involved the development of the soil-adjusted vegetation index (SAVI) and the enhanced vegetation index (EVI), which became operational satellite products on MODIS and VIIRS sensors. He has also studied tropical forest phenology and Amazon forest greening in the dry season, and his work was featured in a *National Geographic* television special entitled "The Big Picture." Currently, he is involved with the Australian Terrestrial Ecosystem Research Network (TERN), helping to produce national operational phenology products; as well as the AusPollen network, which couples satellite sensing to better understand and predict pollen phenology from allergenic grasses and trees.

Contributors

Itiya Aneece
United States Geological Survey
Southwestern Geographic Center
Flagstaff, Arizona

Deepak Bhatnagar
USDA-ARS-SRRC
New Orleans, Louisiana

Trent Biggs
Department of Geography
San Diego State University
San Diego, California

Jin Chen
College of Remote Sensing Science and Engineering
Beijing Normal University
Beijing, China

Thomas E. Cleveland
USDA-ARS-SRRC
New Orleans, Louisiana

Philip Dennison
Department of Geography
University of Utah
Salt Lake City, Utah

Hannes Feilhauer
Department Geographie und Geowissenschaften Abteilung: Institut für Geographie
University of Erlangen-Nuremberg
Erlangen, Germany

Daniel Foley
United States Geological Survey
Southwestern Geographic Center
Flagstaff, Arizona

Chengcheng Gang
College of Life Science
Nanjing University
Nanjing, People's Republic of China

Paul E. Geissler
United States Geological Survey
Astrogeology Science Center
Flagstaff, Arizona

Ioannis Gitas
Laboratory of Forest Management and Remote Sensing, School of Forestry and Natural Environment
University of Thessaloniki
Thessaloniki, Greece

Edward P. Glenn
Department of Soil, Water, and Environmental Science
Environmental Research Lab
Tucson, Arizona

Will M. Grundy
Lowell Observatory
Flagstaff, Arizona

Justin J. Hagerty
United States Geological Survey
Astrogeology Science Center
Flagstaff, Arizona

Yanbo Huang
USDA-ARS-CPSRUStoneville
Mississippi

Alfredo Huete
School of Life Sciences
University of Technology Sydney
Ultimo, New South Wales, Australia

Glynn Hulley
NASA Jet Propulsion Laboratory
California Institute of Technology
Pasadena, California

Jeffrey R. Johnson
Johns Hopkins University Applied Physics Laboratory
Laurel, Maryland

Olga Kalashnikova
NASA Jet Propulsion Laboratory
California Institute of Technology
Pasadena, California

Thomas Katagis
Lab of Forest Management and Remote Sensing
University of Thessaloniki
Thessaloniki, Greece

Werapong Koedsin
Remote Sensing and Geospatial Research Unit
Prince of Songkla University
Phuket, Thailand

Le Kuai
NASA Jet Propulsion Laboratory
California Institute of Technology
Pasadena, California

and

University of California
Los Angeles, California

Tobias Landmann
International Centre of Insect Physiology and Ecology
Nairobi, Kenya

Jianlong Li
College of Life Science
Nanjing University
Nanjing, People's Republic of China

John G. Lyon
Former Professor at the Department of Civil Engineering
The Ohio State University
Columbus, Ohio

Kiril Manevski
Department of Agroecology
Aarhus University
Tjele, Denmark

Michael Marshall
Faculty of Geo-Information Science and Earth Observation
University of Twente
Enschede, The Netherlands

B. B. Maruthi Sridhar
Department of Environmental and Interdisciplinary Sciences
Texas Southern University
Houston, Tennessee

David P. Mayer
United States Geological Survey
Astrogeology Science Center
Flagstaff, Arizona

Ran Meng
Brookhaven National Laboratory
Upton, New York

Tomoaki Miura
Department of Natural Resources and Environmental Management
University of Hawaii at Manoa
Honolulu, Hawaii

Pamela Lynn Nagler
United States Geological Survey
Sonoran Desert Research Station
Southwest Biological Science Center
Tucson, Arizona

Izaya Numata
Geospatial Sciences Center of Excellence
South Dakota State University
Brookings, South Dakota

Aaryn Dyami Olsson
Lab of Landscape Ecology and Conservation Biology
College of Engineering, Forestry and Natural Sciences
Northern Arizona University
Flagstaff, Arizona

Prem Chandra Pandey
Institute of Environment and Sustainable Development
Banaras Hindu University
Varanasi, India

George P. Petropoulos
Department of Geography and Earth Sciences
University of Aberystwyth
Wales, United Kingdom

and

Department of Mineral Resources Engineering
Technical University of Crete
Crete, Greece

Suresh Raina
International Centre of Insect Physiology and Ecology
Kenya and Pollination Services
Kenya

Elijah Ramsey III
United States Geological Survey
Wetland and Aquatic Research Center
Lafayette, Louisiana

Amina Rangoonwala
United States Geological Survey
Wetland and Aquatic Research Center
Lafayette, Louisiana

Dar Roberts
Department of Geography
University of California
Santa Barbara, California

Miaogen Shen
Institute of Tibetan Plateau Research
Chinese Academy of Sciences
Beijing, People's Republic of China

E. Terrence Slonecker
United States Geological Survey
National Civil Applications Center
Reston, Virginia

Laurence A. Soderblom
United States Geological Survey
Astrogeology Science Center
Flagstaff, Arizona

Prashant K. Srivastava
Institute of Environment and Sustainable Development
Banaras Hindu University
Varanasi, India

Natasha Stavros
NASA Jet Propulsion Laboratory
California Institute of Technology
Pasadena, California

Lie Tang
Department of Agricultural and Biosystems Engineering
Iowa State University
Ames, Iowa

Chunliu Tao
Chien-shiung Institute of Technology
Taicang, China

Prasad S. Thenkabail
United States Geological Survey
Western Geographic Science Center
Reston, Virginia

Valerie Thomas
Department of Forest Resources and Environmental Conservation
Virginia Tech, Blacksburg, Virginia

Lei Tian
Department of Agricultural and Biological Engineering
University of Illinois at Urbana Champaign
Urbana, Illinois

Timothy N. Titus
United States Geological Survey
Astrogeology Science Center
Flagstaff, Arizona

Willem J. D. van Leeuwen
School of Geography and Development and School of Natural Resources and the Environment
Office of Arid Lands Studies, Arizona Remote Sensing Center
University of Arizona
Tucson, Arizona

R. Greg Vaughan
United States Geological Survey
Astrogeology Science Center
Flagstaff, Arizona

Sander Veraverbeke
Vrije Universiteit Amsterdam and University of California
Irvine, California

Jin Wu
Department of Environmental and Climate Sciences
Brookhaven National Laboratory
New York

and

School of Biological Sciences
University of Hong Kong
Hong Kong, People's Republic of China

Cai Xueliang
UNESCO-IHE
Delft, The Netherlands

Haibo Yao
Geosystems Research Institute
Mississippi State University
Starkville, Mississippi

Hiroki Yoshioka
Department of Information Science
and Technology
Aichi Prefectural University
Nagakute, Japan

Jingjing Zhang
Anhui University
Hefei, China

Dehua Zhao
College of Life Science
Nanjing University
Nanjing, People's Republic of China

Acronyms and Abbreviations

ΔEOS	Difference between Acquisition Date and End of Season
ADB	Aboveground dry biomass
ADEOS	Advanced Earth Observing Satellite
AERONET	Aerosol Robotic Network
AHI	Airborne Hyperspectral Imager
AIVI	Angular insensitivity vegetation index
ALEXI	Atmosphere-land exchange inverse model
ALI	Advanced Land Imager
ANN	Artificial neural network
ANOVA	Analysis of variance
AOT	Aerosol optical thickness
ARI	Anthocyanin reflectance index
ASD	Analytical Spectral Devices
ASI	Italian Space Agency
ASTER	Advanced Spaceborne Thermal Emission and Reflection Radiometer
ASTER	Assessment Tools for the Evaluation of Risk
AVHRR	Advanced very-high-resolution radiometer
AVIRIS-NG	Airborne Visible/Infrared Imaging Spectrometer—Next Generation
AVIRIS	Airborne Visible Infrared Imaging Spectrometer
AWB	Aboveground wet biomass
BLH	Bottomland hardwood forests
BRDF	Bidirectional reflectance distribution function
C-CAP	Coastal Change Analysis Program
CAI	Cellulose absorption index
CAIs	Calcium- and aluminum-rich inclusions
CAR	Carotene
CASI	Compact Airborne Spectrographic Imager
CC	Chlorophyll content
CCD	Charge-coupled device
CEC	Cation exchange capacity
CFMask	Cloud function of mask
CHL	Chlorophyll
CHRIS	Compact High Resolution Imaging Spectrometer
CIR	Color infrared
CMG	Climate modeling grid
CO_2	Carbon dioxide
CRESDA	Chinese Centre for Resources Satellite Data and Application
CRISM	Compact Reconnaissance Imaging Spectrometer for Mars
CWP	Crop water productivity
DA	Discriminant analysis
DAIS	Digital Airborne Imaging Spectrometer
DAT	Day after treatment
DGVI	Derivative greenness vegetation index
DISORT	Discrete Ordinates Radiative Transfer
DN	Digital number
DVI	Difference vegetation index
DWAB	Dry weight of aboveground biomass

E1	Experiment 1
E2	Experiment 2
ECOSTRESS	ECOsystem Spaceborne Thermal Radiometer Experiment on Space Station
EGS	End of growing season
EJSM	Europa Jupiter System Mission
Elv	Elevation
EM	Electromagnetic
EnMAP	Environmental Mapping and Analysis Program
ENVI	Environment for Visualizing Images
EO-1	Earth Observing-1
EPOXI	Extrasolar Planet Observation eXtended Investigation
ESA	European Space Agency
ET_A	Actual evapotranspiration
ET_C	Transpiration
ETM+	Enhanced Thematic Mapper Plus
EVI2	Two-band enhanced vegetation index
EVI	Enhanced vegetation index
F_{PAR}	Fraction of photosynthetically active radiation
FC1 & FC2	Framing Camera 1 and 2
FLAASH	Fast Line-of-sight Atmospheric Analysis of Spectral Hypercubes
FLEX	FLuorescence Explorer
FOV	Field of view
FTIR	Fourier transform infrared
GA	Genetic algorithm
GAC	Global area coverage
GCC	Green chromatic coordinate
GCOM-C	Global Change Observation Mission—Climate
GDD	Growing degree-day
GIS	Geographic information system
GLI	Global Imager
GLM	General linear model
GOME-2	Global Ozone Monitoring Experiment-2
GOSAT	Greenhouse Gases Observing Satellite
GPP	Gross primary production
GPS	Global Positioning System
GR	Glyphosate-resistant
GRaND	Gamma Ray and Neutron Detector
GSD	Ground sampling distance
H	Sensible heat
HAT	Hour after treatment
HFI	Hyperspectral flower index
HI	Harvest index
HiRISE	High-Resolution Imaging Science Experiment
HN	High nitrogen
HNB	Hyperspectral narrowband
HRI-IR	High-Resolution Instrument Infrared
HRS	Hyperspectral remote sensing
HSI	Hyperspectral imaging
HVI	Hyperspectral narrowband vegetation index
HyspIRI	Hyperspectral Infrared Imager
HyTES	Hyperspectral Thermal Emission Spectrometer

ID Site identification
IFOV Instantaneous field of view
ILTER International long-term ecological research
IPM Intelligent Payload Module
ISA Israel Space Agency
ISRO Indian Space Research Organization
ISS Imaging Science Subsystem
JHU The Johns Hopkins University
JPL Jet Propulsion Laboratory
JPSS Joint Polar Satellite System
KNN K-nearest neighbor
LAD Leaf angle distribution
LAI Leaf area index
LAP Light attenuation profile
LCCS Land Cover Classification System
LDA Linear discriminant analysis
LE Latent heat
LED Light-emitting diode
LEDAPS Landsat Ecosystem Disturbance Adaptive Processing System
LEISA Linear Etalon Imaging Spectral Array
LGS Length of growing season
LHI-4 Leafhopper Index-4
LISCT Lunar International Science Calibration/Coordination Targets
LL R2M Lambda–lambda R2 models
LN Low nitrogen
LOLA Lunar Orbiter Laser Altimeter
LS-SVM Least-squares support vector machine
LULC Land use/land cover
M^3 Moon Mineralogy Mapper
MAB Main asteroid belt
MASCS Mercury Atmospheric and Surface Composition Spectrometer
MBVI Multispectral broadband vegetation index
MERIS MEdium Resolution Imaging Spectrometer
MESMA Multiple endmember spectral mixture analysis
MESSENGER MErcury Surface, Space ENvironment, Geochemistry, and Ranging
METRIC Mapping Evapotranspiration at high Resolution with Internalized Calibration model
MGS Minimum greenness value
MISP Mobile Instrumented Sensor Platform
ML Maximum likelihood
MLC Maximum-likelihood classification
MLR Multiple linear regression
MN Medium nitrogen
MNDVI Modified normalized vegetation index
MODIS Moderate Resolution Imaging Spectroradiometer
MODTRAN MODerate resolution atmospheric TRANsmission
MOLA Mars Orbiter Laser Altimeter
MS Multispectral
MSAVI Modified soil-adjusted vegetation index
MSAVI2 Modified second soil-adjusted vegetation index
MSBB Multispectral broadband

MSI	Mealybug stress index
MTCI	MERIS Total Chlorophyll Index
MTMF	Mixture-tuned matched filtering
MTVI2	Modified Triangular Vegetation Index 2
MVIC	Multi-spectral Visible Imaging Camera
N	Nitrogen
NASA	National Aeronautics and Space Administration
ND	Normalized difference
NDRE	Normalized difference red-edge-based index
NDVI	Normalized difference vegetation index
NDWI	Normalized difference water index
NGC	New General Catalogue
NIMS	Near-Infrared Mapping Spectrometer
NIR	Near-infrared
NIST	National Institute of Standards and Technology
nm	Nanometer
NN	Neural network
NOAA	National Oceanic and Atmospheric Administration
NPP	The Suomi National Polar-orbiting Partnership or Suomi NPP
NPP	Net primary production
NPV	Nonphotosynthetic vegetation
NRC	National Research Council
NRI	Normalized ratio index
OCO-2	Orbiting Carbon Observatory-2
OM	Organic matter
OMEGA	Observatoire pour la Minéralogie, l'Eau, les Glaces et l'Activité (The OMEGA is a visible and near-IR imaging spectrometer to study the Planet Mars, gathering data in 0.38–5.1 micrometer)
OSAVI	Optimized soil-adjusted vegetation index
PAN	Panchromatic
PCA	Principal component analysis
PDS	Planetary Data System
PEN	Phenological Eyes Network
PGS	Peak of growing season
PLS-LDA	Partial least-squares linear discriminant analysis
PLSR	Partial least square regression
PPT	Precipitation
PRI	Photochemical reflectance index
PRISMA	**PR**ecursore **I**per**S**pettrale della **M**issione **A**pplicativa- Hyperspectral Precursor of the Application Mission
PRISMA	PRecursore IperSpettrale della Missione Applicativa
PROBA	Project for On-Board Autonomy
PSF	Point spread function
PSRI	Plant senescence reflection index
PVA	Polytopic vector analysis
PVI	Perpendicular Vegetation Index
QDA	Quadratic discriminant analysis
R_1	Band 1
R_2	Band 2
R^2	Coefficient of determination
REP	Red edge position

RF	Random forest
RMSE	Root-mean-squared error
RMSECV	Root-mean-square error of cross-validation
RS	Remote sensing
RS	Root-to-shoot ratio
RSL	Remote Sensing Laboratories
RT	Radiative transfer
RVI	Remote visual inspection
RVSI	Red-edge vegetation stress index
SAM	Spectral Angle Mapper
SAR	Synthetic aperture radar
SAVI	Soil-adjusted vegetation index
SAVI2	Second soil-adjusted vegetation index
SCCCI	Simplified canopy chlorophyll content index
SDA	Stepwise discriminant analysis
SeaWiFS	Sea-viewing Wide Field-of-view Sensor
SEBAL	Surface Energy Balance Algorithm model
SEBASS	Spatially Enhanced Broadband Array Spectrograph System
SGLI	Second-generation global imager
SGS	Start of growing season
SHALOM	Spaceborne Hyperspectral Applicative Land and Ocean Mission
SIF	Solar-induced chlorophyll fluorescence
SMA	Spectral mixture analysis
SNR	Signal-to-noise ratio
SPOT	Système Pour l'Observation de la Terre
SPSS	Statistical Package for the Social Sciences
SR	Simple ratio
SVM	Support vector machines
SWIR	Shortwave infrared
SWNIR	Shortwave near-infrared
T	Temperature
$\mathbf{T_r}$	Land surface temperature
TCARI	Transformed chlorophyll absorption reflectance index
TDI	Time-delay integration
TFOV	Total field of view
TIR	Thermal infrared
TOA	Top of atmosphere
TOC	Top of canopy
TSAVI	Transformed soil-adjusted vegetation index
TVI	Transformed vegetation index
UAV	Unmanned aerial vehicle
UAV	Unmanned autonomous vehicle
USGS	United States Geological Survey
UTEP	The University of Texas at El Paso
UV	Ultraviolet
UVVS	Ultraviolet-Visible Spectrometer
VARI	Visible atmospherically resistant index
VI	Vegetation index
VIIRS	Visible Infrared Imaging Radiometer Suite
VIMS	Visual and Infrared Mapping Spectrometer
VIR	VNIR/SWIR slit spectrometer

VIRS	Visible-InfraRed Spectrograph
Vis	Vegetation indices
VIS	Visible wavelength region
VIS	Visible
VNIR	Visible and near-infrared wavelength regions
VNIR	Visible, near-infrared
VSL	Vegetation Spectral Library
WDVI	Weighted difference vegetation index
WUE	Water use efficiency
XRD	X-ray diffraction
Y	Crop yield

Section I

Detecting Crop Management Practices, Plant Stress, and Disease

1 Using Hyperspectral Data in Precision Farming Applications

Haibo Yao, Yanbo Huang, Lie Tang, Lei Tian,
Deepak Bhatnagar, and Thomas E. Cleveland

CONTENTS

1.1 INTRODUCTION

1.1.1 Precision Farming

Rather than being managed as a single, uniform unit, a crop field can be handled site specifically based on local field needs. This is the concept behind using precision agriculture for in-field variability management. The goals of precision agriculture can be described as follows and are based on economic, productivity, and environmental considerations:

- Greater yield than traditional farming with the same amount of input;
- The same yields with reduced input;
- Greater yield than traditional farming with reduced input.

The precision agriculture concept has drawn significant attention from farmers and researchers around the world (National Research Council, 1997; Zhang et al., 2002; Hedley, 2015). A complete precision agriculture system can be described in terms of four indispensable parts: (a) field variability sensing and information extraction, (b) decision making, (c) precision field control, and (d) operation and result assessment. The success of any precision agriculture system depends on the correct implementation of these four parts. Among the four parts, the decision-making step is the central component (Stafford, 2000). The decision-making process involves making the right management decisions based on the variability information derived from data collected in the field.

To make sound decisions, the most important step is to obtain accurate information about in-field variabilities. Agricultural engineers devote significant efforts to field variability sensing and information extraction, as well as to precision field control and operation. Sensing and information extraction are crucial parts of the system requiring that the desired information be obtained at the right location at the right time. Sensing and information extraction involve using various sensors to capture data on field conditions. Once the raw data are obtained, appropriate algorithms can be used to extract field information. Sensing either from a close distance (ground) or remotely, such as from airborne or spaceborne sensors, is an import method of field data acquisition (Scotford and Miller, 2005; Larson et al., 2008; McIntyre and Corner, 2016; Skowronek et al., 2017). Agricultural remote sensing typically involves the use of surface reflectance information in the visible (VIS) and near-infrared (NIR) region of the electromagnetic spectrum. It provides a fast and economical way to acquire detailed field data in a short period of time. Remote sensing has thus been used in a broad range of applications in the farming industry. Mulla (2013) reviewed progress made in the previous 25 years on remote sensing in precision agriculture. The article pointed out the potential to collect massive amounts of data from different sensors and platforms for agricultural applications. Another review (Wolfert et al., 2017) further described scenarios in which big data influenced farm operations with its substantial impact on the entire food supply chain.

1.1.2 Hyperspectral Data

Traditionally, agricultural remote sensing has used multispectral broadband imagery. With advances in sensor technology over the past two decades, the introduction of hyperspectral remote sensing imagery to agriculture provided more opportunities for field-level information extraction. One comparison study (Mariotto et al., 2013) with satellite sensors demonstrated that hyperspectral imagery (Hyperion) was advantageous over multispectral broadband imagery (Landsat-7, Advanced Land Imager, Indian Remote Sensing, IKONOS, and QuickBird) in crop productivity modeling. The five crops under investigation were cotton, wheat, corn, rice, and alfalfa. The results showed that hyperspectral-based crop biophysical models explained around 25% greater variability than multispectral broadband-based models. For crop-type identification, hyperspectral data produced much higher accuracy (>90%) than multispectral broadband data (45%–84%).

Figure 1.1 presents a system approach to using hyperspectral imagery for precision agriculture applications. A hyperspectral image has more bands (tens to hundreds or even thousands) with a narrow bandwidth (one to several nanometers) in the same spectral range (e.g., 400–2500 nm) as a multispectral image. In this chapter, *hyperspectral* by default means narrowband spectral data. When presenting hyperspectral imagery, each pixel within an image is typically described as a data vector and the entire image as an image cube. Due to the high data volume of a hyperspectral image, hyperspectral imagery could potentially provide more information for precision agriculture. On the other hand, the increased number of data dimensions in a hyperspectral image also increases the complexity in image processing and might impact accuracy. One example of the influence of data dimensionality on accuracy is the Hughes phenomenon (Hughes, 1968), which shows that classification accuracy decreases as data dimensions increase, especially when a large number of wavebands is involved. To reduce image-processing complexity and to increase image-interpretation accuracy, it is desirable to reduce the original image's dimensionality through a feature reduction process.

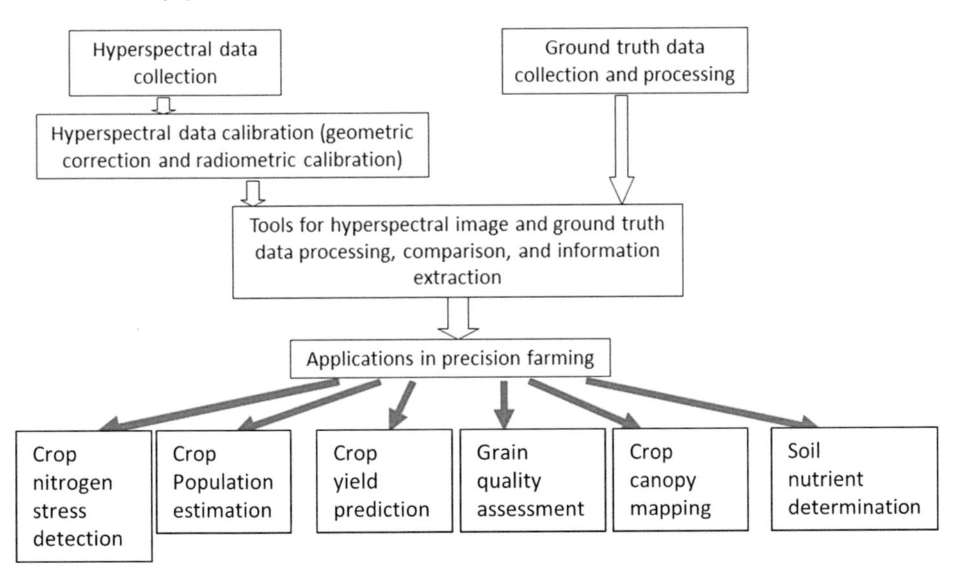

FIGURE 1.1 System diagram of using hyperspectral data in precision farming applications. (Adapted from Yao, H. 2004. Hyperspectral imagery for precision agriculture. PhD dissertation, University of Illinois at Urbana-Champaign, Urbana, IL.)

There are two major types of feature reduction methods (Richard and Jia, 1999): feature selection and feature extraction. The purpose of feature selection is to remove the least effective features (image bands) and select the most effective features. Feature selection consists in evaluating an existing set of features of a hyperspectral image in order to select the most discriminating features and discard the rest. Feature extraction involves transforming a pixel vector into a new set of coordinates in which the basis for feature selection is more evident. Common feature-extraction techniques used in remote sensing include the linear combination of image bands, such as in principal component transformation and canonical analysis, and arithmetic transformation, such as vegetation indices. Vegetation indices and their application in precision farming will be discussed later in this chapter.

1.2 APPLICATIONS OF HYPERSPECTRAL DATA IN PRECISION AGRICULTURE

1.2.1 Precision Farming Management Considerations

Crop production in agriculture has relied heavily on the development and implementation of various technologies. Crop yield can be regarded as the single most important output of crop production systems. Other aspects related to crop production, such as field topography, soil characteristics and fertility, tillage practices, fertilizer application, crop rotation, seeding, weed and pest control, irrigation, and weather, can all be regarded as inputs for crop production systems. Remote sensing (Mulla, 2013) provides field variability information on the manageable inputs in a map-driven approach to precision farming practices. For example, one of the most successful precision farming technologies is variable rate technology (Zhang et al., 2002). This map-driven approach provides a prescription map based on field variability measured by remote sensing. Subsequent variable rate applications of fertilizer, herbicide, or other agricultural chemicals (Hedley, 2015) can be made using the prescription map. In this process, the use of the Global Positioning System (GPS) and a geographic information system (GIS) is also necessary. Additionally, the concept of a "management zone" is an important topic in precision farming. Management zones are smaller sections of a large field where the field properties of interest are regarded as relatively homogeneous. Remote sensing has proved to be quite a useful tool in management zone delineation.

Several issues demonstrate the importance of remote sensing technologies to precision farming, and they are discussed in what follows.

1.2.2 Spatial, Spectral, and Temporal Considerations

Three issues relate to the use of hyperspectral remote sensing imagery in agricultural applications. They are the spatial, spectral, and temporal issues of an image. One of the advantages of airborne or spaceborne remote sensing is the large spatial coverage. Aerial and space remote sensing data can cover a large area in a short period of time. Thus, those techniques can provide a fast, accurate, and economical method for precision applications. Spatial resolution is an important factor that varies dramatically depending on the sensor platform. For proper data interpretation, the spatial resolution of remote sensing data and the ground truth should be matched. Sometimes spatial resampling on one data type is necessary to meet this requirement.

The second issue is image spectral range and resolution. The spectral range normally is in the VIS and NIR regions from 400 to 1000 nm. This is the region where plants show distinct spectral signatures under different conditions. Some applications, such as soil characterizations, extend the spectral region to shortwave infrared, such as from 1000 to 2500 nm. For spectral resolution, agricultural remote sensing has traditionally used multispectral images with a spectral resolution (or bandwidth) of several hundred nanometers. Multispectral images are sometimes called broadband images, with each broadband covering a specific wavelength range such as blue, green, red, or NIR. On the other hand, a hyperspectral image has a bandwidth of one to several nanometers and thus provides significant fine image spectral resolution. Hence, hyperspectral imagery provides the potential for allowing more detailed information extraction in agricultural applications.

The third issue is related to temporal hyperspectral data acquisition. It pertains to the time of acquisition of each image and the time interval between image acquisitions. For example, identifying the temporal relationship between image and yield is helpful for yield estimation and management. The spatial yield pattern does not appear immediately before harvest. Rather, the yield pattern is built up gradually during the growing season. One study found that the spectral reflectance of plants had both a temporal and a spatial aspect (Zwiggelaar, 1998). Because this variation in crop spectral reflectance during the growing season can be related to yield, it could help growers estimate yield during the growing season.

1.2.3 Hyperspectral Narrowband Vegetation Indices

Vegetation indices have been used widely in remote sensing. The most widely known vegetation index is the normalized difference vegetation index (NDVI) calculated by using the red and NIR wavelengths. The use of hyperspectral images makes it possible to build more refined vegetation indices by using distinct narrowbands and improving the indices for the correction of soil background effects (Gong et al., 2003). Many hyperspectral vegetation indices (HVIs) have been developed for different applications (Roberts et al., 2011; Mulla, 2013). The simplest vegetation index is based on individual bands. Filella et al. (1995) used individual image bands located at 430, 550, 680, and 780 nm to build different indices for wheat nitrogen (N) status evaluation. Blackburn (1998) also used individual spectral bands to develop various hyperspectral indices for estimating chlorophyll concentrations.

Vegetation indices can be calculated based on band ratio and combinations of bands. Elvidge and Chen (1994) used narrowbands with a 4 nm bandwidth at 674 and 755 nm to calculate several narrowband indices for the leaf area index (LAI) and percentage green cover and compared the results with the corresponding broadband indices. Hurcom and Harrison (1998) used the NDVI calculated from 677 and 833 nm to measure vegetation cover in a semiarid area. Serrano et al. (2000) used two image bands at 680 and 900 nm to compute vegetation indices, including the NDVI, to estimate the biomass and yield of winter wheat. Broge and Leblanc (2000) calculated narrowband vegetation indices from spectral bands centered at 670 and 800 nm and having a 10 nm bandwidth. Daughtry et al. (2000) used discrete bands at 550, 670, and 801 nm to develop narrowband indices for N stress estimation in corn.

Vegetation indices were also studied based on data from different platforms. Broge and Mortensen (2002) utilized field spectrometer data and spectral bands centered at 550, 650, and 800 nm and having a 10 nm bandwidth to calculate various HVIs. The authors also used an aerial hyperspectral image, the Compact Airborne Spectral Imager (CASI), and chose, based on their sensitivity to chlorophyll, several individual image bands for vegetation index calculation. These indices were used for LAI and N prediction over different types of crops. Another study using CASI images (Haboudane et al., 2002) calculated several vegetation indices using image bands centered at 550, 670, 700, and 800 nm for crop chlorophyll content prediction. The reason for choosing 700 nm is that it is located at the edge between the region where vegetation reflectance is dominated by pigment absorption and the beginning of the red-edge region where reflectance is more affected by vegetation structural characteristics. Hyperspectral imagery acquired from a spaceborne Hyperion sensor was also used for the calculation of vegetation indices (Gong et al., 2003). This study evaluated 12 vegetation indices using 168 bands selected from an image after removal of the water absorption bands and noise bands. These indices were two-band indices and were constructed using all possible two-band combinations.

A common trend in the aforementioned HVIs is the use of individual image bands, where most of the time one specific image band pair is selected based on crop characteristics. One reason for doing this is that it is simple to construct such indices. Another reason is that it is complicated to construct a multiple-narrowband–based index if one follows the traditional construction and comparison approach for all possible solutions. For example, even though many vegetation indices have been designed, the studies in which they were developed generally only tested several indices for result comparison and best index identification. This practice may miss some important indices in the vast vegetation index database. Yao (2004) presented a generic approach to automating the process for vegetation index selection and generation with hyperspectral data. The study first established a collection of available vegetation indices. A genetic algorithm (GA)-based method was then used to select the best vegetation index and spectral band combination for a specific application. The reader is also referred to a much broader discussion on wavebands and indices in other chapters of the first and this edition of *Hyperspectral Remote Sensing of Vegetation*.

1.2.4 Application 1: Soil Management Zoning

Research has been carried out using remotely sensed data for soil property mapping. It is expected that soil surface spectral reflectance could be used for soil constituent and nutrient content discrimination. In a review on remote sensing of soil properties, Ge et al. (2011a) pointed out that many soil properties, such as texture, organic and inorganic carbon content, macro- and micronutrient content, moisture content, cation exchange capacity (CEC), electrical conductivity, pH, and iron, could be quantified successfully to varying extents. As discussed earlier, soil property maps are used to prescribe variable rate applications. For example, a soil pH map is a good resource for decision making in variable rate lime applications (Schirrmann et al., 2011). Based on a simulation study using soil maps derived from spaceborne hyperspectral images, Casa et al. (2012) pointed out that site-specific irrigation could prevent significant water loss better than uniform irrigation. A general concept for utilizing remote sensing in soil nutrient mapping, as recommended by Moran et al. (1997), could be stated as follows: "Measurements of soil and crop properties at sample sites combined with multispectral imagery could produce accurate, timely maps of soil and crop characteristics for defining precision management units." Some ground- and lab-based studies have focused on using reflectance in the VIS and NIR region to determine soil nutrients. It was found that there are different sensitive regions in the electromagnetic spectrum for different soil nutrient properties under controlled lab conditions. Ben-Dor and Banin (1995) used reflectance curves in the infrared region to study six soil properties: clay content, specific surface area, CEC, hygroscopic moisture, carbonate content, and soil organic matter (OM) content. The results showed that the optimum prediction performance of each property required a different number of bands ranging from 25 to 3113. Palacios-Orueta and Ustin (1998) found that the total iron and OM contents were the main factors affecting soil spectral shape and concluded that the levels of iron and OM could be identified from Advanced Visible/Infrared Imaging Spectrometer (AVIRIS) images (NASA, Washington, DC). Thomasson et al. (2001) found that the spectral regions from 400 to 800 nm and from 950 to 1500 nm were sensitive to soil nutrient composition.

The foregoing results indicate that there were different sensitive regions in the image spectral data for different nutrient properties, and such data would be a viable source of information for soil nutrient content classification and mapping. In this case, single hyperspectral imagery would provide the opportunity for different nutrient classifications using various sensitivity ranges. Table 1.1 summarizes some research on soil property sensing with hyperspectral data. With a 79-band hyperspectral image (400–1400 nm) Ben-Dor et al. (2002) were able to build a multiple regression model for each of four soil properties, OM, soil field moisture, soil saturated moisture, and soil salinity, each with different bands. Because the sensitive spectral region is valuable for soil nutrient identification and mapping using hyperspectral images, identification of such sensitive regions remains a major task in hyperspectral remote sensing research.

A more common and traditional method in this application is to explore spectral information using only hyperspectral data. Whiting et al. (2006) summarized spectral processing techniques for soil classification, mineral, moisture, and nutrient determinations and noted that it was possible to use the mineral spectral absorption position and depths to identify mineral contents. Ge and Thomasson (2006) incorporated wavelet analysis using conventional regression methods with field spectrometer measurements for soil property determination. It was found that Ca, Mg, clay, and Zn could be predicted with reasonable R^2 values (>0.5). A different ground-based study (Hu et al., 2016) with 401–2450 nm reflectance data and partial least-squares regression (PLSR) found that the estimation of soil plant-available P and K concentrations was not reliable when soil samples were grouped by CEC and OM, suggesting separate P and K estimation for each field. DeTar et al. (2008) pointed out that some soil properties could be accurately detected using aerial hyperspectral data over nearly bare fields. The best regression R^2 (0.806) was for percentage sand. Other properties, such as silt, clay, chlorides, electrical conductivity, and P, had slightly lower R^2 values (0.66–0.76). Bajwa and Tian (2005) used first derivatives from aerial hyperspectral data (400–1000 nm) and PLSR to model soil fertility factors, including pH, OM, Ca, Mg, P, K, and soil electrical conductivity. They drew the conclusion that certain wavebands explained a high degree of variability in the model. Another study (Hively et al., 2011) included shortwave near-infrared (SWNIR) airborne hyperspectral data (400–2450 nm) in six tilled fields. The PLSR results showed that 13 out of 19 agronomically important elements under investigation had $R^2 > 0.5$, that is, with carbon (0.65), aluminum (0.76), iron (0.75), and silt content (0.79).

Geostatistical techniques were also used with hyperspectral data to incorporate spatial information of soil samples for soil property determination. Examples of the techniques, such as regression-kriging (Ge et al., 2011b) and collocated cokriging (Yao et al., 2014), proved that they could produce better results than regression modeling. Yao (2004) applied two geostatistical approaches, collocated ordinary cokriging and sequential Gaussian cosimulation, to predict soil nutrient factors. It was found that the cosimulation method yielded the best estimation ($R^2 = 0.71$) for K prediction. Figure 1.2 presents an in-field pH map generated from the cosimulation process ($R^2 = 0.58$). It shows that the pH zones can be divided into two regions along the grass waterway located in the middle-left of the

TABLE 1.1
Hyperspectral Applications in Soil Property Sensing

Soil Sensing Application	Hyperspectral Narrowband Wavelength (nm)	Hyperspectral Analysis Technique	R^2	Reference
Aluminum	400–2450	PLS	0.76	Hively et al. (2011)
Ca	414, 574, 1406, 1854, 1166, 1806, 1902, 2158, 2318	Discrete wavelet transform	0.73	Ge and Thomasson (2006)
Carbon	400–2450	PLSR	0.65	Hively et al. (2011)
Chlorides	763, 753, 657, 443	MLR	0.74	DeTar et al. (2008)
Clay	670, 2334, 446, 574, 638, 1086, 406, 2110, 366, 1678, 1966	Discrete wavelet transform	0.56	Ge and Thomasson (2006)
Clay	889, 734, 666, 59	MLR	0.67	DeTar et al. (2008)
Electric conductivity (EC)	763, 753, 647, 443	MLR	0.67	DeTar et al. (2008)
Iron	400–2450	PLS	0.75	Hively et al. (2011)
K	1101–2450	PLSR	0.44	Hu et al. (2016)
K	471	Collocated cokriging	0.84	Yao et al. (2014)
Mg	510, 1406, 1726, 1854, 398, 558, 686, 750, 1422, 1742	Discrete wavelet transform	0.73	Ge and Thomasson (2006)
OM	471	Collocated cokriging	0.74	Yao et al. (2014)
Organic matter	722, 2328, 705, 1678	MLR	0.83	Ben-Dor et al. (2002)
P	1101–2450	PLSR	0.35	Hu et al. (2016)
P	753, 724, 531, 502	MLR	0.70	DeTar et al. (2008)
P	498	Collocated cokriging	0.67	Yao et al. (2014)
PH	639	Collocated cokriging	0.76	Yao et al. (2014)
pH	986, 947, 889, 763	MLR	0.62	DeTar et al. (2008)
Sand	627, 647, 724, 840	Regression	0.81	DeTar et al. (2008)
Silt content	400–2450	PLS	0.79	Hively et al. (2011)
Silt	995, 957, 637, 579	MLR	0.75	DeTar et al. 2008)
Soil field Moisture	739, 1650, 689	MLR	0.65	Ben-Dor et al. (2002)
Soil salinity (EC)	739, 1650, 2166	MLR	0.67	Ben-Dor et al. (2002)
Soil saturated moisture	2085, 2314, 2183, 1563, 1538	MLR	0.76	Ben-Dor et al. (2002)
Zn	1054, 2334, 958, 462, 494, 662, 654, 2158, 2350	Discrete wavelet transform	0.57	Ge and Thomasson (2006)

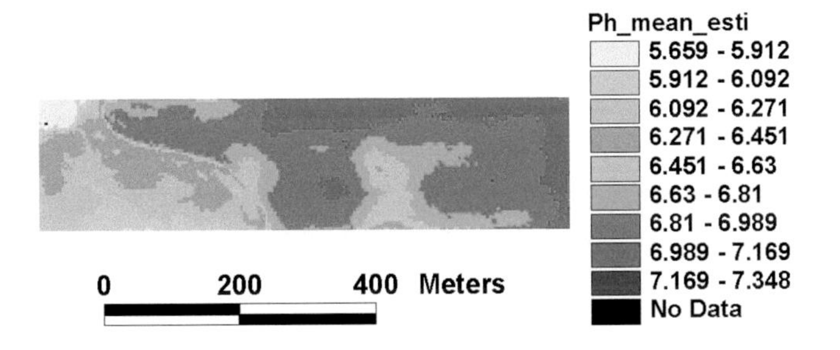

FIGURE 1.2 In-field variability of soil pH as indicated by sequential Gaussian cosimulation method using aerial hyperspectral data. (From Yao, H. 2004. Hyperspectral imagery for precision agriculture. PhD dissertation, University of Illinois at Urbana-Champaign, Urbana, IL.)

field. To the left of the waterway, the soil is acidic with low pH value estimations. To the right of the waterway the soil varies from acidic to basic. This analysis thus provided important information to assist in decision making on variable rate lime applications. Ge et al. (2007) worked on a regression-kriging method to analyze soil sampling data and reflectance measurements. The regression-kriging model R^2 was 0.65 for Na, which was much better than a principal component regression approach. Bilgili et al. (2011) also concluded that cokriging and regression-kriging improved the predictions of soil properties with reflectance data. Ladoni et al. (2010) reviewed statistical techniques including simple regression, the "soil line" approach, principal component analysis, and geostatistics for soil OM estimation using remote sensing data. The study pointed out that remote sensing data could help in the design of a soil sampling strategy.

1.2.5 Application 2: Weed Sensing

Effective weed management is of vital importance for ensuring the profitability of agricultural and horticultural crops. However, weed control has relied heavily on the application of herbicides, leading to increased environmental contamination that has become of concern to the general public and that imposes high costs on producers (Timmermann et al., 2003). This situation calls for more effective use of herbicides, that is, applying a minimal dosage of herbicide to only weeds. Recent years has seen a growing trend toward organic farming, particularly in vegetable crop production. Weed control in organic farming practice excludes the use of synthetic chemicals and often requires mechanical means to control weeds without collateral damage to crops. For both reduced herbicide application and organic farming scenarios, it is self-evident that there must be an effective and reliable weed sensing system that allows for the differentiation of weeds from crops and soils. Numerous sensing technologies have been investigated and developed for weed detection (Thorp and Tian, 2004), among which optical and machine vision systems dominate. Compared with conventional machine vision systems, remote sensing provides a fast and economical means for weed mapping. However, because of the similarities between weeds and crops, as well as soil and vegetation interactions before canopy closure (Thorp and Tian, 2004), the use of remote sensing for weed mapping remains a challenging task.

Hyperspectral imagery has been used from ground-based (Thenkabail et al., 2004; Huang et al., 2016a; Wendel and Underwood, 2016), airborne (Mirik et al., 2013; Skowronek et al., 2017), and spaceborne (McIntyre and Corner, 2016) platforms for weed sensing applications. Table 1.2 presents some studies that used hyperspectral data for weed sensing and mapping. The information enrichment offered by hyperspectral sensors has a direct implication for weed detection, which, however, has long been a very challenging task stemming from the biological complexity of a rather large number of weed species and their similarities to crop plants in the VIS color domain as well as in the morphological feature space. To this end, a large amount of research work in the literature has been found in the area of developing hyperspectral-based weed detection methods. Though the fine spectral resolution of hyperspectral images provides an invaluable source of information for more accurate classification, it is also agreed that the high dimensionality of the data presents challenges in image analysis and classification. Without the development of effective image processing and pattern recognition algorithms, the advantages offered by the rich information in the spectral dimension could not be utilized effectively. Furthermore, these algorithms are often application-specific and require substantial efforts for exploration and testing. To cope with the high dimensionality of hyperspectral data, multivariate analysis and computational intelligence techniques have been widely used and reported in the literature. When processing raw hyperspectral data to delineate the spectral characteristics of weeds, techniques such as principal component analysis (Koger et al., 2003; Thenkabail et al., 2004), wavelet transforms (Koger et al., 2003; Okamoto et al., 2007), and spectral angle mapping (Hestir et al., 2008) are often used. Commonly found classification algorithms include linear or stepwise discriminant analysis (Thenkabail et al., 2004; Nieuwenhuizen et al., 2010; Wendel and Underwood, 2016), spectral mixture analysis (Koger et al., 2003; Hestir et al., 2008),

TABLE 1.2
List of Hyperspectral Applications in Weed Sensing and Mapping

Weed Sensing Application	Hyperspectral Narrowband Wavelength (nm)	Hyperspectral Analysis Techniques	Results	Reference
		Weed mapping		
• Ground based	13–22 bands selected in 400–2500	PCA, LL R2M, SDA, DGVI	>90% accuracy	Thenkabail et al. (2004)
• Ground based	14 bands selected in 400–1000	LDA + MLC	>90% accuracy	Huang et al. (2016a)
• Ground based	20 PCA features in 391–887	LDA, SVM	>90% accuracy	Wendel and Underwood (2016)
• Airborne	25 bands in 509–706	SVM	79% preflower, 91% peak flowering	Mirik et al. (2013)
• Airborne	285 bands in 412–2432	Maxent modeling	75% accuracy	Skowronek et al. (2017)
• Spaceborne	49 bands in 428–917	Matched filter classification	>81% accuracy	McIntyre and Corner (2016)
		Weed spectral characterization		
• Ground based	2151 bands in 350–2500	Wavelet	87% accuracy	Koger et al. (2003)
• Ground based	10 bands selected from 450–900, 900–1650	DA, NN	87% accuracy	Nieuwenhuizen et al. (2010)
• Airborne	72 bands, 409–947	SVM	86%	Karimi et al. (2006)
		Invasive weed species		
• Airborne	126 bands in 450–2500	MTMF	>87%	Glenn et al. (2005)
• Ground based	172 bands in 400–900	MLC	>94%	Samiappan et al. (2017)
		Weed stress characterization		
• Airborne	71 bands in 409–947	GLM	Vary with different growth stage	Goel et al. (2003)
• Ground based	91 bands in 400–850	PLS-LDA	85%	Herrmann et al. (2013)
• Ground based	160 bands in 415–1000	PLS-LDA	96%	Hadoux et al. (2014)
		Weed species identification		
• Ground based	Multispectral bands at 450, 550, and 700	QDA	72%	Piron et al. (2008)

artificial neural networks (ANNs) (Goel et al., 2003; Karimi et al., 2006; Nieuwenhuizen et al., 2010; Eddy et al., 2013), support vector machines (SVMs) (Karimi et al., 2006; Mirik et al., 2013; Wendel and Underwood, 2016), mixture-tuned matched filtering (MTMF) (Glenn et al., 2005), Maxent modeling classification (Skowronek et al., 2017), and matched filter with logistic regression (McIntyre and Corner, 2016).

Using hyperspectral images for vegetation analysis, Thenkabail et al. (2004) identified 22 optimal bands (in the 400–2500 nm spectral range) that best characterize and classify vegetation and agricultural crops. Accuracies of over 90% were attained when classifying shrubs, weeds, grasses, and agricultural crop species. Hyperspectral imaging for weed sensing can be categorized into the following typical application areas: mapping invasive weed species (Glenn et al., 2005; Hestir et al., 2008), weed stress characterization (Goel et al., 2003; Karimi et al., 2006; Herrmann et al., 2013; Hadoux et al., 2014), and weed species identification (Koger et al., 2003; Piron et al., 2008).

For invasive weed species mapping, Hestir et al. (2008) used remotely sensed hyperspectral images to map invasive weeds in wetland systems. They reported a moderate mapping accuracy primarily due to significant spectral variation of the mapped invasive species. Samiappan et al. (2017) applied a hybrid approach to wetland plant and weed species mapping. In this approach, laboratory hyperspectral data were first used to classify different wetland species to identify optimal multispectral bands that could discriminate between species. Second, an unmanned aerial vehicle (UAV) was used to take multispectral broadband data over wetlands to map 11 species. The UAV multispectral broadband data achieved an overall accuracy of 75% in a 4-class situation and a 58% overall accuracy in an 11-class situation.

For weed stress characterization, Goel et al. (2003) conducted research using hyperspectral data acquired by a CASI imager to classify the results of four different weed management strategies in corn fields where three different N application rates were also employed. Satisfactory classification results were obtained when one factor (weed or N) was considered at a time. Ground-based weed detection studies had the advantage of using hyperspectral images with both high spectral and high spatial resolutions. The results provided support for upscale detection algorithms for airborne or spaceborne imagers. One ground-based study (Hadoux et al., 2014) on identifying weeds from wheat plants used image spectra data collected at different times and from different locations in a field. It was found that spectral pretreatment with logarithm transformation combined with partial least-squares linear discriminant analysis (PLS-LDA) produced the best results. Herrmann et al. (2013) found it was possible to separate between weeds (grasses and broadleaf weeds) and wheat with high spectral and spatial resolution ground-based hyperspectral images using PLS-LDA. When hyperspectral images were used to identify seedling cabbages and grass weeds (Deng et al., 2015), a spectral angle mapper (SAM) classifier could successfully separate weeds from soil background and cabbage leaves.

As for weed species identification, Vrindts et al. (2002) used reflectance spectra to classify sugar beets, corn, and seven weed species (Figure 1.3). When tested under controlled laboratory conditions, crops and weeds were separated with greater than 97% accuracy using a limited number of wavelength band ratios. In testing under field conditions, over 90% of crop and weed spectra were

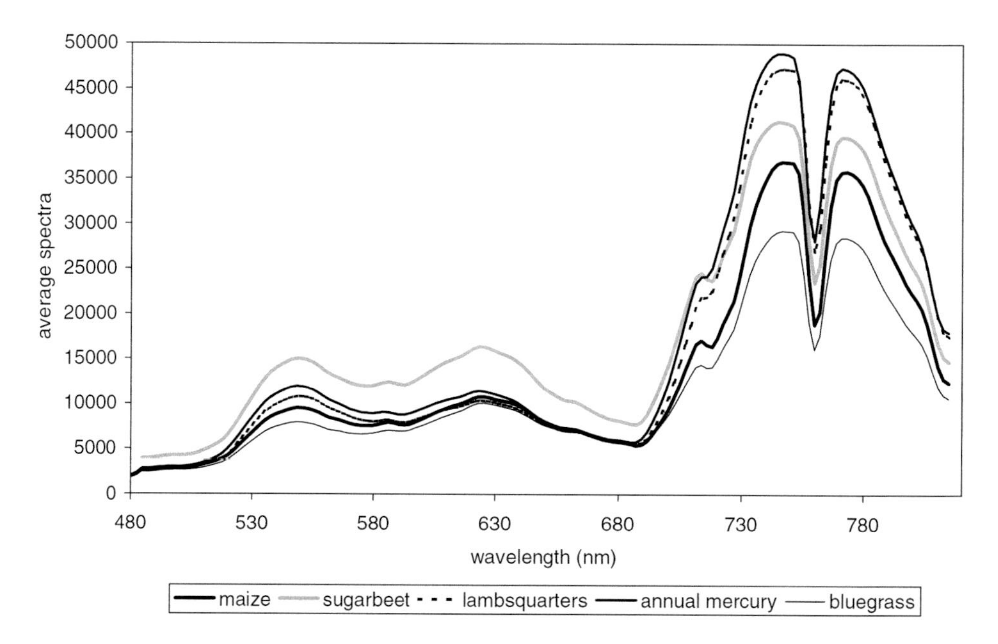

FIGURE 1.3 Reflectance spectra of corn, sugar beet, lambsquarter, annual mercury, bluegrass. (From Vrindts, E. et al. 2002. *Precision Agriculture*, 3, 63–80.)

classified correctly when the model was specific to the prevailing light conditions. Nieuwenhuizen et al. (2010) investigated the use of two spectral sensors that employed spectra of 450–900 and 900–1650 nm for differentiating volunteer potato plants from sugar beets. They found the best classification accuracy was achieved when 10 wavebands in the NIR range were coupled with an ANN algorithm. Zhang and Slaughter (2011) tested the feasibility of identifying six tomato cultivars from two weed species using hyperspectral images in different thermal conditions. The classification accuracy was 92.2% using a global calibration method to cope with four temperature conditions.

1.2.6 Application 3: Crop Herbicide Damage Detection

Detection and assessment of herbicide damage to plants are also important in precision agriculture. One main application of such detection is in the estimation of plant damage due to herbicide drift. Herbicide drift (EPA, 2015) happens when herbicide particles move into off-target areas under weather conditions favorable to drift. It happens more commonly during aerial chemical applications. When this occurs, unwanted plant damage can follow when the herbicide lands on off-target plant surfaces (Figure 1.4). Farmers are interested in determining the severity of unintended injury as early as possible, especially before visible symptoms are observable by the naked eye, in order to take appropriate action to best protect their interests. Thus, it is important to be able to detect the onset of crop injury caused by herbicide drift and preferably be able to determine possible sources and the direction of drift and dosage levels from the drift. The use of hyperspectral data for such an application depends on how the data are collected, that is, whether a fiber optic spectrometer or a hyperspectral imager is used (Yao et al., 2012).

Table 1.3 provides a summary of research on the use of hyperspectral data for herbicide damage detection. Hyperspectral data collected from a fiber optic spectrometer contain information similar to data from a hyperspectral imager. The main difference is that a spectrometer measures only a point that provides no spatial information about the data. Due to their ease of use, visible NIR spectrometers are widely employed in many applications, including herbicide damage assessment (Henry et al., 2004; Huang et al., 2012; Suarez et al., 2017). Since reflectance data are a mixed signal of all the reflectance within a probe's field of view, necessary steps must be taken to ensure that high-quality spectral data are acquired. One data processing method for obtaining pure endmember signals is through spectral unmixing if spectral information from different endmembers is known

FIGURE 1.4 Glyphosate crop injury figure: soybean injured by off-target drift of aerially applied glyphosate.

TABLE 1.3
List of Hyperspectral Applications in Herbicide Damage Detection

Herbicide Damage Application	Hyperspectral Narrowband Wavelength (nm)	Hyperspectral Analysis Technique	Results	Reference
Paraquat injury on corn plants	2,151 bands in 350–2500	Wavelet	Detection with 24 hours after treatment	Henry et al. (2004)
Glyphosate injury on soybean plants	676, 751	NDVI, SAVI, TVI, VNNIR, VNIR	Detection with 24 hours after treatment	Yao et al. (2012)
Glyphosate injury on soybean plants	519, 670, 685, 697	Modified first derivatives	Detection with 4 hours after treatment	Yao et al. (2012)
Glyphosate injury on soybean plants	662, 734	SAVI and DVI	Detection with 24 hours after treatment	Huang et al. (2012)
Glyphosate injury on cotton plants	400–2500	Canonical analysis	Detection 48 h after treatment	Zhao et al. (2014)
2, 4-D injury on cotton plants	400–900	PLS	Detection 2 days after treatment	Suarez et al. (2017)
Dicambo injury on cotton plants	531, 550, 570, 700	ARI, PRI	Detection 24 h after treatment	Huang et al. (2016c)

beforehand. If not, precautions must be taken in the data collection. For example, Huang et al. (2012) avoided the spectral mixture problem by holding the optical fiber directly over the surface of a single leaf. This approach limited the viewing area so reflectance from the entire leaf could not be measured. In a field situation, Huang et al. (2016c) had to exclude some data from analysis due to spectral mixing of crop plants and weeds. Zhang et al. (2010) conducted a study to evaluate the efficacy (or plant damage) of glyphosate in aerial applications. Spectral data were collected with a ground-vehicle-mounted spectrometer. Spectral data were processed using NDVI and analysis of variance (ANOVA). The results indicated at 17 days after treatment (DAT) that the treatment difference was significant, while there was no difference between 1 and 8 DAT. The days after treatment value to observe significant treatment differences was much greater than in other studies. In one study conducted for the purpose of evaluating field glyphosate damage in non-glyphosate-resistant (non-GR) soybean and non-GR cotton plants (Zhao et al., 2014), it was found that herbicide damage could be detected at 72 h after treatment (HAT) with vegetation indices such as NDVI. A physically based leaf radiation transfer model analysis conducted as part of the study further indicated that the damage could be detected at 48 HAT. Huang et al. (2016c) further researched dicamba drift damage on soybean plants in a field study where vegetation-index-based results showed the difficulty of differentiating the dose response of dicamba treatment at 72 HAT. Meanwhile, the difference between treatment groups and control was evident at 24 HAT. A recent work (Suarez et al., 2017) with hyperspectral data on cotton plants applied with the herbicide 2,4-D demonstrated that damage could be identified at 2 DAT.

A hyperspectral imager can provide hyperspectral data with both high spectral and spatial resolutions. It enables spectral data analysis to be carried out only on the pure plant canopy or leaf spectra. A general approach in data preparation is that a segmentation/classification step is used to separate plant pixels from other pixels. Then pure plant spectra can be extracted from the image for later analysis. Yao et al. (2012) implemented this process in a greenhouse-based study to evaluate glyphosate damage to soybean plants. It was found that while the vegetation indices could detect plant injury at 24 HAT, the vegetation indices had difficulty in separating treatment groups with different dosages. Furthermore, a modified derivative analysis developed in the study could potentially detect crop injury at 4 HAT with the first derivatives located at wavelengths of 519, 670,

685, and 697 nm. In addition, the derivatives also demonstrated the ability to differentiate treatment groups with different dosages.

1.2.7 Application 4: Hyperspectral Imagery for Crop Nitrogen Stress Detection

In-season crop nutrient management, such as side dressing applications of N for corn plants, requires crop N content estimation (N stress detection) for the generation of an application map (or prescription map). Many studies have suggested the use of remotely sensed canopy reflectance for crop N detection. Different N levels in plants affect crop chlorophyll concentrations and result in different canopy reflectance spectra (Walburg et al., 1982). For ground-truth estimation, the actual plant N level can be obtained by (1) using as-applied plot N level; (2) chemical measurement of plants tissues, or (3) estimation of the leaf chlorophyll concentration using a field instrument such as the Minolta SPAD meter.

Table 1.4 presents some studies of hyperspectral-based crop N sensing. To estimate corn plant N, Zara et al. (2000) found that the slope of the reflectance spectra between 560 and 580 nm produced the best results for corn N stress detection with AVIRIS aerial hyperspectral images. Three traditional indices (Cassady et al., 2000), NDVI, Photosynthetic Reflectance Index, and Red-Edge Vegetation Stress Index (RVSI), were computed using certain bands from AVIRIS images. It was found that RVSI had the highest correlation with both applied N and measured chlorophyll levels in corn. Boegh et al. (2002) concluded that CASI image green and NIR bands, which are the maximum reflectance bands of chlorophyll, were the most important predictors. Haboudane et al. (2002) worked to develop a combined modeling-based and index-based approach to predicting corn chlorophyll content using CASI imagery. This method used the ratio of an index sensitive to low chlorophyll values to a soil-adjusted index, both calculated from distinct narrow image bands, to build a prediction model. Yao (2004) developed a generic vegetation index generation algorithm and obtained an R^2 of 0.79 for corn plant N estimation. A recent study (Gabriel et al., 2017) indicated that the model R^2 for corn N prediction could be 0.89. In this work, the spatial resolution from the VIS-NIR airborne hyperspectral sensor was 30 $\times$ 30 cm. The results were produced from indices (i.e., Transformed Chlorophyll Absorption Reflectance Index/Optimized Soil-Adjusted Vegetation Index, or TCARI/OSAVI) that combined chlorophyll estimation with canopy structure.

Rice is an important staple food for two-thirds of the world's population. Nguyen et al. (2006) developed regression models for rice plant N estimation with R^2 from 0.76 to 0.87 for validation data. The in-field variation maps from this study are presented in Figure 1.5. Bajwa (2006) stated that PLSR models could explain 47% to 71% of the variability in rice plant N. Additionally, in a 3-year study, a regression model was reported to have an $r = 0.938$ for N estimation in rice (Ryu et al., 2009). Inoue et al. (2012) reported that PLSR prediction of rice N had better results ($R^2 = 0.89$) with 56 selected wavebands (25% of the total) than with all bands. Furthermore, a ratio spectral index (RSI) using derivatives from bands at 740 and 520 nm yielded a prediction R^2 of 0.90. Mahajan et al. (2017) attempted to estimate rice N ($R^2 = 0.80$), phosphorus ($R^2 = 0.69$), and sulfur ($R^2 = 0.73$) content using hyperspectral data collected from an Analytical Spectral Devices (ASD) (ASD Inc., Longmont, CO) field spectrometer with a wavelength range of 350–2500 nm. Unlike other similar studies, this work used reflectance information from the SWNIR region (1000–2500 nm). Research (Onoyama et al., 2015) was also carried out to incorporate both reflectance and growing degree-days (GDD) to account for differences in growing temperature conditions. The PLSR model showed that spectral information could not predict variation in the amount of growth caused by weather variation expressed as GDD.

Similar studies for plant N estimation have been conducted on other crops. Christensen et al. (2004) found that N content in barley could be predicted with 81% accuracy. Xu et al. (2014) worked on malting barley N estimation using an ASD spectrometer and reported a prediction R^2 of 0.82 with a combined model that integrated the first-order derivatives from five wavebands. Zhao et al. (2005) achieved an accuracy of 62.4% in discriminating N in cotton plants with canonical discriminant analysis. Working on cotton, Raper and Varco (2015) used a diode array spectrometer

TABLE 1.4
List of Hyperspectral Applications in Crop Nitrogen Sensing

Nitrogen Sensing Application	Hyperspectral Narrowband Wavelength (nm)	Hyperspectral Analysis Technique	Results	Reference
		Corn		
• Airborne	560, 580	Band slope	$R^2 = 0.97$	Zara et al. (2000)
• Airborne	714, 733, 752	RVSI	$R^2 = 0.82$	Cassady et al. (2000)
• Airborne	550	Regression	$R^2 = 0.78$	Boegh et al. (2002)
• Airborne	550, 670, 700, 800	TCARI/OSAVI	$R^2 = 0.81$	Haboudane et al. (2002)
• Airborne	Selected bands from 400–900	Genetic algorithms selected VIs	$R^2 = 0.79$	Yao (2004)
• Airborne	550, 670, 700, 800	TCARI/OSAVI	$R^2 = 0.89$	Gabriel et al. (2017)
		Rice		
• Airborne	68 bands from 400–1000	PLSR	$R = 0.938$	Ryu et al. (2009)
• Airborne	740 and 520	Ratio of first derivatives	$R^2 = 0.9$	Inoue et al. (2012)
• Ground based	300–1100	PLSR	$R^2 = 0.76, 0.87$	Nguyen et al. (2006)
• Ground based	56 selected bands from 400–1100	PLSR	$R^2 = 0.89$	Inoue et al. (2012)
• Ground based	350–2500	NRI1510	$R^2 = 0.8$	Mahajan et al. (2017)
		Barley		
• Ground based	400–750	PLSR	81% accuracy	Christensen et al. (2004)
• Ground based	496, 499, 689, 797, 882	Regression of first derivatives	$R^2 = 0.82$	Xu et al. (2014)
		Cotton		
• Ground based	400–1000	Canonical discriminant analysis	62.4% accuracy	Zhao et al. (2005)
• Ground based	650, 720, 840	SCCCI	$R^2 = 0.62$	Raper and Varco (2015)
		Potato		
• Airborne	401 to 982	PLSR	$R^2 = 0.79$	Nigon et al. (2015)
• Ground based	710, 750	Ratio index	$R^2 = 0.551$	Jain et al. (2007)
		Wheat		
• Ground based	370, 400	Ratio index	$R^2 = 0.58$	Li et al. (2010)
• Ground based	445, 573, 720, 735	AIVI	$R^2 = 0.84–0.87$	He et al. (2016)

with a wavelength range of 400–850 nm. Their research suggested a simplified canopy chlorophyll content index using bands from red, red-edge, and NIR regions that produced an R^2 of 0.62. Jain et al. (2007) developed a regression model for N estimation in potato plants with R^2 equal to 0.551. Nigon et al. (2015) found potato N estimation could have an R^2 of 0.79. Wheat, a major crop, was also studied widely for its N estimation. Li et al. (2010) studied N content in winter wheat. The R^2 was 0.58 in an experimental field and 0.51 in a farmer's field. Mahajan et al. (2014) worked on simultaneously monitoring nitrogen, phosphorus, sulfur, and potassium content in wheat. The sensor platform was an ASD 350–2500 nm spectrometer. The study incorporated SWNIR bands in the prediction model with an R^2 of 0.47 for phosphorus. He et al. (2016) took a different approach to spectral data collection, where reflectance was measured from different angles of wheat plants.

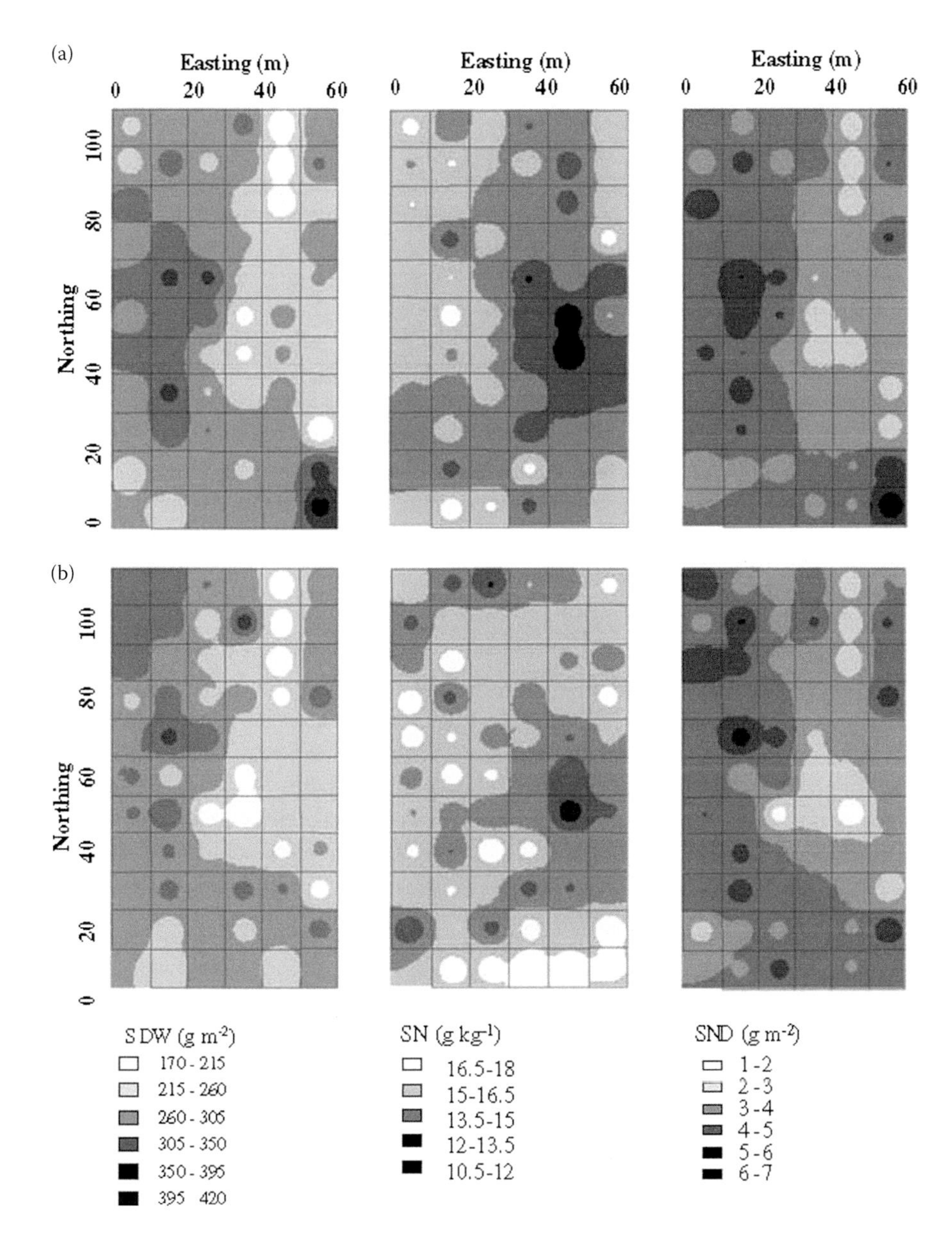

FIGURE 1.5 In-field variation of measured (a) and calculated (b) values by PLSR of shoot dry weight (SDW), shoot N concentration (SN), and shoot N density (SND) in year 2004. The waveband selected by a multiple stepwise linear regression for an N nutrition index (NNI) calculation are 977, 583, 702, and 725 nm. (From Nguyen, H.T. et al. 2006. *Precision Agriculture*, 7, 249–264.)

A novel Angular Insensitivity Vegetation Index (AIVI) based on red-edge, blue, and green bands was introduced. The most consistent prediction ($R^2 = 0.84–0.87$) from AIVI was observed at ±20° (including nadir) with reference to the nadir position.

1.2.8 Application 5: In-Season Crop Yield Estimation

Crop yield prediction or estimation is one of the most important activities in the farming industry. Traditionally, farmers have estimated crop yield for a whole field or for large parts of a field. This approach involves manually counting the number of ears per acre and the number of kernels per ear after the kernel number is established. The total yield of a field can also be predicted with a yield model, using information such as weather, crop type, and up to 170 soil parameters (Villalobos et al., 1996). Utilizing a yield monitor and GPS, crop yields can be measured and stored with detailed location information on the go. In this way, yield maps with both high accuracy and high spatial resolution can be generated (Taylor et al., 2001) to be incorporated into the farm GIS database. For management practices, yield maps can be used to provide input prescriptions for future growing seasons. Although a yield monitor can output yield data with considerable detail in real time, its yield estimation is still a "postharvest" estimation.

One commonly used method for in-season yield monitoring is based on remote sensing images (Ma et al., 1996; Taylor et al., 1997). Different crop yields can cause different crop canopy reflectance. Such reflectance differences can be recorded by images captured from a close distance or by aerial or satellite remote sensing systems. Based on satellite data, regression models (Thenkabail et al., 1994; Hammar et al., 1997) were developed for whole-field yield estimation. Satellite images were also used for an in-field yield variability study (Hayes and Decker, 1996). Table 1.5 below presents some work in using hyperspectral data for crop yield estimation.

TABLE 1.5
List of Hyperspectral Applications in Yield Estimation

Yield Estimation Application	Hyperspectral Narrowband Wavelength (nm)	Hyperspectral Analysis Technique	Results	Reference
		Corn		
• Airborne	Selected bands in 400–900 (including 694, 700, 706)	GAs selected VIs	$R^2 = 0.59$	Yao (2004)
• Airborne	408–947	ANN	$R = 0.76$	Uno et al. (2005)
• Ground based	355–2300	PLSR	$R^2 = 0.92$	Perbandt et al. (2011)
		Sorghum		
• Airborne	457–922	Regression	$R^2 = 0.69, 0.82$	Yang et al. (2004a)
• Airborne	457–922	Spectral unmixing	$R = 0.67, 0.82$	Yang et al. (2007b)
• Airborne	477–844	SAM	$R^2 = 0.60, 0.81$	Yang et al. (2008)
• Airborne	550, 670, 800, 825	MSAVI	$R = 0.79, 0.86$	Yang and Everitt (2012)
		Wheat		
• Spaceborne	400–1000	Crop model based	Yield range matched	Migdall et al. (2009)
• Ground based	350–1050	GA + PLSR	Best estimation at 125 days after planting	Thorp et al. (2017)
• Ground based	630, 750	NDVI	$R = 0.74$	Reyniers et al. (2004)
		Cotton		
• Airborne	457–922	Regression	$R^2 = 0.61, 0.69$	Yang et al. (2004b)
		Onion		
• Ground based	670, 800	TSAVI	$R^2 = 0.67$	Marino and Alvino (2015)
		Rice		
• Ground based	730, 780	NDRE	$r = 0.90$	Kanke et al. (2016)

1.2.8.1 Corn Yield Estimation

Yao (2004) worked on using aerial hyperspectral images for corn yield estimation. The ground-truth yield ranging from 128 to 259 bushels/acre was obtained with a plot harvester for each 50 m^2 experimental plot. The total number of harvested plots was 570. The R^2 value for yield estimation with a vegetation index was 0.59. Uno et al. (2005) also worked on corn yield with aerial hyperspectral imaging. The ground-truth yield was obtained from four 1 m^2 subplots for each 400 m^2 treatment plot (48 total treatment plots). The highest yield prediction using an ANN was $r = 0.76$. The study concluded that there was no clear difference between ANNs and stepwise multiple linear regression models. Perbandt et al. (2011) conducted a study to collect corn plant reflectance data at different angles for yield estimation using a field spectrometer. It was found that the best prediction R^2 (0.92) was with off-nadir measurements rather than with nadir measurements.

1.2.8.2 Grain Sorghum Yield Estimation

Yang et al. (2004a) used aerial hyperspectral images of two Texas fields to estimate grain sorghum yield. The ground-truth yield was collected with a grain combine. It was reported that the sorghum yield significantly correlated with the VIS and NIR bands. Regression analysis using both principal component transformation and stepwise band selection produced similar results. The yield estimation R^2 was from 0.69 to 0.82 for the two fields. In addition, linear spectral unmixing techniques were used to estimate sorghum yield variability using aerial hyperspectral (Yang et al., 2007b) and multispectral broadband (Yang et al., 2007a) images. These techniques were based on the assumption that canopy reflectance is a linear mixture of different spectral components (endmembers) such as soil and sorghum plants. The multispectral broadband image study gave the best correlation coefficient, 0.90. The two-study field with hyperspectral data had the best fraction-based R values (0.67 and 0.82). Yang et al. (2008) further applied a SAM algorithm for sorghum yield estimation. It was concluded that the SAM technique could be used alone or with other vegetation indices for yield estimation with hyperspectral imagery. Yang et al. (2009) also showed that hyperspectral imagery had the potential to improve yield estimation accuracy significantly compared with multispectral imagery. In this application, relative yield variation can be estimated using a modified soil-adjusted vegetation index (MSAVI) image based on one NIR band (800 or 825 nm) and one VIS band (550 or 670 nm) (Yang and Everitt, 2012).

1.2.8.3 Wheat Yield Estimation

In a study using ground-based canopy reflectance for winter wheat yield estimation, it was found that spectral indices, such as infrared/red, the normalized difference (ND), the transformed vegetation index, and the greenness index obtained from flowering to milking stages, gave the best results (Das et al., 1993). The indices were calculated by integrating the reflectance data taken by spectrometer into broadband images. Reyniers et al. (2004) used a line scanner mounted on a tractor for hyperspectral data acquisition. In this study, winter wheat was planted on 60 plots of 12×16 m each. The ground-truth yield data were collected through a plot harvester. Narrowband NDVI calculated at wavelengths of 630 and 750 nm was used for crop coverage measurement. The optically measured crop coverage was positively correlated with grain yield. The correlation coefficient was 0.74 between yield and coverage data. Migdall et al. (2009) also provided a modeling-based approach to simulate winter wheat yield using airborne and spaceborne hyperspectral imagery. Thorp et al. (2017) compared broadband reflectance, narrowband reflectance, and spectral derivatives to estimate durum wheat traits, including LAI, canopy weight, plant N content, grain yield, and grain N content. Reflectance data were collected from a field spectrometer. The study developed a GA to identify the most relevant spectral features for the aforementioned durum wheat trait estimation. Results showed that the GA-based method had the least root-mean-square errors of cross-validation (RMSECV).

1.2.8.4 Other Crops

Rasmussen (1997) used advanced very-high-resolution radiometer (AVHRR) images for millet yield forecasting. The NDVI calculated from the broadband data was used to build a yield regression model. The researcher concluded that millet yield could be measured 1 month before harvest. Yang et al. (2004b) used aerial hyperspectral images to estimate cotton yield on two fields in Texas. Cotton yield ground-truth data were generated using a cotton yield monitor mounted on a cotton picker. To compensate for image resolution and geo-registration errors, both the image data and yield data were aggregated into 8 × 8 m cells. The stepwise regression analysis produced a yield estimation R^2 of 0.61 and 0.69 for the two experimental fields. The yield variation map of one field is presented in Figure 1.6. Pettersson et al. (2006) used a handheld multispectral scanner (400–1000 nm with 10 nm intervals) to collect canopy reflectance data on malting barley. The reflectance data were then used to generate nine vegetation indices. It was reported that all the vegetation indices were significantly correlated with grain yield. The correlation coefficient was 0.9 for the regression models. Marino and Alvino (2015) used field spectrometer measurements to estimate onion yield. For ground-truth establishment, the study applied ordinary kriging, and four productive zones were identified in the field. Among 11 vegetation indices tested in the analysis, it was concluded that the transformed soil-adjusted vegetation index (TSAVI) seemed to be more effective at identifying onion yield spatial variability. In estimating rice yield, Kanke et al. (2016) mentioned that water turbidity affected spectral reflectance when canopy coverage was less than 50%. Results showed that rice grain yield might be more accurately predicted using a normalized difference red-edge-based index (NDRE, $r = 0.90$) and ratio indices rather than red-based normalized and ratio indices.

1.2.8.5 Hyperspectral Imagery for Temporal Yield Analysis

Research has been carried out to establish the temporal relationship between remote sensing imagery and yield. Moran et al. (1997) suggested that remote sensing images from the late growing season had the best results for preharvest crop yield prediction. Vellidis et al. (1999) used an unsupervised classification method and found that a cotton yield pattern could be identified in the early growth stage, within 10 weeks of crop growth, from multispectral broadband color infrared aerial photos. Although the aforementioned multispectral image-based research showed that images obtained from different times and different broadband vegetation indices could be used for yield estimation, problems like when to acquire images and how to properly correlate spectral information from hyperspectral imagery with yield still require much more research.

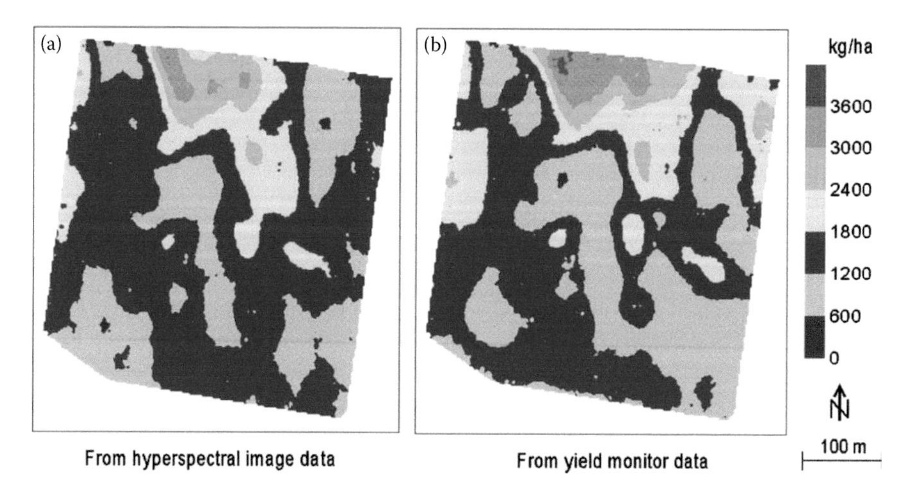

FIGURE 1.6 In-field yield variability maps generated from (a) an airborne hyperspectral image using a nine-band regression model and (b) yield monitor data for a 16 ha cotton field. The nine bands are 499, 546, 601, 702, 717, 738, 771, 778, and 826 nm. (From Yang, C. et al. 2004b. *Precision Agriculture*, 5, 445–461.)

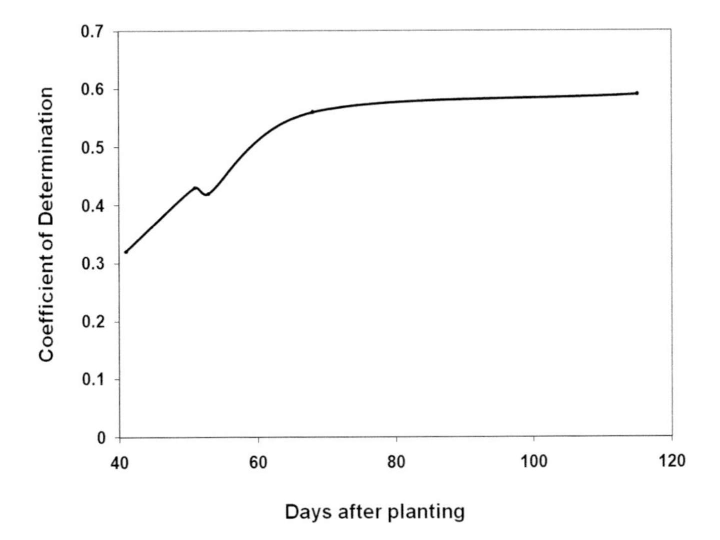

FIGURE 1.7 R^2-values between corn yield and the best calculated vegetation indices, plotted in days after planting (DAP). These indices all included narrowbands either at 694 or 700 nm as selection for red bands. (From Yao, H. 2004. Hyperspectral imagery for precision agriculture. PhD dissertation, University of Illinois at Urbana-Champaign, Urbana, IL.)

The potential for using hyperspectral image data for yield estimation varied based on the date of image data acquisition during the growing season. From a decision-making point of view, it is desirable to know when would be the best time for image data collection. The best time can be determined by estimation accuracy, economic considerations, and other factors. Of these, the ability to accurately estimate crop yield is of utmost importance. Several researchers have worked on using hyperspectral imagery for temporal yield analysis. Goel et al. (2003) showed that the largest correlation between reflectance spectra and crop yield occurred at a tasseling stage. In a study using ground spectrometer data for temporal yield analysis, an ANN model was used (Yang et al., 2002). The researchers concluded that an ANN could be used to predict yield in the early planting stage using ground-spectrometer-based hyperspectral data. Reyniers et al. (2004) pointed out that spectral data from midseason were more related to wheat yield measurements at harvest. Yao (2004) used an index-based approach to estimate corn temporal yield. The aerial hyperspectral image data collected at five different dates were correlated with corn yield. It was found that image bands at around 700 nm (the red-edge region), including 694, 700, and 706 nm, were strong indicators of yield estimation in all five images. The late-season images showed better correlation with the measured corn yield than the early-season images (Figure 1.7). In a recent work, Thorp et al. (2017) found that wheat grain yield could be optimally estimated from ground-based canopy spectral measurements (collected weekly) between 110 and 130 days after planting.

1.2.9 Application 6: Pest and Disease Detection

Changes in plant canopy reflectance due to pest infestation or disease infection can be used for pest and disease detection. Similarly, the concept of precision agriculture is also applicable to plant disease management, which can lead to site-specific pesticide and fungicide applications. Unlike applications discussed earlier, where abiotic plant stress was addressed, detection of pest and disease in the field involves sensing plant biotic stress. Due to the diverse nature of many different pests and diseases, a significant amount of research has been conducted on plant biotic stress detection in recent years (Lowe et al., 2017). According to Mahlein (2016), detection should consider the following factors: (1) early detection of onset of attack, (2) separation of different diseases, (3) differentiation of biotic

TABLE 1.6
List of Hyperspectral Applications in Pest and Disease Detection

Weed Sensing Applications	Hyperspectral Narrowband Wavelength (nm)	Hyperspectral Analysis Technique	Results	Reference
		Pest infestation		
• Aphid on wheat	Multispectral, 675, 800	R800/R675	$R^2 = 0.87$	Mirik et al. (2012)
• Aphid on wheat	350–2500	MLR + wavelet	$R^2 = 0.72$	Luo et al. (2013)
• Aphid on cabbage	694, 742	LS-SVM	>70% accuracy	Zhao et al. (2017)
• Spider mite	57 bands in 459–1002	Spectral unmixing	Infestation detected	Fitzgerald et al. (2004)
• Leaf hopper	550, 691, 715, 761	LHI-4	$R^2 = 0.825$	Prabhakar et al. (2011)
• Planthopper	543, 568, 602	LS-SVM	90% accuracy	Zhao et al. (2012)
• Mealybug	492, 550, 674, 768, 1454	MSIs	$R^2 = 0.82$	Prabhakar et al. (2013)
		Bacterial		
• Bacterial leaf blight	745	Regression	$R^2 = 0.978$	Yang (2010)
		Fungal infection		
• *Fusarium* in wheat	17 bands in 665–675 and 550–560	Head blight index	67% accuracy	Bauriegel et al. (2011a)
• *Fusarium* in wheat	450–700	PLS-LDA	86% accuracy at 8 days after inoculation	Menesatti et al. (2013)
• Gray mold disease in tomato	Selected bands in 400–780	KNN	>94% accuracy	Xie et al. (2017)
• *Cercospora* leaf spot in sugar beet	570, 698, 734	*Cercospora* leaf index	92%	Mahlein et al. (2013)
• Sugar beet rust	513, 570, 704	Sugar beet rust index	87%	Mahlein et al. (2013)
• Powdery mildew in sugar beet	520, 584, 724	Powdery mildew index	85%	Mahlein et al. (2013)
• Yellow rust in wheat	531, 570	NDVI	$R^2 = 0.91$	Huang et al. (2007)
• Leaf rust in wheat	455, 605, 695	Leaf Rust Disease Severity Indices	$R^2 = 0.94$	Ashourloo et al. (2014)

and abiotic stress, and (4) disease severity quantification. Challenges also exist owing to interactions between biotic and abiotic stresses. Here we summarize hyperspectral-based detection applications (Table 1.6) based on stress type, that is, pest infestation such as insect attack and biopathogen infection such as fungi and rust infection.

1.2.9.1 Pest Infestation

In this category, many studies have focused on aphid detection. Mirik et al. (2012) found robust relationships between Russian wheat aphid feeding damage and multispectral broadband vegetation indices. The coefficients of determination (R^2) were from 0.62 to 0.90 for irrigated wheat, from 0.50 to 0.87 for dryland wheat, and from 0.84 to 0.87 for a greenhouse experiment. Luo et al. (2013) used ASD FieldSpec spectroradiometer measurements (350–2500 nm) to quantify leaf aphid density. Results showed that the multivariate linear regression model based on five wavelet features had a better R^2 (0.72) than results based on six sensitivity spectral indices ($R^2 = 0.56$). Another study (Yuan et al., 2014) attempted to discriminate yellow rust (*Puccinia striiformis f. sp. Tritici*), powdery mildew (*Blumeria graminis*), and wheat aphid with field spectrometer measurements at the leaf level during the early grain filling stage. Fisher's linear discriminant analysis indicated an overall accuracy of 0.75. Also at the field level Zhao et al. (2017) conducted a study using a hyperspectral imager to detect aphid infestation on Chinese cabbage plants. The results revealed that reflectance and textural features at 694 and 742 nm were the most important in distinguishing aphid infestation.

Other pests under investigation included spider mite (Fitzgerald et al., 2004), leafhopper (Prabhakar et al., 2011), planthopper (Zhao et al., 2012), and mealybug (Prabhakar et al., 2013). The cotton-plant-feeding strawberry spider mite causes leaf puckering and reddish discoloration in early stages of infestation, followed by leaf drop. Fitzgerald et al. (2004) successfully distinguished between adjacent mite-free and mite-infested cotton field areas by implementing a spectral unmixing process on AVIRIS imagery. Prabhakar et al. (2011) pointed out that there was a significant difference in VIS and NIR regions between healthy and leafhopper-infested plants. Spectral data were collected with a field spectrometer, and the new leafhopper indices showed high coefficients of determination across locations ($R^2 = 0.521–0.825$). Zhao et al. (2012) worked on the detection of injury severity of rice plants caused by brown planthoppers. The results demonstrated that the combination of 16 wavelengths had a classification accuracy of 98% using a least-squares support vector machine (LS-SVM) model.

1.2.9.2 Bacterial

Bacterial leaf blight is a vascular disease of irrigated rice. Its serious infestations might result in up to 50% yield loss. Yang (2010) found that changes in leaf color and appearance were caused by different levels of disease severity, reflected in reflectance spectra differences. The model for infestation area estimation of the highly susceptible cultivar had R^2 equal to 0.978 based on a single spectral band (745 nm).

1.2.9.3 Fungal Infection

Another important disease for many agricultural crops is fungal pathogen infection. Much research has been carried out in this area, including on *Fusarium* in wheat, late blight in tomato plants, fungal pathogens in sugar beets, and rust in wheat.

Infection of *Fusarium* in wheat plants can cause serious yield loss and the production of mycotoxins (Bauriegel et al., 2011a). Researchers have used hyperspectral images for the detection of *Fusarium* infection (head blight) in wheat (Bauriegel et al., 2011a,b). The hyperspectral images were acquired with a push-broom, line-scanning imaging system (400–1000 nm) in ground-based studies. Rather than collecting canopy reflectance commonly seen in other remote sensing research, single wheat ears were imaged (Figure 1.8). The fungal infection was introduced manually, with

FIGURE 1.8 Examples of image classification of *Fusarium*-infected wheat ears. (a) RGB image, (b) grayscale image according to head blight index, (c) classification result (dark gray/red: diseased, light gray/green: healthy). (From Bauriegel, E. et al. 2011a. *Computers and Electronics in Agriculture*, 75, 304–312.)

time-series hyperspectral images subsequently acquired. The study found that the head blight could be detected using the hyperspectral imaging method at 7 days after inoculation. The best time for detection was at the beginning of the medium milk stage, and it was difficult to distinguish at the beginning of ripening. A head blight index, which used bands in ranges of 665–675 nm and 550–560 nm, was developed by Bauriegel et al. (2011a). Menesatti et al. (2013) used both laboratory and in-field conditions for *Fusarium* head blight detection. A PLS-LDA method was used to process the hyperspectral images. The results showed that the onset of infection could be identified with 86% accuracy at 8 days after inoculation in laboratory conditions, while at the same growth stages, the in-field data had a lower accuracy, 63.6%–77%.

Tomato plant late blight caused by the fungal pathogen *Phytophthora infestans* is a serious disease (Zhang et al., 2005). Zhang et al. (2003) used hyperspectral remote sensing for the detection of stress in tomatoes induced by late blight disease on a large scale in the field. The study used an AVIRIS image with 224 bands within a wavelength range of 400–2500 nm acquired during the growing season. The results revealed, based on field reflectance samples measured with a spectrometer, that the NIR region, especially 700–1300 nm, was much more valuable than the VIS region for the detection of the disease. The disease level was divided into four levels, from light symptoms to severe damage. The classification results with the SAM method showed that the late blight diseased tomatoes at stage 3 or above could be distinguished from healthy plants. Meanwhile, the less infected plants at stage 1 or 2 were difficult to distinguish from the healthy plants. Zhang et al. (2005) also analyzed field spectrometer data to study late blight disease. The spectral data showed that the diseased tomatoes could be distinguished from healthy ones before economic damage happened. Xie et al. (2017) used a hyperspectral imager to detect another tomato plant disease, gray mold disease, with spot inoculation in tomato leaves in greenhouse conditions. The detection accuracy for distinguishing healthy from diseased leaves exceeded 94%. As a comparison, early detection accuracy at one day after inoculation was 66.7%.

Sugar beet leaf diseases may cause losses in sugar yield (Rumpf et al., 2010). Mahlein et al. (2010) explored the potential of using hyperspectral data to detect and differentiate three fungal leaf diseases in sugar beets. The research used a spectrometer to measure reflectance spectra (400–1050 nm) of leaves infected with the sugar beet fungal pathogens *Cercospora beticola*, *Erysiphe betae*, and *Uromyces betae*, which caused *Cercospora* leaf spot, powdery mildew, and rust. The vegetation indices evaluated, NDVI, Anthocyanin Reflectance Index, and modified Chlorophyll Absorption Integral, varied in their ability to assess the different diseases at an early stage of disease development, or even before the first symptoms became visible. The conclusion the researchers drew was that the use of spectral vegetation indices for the differentiation of these three sugar beet diseases was possible when using a combination of two or more indices. Rumpf et al. (2010) focused on using SVM classification to differentiate healthy and the aforementioned three types of diseased leaves. The detection accuracy was 97% between healthy and all diseased leaves and 86% in classifying healthy and three individual disease types. Furthermore, Mahlein et al. (2013) used a hyperspectral imager for the detection of these diseases. Four disease-specific indices developed in the research yielded classification accuracies of 89%, 92%, 87%, and 85% for healthy leaves and leaves infected with *Cercospora* leaf spot, sugar beet rust, and powdery mildew, respectively.

There are three types of wheat rust disease: yellow rust, leaf rust, and stem rust (Ashourloo et al., 2014). Their occurrence is a serious threat to wheat yield and grain quality. Huang et al. (2007) used wavelengths of 531 and 570 nm to calculate a normalized vegetation index for yellow rust detection in wheat. The infestation ground truth was interpreted as a disease index through an independent field measurement and assessment. The resulting R^2 was 0.91 between the disease index and the vegetation index generated from airborne data. In another study, Yuan et al. (2013) used a spectrometer to measure leaf reflectance of inoculated leaves for yellow rust detection at joint (early) and grain filling (later) stages. The PLSR models showed a better performance at the later stage, with $R^2 = 0.85$, than the jointing stage ($R^2 = 0.49$). Using a wavelet analysis, Zhang et al. (2014) also found the later stage had better detection of the disease with $R^2 = 0.89$. The early-stage prediction was also improved with $R^2 = 0.69$ using wavelet analysis. Ashourloo et al. (2014) researched wheat leaf rust detection with spectrometer data. It was found that the two Leaf Rust Disease Severity Indices developed could produce an R^2 of 0.94 for disease estimation. Spectral bands used for the Leaf Rust Disease Severity Indices were 455, 605, and 695 nm.

1.2.10 Current Trend: UAV in Precision Agriculture

UAVs provide a unique platform (Figure 1.9) for precision agriculture by mounting portable sensors to remotely sense crop fields at low altitude, from 10 to 100 m typically. In precision agriculture the UAVs that are used are fixed-wing airframes, helicopters, and multirotors (e.g., quadrotor, hexcopter, octocopter). The portable sensors used include digital cameras [such as GoPro action camera (GoPro, San Mateo, CA, USA)], multispectral broadband cameras [such as Tetracam ADC and MCA (Tetracam, Inc., Chatsworth, CA, USA), Micasense RedEdge and Sequoia (Micasense, Seattle, WA, USA)], thermal cameras [such as FLIR Vue Pro 640 (FLIR Systems, Wilsonville, OR, USA) and ICI 9000 (ICI, Beaumont, TX, USA)], hyperspectral cameras [such as cameras developed by BaySpec, Inc. (San Jose, CA, USA) and Rikola Ltd. (Oulu, Finland)], and some other inexpensive Light Detection and Ranging (LIDAR) and Synthetic Aperture Radar (SAR) sensors.

Digital cameras, multispectral broadband cameras, and thermal cameras can be operated as stand-alone and used on any UAVs, from fixed-wing and helicopters to multirotors. Although the body of a portable hyperspectral camera can be small, the camera must attach to a portable computer to work. Also, if the camera is line-scanned, a helicopter or multirotor is a better choice to carry it than a fixed-wing airframe because of the lesser air disturbance with the UAV.

Hyperspectral sensors used on UAVs are frame-based or line-scanned. Frame-based sensors provide full 2D images at every exposure, enabling hyperspectral stereo photogrammetry in UAVs without much geometric distortion. However, frame-based hyperspectral sensors provide much less information than lined-scanned ones. For example, BaySpec, Inc. developed a frame-based snapshot hyperspectral camera, OCI-UAV-2000, and a line-scanned push-broom one, OCI-UAV-1000. In a spectral range of 600–1000 nm, the frame-based camera only provides 20–25 bands, while the line-scanned one provides 100 bands. However, aerial line-scanned hyperspectral sensors always face the issue of geometric distortion, which must be fixed before the data can be used.

FIGURE 1.9 UAV remote sensing over a soybean field.

TABLE 1.7
Multispectral and Hyperspectral UAV Plant Sensing

	Multispectral	Hyperspectral
Spectral bands	Blue, Green, Red, Red Edge, NIR	20–100 bands in 400–1000 nm
Information amount	Low	High
Data amount	Relatively small	At least 5 times more
	Applications	
Detection of weeds	Big weed patches	Small weed patches and species
Detection of pests	Big pest infestations	Small pest infestations
Detection of diseases	Big disease infestations	Small disease infestations
Detection of water deficiency	Very limited	Limited
Detection of crop vigor	Works fairly well	Works well
Crop yield prediction	Works well	Works well

Multispectral and hyperspectral systems in the VIS-NIR spectral range are used to sense plants of weeds and crops in the field. Broadband multispectral cameras can be used to monitor plant growth vigor, while hyperspectral sensors can be used to detect subtle changes in plant spectra in narrowbands to differentiate species and specify stress from pests and diseases. Table 1.7 explains the difference between multispectral and hyperspectral systems in plant sensing in the context of UAVs at low altitude for precision agriculture. The table shows the advantages of hyperspectral UAV systems. With the high resolution of the imagery (5 mm–5 cm per pixel), they might be able to detect small weed patches and even identify weed species. They might identify small pest and disease infestations, but they are still limited in the detection of water stress in comparison with thermal imaging. However, data quantities from hyperspectral sensors are huge compared with data from multispectral broadband sensors. Thus, it is appropriate that, through hyperspectral data processing, specific sensitive bands are identified, and then, based on the identified bands, narrowband multispectral sensors can be developed. Tetracam MCA, Micasense RedEdge, and Sequoia are examples of narrowband multispectral cameras that are used on UAVs.

Digital color and multispectral sensors are widely used on UAVs for precision agriculture (Zhang and Kovacs, 2012; Honkavaara et al., 2013; Huang et al., 2016b, 2017). Hyperspectral sensors need more operational, data processing efforts on UAVs so their application is still limited. Even so, the potential of portable hyperspectral sensors on UAVs is promising in terms of enhancing plant sensing in precision agriculture (Honkavaara et al., 2012; Zhang and Kovacs, 2012; Huang et al., 2017), with numerous industrial developments from BaySpec, Rikola, HeadWall (Headwall Photonics, Inc., Bolton, MA, USA), and Cubert GmbH (Cubert GmbH Real Time Spectral Imaging, Ulm, Germany).

Honkavaara et al. (2013) conducted a study on using a small UAV imaging system to carry a portable frame-based hyperspectral sensor with a customized digital camera for NIR imaging (both are less than 500 g) for biomass estimation on farmland with wheat and barley plots. With the application, a data processing chain was developed for the production of high-density point clouds and hyperspectral reflectance image mosaics, which were used as inputs in the precision agricultural applications.

Zarco-Tejada et al. (2012) conducted a study on the remote detection of water stress in a citrus orchard where UAV airborne data of thermal and hyperspectral imagery were acquired at the time of maximum stress difference among treatments, prior to the rewatering phase. The hyperspectral imagery was acquired at 40 cm resolution and 260 spectral bands in the 400–885 nm spectral range at 6.4 nm full width at half maximum spectral resolution and 1.85 nm sampling interval, enabling the identification of pure crowns for extracting radiance and reflectance hyperspectral spectra from each tree. With hyperspectral imagery a FluorMOD model (Miller et al., 2003; Pedrós et al., 2004) was used to retrieve chlorophyll fluorescence by applying the Fraunhofer line depth (FLD) principle (Damm et al., 2011) using three spectral bands to track stress levels at stomatal conductance and water potential.

Huang et al. (2017) is undertaking a research project to map the distribution of naturally grown glyphosate-resistant (GR) weeds in a soybean field. Hyperspectral plant sensing techniques have been developed to detect GR Palmer amaranth (*Amaranthus palmeri* S. Wats.) and Italian ryegrass [*Lolium perenne* L. ssp. *multiflorum* (Lam.) Husnot] with accuracies of 90% and 80%, respectively (Lee et al., 2014; Reddy et al., 2014). However, hyperspectral weed sensing in the lab or field is time consuming and laborious. For example, in fields, sensors are either mounted on a tractor for imaging or hand-held by a technician to measure plant spectra in certain areas. This manner of hyperspectral data acquisition is tedious, restricted by field conditions, and becomes an obstacle to transferring the research findings to practical use. Application of UAVs is a promising alternative to flying over a crop field to rapidly map the distribution of GR weeds for precision, site-specific weed management. In this project, portable multispectral and hyperspectral cameras were mounted on small UAVs to fly at a very low altitude (10–20 m) to quickly determine the distribution of naturally grown GR weeds, mainly Palmer amaranth and barnyard grass [*Echinochloa crus-galli* (L.) P.Beauv], over a soybean field.

1.3 CONCLUSIONS

Precision farming practices such as variable rate applications of fertilizer and agricultural chemicals require accurate field variability mapping. The value of hyperspectral remote sensing in providing useful information for seven applications of precision farming was elucidated. Such information included (a) soil property detection, (b) weed detection, (c) herbicide drift sensing, (d) N stress detection, (e) crop herbicide damage detection, (f) crop yield estimation, and (g) pest and disease detection. When using remotely sensed hyperspectral data for soil management zone delineation, it was found that there were different sensitive regions in the electromagnetic spectrum (400–1400 nm) for different soil nutrient properties. In addition, the combination of geostatistical techniques and remote sensing data holds great potential for soil nutrient mapping. For selective weed control, canopy reflectance in the spectral region from 450 to 900 nm, with an emphasis on the region from red to NIR, was demonstrated to be important for weed and crop differentiation. Successful classification between broadleaved plants and grasses using reflectance spectra has been reported in the literature. This could potentially lead to smart herbicide application systems that can correctly deliver broadleaf-specific or grass-specific herbicides to correct weed patch targets, resulting in improved weed control efficacy and reduction of herbicide usage. Herbicide drift is a situation where an applied herbicide lands on off-target crops on windy days. Research has shown that crop injury due to herbicide drift can be detected with hyperspectral imaging before visible symptoms are observed by humans. This detection information can be used by farmers to take early action when a drift incident occurs.

Canopy reflectance has also been used in plant N stress sensing and yield estimation, as well as pest and disease detection. In plant N stress sensing applications, the most significant spectral region is the VIS to NIR region. Many vegetation indices have been developed for N stress detection. Among them the most frequently cited index is the NDVI. The prediction accuracy varied among different crops and even for the same crop. For example, the coefficient of determination could be as low as 0.47 or as high as 0.87 for N in rice plants. Such variability could stem from many different sources, such as plant cultivar, planting date and growth stages, local environment, weather conditions, sensor platforms, sensor calibration, and data processing. For crop yield estimation, it is generally regarded that canopy reflectance measured in the middle to late growing season gives the best prediction results. Many studies have used vegetation indices and explored the VIS-NIR region for different crops. The subsequently generated yield map is one of the most important maps for precision farming practices. Lastly, we summarized another potential precision farming application using hyperspectral data, which included research on the detection and assessment of plant biotic stress, such as insect invasion, onset of disease, and fungal pathogen infection. A significant amount of research has been done on the aforementioned plant biotic stress detection in recent years due to the diverse nature of many different agricultural pests and diseases.

In closing, hyperspectral remotely sensed data can be an important source of field variability sensing and information extraction. This is a crucial step for the implementation of precision farming technology, which also consists in management decision making, precision field operation control, and results assessment. A great deal of research has focused on the use of canopy reflectance for N stress detection and crop yield estimation. The red-edge region where the spectrum shifts from the red to the NIR region was found to be important in plant canopy reflectance, as the reflectance differences at these wavelengths reflect the differences in internal leaf tissue (mesophyll) structures. It was also concluded that three aspects of hyperspectral data, including spatial, spectral, and temporal issues, are all important in precision farming applications. With continuous improvement in compact hyperspectral imagers, more UAV-based imaging systems will be available for precision agriculture applications. This development will greatly improve the three aforementioned issues associated with hyperspectral data.

The use of relative reflectance has been reported in many studies. The use of relative reflectance values can minimize the influence of variable lighting conditions and the effects of relative positioning variation between sensor and object. However, outdoor variable lighting (in particular direct sunlight conditions) still poses challenges for developing practicable spectral systems for robust outdoor plant sensing. It became obvious that hyperspectral data obtained from different ground-based, airborne, and spaceborne sensors for precision farming were generally calibrated in different ways. For this reason, it is important that all future studies involving hyperspectral imaging calibrate image sensors to produce data in standard radiometric units in order to establish a calibrating standard across all sensors. More applications of hyperspectral data in precision farming are expected with better data availability and improved data quality.

Finally, one common issue related to hyperspectral remote sensing is the spectral signature mixture of different target components. The implementation of different unmixing techniques would help to extract canopy spectral reflectance information for better data analysis. Another potential direction is the use of geostatistical techniques. We reviewed studies that investigated combining such techniques with soil reflectance for soil nutrient mapping. Similar approaches could be extended to plant reflectance to improve data interpretation.

REFERENCES

Ashourloo, D., Mobasheri, M.R., and Huete, A. 2014. Developing two spectral disease indices for detection of wheat leaf rust (Pucciniatriticina). *Remote Sensing*, 6, 4723–4740.

Bajwa, S.G., and Tian, L.F. 2005. Soil fertility characterization in agricultural fields using hyperspectral remote sensing. *Transactions of the ASAE*, 48, 2399–2406.

Bajwa, S.G. 2006. Modeling rice plant nitrogen effect on canopy reflectance with partial least square regression (plsr). *Transactions of the ASAE*, 49, 229–237.

Bauriegel, E., Giebel, A., Geyer, M., Schmidt, U., and Herppich, W.B. 2011a. Early detection of fusarium infection in wheat using hyper-spectral imaging. *Computers and Electronics in Agriculture*, 75, 304–312.

Bauriegel, E., Giebel, A., and Herppich, W.B. 2011b. Hyperspectral and chlorophyll fluorescence imaging to analyse the impact of fusarium culmorum on the photosynthetic integrity of infected wheat ears. *Sensors*, 11, 3765–3779.

Ben-Dor, E., and Banin, A. 1995. Near-infrared analysis as a rapid method to simultaneously evaluate several soil properties. *Soil Science Society of America Journal*, 59, 364–372.

Ben-Dor, E., Patkin, K., Banin, A., and Karnieli, A. 2002. Mapping of several soil properties using dais-7915 hyperspectral scanner data—A case study over clayey soils in Israel. *International Journal of Remote Sensing*, 23, 1043–1062.

Bilgili, A.V., Akbas, F., and van Es, H.M. 2011. Combined use of hyperspectral VNIR reflectance spectroscopy and kriging to predict soil variables spatially. *Precision Agriculture*, 12, 395–420.

Blackburn, G.A. 1998. Quantifying chlorophylls and caroteniods at leaf and canopy scales: An evaluation of some hyperspectral approaches. *Remote Sensing of Environment*, 66, 273–285.

Boegh, E., Soegaard, H., Broge, N., Hasager, C.B., Jensen, N.O., Schelde, K., and Thomsen, A. 2002. Airborne multispectral data for quantifying leaf area index, nitrogen concentration, and photosynthetic efficiency in agriculture. *Remote Sensing of Environment*, 81, 179–193.

Broge, N.H., and Leblanc, E. 2000. Comparing prediction power and stability of broad-band and hyperspectral vegetation indices for estimation of green leaf area index and canopy chlorophyll density. *Remote Sensing of Environment*, 76, 156–172.

Broge, N.H., and Mortensen, J.V. 2002. Deriving green crop area index and canopy chlorophyll density of winter wheat from spectral reflectance data. *Remote Sensing of Environment*, 81, 45–57.

Casa, R., Castaldi, F., Pascucci, S., and Pignatti, S. 2012. Potential of hyperspectral remote sensing for field scale soil mapping and precision agriculture applications. *Italian Journal of Agronomy*, 7, 331–336.

Cassady, P.E., Perry, E.M., Gardner, M.E., and Roberts, D.A. 2000. Airborne hyperspectral imagery for the detection of agricultural crop stress. In: *Proceeding of SPIE*, 4151, pp. 197–204.

Christensen, L.K., Bennedsen, B.S., Jørgensen, R.N., and Nielsen, H. 2004. Modelling nitrogen and phosphorus content at early growth stages in spring barley using hyperspectral line scanning. *Biosystems Engineering*, 88, 19–24.

Damm, A., Erler, A., Hillen, W., Meroni, M., Schaepman, M.E., Verhoef, W., and Rascher, U. 2011. Modeling the impact of spectral sensor configurations on the FLD retrieval accuracy of sun-induced chlorophyll fluorescence. *Remote Sensing of Environment*, 115, 1882–1892.

Das, D.K., Mishra, K.K., and Kalra, N. 1993. Assessing growth and yield of wheat using remotely-sensed canopy temperature and spectral indices. *International Journal of Remote Sensing*, 14, 3081–3092.

Daughtry, C.S.T., Walthall, C.L., Kim, M.S., Colstoun, E.B.D., and McMurtrey, J.E. 2000. Estimating corn leaf chlorophyll concentration from leaf and canopy reflectance. *Remote Sensing of Environment*, 74, 229–239.

Deng, W., Huang, Y., Zhao, C., and Wang, X. 2015. Identification of seedling cabbages and weeds using hyperspectral imaging. *International Journal of Agricultural and Biological Engineering*, 8, 65–72.

DeTar, W.R., Chesson, J.H., Penner, J.V., and Ojala, J.C. 2008. Detection of soil properties with airborne hyperspectral measurements of bare fields. *Transactions of the ASAE*, 51, 463–470.

Eddy, P.R., Smith, A.M., Hill, B.D., Peddle, D.R., Coburn, C.A., and Blackshaw, R.E. 2013. Weed and crop discrimination using hyperspectral image data and reduced bandsets. *Canadian Journal of Remote Sensing*, 39, 481–490.

Elvidge, C.D., and Chen, Z. 1994. Comparison of broad-band and narrow-band red and near-infrared vegetation indices. *Remote Sensing of Environment*, 54, 38–48.

EPA. 2015. Introduction to pesticide drift. Accessed October 10, 2017, https://www.epa.gov/reducing-pesticide-drift/introduction-pesticide-drift.

Filella, I., Serrano, L., Serra, J., and Penuelas, J. 1995. Evaluating wheat nitrogen status with canopy reflectance indices and discriminant analysis. *Crop Science*, 35, 1400–1405.

Fitzgerald, G.J., Maas, S.J., and Deter, W.R. 2004. Spider mite detection and canopy component mapping in cotton using hyperspectral imagery and spectral mixture analysis. *Precision Agriculture*, 5, 275–289.

Gabriel, J.L., Zarco-Tejada, P.J., López-Herrera, P.J., Pérez-Martín, E., Alonso-Ayuso, M., and Quemada, M. 2017. Airborne and ground level sensors for monitoring nitrogen status in a maize crop. *Biosystems Engineering*, 160, 124–133.

Ge, Y., and Thomasson, J.A. 2006. Wavelet incorporated spectral analysis for soil property determination. *Transactions of the ASAE*, 49, 1193–1201.

Ge, Y., Thomasson, J.A., Morgan, C.L., and Searcy, S.W. 2007. VNIR diffuse reflectance spectroscopy for agricultural soil property determination based on regression-kriging. *Transactions of the ASAE*, 50, 1081–1092.

Ge, Y., Thomasson, J.A., and Sui, R. 2011a. Remote sensing of soil properties in precision agriculture: A review. *Frontiers of Earth Science*, 5, 229–238.

Ge, Y., Thomasson, J.A., Sui, R., and Wooten, J. 2011b. Regression-kriging for characterizing soils with remotesensing data. *Frontiers of Earth Science*, 5, 239–244.

Glenn, N.F., Mundt, J.T., Weber, K.T., Prather, T.S., Lass, L.W., and Pettingill, J. 2005. Hyperspectral data processing for repeat detection of small infestations of leafy spurge. *Remote Sensing of Environment*, 95, 399–412.

Goel, P.K., Prasher, S.O., Patel, R.M., Landry, J.A., Bonnell, R.B., Viau, A.A., and Miller, J.R. 2003. Potential of airborne hyperspectral remote sensing to detect nitrogen deficiency and weed infestation in corn. *Computers and Electronics in Agriculture*, 38, 199–124.

Gong, P., Ru, R.L., and Biging, G.S. 2003. Estimation of forest leaf area index using vegetation indices derived from Hyperion hyperspectral data. *IEEE Transactions on Geoscience and Remote Sensing*, 41, 1355–1362.

Haboudane, D., Miller, J.R., Tremblay, N., Zarco-Tejada, P.J., and Dextraze, L. 2002. Integrated narrow-band vegetation indices for prediction of crop chlorophyll content for application to precision agriculture. *Remote Sensing of Environment*, 18, 416–426.

Hadoux, X., Gorretta, N., Roger, J.-M., Bendoula, R., and Rabatel, G. 2014. Comparison of the efficacy of spectral pre-treatments for wheat and weed discrimination in outdoor conditions. *Computers and Electronics in Agriculture*, 108, 242–249.

Hammar, D., Ferencz, C., Lichtenberger, J., Tarcsai, G., and Ferencz-Arkos, I. 1997. Yield estimation for corn and wheat in the Hungarian great plain using LANDSAT MSS data. *International Journal of Remote Sensing*, 17, 1689–1699.

Hayes, M.J., and Decker, W.L. 1996. Using NOAA AVHRR data to estimate maize production in the United State corn belt. *International Journal of Remote Sensing*, 17, 3189–3200.

He, L., Song, X., Feng, W., Guo, B.-B., Zhang, Y.-S., Wang, Y.-H., Wang, C.-Y., and Guo, T.-C. 2016. Improved remote sensing of leaf nitrogen concentration in winter wheat using multi-angular hyperspectral data. *Remote Sensing of Environment*, 174, 122–133.

Hedley, C. 2015. The role of precision agriculture for improved nutrient management on farms. *Journal of the Science of Food and Agriculture*, 95, 12–19.

Henry, W.B., Shaw, D.R., Reddy, K.R., Bruce, L.M., and Tamhankar, H.D. 2004. Remote sensing to detect herbicide drift on crops. *Weed Technology*, 18, 358–368.

Herrmann, I., Shapira, U., Kinast, S., Karnieli, A., and Bonfil, D.J. 2013. Ground-level hyperspectral imagery for detecting weeds in wheat fields. *Precision Agriculture*, 14, 637–659.

Hestir, E.L., Khanna, S., Andrew, M.E., Santos, M.J., Viers, J.H., Greenberg, J.A., Rajapakse, S.S., and Ustin, S.L. 2008. Identification of invasive vegetation using hyperspectral remote sensing in the California Delta ecosystem. *Remote Sensing of Environment*, 112, 4034–4047.

Hively, W.D., McCarty, G.W., Reeves, J.B., Lang, M.W., Oesterling, R.A., and Delwiche, S.R. 2011. Use of airborne hyperspectral imagery to map soil properties in tilled agricultural fields. *Applied and Environmental Soil Science*, 2011, 13.

Honkavaara, E., Kaivosoja, J., Mäkynen, J., Pellikka, I., Pesonen, L., Saari, H., Salo, H., Hakala, T., Marklelin, L., and Rosnell, T. 2012. Hyperspectral reflectance signatures and point clouds for precision agriculture by light weight UAV imaging system. *ISPRS Annals of Photogrammetry, Remote Sensing and Spatial Information Sciences*, 353–358.

Honkavaara, E., Saari, H., Kaivosoja, J., Pölönen, I., Hakala, T., Litkey, P., Mäkynen, J., and Pesonen, L. 2013. Processing and assessment of spectrometric, stereoscopic imagery collected using a lightweight UAV spectral camera for precision agriculture. *Remote Sensing*, 5, 5006.

Hu, G., K.A. Sudduth, He, D., Brenton Myers, D., and Nathan, M.V. 2016. Soil phosphorus and potassium estimation by reflectance spectroscopy. *Transactions of the ASABE*, 59, 97.

Huang, W., Lamb, D.W., Niu, Z., Zhang, Y., Liu, L., and Wang, J. 2007. Identification of yellow rust in wheat using in-situ spectral reflectance measurements and airborne hyperspectral imaging. *Precision Agriculture*, 8, 187–5197.

Huang, Y., Thomson, S.J., Molin, W.T., Reddy, K.N., and Yao, H. 2012. Early detection of soybean plant injury from glyphosate by measuring chlorophyll reflectance and fluorescence. *Journal of Agricultural Science*, 4, 117–124.

Huang, Y., Lee, M., Thomson, S., and Reddy, K. 2016a. Ground-based hyperspectral remote sensing for weed management in crop production. *International Journal of Agricultural and Biological Engineering*, 9, 98–109.

Huang, Y., Thomson, S.J., Brand, H.J., and Reddy, K.N. 2016b. Development and evaluation of low-altitude remote sensing systems for crop production management. *International Journal of Agricultural and Biological Engineering*, 9, 1–11.

Huang, Y., Yuan, L., Reddy, K.N., and Zhang, J. 2016c. In-situ plant hyperspectral sensing for early detection of soybean injury from dicamba. *Biosystems Engineering*, 149, 51–59.

Huang, Y., Reddy, K.N., Fletcher, R., and Pennington, D. 2017. UAV low-altitude remote sensing for precision weed management. *Weed Technology*, Accepted.

Hughes, G.F. 1968. On the mean accuracy of statistical pattern recongnizers. *IEEE Transactions on Information Theory*, 14, 55–63.

Hurcom, S.J., and Harrison, A.R. 1998. The NDVI and spectral decomposition for semi-arid vegetation abundance estimation. *International Journal of Remote Sensing*, 19, 3109–3125.

Inoue, Y., Sakaiya, E., Zhu, Y., and Takahashi, W. 2012. Diagnostic mapping of canopy nitrogen content in rice based on hyperspectral measurements. *Remote Sensing of Environment*, 126, 210–221.

Jain, N., Ray, S.S., Singh, J.P., and Panigrahy, S. 2007. Use of hyperspectral data to assess the effects of different nitrogen applications on a potato crop. *Precision Agriculture*, 8, 225–239.

Kanke, Y., Tubaña, B., Dalen, M., and Harrell, D. 2016. Evaluation of red and red-edge reflectance-based vegetation indices for rice biomass and grain yield prediction models in paddy fields. *Precision Agriculture*, 17, 507–530.

Karimi, Y., Prasher, S.O., Patel, R.M., and Kim, S.H. 2006. Application of support vector machine technology for weed and nitrogen stress detection in corn. *Computers and Electronics in Agriculture*, 51, 99–109.

Koger, C., Bruce, L.M., Shaw, D.R., and Reddy, K.N. 2003. Wavelet analysis of hyperspectral reflectance data for detecting pitted morningglory (Ipomoea lacunosa) in soybean (Glycine max). *Remote Sensing of Environment*, 86, 108–119.

Ladoni, M., Bahrami, H.A., Alavipanah, S.K., and Norouzi, A.A. 2010. Estimating soil organic carbon from soil reflectance: A review. *Precision Agriculture*, 11, 82–99.

Larson, J.A., Roberts, R.K., English, B.C., Larkin, S.L., Marra, M.C., Martin, S.W., Paxton, K.W., and Reeves, J.M. 2008. Factors affecting farmer adoption of remotely sensed imagery for precision management in cotton production. *Precision Agriculture*, 9, 195–208.

Lee, M.A., Huang, Y., Nandula, V.K., and Reddy, K.N. 2014. Differentiating glyphosate-resistant and glyphosate-sensitive Italian ryegrass using hyperspectral imagery. In: M.S.K.K. Chao (Ed.), *SPIE Proceedings Vol. 9108: Sensing for Agriculture and Food Quality and Safety VI.*

Li, F., Miao, Y., Hennig, S.D., Gnyp, M.L., Chen, X., Jia, L., and Bareth, G. 2010. Evaluating hyperspectral vegetation indices for estimating nitrogen concentration of winter wheat at different growth stages. *Precision Agriculture*, 11, 335–357.

Lowe, A., Harrison, N., and French, A.P. 2017. Hyperspectral image analysis techniques for the detection and classification of the early onset of plant disease and stress. *Plant Methods*, 13, 80.

Luo, J., Huang, W., Yuan, L., Zhao, C., Du, S., Zhang, J., and Zhao, J. 2013. Evaluation of spectral indices and continuous wavelet analysis to quantify aphid infestation in wheat. *Precision Agriculture*, 14, 151–161.

Ma, B.L., Morrison, M.J., and Dwyer, L.M. 1996. Canopy light reflectance and field greenness to assess nitrogen fertilization and yield of maize. *Agronomy Journal*, 88, 915–920.

Mahajan, G.R., Sahoo, R.N., Pandey, R.N., Gupta, V.K., and Kumar, D. 2014. Using hyperspectral remote sensing techniques to monitor nitrogen, phosphorus, sulphur and potassium in wheat (Triticum aestivum L.). *Precision Agriculture*, 15, 499–522.

Mahajan, G.R., Pandey, R.N., Sahoo, R.N., Gupta, V.K., Datta, S.C., and Kumar, D. 2017. Monitoring nitrogen, phosphorus and sulphur in hybrid rice (Oryza sativa L.) using hyperspectral remote sensing. *Precision Agriculture*, 18, 736–761.

Mahlein, A.-K., Steiner, U., Dehne, H.-W., and Oerke, E.-C. 2010. Spectral signatures of sugar beet leaves for the detection and differentiation of diseases. *Precision Agriculture*, 11, 413–431.

Mahlein, A.-K. 2016. Plant disease detection by imaging sensors–parallels and specific demands for precision agriculture and plant phenotyping. *Plant Disease*, 100, 241–251.

Mahlein, A.K., Rumpf, T., Welke, P., Dehne, H.W., Plümer, L., Steiner, U., and Oerke, E.C. 2013. Development of spectral indices for detecting and identifying plant diseases. *Remote Sensing of Environment*, 128, 21–30.

Marino, S., and Alvino, A. 2015. Hyperspectral vegetation indices for predicting onion (Allium cepa L.) yield spatial variability. *Computers and Electronics in Agriculture*, 116, 109–117.

Mariotto, I., Thenkabail, P.S., Huete, A., Slonecker, E.T., and Platonov, A. 2013. Hyperspectral versus multispectral crop-productivity modeling and type discrimination for the HyspIRI mission. *Remote Sensing of Environment*, 139, 291–305.

McIntyre, D.L., and Corner, R.J. 2016. Using EO-1 Hyperion satellite hyperspectral imagery to detect the pasture weed Paterson's curse (Echium plantagineum L.) in Southern Western Australia. In: *Twentieth Australasian Weeds Conference*, pp. 196–199.

Menesatti, P., Antonucci, F., Pallottino, F., Giorgi, S., Matere, A., Nocente, F., Pasquini, M., D'Egidio, M.G., and Costa, C. 2013. Laboratory vs. in-field spectral proximal sensing for early detection of Fusarium head blight infection in durum wheat. *Biosystems Engineering*, 114, 289–293.

Migdall, S., Bach, H., Bobert, J., Wehrhan, M., and Mauser, W. 2009. Inversion of a canopy reflectance model using hyperspectral imagery for monitoring wheat growth and estimating yield. *Precision Agriculture*, 10, 508–524.

Miller, J.R., Berger, M., Alonso, L., Cerovic, Z., Goulas, Y., Jacquemoud, S., Louis, J. et al. 2003. Progress on the development of an integrated canopy fluorescence model. In: *Proceedings of* 2003 *IEEE International Geoscience and Remote Sensing Symposium*. IGARSS, 03, 1, pp. 601–603. Toulouse, France.

Mirik, M., Ansley, R.J., Michels, G.J., and Elliott, N.C. 2012. Spectral vegetation indices selected for quantifying Russian wheat aphid (Diuraphis noxia) feeding damage in wheat (Triticum aestivum L.). *Precision Agriculture*, 13, 501–516.

Mirik, M., Ansley, R.J., Steddom, K., Jones, D., Rush, C., Michels, G., and Elliott, N. 2013. Remote distinction of a noxious weed (Musk Thistle: Carduus Nutans) using airborne hyperspectral imagery and the support vector machine classifier. *Remote Sensing*, 5, 612.

Moran, M.S., Inoue, Y. and Barnes, E.M. 1997. Opportunities and limitations for image-based remote sensing in precision crop management. *Remote Sensing of Environment*, 61, 319–346.

Mulla, D.J. 2013. Twenty five years of remote sensing in precision agriculture: Key advances and remaining knowledge gaps. *Biosystems Engineering*, 114, 358–371.

National Research Council. 1997. *Precision agriculture in the 21st century: Geospatial and information technologies in crop management.* National Academy Press, Washington, DC, 20055.

Nguyen, H.T., Kim, J.H., Nguyen, A.T., Nguyen, L.T., Shin, J.C., and Lee, B.-W. 2006. Using canopy reflectance and partial least squares regression to calculate within-field statistical variation in crop growth and nitrogen status of rice. *Precision Agriculture*, 7, 249–264.

Nieuwenhuizen, A.T., Hofstee, J.W., van de Zande, J.C., Meuleman, J., and van Henten, E.J. 2010. Classification of sugar beet and volunteer potato reflection spectra with a neural network and statistical discriminant analysis to select discriminative wavelengths. *Computers and Electronics in Agriculture*, 73, 146–153.

Nigon, T.J., Mulla, D.J., Rosen, C.J., Cohen, Y., Alchanatis, V., Knight, J., and Rud, R. 2015. Hyperspectral aerial imagery for detecting nitrogen stress in two potato cultivars. *Computers and Electronics in Agriculture*, 112, 36–46.

Okamoto, H., Murata, T., Kataoka, T., and Hata, S.I. 2007. Plant classification for weed detection using hyperspectral imaging with wavelet analysis. *Weed Biology and Management*, 7, 31–37.

Onoyama, H., Ryu, C., Suguri, M., and Iida, M. 2015. Nitrogen prediction model of rice plant at panicle initiation stage using ground-based hyperspectral imaging: Growing degree-days integrated model. *Precision Agriculture*, 16, 558–570.

Palacios-Orueta, A., and Ustin, S.L. 1998. Remote sensing of soil properties in the Santa Monica mountains I. Spectral analysis. *Remote Sensing for Environment*, 65, 170–183.

Pedrós, R., Jacquemoud, S., Goulas, Y., Louis, J., and Moya, I. 2004. A new leaf fluorescence model. In: *2nd International Workshop on Remote Sensing of Vegetation Fluorescence*, Montreal, Canada.

Perbandt, D., Fricke, T., and Wachendorf, M. 2011. Off-nadir hyperspectral measurements in maize to predict dry matter yield, protein content and metabolisable energy in total biomass. *Precision Agriculture*, 12, 249–265.

Pettersson, C.G., Soderstrom, M., and Eckersten, H. 2006. Canopy reflectance, thermal stress, and apparent soil electrical conductivity as predictors of within-field variability in grain yield and grain protein of malting barley. *Precision Agriculture*, 7, 343–359.

Piron, A., Leemans, V., Kleynen, O., Lebeau, F., and Destain, M.-F. 2008. Selection of the most efficient wavelength bands for discriminating weeds from crop. *Computers and Electronics in Agriculture*, 62, 141–148.

Prabhakar, M., Prasad, Y.G., Thirupathi, M., Sreedevi, G., Dharajothi, B., and Venkateswarlu, B. 2011. Use of ground based hyperspectral remote sensing for detection of stress in cotton caused by leafhopper (Hemiptera: Cicadellidae). *Computers and Electronics in Agriculture*, 79, 189–198.

Prabhakar, M., Prasad, Y.G., Vennila, S., Thirupathi, M., Sreedevi, G., Ramachandra Rao, G., and Venkateswarlu, B. 2013. Hyperspectral indices for assessing damage by the solenopsis mealybug (Hemiptera: Pseudococcidae) in cotton. *Computers and Electronics in Agriculture*, 97, 61–70.

Raper, T.B., and Varco, J.J. 2015. Canopy-scale wavelength and vegetative index sensitivities to cotton growth parameters and nitrogen status. *Precision Agriculture*, 16, 62–76.

Rasmussen, M.S. 1997. Operational yield forecasting using AVHRR NDVI data: Reduction of environmental and inter-annual variability. *International Journal of Remote Sensing*, 18, 1059–1077.

Reddy, K.N., Huang, Y., Lee, M.A., Nandula, V.K., Fletcher, R.S., Thomson, S.J., and Zhao, F. 2014. Glyphosate-resistant and glyphosate-susceptible Palmer amaranth (Amaranthus palmeri S. Wats.): Hyperspectral reflectance properties of plants and potential for classification. *Pest Management Science*, 70, 1910–1917.

Reyniers, M., Vrindts, E., and Baerdemaeker, J.D. 2004. Optical measurement of crop cover for yield prediction of wheat. *Biosystems Engineering*, 89, 383–394.

Richard, J.A., and Jia, X. 1999. *Remote sensing digital image analysis: An introduction.* Berlin Heidelberg, Germany Springer-Verlag.

Roberts, D., Roth, K., and Perroy, R. 2011. Hyperspectral vegetation indices. In: P. S. Thenkabail, J. G. Lyon & A. Huete (Eds.), *Hyperspectral Remote Sensing of Vegetation*, pp. 309–327. Baca Raton, FL CRC Press, Taylor & Francis Group.

Rumpf, T., Mahlein, A.-K., Steiner, U., Oerke, E.-C., Dehne, H.-W., and Plümer, L. 2010. Early detection and classification of plant diseases with support vector machines based on hyperspectral reflectance. *Computers and Electronics in Agriculture*, 74.

Ryu, C., Suguri, M., and Umeda, M. 2009. Model for predicting the nitrogen content of rice at panicle initiation stage using data from airborne hyperspectral remote sensing. *Biosystems Engineering*, 104, 465–475.

Samiappan, S., Turnage, G., Hathcock, L., Yao, H., Kincaid, R., Moorhead, R., and Ashby, S. 2017. Classifying common wetland plants using hyperspectral data to identify optimal spectral bands for species mapping using a small unmanned aerial system–a case study. In: *Proceeding of IEEE IGARSS*, Fort Worth, TX IEEE, pp. 2570.

Schirrmann, M., Gebbers, R., Kramer, E., and Seidel, J. 2011. Soil pH mapping with an on-the-go sensor. *Sensors*, 11, 573.

Scotford, I.M., and Miller, P.C.H. 2005. Applications of spectral reflectance techniques in northern European cereal production: A review. *Biosystems Engineering*, 90, 235–2250.

Serrano, L., Filella, I., and Peuelas, J. 2000. Remote sensing of biomass and yield of winter wheat under different nitrogen supplies. *Crop Science*, 40, 723–731.

Skowronek, S., Ewald, M., Isermann, M., Van De Kerchove, R., Lenoir, J., Aerts, R., Warrie, J. et al. 2017. Mapping an invasive bryophyte species using hyperspectral remote sensing data. *Biological Invasions*, 19, 239–254.

Stafford, J.V. 2000. Implementing precision agriculture in the 21st century. *Journal of Agricultural Engineering Research*, 76, 267–275.

Suarez, L.A., Apan, A., and Werth, J. 2017. Detection of phenoxy herbicide dosage in cotton crops through the analysis of hyperspectral data. *International Journal of Remote Sensing*, 38, 6528–6553.

Taylor, J.C., Wood, G.A., and Thomas, G. 1997. Mapping yield potential with remote sensing. *Precision Agriculture—Soil and Crop Modeling*, 2, 713–720.

Taylor, R.K., Kluitenberg, G.J., Schrock, A.D., Zhang, N., Schmidt, J.P., and Havlin, J.L. 2001. Using yield monitor data to determine spatial crop production potential Source. *Transactions of the ASAE*, 44, 1409–1414.

Thenkabail, P.S., Ward, A.D., and Lyon, J.G. 1994. Landsat-5 Thematic Mapper models of soybean and corn crop characteristics. *International Journal of Remote Sensing*, 15, 49–61.

Thenkabail, P.S., Enclona, E.A., Ashton, M.S., and Van Der Meer, B. 2004. Accuracy assessments of hyperspectral waveband performance for vegetation analysis applications. *Remote Sensing of Environment*, 91, 354–376.

Thomasson, J.A., Sui, R., Cox, M.S., and Al-Rajehy, A. 2001. Soil reflectance sensing for determining soil properties in precision agriculture. *Transactions of the ASAE*, 44, 1445–1453.

Thorp, K.R., and Tian, L.F. 2004. A review on remote sensing of weeds in agriculture. *Precision Agriculture*, 5, 477–508.

Thorp, K.R., Wang, G., Bronson, K.F., Badaruddin, M., and Mon, J. 2017. Hyperspectral data mining to identify relevant canopy spectral features for estimating durum wheat growth, nitrogen status, and grain yield. *Computers and Electronics in Agriculture*, 136, 1–12.

Timmermann, C., Gerhards, R., and Kühbauch, W. 2003. The economic impact of site-specific weed control. *Precision Agriculture*, 4, 249–260.

Uno, Y., Prasher, S.O., Lacroix, R., Goel, P.K., Karimi, Y., Viau, A., and Patel, R.M. 2005. Artificial neural networks to predict corn yield from compact airborne spectrographic imager data. *Computers and Electronics in Agriculture*, 47, 149–161.

Vellidis, G., Thomas, D., Wells, T., and Kvien, C. 1999. Cotton yield maps created from aerial photographs. In: *ASAE Meeting Paper No.* 991139, St. Joseph, MI: ASAE.

Villalobos, F.J., Hall, A.J., Ritchie, J.T., and Orgaz, F. 1996. OILCROP-SUN: A development, growth, and yield model of sunflower crop. *Agronomy Journal*, 88, 403–415.

Vrindts, E., Baerdemaeker, J.D., and Ramon, H. 2002. Weed detection using canopy reflection. *Precision Agriculture*, 3, 63–80.

Walburg, G., Bauer, M.E., Daughtry, C.S.T., and Housley, T.L. 1982. Effects of nitrogen nutrition on the growth, yield, and reflectance characteristics of corn canopies. *Agronomy Journal*, 74, 677–683.

Wendel, A., and Underwood, J. 2016. Self-supervised weed detection in vegetable crops using ground based hyperspectral imaging. In: 2016 *IEEE International Conference on Robotics and Automation (ICRA)*, pp. 5128–5135.

Whiting, M.L., Ustin, S.L., and Zerco-Tejada, P. 2006. Hyperspectral mapping of crop and soils for precision agriculture. In: W. Gao & S. L. Ustin (Eds.), *Proceeding of SPIE: Remote Sensing and Modeling of Ecosystems for Sustainability III*, 6298, pp. 62980B–662981.

Wolfert, S., Ge, L., Verdouw, C., and Bogaardt, M.-J. 2017. Big data in smart farming—A review. *Agricultural Systems*, 153, 69–80.

Xie, C., Yang, C., and He, Y. 2017. Hyperspectral imaging for classification of healthy and gray mold diseased tomato leaves with different infection severities. *Computers and Electronics in Agriculture*, 135, 154–162.

Xu, X., Zhao, C., Wang, J., Zhang, J., and Song, X. 2014. Using optimal combination method and *in situ* hyperspectral measurements to estimate leaf nitrogen concentration in barley. *Precision Agriculture*, 15, 227–240.

Yang, C.-C., Prahser, S.O., and Whalen, J. 2002. In-season yield predctionof corn and soybean with hyperspectral imagery. In: *ASAE Meeting Paper No.* 023139, St. Joseph, MI: ASAE.

Yang, C.-M. 2010. Assessment of the severity of bacterial leaf blight in rice using canopy hyperspectral reflectance. *Precision Agriculture*, 11, 61–81.

Yang, C., Everitt, J.H., and Bradford, J.M. 2004a. Airborne hyperspectral imagery and yield monitor data for estimating grain sorghum yield variability. *Transactions of the ASAE*, 47, 915–924.

Yang, C., Everitt, J.H., Bradford, J.M., and Murden, D. 2004b. Airborne hyperspectral imagery and yield monitor data for mapping cotton yield variability. *Precision Agriculture*, 5, 445–461.

Yang, C., Everitt, J.H., and Bradford, J.M. 2007a. Using multispectral imagery and linear spectral unmixing techniques for estimating crop yield variability. *Transactions of the ASAE*, 50, 667–674.

Yang, C., Everitt, J.H., and Bradford, J.M. 2007b. Airborne hyperspectral imagery and linear spectral unmixing for mapping variation in crop yield. *Precision Agriculture*, 8, 279–296.

Yang, C., Everitt, J.H., and Bradford, J.M. 2008. Yield estimation from hyperspectral imagery using spectral angle mapper (SAM). *Transactions of the ASAE*, 51, 729–737.

Yang, C., Everitt, J.H., Bradford, J.M., and Murden, D. 2009. Comparison of airborne multispectral and hyperspectral imagery for estimating grain sorghum yield. *Transactions of the ASAE*, 52, 641–649.

Yang, C., and Everitt, J.H. 2012. Using spectral distance, spectral angle and plant abundance derived from hyperspectral imagery to characterize crop yield variation. *Precision Agriculture*, 13, 62–75.

Yao, H. 2004. Hyperspectral imagery for precision agriculture. PhD dissertation, University of Illinois at Urbana-Champaign, Urbana, IL.

Yao, H., Huang, Y., Hruska, Z., Thomson, S.J., and Reddy, K.N. 2012. Using vegetation Index and modified derivative for early detection of soybean plant injury from glyphosate. *Computers and Electronics in Agriculture*, 89, 145–157.

Yao, H., Tian, L., Wang, G., and Colonna, I. 2014. Estimation of soil fertility using collocated cokriging by combining aerial hyperspectral imagery and soil sample data. *Applied Engineering in Agriculture*, 30, 113–121.

Yuan, L., Zhang, J.-C., Wang, K., Loraamm, R.-W., Huang, W.-J., Wang, J.-H., and Zhao, J.-L. 2013. Analysis of spectral difference between the foreside and backside of leaves in yellow rust disease detection for winter wheat. *Precision Agriculture*, 14, 495–511.

Yuan, L., Huang, Y., Loraamm, R.W., Nie, C., Wang, J., and Zhang, J. 2014. Spectral analysis of winter wheat leaves for detection and differentiation of diseases and insects. *Field Crops Research*, 156.

Zara, P.M., Doraiswamy, P.C., and McMutrey, J. 2000. Assessing variability of nitrogen status in corn plants with hyperspectral remote sensing. In: *Proc. of ASPRS Annual Meeting.*

Zarco-Tejada, P., Gonzalez-dugo, V., and Berni, J.A. 2012. Fluorescence, temperature and narrow-band indices acquired from a UAV platform for water stress detection using a micro-hyperspectral imager and a thermal camera. *Remote Sensing of Environment*, 117, 322–337.

Zhang, C., and Kovacs, J.M. 2012. The application of small unmanned aerial systems for precision agriculture: A review. *Precision Agriculture*, 13, 693–712.

Zhang, H., Lan, Y., Lacey, R., Hoffmann, W.C., Martin, D.E., Fritz, B., and Lopez, J. 2010. Ground-based spectral reflectance measurements for evaluating the efficacy of aerially-applied glyphosate treatments. *Biosystems Engineering*, 107, 10–15.

Zhang, J., Yuan, L., Pu, R., Loraamm, R.W., Yang, G., and Wang, J. 2014. Comparison between wavelet spectral features and conventional spectral features in detecting yellow rust for winter wheat. *Computers and Electronics in Agriculture*, 100, 79–87.

Zhang, M., Qin, Z., Liu, X., and Ustin, S.L. 2003. Detection of stress in tomatoes induced by late blight disease in california, usa, using hyperspectral remote sensing. *International Journal of Applied Earth Observation and Geoinformation*, 4, 295–310.

Zhang, M., Qin, Z., and Liu, X. 2005. Remote sensed spectral imagery to detect late blight in field tomatoes. *Precision Agriculture*, 6, 489–508.

Zhang, N., Wang, M., and Wang, N. 2002. Precision agriculture-a worldwide overview. *Computers and Electronics in Agriculture*, 36, 113–132.

Zhang, Y., and Slaughter, D.C. 2011. Hyperspectral species mapping for automatic weed control in tomato under thermal environmental stress. *Computers and Electronics in Agriculture*, 77, 95–104.

Zhao, D.H., Li, J.L., and Qi, J.G. 2005. Identification of red and nir spectral regions and vegetative indices for discrimination of cotton nitrogen stress and growth stage. *Computers and Electronics in Agriculture*, 48, 155–169.

Zhao, F., Huang, Y., Guo, Y., Reddy, K., Lee, M., Fletcher, R., and Thomson, S. 2014. Early detection of crop injury from glyphosate on soybean and cotton using plant leaf hyperspectral data. *Remote Sensing*, 6, 1538.

Zhao, Y., Xu, X., Liu, F., and He, Y. 2012. A novel hyperspectral waveband selection algorithm for insect attack detection. *Transactions of the ASABE*, 55, 281.

Zhao, Y., Yu, K., Feng, C., Cen, H., and He, Y. 2017. Early detection of aphid (myzus persicae) infestation on Chinese cabbage by hyperspectral imaging and feature extraction. *Transactions of the ASABE*, 60, 1045.

Zwiggelaar, R. 1998. A review of spectral properties of plants and their potential use for crop/weed discrimination in row-crops. *Crop Protection*, 17, 189–1206.

2 Hyperspectral Narrowbands and Their Indices in Study of Nitrogen Content of Cotton Crops

Jianlong Li, Jingjing Zhang, Chunliu Tao, Dehua Zhao, and Chengcheng Gang

CONTENTS

2.1 INTRODUCTION

The use of remote sensing for precision agriculture applications is very popular. Nitrogen is one of the most important fertilizer elements for crop production. Nitrogen is an essential element for crop growth, development and yield formulation. Nitrogen content deficiency will bring a series of changes to crop leaves, such as color, thickness, water content, form and structure. A lack of Nitrogen will directly affect the composition of amino acid, protein, nucleic acid and other materials, which will lead to the reduction of crops photosynthesis capacity and the final yield. Thereby, nitrogen management is a very important management measure in agriculture produce for obtaining high yields and good quality. At the same time, over application of nitrogen will pollute underground water and also gets into streams. To overcome this and other similar problems, precision agriculture (also referred to as precision farming or site-specific farming) has been put forward. Precision nitrogen management is a key content of precision agriculture, which reduces pollution of the water resources and yet resulting in sustained high yields over space and time.

Given the importance of nitrogen in crop growth and yield as well as the need to maintain environmentally acceptable levels of N application, many studies in precision farming are focused on nitrogen application rate and timing for high yield, crop quality and environmental pollution control (Pattey et al. 2001 and Weisz et al. 2001). Under normal conditions, nitrogen fertilizer influences chlorophyll concentration in green leaves. Since chlorophyll, a key indicator of crop physiological status, has a strong absorbance peak in the red spectral region, empirical models of predicting chlorophyll status from spectral reflectance are largely based on red spectra (Broge and Leblanc 2001, Gitelson et al. 2002 and Sims and Gamon 2002). Arrangement of cells within mesophyllic layers of leaves and canopy structure, simulataneously affected with chlorophyll status by nitrogen supply, are the most important factors determining canopy near-infrared (NIR) reflectance (Serrano et al. 2000, Kumar et al. 2001 and Mutanga et al. 2003).

Traditional methods to determine plant tissue nutrient concentrations in a laboratory are time consuming and costly. Remote sensing has a great and realized potential to assess and manage timely crop stress affected by the environment from leaf to landscape scales of crop physiology (Daughtry et al. 2000 and Zarco-Tejada et al. 2000). Recent studies have found close relationships between plant physiological parameters and spectral reflectance. Several studies have documented that N status of field crops' leaves can be assessed by spectral reflectance data of crops' leaf or canopy.

Red–NIR-based VIs could also be used to estimate crop nitrogen stress (Boegh et al. 2002, Strachan et al. 2002 and Hansen and Schjoerring 2003). The red and NIR reflectance data, used to generate the popular VIs and monitor crop growth conditions, are acquired from two kinds of sensors. Broad-band spectral reflectance, currently the popular remotely sensed data (Table 2.1 of

TABLE 2.1
Multiple Comparisons of Mean Values of Three Cotton Variables Observed on July 15, August 14, and October 1 and Seed Cotton Yield under Different Nitrogen Treatments (at 95% Confidence Level)

		Means[a]			
Variables	**Treatments**	**July 14**	**August 15**	**October 1**	**Seed Cotton Yield (kg/hm²)**
LAI (m²/m²)	N0	0.90a	2.23a	1.34a	
	N90	0.97ab	2.75b	1.92b	
	N180	1.06bc	3.12c	2.37c	
N360	1.12c	3.25c	2.33c		
CC (%)	N30	1.25a	0.85a	0.53a	
	N90	1.31ab	1.23b	1.07b	
	N180	1.37ab	1.37bc	1.26c	
	N360	1.41b	1.52c	1.37c	
Aboveground dry Biomass (g/m²)	N0	118.7a	426.8a	636.2a	
	N90	130.2ab	488.7b	816.9b	
	N180	139.4bc	548.6bc	911.7c	
	N360	153.3c	625.2c	1032.3d	
Seed cotton Yield (kg/hm²)	N0				2908.5a
	N90				3914.9b
	N180				4474.9c
	N360				4592.3c

[a] Means within columns followed by the same letter (a–d) are not significantly different based on ANOVA at 95% confidence level ($p \leq 0.05$).

Volume 1), is obtained from the current generation of earth-orbiting satellites carrying multispectral sensors such as the Advanced Spaceborne Thermal Emission and Reflection Radiometer (ASTER), Moderate Resolution Resolution Imaging Spectroradiometer (MODIS) and Landsat-7 Enhanced Thematic Mapper Plus (ETM+). Most of these sensors have several channels among which the red NIR and are the most popular bands. Narrow-band spectral data used to monitor crop condition is generated from imaging sensors such as Airborne Visible/Infrared Imaging Spectrometer (AVIRIS) and Compact High Resolution Imaging Spectrometer (CHRIS) Project for on Board Autonomy (PROBA).

Many spectral vegetation indices (VIs) have been developed in the past three decades to provide more sensitive measurements of plant biophysical parameters and to reduce external noise interferences such as those related to soil and the atmosphere. Some VIs were developed based on narrowband spectral data and others on broadband sensors. Therefore, although the mathematical Equations defining VIs are the same, their calculated values are different, thus affecting their stability in predicting agronomic variables such as total green leaf area index.

Broadband and narrowband based vegetation indices have been compared for their ability to estimate crop agronomic variables such as green vegetation cover, LAI and CCD. In general, the narrowband VIs may be slightly better than their broadband versions for estimating crop variables, although some reported no difference between them.

Hyperspectral narrow-band remote sensing provides a perfect opportunity to characterize and advance the study N content in plants (Pattey et al. 2001 and Strachan et al. 2002). A number of authors (Craig 2001) discussed the superiority of narrow-band/hyperspectral imaging sensors over broad-band/multispectral instruments. Multispectral imaging sensors gather spectral data in large, non-contiguous ranges of the electromagnetic spectrum, thus, a single band represents the average of a relatively large portion of the spectrum. When comparing the predictive powers and stability of broad-band and narrow-band VIs for deriving crop growth variables, there are some other opinions (Broge and Leblanc 2001 and Broge and Mortensen 2002). Besides the indicators of crop growth variables, VIs was applied to detect nitrogen stress (Craig 2001 and Strachan et al. 2002). The selection of optimum wavebands in hyperspectral data has been performed in a number of cases focused mainly on how to improve the correlation between VIs and crop biophysical/biochemical variables. But few studies have been focused on how to increase the sensitivity of the VIs to nitrogen stress.

Our objective was to analyze hyperspectral remote sensing capability in detecting characteristic differences of agricultural crops under different nitrogen application rates and different growing stages. We chose cotton crops to address the overall objective, through three specific sub-objectives, which were to:

1. Identify sensitive hyperspectral wavelengths to different N treatments;
2. Evaluate if the continuum-removal method improves the ability to recognize different N status in the spectral wavebands of absorbing chlorophyll (550–750 nm) at full green canopy coverage period; and
3. Test canonical discriminant analysis.

2.2 MATERIALS AND METHODS

The narrow wavebands located in specific portions of the spectrum have the ability to provide required optimal information sought for a given application. So far, the most common technique to extract information content from spectral measurements is the computation of spectral vegetation indices (VIs). The normalized difference vegetation index (NDVI) and ratio vegetation index (RVI) have been used extensively in correlating remote sensing observations with the characteristics of vegetation. In particular, these vegetation indices were found to be quantitatively and functionally related to several vegetation parameters such as leaf area index (LAI), percent vegetation cover, intercepted photosynthetically active radiation (IPAR), and green biomass.

To enhance their sensitivities to green vegetation spectral signals and to reduce external effects such as noise-related soil and atmospheric influences, many VIs have been developed in the past three decades. These VIs can be divided into four groups: (1) Ratio-based VIs, which is based on the ratio between red and NIR reflectance. The normalized difference vegetation index (NDVI) and ratio vegetation index (RVI) are the most commonly used ratio-based VIs; (2) Orthogonal VIs, which are defined by a line in spectral space for identification at bare soils. The transformed soil-adjusted vegetation index (TSAVI), second soil-adjusted vegetation index (SAVI2) and modified second soil-adjusted vegetation index (MSAVI2) are examples of orthogonal VIs; (3) Derivative VIs: Elvidge and Chen (1995) introduced the first and second-order derivative green VIs. And (4) Atmospheric corrected indices, such as the visible atmospherically resistant index (VARI) (Gitelson et al. 2002). These VIs have been shown to be quantitatively and functionally related with canopy parameters such as the leaf area index (LAI), aboveground biomass, chlorophyll content (CC) and vegetation fraction. Research results indicate these VIs have potential applications in agriculture for forecasting and estimating crop productions, monitoring crop conditions, classifying and mapping crops, and directing precision farming activities (Serrano et al. 2000, Broge and Mortensen 2002 and Strachan et al. 2002).

2.2.1 Experiment Design and Treatment

2.2.1.1 Experiment 1 (E1)

A completely randomized design experiment containing three replicates was conducted in a cotton (Gossypium hirsutum L. cv. Sumian 12) field at Zhejiang University, Zhejiang Province, China (30°4′N, 120°10′E). Treatments included three N application rates of 0, 60 and 120 kg N ha^{-1} (termed LN, MN and HN, respectively). The soil of experiment field is sandy, which contains 0.95 g/kg total-nitrogen, 148.5 mg/kg available-N, 1.21 g/kg available-P, 72.7 mg/kg available-K and 9.96 g/kg organic matters. Each sampling plot consisted of two rows of 0.3 m apart, 3.7 m wide and 5.0 m long ($3.7 \times 5.0 = 17.5$ m^2) with a density of 50,000 plants/ha. Cotton was sown on April 29 directly in fields with north/south row orientation. Phosphorous and potassium fertilizers were supplied in adequate amounts according to the general nutrient status of the field as determined by soil samples: 80 kg/ha P_2O_5 and 160 kg/ha K_2O.

2.2.1.2 Experiment 2 (E2)

A completely randomized design experiment containing three replicates was conducted in a cotton (Gossypium hirsutum L. cv. Sumian 3) field at Zhangjiagang, Jiangsu Province, China (31°50′N, 120°49′E). The soil of experiment field is sandy, which contains 41.6 mg/kg available-N, 47.2 mg/kg total-P, 63.9 mg/kg total-K and 13.2 g/kg organic matters. Treatments included three N application rates of 90, 180 and 360 kg N/ha (termed LN, MN, and HN, respectively). Each sampling plot consisted of two rows of 0.8 m apart, 0.4 m wide and 14 m long ($2.4 \times 14 = 33.6$ m^2) with a density of 45,000 plants/ha. Cotton was sown on April 12 in greenhouse and later transplanted on May 28 to fields with north/south row orientation. Phosphorous and potassium fertilizers were also supplied: 180 kg/ha P_2O_5 and 240 kg/ha K_2O. Irrigation was not used due to the high rainfall (above 1200 mm) and high ground water table of the soil at the study site.

2.2.2 Observed Dates

According to canopy structure and leaf function of cotton plants, the cotton growth cycle was divided into three stages: (1) rapid growth period (early stage when the soil was partially covered by cotton and, therefore, its contribution to spectral signals was significant), (2) full green canopy coverage period (middle stage when the canopy reached almost 100% cover) and (3) senescent period (late stage when cotton bolls were opened and part of the leaves were senesced). Timing of growth stages corresponded to sampling dates of July 15, August 14 and October 1, 2002 in experiment one, and of July 12, August 22 and September 29, 2002 in experiment two, respectively, when agronomic and hyperspectral data were collected.

2.2.3 Canopy Hyperspectral Reflectance Measurements

A 512-channel spectroradiometer (300–1100 nm) by Analytical Spectral Devices™ (FieldSpec FR) was used to acquire cotton canopy spectral data. Noise at both ends of the spectrum limited the useful data range to between 400 and 1000 nm in the analysis. Data were collected on cloudless days with solar elevations ranging from 50° to 55° to minimize external effects from atmospheric conditions and changes in solar position. Prior to the cotton planting, spectral reflectance measurements of the bare soil surface were made. Spectral reflectance was calculated as the ratio of measured radiance to radiance from a white standard reference panel. Immediately after the white standard radiance measurement, two spectral measurements of the cotton canopy were obtained—one with the sensor located directly over the center of two rows on a ridge, the other one with the sensor located directly over the furrow. Then the two spectra were averaged to represent a single mean field spectrum of ridge. The measurements were repeated 10 times for each plot. For experiment 1 (E1), reflectance measurements were obtained three times on July 12, August 22 and September 29 by the spectroradiometer with 25 o field of view and 1.0 m nadir orientation above the canopy which resulted in a sensor field of view of 45 cm diameter. For experiment 2 (E2), reflectance measurements were obtained three times on July 15, August 14 and December 1 by the spectroradiometer with 15 o field of view and 2.3 m nadir orientation above the canopy which resulting in a sensor field of view of 60 cm diameter.

2.2.4 Biomass Measurements

After spectral measurement, 10 cotton samples were selected to analyze biophysical variables immediately in the same place. First, the samples were dried at 70°C in an oven for 48 hours to constant weight, and then dry weight biomass was determined by weight.

2.2.5 Agronomic Variable Measurements

Six cotton plants were harvested on the same days that the canopy spectral measurements were made. Each plant was separated into leaves, branches, and stems and then weighed for leaf biomass calculations (g/m^2). The green leaves from two plants (thus decreasing the workload) were measured with a leaf area meter (CI-203, CID) to estimate the total leaf area per sample plot (1.333 m^2) in order to calculate leaf area index. The leaf area index of cotton was computed as the ratio of green leaf area per sampled area (m^2/m^2). Chlorophyll content was measured from 0.15 g leaf samples that were ground in 3 mL cold acetone/Tris buffer solution (80:20 Vol/Vol, pH = 7.8), centrifuged to remove particulates, and the supernatant diluted to a final volume of 15 mL with additional acetone/Tris buffer. The absorbance of the extract solutions was measured with a U-3000 spectrophotometer at 663, 647 and 537 nm. The chlorophyll concentration was calculated using the following equation (Sims and Gamon 2002):

$$\text{Chla} = 0.01373\text{A}663\text{-}0.000897\text{A}537\text{-}0.0030464\text{A}647$$

$$\text{Chlb} = 0.120405\text{A}647\text{-}0.004305\text{A}537\text{-}0.005507\text{A}663$$

where Ax is the absorbance of the extract solution in a 1 cm path length cuvette at wavelength x. The units for all the equations were micromoles per milliliter (μmol/mL). Canopy chlorophyll density (g/m^2) was computed by multiplying chlorophyll content by total leaf weights.

2.2.6 Data Process and Analysis

The research hypothesis was whether the means of the reflectance between the three treatments were significantly different at each wavelength and was statistically tested using one-way analysis of variance. From this test one can conclude that there are differences between the groups. The statistical tests were done at different time periods (rapid growth period, full green canopy coverage period, senescent period) in order to assess the spectral differences between treatments at different stages of the plants' physiological status. Specially, we tested the utility of the visible absorption feature (R550–R750 nm) to discriminate different levels of nitrogen concentration after continuum removal (defined below). This red absorption feature was selected since it has consistently proved to be an indicator of vegetation condition and is not affected by water absorption in fresh plants. This is in contrast to the mid-infrared bands where chemical absorption is largely masked by water. Continuum removeal normalizes reflectance spectra to allow comparison of individual absorption features from a common baseline (Kokaly 2001). The continuum is a convex hull fitted over the top of a spectrum utilizing straight-line segments that connect local spectra maxima. The continuum is removed by dividing the reflectance value for each point in the absorption pit by the reflectance level of the continuum line (convex hull) at the corresponding wavelength. The first and last spectral data values are on the hull and therefore the first and last bands in the output continuum-removed data file are equal to 1. The output curves have values between 0 and 1, in which the absorption pits are enhanced and the absolute variance removed (Schmidt and Skidmore 2001). At first, this method was applied to identify mineral component in geology, then applied on vegetation science by Kokaly and Clark (1999) to analyze chemical component of several plant dry leaves. In recent years, continuum-removal measure has been applied to vegetation canopies for measuring biochemical contents of plant (Mutanga et al. 2003).

One-way analysis of variance (ANOVA) method is often used to assess significant degree of the spectral reflectance difference between different N treatment during four growth stages with Statistical Product and Service solutions (SPSS 11.0) software.

2.3 RESULTS AND ANALYSIS

2.3.1 Biomass Analysis under Different Nitrogen Treatment

Biomass is an important agriculture parameter in reflecting crops canopy structure. In this study, dry weight of aboveground biomass (DWAB) was selected as an assessment standard. As expected, different nitrogen treatments resulted in significantly different DWABs at experiment 1 and experiment 2 (Figure 2.1). In general, nitrogen increased cotton DWAB. For these two experiments,

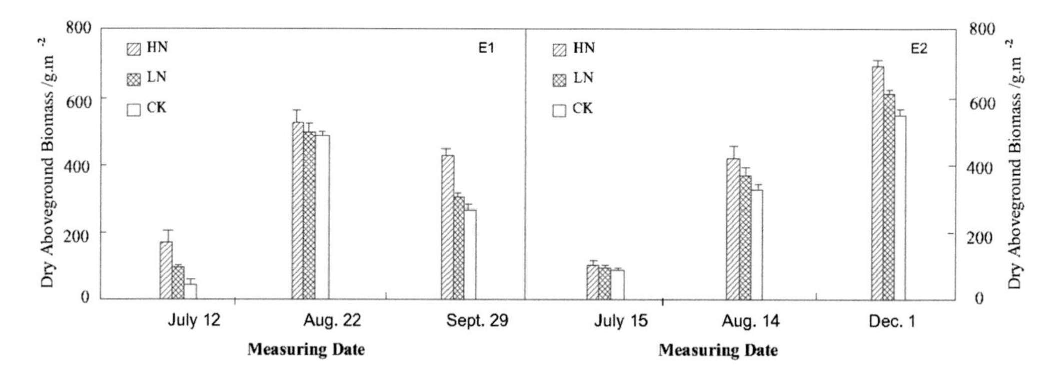

FIGURE 2.1 The variation of cotton biomass under different treatments at experiments 1 and 2 (HN: high nitrogen; LN: low nitrogen).

dry weight of above ground biomass experienced significant difference between LN- and HN-treatments at their respective three observed dates ($p < 0.01$).

2.3.2 Difference of Canopy Spectral Reflectance under Different N Treatments

Figures 2.2 and 2.3 were canopy reflectance spectra and results of one-way ANOVA of canopy reflectance among three N treatments at different wavelengths. In general, canopy spectral reflectance showed no significant difference between nitrogen treatments in visible light. But in near-infrared regions, canopy spectral reflectance showed significant difference between N treatments ($p < 0.05$).

Figure 2.2 showed canopy reflectance spectra under HN, MN and LN treatment decreased in turn in NIR region, especially in full green canopy coverage period, which presented a significant difference between different N treatments. This vigorous growth stage is not only nutritional growth but also the development growth of young cotton buds and bolls. In the senescent period, all of canopy spectra reflectance under three different N applications was lower and spectral reflectance curves present identical trends in different N treatment, because most cotton leaves had fallen off and only a few withered leaves and unpicked cotton bolls were left.

2.3.3 Changes of Normalized Difference Spectra Characteristic

Figures 2.3 and 2.4 were continuum-removed mean reflectance spectra and results of one-way ANOVA of continuum-removed reflectance among three N treatments at the wavelengths of chlorophyll maximal absorptance (550–750 nm) at experiment 1 and experiment 2. Results suggested that the difference of spectral reflectance between 550 and 750 nm was improved by using

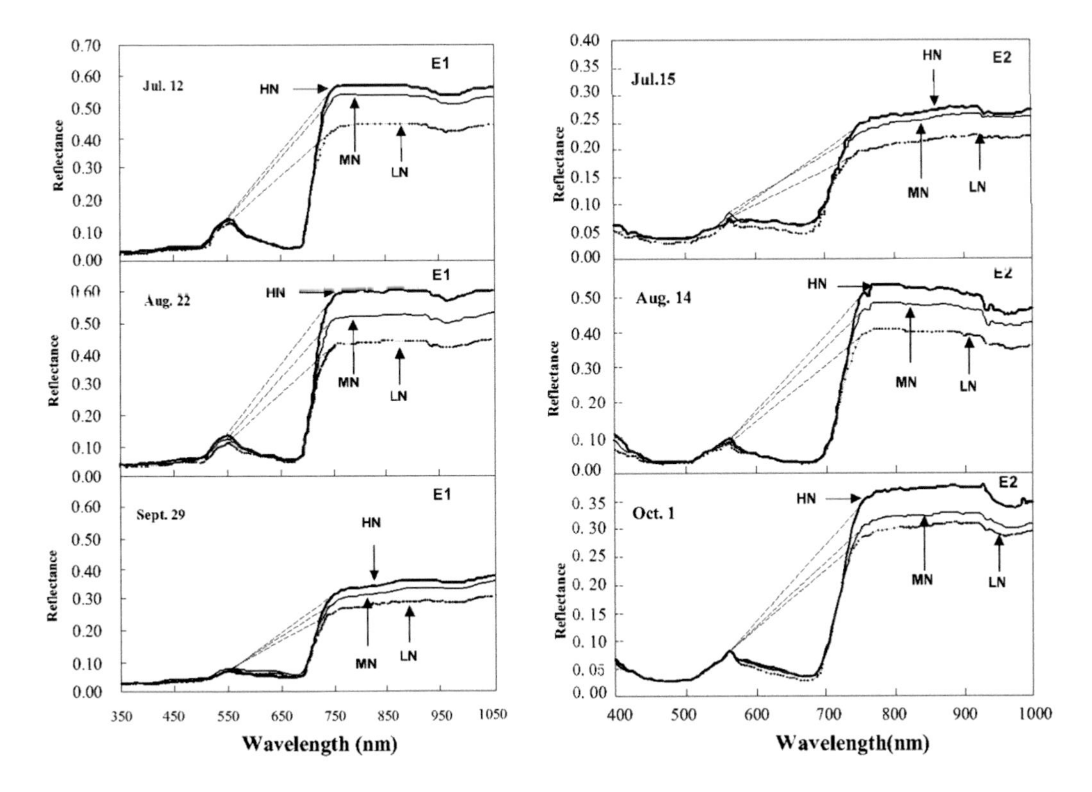

FIGURE 2.2 Mean canopy spectral reflectance with continuum removed line under different treatments (HN: high nitrogen; MN: medium nitrogen; LN: low nitrogen).

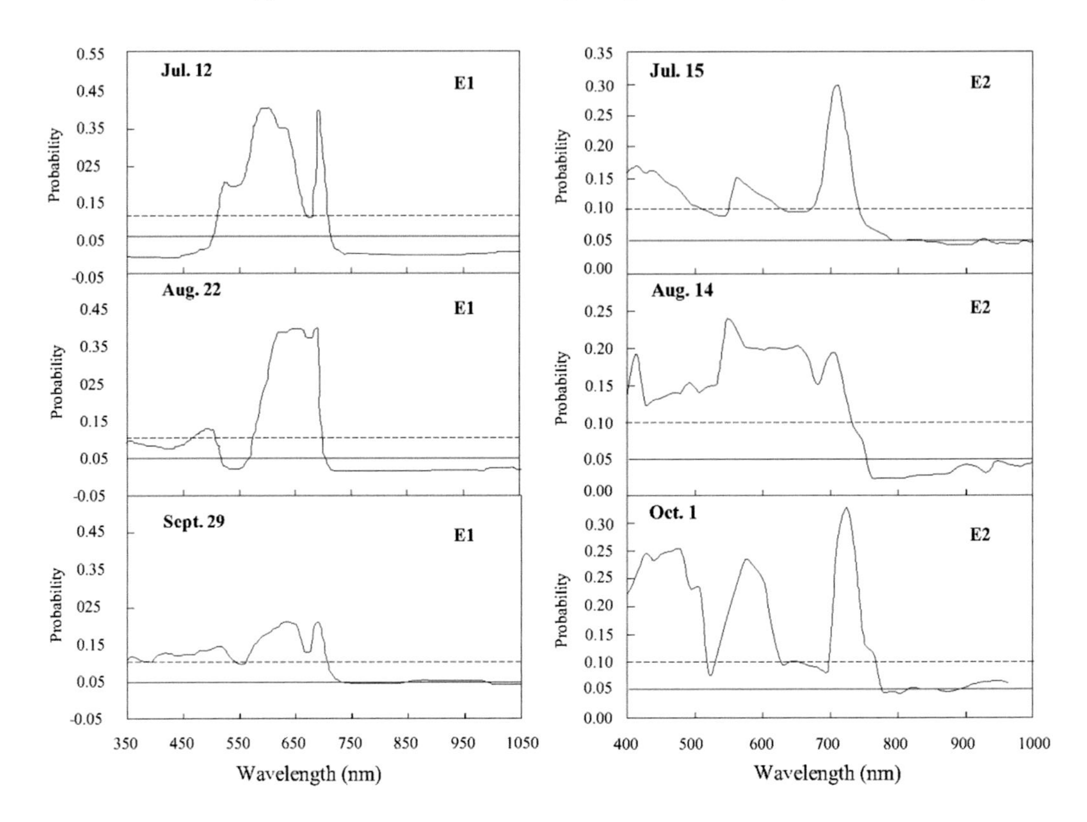

FIGURE 2.3 Results of one-way ANOVA of cotton canopy reflectance among the three N treatments at different wavelength.

continuum-removed technology. Nitrogen could enlarge spectral values between 550 and 750 nm especially in the full green canopy coverage period. One-way ANOVA results showed that there had always been some wavelengths at which the canopy reflectance showed significant difference between N treatments ($p < 0.05$). The most sensitive reflectance to N rate was located at two sides of the chlorophyll maximal absorptance (680 nm).

2.3.4 Multiple Variable Comparison Analysis under Different Treatments

Most of the red-NIR Vis were established for the purpose of estimating plant biophysical/biochemical variables, and have been related to crop variables such as biomass, LAI, and chlorophyll (Broge and Leblanc 2001 and Sims and Gamon 2002 and Serrano et al. 2000 and Kokaly 2001). In general, since N conditions resulted in a significant variation in these variables, which has been proved by this experiment, it is also possible to discriminate canopies grown under different N treatments using these Vis (Mutanga et al. 2003 and Strachan et al. 2002).

As expected, the N fertilizer treatments resulted in broad variations in the three variables (Table 2.1). Generally, variable values increased with the N application rates. With cotton growth, the difference between N treatments was greater. Multiple comparison analysis was used to test if the mean values of the three variables were significantly different between N treatments. The results showed that the difference between N1 (no nitrogen applied) and N360 treatments (the highest N rate) were significant ($P < 0.05$). At middle and late growth stages, the N90 treatments also was significantly different from other treatments ($P < 0.05$). At late stages, each difference in above ground dry biomass (ADB) between two N treatments was significant at the 95% level.

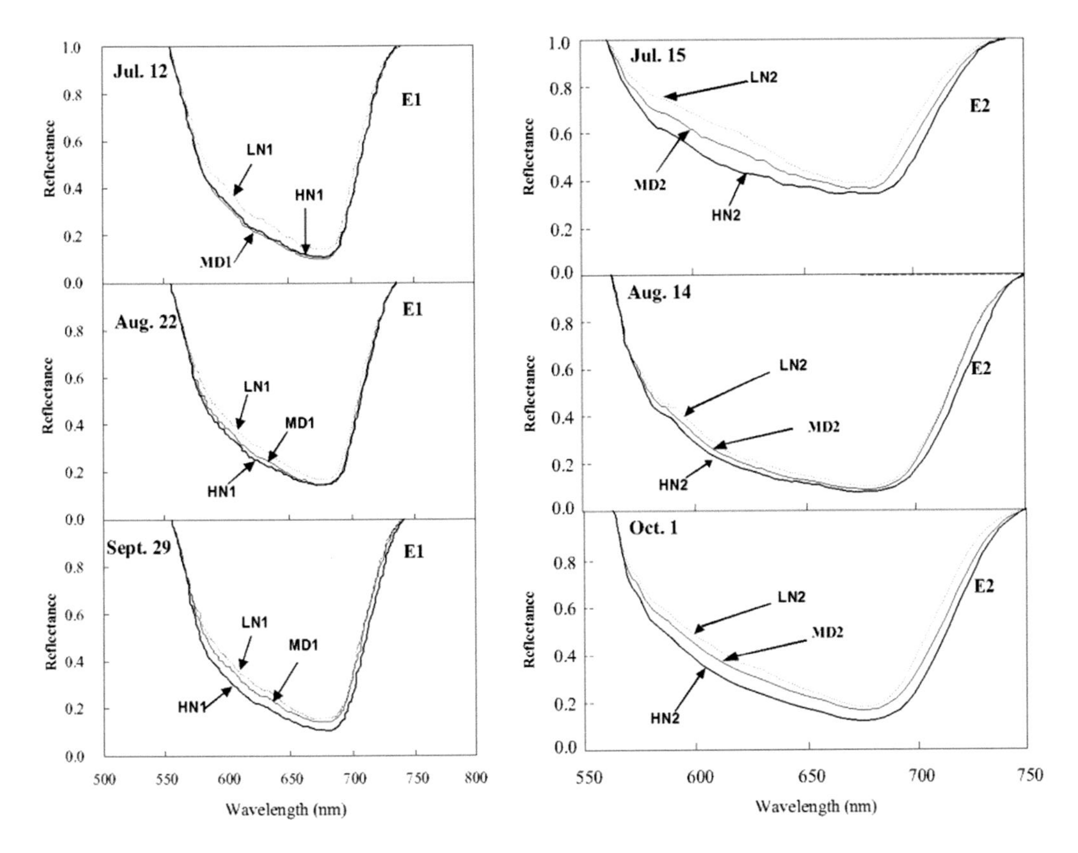

FIGURE 2.4 Continuum removed mean reflectance spectra under different treatments at experiments 1 and 2 (HN: high nitrogen; MN: medium nitrogen; LN: low nitrogen).

2.4 DISCUSSIONS

The above results showed that the nitrogen nutrition difference of crops greatly affected the canopy spectral reflectance (Figures 2.3 and 2.4). In the visible light region, spectral reflectance at around 550 nm showed obvious difference tendencies in statistics, especially at the senescent period, due to the absorption difference of chlorophyll, which was caused by different nitrogen application and lead to the difference of canopy spectral reflectance between different treatments due to the close positive linear correlation between nitrogen and chlorophyll.

Increasing the nitrogen application will increase the content of chlorophyll and biomass of crops, which could decrease the spectral reflectance in the visible light region. But this study has shown that the differences of canopy spectral reflectance were not stable in visible light. The differences could reach 10% relative difference level from spectral band 660–680 nm in the rapid growth period and senescent period, due to little plant material present, partially uncovered soil in early growth stage, and part of the senesced leaves showing in senescent period. But in the full green canopy coverage period the relative difference was not ideal because the canopy coverage in three different treatments could reach almost 100% cover and show no obvious difference between each other.

Here, canopy reflectance showed a stable and significant difference in the near–infrared spectral region, which is sustainable from the rapid growth period to senescent period. The spectral reflectance of plant in the near–infrared is mainly affected by leaf and canopies structure (Serrano et al. 2000). The biomass and LAI of plant is added with the increasing of nitrogen amount, which lead to the evident difference of reflection spectra in different treatment. This study also showed that in near—infrared spectra region the spectral difference between different nitrogen applications could reach

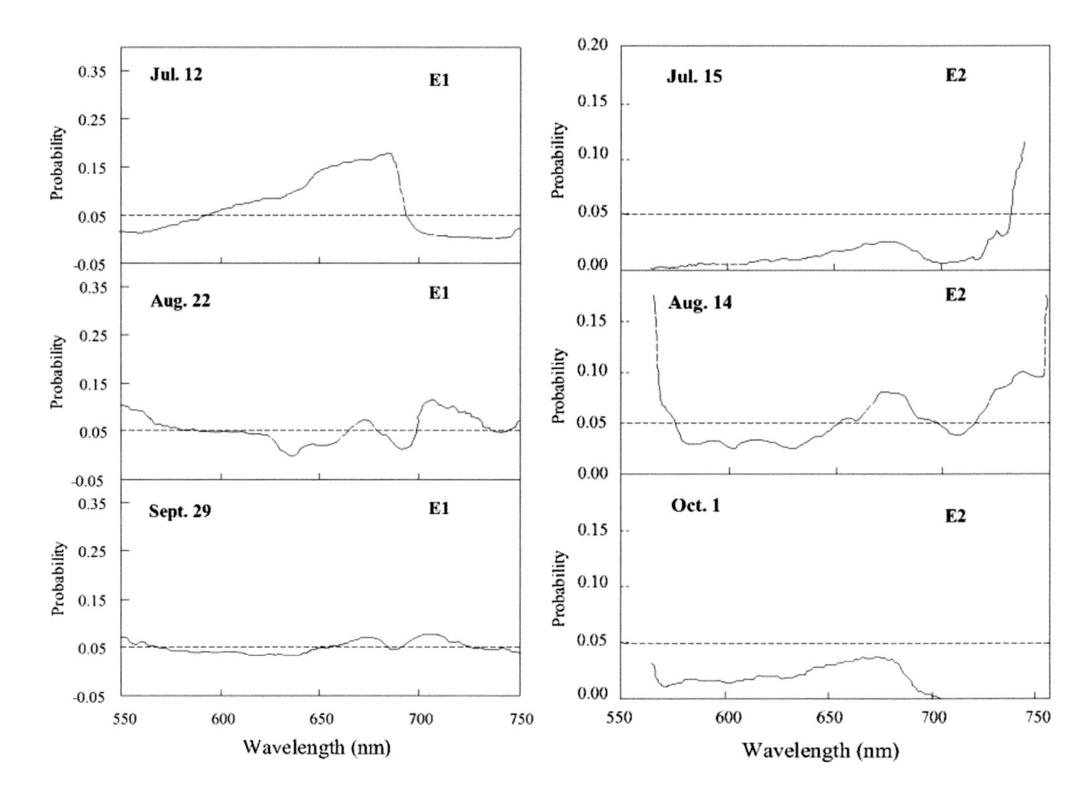

FIGURE 2.5 Results of one-way ANOVA showing wavelengths where continuum removed reflectance differences between three treatments were significant. Horizontal dashed and solid lines show 95% and 90% confidence limits, respectively.

a 5% significant level. The reflected characteristic in near–infrared spectra provided a possibly for distinguishing the nutrition condition of plant nitrogen in different nitrogen treatments.

Results of absorption characteristic in red valley absorption area between 550 and 750 nm in normalized difference method showed that the width and depth of absorption characteristic will increase with nitrogen application rate increase. In general, it is believed that the absorption center of the chlorophyll is 680 nm. This study on two experiments showed the difference of spectra statistics at 680 nm was not significant; the best spectral difference among nitrogen treatments existed on the absorption valley side of slopes (Figures 2.4 and 2.5). The reason might be that higher levels of chlorophyll contents made the red valley wider. Therefore, the normalized difference approach can improve the difference of reflected spectra in the visible light region under different nitrogen treatments; the best spectral band existed around the 620–640 nm and 690–710 nm regions. These results confirmed the research conclusions of Carter (1994) and Mutanga et al. (2003) which indicted the sensitive spectra area exists at the 535–640 nm and 685–700 nm. The reason that the normalized difference approach improved the difference of reflected spectra in different nitrogen treatments, might be that this method can eliminate much of the absolute error by reflectance and can strengthen the absorption valley.

2.5 CONCLUSIONS

Increasing the nitrogen (N) application will increase the content of chlorophyll and biomass of crops, which typically decreases the spectral reflectance in the visible light region. But this study has shown that the difference of canopy spectral reflectance were not stable in visible light and are dependent on factors such as crop growth stages and soil background effects. The difference in reflectance, for example, between 10% difference level from spectral band 660 and 680 nm is as high as 10% or greater during the rapid growth period as a result of significant soil background effects (as a result of <100% canopy cover) or during senescent period, as a result of last loosing chlorophyll (lower reflectance or higher absorption associated with healthier plants with greater canopy cover, plant moisture, and biomass). But in the full green canopy coverage period the difference was not significant because the canopy coverage in three different treatments could all reach almost 100% cover and show no obvious difference between each other. These factors highlight the need to use specific narrowbands from targeted portions of the spectrum to better characterize and study vegetation.

Overall, the results of this study offer a possibility to estimate cotton canopy quality at the field level. The results trigger the need to investigate band depths and slopes, particularly the red edge, and to estimate cotton nutrition quality at the canopy level. The development of models that can manipulate the influence of factors such as the atmosphere, species mix and non-photosynthetic vegetation (standing litter, woody stems, etc.) at different times will be important for further hyperspectral remote sensing of cotton quality. Normalized difference methods can successfully identify the characteristic of cotton canopies without being affected by spectral reflectance of cotton canopies structure and background. These results have an important significance to evaluation of nitrogen content in visible spectral areas.

REFERENCES

Boegh, E., Soegaarda, H., Broge, N., Hasagerc, C.B., Jensenc, N.O., Scheldeb, K., Thomsen, A., 2002. Airborne multispectral data for quantifying leaf area index, nitrogen concentration, and photosynthetic efficiency in agriculture. *Remote Sens. Environ.* 81, 179–193.

Broge, N.H., Leblanc, E., 2001. Comparing prediction power and stability of broad-band and hyperspectral vegetation indices for estimation of green leaf area index and canopy chlorophyll density. *Remote Sens. Environ.* 76, 156–172.

Broge, N.H., Mortensen, J.V., 2002. Deriving crop area index and canopy chlorophyll density of winter wheat from spectral reflectance data. *Remote Sens. Environ.* 81, 45–57.

Carter, G.A., 1994. Ratios of leaf reflectances in narrow wavebands as indicators of plant stress. *Int. J. Remote Sens.* 15, 697–704.

Craig, J.C., 2001. Multi-scale remote sensing techniques for vegetation stress detection. PhD dissertation. University of Florida, Gainesville, Florida.

Daughtry, C.S.T., Walthall, C.L., Kim, M.S., Brown de Colstoun, E., McMurtrey, III J.E., 2000. Estimating corn leaf chlorophyll concentration from leaf and canopy reflectance. *Remote Sens. Environ.* 74, 229–239.

Elvidge C.D., Chen, Z., 1995. Comparison of broad-band and narrow-band red and near-infrared vegetation indices. *Remote Sensing of Environment*, 54, 38–48.

Gitelson, A.A., Kaufman, Y.J., Stark, R., Rundquist, D., 2002. Novel algorithms for remote estimation of vegetation fraction. *Remote Sens. Environ.* 80, 76–87.

Hansen, P.M., Schjoerring, J.K., 2003. Reflectance measurement of canopy biomass and nitrogen status in wheat crops using normalized difference vegetation indices and partial least squares regression. *Remote Sens. Environ.* 86, 542–553.

Kokaly, R.F., 2001. Investigating a physical basis for spectroscopic estimates of leaf nitrogen concentration. *Remote Sens. Environ.* 75, 153–161.

Kokaly, R.F., Clark, R.N., 1999. Spectroscopic determination of leaf biochemistry using band-depth analysis of absorption features and stepwise multiple linear regression. *Remote Sens. Environ.* 67, 267–287.

Kumar, L., Schmidt, K.S., Dury, S., Skidmore, A.K., 2001. Imaging spectrometry and vegetation science. In: Van der Meer, F., De Jong, S.M. (Eds.), *Imaging Spectrometry*. Kluwer Academic Publishers, Dordrecht, The Netherlands, pp. 111–155.

Mutanga, O., Skidmore, A.K., Wieren, S., 2003. Discrimination tropical grass (Cenchrus ciliaris) canopies grown under different nitrogen treatments using spectroradiometry. *J. Photogram. Remote Sens.* 57, 133–272.

Pattey, E., Strachan, I.B., Boisvert, J.B., Desjardins, R.L., McLaughlin, N.B., 2001. Detecting effects of nitrogen rate and weather on corn growth using micrometeorological and hyperspectral reflectance measurements. *Agric. Forest Meteorol.* 108, 85–99.

Schmidt, K.S., Skidmore, A.K., 2001. Exploring spectral discrimination of vegetation types in African rangelands. *Int. J. Remote Sens.* 22, 3421–3434.
Serrano, L., Filella, L., Peñuelas, J., 2000. Remote sensing of biomass and yield of winter wheat under different nitrogen supplies. *Crop Sci.* 40, 723–731.
Sims, D.A., Gamon, J.A., 2002. Relationships between leaf pigment content and spectral reflectance across a wide range of species, leaf structures and developmental stages. *Remote Sens. Environ.* 81, 331–354.
Smil, V., 1997. Global population and nitrogen cycle. *Sci. Am.* 277, 76–81.
Strachan, I.B., Pattey, E., Boisvert, J.B., 2002. Impact of nitrogen and environment conditions on corn as detected by hyperspectral reflectance. *Remote Sens. Environ.* 80, 213–224.
Weisz, R., Crozier, C.R., Heiniger, R.W., 2001. Optimizing nitrogen application timing in no-till soft red winter wheat. *Agron. J.* 93, 435–442.
Zarco-Tejada, P.J., Miller, J.R., Mohammed, G.H., Noland, T.L., Sampson, P.H., 2000. Chlorophyll fluorescence effects on vegetation apparent reflectance: II. Laboratory and airborne canopy-level measurements with hyperspectral data. *Romote Sens. Environ.* 74, 596–608.

3 Analysis of the Effects of Heavy Metals on Vegetation Hyperspectral Reflectance Properties

E. Terrence Slonecker

CONTENTS

3.1 INTRODUCTION

Absolute definitions of "heavy metals" are elusive in modern science. Many different definitions have been proposed. Some are based on density, some on atomic number or atomic weight, and some on chemical properties or toxicity [1]. One definition holds that they are elements with a specific weight higher than 6 g/cm^3 [2]. See Figure 3.1. But no single definition fits well in modern usage. The term "toxic metals" has become to some extent synonymous with heavy metals, but that term is equally problematic because levels of toxicity are highly variable between different metals and vegetation species. At best, heavy metals can be classified as a loosely defined subset of elements that exhibit metallic properties and are toxic to living organisms at some level of concentration or exposure. The term "heavy metals" itself has been criticized as functionally meaningless [1,3].

Metals in the environment, however, are a real concern for a variety of reasons, including their commercial and industrial value, medicinal applications, use in agricultural products, and their toxic effects on human and ecological resources as chemical weapons or as fugitive, uncontrolled, anthropogenic releases into the environment. Some metals, such as selenium, copper, and zinc, are micronutrients that are actually required by most plant and animal life forms in very small doses, while others, such as mercury and lead, are toxic and have no known benefit to living organisms.

Although the toxicity of many heavy metals can vary widely, the term has evolved to have pejorative connotations that make it synonymous with anthropogenic pollution. Heavy metals can occur naturally and can arise from many anthropogenic sources, such as mining and processing of other metals, the smelting of copper, the processing of gold, steel, iron, and coal, the preparation of nuclear fuels, and the production of industrial construction materials. In addition, many computer parts and chips contain heavy metals or involve a production process that results in waste products with heavy metals. Electroplating is a primary source of chromium and cadmium pollution. Arsenic

hydrogen 1 H 1.0079																		helium 2 He 4.0026
lithium 3 Li 6.941	beryllium 4 Be 9.0122												boron 5 B 10.811	carbon 6 C 12.011	nitrogen 7 N 14.007	oxygen 8 O 15.999	fluorine 9 F 18.998	neon 10 Ne 20.180
sodium 11 Na 22.990	magnesium 12 Mg 24.305												aluminium 13 Al 26.982	silicon 14 Si 28.086	phosphorus 15 P 30.974	sulfur 16 S 32.065	chlorine 17 Cl 35.453	argon 18 Ar 39.948
potassium 19 K 39.098	calcium 20 Ca 40.078		scandium 21 Sc 44.956	titanium 22 Ti 47.867	vanadium 23 V 50.942	chromium 24 Cr 51.996	manganese 25 Mn 54.938	iron 26 Fe 55.845	cobalt 27 Co 58.933	nickel 28 Ni 58.693	copper 29 Cu 63.546	zinc 30 Zn 65.39	gallium 31 Ga 69.723	germanium 32 Ge 72.61	arsenic 33 As 74.922	selenium 34 Se 78.96	bromine 35 Br 79.904	krypton 36 Kr 83.80
rubidium 37 Rb 85.468	strontium 38 Sr 87.62		yttrium 39 Y 88.906	zirconium 40 Zr 91.224	niobium 41 Nb 92.906	molybdenum 42 Mo 95.94	technetium 43 Tc [98]	ruthenium 44 Ru 101.07	rhodium 45 Rh 102.91	palladium 46 Pd 106.42	silver 47 Ag 107.87	cadmium 48 Cd 112.41	indium 49 In 114.82	tin 50 Sn 118.71	antimony 51 Sb 121.76	tellurium 52 Te 127.60	iodine 53 I 126.90	xenon 54 Xe 131.29
caesium 55 Cs 132.91	barium 56 Ba 137.33	57-70 *	lutetium 71 Lu 174.97	hafnium 72 Hf 178.49	tantalum 73 Ta 180.95	tungsten 74 W 183.84	rhenium 75 Re 186.21	osmium 76 Os 190.23	iridium 77 Ir 192.22	platinum 78 Pt 195.08	gold 79 Au 196.97	mercury 80 Hg 200.59	thallium 81 Tl 204.38	lead 82 Pb 207.2	bismuth 83 Bi 208.98	polonium 84 Po [209]	astatine 85 At [210]	radon 86 Rn [222]
francium 87 Fr [223]	radium 88 Ra [226]	89-102 **	lawrencium 103 Lr [262]	rutherfordium 104 Rf [261]	dubnium 105 Db [262]	seaborgium 106 Sg [266]	bohrium 107 Bh [264]	hassium 108 Hs [269]	meitnerium 109 Mt [268]	ununnilium 110 Uun [271]	unununium 111 Uuu [272]	ununbium 112 Uub [277]		ununquadium 114 Uuq [289]				

FIGURE 3.1 The Periodic Table showing the elements generally considered heavy metals. Lanthanides and actinides are not shown. (Modified from Shaw, B. et al. In *Heavy Metal Stress in Plants: From Biomolecules to Ecosystems*, 2004; Vol. 2, pp. 84–126.[71])

has been used extensively in pesticides and in wood treating [4] and, because of its toxicity, has been used for years as a base compound for chemical warfare weapons such as Lewisite gas [5].

Hyperspectral remote sensing (HRS), also known as imaging spectroscopy, and, to a greater extent, traditional field and laboratory spectroscopy have a long history of being used to investigate the identification of metals and their effects on vegetation in the environment. However, fugitive metals in the environment do not usually exist in their pure form but rather in a soil-water-vegetation matrix as waste rock materials or sediments or as a result of soil deposition. Besides detecting the minerals themselves, spectroscopy and imaging spectroscopy can also be used to detect the composition and condition of vegetation, which can then be used to interpret mineral deposits or metal composition of the soil in the area of vegetation growth. It has long been acknowledged by scientists that a relationship exists between vegetation, soils, and underlying mineral deposits [6]. In several studies, airborne spectroscopy was used to detect "hidden" mineral deposits through forest-covered areas by revealing subtle variations in the reflected spectrum of vegetation under stress due to the presence of heavy metals [7–10]. In addition, recent scientific literature reflects a growing interest in the spectroscopic identification of environmental hazards, many of which are metals in the soils and vegetation. Another area that has received increased attention in the area of spectroscopy of metal stress in vegetation is that of vegetation indices (VIs). This chapter reviews the scientific background of spectroscopy and imaging spectroscopy with respect to the effects of heavy metals on vegetation reflectance.

3.2 PHYSIOLOGY OF METAL STRESS IN PLANTS

Plants are generally more exposed to pollution risks in the environment because they are stationary and cannot avoid interacting with environmental pollutants such as metals. Plants have evolved various complex strategies for adapting to heavy metal pollution in soil or water media. Plants respond to exposure to heavy metals in several different ways. Metals usually interfere with basic plant metabolism, and enzyme activity is often negatively affected. Metals present in plant tissues can cause plants to form chelate structures, molecules that enclose and isolate metal ions and cause them to lose their functional properties in metabolic cycles such as the citric acid cycle.

Plants generally fall into two categories with respect to strategies for dealing with exposure to heavy metals: *accumulators* uptake metal ions and process them in some manner, storing them in internal tissues or reducing or processing them in biochemical reactions, whereas *excluders* generally

TABLE 3.1
Examples of Visual Symptoms of Metals Stress in Plants

Metal	Characteristics	References
Arsenic	Red/brown necrotic spots on old leaves, yellow/brown roots, reduced growth	[36,37]
Aluminum	Stunted growth, inhibition of root elongation, purple Coloration, curling and yellowing of leaf tips	[72,73]
Cadmium	Brown edges to leaves, chlorosis, necrosis, curled leaves, stunted roots	[74,75]
Copper	Chlorosis, yellow and purple coloration, decreased root growth and leaf biomass	[76–78]
Lead	Dark green leaves, stunted growth, chlorosis, and blackening of root system	[79]
Mercury	Severe stunting of seedlings and roots, chlorosis, reduced biomass	[80]
Nickel	Chlorosis, necrosis, stunting, reduced root and leaf growth	[81]
Selenium	Interveined chlorosis, black spots, bleaching and yellowing of young leaves, pink spots on roots	[17]
Zinc	Chlorosis, stunting, reduced root elongation	[82]

Source: Modified from Shaw, B. et al. In *Heavy Metal Stress in Plants-from Biomolecules to Ecosystems*, 2004; Vol. 2, pp. 84–126.[71]

restrict the uptake of metals by preventing their uptake into plant tissues. This is often accomplished by trapping metal ions in the cell walls of the root tissue.

Whether plants are accumulators or excluders, excess metals in the soil or in plant tissues tend to have negative effects on plant health, growth, and biomass accumulation and can cause visual symptoms at toxic levels. Table 3.1 shows examples of the visual injuries to various flowering plants from metal exposure. These visual symptoms also affect the reflectance characteristics of the typical vegetation spectra. Figure 3.2a and b show increasing visual damage to plant health and the corresponding changes in the blue and red energy absorption troughs at 480 and 680nm, respectively, seen as increasing reflectance and a blue shift.

Excess metal exposure negatively affects photosynthetic processes and typically induces a general "stress" reaction in plants. In some cases, the absorbed metal ion will replace the central magnesium atom in the chlorophyll molecule, which generally causes oxidative stress in the plant. This substitution reduces or prevents photosynthetic light harvesting and results in a breakdown of photosynthesis [11].

Heavy metal exposure can also interfere with plant-water relations. Metals may alter plasma membrane properties, affect enzyme activities, inhibit root growth and elongation, affect osmotic potential, and generally inhibit the ability of the plant to acquire water [12]. This may be manifested as a general drought-stress response but is actually caused primarily by the interference of heavy metals and not simply the lack of water availability.

In general, many different photosynthetic reactions and physiological processes are negatively affected by plant exposure to heavy metals. These vary widely among different species and metals, but in many cases both light and dark photosynthetic reactions are generally inhibited [13].

3.3 BASIC SPECTROSCOPY OF VEGETATION

Spectroscopy is the study of the interaction between energy and matter as a function of either wavelength (λ) or frequency (ν). Historically, spectroscopy referred to the use of visible light dispersed by a prism according to its wavelength and is the parent science to all visible and near-infrared (VNIR) HRS. Dating from the nineteenth century [14], spectroscopic techniques have been used widely in analytical chemistry and astronomy to identify many elemental substances, minerals, and organic compounds.

(a)

(b)

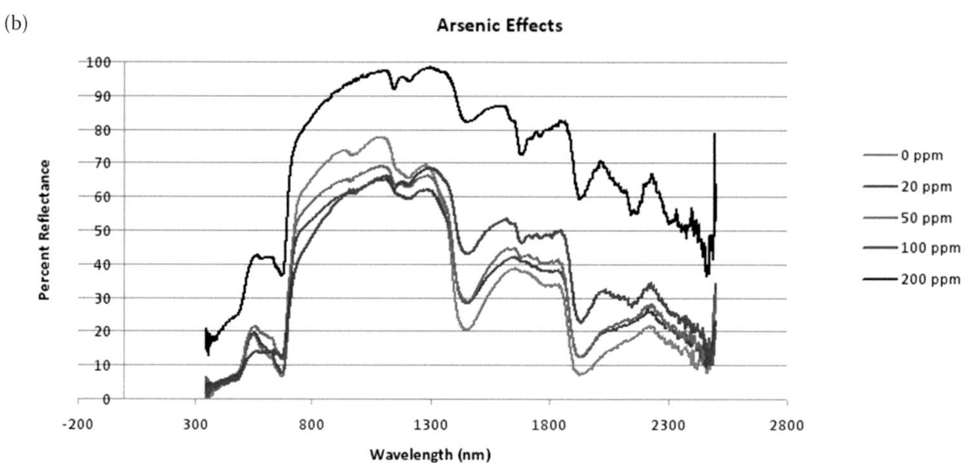

FIGURE 3.2 (a) Visual effects of arsenic stress on *Nephrolepis exaltata* (Boston fern). Ferns are planted in clean sand amended with, from left to right, 0, 20, 50, 100, and 200 ppm sodium arsenate. (From Slonecker, E., *Remote Sensing Investigations of Fugitive Soil Arsenic and Its Effects on Vegetation Reflectance*. George Mason University: Fairfax, Virginia, 2007.[36]) (b) Laboratory reflectance spectra of arsenic-affected ferns in Figure 3.2a above. Spectra were collected with an ASD full-range spectrometer from 15 cm above the canopy of each plant. Note the loss of photosynthetic absorption at 680 nm, causing higher reflectance, the blue shift, and the general increase in reflectance in shortwave infrared (due to loss of water) with increasing soil arsenic. (From Barcelo, J., Poschenrieder, C., *Journal of Plant Nutrition* 1990, *13*, 1–37.[12])

The use of spectral reflectance methods to gain an understanding of photosynthesis and related vegetative processes is a field of scientific study that has been ongoing for decades [15,16]. Laboratory instruments called spectrometers, spectrophotometers, spectrographs, and spectroradiometers are all different names for instruments that essentially use some type of prism to separate light into its component parts and measure the reflectance and absorption of each of those individual component parts from a target surface. Early instruments separated light into the basic colors of the spectrum. Modern instruments separate light into individual nanometers of reflectance energy.

In this review, "hyperspectral" remote sensing technology is afforded the broadest possible definition. The papers reviewed here represent a variety of spectroscopic remote sensing systems and approaches that include individual leaf-level and plant-level analysis under controlled conditions in the laboratory to spectroscopic measurements of plants in the field to overhead aircraft and satellite systems. The common thread is that multiple bands of energy reflectance are recorded and analyzed with spectroscopic methods.

Different spectroscopic collection perspectives also contain inherent advantages and disadvantages that include complications involving the detection and analysis of the reflected energy signal. Outside of a pure laboratory setting, field collections generally involve variable solar lighting, background effects from soil and other materials, and effects from the bidirectional reflection distribution function. Aircraft and especially satellite sensors contain increasingly significant signal noise from atmospheric moisture and constituent gases.

The majority of papers and research studies reviewed here involve spectrometers used in either a laboratory or field setting. Some utilize aircraft and satellite systems, and a few represent a multiscale data collection from the laboratory to field to aircraft or satellite sensor. While the availability and applications of aircraft and satellite systems is growing significantly, and this will be a prime focus area of future research, hyperspectral research in the laboratory and field represents a critical first step in developing and understanding, in repeatable spectral measurement, the effects of heavy metals on plant reflectance.

3.4 SPECTROSCOPY AND IMAGING SPECTROSCOPY OF METAL INTERACTIONS WITH PLANTS

Early spectroscopic analysis of vegetation-metal interactions from both laboratory and aircraft sensors can be traced to the late 1970s and early 1980s, when researchers such as Collins, Milton, and Horler demonstrated repeated shifts in the so-called red edge of typical vegetation reflectance-based stress or enhanced growth caused by excessive exposure to metals in the soil [7,17,18]. This has evolved into a fundamental spectroscopic-plant principle that is still widely used today. The red edge of vegetation reflectance is an area usually centered around 720 nm and represented by the typical sharp rise in reflectance in the 680–760 nm range of the classic vegetation spectral signature. Figure 3.3 shows the classic red edge area of vegetation spectra.

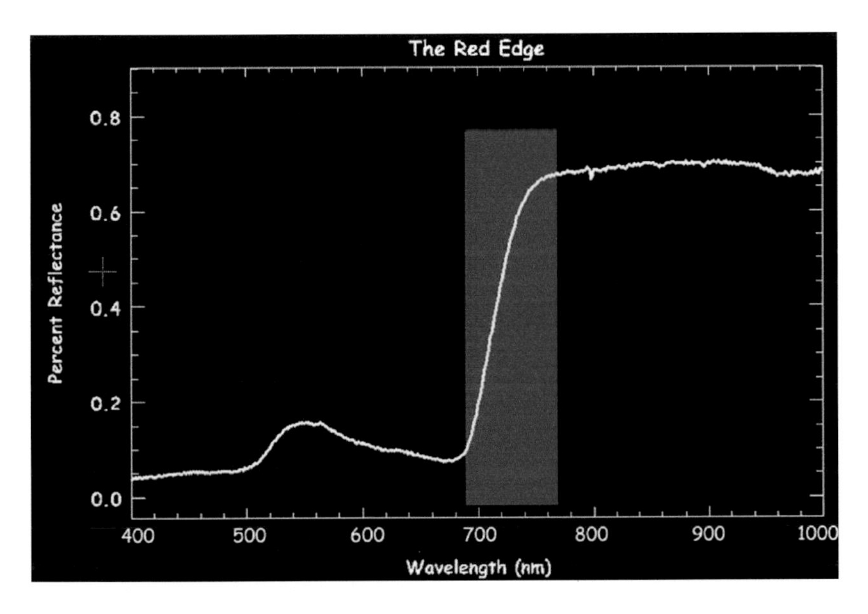

FIGURE 3.3 Red edge. An important region of vegetation spectra is known as the red edge. Much research has focused on measuring shifts in this region corresponding to stress or enhancement of chlorophyll. (From Slonecker, T. et al., *Remote Sensing* 2009, *1*, 644.[37])

Although the general concept of the red edge is easily understood as the area of a sharp rise in reflectance, a variety of definitions and quantitative methods for computing the red edge are found in the literature. Ray [19] defined the red edge as the sharp transition between absorption by chlorophyll in the visible wavelengths and the strong scattering in the NIR from the cellular structure of leaves. The red edge is defined by Horler et al. [18] as the wavelength of maximum $\Delta R/\Delta\lambda$, where R is reflectance and λ is the specific wavelength. Guyot [20] defines the red edge as an inflection in the sharp rise in reflectance between 670 and 760 nm. Although variable in the literature, most modern definitions of the red edge involve the peak of the first derivative [21]. Additional red-edge-related measurements include a ratio of R_{740}/R_{720} and a ratio of first derivative values D_{715}/D_{705} [22].

The general movement of the spectral features in the red edge area is one of the keys to its analytical strength. When plants are healthy and producing more chlorophyll, the red edge tends to shift toward the right to longer wavelengths. This is also usually accompanied by an increase in the absorption trough at 680 nm as the plant absorbs more energy in the photosynthetic process. When a plant is stressed, such as in the case of excessive heavy metals in the soil, the spectra tend to shift toward the left and shorter wavelengths. Stress also tends to produce an increase in reflectance at the 680 absorption through as less light is being utilized for photosynthesis and chlorophyll production. Figure 3.4 shows an example of this stress based on a laboratory experiment with varying levels of copper sulfate in the soil.

Horler [18] studied the feasibility of utilizing a red edge measurement as an indication of plant chlorophyll status. Using derivative reflectance spectroscopy in the laboratory, plant chlorophyll status, and red edge measurements were acquired from single leaves of several different species

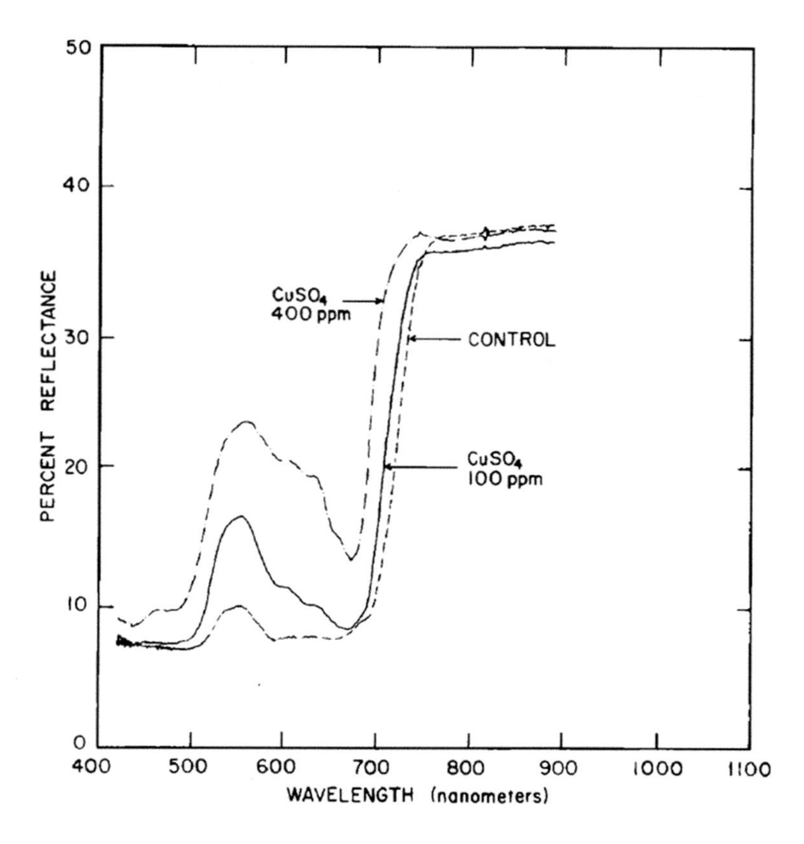

FIGURE 3.4 The "blue" shift in the red edge in laboratory-grown sorghum exposed to different levels of copper sulfate in soil. (From Chang, S., Collins, W., *Economic Geology* 1983, *78*, 723.[10])

under heavy metal stress. By using spectroscopic and laboratory methods to measure the chlorophyll content of the same leaf samples, direct evidence of the red edge–chlorophyll correlation was obtained. Measuring in situ vegetation using a field spectrometer, Ray [19] discovered significant differences in the size and shape of the red edge in different types of arid vegetation and found for a common yellow grass species that there was no chlorophyll "bump" at the green peak and no detectable red edge.

A critical component of spectral analysis of vegetation is the shift in absorption and reflectance features that occur as a result of chemical and nutrient exposures. A general relation between increases in chlorophyll concentration and a "red shift" toward longer wavelengths has been established by several researchers utilizing both laboratory and field spectrographic methods. Gates [23] showed the basic relationship between increased chlorophyll and plant health and the shift of the red edge toward longer wavelengths. Guyot [20] similarly showed that the red edge inflection point shifts to longer red wavelengths as chlorophyll concentrations increase. This general correlation between chlorophyll content and red shift was confirmed by Horler [24] and Baret [25] for different crop species.

More important for this specific research topic, however, is the "blue shift" (i.e., shift toward shorter wavelengths) of the red edge that occurs when vegetation has undergone stress from some mineral or chemical agent. The blue or red shift toward shorter or longer wavelengths, respectively, is one of the keys to detecting stress and growth in all green vegetation. The blue shift is usually accompanied by a general increase in overall reflectance and an increase in the 680 nm absorption feature showing that less light energy is being utilized for photosynthesis.

In some of the first applications comparing field and airborne spectroscopic measurements of metal stress, Collins [8] and Chang and Collins [10] showed a blue shift in the 700–780 nm region of reflectance spectra from conifers affected by metal sulfide. See Figure 3.4. Similar blue shift results have been reported by Schwaller and Tkach from field applications and aerial photographs [26] and Milton in the laboratory [17,27]. In a seminal remote sensing research application using both in situ and airborne measurements, Rock [28] demonstrated a 5 nm blue shift in spruce and fir species in Vermont and Germany as a result of stress caused by airborne pollutant deposition.

Although the underlying physiology is not completely understood, the uptake of heavy metals has the effect of reducing photosynthetic activity and the concentration of chlorophyll. One mechanism of heavy-metal-induced damage in plants that leads to a reduction in photosynthesis involves the in vivo replacement of the central Mg^{2+} ion in the chlorophyll molecule by a heavy metal ion. This replacement is generally toxic to the plant depending on the metal and, at the very least, inhibits the overall ability of the plant to conduct photosynthesis. In general, the magnesium-chlorophyll molecule has a much higher capacity to release electrons than other metals, and replacement by other metals quenches or reduces the ability of the plant to regulate excess light energy and protect the plant from damage [11,29,30].

In another classic paper utilizing both lab and field spectral measurements, Horler [18] studied the effects of heavy metals on the reflectance spectra of plants. Utilizing both natural vegetation growing in known areas of metal concentrations and specific greenhouse experiments, relationships were established between metal stress, total chlorophyll, chlorophyll a/b ratios, and reduced reflectance at specific wavelengths. Controlled experiments with pea plants and other species showed that the general effect of exposure to cadmium (Cd), copper (Cu), lead (Pb), and zinc (Zn) was growth inhibition. Also, the pea plants showed changes in the leaf chlorophyll a/b ratios for exposure to Cd and Cu but showed no changes for Pb and Zn. Metal-treated plants in both controlled and natural environments showed a decrease in reflectance at 850, 1650, and 2,200 nm and an increase at 660 nm. Metal concentration in soil has strong negative correlations to reflectance at 1650 and 2200 nm and strong positive correlations at 660 nm. In general, the ability to measure stress effects from heavy metals is dependent on species, the phase of the growth cycle, and the environment.

Kooistra [31] conducted a study to examine the possibilities for in situ evaluation of soil properties in river floodplains using field reflectance spectroscopy of cover vegetation. Results determined that a combination of field spectroscopy and multivariate calibration leads to a qualitative relation between organic matter and clay content, which are intercorrelated with levels of Cd and Zn. The study indicated the potential for these multivariate methods for mapping soil properties using HRS techniques. Kooistra [32,33] conducted two additional studies to investigate the relation between vegetation reflectance and soil characteristics, including elevated concentrations of the metals Ni, Cd, Cu, Zn, and Pb found in floodplain soils along the Rhine and Meuse Rivers in the Netherlands. These studies obtained high-resolution vegetation reflectance spectra in the visible to NIR using a field radiometer [32]. The relationships were evaluated using simple linear regression in combination with two spectral VIs: the difference vegetation index (DVI) and the red edge position (REP). The R^2 values between metal concentrations and vegetation reflectance ranged from 0.50 to 0.73. The results of the study demonstrated the potential of remote sensing data to contribute to the survey of spatially distributed soil contaminants in floodplains under natural grasslands, using the spectral response of the vegetation as an indicator. Modeling the relationship between soil contamination and vegetation reflectance resulted in similar results for DVI, REP, and the multivariate approach using partial least-squares (PLS) regression [32,33].

Similar studies were conducted by Clevers et al. [34,35] in contaminated floodplains in the Netherlands. Analysis of field spectrometer measurements of reflectance found that REP and the first derivative peaks around 705 and 725 nm were the best predictors of heavy metal contamination. Similarly, Slonecker [36,37] showed the spectral relationship between arsenic uptake and spectral reflectance in arsenic-hyperaccumulating *Pteris* ferns using a PLS regression. Rosso et al. successfully detected plant stress due to metal pollution at the leaf level and reiterated that more investigations need to be undertaken that link their results to canopy-level reflectance [38].

Slonecker [36] used both laboratory spectra and HyMAP imagery spectra of arsenic stress in common lawn grasses to map the distribution of fugitive arsenic and other metals in household lawns in an urban setting. The hyperspectral imagery was processed with a linear spectral unmixing algorithm and mapped with a maximum-likelihood classifier. Classes included grass, arsenic-affected grass, trees, buildings, soil, asphalt, and concrete and showed an overall accuracy of 82.9%. Critical spectral parameters for identifying arsenic stress were located in the green, red, NIR plateau, and water-absorption bands in both laboratory and imagery spectra. Validated against comprehensive ground sampling efforts, final maps of the arsenic-affected grass showed an overall producer's accuracy of 55.8% and an overall user's accuracy of 82.7%. See Figure 3.5.

Gallagher [39] utilized field spectrometry and Ikonos multispectral satellite measurements to assess basal area, plant productivity, and chlorophyll content of gray birch growing in soils containing elevated metals in a New Jersey Brownfields site. Biomass production, measured by a red/green ratio index, showed an inverse relationship ($R^2 = 0.46 - 0.81$) to soil zinc concentration. The relationship was stronger when the total metal levels (TMLs) were higher. Threshold TMLs were established for several species beyond which the normalized difference vegetation index (NDVI) decreased at both the assemblage and individual tree level.

Mars and Crowley [40] utilized AVIRIS and digital elevation model (DEM) data to evaluate hazardous waste contamination in southeastern Idaho, including mine waste dumps, wetland vegetation, and other relevant vegetation types. With the mapped information and the DEM, the delineation of mine dump morphologies, catchment watershed areas above each mine dump, flow directions from the dumps, stream gradients, and the extent of downstream wetlands available for selenium absorption were determined. Compared to ground-truth maps, the AVIRIS imagery correctly identified 76% of all mine waste pixels. Additionally, Mars and Crowley were able to characterize the physical settings of mine dumps and test hypotheses concerning the causes of selenium contamination in the area [40].

Ren et al. [41] found that rice exposed to lead in the soil weakened the photosynthetic process of rice as measured by field spectral measurements. Lead concentrations in rice could be reliably predicted by changes in the normalized band absorption depth, blue shifts in the red edge region, and the distance of the shift.

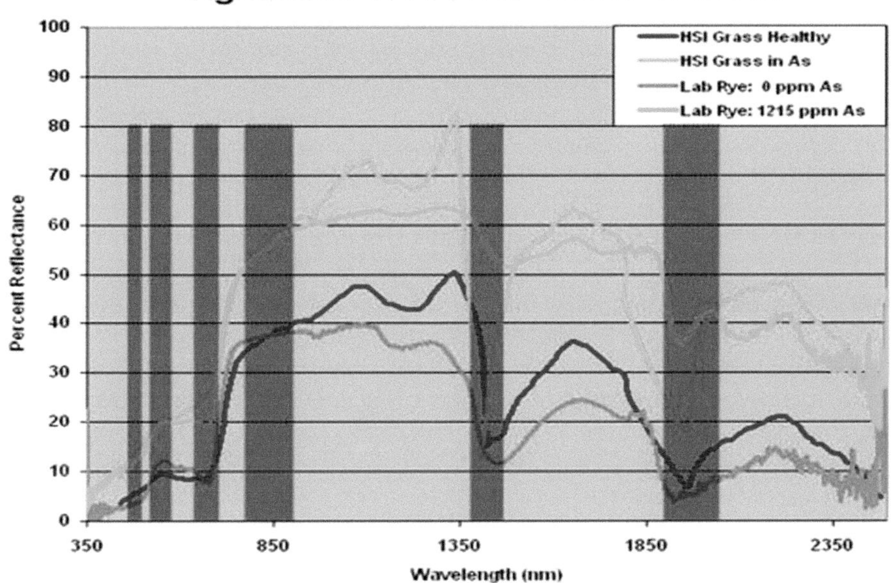

FIGURE 3.5 Healthy and stressed grass signatures from both laboratory and hyperspectral imagery. The same critical areas in the green, red, near-infrared, and shortwave infrared show the patterns of spectral separation between healthy and stressed grass that enable the image processing algorithm to separate, identify, and map arsenic-stressed grasses. (From Slonecker, E., *Remote Sensing Investigations of Fugitive Soil Arsenic and Its Effects on Vegetation Reflectance*. George Mason University: Fairfax, Virginia, 2007.[36])

3.5 VEGETATION INDICES

One area that has received recent attention in the area of spectroscopy of metal stress in vegetation is that of vegetation indices (VIs), which are mathematical manipulations of digital number values of two or more bands of data; they have been a fundamental part of the remote sensing analysis of vegetation for decades. VIs typically stretch or enhance a particular part of the reflected electromagnetic spectrum (EMS) known to relate to specific vegetation qualities such as chlorophyll content, leaf moisture, pigment ratios, and stress level. The search for stressed or unusual growth patterns in cover vegetation, such as potential metal stress patterns, has been enhanced by the use of one or more VIs reported in the scientific literature.

The most widely known and used VI is the NDVI, which is calculated by the following general band formula:

$$\text{NDV} = \frac{\text{NIR} - \text{Red}}{\text{NIR} + \text{Red}}$$

where NIR is the reflectance from the near-infrared band, and R is the reflectance from the red visible band. The NDVI was first proposed by Pearson and Miller in 1972 [42] and has been widely utilized as a general measure of vegetation condition and has both broadband and narrowband formulas for its computation. Although the NDVI has been the most widely used VI, it has clear limitations. The NDVI becomes saturated in areas of multilayered canopy and shows nonlinear relationships with critical vegetation parameters such as the leaf area index (LAI). As a result, substantial efforts have been devoted to developing new indices that improve on the shortcomings of the NDVI [43].

VIs have often been developed for specific purposes and optimized to assess a specific condition or process. Also, the emergence and increasing availability of hyperspectral data and imagery have resulted in a new class of VIs, known as narrowband indices, that capitalize on the increased spectral resolution of hyperspectral data.

For example, Penuelas [44] proposed a structurally insensitive pigment index (SIPI) that incorporates a NIR band (800 nm) to minimize internal leaf structure effects such as increased scattering due to refractive index discontinuities between air and cell walls inside leaves. Gamon [45] developed the Photochemical Reflectance Index (PRI) to estimate the physiological parameters of sunflowers undergoing nitrogen stress. Huete [46] developed a VI that accounts for, and minimizes, the effect of soil background conditions. The soil-adjusted vegetation index (SAVI) equation introduces a soil-brightness-dependent correction factor, L, that compensates for the difference in soil background conditions. NIR is the reflectance from the near-infrared band, and R is the reflectance from the red visible band. Applying a correction for the soil provides more accurate information on the condition of the vegetation itself. The Triangular Vegetation Index (TVI) was developed as a very precise measure of chlorophyll concentration and absorption and depends on very specific narrow wavelengths [47].

Agricultural vegetation applications of both field and airborne hyperspectral data analysis have been conducted by several researchers, showing the promise of this technology in monitoring plant production for food supplies. Strachan [48] and Daughtry [49] both showed that very narrow, crop-specific VIs could be developed and utilized from hyperspectral data and applied to the assessment of agricultural productivity. In general, the use of VIs has seen a significant increase with the development and availability of hyperspectral data. Elvidge and Chen [50], Blackburn [51,52], and Thenkabail et al. [53,54] have demonstrated the effectiveness of narrowband VIs, which continues as one of the most important analytical approaches in the area of spectroscopic analysis of vegetation. Table 3.2 shows the several VIs that are mentioned in this paper along with the spectral calculation and literature source.

VIs have also played an important role in the detection and analysis of stress due to heavy metals (Table 3.3). Reusen et al. [55] successfully mapped heavy metal contamination in Belgium through the expressions of vegetation stress in conifers near abandoned zinc smelting facilities. Utilizing imaging data from an airborne hyperspectral sensor (CASI), they utilized a Spectral Angle Mapper (SAM) classification to build a mask for pine trees and then computed 18 separate VIs of stress. The Edge-Green First derivative Normalized difference (EGFN) VI proved to be the best indicator of zinc stress in the pine trees in the surrounding area [55].

Götze et al. [56] used reflectance spectroscopic methods in both the laboratory and field to quantify and separate heavy metal stress in floodplain vegetation. Testing a series of VIs, they showed that metal stress could be uniquely separated from other forms of stress such as water or nutrient stress. The indices that proved to be most sensitive to the stress from heavy metals in the soil were the normalized pigment chlorophyll index (NPCI), the PRI, the REP, and the continuum removed band depth at 1730 nm (CR1730) [56].

Using both field and laboratory measurements, Slonecker [36] showed that the PRI was sensitive to metal stress in the form of inorganic arsenic, Thorhaug [57] showed that the PRI was sensitive to the effects of low salinity in seagrass health, and Gallagher [39] showed that a red/green ratio index had an inverse relationship with zinc concentrations in gray birch trees.

Several VIs seem to dominate the literature with respect to metal stress in vegetation. The REP described earlier is the most dominant spectral feature used to assess plant stress. It has been used by many researchers to evaluate decreases in plant chlorophyll, biomass, or physiological health with respect to metal stress [7–10,17,24,34,35,39,41,56,58–63].

The PRI was developed by Gamon et al. [45] as a narrowband hyperspectral indicator of changes in the pigment balance of plants due to photosynthetic stress. Originally designed to track diurnal

TABLE 3.2
Vegetation Indices Specifically Referenced in This Paper

Name	Acronym	Formula	Reference
Anthocyanin Reflectance Index	ARI	(1/R550) – (1/R700)	Gitleson et al. [84]
Difference Vegetation Index	DVI	2.4 * MSS7 – MSS5	Richardson and Wiegand [85]
Modified Triangular Vegetation Index 2	MTVI2	1.5[1.2(R800 – R550) – 1.3 (R670 – R550)]/ SQRT[(2 * (R800 + 1)2) – (6 * R800 – 5 * SQRT(R670)) – 0.5]	Haboudane et al. [86]
Moisture Stress Index	MSI	(R1599 – R819)	Hunt and Rock [87]
Normalized Difference Vegetation Index (Broadband)	NDVI	(NIR – RED)/(NIR + RED)	Rouse [88]
Normalized Difference Vegetation Index	NDVI	(R800 – R670)/(R800 + R670)	Sims and Gamon [89] (Narrowband)
Normalized Pigment Chlorophyll Index	NPCI	(R680 – R430)/(R680 + R430)	Peñuelas et al. [44]
Photochemical Reflectance Index	PRI	(R531 – R570)/(R531 + R570)	Gamon et al. [45]
Red Edge Position	REP	R1Dmax: (R1D690 – R1D740)	Curran et al. [90]
Red Edge Vegetation Stress Index	RVSI	((R714 – R752)/2) – R733	Merton [91]
Soil-Adjusted Vegetation Index	SAVI	(1 + 0.5) (R800 – R670)/ (R800 + R670 + 0.5)	Huete [46]
Structure-Insensitive Pigment Index	SIPI	(R800 – R445)/(R800 – R680)	Penuelas et al. [44]
Triangular Vegetation Index	TVI	0.5 * (((120 * (R750 – R550)) – (200 * (R670 – R550)))	Broge and Leblanc 2000 [47]

TABLE 3.3
Some Key Spectral Features and Vegetation Indices Related to Metal Stress in the Literature

Spectral Feature	Metal(s)	Vegetation Type	Sensor	Reference(s)
DVI, REP	Ni, Cd, Cu, Pb, Zn	Floodplain, ryegrass	ASD	[32]
EGFN	Zn	Conifer	CASI	[55]
NDVI	Cr, Pb, Zn, V	Gray birch	ASD	[39]
RGI			Ikonos	
NDVI	Ni, Cd, Cu, Pb, Zn	Rice	Landsat TM	[83]
PRI	General HM	Floodplain	ASD	[56]
PRI	As	Ferns	ASD	[36,37]
REP	Pb	Rice	ASD	[41]
REP	Cu	Peas, maize	PE 554	[61]
	Zn	Sunflower		
REP	General HM	Floodplain Bluegrass, ryegrass	ASD	[34,35]
RVI	Hg	Mustard spinach	ASD	[59]
NDVI, REP				
NPCI, PRI, REP	General HM	Stinging nettles Reed canary grass Meadow foxtail	ASD	[56]
R_{850}	Cd, Cu, Pb, Zn, As	Peas	PE 554	[18]
R_{1650}	Cd, Cu, Pb, Zn, As	Peas	PE 554	[18]
CR_{1730}	General HM	Floodplain	ASD	[56]
R_{2200}	Cd, Cu, Pb, Zn, As	Peas	PE 554	[18]

changes in photosynthetic efficiency, the PRI is sensitive to changes in carotenoid pigments and the epoxidation state of the xanthrophyll cycle. This is a measure of photosynthetic light use efficiency and the rate of carbon dioxide uptake. The PRI measures the relative reflectance on either side of the green maxima around 550 nm and compares reflectance parameters in both the red and green regions simultaneously. Because the change in pigment concentrations due to metal stress in most vascular plants is similar, the PRI has been shown to be a successful indicator of a variety of stress conditions, including stress from soil metals. Slonecker [36] computed a suite of 67 broadband and hyperspectral VIs and used a PLS and stepwise linear regression (SLR) analysis to isolate the best VIs for explaining arsenic stress in Boston ferns and arsenic hyperaccumulating *Pteris* ferns. The results showed for the control Boston ferns that the PRI, along with the Moisture Stress Index, the red edge vegetation stress index and the modified TVI2 provided the best model for explaining the level of arsenic uptake. These indices measure plant stress in one form or another, which generally increases with higher concentrations of soil arsenic. The best indices for the hyperaccumulating *Pteris* ferns were the broadband green index (GI), the sum green index (SGI), and the carotenoid reflectance index (CRI1), all relating to the green part of the spectrum. Although not fully understood, the different indices for stressed and hyperaccumulating species reflect key differences in internal plant physiology [36].

Götze et al. [56] found that four indices were highly correlated between heavy metal content and chlorophyll content. The R^2 values for the NCPI (0.91), PRI (0.75), REP (0.80), and the continuum-removed spectra at 1730 nm (0.74) were all sensitive to metal stress in plants. Although the underlying physiology is not fully understood, the authors speculate that the correlation could be related to lignin or protein production in the plant synthesis. Further, this study shows promising results for using these values to separate heavy metal stress from water and nutrient stress [56].

3.6 EMERGING STATISTICAL METHODS

A wide variety of analytical methods can be noted in a review of the hyperspectral analysis of vegetation and vegetation stress. One of the fundamental issues relates to the fact that the analysis of hyperspectral data presents unique analytical problems for standard multivariate techniques because of the highly correlative and overlapping nature of data. The large numbers of independent variables (>1,500 spectral bands) and the highly correlated nature of those variables stem from the fact that each individual spectral band is only a few nanometers away from the spectral bands above and below it, and the result is that each spectral band records an energy pattern that is similar to its neighboring bands. Highly correlated independent variables create a condition known as collinearity, which violates the assumptions of linear regression. To develop a predictive and effective linear model, variables must be independent. The overall result of a collinearity condition is that correlated independent variables have unstable coefficients, and although the model developed may have a high r^2 value and low residuals, it will perform poorly outside of the immediate data set that was used to develop it.

In recent years, a special statistical technique has emerged that addresses the problems of numerous, highly correlated variables. The technique, known as partial least squares (PLS), was first introduced in 1966 by Swedish mathematician Herman Wold as an exploratory analysis technique in the field of econometrics [64]. It was specifically designed to help researchers in situations of small, nonnormally distributed data sets with numerous but highly correlated explanatory variables. General PLS and all of its variants consist of a set of regression and classification tasks as well as dimension reduction techniques and modeling tools. Sometimes called a "soft" modeling technique, the strength of PLS resides in its relaxation, or "softening," of the distribution, normality, and collinearity restrictions that are inherent in standard multiple linear regression techniques [65,66].

The underlying assumption of all PLS methods is that the observed data are generated by a system or process that is driven by a small number of latent (not directly observed or intuitive) variables. Projection of the observed data to their latent structure by means of PLS is a variation of principal component analysis (PCA). PLS generalizes and combines features from PCA and multiple regression and is similar to canonical correlation analysis in that it can also relate a set of independent variables to a set of multiple dependent response variables and extract latent vectors with maximum correlation [67,68].

The overall goal of PLS processing of laboratory spectral data is the reduction of 2,151 variables (bands 350–2,500 spectrometer data) down to a manageable number of variables (approximately 100) that have a high probability of significance in a predictive model. The PLS regression produces a number of significant factors using a "leave-one-out" cross-validation method [60]. At several stages in the PLS process, diagnostic checks are performed, sometimes graphically, to help isolate variables for deletion in the model that do not have any significant predictive value or are outliers. The end result of a PLS run is a variable importance in projection (VIP) table. The VIP represents the value of each variable in fitting a PLS model for both predictors and responses. The VIP for each factor is defined as the square root of the weighted average times the number of predictors. If a predictor has a relatively small coefficient (in absolute value) and a small value of VIP, then it is a prime candidate for deletion. Variables with VIP values less than 0.8 and outliers are dropped from the variable list. The VIP table results are then typically divided into four to nine groups. The PLS analysis process is then repeated on the individual groups of variables. Typically the process is iterated two to five times until a manageable subset of variables can be identified based on the top VIP scores in each group and some a priori knowledge of the process being modeled. PLS itself can be used to construct a predictive model, but it has some drawbacks. One of the strengths of PLS is its relaxation of collinearity and distribution assumptions, but this can also result in a set of collinear or redundant independent variables. Also, the best combinations of variables are not necessarily reflected in the VIP table values.

In spectral applications, a common practice is to take the final subset of variables and then place them in a SLR model. The stepwise method is a modification of the forward variable selection technique and differs in that variables already in the model do not necessarily stay there. The SLR model computes the F-statistic for each variable and contains parameters for significance levels for variables to *enter* and *stay* in the model. The SLR process computes all possible combinations of linear variables and ends when none of the variables outside the model has significance (p-value) at or below the *entry* level and every variable in the model is significant at the *stay* level. Using these sigma-restricted parameterization and general linear model methods, the SLR process simply regresses all possible combinations of input variables and returns the model with the best regression coefficient and the lowest residuals [36,65].

PLS is also used as an exploratory/data mining and analysis tool in remote sensing. As a relatively new technique, the full utilization of PLS is still evolving, but it is clear that it has a major role to play in several types of spectral, remote sensing analyses due to the large numbers of potential predictive variables and the highly correlated nature of hyperspectral reflectance and hyperspectral imaging data.

3.7 SUMMARY AND CONCLUSIONS

This paper has reviewed the hyperspectral applications of detecting the effects on vegetation of heavy metals in soil. Most spectral applications have been in the form of laboratory or field studies with portable spectrometers, as opposed to hyperspectral imagery applications. But because field spectrometers and HRS instruments essentially measure the same phenomenon at high spatial and spectral resolutions, these studies serve as a form of benchmark for airborne or spaceborne remote sensing development and several studies with airborne or spaceborne HRS instruments, such as AVIRIS [40], CASI [55], and HyMAP [36], have successfully demonstrated, metal-specific vegetation applications of hyperspectral imagery.

The metals involved included a wide range of elements, including general heavy metal contamination, as might be expected in industrial or urban floodplains [31–35,56], and metal-specific applications, such as arsenic [17,36,37,69], lead [41], zinc [55], and selenium [17]. Vegetation targets included general forest canopy, general floodplain, common grasses, and species-specific applications. Hyperspectral methods included standard applications of the NDVI and red edge and newer methods that included VIs such as the PRI, NCPI, and EGFN and a very interesting application of continuum removal and 1730 nm [56,70].

Research on hyperspectral detection of heavy metals and their effects on vegetation is in its infancy. Although much research has been carried out on other forms of vegetation condition such as stress or agricultural productivity, specific attention to metals is currently a primary scientific gap that demands research attention.

One of the direct needs for hyperspectral research is developing the ability to differentiate metal-induced stress from other types of stress such as a drought or nutrient stress. Greenhouse experiments, where stress levels are controlled and then measured with a field spectrometer, could be extremely valuable in determining where metal stress can be reliably and uniquely identified in spectra and for establishing underlying mechanisms causing spectral variation. Götze et al. [56] made a breakthrough in the identification of specific stress agents, and additional work in this area is encouraged.

Further, controlled experiments could be conducted to determine whether stress from specific metals can be uniquely identified using hyperspectral methods. As various metals interact differently with plant biochemistry and photosynthetic processes, it is feasible that stress patterns due to specific metals could be identified and utilized effectively. There could also be specific indicator species that identify the presence of metals in soil, and development of this line of research would have commercial as well as ecological value.

Additional studies that utilize both field and overhead instruments and scale up the spectral responses as a function of spatial scale are needed and represent a critical gap in the current state of the science. Lastly, data-mining efforts, such as those using PLS, that systematically consider thousands or even millions of possible band combinations and compute their statistical relevance against a known data set, would be a valuable approach to teasing out very narrow and specific spectral parameters that are not fully understood.

3.8 FUTURE APPLICATIONS

A better understanding of the spectral response to metals in soil has three primary and valuable applications. First, economic prospecting for metal deposits was one of the early applications and remains just as viable today. Second, metals often hinder agricultural productivity, and a method of monitoring their presence remotely would have immediate application to food production throughout much of the world. Third, the problem of fugitive hazardous wastes in the environment is not one that is likely to diminish in the future. As the global population grows, the need for natural resource exploitation will increase dramatically, along with the negative side effects of mining, industrial byproducts, and both controlled and fugitive wastes. As this review has indicated, there have been numerous successful hyperspectral applications of remote sensing for the location and monitoring of hazardous metals in the environment. Unlike earlier systems, HRS has the potential to identify specific materials based on molecular structure, and although considerable laboratory research continues, overhead aircraft and satellite remote sensing applications are still in their infancy due to complex atmospheric interferences, cost, and data availability. But all of these factors are steadily improving, and there is opportunity for considerable research in the area of hyperspectral monitoring of metal effects on vegetation.

REFERENCES

1. Duffus, J., Heavy metals–a meaningless term. *Pure and Appied. Chemistry* 2002, *74*, 793–807.
2. Alloway, B., *Heavy Metals in Soils*. Springer: 1995.

3. Nieboer, E., Richardson, D., The replacement of the nondescript term [] heavy metals' by a biologically and chemically significant classification of metal ions. *Environmental Pollution Series B, Chemical and Physical* 1980, *1*, 3–26.
4. Nriagu, J. O., *Arsenic in the Environment: Cycling and Characterization*. John Wiley & Sons, Inc.: New York, NY, 1994; Vol. *1*.
5. Albright, R., *Cleanup of Chemical and Explosive Munitions: Locating, Identifying Contaminants, and Planning for Environmental Remediation of Land and Sea Military Ranges and Ordnance Dumpsites*. William Andrew Publishing: 2008; p. 267.
6. Sabins, F., Remote sensing for mineral exploration. *Ore Geology Reviews* 1999, *14*, 157–183.
7. Collins, W., Spectroradiometric detection and mapping of areas enriched in ferric iron minerals using airborne and orbiting instruments. Unpublished PhD dissertation, Columbia University, 120 p. 1978, Remote sensing of crop type and maturity: *Photo-grammetric Engineering and Remote Sensing*, 1976.
8. Collins, W., Chang, S., Kuo, J., *Detection of hidden mineral deposits by airborne spectral analysis of forest canopies*. NASA Contract NSG-5222, Final Report 1981, p. 61.
9. Collins, W., Chang, S., Raines, G., Canney, F., Ashley, R., Airborne biogeophysical mapping of hidden mineral deposits. *Economic Geology* 1983, *78*, 737.
10. Chang, S., Collins, W., Confirmation of the airborne biogeophysical mineral exploration technique using laboratory methods. *Economic Geology* 1983, *78*, 723.
11. Küpper, H., Küpper, F., Spiller, M., Environmental relevance of heavy metal-substituted chlorophylls using the example of water plants. *Journal of Experimental Botany* 1996, *47*, 259.
12. Barcelo, J., Poschenrieder, C., Plant water relations as affected by heavy metal stress: A review. *Journal of Plant Nutrition* 1990, *13*, 1–37.
13. Mysliwa-Kurdziel, B., Prasad, M., Strzalka, K., Photosynthesis in heavy metal stressed plants. *Heavy Metal Stress in Plants: From Biomolecules to Ecosystems* 2004, 198.
14. Rood, J. J., *Modern Chromatics with Application to Art and Industry*. D. Appleton and Company: New York, 1879.
15. Shull, C., A spectrophotometric study of reflection of light from leaf surfaces. *Botanical Gazette* 1929, *87*, 583–607.
16. Willstatter, R., Stoll, A., *Investigations on Chlorophyll, 1913*. Trans. by Schertz and Merz.). Science Press: Lancaster, Pa, 1928; pp. 290–291.
17. Milton, N., Ager, C., Eiswerth, B., Power, M., Arsenic-and selenium-induced changes in spectral reflectance and morphology of soybean plants. *Remote Sensing of Environment* 1989, *30*, 263–269.
18. Horler, D., Barber, J., Barringer, A., Effects of heavy metals on the absorbance and reflectance spectra of plants. *International Journal of Remote Sensing* 1980, *1*, 121–136.
19. Ray, T., Murray, B., Chehbouni, A., Njoku, E. The red edge in arid region vegetation: 340–1060 nm spectra, In *Summaries of the 4th Annual JPL Airborne Geoscience Workshop. Volume 1: AVIRIS Workshop*, R.O. Green, Editor, 1993, 149–152
20. Guyot, G., Baret, F., Jacquemoud, S., Imaging spectroscopy for vegetation studies. *Imaging Spectroscopy: Fundamentals and Prospective Application* 1992, 145–165.
21. Curran, P., Dungan, J., Gholz, H., Exploring the relationship between reflectance red edge and chlorophyll content in slash pine. *Tree Physiology* 1990, *7*, 33.
22. Vogelmann, J., Rock, B., Moss, D., Red edge spectral measurements from sugar maple leaves. *International Journal of Remote Sensing* 1993, *14*, 1563–1575.
23. Gates, D., Keegan, H., Schleter, J., Weidner, V., Spectral properties of plants. *Applied Optics* 1965, *4*, 11–20.
24. Horler, D., Dockray, M., Barber, J., The red edge of plant leaf reflectance. *International Journal of Remote Sensing* 1983, *4*, 273–288.
25. Baret, F., Champion, I., Guyot, G., Podaire, A., Monitoring wheat canopies with a high spectral resolution radiometer. *Remote Sensing of Environment* 1987, *22*, 367–378.
26. Schwaller, M., Tkach, S., Premature leaf senescence; remote-sensing detection and utility for geobotanical prospecting. *Economic Geology* 1985, *80*, 250.
27. Milton, N., Eiswerth, B., Ager, C., Effect of phosphorus deficiency on spectral reflectance and morphology of soybean plants. *Remote Sensing of Environment* 1991, *36*, 121–127.
28. Rock, B., Hoshizaki, T., Miller, J., Comparison of *in situ* and airborne spectral measurements of the blue shift associated with forest decline. *Remote Sensing of Environment* 1988, *24*, 109–127.
29. Küpper, H., Küpper, F., Spiller, M., In situ detection of heavy metal substituted chlorophylls in water plants. *Photosynthesis Research* 1998, *58*, 123–133.

30. Küpper, H., Šetlík, I., Spiller, M., Küpper, F., Prášil, O., Heavy metal-induced inhibition of photosynthesis: targets of *in vivo* heavy metal chlorophyll formation1. *Journal of Phycology* 2002, *38*, 429–441.
31. Kooistra, L., Wehrens, R., Leuven, R., Buydens, L., Possibilities of visible-near-infrared spectroscopy for the assessment of soil contamination in river floodplains. *Analytica Chimica Acta* 2001, *446*, 97–105.
32. Kooistra, L., Salas, E., Clevers, J., Wehrens, R., Leuven, R., Nienhuis, P., Buydens, L., Exploring field vegetation reflectance as an indicator of soil contamination in river floodplains. *Environmental Pollution* 2004, *127*, 281–290.
33. Kooistra, L., Wanders, J., Epema, G., Leuven, R., Wehrens, R., Buydens, L., The potential of field spectroscopy for the assessment of sediment properties in river floodplains. *Analytica Chimica Acta* 2003, *484*, 189–200.
34. Clevers, J., Kooistra, L., Assessment of heavy metal contamination in river floodplains by using the red-edge index. *Chemical Analysis* 2001.
35. Clevers, J., Kooistra, L., Salas, E., Study of heavy metal contamination in river floodplains using the red-edge position in spectroscopic data. *International Journal of Remote Sensing* 2004, *25*, 3883–3895.
36. Slonecker, E., *Remote Sensing Investigations of Fugitive Soil Arsenic and Its Effects on Vegetation Reflectance*. George Mason University: Fairfax, Virginia, 2007.
37. Slonecker, T., Haack, B., Price, S., Spectroscopic analysis of arsenic uptake in Pteris ferns. *Remote Sensing* 2009, *1*, 644.
38. Rosso, P., Pushnik, J., Lay, M., Ustin, S., Reflectance properties and physiological responses of Salicornia virginica to heavy metal and petroleum contamination. *Environmental Pollution* 2005, *137*, 241–252.
39. Gallagher, F., Pechmann, I., Bogden, J., Grabosky, J., Weis, P., Soil metal concentrations and productivity of Betula populifolia (gray birch) as measured by field spectrometry and incremental annual growth in an abandoned urban Brownfield in New Jersey. *Environmental Pollution* 2008, *156*, 699–706.
40. Mars, J., Crowley, J., Mapping mine wastes and analyzing areas affected by selenium-rich water runoff in southeast Idaho using AVIRIS imagery and digital elevation data. *Remote Sensing of Environment* 2003, *84*, 422–436.
41. Ren, H., Zhuang, D., Pan, J., Shi, X., Wang, H., Hyper-spectral remote sensing to monitor vegetation stress. *Journal of Soils and Sediments* 2008, *8*, 323–326.
42. Pearson, R., Miller, L., Remote mapping of standing crop biomass for estimation of the productivity of the shortgrass prairie. In 1972; p. 1355.
43. Carlson, T., Ripley, D., On the relation between NDVI, fractional vegetation cover, and leaf area index* 1. *Remote Sensing of Environment* 1997, *62*, 241–252.
44. Penuelas, J., Gamon, J., Fredeen, A., Merino, J., Field, C., Reflectance indices associated with physiological changes in nitrogen-and water-limited sunflower leaves. *Remote Sensing of Environment* 1994, *48*, 135–146.
45. Gamon, J., Pe uelas, J., Field, C., A narrow-waveband spectral index that tracks diurnal changes in photosynthetic efficiency* 1. *Remote Sensing of Environment* 1992, *41*, 35–44.
46. Huete, A., A soil-adjusted vegetation index (SAVI). *Remote Sensing of Environment* 1988, *25*, 295–309.
47. Broge, N., Leblanc, E., Comparing prediction power and stability of broadband and hyperspectral vegetation indices for estimation of green leaf area index and canopy chlorophyll density. *Remote Sensing of Environment* 2001, *76*, 156–172.
48. Strachan, I., Pattey, E., Boisvert, J., Impact of nitrogen and environmental conditions on corn as detected by hyperspectral reflectance. *Remote Sensing of Environment* 2002, *80*, 213–224.
49. Daughtry, C., Walthall, C., Kim, M., De Colstoun, E., McMurtrey III, J., Estimating corn leaf chlorophyll concentration from leaf and canopy reflectance. *Remote Sensing of Environment* 2000, *74*, 229–239.
50. Elvidge, C., Chen, Z., Comparison of broad-band and narrow-band red and near-infrared vegetation indices. *Remote Sensing of Environment* 1995, *54*, 38–48.
51. Blackburn, G., Quantifying chlorophylls and caroteniods at leaf and canopy scales: An evaluation of some hyperspectral approaches. *Remote Sensing of Environment* 1998, *66*, 273–285.
52. Blackburn, G., Hyperspectral remote sensing of plant pigments. *Journal of Experimental Botany* 2007, *58*, 855.
53. Thenkabail, P., Smith, R., De Pauw, E., Hyperspectral vegetation indices and their relationships with agricultural crop characteristics. *Remote Sensing of Environment* 2000, *71*, 158–182.
54. Thenkabail, P., Smith, R., De Pauw, E., Evaluation of narrowband and broadband vegetation indices for determining optimal hyperspectral wavebands for agricultural crop characterization. *Photogrammetric Engineering and Remote Sensing* 2002, *68*, 607–622.

55. Reusen, I., Bertels, L., Debruyn, W., Deronde, B., Fransaer, D., Sterckx, S., Species Identification and Stress Detection of Heavy-Metal Contaminated Trees. 2003.
56. Götze, C., Jung, A., Merbach, I., Wennrich, R., Gläßer, C., Spectrometric analyses in comparison to the physiological condition of heavy metal stressed floodplain vegetation in a standardised experiment. *Central European Journal of Geosciences* 2010, *2*, 132–137.
57. Thorhaug, A., Richardson, A., Berlyn, G., Spectral reflectance of Thalassia testudinum (Hydrocharitaceae) seagrass: low salinity effects. *American Journal of Botany* 2006, *93*, 110.
58. Choe, E., van der Meer, F., van Ruitenbeek, F., van der Werff, H., de Smeth, B., Kim, K., Mapping of heavy metal pollution in stream sediments using combined geochemistry, field spectroscopy, and hyperspectral remote sensing: A case study of the Rodalquilar mining area, SE Spain. *Remote Sensing of Environment* 2008, *112*, 3222–3233.
59. Dunagan, S., Gilmore, M., Varekamp, J., Effects of mercury on visible/near-infrared reflectance spectra of mustard spinach plants (Brassica rapa P.). *Environmental Pollution* 2007, *148*, 301–311.
60. Guang-yu, C., Xin-hui, L., Su-hong, L., Zhi-feng, Y., Spectral Characteristics of Vegetation in Environment Pollution Monitoring [J]. *Environmental Science and Technology* 2005, *1*.
61. Horler, D., Barber, J., Darch, J., Ferns, D., Barringer, A., Approaches to detection of geochemical stress in vegetation. *Advances in Space Research* 1983, *3*, 175–179.
62. Wickham, J., Chesley, M., Lancaster, J., Mouat, D., *Remote Sensing for the Geobotanical and Biogeochemical Assessment of Environmental Contamination*; DOE/NV/10845–27, Nevada Univ., Reno, NV (United States). Desert Research Inst.: 1993.
63. Xia, L., Shoo-Feng, L., Zheng, L., High spectral resolution data applied to identify plant stress response to heavy metal in mine site [J]. *Science of Surveying and Mapping* 2007, *2*.
64. Wold, H., Estimation of principal components and related models by iterative least squares. *Multivariate Analysis* 1966, *1*, 391–420.
65. Tobias, R., An introduction to partial least squares regression. In *Citeseer*: 1995; pp. 1250–1257.
66. Abdi, H., Partial least squares (PLS) regression. In *Encyclopedia of Social Sciences Research Methods* (eds. M. Lewis–Beck, A. Bryman and T. Futing) 2003; pp. 1–7.
67. Höskuldsson, A., PLS regression methods. *Journal of Chemometrics* 1988, *2*, 211–228.
68. Rosipal, R., Krämer, N., Overview and recent advances in partial least squares. *Subspace, Latent Structure and Feature Selection* 2006, 34–51.
69. Sridhar, B., Han, F., Diehl, S., Monts, D., Su, Y., Spectral reflectance and leaf internal structure changes of barley plants due to phytoextraction of zinc and cadmium. *International Journal of Remote Sensing* 2007, *28*, 1041–1054.
70. Clark, R., Roush, T., Reflectance spectroscopy: Quantitative analysis techniques for remote sensing applications. *Journal of Geophysical Research* 1984, *89*, 6329–6340.
71. Shaw, B., Sahu, S., Mishra, R., Heavy metal induced oxidative damage in terrestrial plants. In *Heavy Metal Stress in Plants-from Biomolecules to Ecosystems*, 2004; Vol. 2, pp. 84–126.
72. Delhaize, E., Ryan, P., Aluminum toxicity and tolerance in plants. *Plant Physiology* 1995, *107*, 315.
73. Roy, A., Sharma, A., Talukder, G., Some aspects of aluminum toxicity in plants. *The Botanical Review* 1988, *54*, 145–178.
74. Jastrow, J., Koeppe, D., *Uptake and Effects of Cadmium in Higher Plants*. John Wiley & Sons: 605 Third Ave., New York, NY 10016 USA, 1980; pp. 607–638.
75. Das, P., Samantaray, S., Rout, G., Studies on cadmium toxicity in plants: A review. *Environmental Pollution* 1997, *98*, 29–36.
76. Kukkola, E., Rautio, P., Huttunen, S., Stress indications in copper-and nickel-exposed Scots pine seedlings. *Environmental and Experimental Botany* 2000, *43*, 197–210.
77. Mocquot, B., Vangronsveld, J., Clijsters, H., Mench, M., Copper toxicity in young maize (Zea mays L.) plants: Effects on growth, mineral and chlorophyll contents, and enzyme activities. *Plant and Soil* 1996, *182*, 287–300.
78. Masarovicová, E., Cicák, A., Štefančík, I., Plant responses to air pollution and heavy metal stress. In *Handbook of Plant and Crop Stress. Marcel Dekker, New York* (ed. M. Pessaraki) Marcel Dekker, Inc.: New York, 1999; pp. 569–598.
79. Sharma, P., Dubey, R., Lead toxicity in plants. *Brazilian Journal of Plant Physiology* 2005, *17*, 35–52.
80. Patra, M., Sharma, A., Mercury toxicity in plants. *The Botanical Review* 2000, *66*, 379–422.
81. Khalid, B., Tinsley, J., Some effects of nickel toxicity on rye grass. *Plant and Soil* 1980, *55*, 139–144.
82. Bonnet, M., Camares, O., Veisseire, P., Effects of zinc and influence of Acremonium lolii on growth parameters, chlorophyll a fluorescence and antioxidant enzyme activities of ryegrass (Lolium perenne L. cv Apollo). *Journal of Experimental Botany* 2000, *51*, 945.

83. Boluda, R., Andreu, V., Gilabert, M., Sobrino, P., Relation between reflectance of rice crop and indices of pollution by heavy metals in soils of Albufera Natural Park (Valencia, Spain). *Soil Technology* 1993, *6*, 351–363.
84. Gitelson, A., Merzlyak, M., Chivkunova, O., Optical Properties and Nondestructive Estimation of Anthocyanin Content in Plant Leaves. *Photochemistry and Photobiology* 2001, *74*, 38–45.
85. Richardson, A., Wiegand, C., Distinguishing vegetation from soil background information (by gray mapping of Landsat MSS data). *Photogrammetric Engineering and Remote Sensing* 1977, *43*, 1541–1552.
86. Haboudane, D., Miller, J., Pattey, E., Zarco-Tejada, P., Strachan, I., Hyperspectral vegetation indices and novel algorithms for predicting green LAI of crop canopies: Modeling and validation in the context of precision agriculture. *Remote Sensing of Environment* 2004, *90*, 337–352.
87. Hunt Jr., E., Rock, B., Detection of changes in leaf water content using near-and middle-infrared reflectances. *Remote Sensing of Environment* 1989, *30*, 43–54.
88. Rouse, J., Monitoring vegetation systems in the Great Plains with ERTS, In *3rd ERTS-1 Symposium*, NASA Goddard Space Flight Center 1974.
89. Sims, D., Gamon, J., Relationships between leaf pigment content and spectral reflectance across a wide range of species, leaf structures and developmental stages. *Remote Sensing of Environment* 2002, *81*, 337–354.
90. Curran, P., Dungan, J., Macler, B., Plummer, S., The effect of a red leaf pigment on the relationship between red edge and chlorophyll concentration. *Remote Sensing of Environment* 1991, *35*, 69–76.
91. Merton, R. Monitoring community hysteresis using spectral shift analysis and the red-edge vegetation stress index, In *Proceedings of the Seventh Annual JPL Airborne Earth Science Workshop.* NASA, Jet Propulsion Laboratory, Pasadena, California, USA 1998.

Section II

Vegetation Processes and Function (ET, Water Use, GPP, LUE, Phenology)

4 Mapping the Distribution and Abundance of Flowering Plants Using Hyperspectral Sensing

Tobias Landmann, Hannes Feilhauer, Miaogen Shen, Jin Chen, and Suresh Raina

CONTENTS

4.1 INTRODUCTION

Flowering is an important phenological response in plants that enables pollination and propagation (Galpaz et al., 2006). Flowering plants and pollination-related traits form spatial patterns on regional scales that are closely linked to environmental (mostly climatic) gradients (Kühn et al., 2006; Pellissier et al., 2010). Further, spatial patterns emerge on local scales but without the strong relation to environmental gradients (Feilhauer et al., 2016; Kohler et al., 2008).

Spatially explicit information on floral patterns may help to monitor the effects of land use (Wesche et al., 2012) and restoration programs (Dixon, 2009), improve the assessment and analysis of ecosystem services and climate change (Von Holle et al., 2010; Schulp et al., 2014), and increase our knowledge of ecosystem functions and plant-pollinator interactions (Burkle and Alarcón, 2011; Steffan-Dewenter et al., 2002).

The concept of pollination syndromes states that flower color and architecture determine which insects will act as pollinators (Müller, 1881; Van der Pijl et al., 1960). According to this concept, bees are, for example, attracted by yellowish and blueish, butterflies by pinkish and reddish, and flies and wasps by whitish and brownish flowers. This specialization is further complicated by flower architecture which is adapted to the morphological and functional features of the respective insect group and by the presence of attractive traits such as UV patterns and floral scents or reward systems like nectar. Pollination syndromes offer the great opportunity to infer directly from spectrally easy-to-map flowering colors the presence or even abundance of specific pollinators. However, the concept has been criticized, and several studies show that the syndromes in general do not hold true and are subject to local variations (Moeller & Geber, 2005; Ollerton et al., 2009; Waser et al., 1996). Still, at least some consistency in pollinator preferences can be assumed (Gong & Huang, 2011) and possibly exploited in combination with remote sensing approaches for ecological studies.

Spatial information about ecosystem-specific floral responses and cycles can also be linked to climate and provide a clear interpretable signal of ecological changes due to climate change and climate variability (Von Holle et al., 2010). Climate effects often alter the ambient temperature, and as a consequence changes in the phenological and floral cycle of plants can occur, although these changes are species and biome specific and generalizations cannot be made over larger areas (Craufurd & Wheeler, 2009). Floral responses to climate changes, often as short-term changes, are, interestingly, expected to also alter changes in interactions between vegetation species, which has ramifications for community structural changes and long-term ecosystem functioning (Dunne et al., 2003).

In this chapter we will give an overview of the spectral properties of flowering, the importance of flower maps, and an overview of challenges and limitations of HS imaging in this regard. Results from important case studies will be compared in terms of the potential of HS imaging to map flowering fractions, site-specific flowering distribution patterns, and specific traits such as flowering color and vegetation-community-associated flowering. The insights from this chapter can help to guide the geospatial mapping community and resource managers to design floral mapping routines and policies for various applications. These applications could pertain to the role of flowering plants for pollination and better management of invasive plant species that can be effectively mapped using their floral signals.

4.2 SPECTRAL FLOWERING RESPONSE

Flowering is a mainly short and subtle response that is essentially associated with plant- and canopy-level decreases in chlorophylls (*a* and *b*), while distinct color changes occur due to alternations in carotenoid pigmentation levels (Ge et al., 2006). The visible (VIS) waveband region (450–680 nm, i.e., the green/blue and red spectral regions) is frequently associated with these flowering-induced pigmentation changes at the leaf or plant canopy level (Chen et al., 2009). Floral-induced changes in carotenoid and chlorophyll levels essentially decrease absorption in the VIS waveband region. Figure 4.1 shows that the spectral response curve for flowering (gray curve and shaded area) for *Solidago gigantea* has a higher reflectance, that is, lower absorption, in the VIS domain than the corresponding "green" leaf reflectance (blackish area with white curve). *S. gigantea* (top right in Figure 4.1) belongs to the *Asteraceae* family and is a yellow flowering weed that is native to North America and a common neophyte in Europe. Since the red waveband spectrum (650–680 nm) is also the maximum chlorophyll absorption area, wavebands centered here provide particularly good separability between flowering and "green" leaved canopies (Sims & Gamon, 2003). In some cases

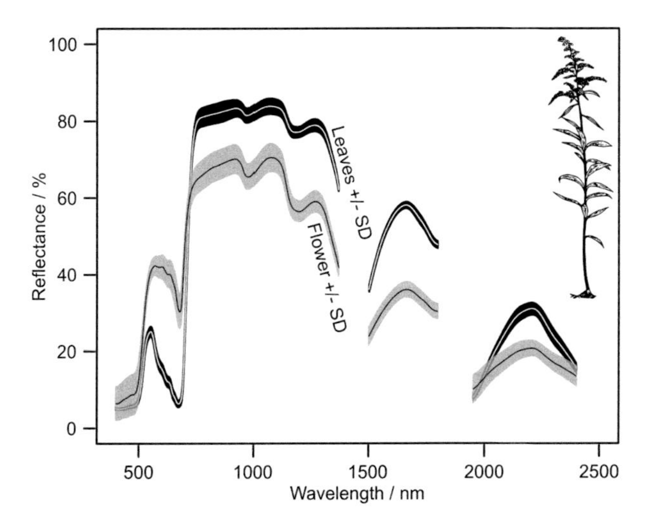

FIGURE 4.1 Mean and standard variation spectral response (from 240 to 2300 nm) between flowers and green leaves of *Solidago gigantea* derived from round-based spectrometer readings taken in the laboratory.

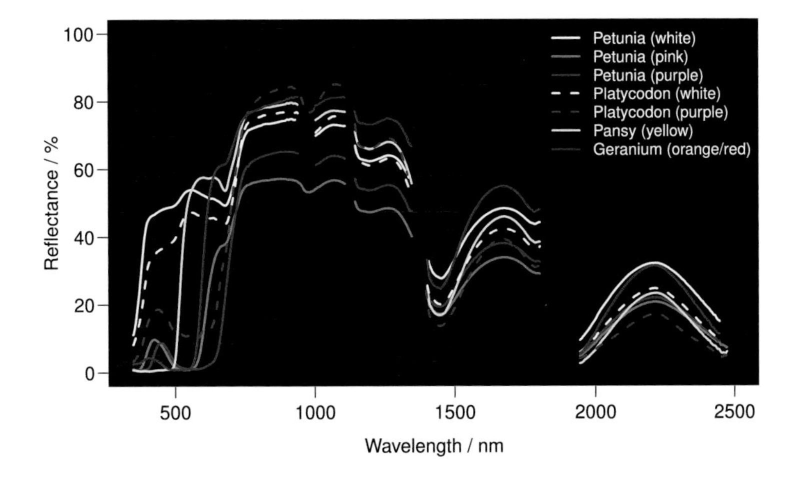

FIGURE 4.2 Effect of flower colors on flower spectrum of Petunia, Platycodon, Pansy, and Geranium plants. Spectral data for this figure were taken from the USGS Spectral Library version 7. Reflectance (%) versus wavelength (nm) is shown.

(specific species), increases in plant carotenoid due to flowering increased absorption in the blue spectral domain (450–480 nm) (Ge et al., 2006).

Flowering generally results in a higher reflectance in the VIS spectrum and lower reflectance in the near-infrared (NIR) and mid-infrared (MIR) waveband regions (Figure 4.1); however, there are also spectral response differences, not considering compaction, atmospheric, and mixing effects, between species and their various flowering colors (Ni & Li, 2000). Figure 4.2 shows various mean spectral response curves for various species (colors) (Kokaly et al., 2017).

4.3 WHY IS IT IMPORTANT TO KNOW ABOUT FLOWERING?

Furthermore, HS-based flowering maps have been proposed in many studies, primarily in the Americas, to better and more effectively manage the spread of invasive species. Invasive species maps are highly desirable for land managers since the spread of invasive species threatens biodiversity and adversely impacts crop and rangeland productivity. Spatially explicit data sets on their spread could aid in their early containment. Invasive species were found in many cases to be distinctly separable at the canopy level from co-occurring vegetation, especially in the maximum flowering season (He et al., 2011; Hestir et al., 2008). Figure 4.3 shows an example of a white flowering noxious weed, *Parthenium hysterophorus*, is distinctly visible and separable from early maize through its white flowering. The differences in phenological stages between the two plants, that is, the flowering stage and the growth stage, combined with different leaf structures, are often utilized in very high pixel resolution optical HS imagery to distinguish invasive weeds from "other" landscape elements and plants (He et al., 2011).

Although HS imagery analysis has been successfully applied (i.e., with permissible accuracies and thematic depth) in identifying invasive species occurrence and spread risk zones over smaller areas, robust and vegetation community-wide HS floral-based mapping routines and general protocols still need to be developed. This is especially true with regard to the assimilation of in situ data and when monitoring methods rely on multiseasonal data metrics for floral-based mapping.

4.4 CHALLENGES AND MAIN ISSUES

In this section, important confounding factors, which often augment each other, are discussed, with solutions for HS-based flowering mapping alluded to in the later part.

FIGURE 4.3 Flowering *Parthenium hysterophorus* in Somaliland between maize plants that are in the initial growth stage.

Field size (when considering large sunflower or poppy fields versus smaller flowering crops) and within-field flowering compaction are the primary influences on spectral variability and the floral-based mappability of larger fields (crops) (Miao et al., 2006). The spectral detection of large flowering fields with no overstory tree cover (i.e., typical poppy seed fields in central Asia) is feasible using spaceborne HS data, even at moderate pixel resolutions (i.e., >30 m) (Wang et al., 2014). However, the use of very-high-resolution (<2 m) multispectral data proved to be of limited use for the mapping of large fields, except when visual interpretation and extensive reference (field) data are used (Tian et al., 2011).

The second profound challenge is spectral mixing of the floral signal with other landscape elements largely owing to the subtleness of the floral signal and the relative large contribution of background reflectance (i.e., soil) and other bidirectional effects (Andrew & Ustin, 2008). In open-landscape-based approaches, these effects are amplified by the high intracanopy variability of flowering, even within individual trees, and small fractional coverage or compaction of flowering plants (Chen et al., 2009). Figure 4.4 illustrates, for a fresh meadow in Germany, that, although some bright flowers are highly conspicuous, their fractional coverage is low, in this case only 4.5% for all species together. Intracanopy variability can result from "other" phenological processes, such as leaf wilting, which occurs alongside flowering within a given plant canopy, and flowering, only occurring within one section of the canopy. Figure 4.5 illustrates this "sectional flowering" for a *Acacia* spp. tree canopy in Kenya.

The third challenge is spectral mixing and wavelength sensitivity as a function of the spatial and spectral characteristics of the instrument or spatial scale used for mapping. When using canopy-level field HS and when flowering compaction is high, the most accurate spectral assessment results can be attained (between 97% and 99% overall accuracies) (Parker Williams & Hunt Jr., 2004). Also, compared to very-high-resolution airborne HS data, field-level HS spectral data showed more sensitivity to flowering response in the blue absorption spectral region (around 460 nm) (Ge et al., 2006) largely due to low spectral mixing effects at this waveband in explicit canopy HS data (Lass & Callihan, 1997). In airborne HS data and when the per-pixel floral fraction and response increase, the visible wavebands (green to red, 525–650 nm) and specifically the red spectra (650 nm) become more important in floral response mapping (Jia et al., 2011). When performing landscape mapping to segregate several flowering stages (pre-, peak, and post

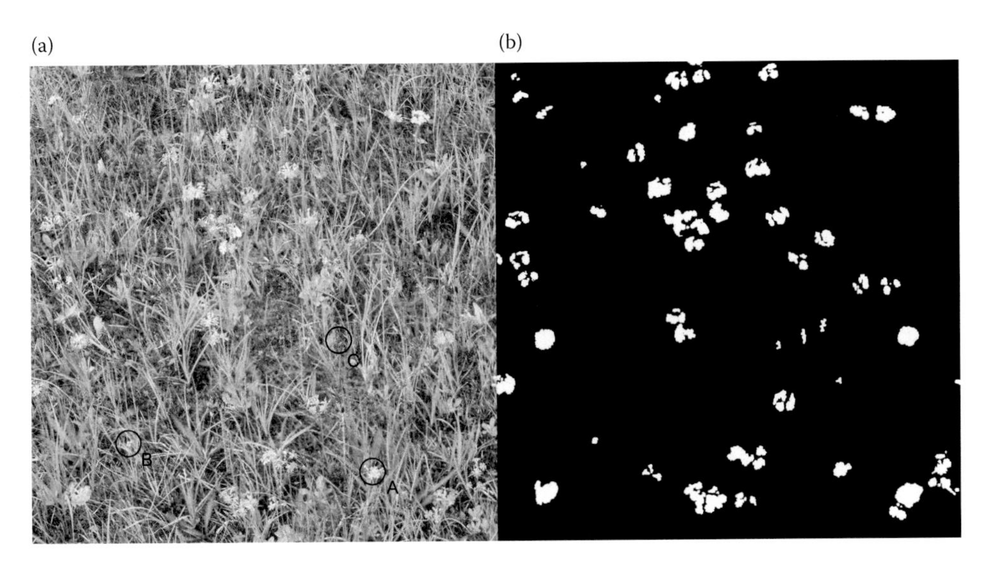

FIGURE 4.4 A typical nutrient-poor fresh meadow in the alpine foothills in Southern Bavaria, Germany, on June 4, 2013 (a) and the corresponding flower cover fractions. (The flower cover fractions were separated with a simple thresholding approach in the RGB color space. The thresholds were determined with regions of interest covering a flower sample and visually cross checked.) (b). Species that contribute to these flower fractions are the pinkish *Primula farinosa* (A), the yellowish *Potentilla erecta* (B), and the blueish *Polygala amarella* (C). Although the flower fraction is very prominent, it only amounts to a total of 4.5% cover.

flowering stages) and densely flowering vegetation communities from nonflowering communities, the spectral sensitivity shifted to more spectral regions, including the red edge (650–690 nm) and the NIR region (730–950 nm) (Miao et al., 2006). Essentially, for both in situ assessments and landscape mapping, the low detectability potential of flowering and spectral mixing is amplified by the nature of the wavelength signal in HS imaging, which is particularly dominated by leaf and canopy reflectance.

FIGURE 4.5 A sectional flowering *Acacia mellifera* tree in a agro-ecological landscape in Kenya. Note the nonflowering canopy sections on the right side and lower canopy parts.

Fourth, landscape flower monitoring and flowering magnitudes are affected by asynchronous floral cycles between and within various vegetation communities (Feilhauer et al., 2016). Asynchronous flowering patterns are often due to small-scale environmental or genetic variability (Almeida-Neto & Lewinsohn, 2004). Because of the short-term floral cycles of many vegetation communities, HS data acquisition must be ideally timed, which can be difficult in areas with extensive cloud cover (Ge et al., 2006).

Thus, one solution to overcoming the mixing effects is to ensure that HS detection takes place at the peak flowering period so that "as pure as possible" flowering pixels can be used as "reference" for wide-area mapping (Lass & Callihan, 1997). Moreover, high-resolution airborne or spaceborne data with a pixel resolution of <1 m are recommended for landscape-based floral mapping. Object-oriented mapping approaches that capture isolated tree canopies or specific vegetation communities could, furthermore, complement pixel-based spectral unmixing to better mitigate the "salt-and-pepper" effect of floral patterns in open landscapes. This would ultimately help to reduce mapping errors due to crown cover geometry and highly variable intracanopy flowering (Figure 4.5) (Abdel-Rahman et al., 2015). In using spectral mixture approaches for preemptively identified flowering vegetation units or clusters, specific reference or dominance areas where flowering is prevalent could be effectively identified.

Using results from pilot studies, the next section describes in more detail the usefulness and possibilities of HS imaging for floral mapping.

4.5 SYNOPSIS OF CURRENT STUDIES

There are currently only a few experimental studies worldwide on flower mapping. Figure 4.6 illustrates the thematic depth of the results (vertical axis) as a function of pixel resolution (data and output) for the most representative and scientifically accredited floral mapping and assessment studies found in the literature (numbered consecutively and itemized on the right). Thematic depth (vertical axis) is point graded according to whether only one species or several species or vegetation communities were assessed or mapped (so one point is given if only one species was considered and two if several were considered) (Figure 4.6). Additionally, another point was added if a hard cover output was produced and two points were added if a soft cover result was produced. Soft cover results allow for the extraction of floral abundances for each pixel or mapping unit that is relatable to the flowering intensity for various vegetation communities or floral intensity differences between various phenological floral stages within the same vegetation community (Landmann et al., 2015).

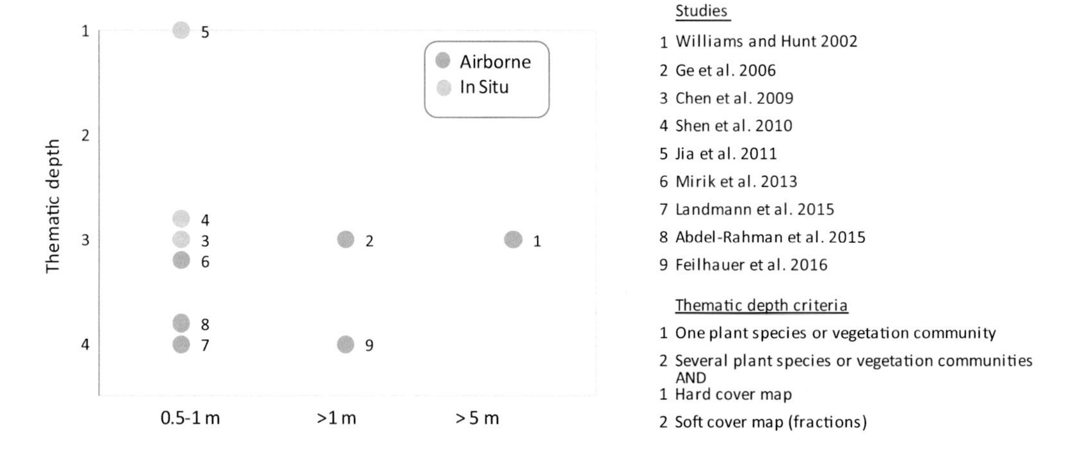

FIGURE 4.6 Two-dimensional illustration using thematic depth versus pixel resolution for various HS-based flower mapping studies numbered 1–9 and deciphered on the right. The thematic depth point scoring scheme is also shown (right, bottom).

In situ canopy studies were considered as having a <1 m "pixel resolution" since the field view of these studies varied significantly but in commonality canopy spectroscopy is spatially explicit (i.e., <1 m field of view). Only three of the nine studies (i.e., studies 2, 5, and 9) made use of the full optical waveband range that is commonly used in spectroscopy (350–2,500 nm). The two outliers (1 and 5) are studies that either mapped only one flowering species, resulting in a low thematic depth, that is, Jia et al. (2011), who only mapped poppy at the canopy level, or when floral mapping results are less spatially explicit due to the pixel size of the input HS data being >5 m (Williams & Hunt, 2002).

The assessment (Figure 4.6) essentially revealed that, for all studies with a pixel resolution <2 m, the thematic depth increased over time, that is, between the earliest study in 2002 and the latest one in 2016. Despite the increase in thematic depth, most of the recent studies did not employ the full spectral waveband region but relied solely on the VIS region (Abdel-Rahman et al., 2015; Mirik et al., 2013). This confirms the suitability of the visible spectrum for differentiating flowering response from other vegetative traits that are "other" leaf-level responses. For all flowering response studies with a <2 m pixel resolution, the use of airborne data (blue dots in Figure 4.6) led to mapping results with greater thematic depth than results attained using in situ spectroscopy data (green dots). In one airborne study that utilized multitemporal 0.6 m AISA/Eagle HS imagery (64 bands with a full width at half maximum of 8–10.5 nm in the spectral range 450–980 nm), short-term flowering phenology could be accurately mapped for various vegetation communities in two different periods (i.e., the pre- and the main flowering periods) (Landmann et al., 2015, study 7 in Figure 4.6).

For the current contribution, a flowering intensity map could also be produced for the two flowering periods using a straightforward spectral-based linear unmixing approach (Figure 4.7) based on the

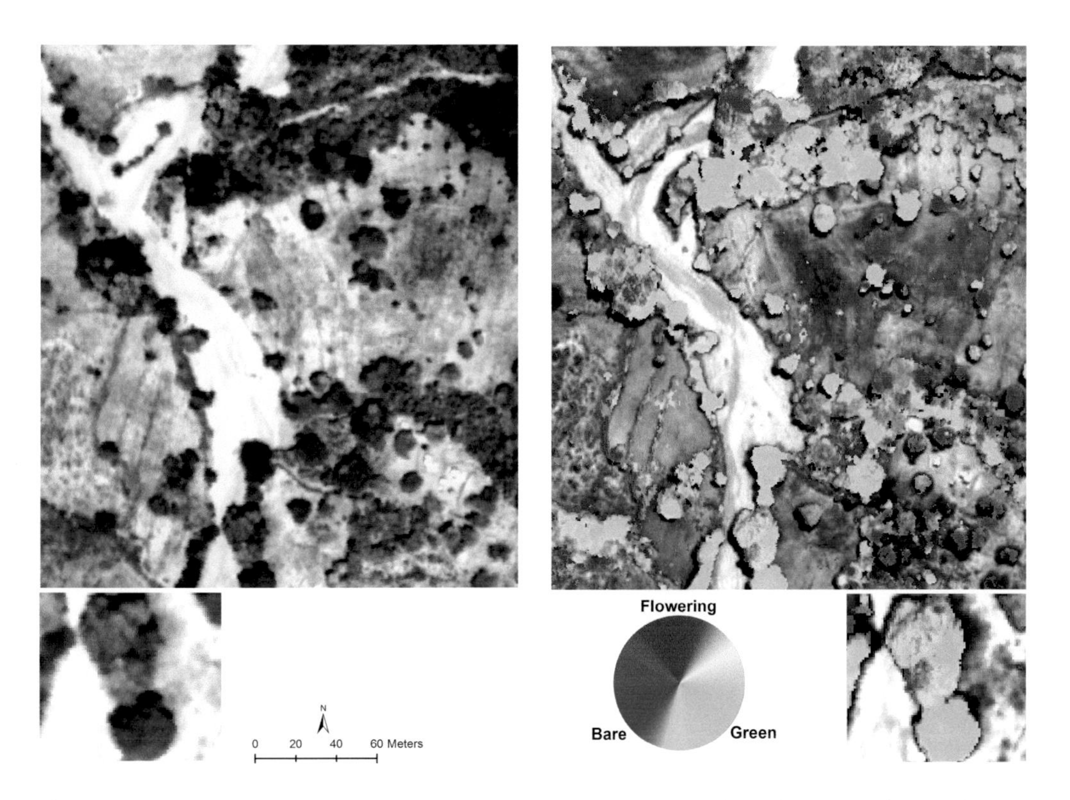

FIGURE 4.7 Left: 0.6 m AISA/Eagle HS imagery captured during peak flowering season for a study area in Kenya; Right: corresponding spectral unmixing result on right using spectral endmember flowering, bare soil, and green vegetation. Flowering is illustrated as reddish shades; more reddish colors exhibit more per-pixel flowering intensity. The zoomed-in image (below) shows a partial flowering tree.

same AISA/Eagle data mentioned earlier. Reddish shades show intense per-pixel floral spectral contributions, and green shades show high spectral abundances of the "green vegetation" spectral endmember that corresponds to green foliage or entirely "green" trees and plants with no flowering. The results of this study showed the already mentioned limitations of HS for flowering mapping over wider areas due to spectral mixing since the authors found that the spectral reference endmember for flowering varied significantly if the area of observation was enlarged. Essentially, in an enlarged area, other phenological effects, such as leaf browning, had "similar to flowering" spectra that led to erroneous results. But on a local scale, if spectral flowering endmembers are carefully selected and verified in the field for indicator trees within specific vegetation communities, even intercanopy flowering intensity gradients can be feasible mapped. The accuracy of the small area mapping results shown in Figure 4.7 varied between 69.6% correctly mapped (for yellow flowering acacia trees) and 97% for white flowering weeds. The intracanopy variability is clearly visible for larger trees and could be accurately mapped (zoomed-in image in Figure 4.7).

In another recent study that also used airborne HS imagery (Abdel-Rahman et al., 2015) (number 8 in Figure 4.6), the authors showed that it was even possible to map various flowering vegetation communities as well as their flowering colors. The greater thematic depth of airborne data is enabled by the greater ground coverage of airborne sensors (as opposed to ground-based and canopy-specific observations), making it possible to effectively map the flowering response from various vegetation communities (or species) or various flowering stages for the same community. A major limitation of airborne data is, however, the cost and thus only being able to capture short phenology cycles of individual plants. This precludes the mapping of other asynchronous flowering vegetation communities and makes it challenging to effectively produce a complete floral cycle map for a given landscape.

Interesting in this regard is that no study has thus far made use of spaceborne HS data for landscape-based floral mapping covering several flowering vegetation communities. This is probably due to the fact that flowering response is a short-term response, i.e., 2–3 weeks in the case of poppy seeds (Jia et al., 2011), and there are currently no operational spaceborne HS sensors available with a high enough temporal revisiting cycle needed for cloud-free detection for specific (i.e., floral) observation periods. For instance, the spaceborne HS imager Hyperion has a revisiting cycle of 16 days, which may not allow for enough data to be collected at critical phenological time periods.

All nine of the mentioned studies (Figure 4.6) make use of an "integrative" approach, defined as the amalgamation of several multisensor, multisource data. In all cases, either ground-based reference data on floral spectral response or flowering compaction or occurrence data were collected. Due to the already mentioned subtleness of the floral response, which is species-, location-, and time-specific, and the use of site-specific reference data in most studies, flowering response mapping is currently performed experimentally and localized rather than operationally.

4.6 CONCLUSIONS AND OUTLOOK

In the context of pollination and climate change effects and landscape ecological studies, floral maps that depict the distribution and abundances and floral cycle of melliferous plants and other species of ecological importance are important information feeds. In HS-based assessments and mapping of floral response, the most important challenges are: the subtleness of flowering, short flowering windows for most species, and inter- and intracanopy or plant flowering variability as a function of ecological gradients and microclimate conditions. However, the case study assessment showed that flower status could be retrieved from in situ and spatially explicit airborne hyperspectral data, especially over smaller areas and for vegetation communities with common plant traits. Dense time series of hyperspectral Earth Observation data may thus offer the opportunity to analyze spatiotemporal dynamics in flowering in future. This applies in particular to a spatial analysis of the effects of asynchronous flowering on pollination (e.g., Thomson & Plowright, 1980). Moreover, this chapter highlighted the application possibilities of floral mapping for the better management of invasive species. Future mapping endeavors might also be able to operationally support narcotics programs that aim to stem the production of poppy seeds.

ACKNOWLEDGMENTS

We gratefully acknowledge the financial support for this research by the following organizations and agencies: German Research Foundation (grant FE 1331/2-1); the European Union, University of Helsinki, the Centre for International Migration and Development (CIM) of the German Development Agency (GIZ); UK's Department for International Development (DFID); Swedish International Development Cooperation Agency (Sida); the Swiss Agency for Development and Cooperation (SDC); and the Kenyan government. The views expressed herein do not necessarily reflect the official opinion of the donors.

REFERENCES

Abdel-Rahman, E. M., Makori, D. M., Landmann, T., Piiroinen, R., Gasim, S., Pellikka, P., & Raina, S. K. 2015. The utility of AISA eagle hyperspectral data and random forest classifier for flower mapping. *Remote Sensing*, *7*, 13298–13318.

Almeida-Neto, M., & Lewinsohn, T. M. 2004. Small-scale spatial autocorrelation and the interpretation of relationships between phenological parameters. *Journal of Vegetation Science*, *15*, 561–568.

Andrew, M. E., & Ustin S. L. 2008. The role of environmental context in mapping invasive plants with hyperspectral image data. *Remote Sensing of Environment*, *112*, 4301–4317.

Burkle, L. A., & Alarcón, R. 2011. The future of plant–pollinator diversity: Understanding interaction networks across time, space, and global change. *American Journal of Botany*, *98*(3), 528–538.

Chen, J., Shen, M., Zhu, X., & Tang, Y. 2009. Indicator of flower status derived from *in situ* hyperspectral measurement in an alpine meadow on the Tibetan Plateau. *Ecological Indicators*, *9*, 818–823.

Craufurd, P., & Wheeler, T. 2009. Climate change and the flowering time of annual crops. *Journal of Experimental Botany*, *60*, 2529–2539.

Dixon, K. W. 2009. Pollination and restoration. *Science*, *325*, 571–573.

Dunne, J. A., Harte, J., & Taylor, K. J. 2003. Subalpine meadow flowering phenology responses to climate change: Integrating experimental and gradient methods. *Ecological Monographs*, *73*, 69–86.

Feilhauer, H., Doktor, D., Schmidtlein, S., & Skidmore, A. K. 2016. Mapping pollination types with remote sensing. *Journal of Vegetation Science*, *27*, 999–1011.

Galpaz, N., Ronen, G., Khalfa, Z., Zamir, D., & Hirschberg, J. 2006. A chromoplast-specific carotenoid biosynthesis pathway is revealed by cloning of the tomato white-flower locus. *The Plant Cell*, *18*, 1947–1960.

Ge, S., Everitt, J., Carruthers, R., Gong, P., & Anderson, G. 2006. Hyperspectral characteristics of canopy components and structure for phenological assessment of an invasive weed. *Environmental Monitoring and Assessment*, *120*(1), 109–126.

Gong, Y. B., & Huang, S. Q. 2011. Temporal stability of pollinator preference in an alpine plant community and its implications for the evolution of floral traits. *Oecologia*, *166*(3), 671–680.

He, K. S., Rocchini, D., Neteler, M., & Nagendra, H. 2011. Benefits of hyperspectral remote sensing for tracking plant invasions. *Diversity and Distributions*, *17*, 381–392.

Hestir, E. L., Khanna, S., Andrew, M. E., Santos, M. J., Viers, J. H., Greenberg, J. A., Rajapakse, S. S., & Ustin, S. L. 2008. Identification of invasive vegetation using hyperspectral remote sensing in the California Delta ecosystem. *Remote Sensing of Environment*, *112*, 4034–4047.

Jia, K., Wu, B., Tian, Y., Li, Q., & Du, X. 2011. Spectral discrimination of opium poppy using field spectrometry. *IEEE Transactions on Geoscience and Remote Sensing*, *49*, 3414–3422.

Kohler, F., Verhulst, J., Van Klink, R., & Kleijn, D. 2008. At what spatial scale do high-quality habitats enhance the diversity of forbs and pollinators in intensively farmed landscapes? *Journal of Applied Ecology*, *45*(3), 753–762.

Kokaly, R. F., Clark, R. N., & Swayze, G. A. 2017 USGS Spectral Library Version 7. Pp. Page, US Geological Survey.

Kühn, I., Bierman, S. M., Durka, W., & Klotz, S. 2006. Relating geographical variation in pollination types to environmental and spatial factors using novel statistical methods. *New Phytologist*, *172*, 127–139.

Landmann, T., Piiroinen, R., Makori, D. M., Abdel-Rahman, E. M., Makau, S., Pellikka, P., & Raina, S. K. 2015. Application of hyperspectral remote sensing for flower mapping in African savannas. *Remote Sensing of Environment*, *166*, 50–60.

Lass, L. W., & Callihan, R. H. 1997. The effect of phenological stage on detectability of yellow hawkweed (Hieracium pratense) and oxeye daisy (Chrysanthemum leucanthemum) with remote multispectral digital imagery. *Weed Technology*, 248–256.

Miao, X., Gong, P., & Swope, S. 2006. Estimation of yellow starthistle abundance through CASI-2 hyperspectral imagery using linear spectral mixture models. *Remote Sensing of Environment*, *101*, 329–341.

Mirik, M., Ansley, R. J., Steddom, K., Jones, D. C., Rush, C. M., Michels, G. J., & Elliott, N. C. 2013. Remote distinction of a noxious weed (Musk Thistle: CarduusNutans) using airborne hyperspectral imagery and the support vector machine classifier. *Remote Sensing*, *5*, 612–630.

Moeller, D. A., & Geber, M. A. 2005. Ecological context of the evolution of self-pollination in Clarkia xantiana: Population size, plant communities, and reproductive assurance. *Evolution*, *59*(4), 786–799.

Müller, H. 1881. Two kinds of stamens with different functions in the same flower. *Nature*, *24*, 307–308.

Ni, W., & Li, X. 2000. A coupled vegetation–soil bidirectional reflectance model for a semiarid landscape. *Remote Sensing of Environment*, *74*, 113–124.

Ollerton, J., Alarcón, R., Waser, N. M., Price, M. V., Watts, S., Cranmer, L., & Rotenberry, J. 2009. A global test of the pollination syndrome hypothesis. *Annals of Botany*, *103*(9), 1471–1480.

Parker Williams, A., & Hunt Jr., E. R. 2004. Accuracy assessment for detection of leafy spurge with hyperspectral imagery. *Journal of Range Management*, *57*, 106–112.

Pellissier, L., Pottier, J., Vittoz, P., Dubuis, A., & Guisan, A. 2010. Spatial pattern of floral morphology: Possible insight into the effects of pollinators on plant distributions. *Oikos*, *119*(11), 1805–1813.

Schulp, C. J. E., Lautenbach, S., & Verburg, P. H. 2014. Quantifying and mapping ecosystem services: Demand and supply of pollination in the European Union. *Ecological Indicators*, *36*, 131–141.

Sims, D. A., & Gamon, J. A. 2003. Estimation of vegetation water content and photosynthetic tissue area from spectral reflectance: A comparison of indices based on liquid water and chlorophyll absorption features. *Remote Sensing of Environment*, *84*, 526–537.

Steffan-Dewenter, I., Münzenberg, U., Bürger, C., Thies, C., & Tscharntke, T. 2002. Scale-dependent effects of landscape context on three pollinator guilds. *Ecology*, *83*(5), 1421–1432.

Thomson, J. D., & Plowright, R. C. 1980. Pollen carryover, nectar rewards, and pollinator behavior with special reference to Diervilla lonicera. *Oecologia*, *46*(1), 68–74.

Tian, Y., Wu, B., Zhang, L., Li, Q., Jia, K., & Wen, M. 2011. Opium poppy monitoring with remote sensing in North Myanmar. *International Journal of Drug Policy*, *22*, 278–284.

Van der Pijl, L. 1960. Ecological aspects of flower evolution. II. Zoophilous flower classes. *Evolution*, *15*(1), 44–59.

Von Holle, B., Wei, Y., & Nickerson, D. 2010. Climatic variability leads to later seasonal flowering of Floridian plants. *PLoS ONE*, *5*, e11500.

Wang, J. J., Zhang, Y., & Bussink, C. 2014. Unsupervised multiple endmember spectral mixture analysis-based detection of opium poppy fields from an EO-1 Hyperion image in Helmand, Afghanistan. *Science of The Total Environment*, *476*, 1–6.

Waser, N. M., Chittka, L., Price, M. V., Williams, N. M., & Ollerton, J. 1996. Generalization in pollination systems, and why it matters. *Ecology*, *77*(4), 1043–1060.

Williams, A. P., & Hunt, E. R. 2002. Estimation of leafy spurge cover from hyperspectral imagery using mixture tuned matched filtering. *Remote Sensing of Environment*, *82*, 446–456.

Wesche, K., Krause, B., Culmsee, H., & Leuschner, C. 2012. Fifty years of change in Central European grassland vegetation: Large losses in species richness and animal-pollinated plants. *Biological Conservation*, *150*(1), 76–85.

5 Crop Water Productivity Estimation with Hyperspectral Remote Sensing

Michael Marshall, Itiya Aneece, Daniel Foley, Cai Xueliang, and Trent Biggs

CONTENTS

5.1 INTRODUCTION

The exchange of carbon dioxide (CO_2) and water vapor between the atmosphere and crops is complex and fundamental to healthy crop development (Bernacchi and VanLoocke, 2015). Carbon assimilation via photosynthesis is the process by which crops use light energy to convert atmospheric CO_2 to carbohydrates (Chapin et al., 2011). Carbohydrates are used to generate additional plant material (biomass), including the harvestable portion known as yield (Y). During carbon assimilation, moisture is lost to the atmosphere via transpiration (ET_C). Assuming the "big leaf" concept for photosynthesis (Dickinson et al., 1998), the regulation of CO_2 and ET_C flux in the crop canopy is determined by the proportion of photosynthesizing canopy area, ratio of internal to atmospheric CO_2 or water vapor, and available solar energy (Steduto et al., 2007). The proportion of photosynthesizing canopy area is measured either as the fraction of photosynthetically active radiation (F_{PAR}) or with the leaf area index (LAI). It is generally assumed to vary with moisture and nutrient availability, temperature extremes, and leaf age (Wang et al., 2014). If too much moisture is transpired during the exchange of CO_2 and water vapor, for example, crops cannot maintain metabolic function or structure, and F_{PAR} declines. Similarly, if crops experience nutrient stress or are exposed to frost or a heat wave, F_{PAR} will be lower. If these conditions persist, F_{PAR} will approach zero, and declines in Y or total crop failure are likely. The ratio of internal to atmospheric CO_2 varies with the type of assimilation (C_3 or C_4). Atmospheric CO_2 is generally assumed constant in either case. Assimilation in C_3 crops (e.g., rice, soy, and wheat) is catalyzed by the RUBISCO enzyme (Collatz et al., 1991). The activity of the enzyme increases with rising temperatures until a given threshold in optimal performance is achieved. Once temperatures exceed an optimal temperature, catalyzation declines.

Assimilation in C_4 crops (e.g., maize) is first catalyzed by PEP carboxylase, which acts to concentrate CO_2 before RUBISCO catalyzation (Collatz et al., 1992). The concentration has the effect of minimizing the temperature dependency, which makes C_4 crops better adapted to hot dry climates. Carbon assimilation increases with increasing F_{PAR}, since crops prioritize light capture (Monteith, 1969). Under cloudy or diffuse light conditions, crops may actually assimilate more CO_2, because more light is able to penetrate deeper into the canopy (Law et al., 2002). Transpiration, on the other hand, increases with the total or net radiation absorbed by the canopy (Monteith, 1965). As more radiation is absorbed by the crop, sensible heat (H) increases and must be offset by an increase in heat generated by ET_C to maintain energy balance (Pieruschka et al., 2010).

Socio-ecological conditions make agroecosystems in dry regions of the world particularly sensitive to CO_2 and water vapor flux (Schlosser et al., 2014). Global climate change is projected to increase evaporative demand, which will lead to more frequent and severe droughts (Cook et al., 2014; Prudhomme et al., 2014). In the near term, Y may increase, because projected increases in corresponding CO_2 emissions and fertilization can act to decrease ET_C. In the long term, however, water stress is projected to overtake the CO_2 offset, leading to Y declines (Gerten et al., 2011; Challinor et al., 2014). Population growth, rapid industrialization and urbanization, water pollution, and the increased use of biofuels will likely put further strain on water resources in this century (Schewe et al., 2014). The Green Revolution, which began to accelerate in the 1960s, consisted of a series of policies and actions that focused on increasing Y to meet the challenges of food insecurity. Among these, agricultural land expansion was the most prevalent. As the amount of arable land dwindles, this option is no longer sustainable. Other measures must be adopted to optimize agricultural production and prevent water scarcity (Porkka et al., 2016).

Increasing crop water productivity (CWP) by maintaining or increasing Y while minimizing water lost to the environment ("more crop per drop") is a key strategy for combating water scarcity in dry regions of the world (Brauman et al., 2013). Increasing CWP is referred to here as a strategy because it is achieved by the coordinated application of several techniques (Ali and Talukder, 2008). Techniques fall into four general categories: physiological, engineering, farm management, and economic (Raza et al., 2012). Physiological methods reduce transpiration while maintaining or increasing Y. This is commonly done by selecting crop types or varieties with a higher tolerance to adverse conditions (Bessembinder et al., 2005). The remaining measures aim to reduce surface runoff and evaporation or increase soil infiltration and drainage. Engineering solutions, such as drip irrigation systems and irrigation channel lining, improve storage and delivery through physical means. They have a high upfront cost, which makes them practical only for farms with access to capital. Deficit irrigation and other irrigation scheduling techniques, mulching, and increasing crop density, are examples of farm management solutions that act to reduce surface runoff and evaporation through biological means. They are more economical than engineering solutions, which makes them more attractive to smallholder farmers in developing countries. Conservation till, soil amendments, and organic matter additions are examples of farm management solutions that increase soil infiltration and drainage by improving soil structure and water holding capacity. Unlike physiological, engineering, or farm management techniques, which can be done on individual farms, economic techniques are intended to incentivize water savings on many farms. As such, they require broad consensus and institutional support. They can be classified as top-down government interventions (e.g., water pricing) or market-based interventions (e.g., virtual water trading) (Liu et al., 2009; Siebert and Döll, 2010).

The cost-effectiveness of these measures remains highly uncertain over both space and time (Evans and Sadler, 2008). Process-based models have been developed to estimate CWP to address these uncertainties (Zwart et al., 2010). These models require extensive input data on current field conditions that are difficult and costly to obtain. Earth observation via satellite remote sensing can partially overcome these challenges, because it captures surface reflectance representing current field conditions over large areas over time at low cost (Chi et al., 2016). CWP has been measured to a limited extent through its components (Y and ET_C) with broadband remote sensing. With broadband remote sensing, a small number of bands or channels known as multispectral broadbands (MSBBs) are used to develop multispectral broadband vegetation indices (MBVIs) that characterize crop properties using surface reflectance in the visible and very near infrared (VNIR) range of the electromagnetic spectrum. Thermal bands are often used for ET_C estimation. MBVIs are often too coarse spectrally to distinguish biophysical properties related to CWP or its components (Ustin et al., 2004). Hyperspectral remote sensing involves hundreds of narrow spectral bands or HNBs that are used to develop hyperspectral vegetation indices (HVIs). HVIs typically provide a higher level of detail that facilitates the characterization of crop-specific properties, such as biomass, moisture, pigments, and lignin/cellulose (woody biomass). Although studies have been conducted on the use of HVIs to estimate biomass, to the authors' knowledge no studies have used HVIs to estimate CWP directly or through its components Y and ET_C. Hyperspectral remote sensing is costlier and more data intensive than broadband remote sensing, which may explain the dearth of studies (Goetz, 2009). In addition, images from past hyperspectral missions (e.g., Hyperion onboard Earth Observing-1) were captured on demand, which resulted in more limited data acquisition.

This chapter presents a methodology that can be used to compare HVIs for CWP estimation. The authors originally compiled available hyperspectral (Hyperion) satellite data for the exercise. Unfortunately, the number of images was too small for any credible interpretation. MBVIs were therefore used instead. Landsat has been widely used to estimate CWP and its components, but analyses tend to focus on the red, VNIR, and thermal regions of the electromagnetic spectrum. The proposed methodology is a data mining approach and therefore covers a wide range of regions of the electromagnetic spectrum. In addition, eddy covariance towers, which measure carbon and moisture flux simultaneously over agricultural fields are used for the first time to demonstrate how CWP models could be developed from hyperspectral remote sensing. The current state of knowledge concerning remote-sensing-based approaches related to CWP is summarized in Section 5.2. Section 5.3 demonstrates the use of multispectral broadband remote sensing to estimate CWP and its components. Section 5.4 provides concluding remarks, including some insights into the weaknesses of the methodology and a path for future work in hyperspectral remote sensing and CWP modeling.

5.2 REMOTE SENSING OF CROP WATER PRODUCTIVITY AND ITS COMPONENTS

CWP is a partial-factor productivity measure that essentially captures the performance of crops or other agricultural output in relation to water applied to the system (Molden et al., 2003). In a cropping system, CWP is expressed as

$$CWP = \frac{\text{Crop yield}\,(\text{kg ha}^{-1}\text{ or \$ ha}^{-1})}{\text{Water input}\,(\text{m}^3\,\text{ha}^{-1})}. \quad (5.1)$$

CWP is often confused with the term water use efficiency (WUE) (Evans and Sadler, 2008). WUE was traditionally defined by irrigation engineers as the ratio of ET_C to water diverted via evaporation, percolation, seepage, or other losses across an irrigation system. WUE has a number of limitations. Efficiency implies that water retained can be expressed as a percentage of the maximum achievable limit, which is not possible. In addition, WUE has been confused with a number of other terms, such as irrigation efficiency and water footprint, which do not consider crop water use. These definitions therefore tend to deemphasize physiological improvements, which may be more practical than engineering, farm management, or economic solutions.

CWP above all is a diagnostic tool to help identify problems and opportunities related to infrastructure, crops, and water management practices. CWP is a useful supplement to irrigation efficiency, which mostly emphasizes engineering solutions. CWP links irrigation efficiency in cropping systems with its desired outcome: optimal crop productivity. Therefore, it helps identify areas where water supplied is not enough or too much. If it is computed periodically through the growing season, it can also be used to determine whether or not water is provided at the right time. All of these can be partly attributed to infrastructure and partly to farm management, for example, as those reflected through the classic head-tail effects in which fields at the head of a canal receive more water than those at the tail. CWP also facilitates the optimization of cropping patterns and tradeoff analysis. For example, Cai et al. (2014) developed a CWP tool for water allocation in a small catchment. With the tool, farmers or farm managers are able to explore a number of options to optimize CWP, including growing a combination of rainfed or irrigated crops.

CWP should not be confused with water footprint either. Water footprint is a concept concerning the resources needed to produce a unit of product. It is frequently used for broad trade and policy analysis, but not for a performance assessment or diagnostic analysis of agricultural systems.

There are several variations of CWP as defined in Equation 5.1 (Ali and Talukder, 2008). The numerator is commonly expressed as grain seed or other agricultural production with a certain moisture content or dry biomass in the case of rainfed systems. CWP expressed with Y is limited to a specific crop type or variety. To assess the performance of multicrop systems, Y of various crops can be converted into monetary ($) terms by multiplying it by market rates. Even crop residue with certain economic value can be valued and tallied. CWP expressed in monetary terms therefore is a way of standardization. This standardization is especially useful if one wants to look at crop diversification and the returns on limited water supply with alternative crop choices. It is however limited to the same agricultural system or similar systems. To enable cross-regional comparisons, climate normalization is required (Bastiaanssen and Steduto, 2017).

The denominator in CWP has an even wider range of interpretations and inputs. In one of the first and most widely cited papers on the subject, Molden (1997) suggested that CWP of an agricultural system can be calculated using, for example, gross inflow, net inflow, irrigation supply, and beneficial consumption. Approaches such as these are used to examine the efficiency of an irrigation system (Roost, 2008). Increasingly, actual water consumed by a crop (ET_C) is used in the denominator to estimate CWP. Actual evapotranspiration (ET_A) may be used in lieu of ET_C because of the difficulty to measure ET_C. ET_A captures not only the water lost from ETc but also water evaporated from the soil or crop surface following a watering event. ET_A or ET_C is used alone in the denominator because the hydrological processes are extremely complex in most cropping systems compared to natural systems. Irrigation, drainage, water reuse, and on-farm practices all contribute to changes in the water cycle, making it difficult to measure the level of water inputs and gauge how efficiently they are used.

The use of remote sensing estimates for CWP assessment marked a significant advance in the development and application of CWP. The advancement was possible because remote sensing data were used to directly estimate and map two important variables: (a) crop biomass/yield and (b) ET_A (Figure 5.1). The remotely sensed data most commonly used to estimate these variables are MBVIs, but hyperspectral narrowbands (HNBs) are also used, though to a much lesser extent. In using remotely sensed data, CWP estimation moved away from the assessment of inputs and outputs that are uncertain, hard to measure, and nonlinearly related to water consumed by a given crop. Moreover, for the first time, CWP could be estimated at the pixel level (typically at 30-m resolution), which helped target interventions. Finally, the use of remote sensing allowed CWP to be estimated at different scales and for different cropping systems, regardless of data availability on rainfall, groundwater, discharge, or irrigation supply.

In the following subsections, remote-sensing-based methods for estimating crop biomass, Y, and ET_A are reviewed.

5.2.1 Crop Biomass and Yield

Y is typically measured as a function of crop biomass and the harvest index (HI). HI is the crop or variety specific fraction of biomass extracted for grain seed or other agricultural production (Hay, 1995). Remote sensing techniques are used to estimate biomass or Y directly, while HI is estimated

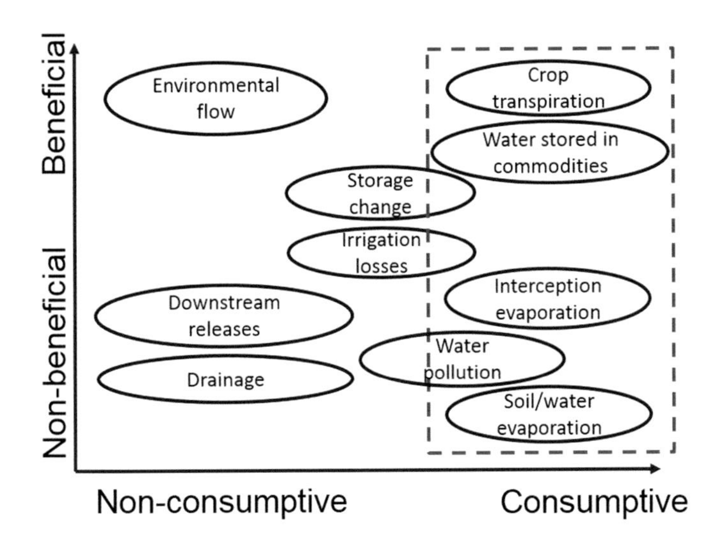

FIGURE 5.1 Remote-sensing-based CWP focuses on consumed water by crop, as opposed to nonconsumptive or inflow/outflow side of the water balance. Similarly, the components beneficial to the grower, consumer, etc. are indicative of productivity, while the non-beneficial components are not.

using non-remote sensing methods (Kemanian et al., 2007). Traditional ways of estimating biomass with field measurements are resource intensive, time consuming, destructive, and difficult to scale up (Feng et al., 2013). MBVIs provide a nondestructive way to estimate biomass or Y and enable such estimation throughout the growing season and over large areas. Common examples include the normalized difference vegetation index (NDVI), the transformed vegetation index (TVI), the simple ratio (SR) index, and the difference vegetation index (DVI) (Johnson, 2016; Mutanga and Skidmore, 2004). These indices incorporate MSBBs in the red and very near-infrared (VNIR) regions of the spectral profile, which correspond with light absorption by pigments in the red region and light scattering due to leaf structural elements in the VNIR region. Highly productive vegetation generally has high absorption in the red region and high reflectance in the VNIR region. Background signatures (such as soil or water) and atmospheric effects can obscure this relationship. Several other popular indices exist to minimize this noise, including the Perpendicular Vegetation Index (PVI), the weighted difference vegetation index (WDVI), the soil-adjusted vegetation index (SAVI), the transformed SAVI (TSAVI), and the modified NDVI (MNDVI). However, these indices often saturate in regions with high biomass and vegetation cover and, thus, are not always effective in estimating biomass throughout the growing season.

Several studies have demonstrated the benefits of using HNBs to estimate aboveground biomass using various indices and methods (Table 5.1). They overcome the saturation problem, resulting in higher correlations overall. Marshall and Thenkabail (2015a), for example, found that VNIR was most correlated with aboveground crop biomass, especially at 794, 845, 865, 943, 963, and 993 nm, followed by the visible at 438, 468, 539, 560, and 631 nm. Feng et al. (2013) also concluded that the 500, 519, 560, 593, 619, 683, 702, and 728 nm wavebands were most important for predicting rice biomass. The Modified Triangular Vegetation Index 2 (MTVI2), calculated using green (550 nm), red (670 nm), and VNIR (800 nm) narrowbands, was able to estimate dry biomass for corn, soybean, and wheat, with R^2 values of 0.95, 0.99, and 0.76, respectively (Liu et al., 2004). Marshall and Thenkabail (2015b) compared the ability of HNBs with MSBBs from MODIS, Landsat, IKONOS, GeoEye-1, and WorldView-2 to successfully estimate aboveground biomass for rice, alfalfa, cotton, and maize. Correlations with biomass were higher with Hyperion HNBs (R^2 from 0.71 to 0.98) than those with the multispectral satellites, with R^2s of 0.10–0.88 for MODIS, 0.32–0.82 for Landsat, 0.50–0.94 for IKONOS, 0.55–0.95 for GeoEye-1, and 0.36–0.87 for WorldView-2. They also found

TABLE 5.1
Summary of the Most Important Hyperspectral Narrowbands (HNBs) for Aboveground Dry (ADB) and Wet (AWB) Biomass per Crop Type and Their Performance versus Multispectral Broadbands (MSBBs)

Crop Type	Variable	HNBs	Variability Explained	Percentage Improvement over Other Studies
Rice, maize, cotton, alfalfa	AWB	794, 845, 865, 943, 963, 933	84%, 59%, 91%, and 86%	12%, 29%, 14%, 6%
Rice, alfalfa, cotton, and maize	AWB	529, 549, 722, 732, 752, 895, 925, 1104	91%, 81%, 97%, and 94%	5%–31% with two-band indices, 3%–33% with three-band indices
Pasture grass	AWB	706, 722, 746, 752, 755, 760	78%, 80%, and 79% using NDVI, SR, and TVI	52% improvement with modified NDVI vs traditional
Maize, soybean, wheat	ADB	550, 670, 800	95%, 99%, and 76%	
Wheat	ADB	874, 900, 955, 1050, 1220, 1225	74% and 78% using NRI and GnyLi	At least 10% improvement over other indices
Rice	AWB and ADB	500, 519, 560, 593, 619, 683, 702, 728	For tillering to elongation, 94% and 93.5% for wet and dry For booting to heading, 89.1% and 78.3% for wet and dry	

Note: In some cases, vegetation indices such as NDVI, SR, TVI, and NRI were used.

that HNB two-band vegetation indices explained 5%–31% more variability in crop biomass than MSBB two-band vegetation indices. Additionally, HNB multiband vegetation indices explained 3%–33% more variability than did MSBB vegetation indices. Similarly, Mariotto et al. (2013) found that the high spatial resolutions of QuickBird and IKONOS images were not sufficient for predicting crop biomass without hyperspectral data.

Several HNBs lie outside the visible and VNIR regions that also produce accurate biomass estimates. For example, Mutanga and Skidmore (2004) found that incorporating red edge bands into traditionally used indices such as NDVI, TVI, and SR, led to higher correlations with biomass of a grass species. The MNDVI using a red edge band at 746 nm and a VNIR band at 755 nm had an R^2 value of 0.78, as opposed to an R^2 value of 0.26 when using a red band at 680 nm and a VNIR band at 833 nm. Similarly, they found that the SR with the highest R^2 value of 0.80 used bands at 706 and 755 nm in the red edge region. The most important bands for the TVI were 752 and 755 nm, also in the red edge region. Marshall and Thenkabail (2015a) also found that the red edge region (centered at 722 nm) was important for biomass estimation. The shortwave infrared region (SWIR) can also improve biomass estimates. For example, Mariotto et al. (2013) found that for predicting crop biomass, 74% of important bands were in the 1051–1331 nm spectral range; 10% in the moisture-sensitive spectral range at 970 nm (VNIR); 10% in the 400–550 nm (blue), 501–600 nm (green), and 760–900 nm (VNIR); and 6% in the red and red edge (630–752 nm) regions. Longer wavelengths have also proved useful. Gnyp et al. (2014), for example, evaluated the normalized ratio index (NRI) and GnyLi, which used a combination of 874 nm (VNIR) and 1225 nm (SWIR) and VNIR (900, 955 nm) and SWIR (1050, 1220 nm), respectively. They explained 74% of the variability in winter wheat biomass, compared to less than 60% with indices from other spectral regions.

Several factors influence which bands or spectral regions are most important for estimating biomass and Y, including crop type, growth stage, crop stress level, and analytical techniques used. The most important bands for estimating biomass vary across crops because of differences in plant architecture and physiology. For example, erectophiles (with vertical leaf angle) such as rice absorb less visible light and scatter more light in the NIR than planophiles (with horizontal leaf angle) such as cotton or alfalfa (Marshall and Thenkabail, 2014). Growth stage is also an important consideration; biomass is often underpredicted at crop growth stages with low biomass (Marshall and Thenkabail, 2014). A more comprehensive review of eco-physiological factors impacting HNB-biomass correlations is presented in Osborne et al. (2002). The statistical method and spectral transformation selected can also impact the strength and significance of the bands correlated with crop biomass. Marshall and Thenkabail (2015a) found that piecewise regression models performed worse than stepwise regression, while Marshall and Thenkabail (2014) found that two-band and multiple band HNB indices selected using sequential search methods were better at predicting biomass than principal component regression. Marshall and Thenkabail (2014) found that correlations between first derivative transforms and biomass were higher than with untransformed spectra, especially at longer wavelengths. The strong correlations were over a much narrower spectral range because the transformation deemphasized background signatures from soil and water in the case of rice. Similarly, second derivatives further deemphasized effects from varying solar illumination, which is especially important for rice in standing water (Marshall and Thenkabail, 2014). However, Marshall and Thenkabail (2015b) found that the untransformed spectra were better in the case of multiband vegetation indices than first derivative spectra.

5.2.2 Evapotranspiration

Total actual evapotranspiration, ET_A, is the second largest term in the water balance equation after precipitation (Katul et al., 2012). ET_C, the water used by crops, is the largest component of ET_A in most locations around the world (Schlesinger and Jasechko, 2014). ET_C rates are most impacted by meteorological conditions, which are the first-order control on the evaporative demand of the atmosphere. ET_C is particularly important during the growing season when vegetation is productive, whereas soil evaporation is important during the initial stages of crop growth, when vegetation cover is low and soil moisture is high. Interception occurs when plant canopies retain rainfall, wetting leaf and stem surfaces, with subsequent evaporation of the intercepted water. The percentage of rainfall that is intercepted decreases with event size, and its importance in the annual water balance depends on the size and distribution of storms and on canopy characteristics. For field crops, interception ranges from 7%–36% of growing-season rainfall in semiarid climates (growing season rainfall 158–275 mm) (Dunne and Leopold, 1978) to 8%–18% of rainfall in humid climates (annual rainfall 1577–1642 mm) (van Dijk and Bruijnzeel, 2001). Interception cannot be measured directly with remote sensing, so it is typically modeled as a function of storm size or relative humidity (Fisher et al., 2008; Mu et al., 2011). Soil evaporation is approximately the same magnitude as evaporation from an open water body when the soil is wet, decreases rapidly with decreasing soil moisture and increasing canopy cover, and is usually insignificant after 5–10 mm of water has evaporated from the soil (Dunne and Leopold, 1978). Soil evaporation is therefore most important in areas with low canopy cover and immediately after rainstorms or in areas receiving frequent irrigation.

In contrast to the extensive research that has evaluated MSBBs and HNBs for crop biomass and yield, considerably less attention has been paid to ET_A estimation (Rodriguez et al., 2011). Biggs et al. (2015) provide a comprehensive review of MSBB methods, which typically use vegetation indices, thermal bands, or a combination of the two to estimate ET_A or its energy equivalent latent energy (LE). Vegetation-index methods include empirical methods that correlate ET_A with a vegetation index, and more complex, process-based models that use vegetation indices to partition ET_A into its components (Wang et al., 2014). NDVI, SAVI, or SAVI's successor, the enhanced vegetation index (EVI) (Huete et al., 2002), are primarily used, because they are sensitive to changes in biomass and F_{PAR}. The Normalized Difference Water Index and Global Vegetation Moisture Index, which rely on the SWIR region sensitive to leaf water absorption, have more recently been suggested to improve estimates of ET_A (Guerschman et al., 2009; Lu and Zhuang, 2010). Thermal methods employ thermal bands to estimate LE as a residual in the energy balance equation and include the Simplified Surface Energy Balance (SSEB) (Senay et al., 2007), Atmosphere-Land Exchange Inverse (ALEXI) (Anderson et al., 1997) model, Mapping Evapotranspiration at high Resolution with Internalized Calibration (METRIC) (Allen et al., 2007), and Surface Energy Balance Algorithm (SEBAL) (Bastiaanssen et al., 1998). Thermal methods estimate sensible heat (H) from the radiometric land surface temperature (T_R), but the relationship between H and T_R is not unique and is difficult to estimate directly from remote sensing. Several thermal methods (METRIC, SEBAL, and related models) use internal calibration for each image to estimate the H-T_R relationship by assuming that LE is zero at the driest, hottest pixels in images and equal to net radiation at the wettest, coldest pixels. Other methods (ALEXI) use the change in remotely sensed temperature during two time periods to constrain an atmospheric boundary layer model (Anderson et al., 2007), which requires the use of geostationary or polar-orbiting satellites with multiple overpasses per day. Unlike METRIC and SEBAL, which are one-source approaches that estimate the combination of evaporation and ET_C, ALEXI uses a vegetation index to separate ET_A into its two primary components (LE and soil heat flux).

ET_A from remote sensing can be compared with point measurements of ET_A from eddy covariance flux towers or Bowen ratio towers and from watersheds where ET_A is calculated as a residual of the water balance. Accuracies of the main methods (vegetation index, thermal, and mixed) are typically in the range of 5%–15% for annual and seasonal ET_A (Biggs et al., 2015). Accuracies typically decrease with decreasing time scale, so remote sensing methods are most accurate for longer time periods (seasonal, annual), where errors average out (Biggs et al., 2015 and references therein).

Several studies have evaluated HVIs for estimating biomass, F_{PAR}, and other biophysical parameters that may constrain ET_A, such as crop nutrient or water stress, but few have been performed to estimate ET_A directly (Rodriguez et al., 2011). Marshall et al. (2016) used ground-based spectroscopy and eddy covariance flux tower data to compare several MBVIs and HVIs for estimating ET_A and its components. In general, MBVIs explained less variability than HVIs: $\Delta R^2 = -0.12$ for ET_A, $\Delta R^2 = -0.17$ for ET_C, and $\Delta R^2 = -0.14$ for soil evaporation. The most highly correlated HVIs for ET_A were in the red edge centered at 672 nm and visible blue (428–478 nm): $R^2 = 0.51$. Several HVIs from the NIR and SWIR were correlated with ET_A as well. Similarly, the most highly correlated HVI for ET_C was also centered on the red edge (672 nm) but performed better with NIR and SWIR channels (722–1050 nm) than the visible blue. The best index for ET_C ($R^2 = 0.68$) was centered at 672 and 733 nm. Correlations between soil evaporation and HVIs were even stronger than for ET_C. Unlike ET_A and ET_C, which were estimated well with HVIs from the literature, soil evaporation was estimated best with channel combinations centered on 743 nm and a range of NIR and SWIR channels (916–1155 nm). The highest correlated index was at 743 and 953 nm ($R^2 = 0.72$). The authors noted that the greatest opportunity for improving ET_A estimates with HNBs was by combining a hyperspectral NDVI for ET_C with the new soil evaporation index. The hyperspectral NDVI could be used to estimate crop biomass and Y as well.

5.3 DEMONSTRATION OF CWP ESTIMATION USING REMOTE SENSING

Data were collected and analyzed to present a methodology and identify MBVIs that should be considered when defining HVIs for CWP estimation when hyperspectral data become more readily available. A variant of CWP was computed for the first time from eddy covariance flux towers for several crop fields in the United States. The towers regularly measure the accumulation of crop biomass and LE over the growing season. Remote sensing data were collected in the footprint of each tower for the analysis. The footprint was defined as the upwind area contributing to the flux measurements. The remote sensing data included MBVIs derived from Landsat-7 ETM+. The Google Earth Engine (https://earthengine.google.com/) cloud computing platform was used to process the large volume of remote sensing data more efficiently. The MBVIs were compared to CWP as two-band difference ratios.

5.3.1 Data and Processing

5.3.1.1 Eddy Covariance Flux Towers

Data from seven eddy covariance flux towers in agricultural areas across the United States were downloaded from the Ameriflux network (http://ameriflux.ornl.gov/). Each station is summarized in Table 5.2. The sites were selected because they had at least 2 years of quality data and at least one near cloud-free satellite image toward the end of the growing season. The stations represent a wide range of C_3 and C_4 crops in rainfed and irrigated fields and altogether span 15 years. Gross primary production (GPP) (μmol CO_2 m^{-2} s^{-1}) and LE (W m^{-2}) were retrieved or derived for each station. The sum of GPP over the growing season (after respiration costs have been discounted) is equal to net primary production (NPP) or total crop biomass. The selected images were sufficiently close to the end of the growing season for us to compute NPP as the sum of GPP from start of season to the date of image retrieval. Start- and end-of-season information was provided by the growers or farm managers. It was assumed, based on Ryan (1991), Gifford (1994), and Waring et al. (1998), that plant respiration was 50% of GPP. GPP was available to download for four stations: US-Ne1, US-Ne2, US-Ne3, and US-Twt. For the other stations, GPP was not available to download. It was instead computed from the carbon flux data using the same approach as at the aforementioned stations, that of the marginal distribution sampling method (Reichstein et al., 2005). The data were available at 30-min or 60-min intervals. They were aggregated to a daily time step over daytime hours (incoming solar radiation >5 W m^{-2}). Daytime averages were computed instead of daily averages because the link between turbulent heat flux and plant response is strong during peak solar hours (Fisher et al., 2008). Finally, a 5-day exponential moving average filter was applied to account for the time lag of

TABLE 5.2
Eddy Covariance Flux Towers Used to Estimate Crop Water Productivity

ID	Site Name	Lat	Long	Years	Crop Type	(R)ainfed (I)rrigated	Elv (m)	PPT (mm)	T (°C)
US-ARM	ARM Southern Great Plains	36.61	−97.49	2006, 2008, 2009, 2010 and 2012	Maize and wheat	R	314	843	15
US-Bo1	Bondville	40.01	−88.29	2004, 2005, and 2007	Maize and soy	R	219	991	11.0
US-Ib1	Batavia	41.86	−88.22	2011	Soy	R	227	929	9.2
US-Ne1	Mead Irrigated	41.17	−96.48	2007	Maize	I	361	887	9.7
US-Ne2	Mead Irrigated	41.16	−96.47	2003, 2008, and 2009	Maize and soy	I	362	887	9.7
US-Ne3	Mead Rainfed	41.18	−96.44	2008	Soy	R	363	887	9.7
US-Twt	Twitchell Island	38.11	−121.65	2013	Rice	I	−5	346	15.9

Note: The site identification (ID), site name, and general bioclimatic information, such as annual rainfall (PPT), elevation (Elv), and average temperature (T), can be found on the Ameriflux website. Only years when a remote sensing image was retrieved and used in the analysis are displayed.

TABLE 5.3
Parameters (HI—Harvest Index, MC—Grain or Seed Moisture Content, and RS—Root-to-Shoot Ratio) Used to Convert Crop Biomass to Yield

Crop Type	RS	MC	HI
Maize	0.18	0.11	0.50
Rice	0.10	0.09	0.40
Soy	0.15	0.1	0.41
Wheat	0.20	0.11	0.37

crop adaptation to new weather conditions. CWP (kg m^{-3}) was computed using Equation 5.1. Crop yield was calculated from NPP using the following equation:

$$\text{Yield} = \text{NPP} * \frac{\text{HI}}{(1+\text{RS})} * \frac{1}{(1-\text{MC})} \tag{5.2}$$

where RS is the root-to-shoot ratio or proportion of belowground to aboveground biomass and MC is the seed moisture content of the grain or seed yield. RS and HI estimates per crop type were taken from Prince et al. (2001), while MC estimates were taken from Lobell et al. (2002) (Table 5.3).

5.3.1.2 Landsat Imagery

Thirteen Landsat images corresponding to the Ameriflux measurements over the time period of 2003–2013 were deemed suitable for analysis. Images were selected based on three criteria: (1) the flux tower footprint was outside the image striping resulting from the failure of the Landsat-7 Scan Line Corrector; (2) the image contained $<20\%$ cloud cover; and (3) the image preceding and closest temporally to the harvest was selected. In general, images for winter crops (wheat) were retrieved in May or June, while for summer crops (maize, soy, and rice) images were retrieved in September or October (Table 5.4). In each case, the images were acquired within 2 weeks of harvest. The raw (Level 1) Landsat data were converted to surface reflectance using the Landsat Ecosystem Disturbance Adaptive Processing System (LEDAPS) atmospheric correction algorithm (Masek et al., 2006). Pixels containing clouds or cloud shadows were flagged using the C function of the Mask (CFMask) algorithm (Foga et al., 2017). Surface reflectance was characterized by blue, green, red, VNIR, and two SWIR bands centered at 485, 560, 660, 830, 1650, and 2215 nm, respectively. These bands are at 30 m spatial resolution. Landsat-7 contains one thermal channel (band 6, at 60 m resolution), which was omitted from the analysis. A 3×3 pixel window in the center of the footprint of each flux tower was averaged and paired with CWP estimates for the analysis.

5.3.2 Analytical Techniques

The MSBBs were converted to MBVIs as two-band difference ratios $[(R_2 - R_1)/(R_1 + R_2)]$ and related to CWP using linear regression. R_1 and R_2 in this case are two given Landsat bands. Band ratioing was selected over more rigorous mining techniques because of the small sample size ($N = 15$). For this reason, the regression was performed on all crops instead of by crop type. The results were first interpreted as λ-λ contour plots (Thenkabail et al., 2000). λ-λ plots are a convenient way to display every possible combination and corresponding R^2 value. The highest performing band combination was then viewed using a scatterplot. Performance was measured with R^2 and

TABLE 5.4
Details on Landsat-7 Enhanced Thematic Mapper Plus (ETM+) Images Used for the Analysis

ID	Crop Type	Image ID	Date Acquired	ΔEOS (Days)
US-ARM	Wheat	LE70280352006159	6/8/2006	13
US-ARM	Maize	LE70280352008261	9/17/2008	8
US-ARM	Wheat	LE70280352009167	6/16/2009	2
US-ARM	Wheat	LE70280352010170	6/19/2010	3
US-ARM	Wheat	LE70280352012144	5/23/2012	5
US-Bo1	Soy	LE70220322004256	9/12/2004	2
US-Bo1	Maize	LE70230322005265	9/22/2005	5
US-Bo1	Maize	LE70230322007271	9/28/2007	5
US-IB1	Soy	LE70230312011282	10/9/2011	4
US-Ne1	Maize	LE70280312007274	10/1/2007	1
US-Ne2	Maize	LE70280312003263	9/20/2003	12
US-Ne2	Soy	LE70280312008261	9/17/2008	8
US-Ne2	Maize	LE70280312009263	9/20/2009	13
US-Ne3	Soy	LE70280312008261	9/17/2008	6
US-Twt1	Rice	LE70440332013258	9/15/2013	8

Note: In each case, the image difference between the acquisition data and end of season (ΔEOS) was less than two weeks.

cross-validated root-mean-squared error (RMSE). The data were initially transformed to account for nonlinearity, but the transformation did not improve the performance of the vegetation indices.

5.3.3 Results

CWP and its components varied considerably across crop types. Rice lost the most water over the growing season (ET_A – 0.52 m), which was offset by relatively high Y (0.83 kg m^{-2}) and led to moderate CWP (1.60 kg m^{-3}). Similarly, maize lost relatively more water over the growing season than other crops ($ET_A = 0.46 \pm 0.05$ m) but produced high Y (1.40 ± 0.46 kg m^{-2}), leading to the highest CWP of all the crops (3.00 ± 0.89 kg m^{-3}). Soy, on the other hand, produced relatively high CWP (1.80 ± 0.48 kg m^{-3}), which was due to low water loss ($ET_A = 0.43 \pm 0.07$ mm) instead of high Y (0.76 ± 0.18 kg m^{-2}). Wheat was the most water efficient ($ET_A = 0.35 \pm 0.05$ m) but had the lowest Y (0.53 ± 0.05 kg m^{-2}). This led to the lowest CWP overall (1.53 ± 0.15 kg m^{-3}).

Figure 5.2 shows how each Landsat combination performed when estimating CWP and its components (Y and ET_A). Figure 5.3 shows the top-performing band combination for each category. Landsat bands 1 (blue) and 3 (red) resulted in the highest correlation and lowest error, while Landsat bands 5 (SWIR1) and 7 (SWIR2) resulted in the lowest correlation and highest error. Landsat band 1 for CWP appeared the most important, because it was in each of the top-performing combinations (with band 3 $R^2 = 0.72$, RMSE = 0.50 kg m^{-3}; band 2 $R^2 = 0.32$, RMSE = 0.70 kg m^{-3}; band 5 $R^2 = 0.31$, RMSE = 0.68 kg m^{-3}; and band 7 $R^2 = 0.30$, RMSE = 0.73 kg m^{-3}). Similarly, Landsat bands 1 and 3 were the highest performing combination for crop yield (Figure 5.2b), with band 1 present in all of the top-performing combinations (with band 3 $R^2 = 0.64$, RMSE = 0.31 kg m^{-2}; band 7 $R^2 = 0.34$, RMSE = 0.38 kg m^{-2}; band 5 $R^2 = 0.33$, RMSE = 0.40 kg m^{-2}; and band 2 $R^2 = 0.28$, RMSE = 0.39 kg m^{-2}). Landsat bands were less effective at estimating ET_A. Landsat

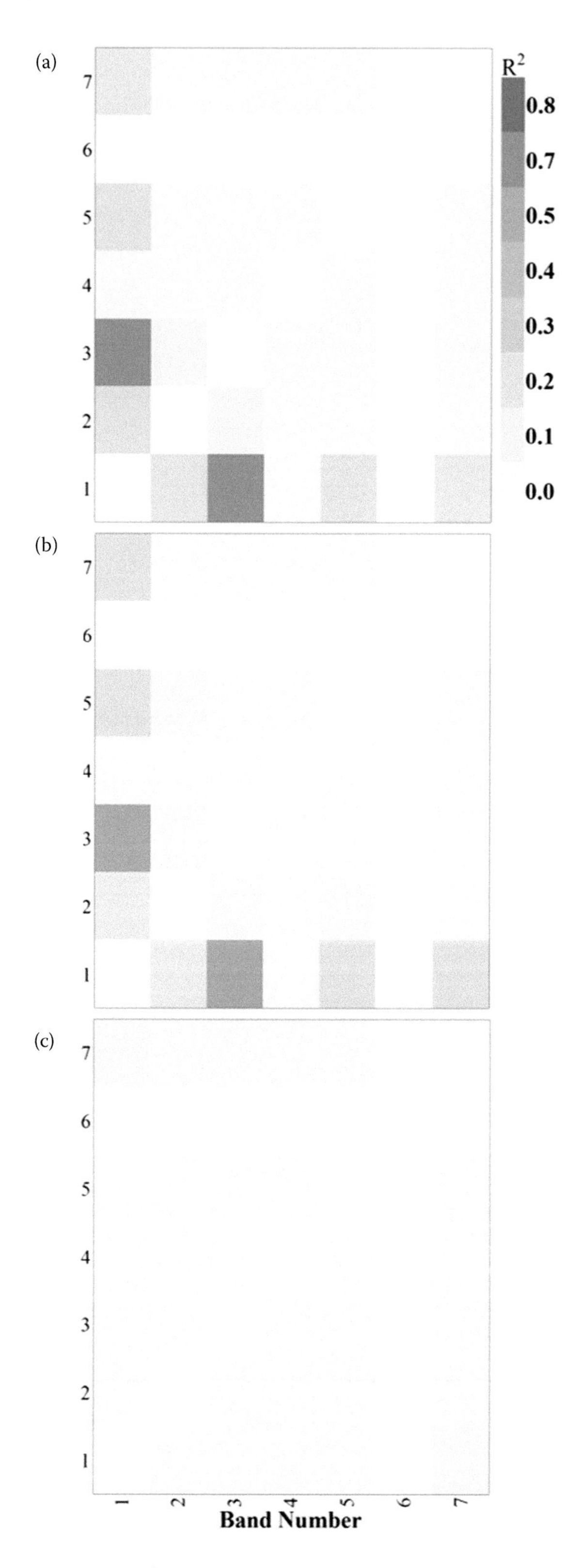

FIGURE 5.2 λ-λ plots expressing R^2 performance of each Landsat band combination with (a) crop yield; (b) evapotranspiration; and (c) CWP.

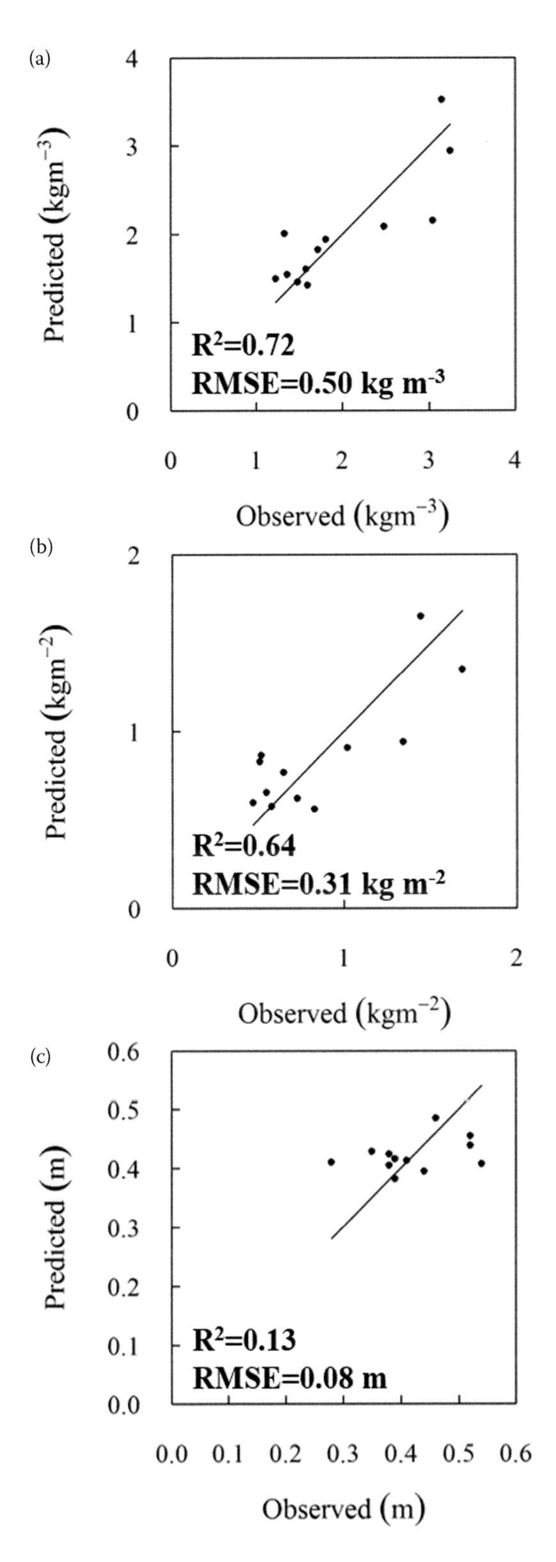

FIGURE 5.3 Predicted versus observed (a) crop yield; (b) evapotranspiration; and (c) CWP. Solid line: perfect correlation ($R^2 = 1.0$). Predicted values were derived from highest performing Landsat bands (1, 3, and 7).

bands 1 and 7 produced the highest performing combination, with $R^2 = 0.13$ and RMSE = 0.08 m (Figure 5.2c).

5.4 DISCUSSION

CWP can vary widely across crop varieties and types (Zwart and Bastiaanssen, 2004). CWP for rice and wheat estimated here were on the upper end of published values (0.6–1.6 and 0.6–1.7 kg m^{-3}), while maize was outside the range of typical values (1.1–2.7 kg m^{-3}), but still less than the maximum value published (3.99 kg m^{-3}). The eco-physiological factors that can account for some CWP variability were reviewed earlier. Methodological differences can also play an important role (Bessembinder et al., 2005). In some studies, yield was estimated just after harvest, while in other studies yield was estimated after some period of drying (MC→0). It was assumed for ease of comparison that variations in GPP were an order of magnitude greater than crop respiration or HI. In this way, respiration became a conservative fraction of GPP summed over the growing season, and HI was constant for each crop type. In reality, respiration and HI can vary considerably across crop varieties and bioclimatic conditions (Turner et al., 2003). The period over which estimates were made can vary among studies, particularly for ET_A. Some studies start estimating ET_A after sowing, while others wait until after emergence. Similarly, some studies estimate ET_A only until physiological maturity, while others estimate ET_A until harvest. During these periods of emergence and senescence, soil evaporation and ET_C can be high, respectively. Finally, instrumentation can be a source of uncertainty. Eddy covariance flux towers are considered the research standard for estimating carbon and moisture flux. Alfieri et al. (2011) found that uncertainties on average for LE and CO_2 flux measured with eddy covariance flux towers were 27 W m^{-2} and 0.10 mg m^{-2} s^{-1}, respectively. These uncertainties increased dramatically in the afternoon and evening hours, when advective conditions are prevalent. Under these conditions, discrepancies with other techniques, such as surface renewal, were high (French et al., 2012).

The results of the broadband analysis, though preliminary, will be invaluable to any future hyperspectral mission, such as HyspIRI (https://hyspiri.jpl.nasa.gov/). The methodology presented here with six spectral bands when applied to HyspIRI with hundreds of spectral bands will undoubtedly lead to more evident CWP spectral features and correlations identified as follows. First, the Landsat blue band proved to be the most effective predictor of CWP and yield. Red was also important, but to a lesser degree. At first glance, this seems counterintuitive, because other studies generally find the strongest correlations with crop biomass and Y to be in the red and VNIR. Further, Johnson (2016) observed that the red-VNIR relationship is strongest midseason during a crop's vegetative stage, when F_{PAR} is high. Landsat imagery for this study was intentionally collected just before harvest to detect the cumulative effects of CWP. At the end of the season, F_{PAR}→0, and the inverse relationship between red and VNIR, which vegetation indices like NDVI exploit, diminishes. The spectra become monotonically increasing through the visible and VNIR, so visible blue and red may be just as effective, if not more so, at estimating yield. Johnson (2016) notes that the advantage of using midseason remote sensing data to predict yield is for forecasting purposes. Forecasting potential was not the objective here, but it should be explored when hyperspectral data become available. Second, Landsat bands 1–5 and 7 were effective at estimating CWP primarily through the Y component. The relationship between these bands and ET_A was poor. This again could be due to late-season effects since crops are typically on a deficit watering regime. Half of the sites analyzed were irrigated. It has been shown that thermal bands, which were omitted from the analysis, are effective at estimating ET_A. The difference in predictability of the two components suggests that CWP cannot be measured directly using the same band combinations. Finally, instead of considering continuous estimates of CWP, it might be more advantageous to use remote sensing to categorize estimates into low, moderate, and high CWP.

HyspIRI will provide global and continuous coverage of surface reflectance with HNBs for the first time. The frequent revisit rate would create the opportunity to compare MBVIs and HVIs at important vegetative and reproductive phases of crop development. A combination of spectral channels, ideally at early stages of crop development, could be used to forecast Y and CWP at the end of the growing season. This would be critical to compare and design preventive strategies to maximize food security and minimize water scarcity.

REFERENCES

Allen, R.G., Tasumi, M., Trezza, R., 2007. Satellite-based energy balance for mapping evapotranspiration with internalized calibration (METRIC)—model. *Journal of Irrigation and Drainage Engineering*, 133: 380–394.

Alfieri, J.G., Kustas, W.P., Prueger, J.H., Hipps, L.E., Chavez, J.L., French, A.N., Evett, S.R., 2011. Intercomparison of nine micrometeorological stations during the BEAREX08 field campaign. *J. Atmos. Ocean. Techol.* 28, 1390–1406.

Ali, M.H., Talukder, M.S.U., 2008. Increasing water productivity in crop production—A synthesis. *Agric. Water Manag.* 95, 1201–1213.

Anderson, M.C., Norman, J.M., Diak, G.R., Kustas, W.P., Mecikalski, J.R., 1997. Two-source time-integrated model for estimating surface fluxes using thermal infrared remote sensing. *Remote Sensing of Environment*, 60: 195–216.

Anderson, M.C., Kustas, W.P., Norman, J.M., 2007. Upscaling flux observations from local to Continental scales using thermal remote sensing. *Agronomy Journal*, 99: 240–254.

Bastiaanssen, W.G.M., Menenti, M., Feddes, R.A., Holtslag, A.A.M., 1998. A remote sensing surface energy balance algorithm for land (SEBAL) 1. Formulation. *Journal of Hydrology*, 212–213: 198–212.

Bastiaanssen, W.G.M., Steduto, P., 2017. The water productivity score (WPS) at global and regional level: Methodology and first results from remote sensing measurements of wheat, rice and maize. *Sci. Total Environ.* 575, 595–611.

Bernacchi, C.J., VanLoocke, A., 2015. Terrestrial ecosystems in a changing environment: A dominant role for water. *Annu. Rev. Plant Biol.* 66, 599–622.

Bessembinder, J.J.E., Leffelaar, P.A., Dhindwal, A.S., Ponsioen, T.C., 2005. Which crop and which drop, and the scope for improvement of water productivity. *Agric. Water Manag.* 73, 113–130. https://doi.org/10.1016/j.agwat.2004.10.004

Biggs, T.W., Petropoulos, G.P., Velpuri, N.G., Marshall, M., 2015. Remote sensing of evapotranspiration from cropland, in: *Remote Sensing Handbook*. Taylor and Francis, London, United Kingdom, pp. 59–99.

Brauman, K.A., Siebert, S., Foley, J.A., 2013. Improvements in crop water productivity increase water sustainability and food security—A global analysis. *Environ. Res. Lett.* 8, 024030.

Cai, X., Kam, S.P., Yen, B.T., Sood, A., Thai, H.C., 2014. CaWAT—A catchment water allocation tool for integrated irrigation and aquaculture development in small watersheds. Presented at the *International Conference on Hydroinformatics*, CUNY Academic Works, New York, NY.

Challinor, A.J., Watson, J., Lobell, D.B., Howden, S.M., Smith, D.R., Chhetri, N., 2014. A meta-analysis of crop yield under climate change and adaptation. *Nat. Clim. Change* 4, 287–291.

Chapin, F.S., Matson, P.A., Vitousek, P., 2011. *Principles of Terrestrial Ecosystem Ecology*. Springer Science & Business Media.

Chi, M., Plaza, A., Benediktsson, J.A., Sun, Z., Shen, J., Zhu, Y., 2016. Big data for remote sensing: Challenges and opportunities. *Proc. IEEE* 104, 2207–2219.

Collatz, G.J., Ball, J.T., Grivet, C., Berry, J.A., 1991. Physiological and environmental regulation of stomatal conductance, photosynthesis and transpiration: A model that includes a laminar boundary layer. *Agric. For. Meteorol.* 54, 107–136.

Collatz, G.J., Ribas-Carbo, M., Berry, J.A., 1992. Coupled photosynthesis-stomatal conductance model for leaves of C4 plants. *Funct. Plant Biol.* 19, 519–538.

Cook, B.I., Smerdon, J.E., Seager, R., Coats, S., 2014. Global warming and 21st century drying. *Clim. Dyn.* 43, 2607–2627.

Dickinson, R.E., Shaikh, M., Bryant, R., Graumlich, L., 1998. Interactive canopies for a climate model. *J. Climate.* 11, 2823–2836.

Dunne, T., Leopold, L.B., 1978. *Water in Environmental Planning*. W. H. Freeman, New York, NY.

Evans, R.G., Sadler, E.J., 2008. Methods and technologies to improve efficiency of water use. *Water Resour. Res.* 44, W00E04.

Feng, H., Jiang, N., Huang, C., Fang, W., Yang, W., Chen, G., Xiong, L., Liu, Q., 2013. A hyperspectral imaging system for an accurate prediction of the above-ground biomass of individual rice plants. *Rev. Sci. Instrum.* 84, 095107.

Fisher, J.B., Tu, K.P., Baldocchi, D.D., 2008. Global estimates of the land–atmosphere water flux based on monthly AVHRR and ISLSCP-II data, validated at 16 FLUXNET sites. *Remote Sens. Environ.* 112, 901–919.

Foga, S., Scaramuzza, P.L., Guo, S., Zhu, Z., Dilley Jr., R.D., Beckmann, T., Schmidt, G.L., Dwyer, J.L., Hughes, M.J., Laue, B., 2017. Cloud detection algorithm comparison and validation for operational Landsat data products—Science direct. *Remote Sens. Environ.* 194, 379–390.

French, A.N., Alfieri, J.G., Kustas, W.P., Prueger, J.H., Hipps, L.E., Chavez, J.L., Evett, S.R. et al. 2012. Estimation of surface energy fluxes using surface renewal and flux variance techniques over an advective irrigated agricultural site. *Adv. Water Resour.* 50, 91–105.

Gerten, D., Heinke, J., Hoff, H., Biemans, H., Fader, M., Waha, K., 2011. Global Water Availability and Requirements for Future Food Production AMSDocument. http://dx.doi.org/10.1175/2011JHM1328.1, last accessed February 19, 2018.

Gifford, R.M., 1994. The global carbon cycle: A viewpoint on the missing sink. *Funct. Plant Biol.* 21, 1–15.

Gnyp, M.L., Bareth, G., Li, F., Lenz-Wiedemann, V.I.S., Koppe, W., Miao, Y., Hennig, S.D. et al. 2014. Development and implementation of a multiscale biomass model using hyperspectral vegetation indices for winter wheat in the North China Plain. *Int. J. Appl. Earth Obs. Geoinformation* 33, 232–242. https://doi.org/10.1016/j.jag.2014.05.006

Goetz, A.F.H., 2009. Three decades of hyperspectral remote sensing of the Earth: A personal view. *Remote Sens. Environ.*, Imaging Spectroscopy Special Issue 113(Supplement 1), S5–S16.

Guerschman, J.P., Van Dijk, A.I.J.M., Mattersdorf, G., Beringer, J., Hutley, L.B., Leuning, R., Pipunic, R.C., Sherman, B.S., 2009. Scaling of potential evapotranspiration with MODIS data reproduces flux observations and catchment water balance observations across Australia. *J. Hydrol.* 369, 107–119.

Hay, R.K.M., 1995. Harvest index: A review of its use in plant breeding and crop physiology. *Ann. Appl. Biol.* 126, 197–216.

Huete, A., Didan, K., Miura, T., Rodriguez, E.P., Gao, X., Ferreira, L.G., 2002. Overview of the radiometric and biophysical performance of the MODIS vegetation indices. *Remote Sens. Environ.* 83, 195–213.

Johnson, D.M., 2016. A comprehensive assessment of the correlations between field crop yields and commonly used MODIS products. *Int. J. Appl. Earth Obs. Geoinformation* 52, 65–81.

Katul, G.G., Oren, R., Manzoni, S., Higgins, C., Parlange, M.B., 2012. Evapotranspiration: A process driving mass transport and energy exchange in the soil-plant-atmosphere-climate system. *Rev. Geophys.* 50, RG3002.

Kemanian, A.R., Stöckle, C.O., Huggins, D.R., Viega, L.M., 2007. A simple method to estimate harvest index in grain crops. *Field Crops Res.* 103, 208–216.

Law, B.E., Falge, E., Gu, L., Baldocchi, D.D., Bakwin, P., Berbigier, P., Davis, K. et al. 2002. Environmental controls over carbon dioxide and water vapor exchange of terrestrial vegetation. *Agric. For. Meteorol., FLUXNET 2000 Synthesis* 113, 97–120.

Liu, J., Miller, J.R., Pattey, E., Haboudane, D., Strachan, I.B., Hinther, M., 2004. Monitoring crop biomass accumulation using multi-temporal hyperspectral remote sensing data, in: IGARSS 2004. *2004 IEEE International Geoscience and Remote Sensing Symposium*. Presented at the IGARSS 2004. 2004 IEEE International Geoscience and Remote Sensing Symposium, pp. 1637–1640, vol. 3.

Liu, J., Zehnder, A.J.B., Yang, H., 2009. Global consumptive water use for crop production: The importance of green water and virtual water. *Water Resour. Res.* 45, W05428.

Lobell, D.B., Hicke, J.A., Asner, G.P., Field, C.B., Tucker, C.J., Los, S.O., 2002. Satellite estimates of productivity and light use efficiency in United States agriculture, 1982–1998. *Glob. Change Biol.* 8, 722–735.

Lu, X., Zhuang, Q., 2010. Evaluating evapotranspiration and water-use efficiency of terrestrial ecosystems in the conterminous United States using MODIS and AmeriFlux data. *Remote Sens. Environ.* 114, 1924–1939.

Mariotto, I., Thenkabail, P.S., Huete, A., Slonecker, E.T., Platonov, A., 2013. Hyperspectral versus multispectral crop-productivity modeling and type discrimination for the HyspIRI mission. *Remote Sens. Environ.* 139, 291–305.

Marshall, M., Thenkabail, P., 2014. Biomass modeling of four leading world crops using hyperspectral narrowbands in support of HyspIRI mission. *Photogramm. Eng. Remote Sens.* 80, 757–772.

Marshall, M., Thenkabail, P., 2015a. Developing *in situ* non-destructive estimates of crop biomass to address issues of scale in. *Remote Sens.* 7, 808–835.

Marshall, M., Thenkabail, P., 2015b. Advantage of hyperspectral EO-1 hyperion over multispectral IKONOS, GeoEye-1, WorldView-2, Landsat ETM+, and MODIS vegetation indices in crop biomass estimation. *ISPRS J. Photogramm. Remote Sens.* 108, 205–218.

Marshall, M., Thenkabail, P., Biggs, T., Post, K., 2016. Hyperspectral narrowband and multispectral broadband indices for remote sensing of crop evapotranspiration and its components (transpiration and soil evaporation). *Agric. For. Meteorol.* 218–219, 122–134.

Masek, J.G., Vermote, E.F., Saleous, N.E., Wolfe, R., Hall, F.G., Huemmrich, K.F., Gao, F., Kutler, J., Lim, T.-K., 2006. A Landsat surface reflectance dataset for North America, 1990–2000. *IEEE Geosci. Remote Sens. Lett.* 3, 68–72.

Molden, D., 1997. Accounting for Water Use and Productivity. IWMI.

Molden, D.J., Murray-Rust, H., Sakthivadivel, R., Makin, I., 2003. A water-productivity framework for understanding and action, in: *Water Productivity in Agriculture: Limits and Opportunities for Improvement.* CABI, pp. 1–18.

Monteith, J.L., 1965. Evaporation and environment. *Symp. Soc. Exp. Biol.* 19, 205–234.

Monteith, J., 1969. Light interception and radiative exchange in crop stands. *Agron. Hortic.—Fac. Publ.*

Mu, Q., Zhao, M., Running, S.W., 2011. Improvements to a MODIS global terrestrial evapotranspiration algorithm. *Remote Sens. Environ.* 115, 1781–1800.

Mutanga, O., Skidmore, A.K., 2004. Narrow band vegetation indices overcome the saturation problem in biomass estimation. *Int. J. Remote Sens.* 25, 3999–4014.

Osborne, S.L., Schepers, J.S., Francis, D.D., Schlemmer, M.R., 2002. Use of spectral radiance to estimate in-season biomass and grain yield in nitrogen- and water-stressed corn. *Crop Sci.* 42, 165–171.

Pieruschka, R., Huber, G., Berry, J.A., 2010. Control of transpiration by radiation. *Proc. Natl. Acad. Sci.* 107, 13372–13377.

Porkka, M., Gerten, D., Schaphoff, S., Siebert, S., Kummu, M., 2016. Causes and trends of water scarcity in food production. *Environ. Res. Lett.* 11, 015001.

Prince, S.D., Haskett, J., Steininger, M., Strand, H., Wright, R., 2001. Net primary production of U.S. Midwest croplands from agricultural harvest yield data. *Ecol. Appl.* 11, 1194–1205.

Prudhomme, C., Giuntoli, I., Robinson, E.L., Clark, D.B., Arnell, N.W., Dankers, R., Fekete, B.M. et al. 2014. Hydrological droughts in the 21st century, hotspots and uncertainties from a global multimodel ensemble experiment. *Proc. Natl. Acad. Sci.* 111, 3262–3267.

Raza, A., Friedel, J.K., Bodner, G., 2012. Improving water use efficiency for sustainable agriculture, in: Lichtfouse, E. (Ed.), *Agroecology and Strategies for Climate Change, Sustainable Agriculture Reviews.* Springer, Netherlands, pp. 167–211.

Reichstein, M., Falge, E., Baldocchi, D., Papale, D., Aubinet, M., Berbigier, P., Bernhofer, C. et al. 2005. On the separation of net ecosystem exchange into assimilation and ecosystem respiration: Review and improved algorithm. *Glob. Change Biol.* 11, 1424–1439.

Rodriguez, J.M., Ustin, S.L., Riaño, D., 2011. Contributions of imaging spectroscopy to improve estimates of evapotranspiration. *Hydrol. Process.* 25, 4069–4081.

Roost, N., 2008. An assessment of distributed, small-scale storage in the Zhanghe irrigation system, China. *Agric. Water Manag.* 95, 685–697.

Ryan, M.G., 1991. Effects of climate change on plant respiration. *Ecol. Appl.* 1, 157–167

Schewe, J., Heinke, J., Gerten, D., Haddeland, I., Arnell, N.W., Clark, D.B., Dankers, R. et al. 2014. Multimodel assessment of water scarcity under climate change. *Proc. Natl. Acad. Sci.* 111, 3245–3250.

Schlesinger, W.H., Jasechko, S., 2014. Transpiration in the global water cycle. *Agric. For. Meteorol.* 189–190, 115–117.

Schlosser, C., Strzepek, K., Gao, X., Gueneau, A., Fant, C., Paltsev, S., Rasheed, B. et al. 2014. *The Future of Global Water Stress: An Integrated Assessment (No. 254), MIT Joint Program on the Science and Policy of Global Change.* Massachusetts Institute of Technology, Cambridge, MA.

Senay, G.B., Budde, M., Verdin, J.P. Melesse, A.M., 2007. A coupled remote sensing and simplified surface energy balance approach to estimate actual evapotranspiration from irrigated fields. *Sensors*, 7, 979–1000.

Siebert, S., Döll, P., 2010. Quantifying blue and green virtual water contents in global crop production as well as potential production losses without irrigation. *J. Hydrol. Green-Blue Water Initiative (GBI)* 384, 198–217.

Steduto, P., Hsiao, T.C., Fereres, E., 2007. On the conservative behavior of biomass water productivity. *Irrig. Sci.* 25, 189–207.

Thenkabail, P.S., Smith, R.B., De Pauw, E., 2000. Hyperspectral vegetation indices and their relationships with agricultural crop characteristics. *Remote Sens. Environ.* 71, 158–182.

Turner, D.P., Urbanski, S., Bremer, D., Wofsy, S.C., Meyers, T., Gower, S.T., Gregory, M., 2003. A cross-biome comparison of daily light use efficiency for gross primary production. *Glob. Change Biol.* 9, 383–395.

Ustin, S.L., Roberts, D.A., Gamon, J.A., Asner, G.P., Green, R.O., 2004. Using imaging spectroscopy to study ecosystem processes and properties. *BioScience* 54, 523–533.

van Dijk, A.I.J.M., Bruijnzeel, L.A., 2001. Modelling rainfall interception by vegetation of variable density using an adapted analytical model. Part 2. Model validation for a tropical upland mixed cropping system. *J. Hydrol.* 247, 239–262.

Wang, L., Good, S.P., Caylor, K.K., 2014. Global synthesis of vegetation control on evapotranspiration partitioning. *Geophys. Res. Lett.* 41, 2014GL061439.

Waring, R.H., Landsberg, J.J., Williams, M., 1998. Net primary production of forests: A constant fraction of gross primary production? *Tree Physiol.* 18, 129–134.

Zwart, S.J., Bastiaanssen, W.G., 2004. Review of measured crop water productivity values for irrigated wheat, rice, cotton, and maize. *Agric. Water Manag.* 69, 115–133.

Zwart, S.J., Bastiaanssen, W.G., de Fraiture, C., Molden, D.J., 2010. WATPRO: A remote sensing based model for mapping water productivity of wheat. *Agric. Water Manag.* 97, 1628–1636.

6 Hyperspectral Remote Sensing Tools for Quantifying Plant Litter and Invasive Species in Arid Ecosystems

Pamela Lynn Nagler, B. B. Maruthi Sridhar, Aaryn Dyami Olsson, Willem J. D. van Leeuwen, and Edward P. Glenn

CONTENTS

6.1 INTRODUCTION: HYPERSPECTRAL REMOTE SENSING OF LANDSCAPE COMPONENTS

Green vegetation can be monitored and distinguished using visible (VIS) and infrared (IR) multiband and hyperspectral remote sensing methods. The problem has been in identifying and distinguishing the nonphotosynthetically active radiation (PAR) landscape components, such as litter and soils, from green vegetation [35–38]. Additionally, distinguishing different species of green vegetation is challenging using the relatively few bands available on most satellite sensors. This chapter focuses both on previously published work by Nagler and others [35–38] that identified hyperspectral remote sensing characteristics that distinguish between green vegetation, soil, and litter (or senescent vegetation), as well as on new research conducted to aid in distinguishing invasive species from the mixed land cover surface.

The main message from the previously published work covered here is that the shortwave infrared (SWIR) wavelength range can be used to distinguish plant litter from soils using the cellulose absorption feature seen at 2100 nm exhibited by litter [1]. A three-band SWIR index that incorporates wavelengths that capture unique absorption differences may prove more useful than the visible to near-infrared (VIS-NIR) range in discriminating plant litter from soils.

Quantifying litter by remote sensing methods is important in constructing carbon budgets of natural and agricultural ecosystems. Distinguishing between plant types is important in tracking the spread of invasive species. Green leaves of different species usually have similar spectra, making it difficult to distinguish between species. However, in this chapter we show that phenological differences between species can be used to detect some invasive species by their distinct patterns of greenness and dormancy over an annual cycle based on hyperspectral data. Both applications require methods to quantify the nongreen cellulosic fractions of plant tissues by remote sensing, even in the presence of soil and green plant cover. We explore these methods and offer three case studies. The first concerns distinguishing surface litter from soil using the cellulose absorption index (CAI), as applied to no-till farming practices where plant litter is left on the soil after harvest. The second involves using different band combinations to distinguish invasive tamarisk from agricultural and native riparian plants on the Lower Colorado River. The third illustrates the use of the CAI and normalized difference vegetation index (NDVI) time-series analyses to distinguish between invasive buffelgrass (*Pennisetum cilliare*) and native plants in a desert environment in Arizona. Together the results show how hyperspectral imagery can be used to inform applications and solve problems that are not amenable to solution by the simple band combinations normally used in remote sensing.

6.1.1 Distinguishing between Green Vegetation, Soil, and Litter Using the CAI in Agricultural Systems

6.1.1.1 Plant Litter

Litter is dead plant material that began as green leaves, stems, or fruits, for example, then fell from the canopy to the surface; it gradually decomposes into soil over time. Senescent or dormant leaves on plants can have the same spectral properties as litter and are classified with litter in this chapter. In agricultural systems, litter is the straw fraction of annual crops left behind after harvest, as in no-till farming. In the context of invasive species, seasonal senescence is a distinguishing feature of some invasive species such as buffelgrass, and methods to detect litter can in theory be used to track the spread of these species.

6.1.1.2 The Importance of Litter to the Soil System

Litter contributes greatly to soil nutrient and energy cycles. The amount and composition of litter change spatially (by region) and temporally (by season) [1,2], and rates of decomposition depend upon species physiology, climate/environmental conditions, and microbial activity [3]. Litter decomposition is more closely correlated with nutrient release than with energy flows [4,5]. Eventually decomposition of litter results in humification and mineralization of the recalcitrant carbon fraction. Initially, worms and other macro animals break the litter into smaller fractions with greater surface area [6]. Then bacteria and fungi break down complex biochemicals to molecular constituents [7]. The remaining litter contains celluloses, hemicelluloses, lignins, and many other materials, including organic nitrogen [5]. Finally, organic material is broken down into CO_2, water, and minerals; nitrogen, phosphorus, calcium, magnesium, and potassium are released [8]. The decomposition of litter contributes to atmospheric CO_2 concentrations and contributes to nitrogen and oxygen cycles [1].

6.1.1.3 Benefits of Litter Left in Agricultural Systems

Part of the interest in quantifying soil litter cover by remote sensing is the recent interest in no-till agriculture, a method used to manage residue cover to protect soils from erosion. McMurtrey et al. [9] cited agricultural statistics in the United States as follows: 330 million acres of arable land is tilled, of which 123 million acres are classified as highly erodible land (HEL); the result is the annual loss of 1.25 billion tons of soil. Leaving organic residues on bare soil also affects water infiltration, evaporation, porosity, and soil temperatures [10]. The decay of litter adds nutrients to the soil, improves soil structure, and facilitates tilling, thereby reducing soil erosion, runoff volumes, sediment transport, and movement of pesticides [11]. Maintaining crop residue on the soil surface is frequently the most cost-effective method of reducing soil erosion and complying with federal regulations [12]. Federal erosion prevention legislation is defined in two acts: the 1985 Food Security Act, specifically the Conservation Compliance Provision (Public Law 99-198), and the 1990 Food, Agriculture, Conservation and Trade Act (Public Law 101-624) [9]. In response to these laws, farmers must implement erosion control practices on highly erodible lands.

6.1.1.4 The Importance of Quantifying Litter

Erosion prediction models, i.e., the Universal Soil Loss Equation and the Water Erosion Prediction Project, incorporate crop residue cover estimates, but there is considerable error in these estimates [13,14]. Usually line-transect methods are used to measure litter cover in the field, but these methods are subject to human error and are time consuming [12,14,15]. More rapid and more accurate spectral measurement techniques are needed to improve litter quantification methods [9,12,13,16].

6.1.1.5 Senescent Leaves in the Life Cycle of Invasive Species

Tracking the spread of invasive species, particularly introduced range grasses, has become a priority goal for lands managers. Many of these species have distinct dormant periods in which leaves are dry and brown and have the same spectral properties as litter. While native plants may also have a dormant period, it is sometimes possible to distinguish between species by their phenology, focusing on the timing of their senescent periods through time series of hyperspectral imagery [61,63]. We explore the use of hyperspectral imagery for this purpose as well as for quantifying plant litter and crop residues in this chapter.

6.1.2 Reflectance Spectra (400–2400 nm) Used in a Lab to Distinguish Green Vegetation, Soil, and Litter in the Landscape

6.1.2.1 Distinguishing between Pure Scenes of Plant Litter and Green Vegetation

Remote sensing studies have typically focused on green vegetation because plant canopies are dominated by green leaf spectral features and because key biophysiological processes, such as photosynthesis, occur in green leaves [1]. Photosynthetically active vegetation has a very distinctive spectral reflectance signature in the landscape, with strong absorption in the VIS bands contrasted with strong reflectance in the NIR (Figure 6.3). Ratio and difference indices based on energy from a band in the VIS (400–700 nm) and in the NIR (700–1100 nm) wavelength ranges were first developed by Jordan [17] to assess spectral features in green vegetation for estimating energy accumulation in plant canopies, biomass, and the leaf area per unit ground, or leaf area index (LAI). The green and nongreen components of a canopy can be separately identified using two wavebands because their spectral reflectance curves have unique shapes. In green leaves, pigment concentrations, water content, and structure affect leaf optical properties [18]. Chlorophylls and other pigments in green vegetation absorb in the blue (450 nm) and red (650 nm) wavelengths, and cell structure and thickness control NIR optical properties; reflectance in the green (550 nm) and NIR wavelengths thereby produces a step-function reflectance curve [19–22]. The ability to discriminate green and nongreen component types by differences in their signatures with these two bands alone is the basis for quantifying vegetation parameters for landscape models [23,24].

6.1.2.2 Distinguishing between Pure Scenes of Plant Litter and Soils

The absorption and scattering properties of leaves change as they senesce and decompose [9]. Previous work by Woolley [18] showed that during senescence, leaves lose moisture and air spaces between cells increase. Spectral changes in the VIS wavelengths occur due to the loss of moisture, pigments, and structure. Celluloses and lignins do not readily compost, resulting in high NIR reflectance [18]. Woolley [18] showed that dried or senescent plant material has higher reflectance than green vegetation at all wavelengths, while Daughtry and Biehl [25] found that litter showed reduced NIR scattering and, thus, lower values for reflectance and transmittance. Unfortunately, soil spectra are generally similar to those of plant litter, making quantifying litter with remote sensing techniques challenging [26–28].

6.1.2.3 Remote Sensing Techniques (and Their Limitations) to Discriminate Litter from Soils and Green Vegetation

To distinguish litter from green leaves and soil, the best waveband regions and resolutions of reflectance spectra must be chosen to distinguish plant litter and soils from an integrated scene. This entails finding differences between litter hyperspectral reflectance data for a variety of species, decay stages, and moisture levels [1,26]. It is important to note wavelength band permutations such as minimum/maximum, greater than/less than, and concave/convex relationships without emphasizing the absolute magnitude of spectral reflectance. For this study, reflectance was measured by sensors that collect data from VIS (400–700 nm), NIR (700–1100 nm), and SWIR (1100–2500 nm) wavelength bands. The appropriate wavelength range and resolution were examined by looking for differences in the spectral curves of the target types; places where litter cellulose and soil minerals absorb energy were seen in the reflectance signatures of nongreen components. We explored methods to develop a robust index that would distinguish plant litter from soils based on differences in their reflectance curve shapes. Several wavebands, including not only the two bands that are commonly used in vegetation studies (VIS and NIR) but also new combinations, were examined to find a diagnostic feature so that an index could be devised to separate plant litter from soils.

6.1.2.3.1 The VIS-NIR Wavelength Range

Although the spectral reflectance of a scene is affected by all included components, such as soil, green vegetation, shadow, surface roughness, and nonsoil residue [29], the spectral reflectance curves of plant litter and soils are often assumed to have the same generally featureless shape in the VIS-NIR (400–1100 nm) [11,12]. There is a problem in discriminating these photosynthetically inactive materials using the VIS-NIR wavelength range because there are generally limited unique features to show differences between the various ground components [30]. An exception to this ability to discriminate contrasting soil and litter signatures is that the VIS and IR wavelength range does depend somewhat on background soil color, for example, if you had an organic or iron-rich soil with overlying yellow-colored litter (van Leeuwen, pers. comm.). Spectral vegetation indices produce values that vary not only within component type but also between types; for instance, the NDVI values for soils (0.08–0.16) and for litter (0.14 to as high as 0.45 for freshly deposited, still green litter) [9] share a common range. Consequently, one difficulty in discriminating plant litter from soils is that wavelengths in the VIS-NIR range do not provide sufficient separability between soils of varying moisture and plant litter of different moisture and ages because their spectral curves are similar and, thus, are indistinguishable at any one wavelength [9].

Reflectance in two wavebands has been of some use in discriminating litter from soil. For example, McMurtrey et al. [9] measured a separation of soil and crop residue NDVI values. The separability or variability of these spectral reflectance curves indicates that the ground components' spectra are not constant, but because their spectra could not be consistently distinguished and were not statistically significant at the LSD (0.05) level, they concluded that the VIS-NIR wavelengths (and thus NDVI) do not produce absorption peaks that can be used to discriminate soils from litter. The two-band index commonly used in vegetation studies (i.e., NDVI) does not provide statistically reliable results for detecting differences among the three classes of photosynthetically inactive plant material, soils, and photosynthetically active green vegetation. Reflectance (400–1100 nm) in three bands has been used in an attempt to distinguish plant litter from soils. McMurtrey et al. [9] found that a third band in the blue (450 nm) range, used in conjunction with the VIS and NIR bands, appeared to capture major differences in the background components, but this band combination was not tested further. This previously published research was undertaken to determine whether key diagnostic differences in nongreen component spectra are revealed in other wavelengths or multiple band combinations.

6.1.2.3.2 Ultraviolet Wavelength Range

Other wavelength regions were investigated. Fluorescence techniques were tested to distinguish plant litter from soils. Daughtry et al. [16] found that plant litter produced greater fluorescence than most soils when illuminated with ultraviolet (320–400 nm) radiation. This method was less ambiguous and better suited for discriminating litter from soils than the VIS-NIR reflectance methods, but several potential problems inhibit the implementation of the fluorescence technique. For instance, (1) excitation energy must be supplied to induce fluorescence, and (2) the fluorescence signal is small relative to normal, ambient sunlight [16].

6.1.2.3.3 SWIR Wavelength Range

Few studies have investigated differences in the SWIR reflectance curves of plant litter and soils [23]; however, several studies have noted spectral features that are unique to each component in the SWIR region [1,31]. A spectral feature common to both litter and soils are two water absorption peaks at 1400 and 1900 nm. In the SWIR spectra of dried plants, a cellulose/lignin absorption peak (a reflectance trough) was noted at 2100 nm [1]. Work with spectral reflectance indicated that the lignocellulose absorption feature at 2100 nm and shoulder peaks at 2000 and 2200 nm were useful for discriminating litter from soils [12]. This feature is absent in the spectra of soils, which show no cellulose absorption but, rather, a clay mineral absorption feature at 2200 nm [28,31].

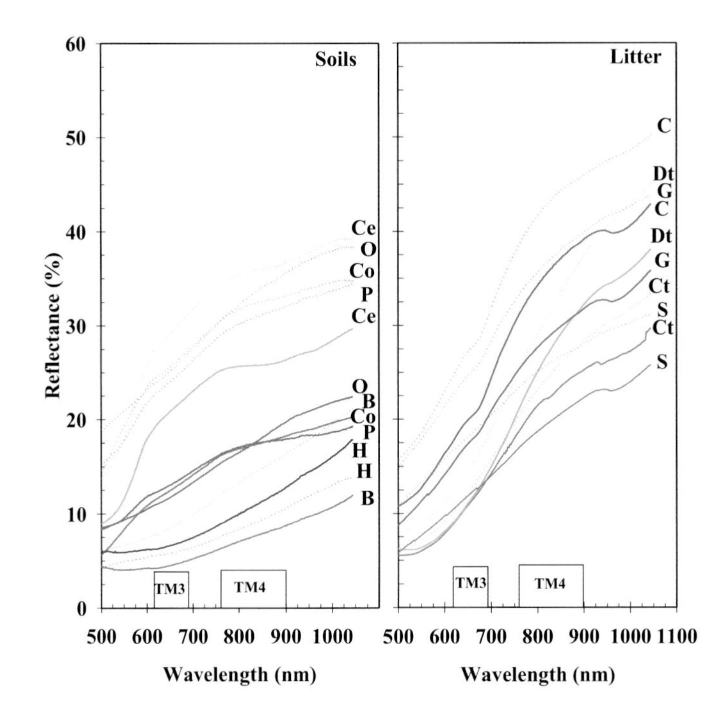

FIGURE 6.1 VIS-NIR spectral reflectance (0.5–1.1 μm) of dry (dashed lines) and wet (solid lines) soils and litter types.

Two different experiments using two instruments were carried out in this work: (1) forest litter and soil samples in the VIS-NIR (400–1100 nm) (Figure 6.1) were first examined with a Spectron Engineering (Wheat Ridge, CO) SE-590 spectroradiometer, and (2) forest litter, crop residue, senescent grass, and soil samples in the VIS-SWIR (400–2500 nm) were then measured with an image replicating imaging spectrometer (IRIS) Mark IV spectroradiometer (Figure 6.2). The litter types were a general representation of the litter surface beneath a canopy. In total, the spectral reflectance of 82 samples of litter (52 forest litter, 24 residues, and 6 grasses) and 7 soils was measured. Five types of litter (coniferous and deciduous forest litter, soybean and corn crop residue, and senescent grasses) were considered for multiple ages.

6.1.2.4 A Diagnostic Feature: Cellulose Absorption Index

The best wavelength range and resolution for discriminating plant litter from soils were defined using three bands in an index called the CAI. Mean spectral reflectance from each sample for three 50 nm wide bands were used to calculate CAI as follows:

$$\mathrm{CAI} = 0.5(\mathrm{R}_{2023\,\mathrm{nm}} + \mathrm{R}_{2215\,\mathrm{nm}}) - \mathrm{R}_{2100\,\mathrm{nm}} \quad (6.1)$$

where $R_{2.0}$, $R_{2.1}$, and $R_{2.2}$ are the wavebands centered at 2023, 2100, 2215 nm, respectively, with a bandwidth of 10 nm. CAI was defined by the relative depth of the spectral absorption at 2100 nm because dry litter exhibited this spectral lignocellulose absorption, as demonstrated first by Elvidge [1]. When contrasted with dry soil spectra from Stoner and Baumgardner [28], the lack of the lignocellulose absorption was evident. The spectral data from soils and plant litter in this work and that of [12] showed that both wet and dry soils and plant litter could be distinguished when their different absorptions at 2100 nm were calculated according to Equation (6.1). The 50 nm bands appeared to be the most useful for discriminating all types of soils from all types of litter. However, the usefulness of the index in discriminating wet soils from wet litter might be improved with smaller bandwidths, which would capture less of the absorption from the water band at 1900 nm.

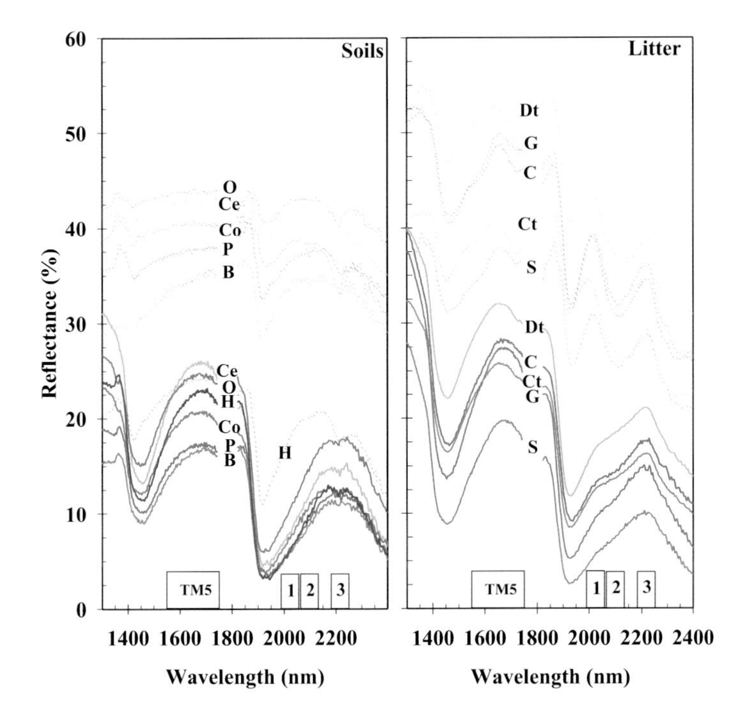

FIGURE 6.2 SWIR spectral reflectance (1.3–2.4 μm) of dry (dashed lines) and wet (solid lines) soils and litters. The soils are Othello (O, o), Cecil (E, e), Codorus (C, c), Portneuf (P, p), Barnes (B, b), and Houston Black clay (H, h). The plant litters are corn (M, m), soybean (S, s), deciduous tree (D, d), coniferous tree (C, c), and grass (G, g).

6.1.2.5 Effects of Water on CAI

To evaluate the effect of sample water content on the index, CAI was plotted as a function of reflectance in the water absorption band (1900–1950 nm) (Figure 6.3). This band around 1900 nm is sensitive to sample moisture content [32]; we use the reflectance in this band to monitor moisture in the plant litter and soil samples in the form of a scatterplot. Wet samples (litter and soils) had reflectance spectra at 1900 nm that were <25%, while dry samples were generally >25%. For wet soils alone, reflectance spectra were <10%, with the exception of the sand, which was very bright even when wet. Dry soil samples had reflectance values in the water absorption band >25%, with the exception of Houston Black clay, which held more moisture when air-dried than other soils.

The presence of water reduced the reflectance of all samples at all wavelengths and made discrimination of litter and soils difficult. Although water absorption dominated the spectral properties of both soils and residues in the SWIR, it was possible to discriminate wet litter from wet soil using the CAI. More than 90% of the wet plant litter samples had positive CAI values. However, five wet litter samples also had negative CAI values. The cellulose absorption feature was negative for three wet deciduous samples and two coniferous samples that were all greater than 1 year old. All five samples were sufficiently decomposed so that the absorption due to cellulose or lignin fibers was easily masked by moisture. Positive CAI values represented the presence of the cellulose spectral feature. In the spectra for all the soils, the cellulose feature was absent and thus produced CAI values between 0 and –5. Although negative CAI values represented an absence of the cellulose spectral feature, they do not necessarily indicate the absence of cellulose, for the lignocellulose feature in plant litter samples was present, but sometimes it was masked by water, and negative CAI values were produced.

The effect of green vegetation on the CAI was also determined in previously published work [36]. Although wet and dry forest litter, crop residues, senesced grass, and soil spectra provided

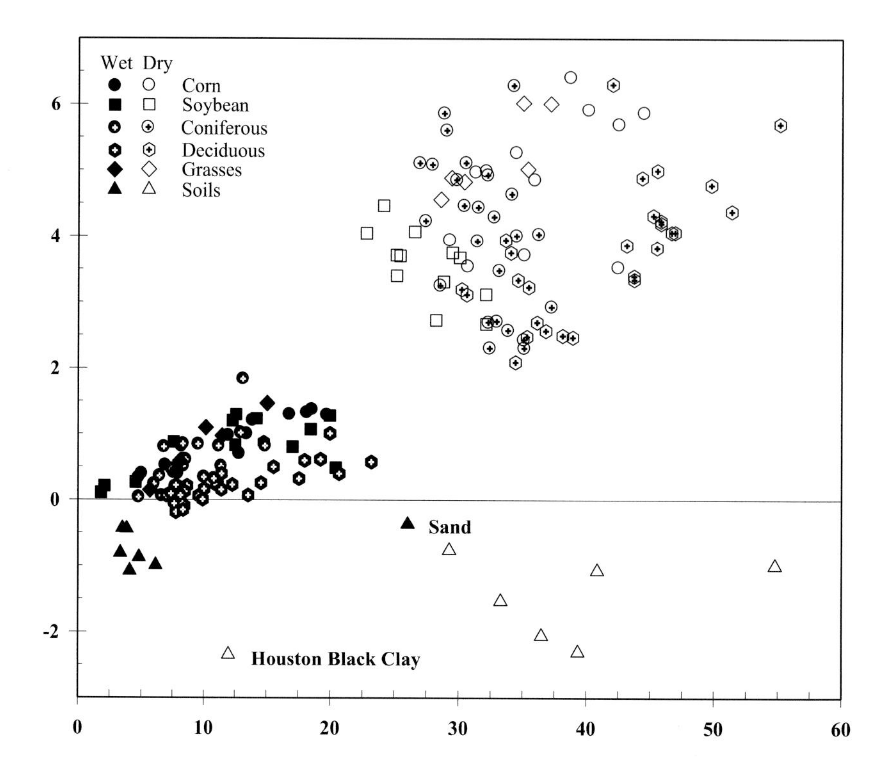

FIGURE 6.3 Plot of CAI as function of reflectance in water absorption band at 1900–1950 nm.

data to represent various ground component spectral reflectance values in the VIS-SWIR range, the original study [36] lacked fresh and dry green-leaf reflectance measurements in this range because the authors were focusing on nonphotosynthetically active targets prior to producing [37]. Lignin, cellulose, and organic compounds found in litter showed distinct absorption peaks in the SWIR range that were not present in the signatures of most green vegetation or yellowed leaves [1]. Hence, it was expected that green vegetation index values would not confound the use of the index in discriminating litter from soils. Information about the spectral behavior of the cellulose feature in the SWIR wavelengths provided a way to index the reflectance values and distinguish most litter from soils, but whether spectra of green vegetation inhibits the usefulness of CAI to distinguish litter from soils was not determined until we looked at mixtures of litter on soils as well as pure green vegetation in the next step of this study (Section 6.1.3 below). With the new mixed study, we repeated the experiment with pure samples and then proceeded with mixtures. Figure 6.4 shows the typical reflectance spectra (400–2400 nm) of pure scenes of black, red, and gray soils and green vegetation (top figure) and spectra of four crop residues and two tree litters (bottom figure).

6.1.2.6 Benefits (and Limitations) of CAI

The spectral resolution requirements (i.e., the spectral, bandwidths, position or center of spectral bands, and number of spectral bands) and sensor proximity to the target (related to the field of view) are some considerations in choosing a sensor for a particular application [33]. A few of the current satellite sensors used are Landsat's Thematic Mapper (TM) and the Moderate Resolution Imaging Spectroradiometer (MODIS). The limited spectral range and resolution of most satellite instruments currently inhibit their use in discriminating nongreen canopy components using the CAI. The bandpasses are too broad for measuring the absorption features of dry plant materials; they lack the ability to spectrally discriminate between plant litter types [1]. However, a high-spectral-resolution sensor such as the Airborne Visible/Infrared Imaging Spectrometer (AVIRIS) can detect reflectance throughout the 400–2500 nm range in continuous narrowbands (10 nm) [34]. Distinguishing plant

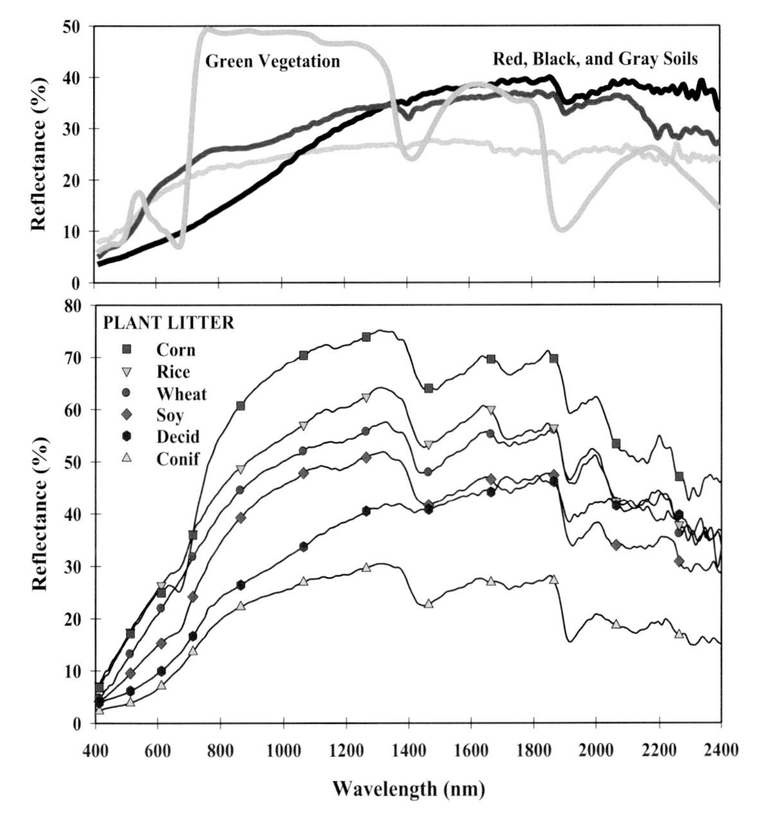

FIGURE 6.4 Typical reflectance spectra (400–2400 nm) of black, red, and gray soils and green vegetation (top figure) and spectra of four crop residues and two tree litters (bottom figure). The symbols in the bottom figure do not represent data sampling points, but rather are placed on the spectral lines for clarity.

litter from soils using AVIRIS and other sensor systems designed with narrow bandwidths in the SWIR region has not yet been tested. Minimizing atmospheric signals due to aerosols and water vapor will be an important prerequisite to quantify the CAI.

6.1.3 Summary of Pure Scenes of Soils and Litter

This study supports research showing that it is not possible to consistently distinguish plant litter from soils using reflectance spectra in the VIS-NIR wavelength range exclusively. The CAI, which we developed using reflectance data from the SWIR wavelength region, is effective at distinguishing litter from soils and from green vegetation and may improve quantification estimates of plant litter in a scene by making them more objective and accurate. Further work could serve to calibrate the CAI for quantifying phytomass to improve estimates of productivity and energy balance. Plant litter and soils, regardless of moisture content, were distinguishable from each other using spectral reflectance data acquired in the VIS-SWIR (400–2500 nm) wavelength region. The developed CAI can be used to successfully discriminate litter from soil.

6.1.4 Mixed Scenes of Plant Litter and Soils

The ability to discriminate plant litter from soils using the CAI allows ground components to be identified using laboratory spectra of pure samples. In the research described for pure scenes, mixed laboratory samples were not measured and discrimination of mixed components using the CAI was not tested. Additionally, the CAI was not tested using field samples.

From the pure scene research it was concluded that the value of this remote sensing method to distinguish crop residues from underlying soils and estimate the quantity of litter in field conditions must be evaluated for field systems like natural canopies or agricultural lands. This can be done by incorporating the experimental CAI values into existing models to derive theoretical estimates of field conditions. However, the hypothetical estimates may not be useful if the CAI is incorporated into a model applied outside its intended use, such as forest systems, where too many unknown variables exist to obtain a fair estimate. Future work to test this methodology in agricultural systems alone is recommended to determine whether the CAI is flexible enough to use in noncanopy field circumstances. If successful, it may then replace the current tedious methods employed in quantifying residues as part of conservation efforts. In this section, we used spectral measurements of laboratory samples with varying percent cover (mixed targets) to obtain CAI values and assess its usefulness as a function of litter weight and litter cover. We employed both (1) photographs of percent cover and (2) SWIR video/images of percent cover to determine whether either the photographs or CAI images were necessary to distinguish varying fractions of litter from underlying soils. The use of SWIR imaging techniques to estimate percent cover and to replace time-consuming SWIR spectral measurement and manipulation techniques is being explored. These can be made using a Vidicon/CCD camera with the three CAI bands; a CAI image of the different fractions of percent residue cover over varying soil backgrounds can be produced.

In the field, soils are rarely completely bare (0% litter cover) or completely covered with plant litter (100% cover), except in some no-till cropping systems. Daughtry [35] varied the moisture content of soil and litter samples but only simulated the effect of mixed scenes; in the present work, the reflectance spectra of wet and dry scenes with different proportions of soil and litter were measured. Figure 6.5 shows the dry (upper graph) and wet (lower graph) reflectance spectra for various amounts of wheat litter on the surface of black soil. As the coverage of plant litter increased in the dry samples, the prominence of the 2100 nm absorption feature also increased. Moisture reduced reflectance and masked the absorption feature at 2100 nm in all the wet, mixed samples. Nagler et al. [36] also showed that discrimination of wet, pure soils from wet, pure litter was possible using the CAI, but here, the wet, mixed samples with >20% litter cover did not show negative CAI values as was seen in the dry, mixed samples. Regardless of moisture, adding wheat litter to the black soil increased reflectance at all wavelengths. On the other hand, adding soybean residue to the gray soil reduced reflectance in the VIS wavelength region but increased reflectance at other wavelengths.

The CAI spectral variable describes the average depth of the cellulose absorption feature at 2100 nm. Positive CAI values indicate the presence of cellulose, so plant litters typically had positive CAI values. Negative CAI values indicate the absence of cellulose. The CAI of soils is typically negative [37,36]. Daughtry et al. [38] observed that in wet samples, absorption by water dominated the reflectance spectra and nearly obscured the differences in their CAI values. The CAI of each mixed scene of plant litter, with seven different levels of cover including the full litter cover, and green vegetation, as well as scenes of pure soils (three types), was plotted as a function of reflectance in the water absorption band at 1910–1950 nm (Figure 6.6). Green vegetation is shown here as being very negative. The mean CAI increased significantly from bare soils (CAI = −0.2) as the amount of plant litter on the soil increased to 100% cover (CAI = 5.2). The plant litter had positive CAI values and the soils had negative values. The CAI of green leaves from Inoue et al. [39] also had large negative values, which indicated that the cellulose absorption feature was obscured by the abundance of water in green leaves. The CAI can be used to distinguish green canopy cover from underlying nongreen landscape components, but it is also possible—given CAI as a function of reflectance in the water absorption band (1910–1950 nm)—to differentiate nongreen components, litter, and soils. A multispectral approach may also be employed; for example, the simple ratio (reflectance in the 760–900 nm band divided by reflectance in 630–690 nm band [30]) could be used to distinguish green vegetation from bare soil, and the CAI could then be used to separate plant litter

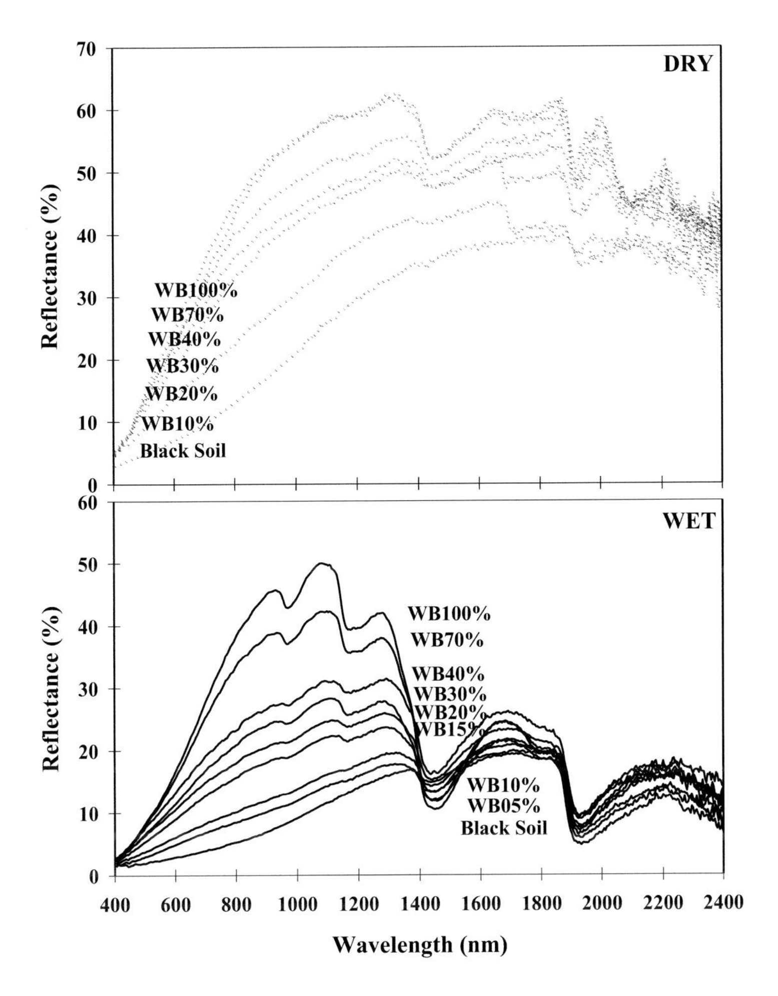

FIGURE 6.5 Mean reflectance spectra for a series of wet and dry mixed scenes of wheat residue with underlying black soil. The litter level, estimated from its weight, of wheat residue in the dry scenes (top figure) were 0% (black soil), 10%, 20%, 30%, 40%, 70%, and 100%. The levels in the wet scenes (bottom figure) were 0% (black soil), 5%, 10%, 15%, 20%, 30%, 40%, 70%, and 100%.

from soil. Thus, the CAI is relevant to situations where it is important to distinguish residues from soils (agricultural systems) and to distinguish green vegetation canopies from underlying nongreen vegetation components (natural systems). In this mixed-scene work, the CAI of all three soils was negative, but as the amount of litter on the soil surface increased, the CAI of the mixed scenes also increased (Figure 6.7). All four residue types showed that mixed scenes with 0% and 10% residue levels by weight and black soil underneath were negative, showing that small amounts of residue on black soil could not be discriminated from bare soil. However, for gray soils, the mixed-scene litter limit varied depending on the litter type. For corn and soybean residues, the mixed scenes with 0% and 10% residue levels and gray soil underneath were negative. With black soil underneath the 0% and 10% residue levels, the mixed scenes were also negative, showing that small amounts of corn and soybean residues on gray and black soil could not be discriminated from bare soil. However, for wheat and rice residue, the mixed scenes with 10% residue level and gray soil underneath were positive, showing that these could be discriminated from bare soil. For red soil, for wheat and soybean residue, the mixed scenes at the 0% and 10% levels were negative, but they were positive with corn and rice residues at these percent cover levels. All four crop residue types had positive CAI values for mixed scenes of more than 20% residue level. These were significantly different from

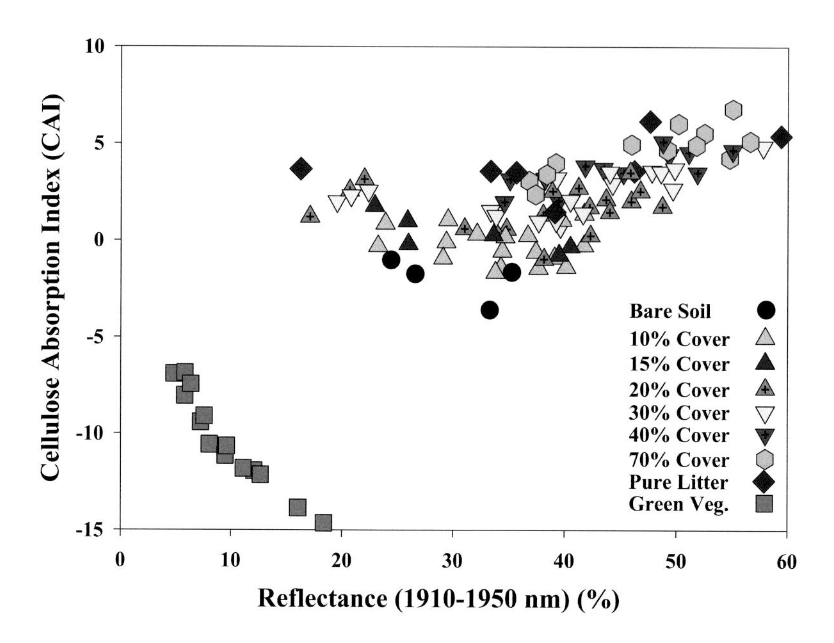

FIGURE 6.6 CAI of all the pure soils, mixed scenes of plant litter with different levels of cover, and green vegetation as a function of percent reflectance (%) in water absorption band (1910–1950 nm).

the CAI values of the soils. For Experiment 2 with the tree litters, both types showed that mixed scenes with 0% litter level for black soil were negative, but that any amount of litter (10%, 15%, and 20% litter level or higher) could be discriminated from black soils using the CAI. The situation was different for gray and red soils. Deciduous, broadleaf tree litter at 10%, 15%, and 20% litter levels had negative CAI values and could not be discriminated from the underlying gray or red soils. The mixed-scene CAI values only became positive at levels greater than 30% litter level. For coniferous tree litter over gray soil, the CAI values were positive for litter levels greater than 10%, showing that this residue could be easily discriminated from a gray background soil. For coniferous tree litter over red soil, the CAI was negative for 10% and 15% residue levels but positive at a 20% residue level.

Residue level by relative weight or relative percent cover (Rel.%C), averaged over three soils, is shown for crop and forest litter levels and is linearly related to the CAI (Figure 6.8). Coniferous tree litter had the most variability and lowest correlation ($R^2 = 0.84$). Deciduous tree litter had the lowest CAI values and a high correlation ($R^2 = 0.98$). Although the discrimination of background soils from litter at low densities or residue levels of less than 10% (crop residues) or less than 20% (tree litters) may be difficult based on these results, the CAI is a very good predictor of the percent of plant litter cover in mixed scenes.

6.1.5 Conclusions for Mixed and Pure Scenes of Soils and Litter

The reflectance spectra of pure and mixed scenes of six plant litter types and three soils were measured, and the CAI was calculated using the spectral feature at 2100 nm. The CAI values of pure plant litter were significantly larger than the CAI value of pure soils. For the mixed scenes, as plant litter cover increased, the CAI increased linearly. The results showed that the CAI was successful in distinguishing fractions of litter from underlying soils in mixed laboratory samples. In some soil types, such as the red soil in this study, a complication arises from using the depth of the cellulose absorption feature at 2100 nm, because the width of the clay mineral absorption feature at 2200 nm matches the minor reflectance peak of cellulose at 2200 nm that was induced by absorptions at 2100 and 2300 nm in plant material, thereby leading to lower CAI values than with either the black or gray soils in this study. Therefore, it is recommended that special attention be paid to the shoulder of the

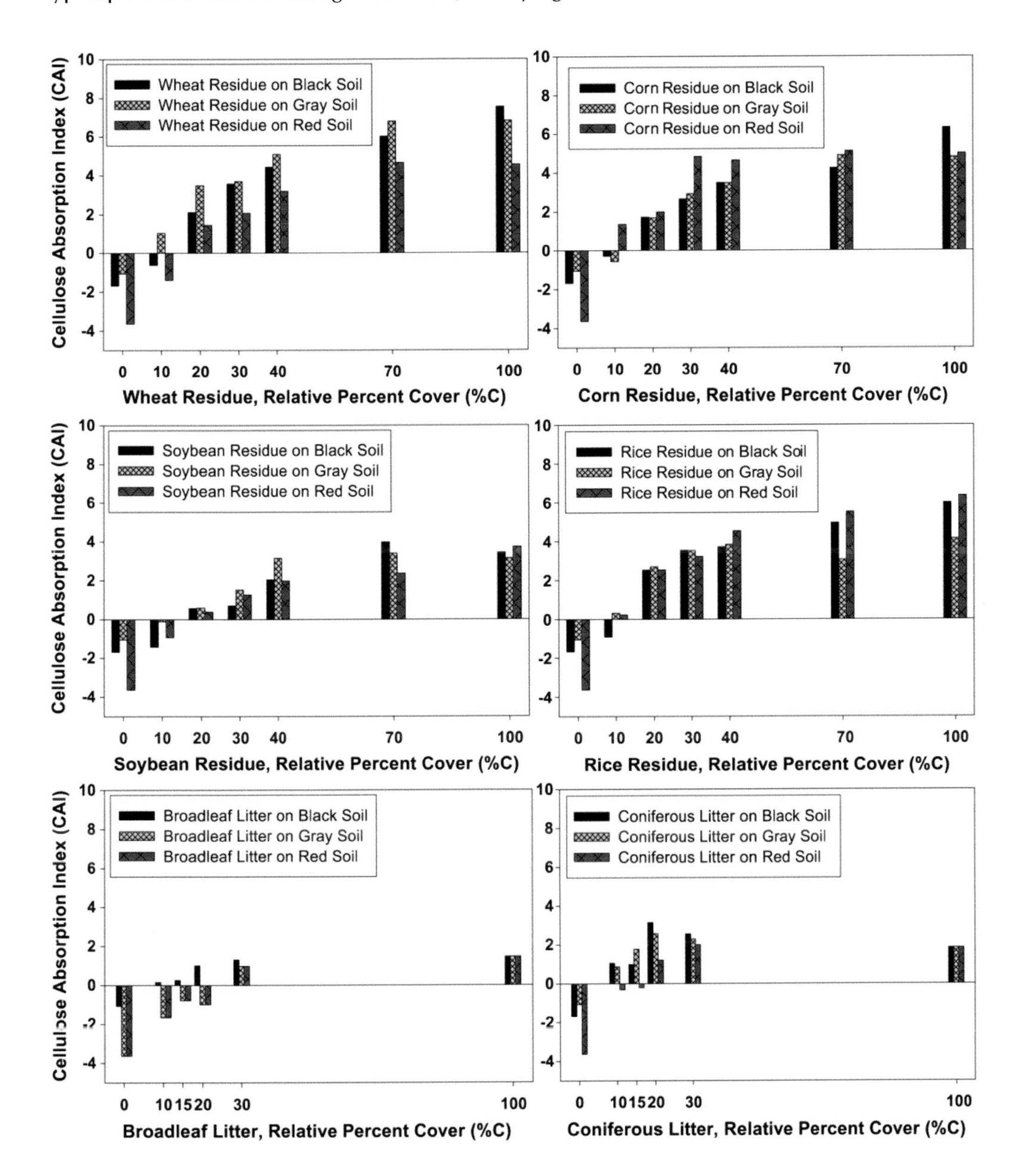

FIGURE 6.7 CAI as function of amount of residue for mixed scenes of varying amounts of each crop residue and tree litter, shown for each of the three soils.

absorption feature at 2200 nm before utilizing the countered absorption and reflectance features to calculate the CAI. Using a two-way analysis of variance for crop residues, soils were found not to be significantly different from one another, although when the statistics were run for tree litters, the red soil was indeed found to be slightly significantly different. Thus, only one soil type in this study inhibited the detection of tree litter or the ability to quantify litter cover. Because the relationship between CAI and litter level did not saturate at low levels of cover in these experiments, this spectral variable was useful over nearly the whole range (>70% cover) of mixed soil–litter scenes and was generally not affected by soil type. Furthermore, the strong linear relationship between crop residues/tree litters and CAI promotes the idea of extrapolating these findings to other residue and litter species, although new experimental data would first have to be obtained.

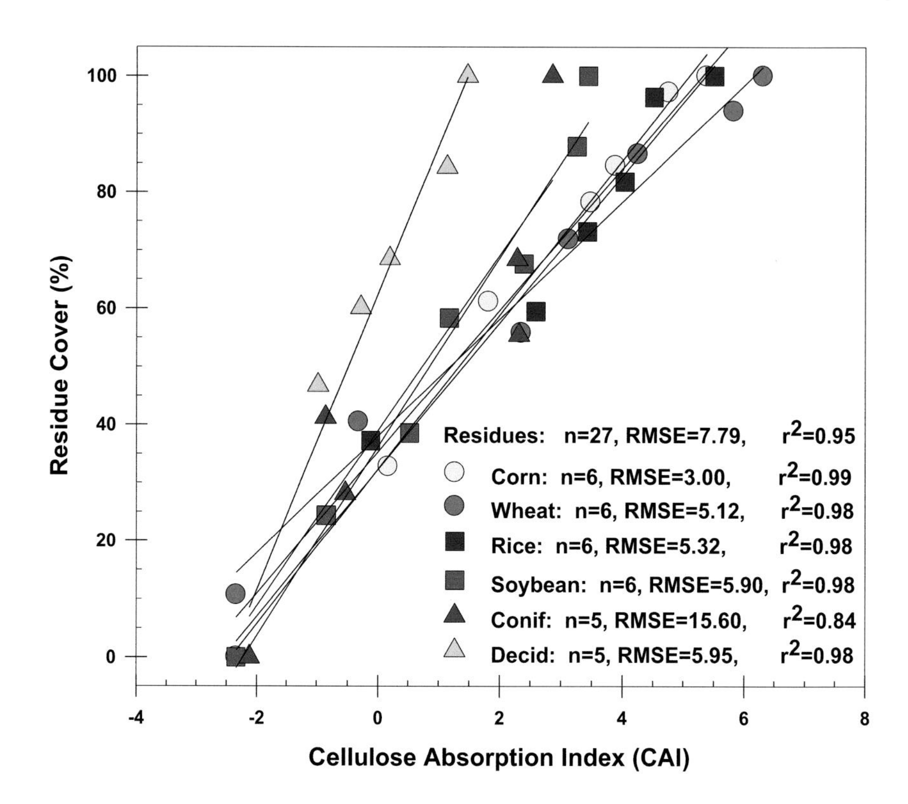

FIGURE 6.8 Crop and forest litter levels estimated by their weight (Rel.% cover) averaged over three soils shown as function of CAI.

The relationships between CAI and percent cover were also determined for each plant litter by image analysis of color slides (%C by video) and residue level by weight (Rel.%C); the polynomial relationship between CAI and the level by weight was more useful than %C by video for all litters, except corn residue. When CAI was regressed with the average percent cover by image analysis (%C by video) and average %C by weight (residue level) for each experiment, crop residues, and tree litters, the crop residues had more robust coefficients of determination (r^2) values than the tree litter across all three soil types. However, %C by video (image analysis) was a more effective method of discriminating crop residues, while %C by weight (residue level) was more effective at discriminating tree litters. The red soil showed a more promising polynomial relationship between CAI and percent cover than the other soils for both methods of estimating percent cover (video and residue level by weight); percent cover was averaged across crop residues and across tree litters. Residue density (g/m^2) can be compared with stacked leaves (weight per unit area), similar to LAI; this may warrant a new study in which the effect of a range of residue densities, all at 100% cover, on CAI is determined. An instrument based on measuring CAI could replace tedious, manual methods of quantifying plant litter cover.

6.2 APPLICATIONS OF HYPERSPECTRAL REMOTE SENSING TO INVASIVE SPECIES: RESEARCH APPROACH AND CASE STUDIES WITH TAMARISK AND BUFFELGRASS

6.2.1 Ecological Importance of Discriminating Invasive Plant Species in the Landscape

Invasive plant species are considered a major threat to global diversity and ecosystem functioning [40]. Plant invasions are known to cause undesirable alterations to native plant and animal populations and their community structure and functioning [40]. They can also cause economic loss [1,81]. Of the myriad invaders of arid lands, riparian shrubs (e.g., *Tamarisk ramosissima* and *Eleagnus* sp.) and fire-promoting grasses (e.g., *Pennisetum* spp., *Schismus* spp., and *Bromus rubens*) elicit the greatest cause for concern.

Tamarisk (*Tamarisk ramosissima*) is an invasive shrub that was introduced into the United States from Asia for ornamental and erosion prevention purposes [41,42]. In due course, tamarisk emerged as a dominant woody species in many riparian sites in the western United States and northern Mexico, where it forms dense, low thickets that displace native vegetation, impede water flow, and increase sedimentation [42]. Tamarisk has greater salt, drought, and fire tolerance and resistance to water stress than native plant species [43,44]. At higher salinities, tamarisk has a clear advantage over native trees. Robinson [42] reported that tamarisk covered about 900,000 acres during the 1960s in the United States and then spread and occupied about 1.5 million acres by 1987 [44]. However, these estimates have not been confirmed by actual surveys, in part due to the difficulty in distinguishing between tamarisk and native species by remote sensing methods.

Invasive grasses threaten desert ecosystems by introducing a grass-fire cycle [45,46]. By increasing the abundance and connectivity of fine fuels, invasive grasses facilitate increases in fire frequency and extent in areas poorly adapted to fire. Invasive grasses recover rapidly following fire, whereas most natives do not. Both C3 and C4 grasses have altered fire frequencies to the detriment of natives in North America. The C3 grasses *Bromus rubens* (Mohave and Sonoran Deserts) and *Schismus arabicus* (Mohave and Sonoran Deserts) are seasonally abundant and depend on winter precipitation, which varies over interannual and multidecadal time frames (e.g., in accordance with phases of El Niño Southern Oscillation and Pacific Decadal Oscillation) [47,48]. C4 grasses are common invaders of subtropical and semiarid systems worldwide and exhibit strong dependence on warm season precipitation, which in North America is coupled tightly with the influence of the North American monsoon. The Sonoran Desert is particularly prone to C4 grass invasion and currently supports various stages of invasion by *Pennisetum ciliare*, *Pennisetum setaceum*, *Eragrostis lehmanniana*, *Melinis repens*, and *Enneapogon cenchroides*, among others. Of particular concern is buffelgrass (*P. ciliare*), which has been planted extensively in pastures in Mexico and is invading southern Arizona but has not yet altered fire regimes in the United States. Monitoring the spread of buffelgrass in the southwestern United States is a key research priority, but at present there are no remote sensing methods to distinguish between buffelgrass and native grasses, cacti, shrubs, and trees.

6.2.1.1 Hyperspectral Reflectance Data as a Monitoring Tool for Invasive Plants

Remote sensing technology has been widely used to monitor weedy and invasive plant species in agricultural and forest environments. Several studies have reported the use of aerial photography [49,50], multispectral airborne digital imagery [51], hyperspectral imagery [52–54], and multispectral satellite imagery like advanced very-high-resolution radiometer sensing [55], Landsat TM [56], and SPOT [57] for detecting the spread of different invasive plant species. The application of remote sensing to the monitoring of invasive plants has proved helpful in assessing the extent of infestations, development of management strategies, and evaluation of control measures for the spread of these unwanted plant populations. Studies have been conducted that included the mapping of different species of tamarisk using remote sensing imagery. Aerial photography was used for mapping *Tamarisk chinensis*, specifically during early winter months [49], when their leaves turn to orange-brown colors prior to the leaf drop. *Tamarisk parviflora* was mapped through the textural analysis of aerial photographs [50], during a time when the trees were without leaves and had pink flowers, making them distinct from other vegetation. A high-spatial-resolution (0.5 m) airborne hyperspectral imager was used to map *Tamarisk* spp. in riparian habitats of Southern California [58] at a time when the trees begin to senesce. A combination of single-band and vegetation indices derived from Landsat Enhanced Thematic Mapper Plus (ETM+) images [59] were also used to map *Tamarisk* spp. MODIS and Advanced Spaceborne Thermal Emission and Reflection (ASTER) data [60] were also used to map *Tamarisk* spp. based on vegetation indices. However, most of these studies depend on the distinct visual characteristics of either tamarisk foliage or flowers seen at only one particular time of year.

The quest to identify invasive grasses in western deserts has led to similar findings, namely, that mapping success is highest when the process is highly manual (e.g., digitization of high-resolution aerial imagery [61–64]) or dependent on a phenological opportunity that inconsistently emerges from year to year.

6.2.2 Reflectance Spectra (0.4–2.4 μm) Used Outdoors in Natural Settings

6.2.2.1 Remote Sensing

This chapter documents two studies that identify the utility of hyperspectral data and imagery to map and monitor invasive species. The first study focuses on the invasive riparian tree salt cedar (*Tamarisk* spp.) in the Lower Colorado River valley, and the second focuses on the invasive C4 grass buffelgrass (*Pennisetum ciliare*) in the Arizona Upland zone of the Sonoran Desert. These studies highlight different methodologies for utilizing hyperspectral remote sensing to identify vegetation in highly dynamic and sensitive environments.

6.2.3 Tamarisk Study

The goal of the tamarisk study was to develop a cost-effective, multiseasonal monitoring approach through satellite remote sensing to identify and map tamarisk and other vegetation types growing in the study area. The specific objectives were (1) to identify the spectral characteristics of the major riparian and agricultural vegetation types in the Lower Colorado River LCR region and (2) to determine whether Landsat TM data could be used to map tamarisk (*Tamarisk ramosissima*) infestations in this region.

6.2.3.1 Description of Study Area and Vegetation

This study was conducted in two areas, the Palo Verde Irrigation District (PVID) and Cibola National Wildlife Refuge (CWR). The study sites are located in Southern California and Northern Arizona, respectively, along the Colorado River. The PVID has a consistent, year-round farming practice growing crops such as alfalfa, cotton, melons, corn, wheat, and other grasses. Among these crops, alfalfa and cotton cover the largest acreage in the study area. Within the riparian areas of the Colorado River and the CWR, there are several native plants, such as honey mesquite (*Prosopis glandulosa*), cottonwood (*Populus fremontii*), quail bush (*Atriplex lentiformis*), arrow weed (*Pulchea sericea*), palo verde (*Parkinsonia microphylla*), and creosote bush (*Larrea tridentata*). Most of these native plant communities have been invaded by and are being replaced by tamarisk. Extensive areas of the CWR are covered with tamarisk, threatening the existence of native plant communities.

6.2.3.2 Spectral Reflectance and Image Analysis

Ground-truth observations from 79 sampling locations across the study area were collected. Observational data recorded were the spectral reflectance measurements of vegetation, type of plant species, plant heights, soil samples, and GPS coordinates for the locations. The study area, overlaid with the sampling locations, is shown in Figure 6.9.

Measurements of the canopy-level spectral reflectance of vegetation at various locations within the study area were designed to coincide with the acquisition of the Landsat TM imagery that was obtained on June 9, 2007. A Fieldspec Pro spectroradiometer (ASD Inc., Boulder, CO, USA) with a spectral range of 350–2500 nm was used to collect field reflectance spectra of the vegetation in the study area. The foreoptics of the spectroradiometer were aligned vertically and placed at 1 m above the surface of the plant canopy, and the instrument was adjusted such that only the reflectance from the targeted area filled the field of view (FOV) of the instrument. The fiber optic input device was held approximately 1 m above the ground, such that the FOV covered a circle of approximately 50 cm

FIGURE 6.9 The Landsat TM image obtained on June 9, 2007, showing PVID and CWR. The CWR is seen toward the southern side (bottom) of the image. Ground-truth sampling locations of the study area are shown as white dots over the dark vegetated surfaces.

in diameter. The calibration spectra of a white spectralon panel (Labsphere Inc., North Sutton, NH) were always acquired before recording of the field spectra. All the spectra were obtained on cloud-free days, with sunlight as the source of illumination. A total of five spectra were collected for each of the sampling locations given in Figure 6.9. The ground-level spectral reflectance acquired at each location was averaged to obtain a mean reflectance spectrum for that location and was plotted to analyze the spectral differences. The leaves of the selected plants were harvested and the reflectance spectra were recorded with a quartz-tungsten-halogen (QTH) lamp as light source. Diffused light from a 100 W Lowel Pro-Light was used to illuminate the dorsal side of leaf surfaces at 45° angles when spectra were collected in the laboratory. The foreoptics were aligned vertically and the height of the foreoptics was adjusted so that only the leaf surface filled the FOV of the instrument. The height of the foreoptics was kept constant throughout the experiment. The same experimental setup was used to obtain the spectra of all the leaf samples. Spectra were collected from five leaf samples of each selected plant type and then averaged to obtain a representative leaf spectrum of the plant. The spectral recording and analysis procedure was similar to that of the canopy-level reflectance as described earlier.

The Landsat TM image obtained on June 9, 2007, was used in this study. The georeferenced and terrain-corrected Landsat TM images were downloaded from the USGS Earth Resources Observation and Science Data Center. The Landsat TM image was processed using ERMapper image processing software, a commercial product of Earth Resources Mapping, Inc. (now part of ERDAS, Redding, CA). Based on the locations of the 79 sampling points, the dark object subtraction (DOS) pixel values corresponding to Landsat TM bands 1–5 and 7 were derived from the June 9, 2007 image. The spectral range of these Landsat TM bands are as follows: band 1: 450–520 nm; band 2: 520–600 nm; band 3: 630–690 nm; band 4: 760–900 nm; band 5: 1550–1750 nm; and band 7: 2080–2350 nm. The dark object of each spectral band is defined as one value less than the minimum digital number found in all the pixels of the image [65]. The detailed procedure for DOS and its effects on the removal of atmospheric haze was given elsewhere [65–67]. From the DOS-corrected digital number (DN) values of the six 30 m resolution Landsat single bands, all the spectral ratio combinations and the NDVI were calculated.

The spectral ratios calculated are as follows: $R_{2,1}$; $R_{3,1}$; $R_{3,2}$; $R_{4,1}$; $R_{4,2}$; $R_{4,3}$; $R_{5,1}$; $R_{5,2}$; $R_{5,3}$; $R_{5,4}$; $R_{7,1}$; $R_{7,2}$; $R_{7,3}$; $R_{7,4}$; $R_{7,5}$ and all their inverse ratios, where R represents the ratio and the numbers represent the Landsat TM band numbers [66]. The NDVI was calculated using the formula NDVI = ((Band4 − Band3)/(Band4 + Band3)) [68]. The spectral ratios and vegetation indices were calculated using MINITAB statistical software (MINITAB Inc., State College, PA, USA).

6.2.3.3 Spectral Characteristics of Riparian and Other Vegetation

6.2.3.3.1 Canopy- and Leaf-Level Spectral Reflectance

The averaged canopy-level spectral reflectance of the various native and invasive plant species obtained in the riparian sites is shown in Figure 6.10, and the spectra of major crop plants obtained in the fields are shown in Figure 6.11. In general, the spectral reflectance of the plants is relatively low in the VIS region (400–700 nm), where light absorption by leaf pigments (primarily due to chlorophyll) is the determining factor. The absorption maxima of leaf pigments occur in the blue and red at 470 and 680 nm, respectively, while the familiar green reflectance peak occurs at 550 nm. In the NIR and middle-IR regions, these pigments are transparent and internal leaf structure and biochemical composition control reflectance. The reflectance spectrum of principal biological interest occurs in the near IR between 700 and 1300 nm, where reflectance is high and absorption is minimal (with two minor water absorption bands at 975 and 1175 nm); beyond 1300 nm, major water absorption bands (at 1450 and 1950 nm) become significant.

Among the spectra of native and invasive plant species, the reflectance of quail bush was higher in the NIR region of 700–1300 nm, followed by tamarisk, mesquite, and arrow weed, respectively (Figure 6.10). The reflectance values of alfalfa and cotton were higher in the entire spectral range from 350 to 2500 nm, compared to that of melons (Figure 6.11).

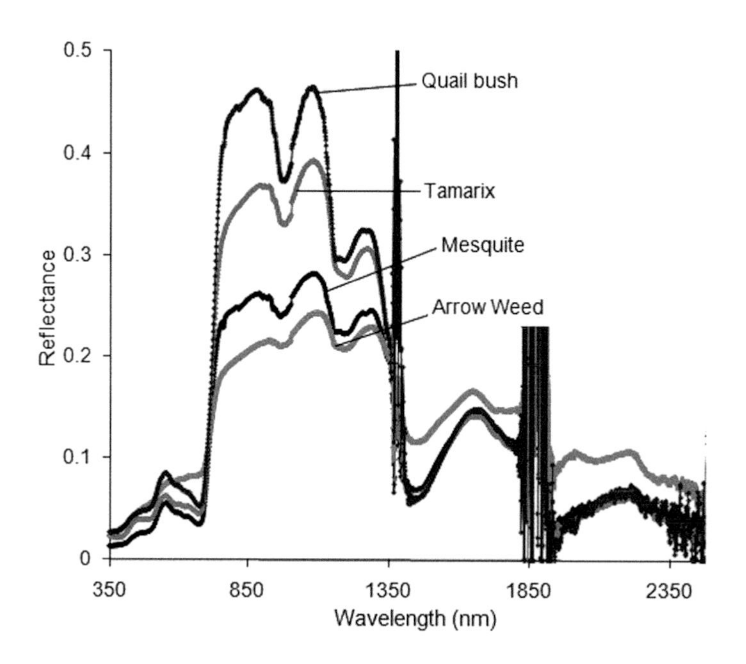

FIGURE 6.10 Average (n = 5) canopy spectral reflectance of native and invasive plant species located in CWR and riparian areas of Colorado River.

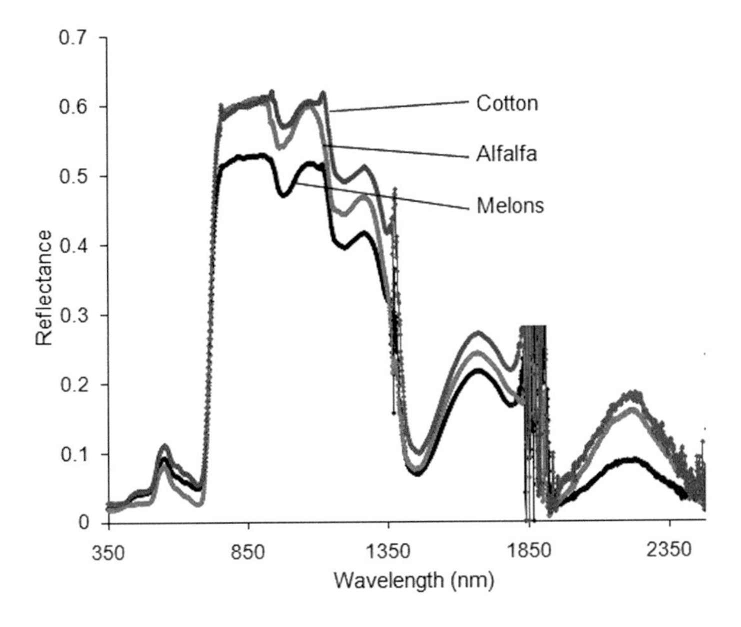

FIGURE 6.11 Average (n = 5) canopy spectral reflectance of agricultural crop plants located in PVID.

Among the leaf-level spectra of native and invasive plant species, the reflectance of cottonwood was higher in the NIR region of 700–1300 nm, followed by quail bush, salt cedar, arrow weed, creosote, and palo verde, respectively (Figure 6.12). The reflectance in the VIS region (400–700 nm) shows a clear chlorophyll peak at 550 nm for cottonwood compared to other plants (Figure 6.12).

The reflectance values of grass and alfalfa were higher in the entire spectral range from 350 to 2500 nm, compared to that of cotton (Figure 6.13). All the plants show a clear chlorophyll peak at 550 nm (Figure 6.13).

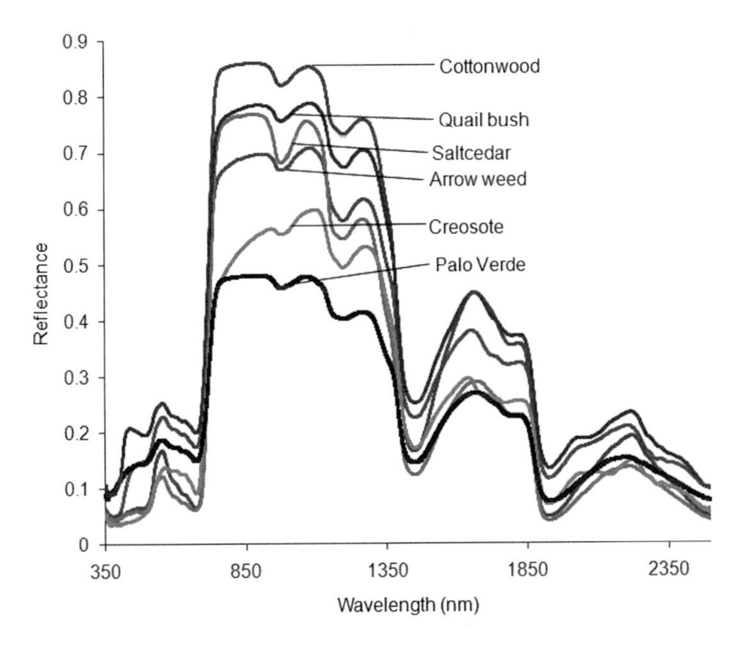

FIGURE 6.12 Average (n = 5) leaf-level spectral reflectance of native and invasive plant species located in CWR and riparian areas of Colorado River.

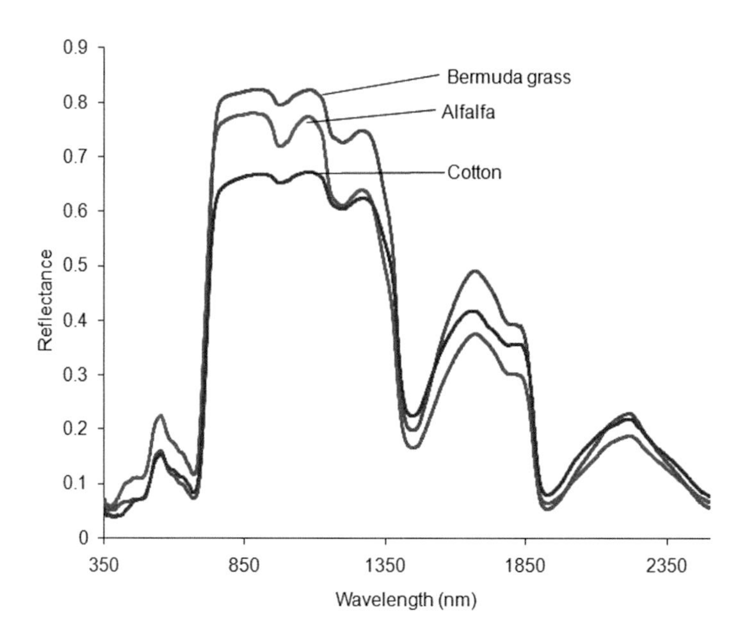

FIGURE 6.13 Average (n = 5) leaf-level spectral reflectance of agricultural crop plants located in PVID.

6.2.3.4 Landsat TM Spectral Ratios and Image Interpretation

Among all the spectral ratios and vegetation indices calculated from the DOS DN values corresponding to the six Landsat TM bands, the NDVI, $R_{1,5}$, and $R_{1,7}$ were chosen as the three best spectral ratios to differentiate major types of vegetation in the region (Figures 6.14 and 6.15a,b). The NDVI values of the alfalfa and melons were significantly ($p < 0.05$) higher, compared to cotton and tamarisk, as shown in Figure 6.14. The dry or senescent grass and soils had NDVI values of less than 0.2 (Figure 6.14). The ratios $R_{1,7}$ and $R_{1,5}$ were significantly ($p < 0.05$) higher for tamarisk compared

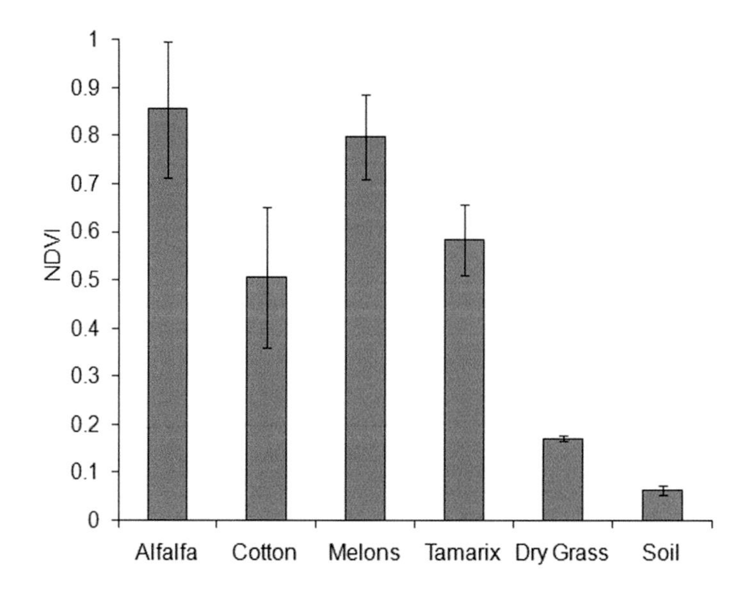

FIGURE 6.14 Differences in NDVI for different vegetation types and soil in Lower Colorado River region. Bars are ± one standard error from 10 replicates.

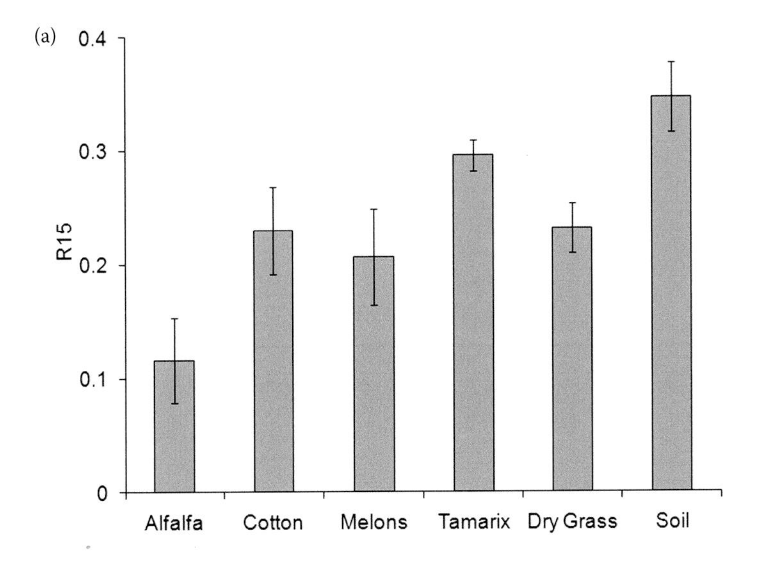

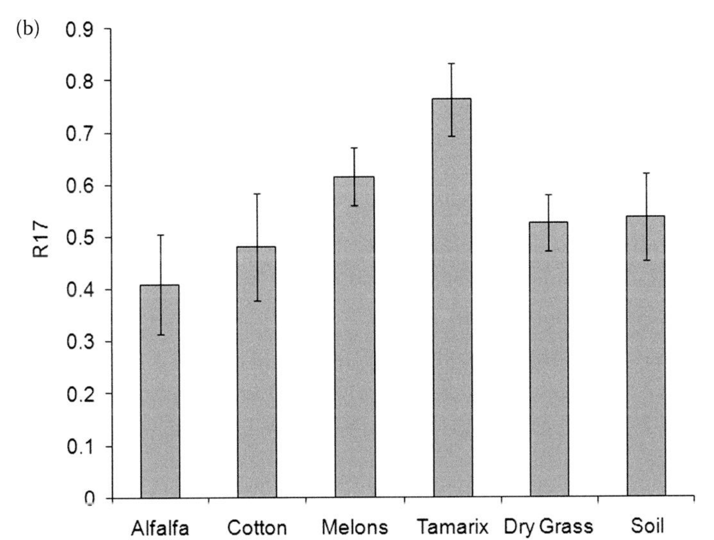

FIGURE 6.15 Differences in $R_{1,5}$ (a) and $R_{1,7}$ (b) for different vegetation types and soil in Lower Colorado River region. Bars are ± one standard error from 10 replicates.

to the rest of the vegetation (Figure 6.15a,b). The alfalfa showed significantly lower values of $R_{1,5}$ ($p < 0.05$) compared to other plants (Figure 6.15b).

The Landsat TM color-composite spectral ratio image, consisting of the band ratios $R_{1,5}$, NDVI, and $R_{1,7}$ assigned to the colors blue, green, and red, respectively, is shown in Figure 6.14. The fully grown alfalfa fields, which were in flowering condition, appear in dark green, and the alfalfa fields that were of medium growth appear in light green (Figure 6.16a). The alfalfa fields in flowering condition were greater than 40 cm in height, while alfalfa fields of medium growth were in the range of 10–40 cm in height. The melon and cotton fields appear in shades of greenish yellow and bluish green, respectively (Figure 6.16a). The tamarisk plants appear in light yellow, and all other vegetation appears in different shades of red (Figure 6.16a). Masks were created to limit image processing to the areas of vegetation, which were identified as pixels with an NDVI of greater than 0.2.

Unsupervised classification images, where the Landsat TM band ratios $R_{1,5}$, NDVI, and $R_{1,7}$ were used as selected inputs for classification, are shown in Figure 6.16b. The different vegetation

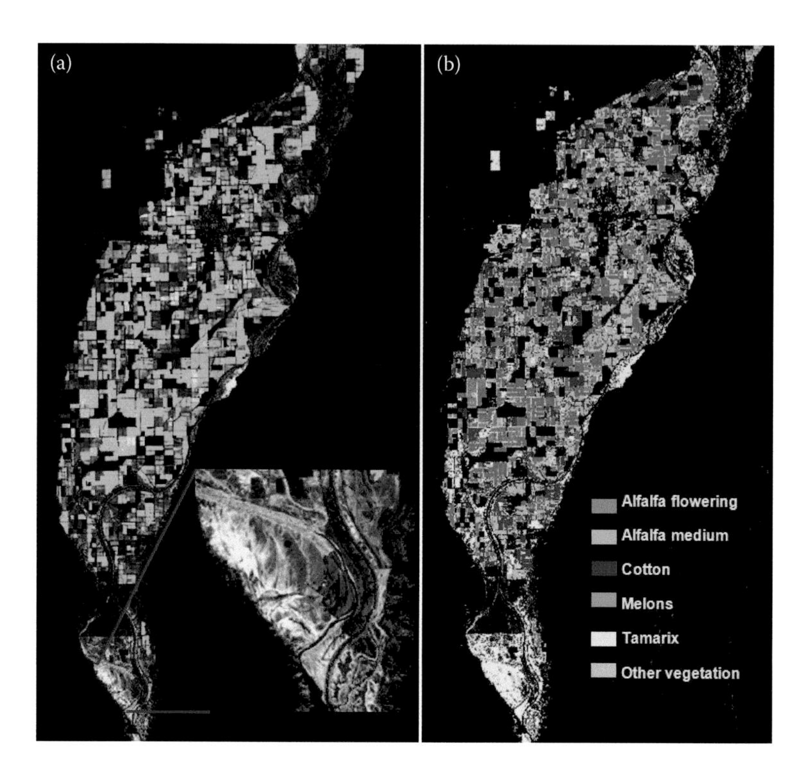

FIGURE 6.16 Landsat TM color-composite spectral-ratio image (NDVI, $R_{1,5}$, and $R_{1,7}$ displayed as BGR, respectively) of Lower Colorado River region (a). The CWR is shown in insert image. Results of unsupervised classification where NDVI, $R_{1,5}$, and $R_{1,7}$ were used as ratio inputs (b).

classes were clearly distinguished in the classified image, and the tamarisk-infested areas of the CWR appear distinctly in the image (Figure 6.16b). An accuracy assessment of the classification was performed using the ground-truth data obtained over 76 sampling locations during the period of Landsat overpass. Each sampling location represents an area of about 30–40 acres. The accuracy results given in Table 6.1 show that an overall accuracy of 76% (kappa = 0.71) was obtained using the three selected band ratios compared to the low accuracies obtained by the Landsat single-band inputs (data not shown). This implies that the selected band ratios are better suited for detecting species with distinct characteristics, such as tamarisk, with its distinctive saline and arid adoptive characteristics.

A visual comparison of all the spectral ratio composite images obtained from July 2007 to June 2008 shows that the tamarisk plants can be clearly mapped separately from other vegetation and are reported elsewhere [69]. High densities of tamarisk in the CWR can be clearly seen in all the images, and the changes in the level of tamarisk density through spring, summer, and fall are distinctly evident, and these results were published elsewhere [69].

6.2.3.5 Discussion and Future Directions

The spectral reflectance of the agricultural plants (alfalfa, cotton, and melons) and tamarisk were higher in the NIR (800–1300 nm) region compared to other plants. The reason for this is that all four of these plant types have higher green cover and plant biomass than the rest of the vegetation types present. Spectral reflectance in the NIR region increases with increased percent green cover, plant biomass [70], and LAI [71]. The spectral reflectance of all the agricultural plants from 1300 to 2500 nm regions was higher than that of tamarisk. It was reported that NIR wavelengths can best distinguish the tamariskfrom other vegetation types [72]. However, they employed the tamarisk spectra from 400 to 900 nm only [72], whereas our study employs spectra covering the wavelength range of 400–2500 nm (Figure 6.10).

TABLE 6.1
Confusion Matrix for Unsupervised Classification of Spectral Ratio (NDVI, $R_{1,5}$, and $R_{1,7}$) Image

Class	Alfalfa Flowering	Alfalfa Medium	Cotton	Melons	Tamarix	Other Vegetation	Total
Alfalfa flowering	22	1	0	0	0	0	23
Alfalfa medium	1	5	7	0	0	0	13
Cotton	0	0	6	0	0	2	8
Melons	2	0	3	6	0	0	11
Tamarix	0	0	1	1	11	0	13
Other vegetation	0	0	0	0	0	8	8
Total	25	6	17	7	11	10	76

Class	User's Accuracy (%)	Producer's Accuracy (%)
Alfalfa flowering	95.65	88
Alfalfa medium	38.46	83.33
Cotton	75	35.29
Melons	54.54	85.71
Tamarix	84.61	100
Other vegetation	100	80

Note: Overall accuracy = 76.32%; kappa coefficient = 0.71.

The leaf-level spectra shows less atmospheric noise in the 1400 and 1900 nm regions compared to the canopy-level reflectance. The minor water absorption bands at 975 and 1175 nm are stronger for the canopy reflectance than for the leaf reflectance. This is due to multipath reflectance in the case of canopy spectra. The canopy-level reflectance differs from the leaf-level spectra because the canopy spectra is affected by factors such as leaf orientation, canopy height, diameter, and leaf density, for example. Variations in leaf area and leaf angle have a dominant effect on canopy reflectance in a full canopy [73]. The interaction of photons with vegetation components in vertical space is known to be highly nonlinear. The scattering behavior is defined by the bidirectional reflectance distribution function and is beyond the scope of this paper.

The three best spectral ratios selected in this study discriminate the different vegetation types based on their biophysical and biochemical properties. The NDVI of the alfalfa and melons was higher compared to the cotton and tamarisk (Figure 6.14). The NDVI [74] has been widely used to estimate vegetation biomass (LAI), photosynthetic activity, and chlorophyll content. The ratios $R_{1,5}$ and $R_{1,7}$ were significantly higher for tamarisk compared to the rest of the vegetation types (Figures 6.15a,b). This can be attributed to the differences in the physical and chemical composition of the tamarisk plants compared to the other vegetation. Tamarisk that has adapted to many different saline soil types is known to secrete a variety of ions such as sodium, chlorine, potassium, calcium, magnesium, and sulfate [75–77] through its salt glands. During the process of evapotranspiration, these salt glands on tamarisk leaves release ions into the transpirational stream, thereby coating all the plant leaves with salt [77]. During the field studies in CWR, we noticed that all the tamarisk plants had a white powdery salt coating over all their leaves. All the salt accumulation on tamarisk leaves makes the leaves appear visually as a bluish green color [78].

The selected Landsat TM spectral ratios (NDVI, $R_{1,5}$, and $R_{1,7}$) emphasize the biophysical and biochemical differences among selected plants, as well as between the selected plants and other vegetation. Assignment of these three ratios to the primary colors green, blue, and red reveals (Figure 6.16a) important information that is not apparent in the single-band image (Figure 6.9) made from the same Landsat TM data set. This methodology is particularly useful in discriminating and mapping vegetation types that do not show strong visible color contrasts. Also, the spectral ratios are more robust and can be applied to multiple satellite overpasses for continuous monitoring of the vegetation.

Application of the spectral-ratio, color-composite technique for all the Landsat TM images obtained from June, 2007 through June, 2008 (Figures 6.14 and 6.16) reveals the seasonal progression of the tamarisk growth. Starting in October the tamarisk gradually loses leaves until January and will remain leafless from January to March, then start producing leaves again from March onward until leaf production peaks in summer. This seasonal progression also reveals that the spectral ratios are robust and can be applied to multiple satellite overpasses to monitor the spread of tamarisk. To map the tamarisk along the Arkansas River in Colorado [59], different combinations of single bands, NDVI, and tasseled cap transformations for each of the six Landsat ETM+ images obtained in April to October. In contrast to the previous study [59], where different Landsat single-band combinations were chosen for different Landsat image analyses, the spectral ratios developed in our study are based on DOS-corrected spectral ratios that are more robust compared to the combination of single spectral bands; hence, the same spectral ratio combination can be used on all dates of Landsat TM data for mapping tamarisk.

In conclusion, spectral-ratio, color-composite Landsat TM images can be used to detect and map tamarisk-infested areas in the Lower Colorado River region. These results show that multispectral and multitemporal Landsat TM data can be a valuable and cost-effective tool with which natural resource managers can develop regional maps depicting where tamarisk infestations occur over large, poorly accessible areas. Generating tamarisk distribution maps over time with Landsat TM data and combining that information with existing GIS data bases can create models for studying several other problems, such as soil salinity, evapotranspiration, forest fire potential, and displacement of native vegetation and wildlife caused by the spread of tamarisk.

6.2.4 Buffelgrass Study

The goal of the buffelgrass study was to identify remote sensing strategies that could be effectively utilized to map and monitor buffelgrass invasion in the Sonoran Desert. The specific objectives were to (1) identify the spectral characteristics of buffelgrass and native vegetation in the Arizona Upland (AU) zone of the Sonoran Desert as they varied throughout a single year and (2) identify optimal timing for discriminating between buffelgrass and native plants.

6.2.4.1 Study Area

This study focused on a sensitive habitat dominated by saguaro cactus (*Carnegiea gigantea*) and palo verde trees (*Parkinsonia microphylla*) in the AU zone of the Sonoran Desert in the Santa Catalina Mountains (Catalinas) just north of Tucson, AZ (Figure 6.17). The lower piedmont of the Catalinas supports some of the most abundant stands of giant saguaro cactus in the world [79]; however, it is currently threatened by buffelgrass invasion. The Catalina Mountains are a sky island, with a forested summit but surrounded by a sea of desert. The saguaro–palo verde association forms a ring at the base of the mountain and has the potential to link high-elevation fuels with urban ignition sources in the suburbs of Tucson.

6.2.4.2 Measurements of Community Composition

Field data collection was performed to (1) characterize the community composition of the habitat buffelgrass has invaded, is invading, and has not yet invaded and (2) measure the canopy-level reflectance of dominant species and cover types found in these communities throughout the year. Ten medium to large buffelgrass patches were identified in the study area at elevations ranging from 883 to 1097 m. Fieldwork occurred between December 2008 and March 2009. At each patch, a transect of contiguous 10 × 10 m plots was oriented such that one end started at the center of the patch and the other extended beyond the patch edge by at least 20 m. Transects were randomly oriented but confined to similar slope, aspect, and geomorphology. Within each plot, species-level projected canopy cover was measured using a point-intercept method along a regular 1 m grid.

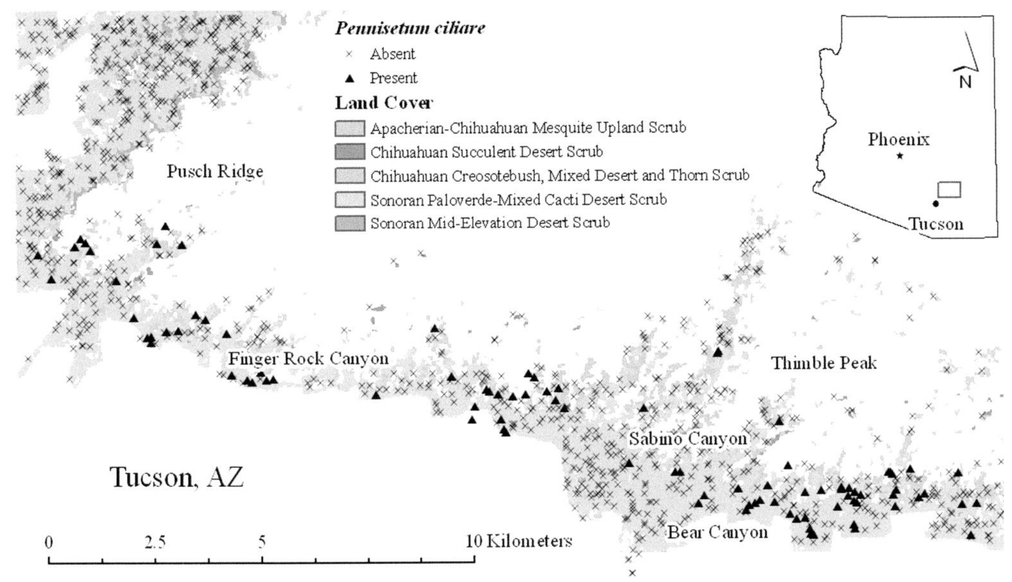

FIGURE 6.17 Buffelgrass study area.

6.2.4.3 Field Spectroscopy of Dominant Cover

On six dates between March and October 2007, we recorded the canopy-level reflectance of each major species found in the AU habitat at Tumamoc Hill, a small hill dominated by saguaros and palo verdes situated about 15–30 km from the field plots. Tumamoc Hill is at a comparable elevation and supports vegetation similar to that found in the Santa Catalina Mountains. Hyperspectral reflectance data at 10 nm intervals between 0.4 and 2.5 μm were measured using an ASD Fieldspec Pro III (Malvern Panalytical, Cambridge, UK) with an 8° FOV held at nadir above the target and adjusted in height to maintain the target in its FOV. Measurements were made between 10 a.m. and noon local time on cloud-free days. The ASD spectrometer was periodically calibrated with a calibrated spectralon reference panel to avoid saturation. Spectralon panel reference measurements were made before and after target measurements to account for solar illumination changes and calculate surface reflectance values. Each reflectance reading was the average of five readings that were taken while the instrument was slightly moved over the target to capture its variability. The same targets were measured on each of the six dates. Throughout the year, approximately 50 targets were measured. The spectra of the most abundant species found in the community characterization step are given in Figure 6.18.

6.2.4.4 Spectral Analysis of Sonoran Desert Vegetation

Spectral separability analysis was performed to assess temporal variations in discriminability between pure buffelgrass and other cover types as well as to distinguish mixed pixels containing different amounts of buffelgrass from pixels without buffelgrass. Due to the low signal-to-noise ratio in the SWIR region of the spectrum, a moving-average filter was applied to the SWIR bands. We calculated Pearson's r correlation of spectral signatures of all targets with buffelgrass for all six acquisition dates. While the landscape we are interested in is highly mixed, combinations of several cover types typically comprise greater than 50% of uninvaded cover: rock, soil, *Encelia farinosa*, *Parkinsonia microphylla*, and *Prosopis glandulosa*. We investigated the discriminability of *P. ciliare* from these dominant cover types in more detail, identifying the wavelengths that generate maximum differentiability for each season. To assess magnitude differences, we calculated the difference between the curves representing the reflectance of each cover type vs. *P. ciliare*.

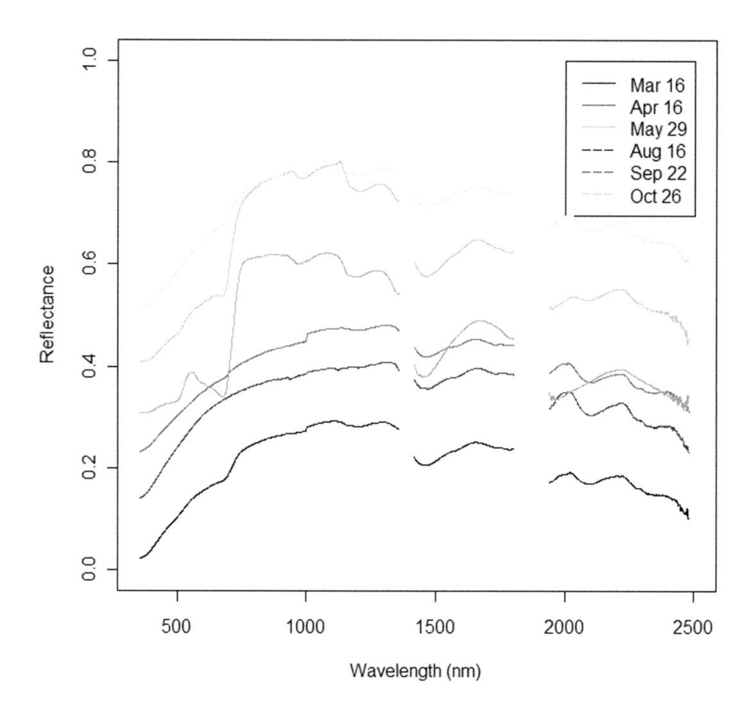

FIGURE 6.18 Reflectance of buffelgrass on six dates in 2007. Curves for acquisitions after March 16, 2007, are offset in sequence of 10% for each subsequent date on *y*-axis to increase visibility of curves.

6.2.4.5 Spectral Separability of Mixed Fractional Cover of Buffelgrass

We utilized plot-level cover measurements and predicted hyperspectral reflectance for all 53 plots for the 6 ASD reflectance data collection dates. We divided plots into low (<5% *P. ciliare* cover), medium (5%–50%), and high (≥50%) *P. ciliare* classes and performed a Student's *t*-test to examine differences between reflectance for high vs. medium and high vs. low cover at each wavelength.

6.2.4.6 Results

Two species comprised 28.4% cover of the low buffelgrass cover plots (48.6% of the total vegetation cover): *Encelia farinosa* and *Parkinsonia microphylla* (Table 6.2) with no other species comprising more than 10% of the vegetation cover. Plots were typically dominated by bare ground with mineral soil [mean = 24.8%, standard deviation (SD) = 8.3%] and rock outcrops (mean = 16.4%, SD = 6.9%) averaging over 40% cover on the uninvaded plots. Therefore, we chose to focus on the relationships between buffelgrass and these four cover types for field-based spectral endmember comparisons. The three plant species (*Pennisetum ciliare*, *Encelia farinosa*, and *Parkinsonia microphylla*) exhibited different spectral responses to phenological changes during the year, varying in their differentiability from each other and from background soil and rock outcrops.

In March, the buffelgrass plants sampled were comprised primarily of senesced blades and rachii from previous years' growth but exhibited some minimal greenup in the form of new leaves arising from the base of the plant. This partial greenup, likely the plants' response to available soil moisture from winter precipitation and rising spring temperatures, was short-lived as the plants became senescent for the hot, arid foresummer preceding the onset of the North American monsoon. By mid-August, the monsoon had manifested in southern Arizona, bringing abundant precipitation that stimulated greenup and leaf elongation in buffelgrass plants. The photosynthetically active phase diminished significantly by September when the plant was yellow, had curled leaves, and had set seed. The plant was almost completely senesced by October and had taken on an orange hue that differed from native plants.

TABLE 6.2
Percent Cover of Species Found in 15 Plots with <5% Buffelgrass Cover on Rocky Slopes in Santa Catalina Mountains

Species	Type	Uninvaded Mean Cover % (SE)
Encelia farinosa	Shrub	16.14 (1.46)
Parkinsonia microphylla	Tree	12.56 (2.24)
Prosopis glandulosa	Tree	3.80 (2.10)
Janusia gracilis	Vine/Shrub	3.20 (1.08)
Jatropha cardiophylla	Shrub	2.98 (1.00)
Lycium berlandieri	Shrub	2.26 (0.68)
Fouquieria splendens	Succulent	1.93 (0.43)
Calliandra eriophylla	Shrub	1.76 (0.52)
Eysenhardtia orthocarpa	Shrub	1.76 (1.08)
Jacquemontia pringlei	Vine/Shrub	1.65 (0.85)
Abutilon incanum	Forb	1.60 (0.60)
Cylindropuntia versicolor	Succulent	1.38 (0.55)
Evolvulus arizonica	Forb	1.32 (0.56)
Opuntia engelmannii	Succulent	1.16 (0.95)
Trixis californica	Shrub	0.88 (0.35)
All succulents		5.5 (1.3)
All grasses		2.1 (0.43)
All forbs		3.1 (1.0)
All shrubs		31.7 (1.7)
All trees		16.7 (2.8)

Spectral characteristics followed buffelgrass phenology and mirrored changes in photosynthetic activity, water content, and cellulose/lignin absorption (Figure 6.18). Absorption at 675 nm was noticeable in March, August, September, and October, although the August absorption was exemplary and the September absorption noteworthy. August was typical of photosynthetically active vegetation across the VIS-NIR, SWIR1, and SWIR2 wavelength regions, exhibiting additional absorption features near 950, 1150, and 1450 nm. SWIR1 and SWIR2 reflectances were comparatively low during August. March and September had somewhat lower SWIR reflectance, likely due to foliar water of green basal leaves (March) or stressed green leaves (September). A cellulose/lignin absorption feature at 2050 nm was one of the most characteristic features of the buffelgrass spectra on five of the six acquisition dates (all dates excluding August 16).

Encelia farinosa followed a bimodal phenological pattern that included spring greenup, floral production, midsummer senescence, and monsoon greenup (Figure 6.19). Greenup had started by March 16 but was close to full spring production by April. April was also characterized by a showy display of yellow composite flowers. Senescence during the arid foresummer resulted in shriveled, desiccated leaves. *Encelia* responded strongly to the summer monsoon and remained photosynthetically active between mid-August and late September. By late October, most leaves had desiccated and dehisced, leaving behind only the skeleton of this mostly <1 m shrub. The spectral reflectance curves show the bimodal growing season and the extended (and more intense) response to the monsoon. This is shown by the chlorophyll feature with a peak at 550 nm and absorption at 675 nm contrasted with strong NIR reflectance and a steep red edge. Additionally, lower SWIR reflectance is demonstrated, as is a strongly peaked reflectance at 1650 nm. Floral reflectance is evident in the April 16 spectra in the form of a shoulder between the 550 nm peak and 675 nm trough. During the arid foresummer, *Encelia* mostly lost the chlorophyll/carotenoid absorption at

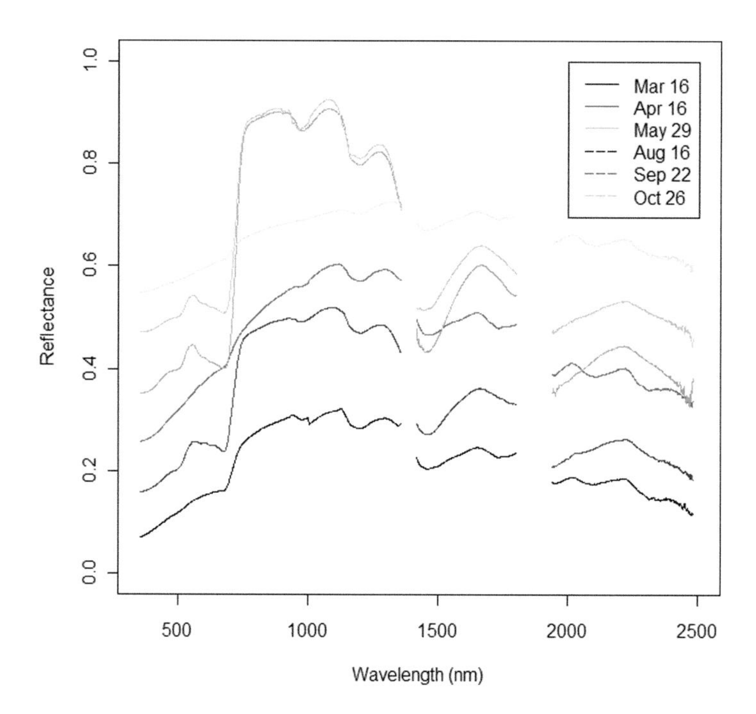

FIGURE 6.19 Reflectance of *Encelia farinosa* on six dates during 2007. Reflectance spectra following March 16 are offset on *y*-axis to increase readability. Note: NIR reflectance (800 nm) for August 16 is actually 10% greater than reflectance on September 22.

675 nm, while the cellulose/lignin absorption feature manifested slightly (more so in May than April). The reflectance of *Encelia* in the NIR in August exceeded 0.6, significantly higher than other vegetation sampled in this study (data not shown). The higher reflectance was mirrored across the VNIR wavelength region with a reflectance of over 0.1 at a strong 675 nm absorption feature and over 0.15 at the 550 nm chlorophyll peak. Other strongly photosynthetic vegetation we sampled in August reflected less than 0.1 at 520 nm and close to 0.05 at 675 nm. This was likely caused by the production of reflective trichomes across the surface of the *Encelia* leaves that act to reduce leaf-level temperatures and increase water use efficiency [80]. Foliar water content of leaves was clearly high in August and September, as evidenced by relatively low SWIR reflectance values. After September, plant deconstruction was rapid, with leaves desiccating, falling off, and blowing away. Remaining leaves in October were brown, but the dominant plant cover was composed of dried stems. This is reflected in the slight absorption feature at 2050 nm.

Parkinsonia microphylla is a deciduous leguminous tree with green photosynthetic bark and small, compound leaves with very small leaflets. The evergreen bark allows it to photosynthesize at all times of the year while the small leaves minimize potential for water loss from transpiration during the hot, dry summer. *Parkinsonia* produces leaves in spring as temperatures increase. Leaf-out occurs in March and April, and floral production is prolific in April. Leaves remain during most summers, although activity is typically suppressed during the arid foresummer because, like many members of Fabaceae, *Parkinsonia* leaves have pulvini at the base of the rachii that retract leaves and minimize their exposure to sun and wind during times of low water availability or humidity. *Parkinsonia* responds to summer precipitation and remains photosynthetically active throughout the summer and fall. *Parkinsonia* reflectance exhibits signs of photosynthetic activity on all six dates (Figure 6.20). While the chlorophyll peak at 550 nm was only apparent from August through late October, the absorption feature at 675 nm was evident on all dates. August and September have the strongest photosynthetic response, as evidenced by the clear chlorophyll absorption feature in the

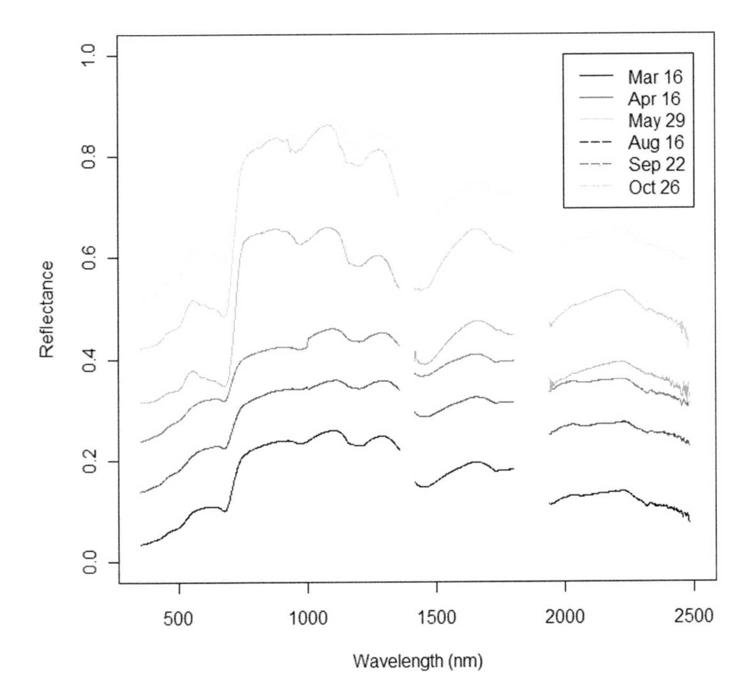

FIGURE 6.20 Spectral reflectance of *Parkinsonia microphylla* on six dates in 2007. Curves following March 16 are sequentially offset by 10% to increase readability.

550–675 nm range and the higher NIR reflectance (750 nm). The slope of the NIR reflectance from 750 to 1250 nm was also noticeably negative in the August spectra. Cellulose absorption features were barely evident in the March, April, and May spectra.

Buffelgrass most strongly resembled *Encelia* on the March 16 acquisition date (Figure 6.19). Only *Encelia* had a cellulose/lignin absorption feature on our first date of the year. Both plants had slight absorption at 675 nm but largely lacked the chlorophyll absorption feature at 550–675 nm. Senesced buffelgrass had a slight orange hue, manifested in the fact that *Encelia* had a lower slope from 450 to 650 nm than buffelgrass. *Parkinsonia microphylla* resembled buffelgrass to a lesser extent but exhibited stronger absorption at 675 nm, strong absorption features in the NIR, a less pronounced cellulose/lignin feature in SWIR2, and overall less reflectance in SWIR1 and SWIR2. Soil and rock differed by virtue of their steep slope from 450 to 700 nm and flat response curves across the SWIR1 and SWIR2. In April, buffelgrass was no longer photosynthetically active and resembled both rock and soil in the VNIR wavelength region (Figure 6.18). In the SWIR2, buffelgrass was the only target with a strong absorption feature at 2050 nm. By late May, all vegetation had slowed photosynthetic activity or stopped altogether. *Encelia* resembled buffelgrass in all parts of the spectrum, including the SWIR2 cellulose/lignin absorption wavelengths. Buffelgrass did lack the absorption feature found at 1200 nm in both *Encelia* and *Parkinsonia*. All plants had greened up by August 16 in response to the monsoon. *Encelia* was exemplary in having increased reflectance across the board. The likely cause of this was the production of reflective trichomes that lowered the leaf surface temperature and increase water use efficiency [80]. *Parkinsonia* and buffelgrass were almost indistinguishable, with only slight differences in the shape of the chlorophyll absorption and slight differences in the overall reflectance in the NIR and depth of the 1200 nm absorption feature. By September, buffelgrass photosynthesis had slowed, differentiating it from both *Parkinsonia* and *Encelia* in the loss of the strong 675 nm absorption and the reappearance of a 2050 nm cellulose/lignin feature. Based on similar SWIR1/SWIR2 reflectance magnitudes, water content between the three plants appeared to be similar. *Encelia* had desiccated by October 26, while *Parkinsonia* remained photosynthetically active. While *Encelia*'s reflectance shape in the SWIR1/SWIR2 regions was similar to that of buffelgrass, buffelgrass reflectance in the SWIR1/SWIR2 was higher overall, possibly due to lower residual water content. Furthermore, the slope from 450 to 750 nm of buffelgrass was quite steep compared to that of *Encelia*. This was also reflected in a strong orange hue to senesced buffelgrass in October that contrasted with that of natives. *Encelia* remains, for example, were mostly gray and light brown.

Overall, buffelgrass exhibited a phenological pattern distinct from that of other dominant native plants and background cover. While some of these differences relate to changes in ecosystem function (e.g., earlier shut-down of photosynthetic activity in the arid foresummer and following the summer monsoon), others relate to changes in ecosystem structure (e.g., the persistence of highly flammable grasses year round is evident in a strong cellulose/lignin absorption feature at 2050 nm that is not reflected in the native species or uninvaded soil background spectra). The spread of buffelgrass across a region that is typically characterized by abundant bare ground surrounding islands of shrubs, trees, cacti, and native grasses results in significant changes in both structure and function. Hyperspectral remote sensing offers the means to both map and monitor invasions and measure the changes in ecosystem structure and function associated with invasions.

6.2.4.7 Conclusions

In summary, the selection of bands in hyperspectral regions is very important for the observation of litter or standing senescent vegetation. The CAI was developed and can be applied to many important resource management questions. Remote sensing technology can help managers and scientists with landscape component discrimination. The phenology of invasive plants that have litter phases can be observed with the CAI. In future research, we will validate the use of the CAI for a range of applications (e.g., agriculture, natural resources, land degradation) and remote sensing platforms with the correct spectral configurations.

The three case studies presented in this chapter show the utility of combining broadband ratios such as NDVI with time-series analyses and narrowband hyperspectral ratios to detect landscape features such as plant litter and invasive species that are otherwise difficult or impossible to resolve with satellite imagery. The CAI makes use of reflectance data in the 400–2500 nm spectral region to distinguish between soil, green vegetation, and cellulose-lignin plant litter, even in mixed scenes, across different soil types, and in soils with varying moisture contents. It was developed to quantify plant litter left on soil as a result of no-till farming. However, the case study with buffelgrass showed that the CAI combined with time-series NDVI imagery could be used to detect the spread of this invasive species based on its phenology. The CAI remains high for most of the year while plants are dormant; then the NDVI signal becomes dominant during the brief summer monsoon season.

In the case of salt cedar, spectral-ratio, color-composite, Landsat TM images can be used to detect and map saltcedar -infested areas in the Lower Colorado River region. These results show that multispectral and multitemporal Landsat TM data can be a valuable and cost-effective tool that natural resource managers can use to develop regional maps depicting where salt cedar infestations occur over large, poorly accessible areas. Generating saltcedar distribution maps over time with Landsat TM data and combining that information with existing GIS databases, one can create models for studying several other problems, such as soil salinity, evapotranspiration, forest fire potential, and displacement of native vegetation and wildlife caused by the spread of saltcedar.

REFERENCES

1. Elvidge, C.D. 1990. Visible and near infrared reflectance characteristics of dry plant materials. *International Journal of Remote Sensing* 11: 1775–1795.
2. van Leeuwen, W.J.D. and A.R. Huete. 1996. Effects of standing litter on the biophysical interpretation of plant canopies with spectral indices. *Remote Sensing of Environment* 55: 123–138.
3. Swift, M.J., O.W. Heal, and J.M. Anderson. 1979. *Decomposition in Terrestrial Ecosystems*. Oxford: Blackwell Scientific Publications.

4. DeAngelis, D.L. 1992. Nutrient interactions of detritus and decomposers. In M. B. Usher, M. L. Rosenzweig and R. L. Kitching (eds). *Population and Community Biology: Dynamics of Nutrient Cycling and Food Webs*. London: Chapman & Hall.
5. Mason, C.F. 1976. *Decomposition*. Southampton: Camelot Press Ltd.
6. Stalfelt. 1972. *Stalfelt's Plant Ecology: Plants, the Soil and Man*. Translated by Jarvis, Marguerite A. Jarvis, Paul G. London: William Clowes and Sons.
7. Richards, B.N. 1974. *Introduction to the Soil Ecosystem*. New York: Longman, Inc.
8. Remmert, H. 1980. *Ecology: A Textbook*. Translated by Brederman-Thorson, Marguerite A. Second ed. Berlin: Springer-Verlag.
9. McMurtrey III, J.E., E.W. Chappelle, C.S.T. Daughtry, and M.S. Kim. 1993. Fluorescence and reflectance of crop residue and soil. *Journal of Soil and Water Conservation* 48(3): 207–213.
10. Skidmore and Siddoway, 1978. Crop residue requirements to control wind erosion. In W. Oschwald (ed). *Crop Residue Management Systems*. ASA Spec. Publ. 31. Mad., WI: ASA, CSSA, and SSSA.
11. Aase, J.K. and D.L. Tanaka. 1991. Reflectances from four wheat residue cover densities as influenced by three soil backgrounds. *Agronomy Journal* 83: 753–757.
12. Daughtry, C.S.T., J.E. McMurtrey III, P.L. Nagler, M.S. Kim, and E.W. Chappelle. 1995. Spectral reflectance of soils and crop residues. In A.M.C. Davies and P. Williams (eds). *Near Infrared Spectroscopy: The Future Waves*. NIR Publications.
13. Ahn, C.W., M.F. Baumgardner, and L.L. Biehl. 1996. Performance of AVIRIS, Adjusted AVIRIS, and Simulated TM data for Classifying Crop Residue.
14. Morrison, J.E. Jr., C. Huang, D.T. Lightle, and C.S.T. Daughtry. 1993. Residue measurement techniques. *Journal of Soil and Water Conservation* 48: 479.
15. Shelton, D.P. and E.C. Dickey. 1995. Conservation tillage. In *Estimating Percent Residue Cover Using the Line-Transect Method*. Conservation Technology Information Center: Cooperative Extension Service.
16. Daughtry, C.S.T., J.E. McMurtrey III, E.W. Chappelle, W.P. Dulaney, J.R. Irons, and M.B. Satterwhite. 1995b. Potential for discriminating crop residues from soil by reflectance and fluorescence. *Agronomy Journal* 87: 165–171.
17. Jordan, C.F. 1969. Derivation of leaf index from quality of light on the floor. *Ecology* 50: 663–666.
18. Woolley, J.T. 1971. Reflectance and transmittance of light by leaves. *Plant Physiology* 47: 656–662.
19. Gates, D.M., H.J. Keegan, J.C. Schleter, and V.R. Weidner. 1965. Spectral properties of plants. *Applied Optics* 4(1): 11–20.
20. Tucker, C.J. 1979. Red and photographic infrared linear combinations for monitoring vegetation. *Remote Sensing of Environment* 8: 127–150.
21. Bracher, G.A. and P.A. Murtha. 1994. Estimation of foliar macro-nutrients and chlorophyll in douglas-fir seedlings by leaf reflectance. *Canadian Journal of Remote Sensing* 20: 102–115.
22. Myneni, R.B., F.G. Hall, P.J. Sellers, and A.L. Marshak. 1995. The interpretation of spectral vegetation indexes. *IEEE Transactions on Geoscience and Remote Sensing* 35(2): 481–486.
23. Goward, S.N. and K.F. Huemmrich. 1992. Vegetation canopy PAR absorptance and the normalized difference vegetation index: An assessment using the SAIL model. *Remote Sensing of Environment* 39: 119–140.
24. Goward, S.N., K.F. Huemmrich, and R.H. Waring. 1994. Visible-near infrared spectral reflectance of landscape components in Western Oregon. *Remote Sensing of Environment* 47: 190–203.
25. Daughtry, C.S.T. and L.L. Biehl. 1985. Changes in spectral properties of detached birch leaves. *Remote Sensing of Environment* 17: 281–289.
26. Kimes, D.S. 1991. Radiative transfer in homogeneous and heterogeneous vegetation canopies. In R.B. Myneni and J. Ross (eds). *Photon-Vegetation Interactions: Applications in Optical Remote Sensing and Plant Ecology*. NY: Springer-Verlag, pp. 339–388.
27. Irons, J.R., R.A. Weismiller, and G.W. Peterson. 1989. Soils reflectance. In G. Asrar (ed). *Theory and Applications of Optical Remote Sensing*. NY: Wiley.
28. Stoner, E.R. and M.F. Baumgardner. 1981. Characteristic variations in reflectance of surface soils. *Soil Science Society of American Journals* 45(6): 1161–1165.
29. Stoner, E.R., M.F. Baumgardner, R.A. Weismiller, L.L. Biehl, and B.F. Robinson. 1980. Extension of laboratory-measured soil spectra to field conditions. *Soil Science Society of American Journals* 44: 572–574.
30. Wiegand, C.L. and A.J. Richardson. 1992. Relating spectral observations of the agricultural landscape to crop yield. *Food Structure* 11: 249–258.
31. Henderson, T.L., M.F. Baumgardner, D.P. Franzmeier, D.E. Stott, and D.C. Coster. 1992. High dimensional reflectance analysis of soil organic matter. *Soil Science Society of American Journals* 56: 865–872.

32. Murray, I. and P.C. Williams. 1988. Chemical principles of near infrared technology. In P. Williams and K. Norris (eds). *Near Infrared Technology in the Agricultural and Food Industries*. St. Paul, MN: Amer. Assoc. Cereal Chemists, pp. 17–34.
33. Sudduth, K.A. and J.W. Hummel. 1993. Soil organic matter, CEC, and moisture sensing with a portable NIR spectrophotometer. *Transactions of the ASAE* 36(6): 1571–1582.
34. Price, J.C. 1992. Variability of high resolution crop reflectance spectra. *International Journal Remote Sensing* 13: 2593–2610.
35. Daughtry, C.S.T. 2001. Discriminating crop residues from soil by shortwave infrared reflectance. *Agronomy Journal* 93: 125–131.
36. Nagler, P.L., C.S.T. Daughtry, and S.N. Goward. 2000. Plant litter and soil reflectance. *Remote Sensing of Environment* 71(2): 207–215.
37. Nagler, P.L., Y. Inoue, E.P. Glenn, A. Russ, and C.S.T. Daughtry. 2003. Cellulose absorption index (CAI) to quantify mixed soil-plant litter scenes. *Remote Sensing of Environment* 87: 310–325.
38. Daughtry, C.S.T., J.E. McMurtrey III, P.L. Nagler, M.S. Kim, and E.W. Chappelle. 1996. Spectral reflectance of soils and crop residues. In A.M.C. Davies and P. Williams (eds). *Near Infrared Spectroscopy: The Future Waves*. NIR Publications. pp. 505–511.
39. Inoue, Y., Morinaga, S., and M. Shibayama. 1993. Non-destructive estimation of water status of intact crop leaves based on spectral reflectance measurements. *Japanese Journal of Crop Science* 62(3): 462–469.
40. Mooney, H.A. and Cleland, E.E. 2001. The evolutionary impact of invasive species. *Procedures of National Academy of Science* 98: 5446–5451.
41. Baum, B.R. 1967. Introduced and naturalized tamarisks in the United States and Canada. *Baileya* 15: 19–25.
42. Robinson, T.W. 1965. Introduction, spread and areal extent of tamarisk (Tamarisk) in the western states. *US Geological Survey Professional Paper 491-A*.
43. Horton, J.S. and Campbell, C.J. 1974. Measurement of phreatophyte and riparian vegetation for maximum multiple use values. *USDA Forest Service Paper RM117.*
44. Brotherson, J.D. and Field, D. 1987. Tamarisk: Impacts of a successful weed. *Rangelands* 9: 110–112.
45. D'Antonio, C.M. and P.M. Vitousek. 1992. Biological invasion by exotic grass, the grass/fire cycle, and global change. *Annual Review of Ecology and Systematics* 23: 63–87.
46. Brooks, M.L., C.M. D'Antonio, D.M. Richardson, J.B. Grace, J.E. Keeley, J.M. diTomaso, R.J. Hobbs, M. Pellant, and D. Pyke. 2004. Effects of invasive plants on fire regimes. *BioScience* 54(7): 677–688.
47. Swetnam, T.W. and J.L. Betancourt. 1990. Fire-Southern oscillation relations in the southwestern United States. *Science* 249: 1017–1021.
48. Brooks, M.L., C.M. D'Antonio, D.M. Richardson, J.B. Grace, J.E. Keeley, J.M. diTomaso, R.J. Hobbs, M. Pelland, and D. Pyke. 2004. Effects of invasive alien plants on fire regimes. *Bioscience* 54(7): 677–688.
49. Everitt, J.H. and C.J. Deloach. 1990. Remote sensing of Chinese Tamarisk (*Tamarisk chinensis*) and associated vegetation. *Weed Science* 38: 273–278.
50. Ge, S., R. Carruthers, P. Gong, and A. Herrera. 2006. Texture analysis for mapping *Tamarisk parviflora* using aerial photographs along the Cache creek, California. *Environmental Monitoring and Assessment* 114: 65–83.
51. Akasheh, O.Z., C.M.U. Neale, and H. Jayanthi. 2008. Detailed mapping of riparian vegetation in the middle Rio Grande River using high resolution multi-spectral airborne remote sensing. *Journal of Arid Environments* 72: 1734–1744.
52. Underwood, E., S. Ustin, and D. DiPietro. 2003. Mapping native plants using hyperspectral imagery. *Remote Sensing of Environment* 46: 150–161.
53. Narumalani, S. et al. 2006. A comparative evaluation of ISODATA and spectral angle mapping for the detection of tamarisk using airborne hyperspectral imagery. *Geocarto International* 21: 59–66.
54. Narumalani, S., D.R. Mishra, R. Wilson, P. Reece, and A. Kohler. 2009. Detecting and mapping four invasive species along the floodplain of North Platte River, Nebraska. *Weed Technology* 23: 99–107.
55. Peters, A.J., B.C. Reed, M.D. Eve, and K.C. McDaniel. 1992. Remote sensing of broom snake weed (*Gutierrezia sarothrae*) with NOAA-10 spectral image processing. *Weed Technology* 6: 1015–1020.
56. Dewey, S.A., K.P. Price, and D. Ramsey. 1991. Satellite remote sensing to predict potential distribution of dyers woad (*Isatis tinctoria*). *Weed Technology* 5: 479–484.
57. Everitt, J.H., D.E. Escobar, R. Villarreal, M.A. Alaniz, and M.R. Davis. 1993. Canopy light reflectance and remote sensing of shin oak (*Quercus havardii*) and associated vegetation. *Weed Science* 41: 291–297.
58. Hamada, Y., D.A. Stow, L.L. Coulter, J.C. Jafolla, and L.W. Hendricks. 2007. Detecting Tamarisk species (*Tamarisk* spp.) in riparian habitats of Southern California using high spatial resolution hyperspectral imagery. *Remote Sensing of Environment* 109: 237–248.

59. Evangelista, P.H., T.J. Stohlgren, J.T. Morisette, and S. Kumar. 2009. Mapping invasive tamarisk (*Tamarisk*): A comparison of single-scene and time series analyses of remotely sensed data. *Remote Sensing* 1: 519–533.
60. Dennison, P.E., P.L. Nagler, K.R. Hultine, E.P. Glenn, and J.R Ehleringer. 2009. Remote monitoring of tamarisk defoliation and evapotranspiration following tamarisk leaf beetle attack. *Remote Sensing of Environment* 113: 1462–1472.
61. Franklin, K.A., K. Lyons, P.L. Nagler, D. Lampkin, and E.P. Glenn. 2006. Buffelgrass (*Pennisetum ciliare*) land conversion and productivity in the plains of Sonora, Mexico. *Biological Conservation* 127: 62–71.
62. Brenner, J.C. 2010. What drives the conversion of native rangeland to buffelgrass (*Pennisetum ciliare*) pasture in Mexico's Sonoran Desert?: The social dimensions of a biological invasion. *Human Ecology* 38(4): 495–505.
63. Olsson, A. 2010. Ecosystem transformation by buffelgrass: Climatology of invastion, effects on Arizona Upland diversity, and remote sensing tools for managers. Dissertation for Arid Lands Resource Sciences, University of Arizona.
64. Huang, C.-Y. and E.L. Geiger. 2008. Climate anomalies provide opportunities for large-scale mapping of non-native plant abundance in desert grasslands. *Diversity and Distributions* 14: 875–884.
65. Vincent, R.K., X. Qin, R.M.L. McKay, J. Miner, K. Czajkowski, J. Savino, and T. Bridgeman. 2004. Phycocyanin detection from LANDSAT TM data for mapping cyanobacterial blooms in Lake Erie. *Remote Sensing of Environment* 89: 381–392.
66. Vincent, R.K. 1997. *Fundamentals of Geological and Environmental Remote Sensing*. Upper Saddle River, NJ: Prentice Hall.
67. Maruthi Sridhar, B.B., R.K. Vincent, J.D. Witter, and A.L. Spongberg. 2009. Mapping the total phosphorus concentration of biosolid amended surface soils using LANDSAT TM data. *Science of the Total Environment* 407: 2894–2899.
68. Rouse, J.W. et al. 1974. *Monitoring the vernal advancement and retrogradation (green wave effect) of natural vegetation*, Type III Final Report, NASA Goddard Space Flight Center, Green belt, Maryland.
69. Maruthi Sridhar, B.B., R.K. Vincent, W.B. Clapaham, S.I. Sritharan, J. Osterberg, C.M.U. Neale, and D.R. Watts. in press. Mapping tamarisk (*Tamarisk ramosissima*) and other riparian and agricultural vegetation in the Lower Colorado River region using multi spectral LandsatTM imagery. *GeoCarto International*.
70. Gamon, J.A., C.B. Field, K.L. Goulden, A.E. Griffin, G. Hartley, G. Joel, J. Penuelas, and R. Vallentini. 1995. Relationships between NDVI canopy structure, and photosynthesis in three California vegetation types. *Ecological Applications* 4: 28–41.
71. Qi, J., F. Cabot, M.S. Moran, and G. Dedieu. 1995. Biophysical parameter estimations using multidirectional spectral measurements. *Remote Sensing of Environment* 54: 71–83.
72. Randquist, B.C. and D.A. Brookman. 2007. Spectral characterization of the invasive shrub tamarisk (*Tamarisk* spp.) in North Dakota. *Geocarto International* 22: 63–72.
73. Asner, G.P. 1998. Biophysical and biochemical sources of variability in canopy reflectance. *Remote Sensing of Environment* 64: 234–253.
74. Curran, P.J. 1989. Remote sensing of foliar chemistry. *Remote Sensing of Environment* 30: 271–278.
75. Waisel, Y. 1961. Ecological studies on *Tamarisk aphylla* (L.) Karst. III. The salt economy. *Plant and Soil* 13: 356–364.
76. Berry, W.L. 1970. Characteristics of salts secreted by *Tamarisk aphylla*. *American Journal of Botany* 57: 1226–1230.
77. Storey, R. and W.W. Thompso. 1994. An X-ray microanalysis study of the salt glands and intercellular calcium crystals of tamarisk. *Annals of Botany* 73: 307–313.
78. Nagler, P.L., E.P. Glenn, T.L. Thompson, and A. Huete. 2004. Leaf area index and normalized difference vegetation index as predictors of canopy characteristics and light interception by riparian species on the Lower Colorado River. *Agriculture and Forest Meteorology* 125: 1–17.
79. Whittaker, R.H. and W.A. Niering. 1973. Vegetation of the Santa Catalina Mountains, Arizona. V. Biomass, production, and diversity along the elevation gradient. *Ecology* 56(4): 771–790.
80. Ehleringer, J.R. and O. Björkman. 1978. Pubescence and leaf spectral characteristics in a desert shrub, *Encelia farinosa*. *Oecologia* 36(2): 151–162.
81. Pimentel, D. et al. 2001. Economic and environmental threats of alien plant, animal, and microbe invasions. *Agriculture, Ecosystems & Environment* 84(1): 1–20.

7 Hyperspectral Applications to Landscape Phenology

Alfredo Huete, Werapong Koedsin, and Jin Wu

CONTENTS

7.1 INTRODUCTION

Phenology is the study of annually recurring biological life cycle events and the drivers and controls of their periodicity. Phenology is a characteristic property of ecosystem functioning and influence local to global biogeochemical and hydrological processes, including photosynthesis, water cycling, and the energy balance (Yang et al., 2014). Plant growth cycles are driven by abiotic and biotic factors, including climate (temperature, radiation, precipitation), soil moisture and nutrients, topography, species composition, plant functional type, and plant- and leaf-scale phenology (Figure 7.1). Shifts in phenology thereby depict plants' integrated response to climate and environmental changes and have become an important source of information on how plants are responding to climate change (de Keyzer et al., 2017).

Satellite data, with their synoptic views, repetitive sampling, and high spectral resolution, offer numerous opportunities to advance the study of phenology. Thus far, satellite products have primarily contributed to studies of "landscape phenology" (Friedl et al., 2006), defined as the aggregate seasonal vegetation patterns detected by satellites. Landscape or satellite phenology is distinct from traditional definitions of phenology that commonly involve in situ observations of individual plants and biological life cycle events, such as budbreak, flowering, pollination, and fruiting.

Investigations of hyperspectral vegetation phenology are very limited and have yet to be exploited, as remotely sensed seasonal vegetation dynamics have rarely been explored at higher spatial and spectral resolutions (Dennison and Roberts, 2003). Yet phenological life cycle events, such as flowering, leaf onset, and litterfall, can be quite dramatic visually and, thus, will strongly alter canopy optical properties. The measurement and detection of finer spatial-, spectral-,

FIGURE 7.1 Phenology of a temperate deciduous forest at Harvard Forest, as viewed through a phenocam. (a) Summer. (b) Autumn. (From http://phenocam.sr.unh.edu/webcam/sites/harvard/)

and temporal-scale phenology signals remain a challenge and offer unique opportunities for hyperspectral remote sensing methods to contribute to phenology studies.

In this chapter, we review current knowledge on phenology optical signals at the leaf, canopy, and landscape scales, provide an overview of current and potential hyperspectral applications to assess life cycle events and determine phenophases, and discuss the challenges and limitations of hyperspectral sensing in phenology applications.

7.2 SATELLITE REMOTE SENSING OF PHENOLOGY

Satellite phenology involves a composite measure of seasonal canopy foliage dynamics, encompassing leaf-on, leaf-off, species foliage composition, and the abundance and structural arrangement of leaves in the canopy. Higher temporal frequency satellite data from coarse (>250 m) and moderate (30 m) resolution sensors enable the retrieval of phenological profiles and generation of phenology products. Coarse satellite data can be from various sensors, such as the Advanced Very High Resolution Radiometer (AVHRR), Moderate Resolution Imaging Spectroradiometer (MODIS), or the MEdium Resolution Imaging Spectrometer (MERIS), while Landsat and Sentinel-2 offer moderate-resolution measures of vegetation seasonality (Clark, 2017). Spectral indices, including vegetation indices (VIs), leaf area index (LAI), and chlorophyll index products, are most often used to depict continuous changes in vegetation growth over phenological cycles.

7.2.1 Phenological Profile Metrics

The landscape growing season profiles are quantified with the use of time-based metrics in order to consistently retrieve measures of phenology in space and time and enable long-term and regional to global phenology analyses. Commonly used phenology metrics are graphically presented in Figure 7.2 and include event-based metrics that depict the start of growing season (SGS), end of growing season (EGS), length of growing season (LGS), and peak of growing season (PGS) and additionally may include metrics representing minimum greenness value, rate of greenup, and rate of drying or curing (Zhang et al., 2003).

In addition to the "timing" of phenophases, one can also retrieve seasonal measures of magnitude, such as (1) the seasonal amplitude of greenness values, (2) peak greenness value, (3) minimum greenness value, and (4) integrals over the growing season. All phenology metrics can be output in image form for further spatial-temporal analysis (Figure 7.2).

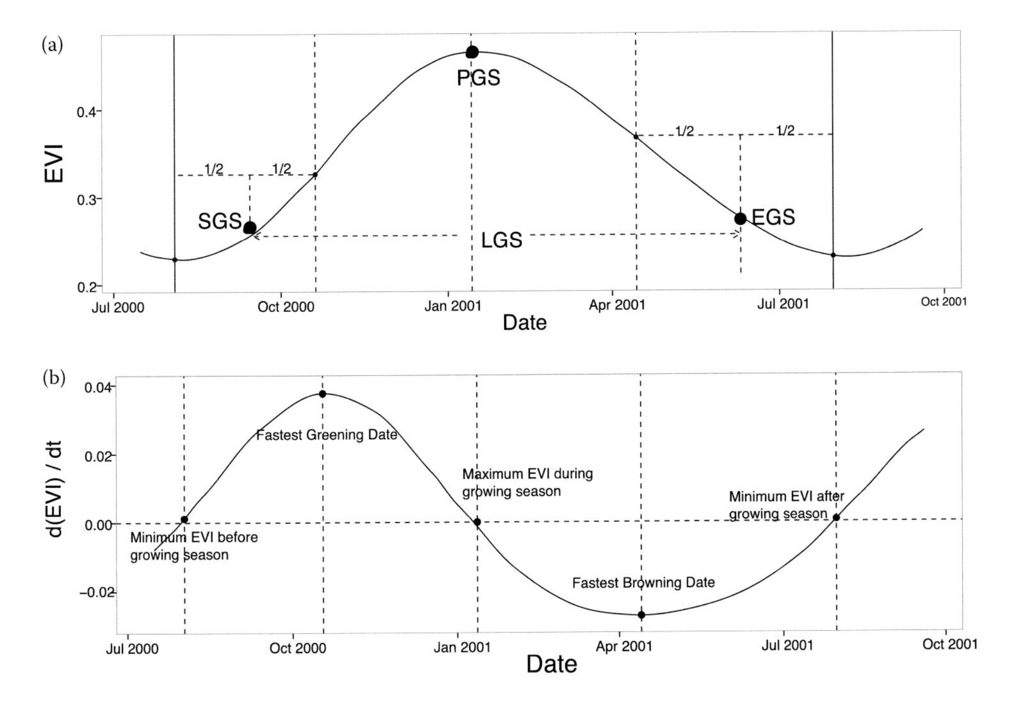

FIGURE 7.2 Phenology metrics derived from MODIS EVI seasonal profile (a) and its first derivative (b), displaying start of growing season (SGS), end of growing season (EGS), length of growing season (LGS), and peak of growing season (PGS). The first derivative shows maximum rates of greening and browning as well as minimum EVI base values pre- and postseason. (Modified from Ma, X. et al. 2013. *Remote Sensing of Environment* 139, 97–115.)

7.2.2 Coarse-Resolution Landscape Phenology

Numerous landscape phenology studies and applications have been reported in the literature that use foliage-based measures of greenness, such as VIs, and encompass a wide range of biomes from tropical and urban areas to northern latitudes (Myneni et al., 1997; Tucker et al., 2001; Huete et al., 2006; Pau et al., 2010). Spectral mixture analysis (SMA) has also been used to retrieve subpixel vegetation spectral features at seasonal time scales. Anderson et al. (2011) applied SMA to MODIS time series data to better interpret canopy phenology in Amazon rainforests through the simultaneous retrieval of seasonal greening and browning patterns. Khwarahm et al. (2017) employed the MERIS Total Chlorophyll Index (MTCI) time series data to estimate the flowering phenophase of birch and grass species and predict pollen release dates based on retrievals of two phenological variables, the start of season and peak of season.

7.2.3 Moderate-Resolution Landscape Phenology

Kauth and Thomas (1976) originally used Landsat Multispectral Scanner (MSS) data to trace growing season dynamics of crops over large agricultural regions. They developed an orthogonal "tasseled cap" transform, analogous to a spectral mixture model, to extract greenness, yellowness, brightness, and wetness biophysical features for characterizing crop phenophases. More recently, fusion techniques have been developed to combine moderate spatial resolution (30 m) Landsat data with higher temporal frequency MODIS data to improve phenology products and applications (Roy et al., 2008). The two Sentinel-2 satellite sensors, as part of the Copernicus program, are expected to further advance phenology studies by offering 5 day repeat acquisitions with 10 m resolution capabilities and a 20 m red edge band (Clark, 2017; Vrieling et al., 2018).

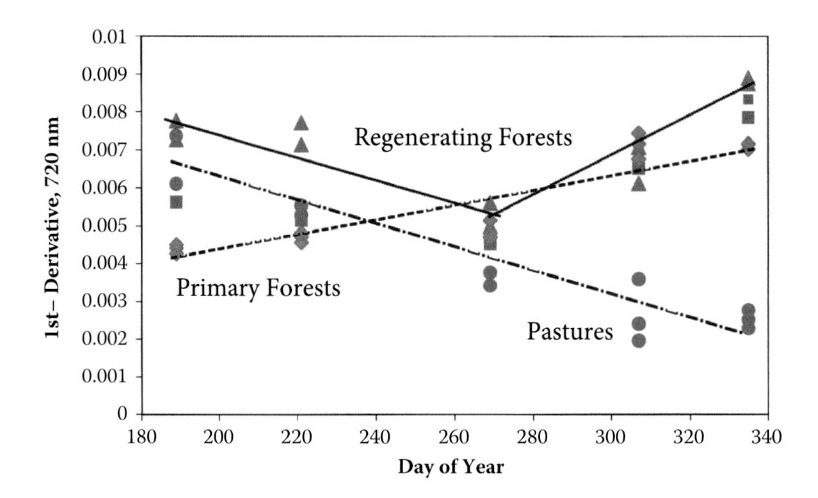

FIGURE 7.3 Dry season phenological profiles of primary forest, regenerating forest, and pasture, depicted by first derivative peak values, at 720 nm across the 2001 and 2002 EO-1 Hyperion acquisition dates. Sites near Tapajos flux tower site. (Modified from Huete, A. et al. 2008. In Kalacska, M., Sanchez-Azofeifa, G. (Ed.), *Hyperspectral Remote Sensing of Tropical and Sub-Tropical Forests* (pp. 233–259). Boca Raton: CRC Press.)

Hyperspectral moderate-resolution applications to landscape phenology have been largely carried out from airborne platforms, such as the Airborne Visible Infrared Imaging Spectrometer (AVIRIS) and the spaceborne EO-1 Hyperion sensor. Hyperspectral data add increased spectral fidelity and enable a wide range of hyperspectral tools, including higher-order derivative analysis (Tsai and Philpot, 1998), band ratioing and continuum removal, lignocellulose absorption index (Elvidge, 1990), mixture modeling (Asner and Lobell, 2000), chlorophyll indices (Blackburn, 1998), and spectral physiological indices such as the photochemical reflectance index (PRI) (Gamon et al., 1997).

Dennison and Roberts (2003) used multiple endmember spectral mixture analysis (MESMA) from AVIRIS to explore seasonal changes of key species in a chaparral woodland biome in California, USA. They found significant seasonal spectral changes, consistent with increases in nonphotosynthetic vegetation (NPV) during dry seasonal periods; however, such variations in NPV inhibited the discrimination between the chaparral species. Knowledge of phenological optical variations were deemed important for discriminating chaparral species.

Miura et al. (2003) reported clear differences in diagnostic vegetation absorption features, such as the red edge and lignocellulose absorptions, enabling the biochemical characterization and discrimination of Brazilian cerrado physiognomies (grass, shrub, tree) with moderate-resolution EO-1 Hyperion imagery. Huete et al. (2008) found unique phenological patterns of pastures, secondary forests, and intact forests in Amazon tropical forests from seasonal Hyperion images. The red edge first derivatives showed different phenology responses during the more sunlit dry season, such that intact forests increased in greenness, pastures and disturbed forests decreased in greenness, and regenerating forests showed initial declines in the early dry season followed by increasing greenness during the latter part of the dry season (Figure 7.3). The mixed response from regenerating forests was attributed to the mixed functional types (herbaceous understory and developing forest layers), in which the shallow-rooted, drying herbaceous layer drove greenness declines, after which the herbaceous layer remained stable, revealing the increasing greenness of the forest (Huete et al., 2008).

7.2.4 Phenology—Biophysical Variable Dependencies

Many studies have shown significant differences in the retrieved phenology metrics from variations in biophysical variables used to trace the growing season profile. Croft et al. (2014) investigated the spatiotemporal variability of chlorophyll content and LAI across a growing season and showed the SGS for LAI to be 1 month earlier than SGS based on chlorophyll content, while the length of the growing season varied by 2 months, as the LAI-measure of the EGS was 20–30 days later then the chlorophyll-measure of EGS. Yang et al. (2014) found a mismatch between camera-derived phenological metrics based on canopy greenness (green chromatic coordinate, or GCC) and the total chlorophyll concentration with the seasonal peak of GCC approximately 20 days earlier than the peak of the total chlorophyll concentration.

7.2.5 Solar-Induced Chlorophyll Fluorescence

Some of the solar radiation absorbed by chlorophyll in photosynthetic plants can be dissipated as heat or reradiated at longer wavelengths in the 650–850 nm region. This reemitted NIR light is known as solar-induced chlorophyll fluorescence (SIF) and, as a byproduct of photosynthesis, represents a more direct measure of photosynthesis (Guanter et al., 2012). SIF is an independent measurement that is more dynamic than greenness and chlorophyll measures, as it will respond more quickly to environmental stress (Porcar-Castell et al., 2014) and yield photosynthetic phenological profiles that are distinct to VI phenological profiles (Song et al., 2018).

The first global maps of SIF time series data were derived using data from the Greenhouse Gases Observing Satellite (GOSAT) (Guanter et al., 2012). The Global Ozone Monitoring Experiment-2 (GOME-2) instrument on board the MetOp-A platform (Joiner et al., 2013) offers SIF data at a finer resolution than GOSAT, yet still fairly coarse (40 km), and more recently, the Orbiting Carbon Observatory-2 (OCO-2) mission has nonimage SIF data with dimensions of 1.29×2.25 km (Sun et al., 2017).

7.3 LEAF- AND PLANT-SCALE PHENOLOGY

Fundamentally, landscape phenology encompasses the aggregate phenology of individual leaves, leaf demography and turnover, plant species composition, and canopy structural changes. Multiple species impart a spatial aspect when interpreting temporal phenological profiles, and many difficulties remain in the interpretation of the dominant sources of phenological variability across scales, from leaves to canopies and landscapes (Wu et al., 2016a,b; de Keyzer et al., 2017).

7.3.1 Leaf Age and Ontogeny

Leaves undergo structural and biochemical changes during their life cycle that generate complex changes in leaf spectroscopy associated with leaf age, ontogeny (development), chemical composition, cell structure, senescence, and physiology (Figure 7.4).

Variations in pigment content (chlorophyll, carotenoids, and anthocyanins) result in leaf color variations that provide information about the leaf phenology. In most dicotyledonous plants, expanding leaves show increases in area, thickness, chlorophyll content, and photosynthesis rates before leaves fully expand; however, certain anthocyanin-producing rainforest trees and African savanna trees show a contrasting ontogenetic pattern of "delayed" greening, with maximum levels of chlorophyll delayed until sometime after the leaves are fully expanded (Choinski et al., 2003). Young leaves of many tropical and subtropical tree species exhibit red coloration caused by the accumulation of anthocyanins (purple/red pigments) that protect young leaves from too much radiation (Sims and Gamon, 2002). There are also plant species with a transient red-coloration phase during the early stages of leaf expansion, such as in certain oak tree species, eucalyptus species, and *Corymbia gummifera* in Australia. Generally, as young leaves expand and develop, there is greater absorptance of visible light, accompanied by lower transmittance and near constant reflectance (Choinski et al., 2003).

Guyot (1990) reported that leaf optical properties of annual plants and deciduous trees changed significantly only during juvenile stages and senescence, remaining constant for most of their life cycle. During leaf senescence, chlorophyll pigments decline more rapidly than carotenoids (yellow

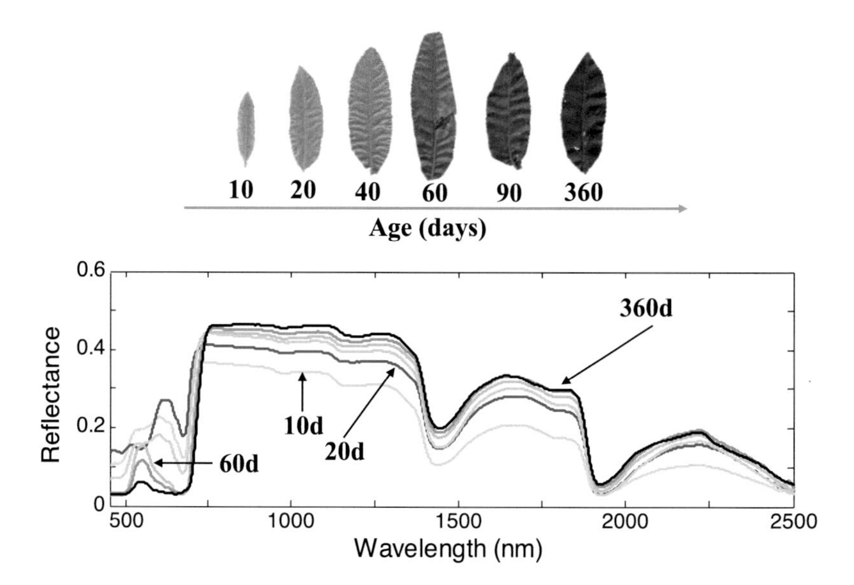

FIGURE 7.4 Leaf reflectance spectra of Amazon tree species *Endopleura uchi* over a 1 year lifespan, as measured with an Analytical Spectral Devices © (ASD) hand-held spectroradiometer. (Courtesy Jin Wu.)

pigments) (Gitelson et al., 2003). Sims and Gamon (2002) analyzed young, mature, and senesced leaves from a wide range of plant functional types and demonstrated how various spectral indices could be used to quantify pigments levels and leaf color relationships. To improve upon spectral index relationships with pigment content across confounding variations in leaf structural characteristics, they developed a set of modified spectral indices, based on use of the 705 nm red edge. They found winter deciduous and evergreen functional type groups produced good correlations between SR705 and chlorophyll, whereas annuals and drought deciduous groups benefitted by use of the modified set of spectral indices, which accounted for highly reflective epidermal surfaces found in these functional groups.

Albert et al. (2018) investigated leaf chemistry seasonal variations in dominant tree species in the Amazon rainforest and reported strong leaf age dependencies of chlorophyll a, b ratios (Chl a:b), and this ratio was found to be approximately 30% higher in young leaves than in mature leaves and 37% higher in mature leaves compared with old leaves. Total chlorophyll was 90% higher in mature than in young leaves, but older leaves had 15% more chlorophyll than mature leaves. A leaf demography analysis showed that leaf age composition varied greatly during the dry season, with new leaves produced before old leaves dropped. This was consistent with the leaf demography–ontogeny hypothesis that evergreen rainforest trees exchange old leaves (with low photosynthetic capacity) for recently mature leaves (with high photosynthetic capacity) during the dry season while maintaining high LAI (Wu et al., 2016b). This demonstrated the importance of leaf-scale phenology in explaining forest-scale patterns of dry season photosynthesis.

Chavana-Bryant et al. (2017) developed a predictive, partial least-squares regression model based on leaf spectra measurements of 1099 leaves from 12 Amazonian canopy trees in southern Peru. They found leaf aging to be a fundamental driver of changes in leaf traits, thereby regulating ecosystem processes and remotely sensed canopy dynamics. Their spectra-based model was more accurate in predicting leaf age compared with trait-based models, and they found narrowband spectral indices, including NDVI, EVI, normalized difference water index (NDWI), and PRI were all leaf-age-dependent. Their study, as well as others (Wu et al., 2017), have highlighted the importance of leaf age in canopy spectroscopy and satellite-derived phenology.

Thus, at the tree crown scale, plant functional traits, leaf age composition, and crown structure jointly determine canopy spectroscopy, while at the landscape scale, leaf age demography, relative abundance of deciduous trees over evergreen trees, and understory dynamics dominate satellite phenological observations (Wu et al., 2016a, 2017).

7.4 DETECTING LIFE CYCLE EVENTS

Although there are strong spectral contrasts across plant phenophases associated with leaf coloration changes, budbreak, onset of new leaves, and flowering, satellite retrieval of such key life cycle events remains quite challenging, even using hyperspectral data. Carvalho (2013) showed how hyperspectral data could be used to discriminate plant phenological stages and found seasonal changes in both leaves and flower spectra. Visible red wavelengths (650–680 nm) provide particularly good separability between flowering and green-leaved canopies (Sims and Gamon, 2002). Ge et al. (2006) reported species-specific increases in plant carotenoids due to flowering with associated increases in blue spectral absorption (450–480 nm).

Flowering status is an important ecological indicator for monitoring plant phenology. Detection of flowering plants is important in pollination studies. Chen et al. (2009) studied flowering times and flower coverage on the Tibetan Plateau as an indicator of ecological processes. Using in situ hyperspectral data, they developed a hyperspectral flower index (HFI) to detect the yellow flowering color of *Halerpestes tricuspis* in an alpine meadow, with the best detection reported when flower coverage was greater than 0.10. Khwarahm et al. (2017) used SGS and PGS phenology information derived from MERIS MTCI to generate a predictive model that enabled mapping of birch and grass flowering across the U.K.

Certain invasive plant species can be detected using their floral signals, and studies have shown invasive species to be discernable from co-occurring vegetation during peaks in the flowering season (Hestir et al., 2008). However, there remain important limitations in the use of hyperspectral data for flowering mapping, including (1) limitations of hyperspectral data for flower mapping over large areas due to spectral mixing and confusion with leaf browning; (2) short-term flowering response cycles, in which 16 day satellite revisit cycles, such as from Hyperion, are insufficient to capture critical flowering periods; for example, Jia et al. (2011) showed that the duration of flowering in poppy seeds was only 2–3 weeks; and (3) the subtleness of the floral signal in many species, particularly grasses. These limitations make it challenging to effectively produce complete floral cycle maps for a given landscape.

Yang et al. (2014) found that senescing leaf redness in deciduous trees was significantly correlated with spectral indices of anthocyanin concentration, enabling new ways to quantify vegetation senescence remotely. On the other hand, confusion can arise during the plant drying phenophase of grasses and dry region plants, in which leaves are neither fully green nor senescent, but fall along a continuum.

7.5 PHENOLOGY SPECIES DISCRIMINATION

Several studies have utilized the unique spectral-phenology patterns of land-cover types to generate land-use/land-cover classification maps from broadband and hyperspectral data (Somers and Asner, 2012, 2013; Dudley et al., 2015). Phenology differences are effective for identifying plant species, and knowledge of plant phenology can help resolve optimal temporal windows to aid in the spectral discrimination of various species and functional types (Bradley, 2014).

Cole et al. (2014) used field spectroscopy to establish a spectral library of phenology variations across key moorland species to determine suitable temporal windows for monitoring upland peatland systems. They used narrowband hyperspectral indices, red edge position (REP) and PRI, during greenup and the cellulose absorption index (CAI) and plant senescence reflection index (PSRI) during senescence. By analyzing spectral response changes with phenology and degree of spectral variation among species, they found the months of April and July to be most suitable for species separability. Ghiyamat and Shafri (2010) used hyperspectral analysis of the chemical composition of foliage and tree phenology to study forest biodiversity and further applied wavelet transforms to identify tree species.

Dillen et al. (2012) showed leaf traits to vary significantly between two co-occurring tree species, red oak (*Quercus rubra*) and paper birch (*Betula papyrifera*), throughout a growing season. Interactions between paper birch and red oak species partly contributed to their contrasting spring phenologies, in which paper birch was characterized by much earlier early bud break and rapid leaf expansion.

Challenges remain on how to use multidate hyperspectral data for the detection of coastal and marine species that present different types of phenological information. Hestir et al. (2015) conducted a comprehensive overview of the potential and need for the global mapping satellite mission (Hyperspectral Infrared Imager, HyspIRI) for measuring and mapping freshwater aquatic ecosystems. Their study found that the temporal resolution of HyspIRI was suitable for characterizing the growing season of wetland species phenology in temperate regions and would allow the identification and mapping of macrophyte communities but was found to be unsuitable for tracking algal blooms.

Phenology-based spectral discrimination methods can be applied to invasive species detection, weeds, as well as discerning intra-species variations. Differences in phenological stages between two plants are often utilized in hyperspectral imagery to discern invasive weeds from other landscape elements and plants. Hyperspectral detection of invasive species relies on knowledge of key phenology periods for optimal discrimination of spectral differences between species and the target of interest. Bradley (2014) reviews when the phenology approach is most useful or most effective for identifying invasive species. Ge et al. (2006) were able to identify and map the different flowering stages of the invasive weed, yellow starthistle (*Centaurea solstitialis*), through spectral characterization and modeling of canopy components (stems, buds, opening flowers, and postflowers), using hyperspectral data.

Gómez-Casero et al. (2010) were able to discriminate wild oat and the weed canary grass based on hyperspectral data applied only during late-season phenology stages. Hestir et al. (2008) found that both the invasive pepperweed and water hyacinth exhibited significant spectral variation related to plant phenology, and that their discrimination based on phenology influences was moderately successful. In a separate case study using airborne hyperspectral data (HyMap), Hestir et al. (2008) were successful in identifying invasive vegetation based on the unique phenologies of three separate species, however in multiple-date applications the accuracy was variable across dates, hindering the determination of phenological stages for each species.

7.6 INFLUENCE OF SPECIES DIVERSITY ON PHENOLOGY

Plant species biodiversity has a large influence on satellite-observed landscape phenological profiles. Species composition with varying phenologies and leaf demographies aggregate to produce a larger-scale "mixed" phenology that is recorded by satellite sensors. Whereas species phenology variations may be useful in discriminating species, the aggregate signal from multiple species can confuse the interpretation of phenology. This is referred to as the spatiotemporal duality of phenological data (de Keyzer et al., 2017), which causes ambiguities to arise from species' presence.

Climate as well as land-cover changes can influence phenology. Thus, a shift in species dominance can alter satellite-measured phenology and be confused with a climate-change-induced phenology shift. Exotic species or the emerging presence of an invasive species can gain a foothold in a landscape and lead to an apparent shift in phenology (Wilsey et al., 2018). Compositional shifts in C_3 and C_4 species, attributed to elevated CO_2 levels, will impact the phenology of grassland areas and LGS. Wilsey et al. (2018) reported that native (C_4) to exotic (C_3) plant conversions in central U.S. grasslands have led to highly altered phenology patterns, as the senescence phenophase arose on average 36 days later in exotic than native-dominated grasslands. They noted that this conversion would have to be considered in developing estimates of how global change will affect phenology in

FIGURE 7.5 Phenocam images showing dry season phenology dynamics at individual tree crown and leaf scales, while whole canopy maintains an evergreen appearance year round. (From 2013–2014, Central Amazonian rainforests (ATTO) (Lopes et al., 2016). (Data courtesy of Bruce Nelson, INPA.)

locations where exotics are present, and particularly in cases where their abundance is increasing concurrently with climate change.

A highly diverse tropical forest tree canopy can also result in a range of "apparent" phenology responses, depending on the extent to which dominant species are synchronous or asynchronous with the same or common environmental factors (Figure 7.5). In a regenerating forest, the canopy phenological profile represents mixtures of functionally diverse, herbaceous, and developing tree layers (Figure 7.3), each with unique phenological responses to dry and wet seasonal periods (Huete et al., 2008).

Marine and coastal ecosystems are among the most highly species diverse and productive of Earth's ecosystems, encompassing mangrove, seagrass meadows, algae, coral reef, and plankton biologic communities. The phenology of the diverse life forms that grow in these habitats often change rapidly, from day to day, in wet and dry seasons and over short distances. A quantitative study of phenology is required to provide the information that can be used to assess their complex drivers in order to protect and manage wetlands and other coastal resources (Osterman et al., 2016).

It remains challenging to use multidate hyperspectral data for monitoring the phenology and changes in plant communities and species composition in marine and coastal areas. The spectral properties of coastal and marine plants are a combination of the optical properties of individual vegetative components, effects of growth forms, density, height, tidal stage, and soil type, as well as water chemistry and turbidity variations. Hyperspectral remote sensing is an important technique to discriminate the main vegetative types, such as mangrove species (Kamal and Phinn, 2011; Koedsin and Vaiphasa, 2013) and wetland species (Hirano et al., 2003; Pengra et al., 2007; Jollineau and Howarth, 2008; Adam et al., 2010; Turpie et al., 2015). Due to the complexity of coastal ecosystems, there is a need for accurate spatial, spectral, radiometric, and temporal image data and information extraction techniques to process high-dimensional hyperspectral remote sensing data.

7.7 CHALLENGES AND FUTURE DIRECTIONS

In this chapter, we explored the optical phenology patterns of variability from leaf to landscape scales and focused on the use of hyperspectral remote sensing to improve our understanding and ecological interpretation of phenology patterns and responses to biotic and abiotic drivers. An understanding of phenology is a prerequisite to interannual studies and predictive modeling of land surface responses to climate change.

High-temporal-frequency measurements are critical in obtaining sufficient amounts of cloud-free data and achieve greater sensitivity to phenological timing retrievals. Improved temporal fidelity would be useful for more accurate retrieval of the timing of budbreak, flowering, and peak greenness, which even coarse-resolution, high-temporal-frequency satellites may not capture because of compositing methods that aggregate data from weekly to monthly intervals. Flowering events and peak greenness periods are short-lived and may be missed entirely in current phenology products. Improved timing accuracies would be needed for many applications, for example, flowering maps and pollen alerts.

Advancements in terms of more accurate phenology assessments require information at adequate spectral, spatial, and temporal resolutions. Phenology studies at finer spatial and spectral resolutions are still relatively scarce, yet they are valuable for understanding species mixture effects on landscape phenology and serve to resolve ambiguities arising from the spatiotemporal duality of phenology signals. Hyperspectral spatial domains are important for discriminating key species and for better understanding the presence and influence of mixed species. Only in realizing the spectral-spatial detail present in fine-resolution hyperspectral data can one effectively interpret and characterize phenological variations in coarser-scale satellite imagery. Lastly, hyperspectral data have the capability of providing biophysical/biochemical information to existing greenness-based phenological profiles for better interpretation of phenology signals and their drivers and controls. Hyperspectral remote sensing provides opportunities to advance phenological science by improving upon the detection of life cycle events, species detection, and biophysical retrievals, thereby bridging the many gaps limiting phenological assessments in coarse-resolution satellite data.

Future directions to advance phenology studies involve the linking of satellite imagery with ground-based sensors that measure plant- and canopy-level processes, such as the expanded use of in situ sensor platforms capable of long-term ecosystem monitoring and analysis. One example is the Mobile Instrumented Sensor Platform (MISP) deployed in the Arctic tundra (Healey et al., 2014). This versatile robotic sensor system can detect and monitor plant-to-ecosystem changes and shifts in productivity, along with species composition, phenology, structure, and function over seasonal, interannual, and decadal time scales.

Unmanned autonomous vehicle (UAV) technology allows flexibility in providing timely data and the level of resolution can be selected to meet complex phenological requirements involving unique spatial, spectral, and temporal resolutions. There have been advances in research and development of hyperspectral sensors specifically designed or adapted for UAV payloads. UAV essentially has the capability of effectively filling in current observational gaps in environmental remote sensing applications, including phenology. They provide crucial information needed for monitoring coastline change, wetland mapping, ecosystem monitoring, and the phenology of coastal and marine plants (Klemas, 2015).

A study by Berra et al. (2016) showed that the GCC data derived from UAV RGB images could be used to estimate individual tree phenology parameters (such as start of season) in tropical situations. UAVs with hyperspectral cameras have been used to successfully map mangroves, wetland vegetation communities, and other aquatic vegetation (Husson et al., 2013; Tian et al., 2017; Cao et al., 2018), but there have yet to be UAV studies utilizing multidate hyperspectral data in coastal applications. These research gaps can ideally test the capabilities of multitemporal hyperspectral data from UAVs for monitoring coastal vegetation phenology.

The Phenological Eyes Network (PEN), established in 2003, is a network of long-term ground observation sites to validate remotely sensed ecological dynamics, in particular, seasonal/phenological changes in vegetation (Nasahara and Nagai, 2015). This includes an automatic digital fish-eye camera, a hemispherical spectroradiometer, and a sun photometer. There are over 30 PEN sites, many of which are integrated with FluxNet or International Long Term Ecological Research (ILTER) sites. The collection of in situ, diurnal, and daily hyperspectral data is of immense value in understanding the capabilities of hyperspectral data in all aspects of phenological science, from leaf and species to individual tree crowns and plant canopy communities. These cross-site long-term monitoring infrastructures facilitate collaborations between remote sensing scientists and ecologists.

REFERENCES

Adam, E., Mutanga, O., Rugege, D. 2010. Multispectral and hyperspectral remote sensing for identification and mapping of wetland vegetation: A review. *Wetlands Ecology and Management* 18, 281–296.

Albert, L. P., Wu, J., Prohaska, N., de Camargo, P. B., Huxman, T. E., Tribuzy, E. S., ... Saleska, S. R. 2018. Age-dependent leaf physiology and consequences for crown-scale carbon uptake during the dry season in an Amazon evergreen forest. *New Phytologist* Mar 4. http://doi.org/10.1111/nph.15056

Anderson, L. O., Aragão, L. E. O. C., Shimabukuro, Y. E., Almeida, S., Huete, A. 2011. Fraction images for monitoring intra-annual phenology of different vegetation physiognomies in Amazonia. *International Journal of Remote Sensing* 32(2), 387–408. https://doi.org/10.1080/01431160903474921

Asner, G. P., Lobell, D. B. 2000. A biogeophysical approach for automated SWIR unmixing of soils and vegetation. *Remote Sensing of Environment* 74, 99.

Berra, E. F., Gaulton, R., Barr, S. 2016. Use of a digital camera onboard a UAV to monitor spring phenology at individual tree level. In *2016 IEEE International Geoscience and Remote Sensing Symposium (IGARSS). Presented at the 2016 IEEE International Geoscience and Remote Sensing Symposium (IGARSS)*, pp. 3496–3499. https://doi.org/10.1109/IGARSS.2016.7729904

Blackburn, G. A. 1998. Quantifying chlorophylls and carotenoids at leaf and canopy scales: An evaluation of some hyperspectral approaches. *Remote Sensing of Environment* 66, 273–285.

Bradley, B. A. 2014. Remote detection of invasive plants: A review of spectral, textural and phenological approaches. *Biological Invasions* 16(7), 1411–1425. http://doi.org/10.1007/s10530-013-0578-9

Cao, J., Leng, W., Liu, K., Liu, L., He, Z., Zhu, Y. 2018. Object-based mangrove species classification using unmanned aerial vehicle hyperspectral images and digital surface models. *Remote Sensing* 10, 89. https://doi.org/10.3390/rs10010089

Carvalho, S., Schlerf, M., van der Puttena, W. H., Skidmore, A. K. 2013. Hyperspectral reflectance of leaves and flowers of an outbreak species discriminates season and successional stage of vegetation. *International Journal of Applied Earth Observation and Geoinformation* 24(1), 32–41. https://doi.org/10.1016/j.jag.2013.01.005

Chavana-Bryant, C., Malhi, Y., Wu, J., Asner, G. P., Anastasiou, A., Enquist, B. J., ... Gerard, F. F. 2017. Leaf aging of Amazonian canopy trees as revealed by spectral and physiochemical measurements. *New Phytologist* 214(3), 1049–1063. http://doi.org/10.1111/nph.13853

Chen, J., Shen, M., Zhu, X., Tang, Y. 2009. Indicator of flower status derived from *in situ* hyperspectral measurement in an alpine meadow on the Tibetan Plateau. *Ecological Indicators* 9(4), 818–823. http://doi.org/10.1016/j.ecolind.2008.09.009

Choinski Jr, J. S., Ralph, P., Eamus, D. 2003. Changes in photosynthesis during leaf expansion in *Corymbia gummifera*. *Australian Journal of Botany* 51(1), 111–118. http://dx.doi.org/10.1071/BT02008

Clark, M. L. 2017. Comparison of simulated hyperspectral HyspIRI and multispectral Landsat 8 and Sentinel-2 imagery for multi-seasonal, regional land-cover mapping. *Remote Sensing of Environment* 200, 311–325.

Cole, B., McMorrow, J., Evans, M. 2014. Spectral monitoring of moorland plant phenology to identify a temporal window for hyperspectral remote sensing of peatland. *ISPRS Journal of Photogrammetry and Remote Sensing* 90, 49–58. http://doi.org/10.1016/j.isprsjprs.2014.01.010

Croft, H., Chen, J. M., Zhang, Y. 2014. Temporal disparity in leaf chlorophyll content and leaf area index across a growing season in a temperate deciduous forest. *International Journal of Applied Earth Observation and Geoinformation* 33(1), 312–320. http://doi.org/10.1016/j.jag.2014.06.005

de Keyzer, C. W., Rafferty, N. E., Inouye, D. W., Thomson, J. D. 2017. Confounding effects of spatial variation on shifts in phenology. *Global Change Biology* 23(5), 1783–1791. http://doi.org/10.1111/gcb.13472

Dennison, P. E., Roberts, D. A. 2003. The effects of vegetation phenology on endmember selection and species mapping in southern California chaparral. *Remote Sensing of Environment* 87(2–3), 295–309. http://doi.org/10.1016/j.rse.2003.07.001

Dillen, S. Y., de Beeck, M. O., Hufkens, K., Buonanduci, M., Phillips, N. G. 2012. Seasonal patterns of foliar reflectance in relation to photosynthetic capacity and color index in two co-occurring tree species, Quercus rubra and Betula papyrifera. *Agricultural and Forest Meteorology* 160, 60–68. http://doi.org/10.1016/j.agrformet.2012.03.001

Dudley, K. L., Dennison, P. E., Roth, K. L., Roberts D. A., Coates, A. R. 2015. A multi-temporal spectral library approach for mapping vegetation species across spatial and temporal phenological gradients. *Special Issue Hyperspectral Infrared Imager HyspIRI* 167, 121–134.

Elvidge, C. D. 1990. Visible and near-infrared reflectance characteristics of dry plant materials. *International Journal of Remote Sensing* 11, 1775.

Friedl, M., Henebry, G., Reed, B., Huete, A., White, M., Morisette, J., Nemani, R., Zhang, X., Myneni, R. 2006. Land Surface Phenology. A Community White Paper requested by NASA. April 10. ftp://zeus.geog.umd.edu/Land_ESDR/Phenology_Friedl_whitepaper.pdf

Gamon, J. A., Serrano, L., Surfus, J. S. 1997. The photochemical reflectance index: An optical indicator of photosynthetic radiation use efficiency across species, functional types, and nutrient levels. *Oecologia* 112, 4921.

Ge, S., Everitt, J., Carruthers, R., Gong, P., Anderson, G. 2006. Hyperspectral characteristics of canopy components and structure for phenological assessment of an invasive weed. *Environmental Monitoring and Assessment* 120(1–3), 109–126. http://doi.org/10.1007/s10661-005-9052-1

Ghiyamat, A., Shafri, H. Z. M. 2010. A review on hyperspectral remote sensing for homogeneous and heterogeneous forest biodiversity assessment. *International Journal of Remote Sensing* 31(7), 1837–1856. http://doi.org/10.1080/01431160902926681

Gitelson, A. A., Gritz, Y., Merzlyak, M. N. 2003. Relationships between leaf chlorophyll content and spectral reflectance and algorithms for non-destructive chlorophyll assessment in higher plant leaves. *Journal of Plant Physiology* 160(3), 271–282. http://www.ncbi.nlm.nih.gov/pubmed/12749084

Guanter, L. et al. 2012. Retrieval and global assessment of terrestrial chlorophyll fluorescence from GOSAT space measurements. *Remote Sensing of Environment* 121, 236–251.

Guyot, G. 1990. Optical properties of vegetation canopies. In *Applications of Remote Sensing in Agriculture*. Sevenoaks, Kent: Butterworths. https://www.cabdirect.org/cabdirect/abstract/19920753700

Gómez-Casero, M. T., Castillejo-González, I. L., García-Ferrer, A., Peña-Barragán, J. M., Jurado-Expósito, M., García-Torres, L., López-Granados, F. 2010. Spectral discrimination of wild oat and canary grass in wheat fields for less herbicide application. *Agronomy for Sustainable Development* 30(3), 689–699. http://doi.org/10.1051/agro/2009052

Healey, N. C., Oberbauer, S. F., Ahrends, H. E., Dierick, D., Welker, J. M., Leffler, A. J., ... Tweedie, C. E. 2014. A mobile instrumented sensor platform for Long-Term terrestrial ecosystem analysis: An example application in an Arctic tundra ecosystem. *Journal of Environmental Informatics* 24(1), 1–10. http://doi.org/10.3808/jei.201400278

Hestir, E. L., Khanna, S., Andrew, M. E., Santos, M. J., Viers, J. H., Greenberg, J. A., ... Ustin, S. L. 2008. Identification of invasive vegetation using hyperspectral remote sensing in the California Delta ecosystem. *Remote Sensing of Environment* 112(11), 4034–4047. http://doi.org/10.1016/j.rse.2008.01.022

Hestir, E. L. et al. 2015. Measuring freshwater aquatic ecosystems: The need for a hyperspectral global mapping satellite mission. *Special Issue Hyperspectral Infrared Imager Hyspiri* 167, 181–195.

Hirano, A., Madden, M., Welch, R. 2003. Hyperspectral image data for mapping wetland vegetation. *Wetlands* 23, 436–448. https://doi.org/10.1672/18-20

Huete, A. R., Didan, K., Shimabukuro, Y. E., Ratana, P., Saleska, S. R., Hutyra, L. R., ... Myneni, R. 2006. Amazon rainforests green-up with sunlight in dry season. *Geophysical Research Letters*, 33(6), 2–5. https://doi.org/10.1029/2005GL025583

Huete, A., Kim, Y., Ratana, P., Didan, K., Shimabukuro, Y., Miura, T. 2008. Assessment of phenologic variability in Amazon tropical rainforests using hyperspectral hyperion and MODIS satellite data. In Kalacska, M., Sanchez-Azofeifa, G. (Ed.), *Hyperspectral Remote Sensing of Tropical and Sub-Tropical Forests* (pp. 233–259). Boca Raton: CRC Press.

Husson, E., Hagner, O., Ecke F., Schmidtlein S. 2013. Unmanned aircraft systems help to map aquatic vegetation. *Applied Vegetation Science* 17, 567–577. https://doi.org/10.1111/avsc.12072

Jia, K., Wu, B., Tian, Y., Li, Q., Du, X. 2011. Spectral discrimination of opium poppy using field spectrometry. *IEEE Transactions on Geoscience and Remote Sensing* 49(9), 3414–3422. http://doi.org/10.1109/TGRS.2011.2126582

Joiner, J. et al. 2013. Global monitoring of terrestrial chlorophyll fluorescence from moderate spectral resolution near-infrared satellite measurements: Methodology, simulations, and application to GOME-2. *Atmospheric Measurement Techniques* 6(2), 3883–3930.

Jollineau, M. Y., Howarth, P. J. 2008. Mapping an inland wetland complex using hyperspectral imagery. *International Journal of Remote Sensing* 29, 3609–3631.

Kamal, M., Phinn, S. 2011. Hyperspectral data for mangrove species mapping: A comparison of pixel-based and object-based approach. *Remote Sensing* 3, 2222–2242.

Kauth, R. J., Thomas, G. S. 1976. The tasseled cap. A graphic description of the spectral-temporal development of agricultural crops as seen by Landsat. *2nd International Symposium on Machine Processing of Remotely Sensed Data*, Purdue University, West Lafayette.

Khwarahm, N. R., Dash, J., Skjoth, C. A., Newnham, R. M., Adams-Groom, B., Head, K., … Atkinson, P. M. 2017. Mapping the birch and grass pollen seasons in the UK using satellite sensor time-series. *Science of the Total Environment* 578, 586–600. http://doi.org/10.1016/j.scitotenv.2016.11.004

Klemas, V. V. 2015. Coastal and environmental remote sensing from unmanned aerial vehicles: An overview. *Journal of Coastal Research* 315, 1260–1267. https://doi.org/10.2112/JCOASTRES-D-15-00005.1

Koedsin, W., Vaiphasa, C. 2013. Discrimination of tropical mangroves at the species level with EO-1 Hyperion data. *Remote Sensing* 5, 3562–3582.

Lopes, A. P., Nelson, B. W., Wu, J., de Alencastro Graça, P. M. L., Tavares, J. V., Prohaska, N., Martins, G. A., Saleska, S. R. 2016. Leaf flush drives dry season green-up of the Central Amazon. *Remote Sensing of Environment* 182, 90–98.

Ma, X. et al. 2013. Spatial patterns and temporal dynamics in savanna vegetation phenology across the north Australian tropical transect. *Remote Sensing of Environment* 139, 97–115. http://doi.org/10.1016/j.rse.2013.07.030

Miura, T. et al. 2003. Discrimination and biophysical characterization of Brazilian Cerrado physiognomies with EO-1 hyperspectral Hyperion. In Green, R. E. (Ed.), *12th JPL Airborne Earth Science Workshop.* NASA, Jet Propulsion Laboratory, California Institute of Technology, 207.

Myneni, R. B., Keeling, C. D., Tucker, C. J., Asrar, G., Nemani, R. R. 1997. Increased plant growth in the northern high latitudes from 1981 to 1991. *Nature* 386, 698–702.

Nasahara, K. N., Nagai, S. 2015. Review: Development of an *in situ* observation network for terrestrial ecological remote sensing: The Phenological Eyes Network (PEN). *Ecological Research* 30(2), 211–223. http://doi.org/10.1007/s11284-014-1239-x

Osterman, S. N. et al. 2016. A spaceborne visible-NIR hyperspectral imager for coastal phenology. In *Sensors, Systems, and Next-Generation Satellites XX. International Society for Optics and Photonics*, 10000, 100001T-10000–13.

Pau, S., Okin, G. S., Gillespie, T. W. 2010. Asynchronous response of tropical forest leaf phenology to seasonal and el Niño-driven drought. *Plos One* 5(6), e11325. http://doi.org/10.1371/journal.pone.0011325

Pengra, B. W., Johnston, C. A., Loveland, T. R. 2007. Mapping an invasive plant, Phragmites australis, in coastal wetlands using the EO-1 Hyperion hyperspectral sensor. *Remote Sensing of Environment* 108, 74–81.

Porcar-Castell, A., Tyystjärvi, E., Atherton, J., van der Tol, C., Flexas, J., Pfündel, E. E., Moreno, J., Frankenberg, C., Berry, J. A. 2014. Linking chlorophyll a fluorescence to photosynthesis for remote sensing applications: Mechanisms and challenges. *Journal of Experimental Botany* 65(15), 4065–4095.

Roy, D. P., Ju, J., Lewis, P., Schaaf, C., Gao, F., Hansen, M., Lindquist, E. 2008. Multi-temporal MODIS–Landsat data fusion for relative radiometric normalization, gap filling, and prediction of Landsat data. *Remote Sensing of Environment* 112(6), 3112–3130. http://www.sciencedirect.com/science/article/pii/S0034425708001065

Sims, D., Gamon, J. 2002. Relationships between leaf pigment content and spectral reflectance across a wide range of species, leaf structures and developmental stages. *Remote Sensing of Environment*, 81(2–3), 337–354. http://doi.org/10.1016/S0034-4257(02)00010-X

Somers, B., Asner, G. P. 2012. Hyperspectral time series analysis of native and invasive species in Hawaiian rainforests. *Remote Sensing* 4(9), 2510–2529.

Somers, B., Asner, G. P. 2013. Invasive species mapping in Hawaiian rainforests using multi-temporal Hyperion spaceborne imaging spectroscopy. *IEEE Journal of Selected Topics in Applied Earth Observations and Remote Sensing* 6(2), 351–359, Apr. 2013.

Song, L., Guanter, L., Guan, K., You, L., Huete, A., Ju, W., Zhang, Y. 2018. Satellite sun-induced chlorophyll fluorescence detects early response of winter wheat to heat stress in the Indian Indo-Gangetic Plains. *Global Change Biology May 10.* http://doi.org/10.1111/gcb.14302

Sun, Y., Frankenberg, C., Wood, J. D., Schimel, D. S., Jung, M., Guanter, L., … Yuen, K. 2017. OCO-2 advances photosynthesis observation from space via solar-induced chlorophyll fluorescence. *Science* 358(6360).

Tian, J., Wang, L., Li, X., Gong, H., Shi, C., Zhong, R., Liu, X. 2017. Comparison of UAV and WorldView-2 imagery for mapping leaf area index of mangrove forest. *International Journal of Applied Earth Observation and Geoinformation* 61, 22–31. https://doi.org/10.1016/j.jag.2017.05.002

Tsai, F., Philpot, W. 1998. Derivative analysis of hyperspectral data. *Remote Sensing of Environment* 66, 41.

Tucker, C. J., Slayback, D., Pinzon, J. E., Los, S. O., Myneni, R. B., Taylor, M. G. 2001. Higher northern latitude normalized difference vegetation index and growing season trends from 1982 to 1999. *International Journal of Biometeorology* 45(4), 184–190.

Turpie, K. R., Klemas, V. V., Byrd, K., Kelly, M., Jo, Y.-H. 2015. Prospective HyspIRI global observations of tidal wetlands. *Remote Sensing of Environment, Special Issue on the Hyperspectral Infrared Imager (HyspIRI)* 167, 206–217. https://doi.org/10.1016/j.rse.2015.05.008

Vrieling, A., Meroni, M., Darvishzadeh, R., Skidmore, A. K., Wang, T., Zurita-Milla, R., … Paganini, M. 2018. Vegetation phenology from Sentinel-2 and field cameras for a Dutch barrier island. *Remote Sensing of Environment March*, 0–1. http://doi.org/10.1016/j.rse.2018.03.014

Wilsey, B. J., Martin, L. M., Kaul, A. D. 2018. Phenology differences between native and novel exotic-dominated grasslands rival the effects of climate change. *Journal of Applied Ecology* 55(2), 863–873. http://doi.org/10.1111/1365-2664.12971

Wu, J., Albert, L. P., Lopes, A. P., Restrepo-Coupe, N., Hayek, M., Wiedemann, K. T., … Christoffersen, B. 2016a. Leaf development and demography explain photosynthetic seasonality in Amazon evergreen forests. *Science* 351(6276), 972–976. http://doi.org/10.1126/science.aad5068

Wu, J., Guan, K., Hayek, M., Restrepo-Coupe, N., Wiedemann, K. T., Xu, X., … Saleska, S. R. 2016b. Partitioning controls on Amazon forest photosynthesis between environmental and biotic factors at hourly to interannual timescales. *Global Change Biology Sep 19*. http://doi.org/10.1111/gcb.13509

Wu, J. et al. 2017. Convergence in relationships between leaf traits, spectra and age across diverse canopy environments and two contrasting tropical forests. *New Phytologist* 214, 1033–1048.

Yang, X., Tang, J., Mustard, J. F. 2014. Beyond leaf color: Comparing camera-based phenological metrics with leaf biochemical, biophysical, and spectral properties throughout the growing season of a temperate deciduous forest. *Journal of Geophysical Research: Biogeosciences* 119(3), 181–191. http://doi.org/10.1002/2013JG002460

Zhang, X., Friedl, M., Schaaf, C. B., Strahler, A. H., Hodges, J. C. F., Gao, F., … Huete, A. 2003. Monitoring vegetation phenology using MODIS. *Remote Sensing of Environment*, 84(3), 471–475. https://doi.org/10.1016/S0034-4257(02)00135-9

Section III

Land Cover, Forests, and Wetland and Urban Applications Using Hyperspectral Data

8 The Use of Hyperspectral Earth Observation Data for Land Use/Cover Classification

Present Status, Challenges, and Future Outlook

Prem Chandra Pandey, Kiril Manevski, Prashant K. Srivastava, and George P. Petropoulos

CONTENTS

8.1 INTRODUCTION

Land use/land cover (LULC) are essential variables of the Earth's system that are intimately connected with anthropogenic and physical environments (Chatziantoniou et al., 2017; Otukei and Blaschke, 2010). Explicit spatial information on land cover, that is to say, where things are, is useful for policy decisions on agricultural resource management, environmental and ecological protection, and native habitat mapping and restoration (Ireland and Petropoulos, 2015; Sanchez-Hernandez et al., 2007). In addition, thematic maps of land use/land cover (LULC) are also linked to systems for monitoring desertification and land degradation—key processes evident in many areas on the Earth experiencing ongoing climate change (Lamine et al., 2017; Singh et al., 2016). At local and regional scales, knowledge of LULC represents a basic dimension of resources available in any geographic unit (Kavzoglu and Colkesen, 2009). At a larger scale, LULC information is of key importance in delineating the broad patterns of climate and vegetation that form the environmental context for human activities.

To monitor LULC and their interrelationships, remote sensing techniques have been employed at various scales. Remote sensing measures reflectance in a spectrally contiguous manner from the visible (VIS) to the shortwave infrared (SWIR) parts of the electromagnetic spectrum (400–2500 nm). However, the "contiguous" manner differs markedly between remote sensing technologies. Since the 1960s–1970s, LULC classification involved routine mapping from "less contiguous," that is, multispectral remote sensing data (Petropoulos et al., 2012a). Since the launch of the Landsat missions, the research community has seen a wide range of spaceborne multispectral sensors put in orbit for applications in various research and application themes. Such remote sensing applications have not looked back after their successful implementation in LULC applications and other purposes thanks to their capacity to provide repetitive data over large areas and synoptic coverage even for inaccessible locations at different spatial and temporal scales, often free of charge. Although multispectral images often come with high spatial resolution, their limited spectral resolution, that is, few to several spectral bands, may hamper target identification and detailed characterization of LULC. This limitation may be overcome by the use of hyperspectral remote sensing. Hyperspectral sensors are delicate tools that possess hundreds of bands between 400 and 2500 nm in a virtually continuous manner, aiming to resolve target-specific and detailed spectral responses—a task considered difficult for any multispectral sensor. For instance, vegetation cover and plant species have varying degrees of spectral characteristics depending upon biochemical pigments, intracellular spaces, water contents, diversity developmental stages, and phenological phenomena, while being governed at times by the surrounding background. At a lower canopy level, where spectral predominance is driven by soil backgrounds, the examination of plant species is even more troublesome and often unreliable (Curran, 2001). Thus, it follows that multispectral imagery is preferred for broader, meaning community-level, vegetation cover analysis, whereas detailed studies involving greater discrimination, for example, deciduous vs. evergreen vegetation, are better served by hyperspectral remote sensing (Thenkabail et al., 2004a).

This chapter provides information pertaining to LULC mapping from a view of hyperspectral remote sensing across scales. First, current advancements in spaceborne hyperspectral LULC studies and methods are presented, emphasizing the classification algorithms and associated factors. In this context, case studies with critical reviews are also included. Then field-scale hyperspectral remote sensing and the basics of detailed plant species discrimination are discussed, alongside the main influencing factors and the major methods for data processing. The main focus is on vegetation land cover as the most complex in terms of spectral response and variability. This is followed by a brief critical overview of the recent rapidly increased use of unmanned aerial vehicle (UAV) platforms in remote sensing of LULC. Finally, the chapter concludes with summary information and puts forward the challenges associated with more accurate estimation of land cover using hyperspectral data.

8.2 MULTISPECTRAL VERSUS HYPERSPECTRAL REMOTE SENSORS

Since decades ago, multispectral images have been in use for LULC mapping. There is an extensive literature describing the use of Landsat data in LULC mapping and other objectives such as tracking seasonal droughts on a large scale (Goerner et al., 2009; Thenkabail et al., 2004a; Zhang et al., 2017). Landsat satellites were launched by NASA and so far there are eight, the last (Landsat 8) enabling processing and analysis of multispectral and thermal data at 30 and 100 m spatial resolutions, respectively (Ding et al., 2014). Another well-known multispectral remote sensing platform in space is the Moderate Resolution Imaging Spectroradiometer (MODIS) (500–1000 m spatial resolution), which has also been used extensively in large-scale LULC mapping (Garcia-Mora et al., 2012). Data from these sensors are often considered too coarse for detailed LULC mapping as many Earth targets typically contain features that are smaller than the spatial resolution of these multispectral images (Lu and Weng, 2004). To resolve the spatial resolution issue, high-spatial-resolution images were introduced with advancements in technology, and to provide accurate results according to user requirements, high-spatial-resolution images were incorporated into mapping, such as by IKONOS, QuickBird, and WorldView1 and 2. One of the most recent sensors launched by the European Space Agency is Sentinel-2, a multispectral device with 13 bands from 443 to 2190 nm (Table 8.1) and a 10 day repeat cycle. The three red edge bands seem especially attractive, as this part of the spectrum is known to contain certain information about fine differences in plant pigments; higher chlorophyll content can indicate higher canopy density or complex community structure or higher nitrogen content in plant tissue (Alvarez-Añorve et al., 2008). Laurin et al. (2016) showed high potential for ecological monitoring using simulated Sentinel-2 data for tropical rainforests in West Africa. Despite its potential in terms of good spatial and spectral resolution, studies involving the actual use of Sentinel-2 for agro-environmental investigations are very limited because the satellite only started providing data in late 2015, so researchers have yet to investigate the actual potential of Sentinel-2 for drought stress detection and management.

As already mentioned, the spatial resolution interacts with landscape features, and the issue of "mixed pixels" arises, where several land-use features are contained in a single pixel of an image. Such a mixture becomes especially prevalent in, for example, residential regions where pavement, buildings, trees, lawns, roads, parking lots, and asphalt can all occur within one pixel. Natural or seminatural landscapes composed of mixtures of agricultural crops or various vegetation types such as trees, shrubs, and grasses with varying botanical descriptions are also common in mixed pixels, which affects the LULC mapping efficiency of multispectral remote sensing datasets, despite their high spatial resolution (Cracknell, 1998; Fisher, 1997). This problem might be mitigated by the introduction of hyperspectral imaging systems.

Although both multi- and hyperspectral imaging systems have advantages and disadvantages, hyperspectral imaging systems have proven to have huge potential for better discrimination between different LULCs and for mapping fragmented landscapes compared to multispectral images. Hyperspectral imaging systems, such as Hyperion, HyspIRI, and EnMap, are able to record reflected light from land cover in numerous narrow, virtually continuous spectral bands, thereby acquiring vast amounts of spectral information observed from higher altitudes (Ben-Dor et al., 2013). Hyperspectral sensors capture data by spatial and spectral scanning, nonscanning, or snapshots and spatiospectral scanning. This approach has been demonstrated by several researchers and scientists (Petropoulos et al., 2011, 2012b; Zomer et al., 2009). Examples of varying contrasts in spectral, spatial, and temporal characteristics between several spaceborne multispectral and hyperspectral imaging techniques are presented in Table 8.1.

NASA launched the first successful civilian hyperspectral satellite sensor, Hyperion EO-1, in Earth's orbit in November 2000 (Pearlman, 2003). The Compact High Resolution Imaging Spectrometer (CHRIS) followed this successful launch in 2001 on the European Space Agency's (ESA) (PROBA) platform (Barnsley et al., 2004). Since the launch of NASA's Hyperion on the EO-1 platform, hyperspectral imaging has seen a boost in development and placement in Earth's orbit for observation and analysis. The Hyperion EO-1 sensor was mainly launched to prepare the mineral spectral library and mineral mapping by the U.S. Geological Survey (USGS) (Vorovencii, 2009), whereas the ESA's CHRISPROBA was used to gather bidirectional reflectance distribution function (BRDF) information for enhanced knowledge of spectral reflectance. The Hyperion EO-1 sensor has a ground-coverage field of view (FOV) providing a 7.5 km swath, while the CHRIS sensor is a high-spatial-resolution hyperspectral spectrometer (18 m at nadir) with a 14 km swath (Ben-Dor et al., 2013).

During the period 2013–2016, the development of the PRecursore IperSpettrale della Missione Applicativa (Hyperspectral Precursor of the Application Mission) (PRISMA) by the Italian Space Agency (ASI), Germany's Environmental Mapping and Analysis Program (EnMAP), and NASA's HyspIRI (Hyperspectral Infrared Imager) enabled commercial and research themes to make progress compared to multispectral imaging. Two instruments were mounted on HyspIRI in low orbit with

TABLE 8.1
Comparison of Several Existing Multi- and Hyperspectral Sensors in Their Spectral Properties and Land Cover Application

Sensor Name	Band Number	Band Name	Range or Spectral Resolution (μm)	Spatial Resolution (m) and Band Key Features	Land Cover (LC) Mapping Applications (According to Xie et al., 2008 and Poursanidis and Chrysoulakis, 2017)
				Multispectral Sensors	
Sentinel 2	13	Coastal aerosol	0.42–0.46	60	Spatial planning, agro-environmental monitoring, water monitoring, forest and vegetation monitoring, land carbon, natural resource monitoring, global crop monitoring; four bands at 10 m spatial resolution aim for continuity with compatible missions such as SPOT-5 or Landsat-8 and address user requirements, in particular for basic land-cover classification; the six bands at 20 m spatial resolution aim for enhanced land-cover classification and for retrieval of geophysical parameters; bands at 60 m are dedicated mainly to atmospheric corrections and cirrus-cloud screening.
		Blue	0.42–0.55	10	
		Green	0.52–0.59	10	
		Red	0.63–0.69	10	
		Vegetation red edge	0.69–0.72	20	
		Vegetation red edge	0.73–0.75	20	
		Vegetation red edge	0.76–0.80	20	
		NIR	0.73–0.96	10	
		Narrow NIR	0.84–0.88	20	
		Water vapor	0.92–0.96	60	
		SWIR (cirrus)	1.36–1.40	60	
		SWIR	1.52–1.700	20	
		SWIR	2.01–2.370	20	
Landsat ETM+	7	Blue	0.45–0.52	30 m multispectral, 15 m panchromatic; single scene is approximately 170 × 183 km; temporal resolution is 16 days; launched 1999 (Landsat 1 in 1972)	Regional-scale vegetation mapping, usually at community level; blue: green reflectance of healthy vegetation; red is important for discriminating among different kinds of vegetation; NIR is especially responsive to amount of vegetation biomass present in scene.
		Green	0.52–0.60		
		Red	0.63–0.69		
		NIR	0.76–0.90		
		Mid-IR	1.55–1.75		
		Thermal IR	10.4–12.5		
		Mid-IR	2.08–2.35		

(Continued)

TABLE 8.1 (*Continued*)
Comparison of Several Existing Multi- and Hyperspectral Sensors in Their Spectral Properties and Land Cover Application

Sensor Name	Band Number	Band Name	Range or Spectral Resolution (μm)	Spatial Resolution (m) and Band Key Features	Land Cover (LC) Mapping Applications (According to Xie et al., 2008 and Poursanidis and Chrysoulakis, 2017)
SPOT 5	5	Pan	0.48–0.71	10 m; single scene is 60 × 60 km; launched 2002 (first SPOT in 1986).	Regional scale usually capable of mapping vegetation at community level or species level or global/national/regional scale mapping land-covertypes (e.g., urban area, classes of vegetation, water area).
		Green	0.50–0.59		
		Red	0.61–0.68		
		NIR	0.78–0.89		
		SWIR	1.58–1.75		
Ikonos	4	Pan	0.45–0.90	0.82 m panchromatic; 3.2 m multispectral;single scene is 11 × 11 km; temporal resolution is 3–5 days; launched 2000.	Local to regional scale land-covermapping at higher level (e.g., vegetation species) or used to validate other classification results.
		Blue	0.44–0.51		
		Green	0.51–0.59		
		Red	0.63–0.69		
		NIR	0.75–0.85		
				Hyperspectral Sensors	
AVIRIS	224	0.4–2.5	0.1	Airborne sensor since 1998; depending on satellite platform and latitude of data collected, spatial resolution ranges from a few to dozens of meters; single scene from several to dozens of kilometers.	At local to regional scale usually capable of mapping vegetation at community level or species level; as images are carried out as one-time operations, data are not readily available as they are obtained on an "as-needed" basis.
Hyperion	220	0.4–2.5 μm	0.1	30 m; data available since 2003.	At regional scale capable of mapping vegetation at community level or species level.
CHRIS Proba	Up to 63	0.415–1.05 μm	0.125	15–30 m; data available since 2001.	Applications in imaging spectrometry and production of laboratory-like reflectance spectra for each pixel in an image.

an Intelligent Payload Module (IPM) to enable data subsets to be broadcast directly (Knox et al., 2010). HyspIRI measures the spectral range 380–2500 nm from VIS to IR in 10 nm contiguous bands, while a second multispectral instrument measures from 3 to 12 um in the mid-IR to thermal IR(TIR) range. The VIS-SWIR/TIR spectra have a spatial resolution of 60 m at nadir with a revisit period of 19 days for VIS-SWIR and 5 day revisit time period for TIR. Subsequently, NASA's Moon Mineralogy Mapping (M3) project, in collaboration with the Indian Space Research Organization (ISRO), was launched to map the Moon's surface (Pieters et al., 2009). This was followed by the HyspIRI satellite mission discussed earlier (https://hyspiri.jpl.nasa.gov/). In 1980–1983, NASA started operating several missions, including a thermal hyperspectral mission known as the thermal infrared multispectral scanner (TIMS) (Kahle and Goetz, 1983). The thermal spectral range has been shown to be an important and promising region for acquiring mineral-based information using TIMS and ASTER sensors.

ISRO's Moon mission Chandrayan-1 launched in 2008 and carried a Hypespectral Imager (HySI) hyperspectral sensor that was useful in delivering information about the mineral composition of the lunar surface, mineral mapping, and water molecules present on the surface of the Moon. Recently, ISRO announced plans to launch the Hyperspectral Imaging Satellite (HySIS), which will be developed using a critical chip called the Optical Imaging Detection Array. HySIS has 55 spectral bands, which can be used in a range of research themes such as crop and environmental monitoring, oil exploration, and mineral mapping. The EnMAP satellite mission (2015) uses a wide range of ecosystem parameters including agriculture, forestry, soil and geological environments, coastal zones, and inland waters, while HyspIRI will provide products to understand ecosystems and deliver important information on natural disasters such as volcanic eruptions, wildfires, and drought. On June 23, 2017, a small and lightweight hyperspectral camera was successfully launched into space on the Aalto-1 nanosatellite, measuring a wavelength range of 500–900 nm. It is a unique hyperspectral tunable spectral imager operating in space due to its miniature size, representing half a CubeSat unit (0.5 U) in size, or 5 × 10 × 10 cm. It is a scalable sensing technology that will offer opportunities for new nanoscale satellite-based services for land cover or other research domains.

8.2.1 Planned Hyperspectral Remote Sensing Missions

Recognizing the importance of hyperspectral imaging, several countries are planning space missions in the near future. The Spaceborne Hyperspectral Applicative Land and Ocean Mission (SHALOM) is a joint mission between the Israeli and Italian space agencies that was unveiled on June 16, 2009, to support research in both countries (Feingersh and Dor, 2015). A SHALOM hyperspectral satellite will be launched to perform a joint study of the feasibility of development, launch, and operation of commercial satellites. SHALOM is designed with a spatial resolution of 10 and 2.5 m for hyperspectral and panchromatic images, respectively. Hyperspectral images will have a 2 day revisit time covering 200 km^2 daily with 10 m spectral resolution and 241 spectral bands at a range of 400–2500 nm. The main use of SHALOM will be to investigate and monitor environmental quality, crises, the search for mineral and natural resources, monitor water bodies, and facilitate precision agriculture activities in Israel and Italy (Ben Dor et al., 2014).

The Italian Space Agency scheduled the launch of the PRecursore Iper Spettrale della Missione Applicativa (Hyperspectral Precursor of the Application Mission) (PRISMA) satellite on May 30, 2018. PRISMA has integrated a hyperspectral sensor with a medium-resolution (panchromatic photo camera) sensitive to all colors. PRISMA is a push-broom sensor covering a spectral range of 400–2500 nm with a swath of 30–60 km and a ground sampling distance (GSD) of 20–30 m (2.5–5 m for panchromatic images). This feature will enable PRISMA to detect the chemical-physical composition of geometric features of objects it observes.

The ESA plans to launch the FLuorescence Explorer (FLEX) satellite as the eighth Earth explorer space mission in 2022 (Colombo et al., 2016; ESA, 2015) to observe vegetation fluorescence (Fletcher, 2015). FLEX will contain an array of three instruments to measure parameters such as fluorescence, hyperspectral reflectance, and canopy temperature and the features related to them. The FLEX mapper, with a 300 m resolution, will be used to monitor croplands and forests due to its unique capability of fluorescence detection, an indicator of photosynthesis in both healthy and physiologically perturbed vegetation (Jonathan, 2015). The FLEX mapper will provide an opportunity to researchers to assess canopy fluorescence at the global level from space to monitor global steady-state chlorophyll fluorescence in terrestrial vegetation (Colombo et al., 2016).

The China Centre for Resources Satellite Data and Application (CRESDA) has developed two small satellites to monitor environmental resources. HJ-1A is a hyperspectral sensor, while HJ-1B is an IR camera. HJ-1A and HJ-1B (*huanjing* = environment) satellites are mounted on the Huan Jing satellite constellation consisting of two more satellites with instruments HJ-1C. These two sensors deliver 3–100 m spatial resolution images.

8.3 SPACEBORNE HYPERSPECTRAL REMOTE SENSING STUDIES OF LAND USE/COVER

The superior capacity of hyperspectral systems to segregate and identify several ground features compared with traditional multispectral systems has been demonstrated by several researchers (Zhang and Ma, 2009). These days, hyperspectral imaging systems are viewed as being among the most noteworthy Earth observation information sources (Du et al., 2010) and therefore are used in various applications of land cover assessment (Li et al., 2010). Generally, hyperspectral imaging system data are preferred over multispectral data sources for digital image classification for land-cover thematic maps (Chintan et al., 2004). A complete and extensive review of the classification approaches used for remote sensing datasets, including methods suitable for hyperspectral data analysis, can be found in Lu and Weng (2007). In general, a commonly utilized land-cover classification group includes three key sets, such as pixel-based, subpixel, and object-based classification techniques. These categories of classification groups are discussed next.

8.3.1 Pixel-Based Classification

Methods belonging to pixel-based classification approaches perform classification by assigning each pixel to the target features of the land cover. This can be accomplished using either supervised or unsupervised classifiers. Supervised classifiers utilize samples of known features of each land cover class, known as "training sites," to classify pixels of unknown characteristics of digital images (Campbell, 1996). In contrast, unsupervised classifiers aggregate pixels with comparative spectral values into unique groups in accordance with predefined statistical criteria whereby the classifier joins and reassigns spectral groups into more informative feature classes. Supervised classifiers are further broadly subdivided into parametric and nonparametric classifiers. In contrast with nonparametric classifiers, for example, artificial neural networks (ANNs), support vector machines (SVMs) (Vapnik, 1995), and random forests (RFs) (Clark and Kilham, 2016; Guidici and Clark, 2017), parametric pixel-based classifiers, for example, maximum likelihood (ML)classifiers (Harris, 1998) require prior knowledge and learning about cover features with respect to the statistical distribution of the information to be characterized for the distinctive classes utilized, information that is frequently hard to come by. A pixel-based classification approach that is the most popular among all approaches, especially when executed with hyperspectral sensors, is the Spectral Angle Mapper (SAM) (Kruse et al., 1993). SAM is a supervised classification approach based on the calculation of a spectral angle corresponding to a reference source and the target feature spectral characteristics in question. SAM's popularity is due to its quick, easy, and reliable execution as well as implementation for the spectral assessment of digital image spectra with respect to the reference spectra. It is a capable classifier algorithm since it overcomes shade influence and suppresses its impact on spectra to emphasize target feature reflectance (de Carvalho and Meneses, 2000). Despite the capability of pixel-based strategies, most such methods require making assumptions with respect to the likelihood distribution of the training samples (for instance, in ML), which may not generally agree or coincide with reality. Other classifiers, for example ANNs, may require critical measures of effort in terms of the adjustment and calibration of neural nets before acquiring an acceptable accuracy of results.

It is generally accepted that such classification methods do not utilize the concept of the spatial dimension of digital images, for example, textural or relevant information exhibited in remotely sensed datasets (Yan et al., 2006). Last but not least, a key advantage of both classification approaches is that their use is not limited by the so-called Hughes phenomenon, also referred to as the "curse of dimensionality" (Hughes, 1968). The Hughes phenomenon occurs when training sites or reference sample numbers are inadequate compared with the number of input target features (and therefore of classifier parameters). This causes a decline in the accuracy of results as a component of the information dimensionality increase with an increased number of spectral channels of digital images (Dalponte et al., 2009; Zhang and Ma, 2009). Pandey et al. (2014) demonstrated methods to reduce the Hughes phenomenon utilizing segmented principal component analysis (PCA) techniques that segregate spectral ranges into several informative discrete spectral groups. Highly informative principal component bands of each group (first three only) were considered for the further processing and incorporation of information as much as possible, giving rise to enhanced visualization of the resultant images. The resultant images were used for classification that generated improved results compared to individual images. This technique reduces the Hughes phenomenon because it separates full spectral ranges into several groups. Despite the fact that this data problem is not looked on account of the several classification approaches, the type of technique used for analysis and assessment can be very useful with any type of classifiers used in LULC classification with hyperspectral imaging.

8.3.2 Object-Based Classification

The early period of the 1980s saw the development of new techniques and methods, such as object-based classification illustrated by de Kok et al. (1999). These methods depend on the idea that information required to represent target features in a digital image is represented in a meaningful object versus the pixel-based methods, where it is represented in a single pixel. The initial part of the object-based method was based on image segmentation techniques to isolate images into regions where each is homogeneous and no two adjacent areas are homogeneous (Pal and Pal, 1993). In the following stage, the segmented image is utilized in conjunction with textural and contextual information, in addition to spectral information, to generate a thematic classified LULC map. A hierarchy of levels of segmentation, such as a lower level and a higher level, can be achieved and be produced for a few large objects and several smaller objects belonging to their categories. This method incorporates the statistical values of objects from digital images, such as minimum, maximum, and standard deviation, along with their spectral characteristics for each spectral channel of the number of pixels used for a given target object. Furthermore, physical factors, such as shape, size, tone, texture, compactness, and other characteristics, are portrayed for each feature and depicted for the spatial features of the target object and can be utilized as part of the classification process to support the object's perception and discrimination (Bock et al., 2005). In this way, the incorporation of spectral and spatial information of target features represents a fundamental preferred attribute and advantage of object-based classifiers. Although spectral information forms the basis of hyperspectral remote sensing image classification and interpretation (Liu and Li, 2013), spectral information alone is not useful for classification and mapping, as demonstrated by Bai et al. (2017). To improve the accuracy of land-cover-classification results, other parameters should be combined along with spectral information. Therefore, Liu and Li (2013) demonstrated an idea to incorporate textural properties with spectral data to improve the accuracy and improved land-cover classification.

The authors used ANN classifiers on textural products for land-cover mapping and presented them as better classifiers compared to only spectral data.

Above all, the error such as "salt and pepper" effect and "edge" display in the thematic image generated by pixel based classifiers, are reduced in the object based classification technique. Likewise it ought to be noticed that in contrast with pixel-based classifiers, object based classification by and large requires higher analyst skills and level of expertise in order to be executed. Likewise, for the most part, the precision of segmentation image specifically influences the execution of the object based classification. A few researches have demonstrated that better segmentation results can prompt object-based image classification and has experienced extraordinary results as compared to pixel-based classification (e.g., Yan et al., 2006).

8.3.3 Spectral Unmixing

This is an altogether different technique for classification that partitions pixels into several spectral segments known as endmembers, and additional information about the general spectral response of the pixels is mandatory (Hostert et al., 2003; Okin et al., 2001; Zhang et al., 2006). Commonly, the endmembers chosen must be less than the spectral channels of the instrument or sensors for this method to be used. Additionally, all endmembers available in a digital image ought to be used with a specific end goal to obtain consistent and reliable products. The spectral unmixing outcome is one for each particular endmember, with pixel values ranging from zero to one, demonstrating the fraction of the first image attributed to the specific endmembers. Subpixel classification techniques are largely separated into linear and nonlinear unmixing methods, depending upon whether each pixel's reflectance is a linear or a nonlinear combination of all target features within the pixel (Plaza et al., 2009; Small, 2001). Hyperspectral information is used in linear methods because it has substantially more spectral channels. Mixing methods are additionally useful for smaller or rare target features such as the identification and discrimination of alien or intrusive species. An overview of linear mixing methods and distinctive limitations can be found in Miao et al. (2006).

8.3.4 Spaceborne Hyperspectral Applications for Land Use/Cover Mapping

In recent decades, several airborne and satellite hyperspectral remote sensing sensors have been launched. The current availability of such remote sensors has additionally supported the improvement of a few techniques for assessment and analysis provided by such sensing systems. Hyperion is a spaceborne satellite hyperspectral sensor mounted on board the Earth Observer-1 (EO-1) satellite, launched by NASA under its New Millennium Program around 2000 (Tables 8.1 and 8.2). It has a total of 242 spectral channels and acquires images at a 30 m spatial resolution and around 10 nm spectral resolution (70 spectral channels are part of the VIS to near-infrared (VIS-NIR) and 172 channels form the shortwave infrared (SWIR) part (Han et al., 2002). The Hyperion sensor is considered the first main "real and genuine" spaceborne hyperspectral remote sensor instrument in Earth's orbit. The public access and availability of Hyperion data have opened one of several opportunities for remote sensing research/studies to explore its potential use for assessment and investigation in LULC mapping, as well as other related fields with future prospects.

Several studies have observed and analyzed the use of Hyperion data with different pixel-based classification approaches to land-cover mapping (Goodenough et al., 2003; Pignatti et al., 2009; Walsh et al., 2008). Moreover, both linear and nonlinear unmixing classification approaches integrated with Hyperion sensors for LULC mapping have additionally been explored in a few studies (e.g., Pignatti et al., 2009; Walsh et al., 2008). Others have likewise investigated the use of object-based approaches to assessing LULC mapping and classification (Walsh et al., 2008;

TABLE 8.2
Brief List of Hyperspectral Imaging Space Mission Accomplished and Planned for Launch in Near Future

Mission Sensors	Platform	Spectral Range (Resolution, nm)	Spatial Resolution (m)	Channels Number	Revisit Time (Days)	Organization/Nation	Launch Year
Hyperion	EO-1	400–2500	30	220	200	NASA	Nov. 2000
CHRIS	PROBA	400–1050	**18** to 36	150	2 (midlatitudes)	ESA	22 Oct. 2001
EnMap		420–2450	30	232	23 & 4 (across track $\pm 30°$)	DLR Germany	2015
HyspIRI		380–2500	60	217	**VSWIR:** 19 **TIR:** 5	NASA	
HySIS		400–1200	30	55	5/19	ISRO India	
SumbandilaSat/MSI		440–2350	15	200	–	South Africa	17 Sep. 2009
VENUS		415–910	5.3	12	2	CNES/Israel	2016
SHALOM	MBT SPACE	400–2500	10	241	2	Israel Space Agency	16 June 2019[a]
PRISMA		400–2500	20–30	237	29	Italian Space agency	30 May 2018[a]
FLEX	Earth Explorer		300		28	ESA	2022[a]
HySI	Chandrayaan-1	400–950 (15)	30	64		ISRO India	2008
HJ-1A/HJ-1B	CAST	450–950 (5)	100	128	31	China	2008
Hero(CASI)		400–2500	30	>200	3		

[a] To be launched in 2022.

Wang et al., 2010). For instance, Walsh et al. (2008) analyzed three different classifiers, such as Spectral Angle Mapper (SAM), spectral unmixing, and object-based classification integrated with a Hyperion sensor, and in particular compared their results and robustness with Hyperion data for mapping intrusive/alien plant species in Ecuador, and they further announced that the object-based arrangement outperformed the other two classification approaches. Wang et al. (2010) utilized an object-based classification approach to Hyperion sensors taken for a test field site in China and demonstrated accuracy results from 72% to 88% in classified maps, subject to the number of reference classes taken and the total number of classes outlined therein. However, a literature review revealed that a couple of studies reported a comparative assessment of the performance of different classifier algorithms using a hyperspectral imaging system for land-cover mapping (Petropoulos et al., 2012a). For instance, Pal and Mather (2005) analyzed various pixel-based classifiers for comparative results and accuracy, including SVMs and ANNs from the multispectral LANDSATETM+ and Digital Airborne Imaging Spectrometer (DAIS) airborne hyperspectral sensors, respectively, for two study sites located in Spain and the UK. Some authors additionally obtained higher classification accuracy results from SVMs compared to two classifiers (ANN and SAM), with an improved accuracy of up to approximately 2%–5%. Karimi et al. (2006) assessed the efficiency of SVMs and an ANN for classifying hyperspectral imagery over an agricultural field, utilizing information from an airborne hyperspectral imaging system flown over test sites in Canada. The authors likewise found the SVMs to outperform the ANN classifier, by around 15% and 0.114 in overall accuracy and kappa coefficient, respectively.

In another study, Pal (2006) explored the use of SVMs with hyperspectral imagery from the DAIS airborne sensor at a test site in Spain and revealed that the overall accuracy of the SVMs exceeded 91%. Koetz et al. (2008) inspected the integrated use of hyperspectral and LiDAR-derived information with SVMs for plant fuel type mapping for a test site in France. Some authors revealed overall accuracy and kappa coefficient of 69.15% and 0.645 separately when the SVM was used with the hyperspectral data, which increased by 6.3% and 0.115 respectively when the two datasets were integrated for the same case study. For example, recently Clark and Kilham (2016) conducted a study on a RF algorithm for three independent variables, reflectance, MNF (Minimum Noise Fraction), matrices with simulated multitemporal (two seasonal imagery) HyspIRI imagery in 2016 at a test site in San Francisco, California (Figure 8.1), to explore the algorithm's ability to classify HypsIRI imagery according to the international Land Cover Classification System (LCCS) in two levels of classification complexity. They demonstrated the robustness of multitemporal matrices (0.9%–3.1%) and hyperspectral matrices (16.4%–21.8%) and concluded that they were superior to other two variables, MNF and reflectance used in the study for classification purposes depending upon pixel or polygon scales for the reported analysis. Their outcomes were provided keeping in mind the future launch of the hyperspectral satellites with enhanced spatial, spectral, and temporal resolutions, which in turn will require a sophisticated and accurate classification algorithm for land-cover mapping results particularly relevant to regional and global scales.

In another recent study, Guidici and Clark (2017) investigated the multiseasonal land-cover classification using simulated HyspIRI imagery with SVM, RF, and neural network (NN) algorithms for more heterogeneous landscape in test sites located in the San Francisco area in 2015 (site similar to that used previously in a study by the same author and investigated by Clark and Kilham 2016). Their outcomes demonstrated an improved classification accuracy of multiseasonal imagery by 2.0% (SVM), 1.9% (NN), and 3.5% (RF) compared to single-season imagery (Figure 8.2). A NN has the potential to provide improved overall classification accuracy (89.9%), similar to SVM-generated classification land-cover results (89.5%), while both outperformed the RF with a difference of almost 7% in land-cover accuracy. The researchers provided insight into hyperspectral imaging with the appropriate use of classification algorithms for improved classification accuracy and an ability to interpret distinct target features in the spatial-spectral-temporal domain in remote sensing products for analysis and visualization.

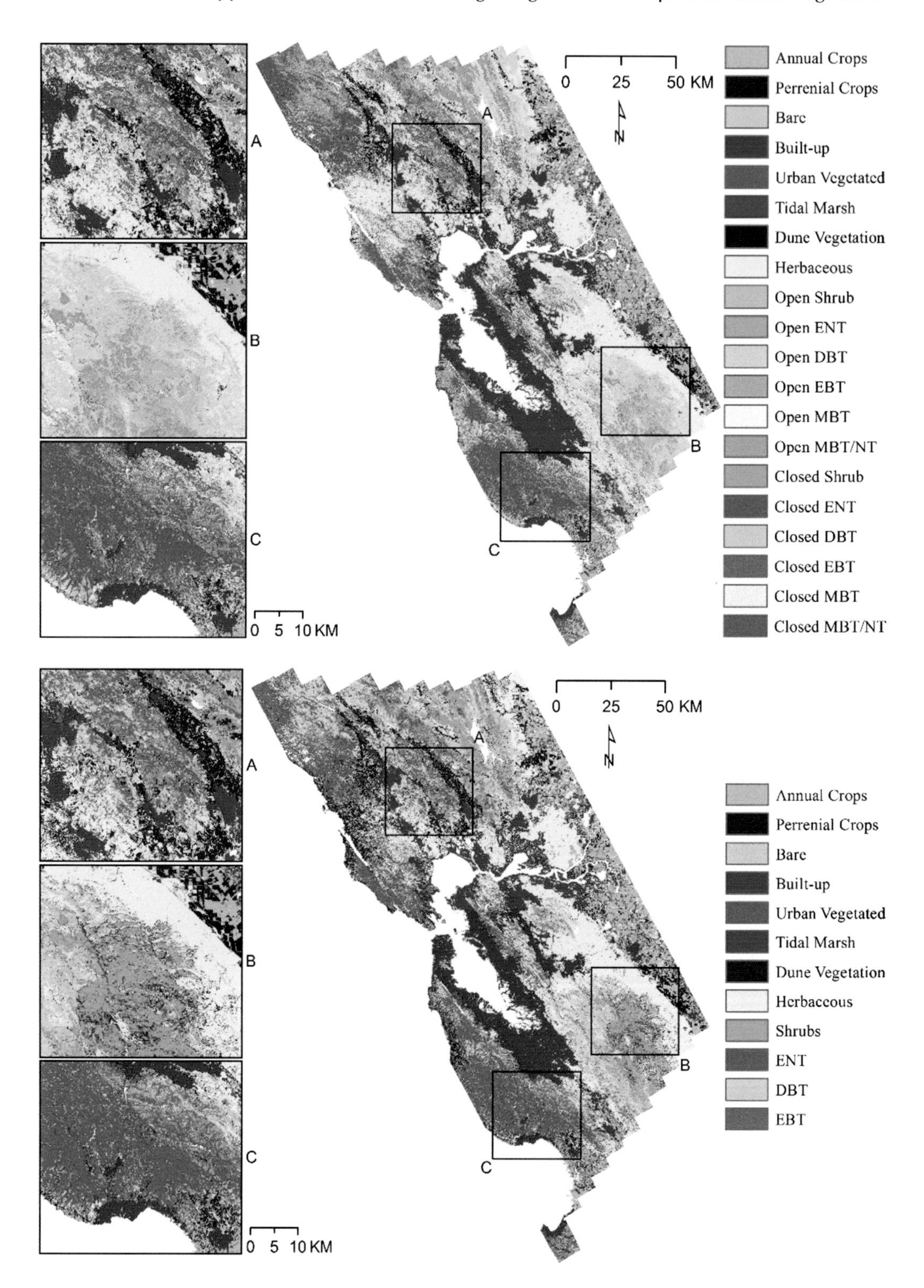

FIGURE 8.1 HyspIRI land-cover classification with multitemporal hyperspectral matrices for 20 LCCS classes in (upper part) and 12 LCCS classes (lower part). (Adapted from Clark, M. and Kilham, N., 2016. *ISPRS J Photogramm Remote Sens*, 119: 228–245.)

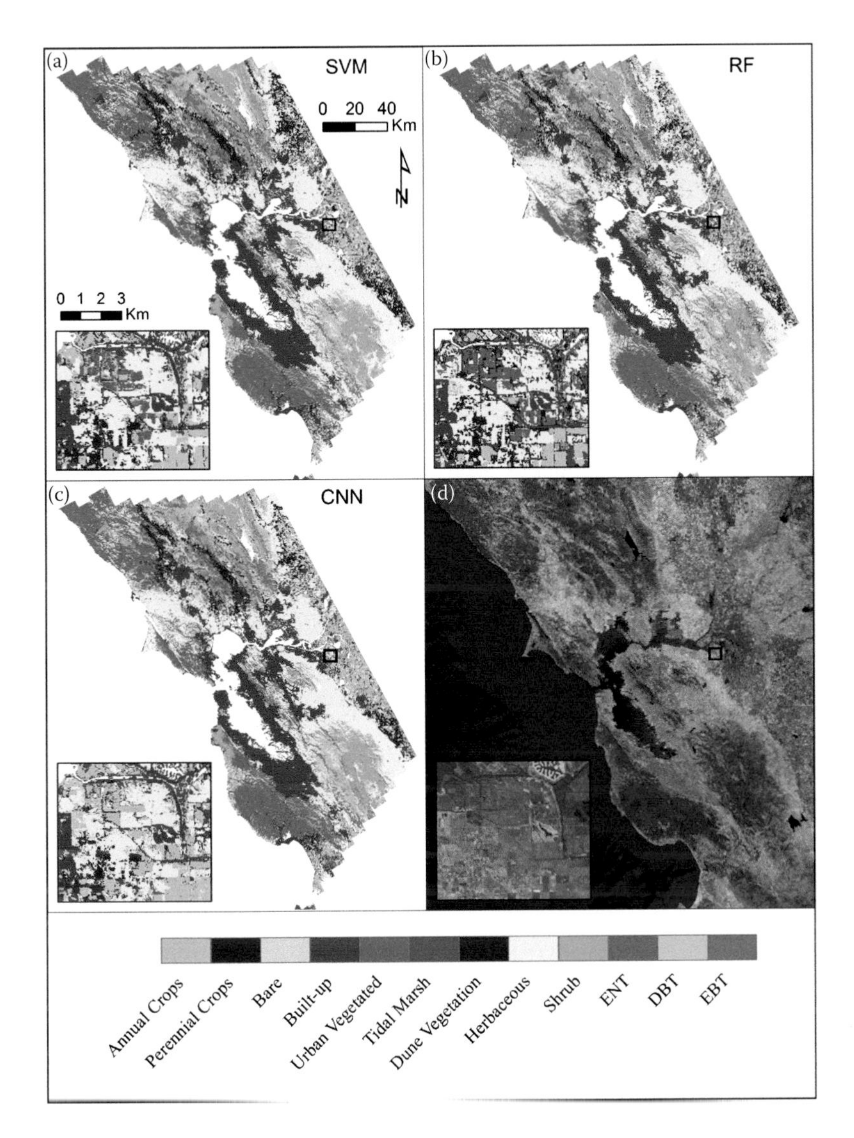

FIGURE 8.2 HyspIRI land-cover classification maps using (a) SVM, (b) RF, (c) NN classification maps. White areas indicate pixels that were not classified (e.g., water, clouds, no data), and (d) Natural color mosaic of imagery of study site from June 2015. (Adapted from Guidici, D. and Clark, M.L., 2017. *Remote Sensing*, 9(6): 629.)

8.4 FIELD HYPERSPECTRAL REMOTE SENSING OF LAND USE/COVER

Hyperspectral remote sensing at the field scale, that is, field spectroradiometry, uses spectral reflectance as a key variable. The solar irradiance, or the spectral flux that reaches a detector per surface unit, dissipates toward the SWIR (1300–2500 nm) from the VIS wavelengths (100–2500 nm). This is because atmospheric constituent gases, such as H_2O, CO_2, O_2, O_3, CH_4, and N_2O, selectively scatter and absorb incident radiation, thereby forming atmospheric "blinds," while allowing some solar radiation to pass through atmospheric "windows" (Avery and Berlin, 1992). Water vapors are the major atmospheric blinds with the strongest absorption at around 1450, 1400, and 1900 nm, thus covering significant portions of the near-infrared (NIR) (700–1300 nm) and SWIR compared to other atmospheric constituents (Avery and Berlin, 1992; Price, 1998). Further, the reflection of solar irradiance from land cover is radiance, a unit widely used in remote sensing as it is independent of sensor characteristics. Moreover, a proportion of the radiance reflected by a target to that reflected by a perfect (lossless) and diffuse (isotropic) standard surface illuminated under similar conditions is reflectance. Because spectral reflectance is not a characteristic that is contingent on targets, it is referred as a "spectral reflectance factor" $R(\lambda)$ (Milton et al., 2009). $R(\lambda)$ is dimensionless, having values from 0 to 1, yet may achieve values beyond1, particularly for specular surfaces, for example, snow.

Typical spectral reflectances have been described for many land covers such as soil, water, and snow (Avery and Berlin, 1992). Vegetation reflectance is much more complex in its pattern and variability. Plants use mostly the red and blue bands in the VIS spectrum for photosynthesis driven by selective absorption by pigments, leaving the reflected mid-VIS to give leaves their green visual color. The most important pigment, chlorophyll-a (Chl-*a*), absorbs light at approximately 430 and 680 nm, whereas Chl-*b* uses 480 and 660 nm and β-carotene from 400 to 550 nm with two local maxima at 470 and 515 nm. Further, the red edge (700–800 nm) is the region of rapid change in reflectance before the NIR. Maximum reflectance is reached in the NIR due to internal light scattering in the sponge mesophyll and also due to external scattering within the canopy. In the SWIR, water dynamics and absorption are prominent, with absorption zones at about 1450–1530 nm and 1900–2000 nm. It should be mentioned that for vegetation, as for any land cover, there is a directional reliance on the Sun's position (anisotropy), which is explicitly described by the bidirectional reflectance distribution function (BDRF) (Schaaf, 2009). BDRF is a four-dimensional capacity measured in units of inverse steradians [sr^{-1}]; the fundamental idea and importance of discussing the BRDF and the directional issues of radiation and reflectance are covered in Nicodemus et al. (1977).

Field spectroradiometry provides nondestructive and repeatable spatial measurements in a wide spectral range with high precision and accuracy in user-friendly computer formats and compact and portable spectroradiometer designs (Milton et al., 2009; Nidamanuri and Zbell, 2011). Nonetheless, measuring land cover such as vegetation on a field is challenging because it involves a consideration of influencing factors. Consequently, the main factors affecting field spectroradiometry are presented next, followed by a discussion of advancements in compiling spectral data. Finally, a summary of the statistical tools utilized in vegetation land-cover discrimination using field spectroradiometry data is given.

8.4.1 Factors Affecting Field Spectroradiometric Measurements

Spectroradiometers are devices that measure relative spectral radiation over a specified wavelength range and calibrated to output spectral measurements (reflectance and transmittance) in absolute units (e.g., energy flux density in W m^{-2} nm^{-1}) for natural and synthetic surfaces and materials. Spectroradiometers can be used in both the lab and the field. However, whereas the environment in the lab is more controllable, several important factors need to be considered in the field prior to and during field spectroradiometric measurements.

First, the instrument-specific "behavior" should be assessed with respect to the environment (e.g., heating) and calibration. Before sampling, the device is normally switched on to warm up for no less than 30 min so as to decrease its temperature sensitivity (known as "step"), often depicted as a sudden deviation in the target reflectance at a particular wavelength (Figure 8.3). The sensitivity drift can sometimes be remedied in the field by holding the fiber-optic cable with the sensor away from the object and permitting the FOV of the strands to overlap (ASD, 2009). Also, the amount of usable information in the spectra depends, among other things, on high signal to- noise ratio (SNR), which in turn increases with the number of averaged spectra (ASD, 2009; Milton et al., 2009). Instrument calibration for the Sun's characteristics (illumination, atmospheric and scattering impacts) is typically done by measuring a surface with a known reflectance of nearly 100% (white board). The most widely recognized is the material Spectralon®, a white reference reflector with 99% reflectance made by Labsphere Inc. of North Sutton, NH, USA. Periodic reflectance estimations of the white board can provide an estimate of how "perfect" a diffuse reflector is amid field measurements.

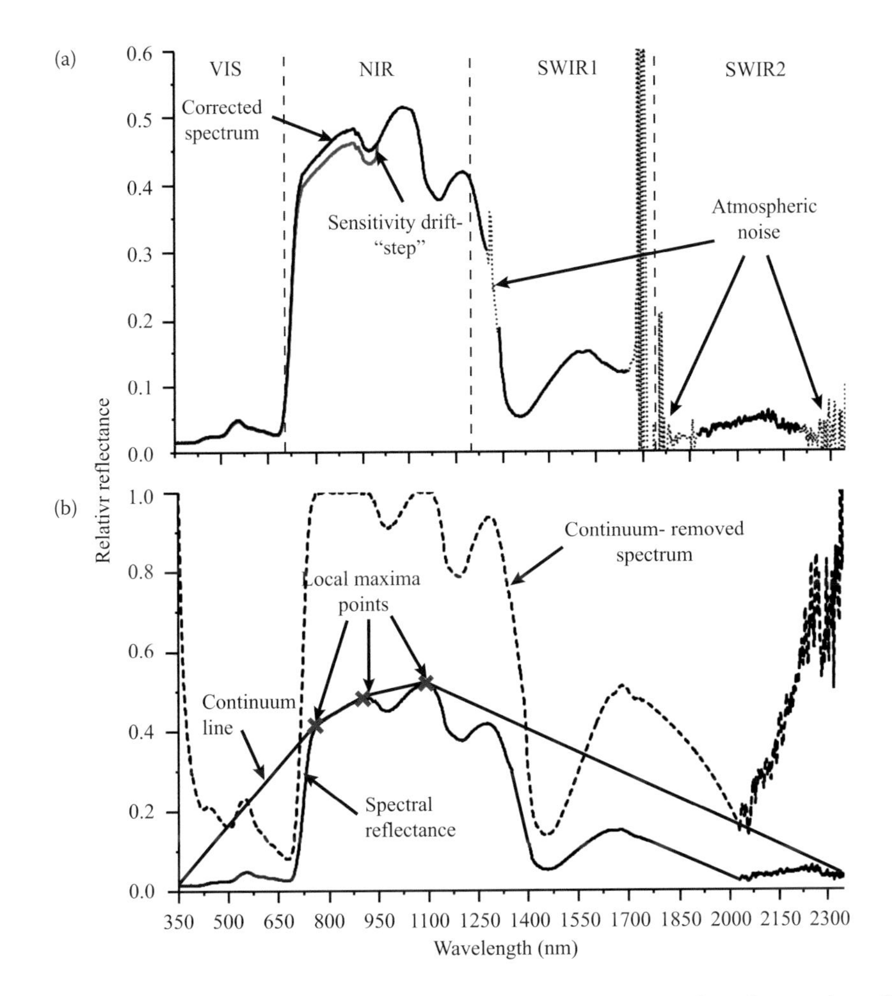

FIGURE 8.3 (a) The four spectral regions from VIS to SWIR and preprocessing of vegetation reflectance spectrum; the sensitivity drift effect from 350 to 1000 nm (red line) is corrected (full line); the spectral regions around 1400, 1950, and 2400 nm (dashed line) are removed due to excessive atmospheric noise; adapted from Petropoulos et al. (2014). (b) Vegetation reflectance spectrum, its continuum with local maxima, and the continuum-removed reflectance spectrum. (Adapted from Manevski, K. et al., 2017. *Sensitivity Analysis in Earth Observation Modelling*. Elsevier, pp. 103–121.)

Second, atmospheric gases and the Sun's illumination variation may significantly influence reflectance measurements. As discussed earlier, atmospheric gases absorb energy at longer wavelengths, and this is somewhat compensated for by reduced spectral resolution in the SWIR wavelengths during instrument manufacture (Salisbury, 1998). So as to further diminish these environmental factors, the smallest nadir position of the sensor is normally utilized to avoid BRDF (Manakos et al., 2010; Pfitzner et al., 2006). Illumination factors, such as background radiance from the surroundings, are typically small and insignificant sources of error, except if a large background object or something close enough to the land cover obscures a large portion of the solid angle viewed by the target. Clouds, shades, and wind attenuate solar irradiance and the target's reflectance (Chang et al., 2005). Therefore, clouds should be avoided as much as possible, and measurements are performed during high Sun position, for example, between 11 a.m. and 2 p.m., when the Sun's height varies by less than 30°. At higher latitudes, directional estimations with off-nadir edges and a wider FOV are occasionally used (Eklundh et al., 2011). Wind affects the spectral response of land cover as the cover, for example vegetation canopy, moves and changes its shadow. Nowadays it is minimized by fast scanning time (100 ms for a complete 350–2500 nm spectrum) of spectroradiometers (ASD, 2009). In addition, replicated measurements further minimize both the wind and illumination synergetic effects and instrumentation factors (Hemmer and Westphal, 2000).

Third, each measurement essentially aims to acquire as pure signals from the land cover of interest as possible, which may contain mixed covers if appropriate FOV adjustments to height are not considered. The FOV ought to be sufficiently wide to cover a homogeneous species distribution. In the meantime, the FOV also ought to limit the underlying soil contribution to the spectral response (Ben-Dor et al., 2009). Several relative readings can be taken along different directions over the canopy followed by white reference reading in order to justify the variation in output (Manakos et al., 2010; Manevski et al., 2012).

8.4.2 Development and Use of Field Spectral Libraries

Extracting field reflectance spectra is crucial for spectral discrimination between land covers if variables that influence the measurements are considered. That information becomes part of spectral libraries that are composed and stored together as spectral signatures with associated metadata (Hueni et al., 2009; Rasaiah et al., 2014). The spectral signatures may represent geo-bio-chemo-physical dynamics pertaining to the land cover and its environment, such as, for example, shadow, moist surface, and the phenology of vegetation, or may aim for a "pure," that is, reference spectra. In both cases, extracting only the necessary information and increasing the uniqueness of the reflectance requires a clear approach to developing a spectral library, and it is essential for guaranteeing sufficient information quality (Milton et al., 2009; Salvaggio et al., 2005). Most studies focus on the use of unaltered field reflectance spectra.

Field reflectance without postcollection quantitative/mathematical manipulations is unaltered reflectance (Figure 8.3a). Such reflectance is widely used in land-cover discrimination in a holistic approach, that is to say, taking into consideration land cover and its surroundings, such as micro-pedo-climate, as an integral part of the spectral signature (Rao et al., 2007; Schmidt and Skidmore, 2003). Hence, unaltered spectral libraries are suitable for solving the spectral variation between pixels on hyperspectral imagery, especially for more homogeneous landscapes or when the FOV matches the image pixel size (Artigas and Yang, 2006; Rao et al., 2007). Spectral unmixing in hyperspectral imagery is also used when the unaltered library contains "reference spectra" to outline distinctive features at a specific spatial scale (Zhang et al., 2003). Separating the influence of canopy structure or soil background from the spectra or detecting differences in vegetation due to various biophysical properties (e.g., canopy thickness, leaf water, or chlorophyll content) can be further enhanced with field hyperspectral libraries by the use of vegetation indices (Castro-Esau et al., 2004; Price, 1994).

One of the most common numerical manipulations of unaltered reflectance is the continuum removal, which is essentially normalization by a continuum line with a high value of associated spectral local maxima utilizing straight line segments, as indicated by the following equation:

$$R_{(\lambda)cr} = \frac{R_{(\lambda)}}{C_{(\lambda)}}, \tag{8.1}$$

where $R_{(\lambda)cr}$ is the continuum-removed spectral reflectance, $R_{(\lambda)}$ is unaltered (raw) spectral reflectance, and $C_{(\lambda)}$ is the continuum line spectral value (Figure 8.3b). The continuum removal technique was initially used in remote sensing for mineral mapping and rock identification studies, and an increasing number of studies use them also on vegetation spectra (Manevski et al., 2011; Prasad and Gnanappazham, 2015; Psomas et al., 2005). It should be kept in mind that applying continuum removal over wide spectral ranges may not necessarily result in better shape characterization because low local maxima can be missed by the continuum line, such as the missed local maxima at about 1300 nm in Figure 8.3b. Moreover, detecting local maxima in noisy bands may introduce artificial reflection/absorption features. Algorithms such as those that are ENVI enabled (ENVI, 2009) can characterize local maxima considering the aforementioned drawbacks.

Similarly to the continuum removal, derivatives essentially represent a change of power in the reflectance, and first- and second-order derivatives are often applied on spectral libraries. The first derivative is calculated for a wavelength halfway between two wavelengths, while the second resolves three firmly closed spaced wavelength reflectance values (Petropoulos et al., 2014). Derivatives highlight sudden changes in spectra, as illustrated for the red edge inFigure 8.4. Derivatives are being utilized to examine various vegetation covers at the leaf scale or to observe vegetation status over various soil backgrounds (Frank and Menz, 2003; Kochubey and Kazantsev, 2012). Derivatives may also portray vegetation cover status in relation to chlorophyll content (Castro-Esau et al., 2004; Kochubey and Kazantsev, 2007), where broadband indices from unaltered reflectance (e.g., NIR/R reflectance ratio) are not that sensitive. Morrey (1968) describes in detail these derivatives and their advantages.

A number of spectral libraries exist that are designed to hold pure spectra data of different land covers. Research is being carried out for the purpose of creating widely used spectral libraries of land cover for several target features (e.g., wetlands, deserts, soil, vegetation species) (Hueni et al.,

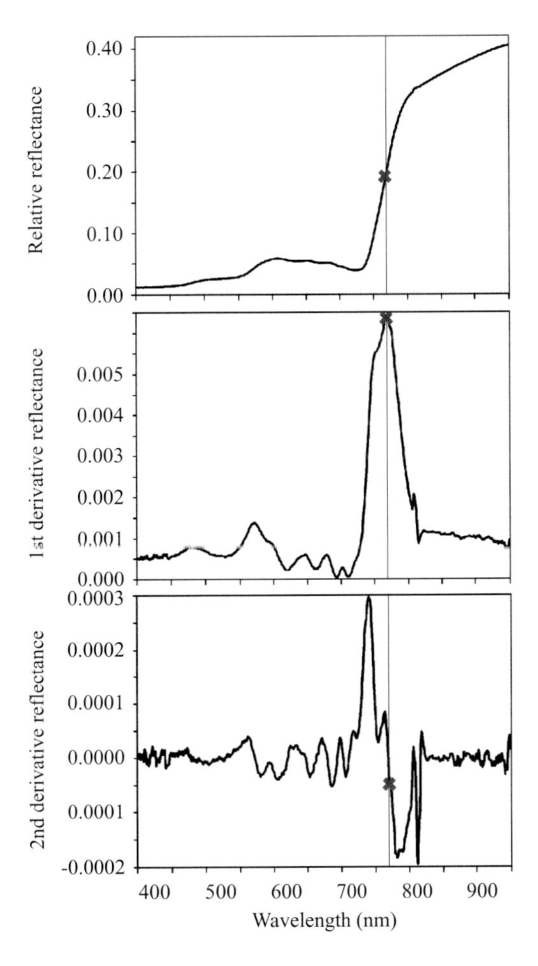

FIGURE 8.4 Unaltered vegetation reflectance (upper graph) and its first (middle graph) and second (lower graph) derivatives in 350–900 nm spectrum. The red edge has an inflection point (point where slope is at its maximum) at 730 nm. This corresponds to the maximum in its first derivative and to local minimum in the second derivative (red squares). (Adapted from Petropoulos, G.P. et al., 2014. *Scale Issues in Remote Sensing*, 285–320.)

2009; Milton et al., 2009). The lack of standardized procedures and metadata collection has for years been reported as a limiting factor in field spectroradiometry, in relation to data transferability and interpretability (Pfitzner et al., 2006; Rasaiah et al., 2015b). However, recently efforts have been made to standardize designs and methods for the collection of spectral data and associated metadata (Rasaiah et al., 2014, 2015a,b). One of the most comprehensive spectral libraries is offered by the USGS (http://speclab.cr.usgs.gov/) and integrates the ASTER spectral library, the Johns Hopkins University (JHU) spectral library, and the Jet Propulsion Laboratory (JPL) spectral library (Baldridge et al., 2009). This library offers a full compilation of spectral references for several land-cover features, such as minerals, vegetation, and artificial materials, measured using several advanced spectroradiometers, such as the Beckman 5270 (200–3000 nm), the ASD field spectroradiometer (350–2500 nm), the Nicolet Fourier-transform infrared (FTIR) interferometer spectrometer (1,300–15,000 nm), and the NASA Airborne Visible/Infrared Imaging Spectrometer (AVIRIS) (400–2500 nm). The SPECCHIO online spectral database system is also an open-access spectral library from the Remote Sensing Laboratories (RSL) of the University of Zurich (http://www.specchio.ch/) that contains a rich source of metadata with reference spectra and field-measured spectral information so as to ensure its durability and shareability in the scientific community (Hueni et al., 2009).

8.4.3 Statistical Approaches in Field Spectral Discrimination

Statistical approaches suitable for analyzing field spectral data can be parametric and nonparametric, depending on whether or not the data follow a normal, or Gaussian, distribution. Studies have shown that vegetation spectral reflectance is quite variable and can follow a Gaussian or other distribution (e.g., Manevski et al., 2011, 2012). Parametric tests are often considered robust, particularly when the number of samples is large (Robson, 1994). Nonparametric techniques make no assumption regarding data distribution and are less sensitive to off-scale values (Manevski et al., 2017). Thus, both approaches have their own particularities, qualities, and shortcomings. When coupled with land-cover spectral data, they should lessen wavelengths to the most significant for spectral discrimination without the loss of vital data (Ben-Dor et al., 2009; Thenkabail et al., 2004b). Hence, dimension reduction is an important task of statistical methods. These include ANOVA (Adam and Mutanga, 2009; Manevski et al., 2011, 2012; Vaiphasa et al., 2005), correlation analysis (Mariotti et al., 1996), linear discriminant analysis (Abdel-Rahman et al., 2010; Clark et al., 2005), and canonical discriminant analysis (van Aardt and Wynne, 2001), among others. In relation to the number of independent variables (factors) used to explain reflectance variability, the approaches can be grouped into univariate and multivariate methods.

Univariate techniques utilize one independent factor (land cover or plant species) in detecting differences for one dependent variable (e.g., reflectance). The single-factor ANOVA investigates the difference between groups as a deviation of each group's mean from the "grand mean" (Robson, 1994). Comparable univariate methods for vegetation discrimination in light of parametric assumptions involve *t*-tests (Jacobsen et al., 1995; Vaiphasa, 2006), analysis of covariance, linear regressions, and generalized linear models.

As opposed to the univariate approach, multivariate statistical approaches make no assumptions about a Gaussian distribution and may result in better descriptions of combinations of more dependable factors that fulfill particular mathematical criteria as a clarification of more autonomous factors. Such strategies are utilized both in data dimensionality reduction and for illustrative purposes. PCA is a two-step multivariate technique that first decomposes the X-matrix (in this case, the reflectance) and then fits a multiple linear regression model, using the principal components instead of the original *X*-variables as predictors (Castro-Esau et al., 2004). Similar to PCA, partial least squares (PLS) also takes dependent variables into account during the calculation of the principal components (it simultaneously models X- and Y-matrices in order to find the "latent" variables of X that will better predict "latent" variables for Y). Canonical discriminant analysis is another data reduction method used to clarify the factors (such as wavelength) that separate the best among types (plant species); it outlines variation among classes, much in the way that PCA compresses the aggregate variation (Dimitrakopoulos, 2001). Other techniques include statistical distances that calculate the distance between probability functions for two classes, thereby suggesting the probable success of between-class discrimination. Such distance measures include the Euclidean, Jeffreys-Matusita, and Bhattacharyya distance (Ouyang et al., 2013; Schmidt and Skidmore, 2003).

8.4.4 Field Spectroradiometry Applications for Land Use/Cover Mapping

Since the 1990s, an increasing number of studies have reported land-cover discrimination results using field spectroradiometry, focusing primarily on vegetation land cover. Prasad et al. (2015) used multiple statistical approaches for the discrimination of mangrove species in India using unaltered and continuum-transformed field and laboratory spectral data. Based on their extensive parametric and nonparametric and spectral distance analyses, the authors identified green, red edge, and SWIR regions as important for vegetation identification and concluded that the continuum removal of additive inverse spectra gave better discrimination than continuum-removed spectra. Artigas and Yang (2006) discriminated saltmarsh species in the USA within the VIS-NIR spectrum using the Mann-Whitney U-test and suggested that the orange and red bands in the VIS, as well as selected bands in the NIR, were important. Psomas et al. (2005) have carried out similar tests for discriminating dry-mesic meadow species in Switzerland. Apart from band identification, a noteworthy outcome of these studies was the seasonal variability of the vegetation spectral signatures. Manevski et al. (2012) utilized the Mann-Whitney U-test and successfully discriminated typical Mediterranean vegetation and emphasized the seasonal impact of vegetation and its importance in discrimination results. This seasonal variability, together with inherent differences between various vegetation land covers and even within the same vegetation but different functional groups (e.g., tree or shrub), was the main reason for the high but "parallel" variability of the vegetation spectral reflectance, and this has implications for the choice of statistical method (Figure 8.5).

Field spectroradiometry has likewise been utilized for assessing and mapping vegetation quality, for example, nitrogen content. Mutanga et al. (2003) spectrally discriminated tropical grass under various nitrogen treatments using parametric ANOVA tests, coupled with both unaltered and continuum-removed spectra, and emphasized the need to further map field-scale variation using hyperspectral imaging. Lacar et al. (2001) investigated the potential to discriminate between four grape vines (Cabernet Sauvignon, Merlot, Semillon, and Shiraz) in southern Australia using one-way ANOVA, combined with Tukey posthoc tests, and emphasized the red edge (~720 nm) trailed by the green and its wings in the VIS as important for their study. In addition to the importance of the red edge for vegetation cover identification, Schmidt and Skidmore (2003) discriminated numerous distinctive Dutch saltmarsh plant species utilizing nonparametric statistics; while no single band was found to discriminate between all the species, the authors still found the NIR bands around 770 nm to be sufficiently powerful; they also demonstrated that the red edge can lose statistical importance for vegetation discrimination studies when large spectral variability is present in the data. In their comparative study, Manevski et al. (2011, 2012) concluded that the use of unaltered reflectance narrows the statistical difference between plants to bands in the VIS and SWIR spectrum but weakens the difference in the NIR spectrum compared to continuum-removed reflectance. Despite the substantial dimension reduction, their discrimination results, as for many other studies, still left numerous wavelengths relevant for discrimination of the studied species. Adam and Mutanga (2009) have proposed a hierarchical method to discriminate plant species that at first includes one-way ANOVA, followed by classification and regression trees and spectral distance analysis. This analysis yielded eight bands that are considered to be practical for upscaling to airborne or spaceborne sensors for mapping the studied vegetation.

To sum up, the general trend so far points to continuum-removed reflectance as being a more powerful input to a nonparametric analysis for the discrimination of vegetation land cover at the field scale, when compared with unaltered reflectance and parametric analysis. However, it should be emphasized that the number of observations may play an important role for discrimination result

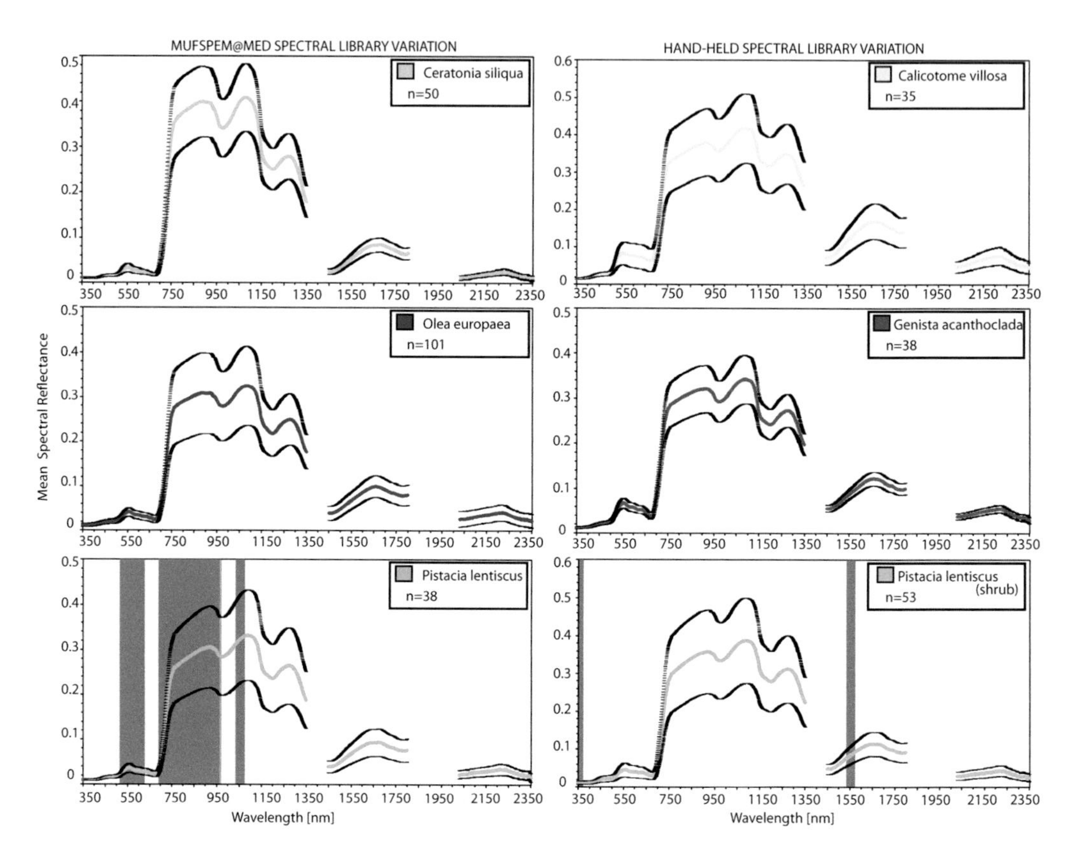

FIGURE 8.5 Mean reflectance spectrum of common Mediterranean vegetation obtained with aerial-lift mobile platform (left) and handheld (right). Spectra are flanked by standard deviation of mean (black lines) at 95% confidence level. Wavelengths where the variances between three plants compared in each spectral library were homogeneous are shaded gray on lower graphs. Spectral regions around 1400, 1940, and 2400 nm were removed due to atmospheric noise. (Adapted from Manevski, K. et al., 2012. *IEEE J-Stars*, 5(2): 604–616; with permission.)

outputs and should be considered when utilizing spectral libraries for discrimination and land-cover classification purposes (Manevski et al., 2017). Last but not least, the synergistic use of field spectroradiometry and spaceborne hyperspectral imaging for land-cover mapping has not been as widely investigated. Land-cover discrimination studies using field spectroradiometry have rapidly increased in number, but these results are seldom applied further on hyperspectral imaging data (van Aardt and Wynne, 2001, 2007). Rao et al. (2007) illustrated the use of field spectroradiometry information of agricultural sites in India, for Hyperion data classification at different spatial scales (canopy level and pixel level), thereby achieving notable overall accuracies of 86%–89%. In any case, the similarity and inconsistency of the spectral characteristics of land cover definitely raises the issue of high mapping vulnerability in the pixel-based analysis of heterogeneous and fragmented landscapes (Xie et al., 2008). Apart from further improving field spectroradiometry studies, further research should focus on upscaling field studies and the validation of hyperspectral imagery.

8.5 UNMANNED AERIAL VEHICLES AND HYPERSPECTRAL REMOTE SENSING FOR LAND USE/COVER MAPPING

In recent years, many advanced platforms, in terms of small size, light weight, and ease of operation, have been developed to carry sensors amid rising demand for improved spatial, spectral, and temporal resolutions. These include primarily UAVs, commonly known as drones. Each platform has its own advantages, and disadvantages, so the choice of platform, and thus data resolution, depends on the study objective, but synergies are ongoing and expected to further enhance results (Pádua et al., 2017). UAVs' greatest advantage over field- and spaceborne remote sensing is the reduced altitude of their flight, which greatly decreases costs and improves data resolution, allowing for higher monitoring frequencies. Also, the thickness of the atmosphere is much smaller compared to spaceborne platforms, which mitigates atmospheric effects. However, their disadvantage is that they require more flights to cover large areas due to their reduced flight time (Gago et al., 2015). As for other data, a possible solution is coupling UAV data with field- or spaceborne remote sensing data. The use of UAV data in hyperspectral remote sensing of LULC is rather recent (Huang et al., 2017; Natesan et al., 2017). As these data are characterized by a very high spatial resolution, they are expected to attract ever more attention since UAVs are expected to be increasingly capable of handling the relatively large spectral variability within and also between plots/fields, as demonstrated by some studies (Yuan et al., 2017). However, some studies report limited classification results, despite increased spectral discrimination between various LULC targets from UAV image-derived endmember pixels, possibly due to extensive processing of these data that involves corrections for radiative and geometric distortions and noise removal, for example (Hunt et al., 2017; Mitchell et al., 2012).

It is noteworthy that, regardless of the remote sensing platform for data acquisition (field-, air-, or spaceborne), hyperspectral data per second consist of hundreds of bands that cover the electromagnetic spectrum, thus comprising a hyperspectral data cube. While a single cube is highly manageable in terms of data processing, many LULC studies involve an ensemble of cubes for multiple purposes during multiple steps and so have the characteristics of "big data": large volume, from terabytes (TB = 1024 GB) to petabytes (PB = 1024 TB) and sometimes exabytes (EB = 1024 PB), multitemporality (collected on different dates) or multiresolution (collected on different scales), and near-real processing time (data should be analyzed in a reasonable time to achieve a given task) (Chi et al., 2016). Therefore, LULC studies using hyperspectral data require an increasingly practical handling of big data of massive volumes, machine learning, and artificial intelligence using supercomputers, which remains challenging. More recently, cloud computing platforms have been developed and optimized for data-intensive loads, parallel file systems, and database management and multilevel data locality (Ma et al., 2015).

8.6 CONCLUDING REMARKS

The rapid development of remote sensing technology makes it increasingly feasible to derive land-cover information from hyperspectral Earth observation data acquired at different observational scales. Numerous studies have demonstrated that hyperspectral imaging from satellite platforms is capable of producing accurate maps of land cover. Validation methods based on error matrix statistics have generally shown overall accuracies of 85% and kappa coefficients of 0.80 or higher. These are generally regarded as very satisfactory for many practical applications requiring regional land-cover maps. However, progress in field-scale remote sensing, such as field spectroradiometry in hyperspectral imaging analysis of LULC, has not yet been fully realized in practice. Many studies suggest that, indeed, prior to any work that implies land- cover analysis from remotely sensed air- and spaceborne imagery at the species level, knowledge of species spectral separability is useful and of vital importance. Yet, this is not a trivial task because the comparison of spectral signatures obtained from the field with those from satellite hyperspectral remote sensing sensors is hindered by various factors, such as the different physical setups of sensors and measurement environments, the latter being especially variable over time and space on the Earth's surface. Also, scale-related factors must be taken into account to minimize the remote measurement discrepancy between what is actually observed at the ground level and what is perceived from remotely sensed imagery before data can be of use.

Although a novel technology, field spectroradiometry in LULC studies, especially for vegetation, have led to remarkable achievements and have proven the significant capability of this technology in land-cover analysis over the past 15 years. Vegetation land cover, both agricultural and natural, remains the focus in most studies, as the spectral characteristics of plants are dynamically influenced by many internal and external factors. Thus, both multi- and univariate statistics are employed and have proven their ability to discriminate between vegetation land-cover reflectance, as determined by reflectance ANOVA, in VIS bands, and especially in the NIR spectrum, including the red edge. Single bands yielded by field-scale dimension-reduction studies for discrimination between targets of a certain type of land cover, such as vegetation species, are reported within the NIR, but findings must be confirmed on satellite imagery in future investigations. In addition, experiences gained through vegetation discriminant analyses at the field scale so far indicate an increasing importance of the SWIR spectrum, considering its ability to characterize water content in vegetation cover. Also, the link between remote sensing datasets acquired at different spatial scales is present but needs to be implemented in order to improve LULC mapping.

Finally, the use of UAVs have become increasingly popular in hyperspectral remote sensing of LULC because they offer both increased spatial and spectral resolution. All in all, UAVs can be used to improve estimation of land cover from hyperspectral remote sensing data. It is important to integrate knowledge on the spectral properties of land-cover targets and the factors affecting spectral variations across scales. In this framework, the synergy between field spectroradiometry, hyperspectral imaging, and processing techniques is and will be further investigated for various types of land cover.

ACKNOWLEDGMENTS

The authors wish to thank the editor for his useful comments on the chapter manuscript. Dr. Pandey's contribution was supported by the Scientific Engineering and Research Board (NPDF/2016/002487), India. Dr. Petropoulos' contribution was supported by the UK research council award (STFC ST/N006836/1) and by the EU Marie Curie Project ENViSIoN-EO.

REFERENCES

Abdel-Rahman, E.M., Ahmed, F.B. and van den Berg, M., 2010. Estimation of sugarcane leaf nitrogen concentration using in situ spectroscopy. *Int J Appl Earth Obs*, 12: S52–S57.

Adam, E. and Mutanga, O., 2009. Spectral discrimination of papyrus vegetation (*Cyperus papyrus* L.) in swamp wetlands using field spectrometry. *ISPRS J Photogramm*, 64(6): 612–620.

Alvarez-Añorve, M., Quesada, M. and de la Barrera, E., 2008. Remote sensing and plant functional groups physiology, ecology, and spectroscopy in tropical systems, Chapter 2. In: M. Kalacska and G.A. Sanchez-Azofeifa (Editors), *Hyperspectral Remote Sensing of Tropical and Sub-Tropical Forests*. CRC Press.

Artigas, F.J. and Yang, J.S., 2006. Spectral discrimination of marsh vegetation types in the New Jersey Meadowlands, USA. *Wetlands*, 26(1): 271–277.

ASD, A.S.D.I., 2009. *Technical Guide* (available upon request to ASD Inc.), Boulder, USA.

Avery, T.E. and Berlin, G.L., 1992. *Fundamentals of Remote Sensing and Airphoto Interpretation*. New York, NY: MacMillan Publishing Company.

Bai, X., Sharma, R.C., Tateishi, R., Kondoh, A., Wuliangha, B. and Tana, G., 2017. A Detailed and High-Resolution Land Use and Land Cover Change Analysis over the Past 16 Years in the Horqin Sandy Land, Inner Mongolia. *Mathematical Problems in Engineering*, 2017.

Baldridge, A.M., Hook, S.J., Grove, C.I. and Rivera, G., 2009. The ASTER spectral library version 2.0. *Remote Sens Environ*, 113(4): 711–715.

Barnsley, M.J., Settle, J.J., Cutter, M.A., Lobb, D.R. and Teston, F., 2004. The PROBA/CHRIS mission: A low cost smallsat for hyperspectral multiangle observations of the earth surface and atmosphere. *IEEE Trans. Geosci Remote Sens*, 42(7): 1512–1519.

Ben-Dor, E. et al., 2009. Using imaging spectroscopy to study soil properties. *Remote Sens Environ*, 113: S38–S55.

Ben Dor, E., Kafri, A. and Giancarlo, V., 2014. SHALOM: Spaceborne Hyperspectral Applicative Land And Ocean Mission: A Joint Project of ASI-ISA An Update For 2014. *A special session on hyperspectral sensor in orbit.* IGARSS July 2014, Quebec.

Ben-Dor, E., Schläpfer, D., Plaza, A.J. and Malthus, T., 2013. Hyperspectral remote sensing. In: *Airborne Measurements for Environmental Research: Methods and Instruments*, pp. 413–456.

Bock, M., Xofis, P., Mitchley, J., Rossner, G. and Wissen, M., 2005. Object-oriented method for habitat mapping at multiple scales—Case studies from northern Germany and Wye Downs, UK. *J Nat Conserv*, 13: 75–89.

Campbell, J.B. 1996. *Introduction to Remote Sensing*. New York: The Guilford Press, p. 622, ISBN-9781609181765.

Castro-Esau, K.L., Sanchez-Azofeifa, G.A. and Caelli, T., 2004. Discrimination of lianas and trees with leaf-level hyperspectral data. *Remote Sens Environ*, 90(3): 353–372.

Chang, J., Clay, S.A. and Clay, D.E., 2005. Clouds influence precision and accuracy of ground-based spectroradiometers. *Commun Soil Sci Plan*, 36(13–14): 1799–1807.

Chatziantoniou, A., Petropoulos, G.P. and Psomiadis, E., 2017. Co-Orbital Sentinel 1 and 2 for LULC mapping with emphasis on wetlands in a Mediterranean setting based on machine learning. *Remote Sens*, 9(12): 1259.

Chi, M.M. et al., 2016. Big data for remote sensing: Challenges and opportunities. *P IEEE*, 104(11): 2207–2219.

Chintan, A.S., Arora, M.K. and K.V. Pramod., 2004. Unsupervised classification of hyperspectral data: An ICA mixture model based approach. *International Journal of Remote Sensing*, 25: 481–487.

Clark, M. and Kilham, N., 2016. Mapping of land cover in northern California with simulated hyperspectral satellite imagery. *ISPRS J Photogramm Remote Sens*, 119: 228–245.

Clark, M.L., Roberts, D.A. and Clark, D.B., 2005. Hyperspectral discrimination of tropical rain forest tree species at leaf to crown scales. *Remote Sens Environ*, 96(3–4): 375–398.

Colombo, R., Meroni, M. and Rossini, M., 2016. Development of fluorescence indices to minimisethe effects of canopy structural parameters. *Annali di Botanica*, 6: 77–83.

Cracknell, A., 1998. Synergy in remote sensing-what's in a pixel? *Int J Remote Sens*, 19(11): 2025–2047.

Curran, P.J., 2001. Imaging spectrometry for ecological applications. *Int J Appl Earth Obs*, 3(4): 305–312.

Dalponte, M., Bruzzone, L. and Gianelle, D., 2009. Fusion of hyperspectral and LIDAR remote sensing data for the estimation of tree stem diameters, pp. II1008–II1011.

De Carvalho, O.A. and Meneses, P.R., 2000. Spectral Correlation Mapper (SCM); An Improvement on the Spectral Angle Mapper (SAM). *Summaries of the 9th JPL Airborne Earth Science Workshop, JPL Publication 00-18*, 9 p.

De Kok, Schneider, T. and Ammer, U., 1999. Object-based classification and applications in the alpine forest environment. *In fusion of sensor data, knowledge sources and algorithms: Proceedings of the joint ISPRS/EARSEL workshop*, 3–4 June 1999, Valladolid, Spain. International of Archives Photogrammetry and Remote Sensing, 32: 7-4-3 W6

Dimitrakopoulos, A.P., 2001. A statistical classification of Mediterranean species based on their flammability components. *Int J Wildland Fire*, 10(2): 113–118.

Ding, Y., Zhao, K., Zheng, X. and Jiang, T., 2014. Temporal dynamics of spatial heterogeneity over cropland quantified by time-series NDVI, near infrared and red reflectance of Landsat 8 OLI imagery. *Int J Appl Earth Obs Geoinf*, 30(Suppl C): 139–145.

Du, P., Tan, K. and Xing, X., 2010. Wavelet SVM in reproducing kernel Hilbert space for hyperspectral remote sensing image classification. *Optics Communications*, in press.

Eklundh, L., Jin, H., Schubert, P., Guzinski, R. and Heliasz, M., 2011. An opticalsensor network for vegetation phenology monitoring and satellite data calibration. *Sensors-Basel*, 11(8): 7678–7709.

ENVI, 2009. *ENVI User's Guide*. ENVI Software.

ESA. 2015. Report for mission selection: FLEX, *ESA SP-1330/2 (2 volume series)*, European Space Agency, noordwijk, The Netherlands.

Feingersh, T. and Dor, E.B., 2015. SHALOM–A commercial hyperspectral space mission. In: S.-E. Qian (Editor), *Optical Payloads for Space Missions*. John Wiley & Sons Ltd, Chichester, UK, p. 247.

Fisher, P., 1997. The pixel: a snare and a delusion. *Int J Remote Sens*, 18(3): 679–685.

Fletcher, F. 2015. Carbonsat FLEX-report for mission selection. An earth explorer to observe vegetation fluorescence. European Space Agency, Noordwijk, the Netherlands. Available online: http://esamultimedia.esa.int/docs/EarthObservation/SP1330-2_FLEX.pdf

Frank, M. and Menz, G., 2003. Detecting Seasonal Changes in a Semi-arid Environment Using Hyperspectral Vegetation Indices. *Proceedings of the 3rd EARSel Workshop on Imaging Spectrometry*, pp. 504–512.

Gago, J. et al., 2015. UAVs challenge to assess water stress for sustainable agriculture. *Agric Water Manage*, 153: 9–19.

Garcia-Mora, T.J., Mas, J.F. and Hinkley, E.A., 2012. Land cover mapping applications with MODIS: a literature review. *Int J Digit Earth*, 5(1): 63–87.

Goerner, A., Reichstein, M. and Rambal, S., 2009. Tracking seasonal drought effects on ecosystem light use efficiency with satellite-based PRI in a Mediterranean forest. *Remote Sens Environ*, 113(5): 1101–1111.

Goodenough, D.G., Dyk, A., Niemann, K.O., Pearlman, J.S., Chen, H., Han, T., Murdoch, M. and West, C., 2003. Processing Hyperion and ALI for forest classification. *IEEE transactions on geoscience and remote sensing*, 41(6): 1321–1331.

Guidici, D. and Clark, M.L., 2017. One-dimensional convolutional neural network land-cover classification of multi-seasonal hyperspectral imagery in the San Francisco Bay Area, California. *Remote Sens*, 9(6): 629.

Han, T., Goodenough, D.G., Dyk, A. and Love, L., 2002. Detection and correction of abnormal pixels in Hyperion images. *Proceedings of the IEEE International Geoscience and Remote Sensing Symposium (IGARSS'02).* Toronto (Ontario): Canada, vol. 3, pp. 1327–1330.

Harris, J.W. and Stocker, H., 1998. Maximum Likelihood Method. *§21.10.4 in Handbook of Mathematics and Computational Science.* New York: Springer-Verlag, p. 824.

Hemmer, T.H. and Westphal, T.L., 2000. Lessons learned in the post-processing of field spectroradiometric data covering the 0.4 to 2.5 im wavelength region. *Proceedings of SPIE. The International Society for Optical Engineering*, pp. 249–260.

Hostert, P., Roder, A. and Hill, J., 2003. Coupling spectral unmixing and trend analysis for monitoring of long-term vegetation dynamics in Mediterranean rangelands. *Remote Sens Environ*, 87(2–3): 183–197.

Huang, J. et al., 2017. Juvenile tree classification based on hyperspectral image acquired from an unmanned aerial vehicle. *Int J Remote Sens*, 38(8–10): 2273–2295.

Hueni, A., Nieke, J., Schopfer, J., Kneubuhler, M. and Itten, K.I., 2009. The spectral database SPECCHIO for improved long-term usability and data sharing. *Comput Geosci-UK*, 35(3): 557–565.

Hughes, R.F., 1968. On the mean accuracy of statistical pattern recognizers. *IEEE Trans. Information Theory, IT*-14(1): 55–63.

Hunt, E.R. et al., 2017. Monitoring nitrogen status of potatoes using small unmanned aerial vehicles. *Precis Agric.*

Ireland, G. and Petropoulos, G.P., 2015. Exploring the relationships between post-fire vegetation regeneration dynamics, topography and burn severity: A case study from the Montane Cordillera Ecozones of Western Canada. *Applied Geography*, 56: 232–248.

Jacobsen, A., Broge, N.H. and Hansen, B.U., 1995. Monitoring Wheat Fields and Grasslands using Spectral Reflectance Data.*79 International Symposium on Spectral Sensing Research (ISSSR)*, Victoria, Australia.

Jonathan, A., 2015. Flex satellite will map Earth's plant glow. *BBC News*. Retrieved. 2015. 11–20.

Kahle, A.B. and Goetz, A.F., 1983. Mineralogic information from a new airborne thermal infrared multispectral scanner. *Science*, 222(4619): 24–27.

Karimi, Y., Prasher, S., Patel, R. and Kim, S., 2006. Application of support vector machine technology for weed and nitrogen stress detection in corn. *Computers and Electronics in agriculture*, 51(1): 99–109.

Kavzoglu, T. and Colkesen, I., 2009. A kernel functions analysis for support vector machines for land cover classification. *Int J Appl Earth Observ Geoinf*, 11(5): 352–359.

Knox, R.G. et al., 2010. PS 79-114: The Hyperspectral Infrared Imager (HyspIRI) mission: A new capability for global ecological research and applications. *The 95th ESA Annual Meeting.*

Kochubey, S.M. and Kazantsev, T.A., 2007. Changes in the first derivatives of leaf reflectance spectra of various plants induced by variations of chlorophyll content. *J Plant Physiol*, 164(12): 1648–1655.

Kochubey, S.M. and Kazantsev, T.A., 2012. Derivative vegetation indices as a new approach in remote sensing of vegetation. *Front Earth Sci-Prc*, 6(2): 188–195.

Koetz, B., Morsdorf, F., Van der Linden, S., Curt, T. and Allgöwer, B., 2008. Multi-source land cover classification for forest fire management based on imaging spectrometry and LiDAR data. *Forest Ecology and Management*, 256(3): 263–271.

Kruse, F.A., Lefkoff, A., Boardman, J., Heidebrecht, K., Shapiro, A., Barloon, P. and Goetz, A., 1993. The spectral image processing system (SIPS)—interactive visualization and analysis of imaging spectrometer data. *Remote sensing of environment*, 44(2-3): 145–163.

Lacar, F.M., Lewis, M.M. and Grierson, I.T., 2001. Use of hyperspectral reflectance for discrimination between grape varieties. *Geoscience and Remote Sensing Symposium, 2001. IGARSS '01. IEEE 2001 International*, vol.6, pp. 2878–2880.

Lamine, S. et al., 2017. Quantifying land use/land cover spatio-temporal landscape pattern dynamics from Hyperion using SVMs classifier and FRAGSTATS®. *Geocarto Int*, 1–17.

Laurin, G.V. et al., 2016. Discrimination of tropical forest types, dominant species, and mapping of functional guilds by hyperspectral and simulated multispectral Sentinel-2 data. *Remote Sens Environ*, 176: 163–176.

Li, D-C., C-W., Liu, 2010. A class possibility based kernel to increase classification accuracy for small data sets using support vector machines. *Expert Systems with Applications*, 37, 3104–3110.

Liu, J. and Li, J., 2013. Land-use and land-cover analysis with remote sensing images, *Information Science and Technology (ICIST)*, 2013 *International Conference on. IEEE*, pp. 1175-1177.

Lu, D. and Weng, Q., 2004. Spectral mixture analysis of the urban landscape in Indianapolis with Landsat ETM+ imagery. *Photogramm Eng Remote Sens*, 70(9): 1053–1062.

Lu, D. and Weng, Q., 2007. A survey of image classification methods and techniques for improving classification performance. *International Journal of Remote Sensing*, 28(5): 823–870.

Ma, Y. et al., 2015. Remote sensing big data computing: Challenges and opportunities. *Future Gener Comput Syst*, 51: 47–60.

Manakos, I., Manevski, K., Petropoulos, G.P., Elhag, M. and Kalaitzidis, C., 2010. Development of a spectral library for Mediterranean land cover types. In: R. Reuter (Editor), *30th EARSeLSymp.: Remote Sensing for Science, Education and Natural and Cultural Heritage*. EARSeL, Chania, Greece, pp. 663–668.

Manevski, K., Jabloun, M., Gupta, M. and Kalaitzidis, C., 2017. Chapter 6—Field-Scale Sensitivity of Vegetation Discrimination to Hyperspectral Reflectance and Coupled Statistics. In: G.P. Petropoulos and P.K. Srivastava (Editors), *Sensitivity Analysis in Earth Observation Modelling*. Elsevier, pp. 103–121.

Manevski, K., Manakos, I., Petropoulos, G.P. and Kalaitzidis, C., 2011. Discrimination of common Mediterranean plant species using field spectroradiometry. *Int J Appl Earth Obs*, 13(6): 922–933.

Manevski, K., Manakos, I., Petropoulos, G.P. and Kalaitzidis, C., 2012. Spectral Discrimination of Mediterranean Maquis and Phrygana Vegetation: Results From a Case Study in Greece. *IEEEJ-Stars*, 5(2): 604–616.

Mariotti, M., Ercoli, L. and Masoni, A., 1996. Spectral properties of iron-deficient corn and sunflower leaves. *Remote Sens Environ*, 58(3): 282–288.

Miao, X., Gong, P., Swope, S., Pu, R., Carruthers, R., Anderson, G., Heaton, J. and Tracy, C.R. 2006. Estimation of yellow starthistle abundance through CASI-2 hyperspectral imagery using linear spectral mixture models, *Remote Sensing of Environment*, 101:329–341.

Milton, E.J., Schaepman, M.E., Anderson, K., Kneubuhler, M. and Fox, N., 2009. Progress in field spectroscopy. *Remote Sens Environ*, 113: S92–S109.

Mitchell, J.J. et al., 2012.Unmanned aerial vehicle (UAV) hyperspectral remote sensing for dryland vegetation monitoring.*2012 4th Workshop on Hyperspectral Image and Signal Processing: Evolution in Remote Sensing (WHISPERS)*, pp. 1–10.

Morrey, J.R., 1968. On determining spectral peak positions from composite spectra with a digital computer. *Anal Chem*, 40(6): 905–914.

Mutanga, O., Skidmore, A.K. and van Wieren, S., 2003. Discriminating tropical grass (Cenchrusciliaris) canopies grown under different nitrogen treatments using spectroradiometry. *ISPRS J Photogramm*, 57(4): 263–272.

Natesan, S., Benari, G., Armenakis, C. and Lee, R., 2017. Land cover classification using a UAV-borne spectrometer. *Int Arch Photogramm Remote Sens Spatial Inf Sci*, XLII-2/W6: 269–273.

Nicodemus, F.F., Richmond, J.C., Hsia, J.J., Ginsberg, I.W. and Limperis, T.L., 1977. Geometrical considerations and nomenclature for reflectance (Vol. Monograph 160). Washington D.C. 20402: National Bureau of Standards.

Nidamanuri, R.R. and Zbell, B., 2011. Transferring spectral libraries of canopy reflectance for crop classification using hyperspectral remote sensing data. *Biosyst Eng*, 110(3): 231–246.

Okin, G.S., Roberts, D.A., Murray, B. and Okin, W.J., 2001. Practical limits on hyperspectral vegetation discrimination in arid and semiarid environments. *Remote Sens Environ*, 77(2): 212–225.

Otukei, J.R. and Blaschke, T., 2010. Land cover change assessment using decision trees, support vector machines and maximum likelihood classification algorithms. *Int J Appl Earth Observ Geoinf*, S12: S27–S31.

Ouyang, Z.T. et al., 2013. Spectral Discrimination of the Invasive Plant Spartina alterniflora at Multiple Phenological Stages in a Saltmarsh Wetland. *Plos One*, 8(6).

Pádua, L. et al., 2017. UAS, sensors, and data processing in agroforestry: a review towards practical applications. *Int J Remote Sens*, 38(8–10): 2349–2391.

Pal, N.R. and Pal, S.K., 1993. A review of image segmentation techniques. *Pattern Recognit*, 26: 1277–1294.

Pal, M., 2006. Support vector machine-based feature selection for land cover classification: A case study with DAIS hyperspectral data. *Int J Remote Sens*, 27(14): 2877–2894.

Pal, M. and Mather, P.M., 2005. Some issues in the classification of DAIS hyperspectral data. *Int J Remote Sens*, 27: 2895–2916.

Pandey, P.C., Tate, N.J. and Balzter, H., 2014. Mapping tree species in coastal Portugal using statistically segmented principal component analysis and other methods. *IEEE Sens J*, 14(12): 4434–4441.

Pearlman, J.S., Barry, P.S., Segal, C.C., Shepanski, J., Beiro, D. and Carman, S.L., 2003. Hyperion a space based imaging spectrometer. *IEEE Trans. Geosci. Remote Sens.*, 41(6), 1160–1173.

Petropoulos, G.P., Kalaitzidis, C. and Vadrevu, K.P., 2011. Support vector machines and object-based classification for obtaining land use/cover cartography from hyperion hyperspectral imagery. *Comput Geosci*, 41: 99–107.

Petropoulos, G.P., Arvanitis, K. and Sigrimis, N., 2012a. Hyperion hyperspectral imagery analysis combined with machine learning classifiers for land use/cover mapping. *Expert Syst Appl*, 39(3): 3800–3809.

Petropoulos, G.P., Vadrevu, K.P. and Kalaitzidis, C., 2012b. Spectral angle mapper and object-based classification combined with hyperspectral remote sensing imagery for obtaining land use/cover mapping in a Mediterranean region. *Geocarto Int*, 1–16.

Petropoulos, G.P., Manevski, K. and Carlson, T.N., 2014. Hyperspectral remote sensing with emphasis on land cover mapping: From ground to satellite observations. *Scale Issues Remote Sens*, 285–320.

Pfitzner, K., Bollhöfer, A. and Carr, G., 2006. A standard design for collecting vegetation reference spectra: Implementation and implications for data sharing. *J Spatial Sci*, 51(2): 79–92.

Pieters, C. et al., 2009. Character and spatial distribution of OH/H2O on the surface of the Moon seen by M3 on Chandrayaan-1. *Science*, 326(5952): 568–572.

Pignatti, S., Cavalli, R.M., Cuomo, V., Fusilli, V., Pascucci, S., Poscolieri, M. and Santini, F., 2009. Evaluating Hyperion capability for land cover mapping in a fragmented ecosystem: Pollino National Park, Italy, *Remote Sens Environ*, 113(3): 622–634.

Prasad, K.A. and Gnanappazham, L., 2015. Multiple statistical approaches for the discrimination of mangrove species of Rhizophoraceae using transformed field and laboratory hyperspectral data. *Geocarto Int*, 1–22.

Plaza, A., Benediktsson, J.A., Boardman, J.W., Brazile, J., Bruzzone, L., Camps-Valls, G., Chanussot, J., Fauvel, M., Gamba, P. and Gualtieri, A., 2009. Recent advances in techniques for hyperspectral image processing. *Remote Sensing of Environment*, 113: S110–S122.

Plaza, J., Plaza, A., Perez, R. and Martinez P., 2009. On the use of small training sets for neural network-based characterisation of mixed pixels in remotely sensed hyperspectral images, *Pattern Recognition*, 42: 3032–3045.

Poursanidis, D. and Chrysoulakis, N., 2017. Remote sensing, natural hazards and the contribution of ESA sentinels missions. *Remote Sens Appl Soc Environ*, 6: 25–38.

Prasad, K.A. and Gnanappazham, L., 2015. Multiple statistical approaches for the discrimination of mangrove species of Rhizophoraceae using transformed field and laboratory hyperspectral data. *Geocarto Int*, 1–22.

Price, J.C., 1994. How unique are spectral signatures? *Remote Sens Environ*, 49(3): 181–186.

Price, J.C., 1998. An approach for analysis of reflectance spectra. *Remote Sens Environ*, 64(3): 316–330.

Psomas, A., Zimmermann, N.E., Kneubühler, M., Kellenberger, T. and Itten, K., 2005. Seasonal variability in spectral reflectance for discriminating grasslands along a dry- mesic gradient in Switzerland. *4th EARSeL Workshop on Imaging Spectroscopy. New Quality in Environmental Studies*. EARSeL Warsaw, Poland, pp. 655–666.

Rao, N.R., Garg, P.K. and Ghosh, S.K., 2007. Development of an agricultural crops spectral library and classification of crops at cultivar level using hyperspectral data. *Precis Agric*, 8(4–5): 173–185.

Rasaiah, B.A., Jones, S.D., Bellman, C. and Malthus, T., 2014. Critical metadata for spectroscopy field campaigns. *Remote Sens*, 6(5): 3662–3680.

Rasaiah, B.A., Bellman, C., Jones, S.D., Malthus, T.J. and Roelfsema, C., 2015a. Towards an interoperable field spectroscopy metadata standard with extended support for marine specific applications. *Remote Sens*, 7(11): 15668–15701.

Rasaiah, B.A., Jones, S.D., Bellman, C., Malthus, T.J. and Hueni, A., 2015b. Assessing Field Spectroscopy Metadata Quality. *Remote Sens*, 7(4): 4499–4526.

Robson, C., 1994. *Experiment, Design and Statistics in Psychology*. Penguin Books.

Salisbury, J.W., 1998. *Spectral Measurements Field Guide*. USA: Defense Technical Information Center.

Salvaggio, C., Smith, L.E. and Antoine, E.J., 2005. Spectral signature databases and their application/misapplication to modeling and exploitation of multispectral/hyperspectral data. *Algorithms and Technologies for Multispectral, Hyperspectral, and Ultraspectral Image*. SPIE, pp. 531–541.

Sanchez-Hernandez, C., Boyd, D.S. and Foody, G., 2007. Mapping specific habitats from remotely sensed imagery: Support vector machine and support vector data description based classification of coastal saltmarsh habitats. *Ecological Inform*, 2: 83–88.

Schaaf, C.B., 2009. Albedo and reflectance anisotropy: Assessment of the status of the development of the standards for the Terrestrial Essential Climate Variables. *Global Terrestrial Observing System.*

Schmidt, K.S. and Skidmore, A.K., 2003. Spectral discrimination of vegetation types in a coastal wetland. *Remote Sens Environ*, 85(1): 92–108.

Singh, S.K. et al., 2016. Landscape transform and spatial metrics for mapping spatiotemporal land cover dynamics using Earth Observation datasets. *Geocarto Int*, 1–16.

Small, C., 2001. Estimation of urban vegetation abundance by spectral mixture analysis. *International Journal of Remote Sensing*, 22: 1305–1334.

Thenkabail, P.S., Enclona, E.A., Ashton, M.S., Legg, C. and De Dieu, M.J., 2004a. Hyperion, IKONOS, ALI, and ETM plus sensors in the study of African rainforests. *Remote Sens Environ*, 90(1): 23–43.

Thenkabail, P.S., Enclona, E.A., Ashton, M.S. and Van der Meer, B., 2004b. Accuracy assessments of hyperspectral waveband performance for vegetation analysis applications. *Remote Sens Environ*, 91(3–4): 354–376.

Vaiphasa, C., 2006. Consideration of smoothing techniques for hyperspectral remote sensing. *ISPRS J Photogramm*, 60(2): 91–99.

Vaiphasa, C., Ongsomwang, S., Vaiphasa, T. and Skidmore, A.K., 2005. Tropical mangrove species discrimination using hyperspectral data: A laboratory study. *Estuar Coast Shelf S*, 65(1–2): 371–379.

van Aardt, J.A.N. and Wynne, R.H., 2001. Spectral separability among six southern tree species. *Photogramm Eng Rem S*, 67(12): 1367–1375.

Van Aardt, J.A.N. and Wynne, R.H., 2007. Examining pine spectral separability using hyperspectral data from an airborne sensor: An extension of field-based results. *Int J Remote Sens*, 28(1–2): 431–436.

Vapnik, V., 1995. *The nature of statistical learning theory*, New York, NY: Springer-Verlag,

Vorovencii, I., 2009. The hyperspectral sensors used in satellite and aerial remote sensing. *Bulletin of the Transilvania University of Braşov* Vol, 2: 51–56.

Walsh, S. J., et al. 2008. QuickBird and Hyperion data analysis of an invasive plant species in the Galapagos Islands of Ecuador: Implications for control and land use management. *Remote Sensing of Environment*, 112: 1927–1941.

Wang, J., Chen, Y., He, T., Lv C. and Liu, A., 2010. Application of geographic image cognition approach in land type classification using Hyperion image: A case study in China. *International Journal of Applied Earth Observation and Geoinformaiton*, 12S: S212–S222

Xie, Y.C., Sha, Z.Y. and Yu, M., 2008. Remote sensing imagery in vegetation mapping: A review. *J Plant Ecol-UK*, 1(1): 9–23.

Yan, G., et al., 2006. Comparison of pixel-based and object-oriented image classification approaches—a case study in a coal fire area, Wud, Inner Mongolia, China. *International Journal of Remote Sensing*, 27(18), 4039–4055.

Yuan, H. et al., 2017. Retrieving soybean leaf area index from unmanned aerial vehicle hyperspectral remote sensing: analysis of RF, ANN, and SVM regression models. *Remote Sens*, 9(4): 309.

Zhang, J.K., Rivard, B., Sanchez-Azofeita, A. and Castro-Esau, K., 2006. Intra and inter-class spectral variability of tropical tree species at La Selva, Costa Rica: Implications for species identification using HYDICE imagery. *Remote Sens Environ*, 105(2): 129–141.

Zhang, M., Qin, Z., Liu, X. and Ustin, S.L., 2003. Detection of stress in tomatoes induced by late blight disease in California, USA, using hyperspectral remote sensing. *Int J Appl Earth Obs*, 4(4): 295–310.

Zhang, R. and Ma, J., 2009. Feature selection for hyperspectral data based on recursive support vector machines. *International Journal of Remote Sensing*, 30, 14, 3669–3677.

Zhang, L.F., Jiao, W.Z., Zhang, H.M., Huang, C.P. and Tong, Q.X., 2017. Studying drought phenomena in the Continental United States in 2011 and 2012 using various drought indices. *Remote Sens Environ*, 190: 96–106.

Zomer, R.J., Trabucco, A. and Ustin, S.L., 2009. Building spectral libraries for wetlands land cover classification and hyperspectral remote sensing. *J Environ Manage*, 90(7): 2170–2177.

9 Hyperspectral Remote Sensing for Forest Management

Valerie Thomas

CONTENTS

9.1 INTRODUCTION

Forests cover about 30% of the Earth's terrestrial surface (FAO, 2016), play a significant role in the climate system, and are integral to numerous ecosystem, cultural, and economic services (Figure 9.1). As such, there is considerable interest in sustainable management to conserve forest resources while balancing the competing interests for their use. Forest management is a multifaceted endeavor, defined as the "application of biological, physical, quantitative, managerial, economic, social, and policy principles to the regeneration, management, utilization, and conservation of forests to meet specified goals and objectives while maintaining the productivity of the forest" (Helms, 1998, p. 71). In practical terms, this often means some type of silviculture practice, protection activity, or forest regulation.

Given the large, and often remote, land area covered by forests, remote sensing technologies have been widely adopted as part of operational forest management portfolios, mainly to monitor the location, type, and amount of forests, as well as changes in them over time. To date, the use of remote sensing for forest management has been largely driven by aerial photography and multispectral satellite imagery (such as Landsat or SPOT) and (more recently) lidar. Hyperspectral technology may allow us to expand our remote characterizations to examine species and biodiversity, forest health and condition, stand structure, threats (such as stress or invasive species), and forest ecosystem function (Carter, 1994; Goetz, 1995; Lichtenhaler et al., 1996; Martin and Aber, 1996; Green et al., 1998; Merton, 1998; Ustin and Trabucco, 2000; Curran, 2001; Rocchini et al., 2010; Treitz et al., 2010; Féret and Asner, 2014; Stein et al., 2014; Thenkabail et al., 2014; Somers et al., 2015; Asner et al., 2017). Indeed, most of the vegetation applications of hyperspectral remote sensing discussed in other chapters of this book could be applicable to forest management. However, despite the many potential benefits of this technology, the use of hyperspectral imagery for operational forest

FIGURE 9.1 Ecosystem, economic, and cultural services of forests.

management applications has been extremely limited. Most of the work has been in the realm of scientific research or pilot/demonstration studies that show the potential value of hyperspectral data in forest conservation, monitoring, and inventory. The reluctance by practicing forest managers to adopt adoption hyperspectral remote sensing on a large scale stems from (1) the size and complexity of both the hyperspectral data sets and forest ecosystems and (2) lack of repeated hyperspectral observations for large regions (in other words, lack of accessible satellite data).

9.2 COMPLEXITIES OF FOREST ECOSYSTEMS

Many factors affect the reflectance of all vegetation and confound remote sensing analysis. These include phenology, insolation, illumination geometry, soil characteristics, the nutrient regime, hydrology, and spectral similarities between many species. In fact, it has been demonstrated that the inherent within-species variability in reflectance, combined with spectral similarities across species, has made some species indistinguishable using reflectance alone (Hoffbeck and Landgrebe, 1996). Analysis of forest ecosystems is further confounded by highly variable three-dimensional structures, particularly in mixed-wood canopies (Figure 9.2). Forests often have complex horizontal and vertical species mixtures (causing layering in the canopy), complex canopy architecture, variable height and biomass, and within- and across-species variability in leaf area and in foliar biochemistry. Variability in leaf morphology and foliar biochemistry has also been demonstrated within a single canopy, particularly between sunlit leaves near the top of the canopy and shaded leaves below (Gholz et al., 1991; Vose et al., 1994; Demerez et al., 1999; O'Neil et al., 2002). Complex canopy architecture, particularly the size and location of gaps within the canopy, also influences the light regime (Hardy et al., 2004) and can influence the rate of photosynthesis and absorption of light (Todd et al., 2003; Thomas et al., 2006a).

When analyzing remote sensing data, the inherent heterogeneity of a forest ecosystem can be further magnified by the pixel resolution of the sensor. If the size of the pixel is greater than the size of a single tree canopy, the reflectance signal may contain mixed effects of shadow, nonleaf reflectance, or species mixtures, for example (Woodcock and Strahler, 1987; St-Onge and Cavayas,

FIGURE 9.2 Complex forest canopy architecture. (a) Evidence of variable species, canopy height, and layering in a boreal mixed-wood forest. (b) Below-canopy variability of a boreal mixed-wood forest.

1995; Curran and Atkinson, 1999; Treitz and Howarth, 2000; Treitz, 2001). This has resulted in considerable research efforts into canopy reflectance modeling in an attempt to quantify or remove some of the nonleaf reflectance from signals. Of particular relevance is the body of research in geometric modeling, which assumes that a forest is comprised of multiple objects with quantifiable dimensions, shapes, and arrangements. The simplest of these models, originally developed for crop canopies, are designed to simulate the effect of shadow on canopy reflectance, where the fractions of vegetation, soil, and their shadowed components are quantified (e.g., Jahnke and Lawrence, 1965; Terjung and Louie, 1972; Jackson et al., 1979).

More complex models combine canopy structural geometry and the principles of radiative transfer within crowns, such as the geometric optical-radiative transfer model (GORT) (Li et al., 1995). These models, referred to as hybrid radiative transfer models, have been used to characterize canopy structure and foliar biochemistry in a variety of heterogeneous forest environments (Kuusk, 1998; Dawson et al., 1999; Demerez and Gastellu-Etchegorry, 2000; Hu et al., 2000; Gastellu-Etchegorry and Bruniquel-Pinel, 2001; Zarco-Tejada et al., 2001; Gemmell et al., 2002; Kimes et al., 2002; Kötz et al., 2004).

Another approach to directly characterizing/measuring/accounting for canopy heterogeneity is the fusion of light detection and ranging (lidar) with hyperspectral data, which has also been shown to provide better information on canopy structure (lidar), biochemistry (hyperspectral), and function (fused). Lidar has been shown to be the ideal technology for characterizing canopy structure, including height, crown shape, leaf area, biomass, and basal area (e.g., Næsset, 1997; Magnussen and Boudewyn, 1998; Means et al., 2000; Lim and Treitz, 2004; Hopkinson et al., 2005; Thomas et al., 2006b). Unlike hyperspectral technology, lidar has been adopted in many places for operational use in forest inventories (e.g., Næesset, 2007; Woods et al., 2011). Fused lidar and hyperspectral data have improved our ability to characterize structure and biochemical variables (e.g., Koetz et al., 2006; Asner et al., 2007, 2011, 2015; Thomas et al., 2008) and have yielded substantial information about species, ecosystem biodiversity, and function (Varga and Asner, 2008; Colgan et al., 2012; Alonzo et al., 2014; Asner et al., 2017).

9.3 FOREST MANAGEMENT APPLICATIONS OF HYPERSPECTRAL REMOTE SENSING

A number of applications for hyperspectral remote sensing can directly benefit forest management programs that have national and local implementations. These include, but are not limited to,

TABLE 9.1
Selected Forest Inventory Variables That Have Been Mapped with Remote Sensing

Forest Characteristic	Definition	Citation
Mean dominant height (m)	Mean height of 100 largest trees in a hectare	Brack and Marshall (1998), Garcia (1998)
Quadratic mean diameter at breast height (cm)	$\sqrt{\frac{\sum \text{DBH}_i^2}{n}}$	Curtis and Marshall (2000)
Total above ground biomass (Mg/ha)	Calculated with allometric model of the form $b \times \text{DBH}^2 \times$ height	Alemdag (1983, 1984)
Crown closure (%)	Ground covered by tree crowns when viewed from above	Avery and Burkhart (2002)
Total crown projected area (m^2/m^2)	$\frac{\sum \pi(\text{crown width})^2}{\text{plot area}}$	Avery and Burkhart (2002), Thomas et al. (2006a,b)
Basal area (m^2/ha)	$\frac{\pi}{40{,}000}\sqrt{\frac{\sum \text{DBH}_i^2}{\text{plot area}}}$	Avery and Burkhart (2002)
Stem density (#/ha)	Number of trees per hectare	Niemann and Goodenough (2003)

information to supplement forest inventories (e.g., species and biophysical variables), improved estimation of biomass and carbon (of interest for a variety of purposes, including climate modeling), and mapping of wildfire fuels for forest fire risk detection.

9.3.1 Forest Inventories

Forest inventories involve a systematic collection of data about forest stands that are used for a variety of local and national purposes. Many countries have their own well-developed and long-standing system of standardized data collection, which is usually very manually intensive. For example, in the United States, the Forest Inventory and Analysis national program has been run by the U.S. Department of Agriculture Forest Service since 1930. Finland has a National Forest Inventory program that has been operating since 1921. Although the specific methods of data collection vary by country, most involve a systematic sampling of trees to attain representative data to characterize a given forest stand. At minimum, this includes species, height, and diameter at breast height (DBH). From these, stand basal area can be calculated and allometric equations can be used to model biomass, volume, and carbon (Table 9.1). At the stand level, some measurement is also usually made of stand age and quality. Despite the complexity of forest environments discussed earlier, hyperspectral remote sensing can provide information that benefits forest inventories, particularly for improved species classifications and mapping certain biophysical variables.

9.3.1.1 Forest Species Mapping (and Species Richness)

Numerous authors have demonstrated the potential of hyperspectral remote sensing to improve species mapping. A number of authors have used hyperspectral data and indices known to be related to foliar pigments to discriminate between canopy species across landscapes (Wessman et al., 1988; Fuentes et al., 2001; Clark et al., 2003; Kokaly et al., 2003; Townsend et al., 2003; Plourde et al., 2007). Others have developed techniques to use fused lidar and hyperspectral data for species mapping, which enables the spectral and structural characteristics of canopies to be examined concurrently (Asner et al., 2008a,b; Thomas et al., 2009).

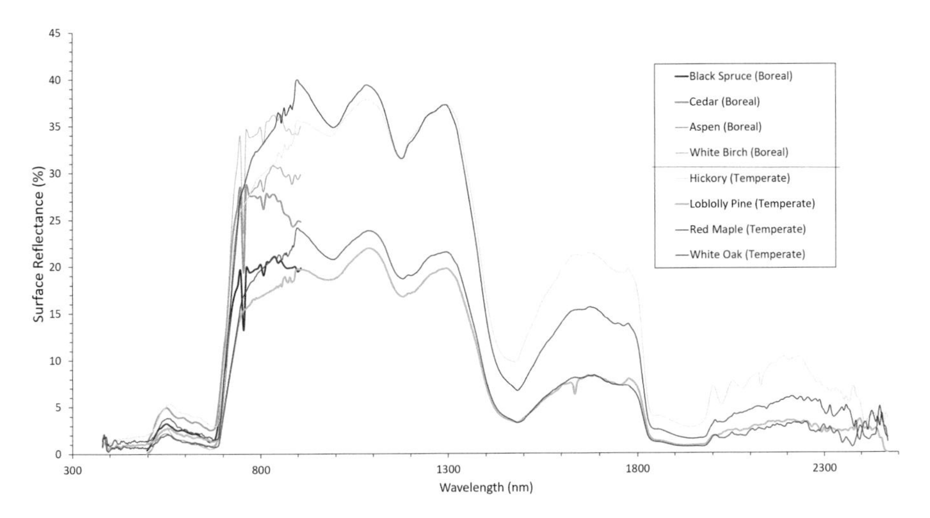

FIGURE 9.3 Spectral signatures for selected boreal and temperate forest species. The boreal species signatures were acquired from a CASI hyperspectral sensor with 56 bands from 498 to 916 nm. The temperate forest signatures were acquired with the European Space Agency's HyPlant instrument, with 629 bands from 380 to 2,537 nm.

Asner et al. (2008a,b) made significant contributions toward developing more robust techniques for species mapping using a combination of hyperspectral and lidar data. They developed techniques to identify invasive species in Hawaiian forests, with the hope of eventually being able to map forest spread over time. Their approach, which builds upon previous work by Blackburn (2002), is to develop masks of sunlit tree crowns at predetermined heights that are clearly visible to an airborne or satellite sensor, thereby eliminating the confounding effect of gaps, shadows, and canopy architecture. Superimposing this mask on a hyperspectral image, Asner et al. then used a two-stage spectral mixture analysis approach to determine (1) the fraction of live and dead canopy within the sunlit areas and (2) the fraction of individual species within the sunlit crown.

There have also been numerous studies in which the spectral reflectance of the hyperspectral data has been matched directly to the species of interest. Spectral libraries have been developed for many species, under a variety of phenological conditions. This has been done at the leaf scale, using a portable handheld spectroradiometer (Clark et al., 2005), and at the canopy scale, by locating pure pixels (i.e., pixels that contain only one species, also referred to as an endmember) within a hyperspectral image (Kokaly et al., 2003) (Figure 9.3).

Spectral matching algorithms, such as the Spectral Angle Mapper (SAM), can be used to match spectral patterns of pixels in an image to endmembers by determining the cosine of the angle in spectral space, where a low angle implies a close match (Buddenbaum et al., 2005) (Figure 9.4). This approach assumes that the spectral response in a pixel is pure, and it tends to work better at higher spatial resolutions where there is less mixture of species within a pixel or for broader classes that include multiple species with a similar spectral response. In cases where there is a known mixture of desired classes within pixels, spectral mixture analysis has been used to quantify the fraction of each species within pixels (Darvishefat et al., 2002).

In recent years, research has emerged that embraces the complexity of heterogeneous forest or other ecosystems as a characteristic that informs us of its biodiversity. The "spectral diversity" or "spectral variation" hypothesis argues that biodiversity should be correlated with spectral heterogeneity, and numerous authors have supported this for a range of different ecosystem types (Carlson et al., 2007; Rocchini et al., 2010; Féret and Asner, 2014; Warren et al., 2014; Heumann et al., 2015). Asner et al. (2017) have used hyperspectral maps of canopy biochemistry and leaf traits to model functional diversity in the Amazon rainforest. Further, the Carnegie Airborne Observatory is developing a

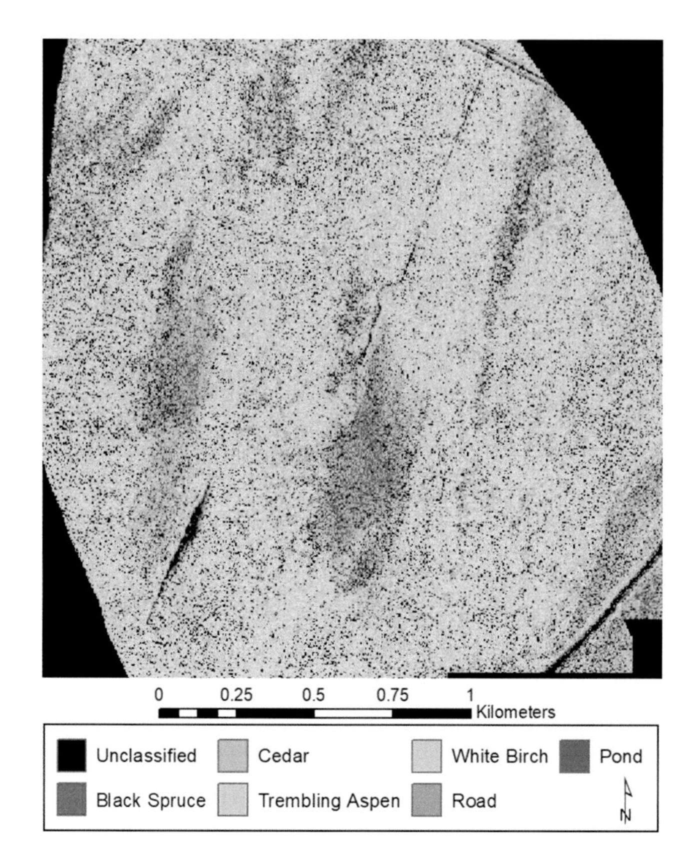

FIGURE 9.4 SAM classification of boreal forest using boreal spectral signatures shown in Figure 9.3 as endmembers.

spectronomic database that links spectral signatures, canopy biochemistry, and phylogenetics for tropical forests (https://cao.carnegiescience.edu/spectranomics). Again, most of this biodiversity work is at the research level and has not yet been successfully implemented for real operational forest management scenarios. However, there are clear forest management applications for these approaches, particularly in the context of forest biodiversity conservation at global and regional scales.

9.3.1.2 Forest Biophysical Variables

Numerous forest biophysical variables are of interest for forest inventories, including leaf area index (LAI), crown closure or crown gaps, tree or canopy height, stem density, crown depth, biomass or volume, diameter at breast height (DBH), and amount of dead trees in a stand (Olthof and King, 1998; Treitz and Howarth, 1999; Sampson, 2000). Although there has been considerable effort to monitor these variables using remote sensing, most of the work to date has been through the use of broadband sensors, such as Landsat, or through the use of active systems, such as synthetic aperture radar (SAR) and lidar. Research has generally followed three paths: (1) the use of broadband indices, such as the calculation of the normalized difference vegetation index (NDVI) for the prediction of LAI (e.g., Peterson et al., 1987; Herwitz et al., 1989; Spanner et al., 1990; Curran et al., 1992; Shippert et al., 1995; Chen and Cihlar, 1996; Green et al., 1997; White et al., 1997; Wang et al., 2005); (2) the analysis of spatial statistics, texture, or multivariate statistical techniques (e.g., Yuan et al., 1991; Hershey et al., 1998; Olthof and King, 1998; Davison et al., 1999; Phinn et al., 1999; Seed et al., 1999; Pellikka et al., 2000); or (3) the use of active-sensor metrics to develop predictive regression models (e.g., Næsset, 1997; Magnussen and Boudewyn, 1998; Means et al., 2000; Lim and Treitz, 2004; Hopkinson et al., 2005; Thomas et al., 2006b).

TABLE 9.2
Logarithmic Bivariate Models to Predict Field Metrics from the Mean Derivative Chlorophyll Index

Field Metric versus DCI (n = 24)	r^2	r^2_{adj}	RMSE	Plot Mean
Mean dominant height (m)	0.59	0.57	2.16	20.4
Quadratic diameter at breast height (cm)	0.55[a]	0.53	3.51	19.3
Total aboveground biomass (mg/ha)	0.63	0.61	1740	4426.5
Crown closure (%)	0.35	0.32	10.8	36.6
Total projected crown area (m^2/m^2)	0.75[a]	0.74	0.48	1.19
Basal area (m^2/ha)	0.64[a]	0.62	0.08	26.9
Leaf area index	0.63	0.61	0.18	2.3
Stem density (#/ha)	No significant relationship.			

Note: All models are significant.
[a] Residuals not normally distributed.

Hyperspectral research in this area has generally focused either on the use of narrowband indices, partial regression analysis, spectral mixture analysis, or canopy reflectance modeling to predict biophysical variables of interest. In the case of canopy reflectance modeling, the models predict reflectance based on canopy characteristics (e.g., foliar biochemistry, water, leaf morphology, and transmittance characteristics), solar-target-sensor geometry, and soil information. For example, Chen et al. (1999) used the canopy reflectance model 4-Scale to improve hyperspectral predictions of LAI and crown closure in a boreal forest ecosystem. When reflectance is the input, these models can be inverted to predict the desired structural information (referred to as inversion modeling). This was demonstrated in a coniferous ecosystem for LAI, canopy cover, water content, and dry material (Schaepman et al., 2005). Fused lidar and hyperspectral data have also been valuable for the prediction of canopy chemistry and leaf mass in numerous environments. For example, Gökkaya et al. (2015a,b) used fused hyperspectral and lidar data to examine canopy nitrogen (N), phosphorus (P), potassium (K), calcium (Ca), and magnesium (Mg) in a boreal environment, and Asner et al. (2016, 2017) mapped foliar N, P, Ca, phenols, lignin, water content, and leaf mass area (LMA) in the Andes-Amazon region with a fused lidar-hyperspectral approach.

Although there is a known sensitivity of hyperspectral reflectance and indices to structure and leaf area/morphology, the use of hyperspectral data by themselves to predict forest biophysical variables has been relatively limited, especially now that many platforms fuse lidar with hyperspectral data. Notable exceptions to this include Gong et al. (2003a), Lee et al. (2004), and Schlerf et al. (2005) for LAI or volume prediction. Thomas et al. (2011) used hyperspectral indices to predict LAI and clumping and found strong relationships between the derivative chlorophyll index (DCI) and clumping (Table 9.2). The DCI is calculated as D705/D722, where D = derivative of reflectance (Zarco-Tejada et al., 2002). Clumping is a variable that describes nonrandom foliage distribution within canopies (Chen et al., 2005) and can be thought of as a descriptor of canopy gap distribution. It partially drives processes that are strongly affected by the canopy light regime, such as photosynthesis, evapotranspiration, and the distribution of canopy foliar nutrients (Stenberg, 1998; Chen et al., 1999; Alt et al., 2000; Dreccer et al., 2000; Bernier et al., 2001; Palmroth and Hari, 2001; Baldocchi et al., 2002; Liu et al., 2002; Thomas et al., 2006a, 2009). Thomas et al. (2011) demonstrated strong relationships between clumping and several canopy nutrients, including N, chlorophyll, carotenoids, P, and Mg.

The work of Schlerf et al. (2005) and Thomas et al. (2011) suggests that some hyperspectral indices may provide direct insight into volume, canopy height, aboveground biomass/carbon, and

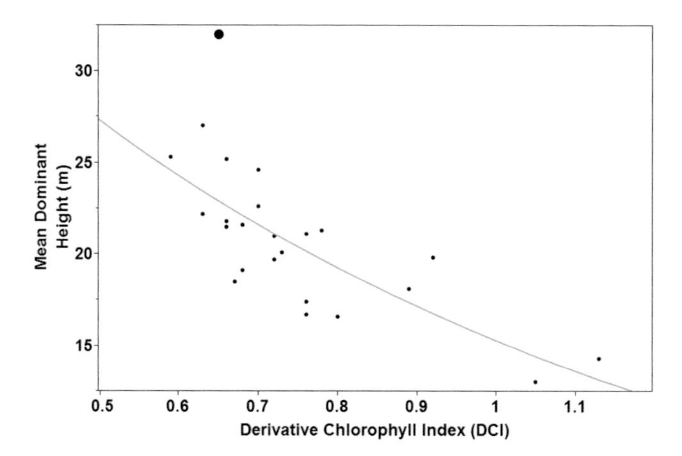

FIGURE 9.5 Logarithmic relationship between hyperspectral DCI and mean dominant height. Regression statistics shown in Table 9.2.

crown closure. Expanding upon the work of Thomas et al. (2011), the relationship between airborne hyperspectral indices and plot-based stand variables was assessed using simple bivariate regression. For most variables, the relationship was shown to be logarithmic (Figure 9.5) and significant, with 50%–75% of the variance explained by the models (Table 9.1). As expected, given the relationship to clumping, crown closure was well predicted by the DCI. These results suggest that hyperspectral data are currently being underutilized for the purpose of stand inventories and may offer more than just species or foliar biochemistry. An example of the spatial distribution of DCI can be seen in Figure 9.6a. Application of the logarithmic bivariate model for mean dominant height (Figure 9.6b) shows that the spatial distribution is related to the species distribution at the site (refer back to Figure 9.4). Most of the tall trees are trembling aspen, and the large patches of lower heights correspond to patches of black spruce at the site.

Goodenough et al. (2006, 2008) demonstrated the ability to map biomass and carbon from Airborne Visible/Infrared Imaging Spectrometer (AVIRIS) data using partial least-squares regression analysis on reflectance data and the first and second derivatives of reflectance ($r^2 = 0.82$). In their work, they also compared the performance of two airborne hyperspectral sensors with significantly different signal-to-noise ratios. The AVIRIS sensor, designed and operated by NASA, is known to produce high-quality images with a high signal-to-noise ratio. These data were compared to imagery from the Airborne Imaging Spectrometer for Applications (AISA) sensor, which had very comparable signal-to-noise ratios in the visible and near-infrared wavelengths, but significantly lower values in the shortwave infrared (SWIR) wavelengths: 3.43 versus 4.50 relative signal-to-noise values for AISA SWIR and AVIRIS SWIR, respectively (Goodenough et al., 2008). They attribute the superior performance of AVIRIS to the higher signal-to-noise ratio, as well as to solar geometry effects in the AISA data. Their findings highlight one of the limiting factors on the use of hyperspectral imaging for operational forest management applications. That is, significant preprocessing of hyperspectral data is necessary to derive high-quality reflectance data, which requires considerable user expertise.

9.3.2 Carbon Exchange

Forests are the largest terrestrial carbon stores and make a significant contribution to the global carbon cycle, which is widely felt to play a fundamental role in regulating the climate of the Earth. In North America, the combined boreal and temperate forests cover a significant portion of our landscape and, in addition to the major role they play in the carbon cycle, also have a significant influence on the water and energy balance, animal habitats, and the economic functioning of many regions. Despite the importance of our forests in climate change and other processes, our understanding of and ability

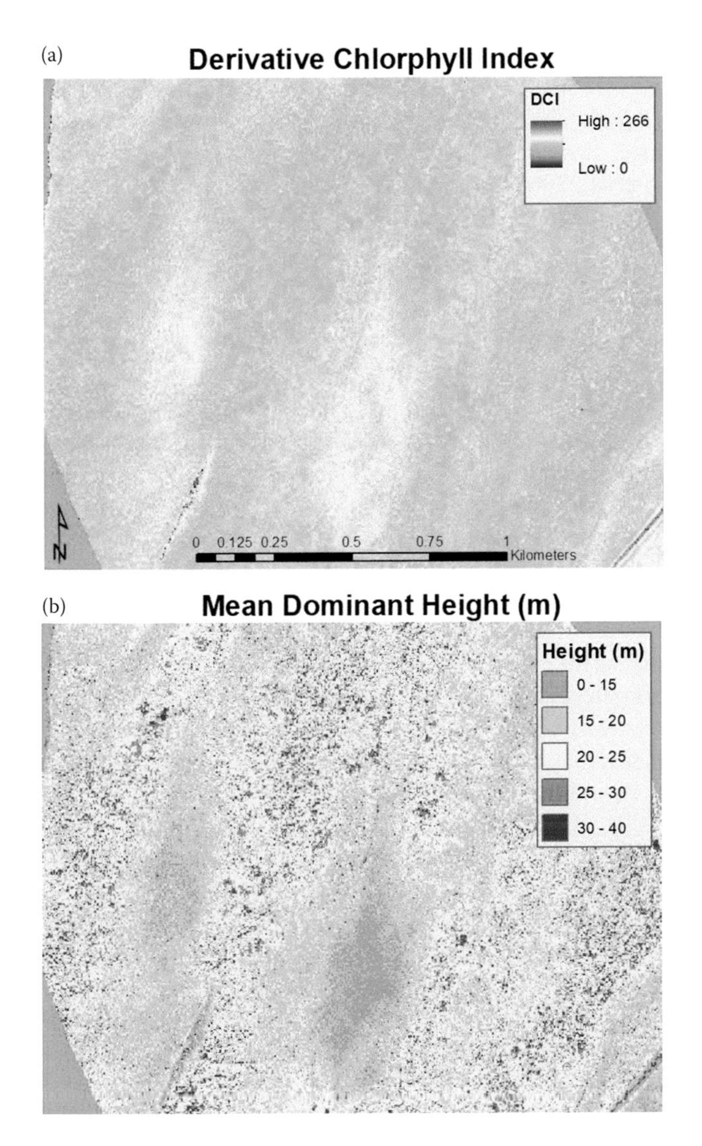

FIGURE 9.6 (a) Spatial patterns of DCI across a boreal mixed-wood forest in northern Ontario, Canada. (b) Mean dominant height mapped using model described in Table 9.2 and Figure 9.5.

to model and predict the carbon cycle remains weak in some areas, and there are numerous ongoing research efforts to improve our understanding of forest carbon dynamics. Part of the challenge lies in the fact that carbon sequestration processes occur at many spatial and temporal scales and it is difficult to envision a measurement/modeling scheme that could adequately represent this variability.

Carbon cycling is often described in terms of net ecosystem productivity (NEP) or net primary productivity (NPP). NEP and NPP are fundamental ecological concepts that not only are critical to regional-, national-, and global-scale carbon budgets and climate models, but also indicate terrestrial land surface conditions and ecological processes (Ciais et al., 2005; Kashian et al., 2006; Bo et al., 2007). Unfortunately, there are significant problems and inaccuracies with current global NEP and NPP models, often resulting in poor correlation between measured NEP at micrometeorological flux stations and national- or global-scale model predictions. A large number of national- and global-scale carbon models calculate NPP from the fraction of photosynthetically active radiation absorbed by the canopy (fPAR) and a light use efficiency (LUE) term that is considered constant for a given vegetation type and is typically biome-specific (Running et al., 1999; Ahl et al., 2004). However, growing evidence suggests that LUE varies according to many factors across and within ecosystems, including ecosystem type, stand age, species composition, nutrient availability, and vegetation stress (e.g., Goetz and Prince, 1996, 1998; Medlyn, 1998; Turner et al., 2003a,b; Ahl et al., 2004; Drolet et al., 2005; Jenkins et al., 2007). To improve global terrestrial models of NEP, it is essential to develop a deeper understanding of the factors that control LUE.

Research at satellite and airborne scales suggest that hyperspectral remote sensing has much to offer for the improved accuracy of global carbon exchange models, though it is underutilized in this context. These data have a direct role to play in improved species mapping (e.g., Wessman et al., 1988; Fuentes et al., 2001; Clark et al., 2003; Kokaly et al., 2003; Townsend et al., 2003; Plourde et al., 2007; Asner et al., 2008a,b, 2017; Thomas et al., 2009), characterization of vegetation structure (e.g., Schlerf et al., 2005; Thomas et al., 2011), quantification of aboveground carbon and biomass (e.g., Schlerf et al., 2005), mapping of fPAR and LUE (Thomas et al., 2009), nutrient availability (e.g., Zarco-Tejada et al., 2001, 2004; Gökkaya et al., 2015a,b; Asner et al., 2016, 2017), vegetation stress, vegetation water content, and wildfire regimes. As shown by many researchers, this is particularly true when such data are fused with other lidar data.

9.3.3 Wildfire Fuel

Another forest management application where hyperspectral remote sensing can make a direct contribution is the mapping of forest fire risk, particularly with regard to the type and quantity of fire fuel, which can affect the behavior and intensity of fires. This topic is attracting increasing attention and urgency, with observed increases in the frequency and intensity of fires due to climate change (Tymstra et al., 2007; Flannigan et al., 2009) and exurban encroachment into forested areas (Syphard et al., 2007a,b). There is an identified need for a better understanding of the spatial distribution of different wildfire fuel types and amounts, as well as areas that have been previously burned (Ustin et al., 2004; Jia et al., 2006; Varga and Asner, 2008).

Wildfire fuels include all dead or living vegetation that can be ignited (Miller, 2001). Their arrangement and availability, including the structure of the canopy itself, will determine whether the burn will be a ground-surface or crown fire (Miller, 2001; Cruz et al., 2003). Hyperspectral remote sensing can be used to map the spatial distribution of wildfire fuel properties, providing much-needed information on the type and arrangement of a fire, as well as the location and severity of previous burns. Most of the work in this area has focused on the use of narrowband vegetation indices or radiative transfer modeling to map fuel properties within the canopy. This may include canopy moisture content (Jacquemoud and Ustin, 2003; Li et al., 2005; Riano et al., 2005), amount of dry or senescent carbon or degree of fuel curing (Melillo et al., 1982; Fourty et al., 1996; Serrano et al., 2002), canopy stress detection and foliar biochemistry (Ustin and Trabucco, 2000; Smith et al., 2003), and relative amount of photosynthetic versus nonphotosynthetic materials (Varga and Asner, 2008).

Another approach to fire fuel mapping follows the species classification work discussed earlier. Some researchers have collected the field reflectance spectra of a number of fuel attributes and used those spectra as endmembers (i.e., pure spectra). The researchers have attempted to quantify the relative fraction of endmembers at the subpixel level using spectral mixture analysis (e.g., Ustin et al., 2004; Green et al., 1998; Clark et al., 2003; Jia et al., 2006) and at the pure-pixel level using spectral matching algorithms (e.g., Jia et al., 2006).

Although fire fuel attribute mapping is really in the research phase, some attempts have been made to move toward operational capabilities in this area. To assist forest planners, ENVI, one of the software packages used to analyze hyperspectral data, has combined a number of vegetation indices that detect canopy water content, greenness, and dry carbon into a "Fire Fuel Tool," which produces a map of relative fire risk based on a combination of these three attributes (RSI, 2009). The tool requires limited user expertise, both in its operation and interpretation of results, setting the stage for the adoption of the technique by users beyond the research community.

In addition to wildfire fuel mapping, hyperspectral remote sensing has also been used as an input to fire behavioral models. Fire behavior is largely controlled by topography, weather, and the availability of fuel (Miller, 2001). Topography and weather are generally well known and accurately characterized across space and time. Fire behavior models range in complexity from simple empirical approaches (Andrews, 1986; Finney, 1998) to complex physics models based upon thermodynamics and the conservation of mass, momentum, and energy (e.g., Linn, 1997; Dupuy and Larini, 1999; Porterie et al., 2000; Grishin, 2001a,b). For many of these models, fire fuel attributes are averaged over large areas or broad ecological types. With advances in the ability to map complex wildfire fuel characteristics from hyperspectral data, fuel maps can now be directly input into some models, for example FIRETEC (Linn, 1997), to provide more accurate simulations of fire behavior (e.g., Bossert et al., 2000; Dennison et al., 2000).

9.4 POTENTIAL FUTURE APPLICATIONS

9.4.1 Fusion with Lidar

It is clear that a highly valuable form of data fusion is the integration of lidar data with hyperspectral data. Such integration can provide insight into both the structure and function of a forest environment and serve as an indicator of many types of risks and processes. This type of data fusion has the potential to advance research in numerous areas, including the characterization of height, aboveground biomass and carbon, bidirectional reflectance modeling, species mapping, geometric optical and radiative transfer modeling, ecosystem modeling, prediction of canopy biochemistry and photosynthesis parameters, and modeling of canopy photosynthesis and carbon exchange.

As noted in Section 9.2, the fusion of lidar with hyperspectral data has improved our ability to characterize structure and biochemical variables (e.g., Koetz et al., 2006; Asner et al., 2007, 2011, 2015; Thomas et al., 2008). Although lidar is typically used by itself to characterize forest height and biomass, some work has shown that the fusion of hyperspectral reflectance with lidar data can also improve estimates of aboveground biomass (Laurin et al., 2014). This has implications for future work in the modeling of carbon stocks and fluxes, particularly in light of emerging international programs such as Reducing Emissions from Deforestation and Forest Degradation (REDD) and REDD+ (http://www.fao.org/redd/en/), which provide financial incentives to reduce deforestation and forest degradation in tropical forests. The fusion of lidar and hyperspectral data may allow for significant improvements in our ability to accurately monitor forest biomass and carbon, which could help satisfy the measurement, reporting, and verification (MRV) requirements of REDD+.

Thomas et al. (2008) used fused lidar and hyperspectral data to scale estimates of chlorophyll from leaf to canopy, to improve species classifications, and to model the spatial variability of the fraction of photosynthetically active radiation absorbed by the canopy (fPAR) for a boreal mixed-wood ecosystem (Thomas et al., 2006a). These data have been incorporated into models of photosynthesis and carbon exchange that, in conjunction with meteorological data collected continuously at a flux tower site, allow for the analysis of variability within the footprint of the flux tower (Thomas et al., 2009). Thus, they provide insight into the importance of canopy architecture on canopy function.

The fusion of lidar and hyperspectral data can also be used to model the vertical light profile of a canopy and its relationship to foliar biochemistry and photosynthesis parameters through the use of GORTs. A number of studies at the leaf scale have successfully predicted foliar biochemistry based on leaf reflectance and transmittance using leaf optical models (e.g., PROSPECT) (Jacquemoud et al., 1996; Baret and Fourty, 1997; Fourty and Baret, 1998). At the canopy scale, radiative transfer (RT) models are used to model the interaction of solar radiation with vegetation elements and can describe the spectral reflectance of a forest stand. The simplest of these models assume that canopies are homogeneous and can be described by simple geometric shapes (e.g., SAIL).

More complex models, such as GORT models, attempt to represent canopy geometry by modeling gaps and shadows within the canopy (e.g., GeoSAIL, GORT). By coupling radiative transfer models with leaf optical models (e.g., PROSPECT-SAIL, PROSPECT-GeoSAIL) and running backwards simulations, numerous researchers have predicted canopy biochemistry from remotely sensed reflectance data (Jacquemoud et al., 2000; Zarco-Tejada et al., 2004; Malenovský et al., 2006). Unfortunately, due to the complexity of forest canopy architecture and the confounding factors highlighted earlier, validation of these inversion model predictions with leaf samples has been challenging. Despite that, significant advances have been made incorporating detailed information about the vertical and horizontal distribution of canopy available from airborne lidar data directly to the geometric-optical RT models. Some work has been done to use radiative transfer models to simulate lidar waveforms and to predict tree height, spacing, gap distribution, foliar chlorophyll, and water content from lidar data (e.g., Peterson et al., 2001; Kotchenova et al., 2003; Koetz et al., 2006, 2007). More recently, the Discrete Anisotropic Radiative Transfer (DART 5) model has been applied with lidar and hyperspectral data to natural and urban landscapes (Gastellu-Etchegorry et al., 2015), and this remains an exciting area for future research.

9.4.2 The Need for Repeated Global Measurements

One of the major limitations to the use of hyperspectral remote sensing in a forest management context is the lack of large-area measurements taken with consistent sensor settings and repeated over time, such as would be possible with satellite-based hyperspectral sensors (Table 9.3). This has resulted in work being done from airborne platforms, in localized contexts, primarily for scientific objectives.

Satellite-based hyperspectral remote sensing offers the potential for more accurate derivations of geochemical, biochemical, and biophysical variables that drive regional and global ecosystem process models. For instance, satellite hyperspectral data could be used to improve the accuracy of maps of species (or communities of species), to derive foliar pigment and nutrient concentrations for large areas, and to estimate LAI, fPAR, and parameters of photosynthesis. Satellite hyperspectral data, acquired continuously, would provide this information across landscapes continually altered by disturbance and recovery processes, land-use management and change, and climate change. This would improve our understanding of biogeochemical cycling and the impact of land management decisions on these processes.

Although several satellite-based hyperspectral sensors have been proposed in the past two decades, very few sensors have been launched, and none designed to provide global mapping. NASA's Hyperion sensor provided 220 bands of data across visible, near-infrared, and SWIR wavelengths (i.e., 0.4–2.5 μm) at 30 m resolution (Folkman et al., 2001). Hyperion was part of the Earth Observing (EO-1) mission, originally intended to validate several instrument and spacecraft bus technologies. The satellite was launched in 2000, with a design life of 18 months. It ran out of fuel in 2017. Hyperspectral data were not collected continuously from this sensor but were acquired according to user requests. This made large-area forest management applications challenging, because a single scene is relatively small (i.e., 7.7 km wide by 42 or 185 km long), and the dates of acquisition could conflict with other users, in which case acquisitions would be prioritized. The other satellite-based

TABLE 9.3
Selected Current and Future Hyperspectral Satellite Missions

Sensor	Agency/Nation	Spectral Range (Bands)	Spatial Resolution (m)	Launch (Lifespan)
Hyperion	NASA	400–2500 nm (220 bands)	30	2000 (16+ years)
CHRIS	ESA	400–1050 nm (19 bands)	17	2001 (still operating)
		Reprogrammable (63 bands)	34	
EnMap	Germany	420–2450 nm	30	2019 (5+ years)
Hyspiri	NASA	380–2500 nm	60	In study stage
PRISMA	Italy	400–2500 nm (240 bands)	30	2018 (5 years)

option is the CHRIS sensor on the European Space Agency's Proba-1 mission. CHRIS has a higher spatial resolution (17 m) but fewer bands (up to 62). It was launched in 2001, with a 2-year lifespan, and is still in operation (ESA, 2017).

Research success at the satellite scale lends credence to the potential use of satellite hyperspectral for forest management in ways that are simply not possible with the existing suite of technologies. In forest contexts, Hyperion has been used for improved mapping of forest species (Goodenough et al., 2003), species communities (Thenkabail et al., 2004), various invasive species (Asner et al., 2006), crown closure and LAI (Gong et al., 2003a,b; Pu et al., 2005), drought stress (Asner et al., 2004), biomass and carbon mapping (Asner et al., 2004; Thenkabail et al., 2004), and canopy nitrogen (Smith et al., 2003). Similarly, CHRIS has been used to map forest successional stages (Galvão et al., 2009), canopy structure and heterogeneity (Koetz et al., 2005), chlorophyll fluorescence (Raddi et al., 2005), foliar biochemistry and water content (Kneubühler et al., 2008), and wildfires (Valencia et al., 2005).

Several satellite-based hyperspectral sensors are tentatively being planned in different countries, but these missions often experience launch delays or are designed to demonstrate the technology involved rather than actual operation. None of the sensors described in the previous edition of this book (EnMap, Hyspiri, and PRISMA from Germany, the United States, and Italy, respectively) were launched according to original plans. Germany does plan to launch the EnMAP mission by 2019 (2014 in the earlier edition of this book) (EnMap, 2017), which, if the launch is successful, will provide repeated global coverage, enabling forest planners to study the impacts of management decisions on the function of biogeochemical cycling in a way never before possible.

9.4.3 Hyperspectral Sensors on Unmanned Aerial Vehicles

Hyperspectral sensors have been successfully miniaturized for unmanned aerial vehicles (UAVs), but they have not been widely used in forests to date. There are certain forest applications for which hyperspectral UAVs are ideally suited. These include multitemporal mapping over constrained areas, such as monitoring water stress over a citrus orchard (Zarco-Tejada et al., 2012; Gonzalez-Dugo et al., 2013), or monitoring stress and damage events in targeted areas, such as bark beetle damage at the tree level (Näsi et al., 2015) or stress in olive orchards (Calderón et al., 2013). UAV technologies are evolving, as are the regulatory frameworks for their operation (which vary by country). It is likely that this area will see significant advances over the next 5–10 years.

9.5 CONCLUSIONS

Although the applications for hyperspectral remote sensing in a forest management context are numerous, the technology is currently being underutilized for this purpose. It is evident that hyperspectral imaging is part of the solution to global monitoring and predictions for biogeochemical cycling, forest biophysical variables, and forest physiology and function. Many successes have been demonstrated at the laboratory and local scales, but consistent, repeatable results are necessary at national and global scales (i.e., through the use of satellite-based technologies) before the full potential of this technology can be realized for forest management and policy. With rising social, political, and scientific concerns surrounding land-use change, global forest loss, forest degradation, and the role of the world's forests in the global carbon cycle and climate, satellite-based hyperspectral data will likely become a critical tool for informing global modeling efforts and enhancing our understanding of the function of remote boreal and tropical ecosystems.

REFERENCES

Ahl, D., S.T. Gower, D.S. Mackay, S.N. Burrows, J.M. Norman, and G.R. Diak, 2004. Heterogeneity of light use efficiency in a northern Wisconsin forest: Implication for modeling net primary production using remote sensing. *Remote Sensing of Environment*, 93: 168–178.

Alemdag, I.S., 1983. Mass equations and mechantability factors for Ontario softwoods. *Canadian Forest Service Petawawa National Forest Institute Information Report*, PI-X-23.

Alemdag, I.S., 1984. Total tree and merchantable stem biomass equations for Ontario hardwoods. *Canadian Forest Service Petawawa National Forest Institute Information Report*, PI-X-46.

Alonzo, M. B. Bookhagen, and D.A. Roberts, 2014. Urban tree species mapping using hyperspectral and lidar data fusion. *Remote Sensing of Environment*, 148: 70–83.

Alt., C., Stutzel, H., and H. Kage, 2000. Optimal nitrogen content and photosynthesis in cauliflower (*Brassica oleracea* L. *botrytis*). Scaling leaf to plant. *Annals of Botany*, 85: 779–787.

Andrews, P. L., 1986. *BEHAVE: Fire behavior prediction and fuel modeling system—BURN Subsystem, Part 1.* Gen. Tech. Rep. INT-194. Ogden, UT: U.S. Department of Agriculture, Forest Service Intermountain Forest and Range Experiment Station. 130 p.

Asner, G.P., D. Nepstad, G. Cardinot, and D. Ray, 2004. Drought stress and carbon uptake in an Amazon forest measured with spaceborne imaging spectroscopy. *PNAS*, 101(16): 6039–6044.

Asner, G.P., R.E. Martin, K.M. Carlson, U. Rascher, and P.M. Vitousek, 2006. Vegetation-climate interactions among native and invasive species in Hawaiian rainforest. *Ecosystems*, 9: 1106–1117.

Asner, G.P., D.E. Knapp, T. Kennedy-Bowdoin, M.O. Jones, R.E. Martin, J.W. Boardman, and C.B. Field, 2007. Carnegie airborne observatory: In-flight fusion of hyperspectral imaging and waveform light detection and ranging for three-dimensional studies of ecosystems. *Journal of Applied Remote Sensing*, 1(1): 013536.

Asner, G.P., M.O. Jones, R.E. Martin, D.E. Knapp, and R.F. Hughes, 2008a. Remote sensing of native and invasive species in Hawaiin forest. *Remote Sensing of Environment*, 112: 1912–1926.

Asner, G.P., D.E. Knapp, T. Kennedy-Bowdoin, M.O. Jones, R.E. Martin, J. Boardman, and R.F. Hughes, 2008b. Invasive species detection in Hawaiian rainforests using airborne imaging spectroscopy and LiDAR. *Remote Sensing of Environment*, 112: 1942–1955.

Asner, G. P., R.E. Martin, D.E. Knapp, R. Tupayachi, C. Anderson, L. Carranza, P. Martinez, M. Houcheime, F. Sinca, and P. Weiss, 2011. Spectroscopy of canopy chemicals in humid tropical forests. *Remote Sensing of Environment*, 12(115): 3587–3598.

Asner, G.P., R.E. Martin, C. Anderson, and D.E. Knapp, 2015. Quantifying forest canopy traits: Imaging spectroscopy versus field survey. *Remote Sensing of Environment*, 158: 15–27.

Asner, G.P., D.E. Knapp, C.B. Anderson, R.E. Martin, and N. Vaughn, 2016. Large-scale climatic and geophysical controls on the leaf economics spectrum. *PNAS*, 113(28): E4043–E4051.

Asner, G.P., R.E. Martin, D.E. Knapp, R. Tupayachi, C.B. Anderson, F. Sinca, N.R. Vaughn, and W. Llactayo, 2017. Airborne laser-guided imaging spectroscopy to map forest trait biodiversity and guide conservation. *Science*, 355(6323): 385–389.

Avery, T.E., and H.E. Burkhart, 2002. *Forest Measurements*. 5th ed. McGraw Hill, New York, NY. 456 p.

Baldocchi, D.D., K.B. Wilson, and L. Gu, 2002. How the environment, canopy structure and canopy physiological functioning influence carbon, water and energy fluxes of a temperate broad-leaved deciduous forest—An assessment with the biophysical model CANOAK. *Tree Physiology*, 22: 1065–1077.

Baret, F., and T. Fourty, 1997. Estimation of leaf water content and specific leaf weight from reflectance and transmittance measurements. *Agronomie*, 17: 455–464.

Bernier, P.Y., F. Raulier, P. Stenberg, and C.H. Ung, 2001. Importance of needle age and shoot structure on canopy net photosynthesis of balsam fir (*Abies balsamea*): A spatially inexplicit modeling analysis. *Tree Physiology*, 21: 815–830.

Blackburn, G.A., 2002. Remote sensing of forest pigments using airborne imaging spectrometer and LIDAR imagery. *Remote Sensing of Environment*, 82: 311–321.

Bo, T., C. MingKui, L. KeRang, G. FengXue, J. JinJun, H. Mei, and Z. LeiMing, 2007. Spatial patterns of terrestrial net ecosystem productivity in China during 1981–2000. *Science in China Series D:Earth Sciences*, 50(5): 745–753.

Bossert, J.E., R.R. Linn, J.M. Reisner, J.L. Winterkamp, P. Dennison, and D. Roberts, 2000. Coupled atmosphere–fire behavior model sensitivity to spatial fuels characterization. *Third Symposium on Fire and Forest Meteorology of the American Meteorological Society*, January 9–14, 2000, Long Beach, CA, pp. 21–26.

Brack, C.L., and P. Marshall, 1998. Sequential sampling with systematic selection for estimating mean dominant height. *Australian Forestry*, 61: 253–257.

Buddenbaum, H., M. Schlerf, and J. Hill, 2005. Classification of coniferous tree species and age classes using hyperspectral data and geostatistical methods. *International Journal of Remote Sensing*, 26(24): 5453–5465.

Calderón, R., J.A. Navas-Cortés, C. Lucena, and P.J. Zarco-Tejada, 2013. High-resolution airborne hyperspectral and thermal imagery for early detection of *verticillium* wilt of olive using fluorescence, temperature and narrow-band spectral indices. *Remote Sensing of Environment*, 139: 231–245.

Carlson, K.M., G.P. Asner, R.F. Hughes, R. Ostertag, and R.E. Martin, 2007. Hyperspectral remote sensing of canopy biodiversity in Hawaiian lowland rainforests. *Ecosystems*, 10: 536–549.

Carter, G.A., 1994. Ratios of leaf reflectances in narrow wavebands as indicators of plant stress. *International Journal of Remote Sensing*, 15: 697–703.

Chen, J.M., and Cihlar, J., 1996. Retrieving leaf area index of boreal conifer forests using landsat TM images. *Remote Sensing of Environment*, 55: 153–162.

Chen, J.M., S.G. Leblanc, J.R. Miller, J. Freemantle, S.E. Loechel, C.L. Walthall, K.A. Innanen, and H.P. White, 1999. Compact airborne spectrographic imager (CASI) used for mapping biophysical parameters of boreal forests. *Journal of Geophysical Research*, 104: 927–945.

Chen, J.M., C.H. Menges, and S.G. Leblanc, 2005. Global mapping of foliage clumping index using multi-angular satellite data. *Remote Sensing of Environment*, 97: 447–457.

Ciais, P., M. Reichstein, N. Viovy, A. Granier, J. Ogee, V. Allard, M. Aubinet et al. 2005. Europe-wide reduction in primary productivity caused by the heat and drought in 2003. *Nature*, 437: 529–533.

Clark, M.L., D.A. Roberts, and D.B. Clark, 2005. Hyperspectral discrimination of tropical rain forest tree species at leaf to crown scales. *Remote Sensing of Environment*, 96: 375–398.

Clark, R.N., G.A. Swayze, K.E. Livo, R.F. Kokaly, S.J. Sutley, J.B. Dalton, R.R. McDougal, and C.A. Gent, 2003. Imaging spectroscopy: Earth and planetary remote sensing with the USGS Tetracorder and expert systems. *Journal of Geophysical Research*, 108(E12): 5131–5146.

Colgan, M.S., C.A. Baldeck, J.-B. Féret, and G.P. Asner, 2012. Mapping savanna tree species at ecosystem scales using support vector machine classification and BRDF correction on airborne hyperspectral and LiDAR data. *Remote Sensing*, 4(11): 3462–3480.

Cruz, M., M. Alexander, and R. Wakimoto, 2003. Assessing canopy fuel stratum characteristics in crown fire prone fuel types of western North America. *International Journal of Wildland Fire*, 12: 39–50.

Curran, P.J., J.L. Dungan, and H.L. Gholz, 1992. Seasonal LAI in slash pine estimated with Landsat TM. *Remote Sensing of Environment*, 39: 3–13.

Curran, P.J., and P.M. Atkinson, 1999. Issues of scale and optimal pixel size, In *Spatial Statistics for Remote Sensing*. A. Stein, F. Van Der Meer, and B. Gorte (eds), Dordrecht: Kluwer Academic Publishers, pp. 115–133.

Curran, P.J., 2001. Imaging spectrometry for ecological applications. *International Journal of Applied Earth Observation and Geoinformation*, 3: 305–312.

Curtis, R.O., and D.D. Marshall, 2000. Why quadratic mean diameter? *Western Journal of Applied Forestry*, 15(3): 137–139.

Darvishefat, A., T. Kellenburger, and K. Itten, 2002. Application of hyperspectral data for forest stand mapping. *Symposium on Geospatial Theory, Processing, and Applictions, ISPRS Commission IV, Symposium 2002*, Ottawa, Canada, July 9–12, 2002. IAPRS, Vol. XXXIV, part 4.

Davison, D., S. Achal, S. Mah, R. Gauvin, M. Kerr, A. Tam, and S. Preiss, 1999. Determination of tree species and tree stem densities in Northern Ontario forests using airborne CASI data. *Proceedings of the Fourth International Airborne Remote Sensing Conference and Exhibition II*, 21–24 June 1999, Ottawa, ON, pp. 187–196.

Dawson, T.P., P.R.J. North, and P.J. Curran, 1999. The propagation of foliar biochemical absorption features in forest canopy reflectance: A theoretical analysis. *Remote Sensing of Environment*, 67: 147–159.

Demerez, V., J.P. Gastellu-Etchegorry, E. Mougin, G. Marty, C. Proisy, E. Duferene, and V.L.E. Dantec, 1999. Seasonal variation of leaf chlorophyll content of a temperate forest. Inverstion of the PROSPECT model. *International Journal of Remote Sensing*, 20: 879–894.

Demerez, V., and J.P. Gastellu-Etchegorry, 2000. A modeling approach for studying forest chlorophyll content. *Remote Sensing of Environment*, 71: 226–238.

Dennison, P.E., D.A. Roberts, and J.C. Regelbrugge, 2000. Characterizing chaparral fuels using combined hyperspectral and synthetic aperture radar data. In *Proc. 9th AVIRIS Earth Science Workshop*. Feb. 23–25, 2000, Pasadena, CA, vol. 6, pp. 119–124.

Dreccer, M.F., M. van Oijen, A.H. Schapendonk, C.S. Pot, and R. Rabbinge, 2000. Dynamics of vertical leaf nitrogen distribution in a vegetative wheat canopy. Impact on canopy photosynthesis. *Annals of Botany*, 86: 821–831.

Drolet, G.G., K.F. Huemmrich, F.G. Hall, E.M. Middleton, T.A. Black, A.G. Barr, and H.A. Margolis, 2005. A MODIS-derived photochemical reflectance index to detect inter-annual variations in the photosynthetic light-use efficiency of a boreal deciduous forest. *Remote Sensing of Environment*, 98: 212–224.

Dupuy, J.L., and M. Larini, 1999. Fire spread through a porous forest fuel bed: A radiative and convective model including fire-induced flow effects. *International Journal of Wildland Fire*, 9: 155–172.

EnMap, 2017. http://www.enmap.org/?q=mission (last cited: December 15, 2017).

ESA, 2017. Proba-1 images Calanda reservoir. ESA Mission News, Earth Online. https://earth.esa.int/web/guest/missions/mission-news/-/article/proba-1-images-calanda-reservoir [last cited: December 15, 2017].

FAO, 2016. *Global Forest Resources Assessment 2015. How are the World's Forests Changing?* 2nd edition. FAO Forestry paper, 45pp, FAO, Rome. Available on-line at: http://www.fao.org/3/a-i4793e.pdf

Féret, J.-B., and G.P. Asner, 2014. Mapping tropical forest canopy diversity using high-fidelity imaging spectroscopy. *Ecological Applications*, 24(6): 1289–1296.

Finney, M.A., 1998. *FARSITE: Fire Area Simulator—Model Development and Evaluation*. Res. Pap. RMRSRP-4. Fort Collins, CO: U.S. Department of Agriculture, Forest Service, Rocky Mountain Research Station. 47 p.

Flannigan, M., B. Stocks, M. Turetsky, and M. Wotton, 2009. Impacts of climate change on fire activity and fire management in the circumboreal forest. *Global Change Biology*, 15: 549–560.

Folkman, M.A., P. Lee, P.J. Jarecke, S.L. Carman, and J. Pearlman, 2001. EO-1/Hyperion hyperspectral imager design, development, characterization, and calibration. *Proc. SPIE*, 4151: 2000.

Fourty, T., F. Baret, S. Jacquemoud, G. Schmuck, and J. Verdebout, 1996. Leaf optical properties with explicit description of its biochemical composition: Direct and inverse problems. *Remote Sensing of Environment*, 56: 104–117.

Fourty T., and F. Baret, 1998. On spectral estimates of fresh leaf biochemistry. *International Journal of Remote Sensing*, 19: 1283–1297.

Fuentes, D.A., J.A. Gamon, H.L. Qui, D.A. Sims, and D.A. Roberts, 2001. Mapping Canadian boreal forest vegetation using pigment and water absorption features derived from the AVIRIS sensor. *Journal of Geophysical Research—Atmospheres*, 106: 33565–33577.

Galvão, L.S., F.J. Ponzoni, V. Liesenberg, and J.R. dos Santos, 2009. Possibilities of discriminating tropical secondary succession in Amazônia using hyperspectral and multiangular CHRIS/PROBA data. *International Journal of Applied Earth Observation and Geoinformation*, 11: 8–14.

Garcia, O. 1998. Estimating top height with variable plot sizes. *Canadian Journal of Forest Research*, 28(10): 1509–1517.

Gastellu-Etchegorry, J.P., and V. Bruniquel-Pinel, 2001. A modeling approach to assess the robustness of spectrometric predictive equations for canopy chemistry. *Remote Sensing of Environment*, 76: 1–15.

Gastellu-Etchegorry, J.-P., T. Yin, N. Lauret, T. Cajgfinger, T. Gregoire, E. Grau, J.-B. Feret et al. 2015. Discrete anisotropic radiative transfer (DART 5) for modeling airborne and satellite spectroradiometer and LIDAR acquisitions of natural and urban landscapes. *Remote Sensing*, 7(2): 1667–1701.

Gemmell, F., J. Varjo, M. Strandstrom, and A. Kuusk, 2002. Comparison of measured boreal forest characteristics with estimates from TM data and limited ancillary information using reflectance model inversion. *Remote Sensing of Environment*, 81: 365–377.

Gholz, H.L., S.A. Vogel, W.P. Cropper, K. McKelvey, K.C. Ewel, R.O. Teskey, and P.J. Curran, 1991. Dynamics of canopy structure and light interception in *Pinus elliottii* stands, North Florida. *Ecological Monographs*, 61: 33–51.

Goetz, A.F.H., 1995. Imaging spectrometry for remote sensing: Vision to reality in 15 years, *Proceedings of the SPIE International Society for Optical Engineers*, 2480, 12p.

Goetz, S.J., and S.D. Prince, 1996. Remote sensing of net primary production in boreal forest stands. *Agricultural and Forest Meteorology*, 78: 149–179.

Goetz, S.J., and S.D. Prince, 1998. Variability in carbon exchange and light utilization among boreal forest stands: Implications for remote sensing of net primary production. *Canadian Journal of Forest Research*, 28: 375–389.

Gökkaya, K., V.A. Thomas, T.L. Noland, J.H. McCaughey, I. Morrison, and P.M. Treitz, 2015a. Prediction of macronutrients at the canopy level using spaceborne imaging spectroscopy and LiDAR data in a mixedwood boreal forest. *Remote Sensing*, 7(7): 9045–9069.

Gokkaya, K., V. Thomas, T. Noland, H. McCaughey, I Morrison, and P. Treitz, 2015b. Mapping continuous forest type variation by means of correlating remotely sensed metrics to canopy nitrogen to phosphorus ratio in a boreal mixedwood forest. *Applied Vegetation Science*, 18: 143–157.

Gong, P., R. Pu, G.S. Biging, and M.R. Larrieu, 2003a. Estimation of forest leaf area index using vegetation indices derived from Hyperion hyperspectral data. *IEEE Transactions on Geoscience and Remote Sensing*, 41: 1355–1362.

Gong, P., R. Pu, G.S. Biging, and M.R. Larrieu, 2003b. Extraction of red edge optical parameters from Hyperion data for estimation of forest leaf area index. *IEEE Transactions on Geoscience and Remote Sensing*, 41: 916–921.

Gonzalez-Dugo, V., P. Zarco-Tejada, E. Nicolás, P.A. Nortes, J.J. Alcarón, D.S. Intrigiolo, and E. Fereres, 2013. Using high resolution UAV thermal imagery to assess the variability in the water status of five fruit tree species within a commercial orchard. *Precision Agriculture*, 14(6): 660–678.

Goodenough, D.G., A. Dyk, K.O. Niemann, J.S. Pearlman, H. Chen, T. Han, M. Murdoch, and C. West, 2003. Processing Hyperion and ALI for Forest Classification. *IEEE Transactions on Geoscience and Remote Sensing*, 41: 1321–1331.

Goodenough, D.G., J.Y. Li, G.P. Asner, M.E. Schaepman, S.L. Ustin, and A. Dyk, 2006. Combining hyperspectral remote sensing and physical modeling for applications in land ecosystems. *IEEE International Geoscience and Remote Sensing Symposium (IGARSS)*, Denver, Colorado, 5p.

Goodenough, D.G., K.O. Niemann, A. Dyk, G. Hobart, P. Gordon, M. Loisel, and H. Chen, 2008. Comparison of AVIRIS and AISA airborne Hyperspectral sensing for above-ground forest carbon mapping. *IGARSS 2008*, II: 129–132.

Green, E.P., P.J. Mumbyb, A.J. Edwards, C.D. Clark, and A.C. Ellis, 1997. Estimating leaf area index of mangroves from satellite data. *Aquatic Botany*, 58: 11–19.

Green, R.O., M.L. Eastwood, and O. Williams, 1998. Imaging spectroscopy and the airborne visible/infrared imaging spectrometer (AVIRIS). *Remote Sensing of Environment*, 65: 227–240.

Grishin, A.M., 2001a. Heat and mass transfer and modeling and prediction of environmental catastrophes. *Journal of Engineering Physics and Thermophysics*, 74: 895–903.

Grishin, A.M., 2001b. Conjugate problems of heat and mass exchange and the physicomathematical theory of forest fires. *Journal of Engineering Physics and Thermophysics*, 74: 904–911.

Hardy, J.P., R. Melloh, G. Koenig, D. Marks, A. Winstral, J.W. Pomeroy, and T. Link, 2004. Solar radiation transmission through conifer canopies. *Agricultural and Forest Meteorology*, 126: 257–270.

Helms, J.A., (editor). 1998. *Terminology of Forest Science, Technology, Practice, and Products*. The dictionary of forestry. Society of American Foresters, Bethesda, MD. 224p.

Hershey, R.R., W.H. McWilliams, and G.C. Reese, 1998. Utilizing the spatial structure available: Creating maps of forest attributes from forest inventory data. *Proceedings of the First International Conference on Geospatial Information in Agriculture and Forestry I*, 1–3 June 1998, Lake Buena Vista, FL, pp. 64–71.

Herwitz, S.R., D.L. Peterson, and J.R. Eastman, 1989. Thematic mapper detection of change in the leaf area index of closed canopy pine plantations in Central Massachusetts. *Remote Sensing of Environment*, 29: 129–140.

Heumann, B.W., R.A. Hackett, and A.K. Monfils, 2015. Testing the spectral diversity hypothesis using spectroscopy data in a simulated wetland community. *Ecological Informatics*, 25: 29–34.

Hoffbeck, J.P., and D.A. Landgrebe, 1996. Classification of remote sensing images having high spectral resolution. *Remote Sensing of Environment*, 57: 119–126.

Hopkinson, C., L.E. Chasmer, G. Sass, I.F. Creed, M. Sitar, W. Kalbfleisch, and P. Treitz, 2005. Vegetation class dependent errors in lidar ground elevation and canopy height estimates in a boreal wetland environment. *Canadian Journal of Remote Sensing*, 31: 191–206.

Hu, B.X., K. Inannen, and J.R. Miller, 2000. Retrieval of leaf area index and canopy closure from CASI data over the BOREAS flux tower sites. *Remote Sensing of Environment*, 74: 255–274.

Jackson, R.D., R.T. Reginato, P.J. Printer, and S.B. Idso, 1979. Plant canopy information extraction from composite scene reflectance of row crops. *Applied Optics*, 18: 3775–3782.

Jacquemoud, S., S.L. Ustin, J. Verdebout, G. Schmuck, G. Andreoli, and B. Hosgood, 1996. Estimating leaf biochemistry using the PROSPECT leaf optical properties model. *Remote Sensing of Environment*, 56: 194–202.

Jacquemoud, S., C. Bacour, H. Poilvé, and J.P. Frangi, 2000. Comparison of four radiative transfer models to simulate plant canopies reflectance—Direct and inverse mode. *Remote Sensing of Environment*, 74: 471–481.

Jacquemoud, S., and S. L. Ustin, 2003. Application of radiative transfer models to moisture content estimation and burned land mapping. *Proc. 4th International Workshop on Remote Sensing and GIS Applications to Forest Fire Management*, Ghent, Belgium, pp. 3–12.

Jahnke, L.S., and D.B. Lawrence, 1965. Influence of photosynthetic crown structure on potential productivity of vegetation, based primarily on mathematical models. *Ecology*, 46: 319–326.

Jenkins, J.P., A.D. Richardson, B.H. Braswell, S.V. Ollinger, D.Y. Hollinger, and M.-L. Smith, 2007. Refining light use efficiency calculations for a deciduous forest canopy using simultaneous tower based carbon flux and radiometric measurements. *Agriculture and Forest Meteorology*, 143: 64–79.

Jia, G.J., I.C. Burke, M.R. Kaufmann, A.F.H. Goetz, B.C. Kindel, and Y. Pu, 2006. Estimates of forest canopy fuel attributes using hyperspectral data. *Forest Ecology and Management*, 229: 27–38.

Kashian, D., W.H. Romme, D.B. Tinker, M.G. Turner, and M.G. Ryan, 2006. Carbon storage on landscapes with stand-replacing fires. *Bioscience*, 56(7): 598–605.

Kimes, D., J. Gastellu-Etchegorry, and P. Esteve, 2002. Recovery of forest canopy characteristics through inversion of a complex 3D model. *Remote Sensing of Environment*, 79: 320–328.

Kneubühler, M., B. Koetz, S. Huber, N.E. Zimmermann, and M.E. Schaepman, 2008. Spectro-directional CHRIS/Proba data over two Swiss test sites for improved estimation of biophysical and chemical variables – Five years of activities. *The International Archives of the Photogrammetry, Remote Sensing and Spatial Information Sciences*, Beijing, Vol. XXXVII. Part B7. 6 pp.

Koetz, B., M. Kneubühler, J.L. Widlowski, F. Morsdorf, M. Schaepman, and K. Itten, 2005. Assessment of canopy structure and heterogeneity from multi-angular CHRIS-PROBA data. *The 9th Int. Symposium on Physical Measurements and Signatures in Remote Sensing (ISPMSRS)*, Beijing, China, Vol. XXXVI, pp. 73–78.

Koetz, B., F. Morsdorf, G. Sun, K.J. Ranson, K. Itten, and B. Allgöwer, 2006. Inversion of a lidar waveform model for forest biophysical parameter estimation. *IEEE Geoscience and Remote Sensing Letters*, 3(1): 49–53.

Koetz, B., G. Sun, F. Morsdorf, K.J. Ranson, M. Kneubühler, K. Itten, and B. Allgöwer, 2007. Fusion of imaging spectrometer and LIDAR data over combined radiative transfer models for forest canopy characterization. *Remote Sensing of Environment*, 106: 449–459.

Kokaly, R.F., D.G. Despain, R.N. Clark, and K.E. Livo, 2003. Mapping vegetation in Yellowstone National Park using spectral feature analysis of AVIRIS data. *Remote Sensing of Environment*, 84: 437–456.

Kotchenova, S.Y., N.V. Shabanov, Y. Knyazikhin, A.B. Davis, R. Dubayah, and R.B. Myneni, 2003. Modeling lidar waveforms with time-dependent stochastic radiative transfer theory for remote estimations of forest structure. *Journal of Geophysical Research*, 108(D15, 4484, ACL 12): 1–13.

Kötz, B., M. Schaepman, F. Morsdorf, P. Bowyer, K. Itten, and B. Allgöwer, 2004. Radiative transfer modeling within a heterogeneous canopy for estimation of forest fire fuel properties. *Remote Sensing of Environment*, 92: 332–344.

Kuusk, A., 1998. Monitoring of vegetation parameters on large areas by the inversion of a canopy reflectance model. *International Journal of Remote Sensing*, 19: 2893–2905.

Laurin, G.V., Q. Chen, J.A. Lindsell, D. A. Coomes, F. Del Frate, L. Guerriero, F. Pirotti, and R. Valentini, 2014. Above ground biomass estimation in an African tropical forest with lidar and hyperspectral data. *ISPRS Journal of Photogrammetry and Remote Sensing*, 89: 49–58.

Lee, K.-S., W.B. Cohen, R.E. Kennedy, T.K. Maiersperger, and S.T. Gower, 2004. Hyperspectral versus multispectral data for estimating leaf area index in four different biomes. *Remote Sensing of Environment*, 91: 508–520.

Li, J.Y., D.G. Goodenough, and A. Dyk, 2005. Mapping relative water content in Douglas-Fir with AVIRIS and a canopy model. *Proceedings of IGARSS 2005*, Seoul, Korea, vol. V, pp. 3572–3574.

Li, X., A.H., Strahler, and C.E. Woodock, 1995. A hybrid geometric optical radiative transfer approach for modeling albedo and directional reflectance of discontinuous canopies. *IEEE Transactions on Geosciences and Remote Sensing*, 33: 466–480.

Lichtenhaler, H.K., M. Lang, M. Sowinska, F. Heisel, and J.A. Mieh, 1996. Detection of vegetation stress via a new high resolution fluorescence imaging system. *Journal of Plant Physiology*, 148: 599–612.

Lim, K.S., and P.M. Treitz, 2004. Estimation of above ground forest biomass from airborne discrete return laser scanner data using canopy-based quantile estimators. *Scandinavian Journal of Forest Research*, 19: 558–570.

Linn, R.R., 1997. *A Transport Model for Prediction of Wildfire Behavior.* Los Alamos National Laboratory Scientific Report, LA 13334-T.

Liu, J., J.M. Chen, J. Chilar, and W. Chen, 2002. Remote sensing-based estimation of net primary productivity over Canadian landmass. *Global Ecology and Biogeography*, 11: 115–129.

Magnussen, S., and P. Boudewyn, 1998. Derivations of stand heights from airborne laser scanner data with canopy-based quantile estimators. *Canadian Journal of Forest Research*, 28: 1016–1031.

Malenovský, Z., J. Albrechtova, Z. Lhotákova, R. Zurita-Milla, J.G.P.W. Clevers, M.E. Schaepman, and P. Cudlín, 2006. Applicability of the PROSPECT model for Norway spruce needles. *International Journal of Remote Sensing*, 27: 5315–5340.

Martin, M.E., and J.D. Aber, 1996. Estimating canopy characteristics as inputs for models of forest carbon exchange by high spectral resolution remote sensing. In *The Use of Remote Sensing in the Modeling of Forest Productivity.* H.G. Gholz, K. Nakane, and H. Shimoda (eds), Kluwer Academic, Dordrecht, The Netherlands, pp. 61–72.

Means, J.E., S.A. Acker, J.F. Brandon, M. Renslow, L. Emerson, and C.J. Hendrix, 2000. Predicting forest stand characteristics with airborne scanning LiDAR. *Photogram. Eng. Remote Sensing*, 66: 1367–1371.

Medlyn, B.E., 1998. Physiological basis of the light use efficiency model. *Tree Physiology*, 18: 167–176.

Melillo, J.M., J.D. Aber, and J.F. Muratore, 1982. Nitrogen and lignin control of hardwood leaf litter decomposition dynamics. *Ecology*, 63: 621–626.

Merton, R., 1998. Monitoring community hysteresis using spectral shift analysis and the red-edge vegetation stress index. *Proceedings of the Seventh Annual JPL Airborne Earth Science Workshop*, January 12–16, 1998, NASA, Jet Propulsion Laboratory, Pasadena, CA.

Miller, M., 2001. *Fire Effects Guide, Chapter III – Fuels*. NEES 2394. National Wildfire Coordinating Group. National Interagency Fire Center, Boise, http://www.nwcg.gov/pms/RxFire/FEG.pdf [cited Sept. 30, 2010].

Næsset, E., 1997. Determination of mean tree height of forest stands using airborne laser scanner data. *ISPRS Journal of Photogrammetry and Remote Sensing*, 52: 49–56.

Næsset, E., 2007. Airborne laser scanning as a method in operational forest inventory: Status of accuracy assessments accomplished in Scaninavia. *Scandinavian Journal of Forest Research*, 22: 433–442.

Näsi, R., E. Honkavaara, P. Lyytikäinen-Saarenmaa, M. Blomqvist, P. Litkey, T. Hakala, N. Viljanen, T. Kantola, T. Tanhuanpää, and M. Holopainen, 2015. Using UAV-based photogrammetry and hyperspectral imaging for mapping bark beetle damage at tree-level. *Remote Sensing*, 7(11): 15467–15493.

Niemann, K.O., and D.G. Goodenough, 2003. Estimation of foliar chemistry of western hemlock using hyperspectral data. Chapter 17. In *Remote Sensing of Forest Environments. Concepts and Case Studies*. M. Wulder and S.E. Franklin (eds). pp. 447–467.

Olthof, I., and D.J. King, 1998. Determination of soil property and forest structure relations with airborne digital camera images spectral and spatial information. *Proceedings of the 19th Canadian Symposium on Remote Sensing*, May 1997, Ottawa, ON, pp. 103–106.

O'Neil, A.L., J.A. Kupiec, and P.J. Curran, 2002. Biochemical and reflectance variation throughout the canopy of a Sitka spruce plantation. *Remote Sensing of Environment*, 80: 134–142.

Palmroth, S., and P. Hari, 2001. Evaluation of the importance of acclimation of needle structure, photosynthesis, and respiration to available photosynthetically active radiation in a Scots pine canopy. *Canadian Journal of Forest Research*, 31: 1235–1243.

Pellikka, P.K.E., E.D. Seed, and D.J. King, 2000. Modelling deciduous forest ice storm damage using CIR aerial imagery and hemispheric photography. *Canadian Journal of Remote Sensing*, 26: 394–405.

Peterson, B., W. Ni-Meister, J. Blair, M. Hofton, P. Hyde, and R. Dubayah, 2001. Modeling lidar waveforms using a radiative transfer model. *International Archives of Photogrammetry, Remote Sensing and Spatial Information Sciences*, 34: 121–124.

Peterson, D.L., M.A. Spanner, S.W. Running, and K.B. Teuber, 1987. Relationship of thematic mapper simulator data to leaf area index of temperate coniferous forests. *Remote Sensing of Environment*, 22: 323–341.

Phinn, S.R., P. Scarth, and D. Mitchell, 1999. Estimation of forest structural parameters for forestry and koala habitat monitoring in South-East Queensland, Australia. *Proceedings of the Fourth International Airborne Remote Sensing Conference and Exhibition II*, 21–24 June 1999, Ottawa, ON, pp. 179–186.

Plourde, L.C., S.V. Ollinger, M.-L. Smith, and M.E. Martin, 2007. Estimating species abundance in a northern temperate forest using spectral mixture analysis. *Photogrammetric Engineering & Remote Sensing*, 73: 829–840.

Porterie B., D. Morvan, J.-C. Loraud, and M. Larini, 2000. Firespread through fuel beds: Modeling of wind-aided fires and induced hydrodynamics. *Physics of Fluids*, 12: 1762–1782.

Pu, R., Q. Yu, P. Gong, and G.S. Biging, 2005. EO-1 Hyperion, ALI, and Landsat 7 ETM+ data comparison for estimating forest crown closure and leaf area index. *International Journal of Remote Sensing*, 26: 457–474.

Raddi, S., S. Cortes, I. Pippi, and F. Magnani, 2005. Estimation of vegetation photochemical processes: An application of the photochemical reflectance index at the San Rossore test site. *Proc. of the 3rd ESA CHRIS/Proba Workshop*, 21–23 March, ESRIN, Frascati, Italy (ESA SP-593, June 2005).

Riano, D., P. Vaughan, E. Chuvieco, P.J. Zarco-Tejada, and S.L. Ustin, 2005. Estimation of fuel moisture content by inversion of radiative transfer models to simulate equivalent water thickness and dry matter content: Analysis at leaf and canopy level. *IEEE Transactions on Geoscience and Remote Sensing*, 43: 819–825.

Rocchini, D., N. Balkenhol, G.A. Carter, G.M. Foody, T.W. Gillespie, K.S. He, S. Kark et al. 2010. Remotely sensed spectral heterogeneity as a proxy of species diversity: Recent advances and open challenges. *Ecological Informatics*, 5(5): 318–329.

RSI, 2009. Fire Fuel Tool. ENVI User's Guide, Version 4.7. ITT Visual Information Solutions, http://www.ittvis.com

Running, S.W., D.D. Baldocchi, D.P. Turner, P.S. Bakwin, and K.A. Hibbard, 1999. A global terrestrial monitoring network integrating tower fluxes, flask sampling, ecosystem modeling and EOS satellite data. *Remote Sensing of Environment*, 70: 108–128.

Sampson, P.H., 2000. Forest Condition Assessment: An Examination of Scale, Structure, and Function Using High Spatial Resolution Remote Sensing Data, *M.Sc. thesis*, York University, Toronto, Canada, 157 p.
Schaepman, M.E., B. Koetz, G. Schaepman-Strub, and K.I. Itten, 2005. Spectrodirectional remote sensing for the improved estimation of biophysical and chemical variables: Two case studies. *International Journal of Applied Earth Observation and Geoinformation*, 6: 271–282.
Schlerf, M., C. Atzberger, and J. Hill, 2005. Remote sensing of forest biophysical variables using HyMap imaging spectrometer data. *Remote Sensing of Environment*, 95: 177–194.
Seed, E.D., D.J. King, and P.K.E. Pellikka, 1999. Multivariate analysis of low cost airborne CIR imagery for the determination of forest canopy structure. *Proceedings of the Fourth International Airborne Remote Sensing Conference and Exhibition II*, June 21–24, 1999, Ottawa, ON, pp. 139–146.
Serrano, L., J. Penuelas, and S.L. Ustin, 2002. Remote sensing of nitrogen and lignin in Mediterranean vegetation from AVIRIS data: Decomposing biochemical from structural signals. *Remote Sensing of Environment*, 81: 355–364.
Shippert, M.M., D.A. Walker, N.A. Auerbach, and B.E. Lewis, 1995. Biomass and leaf-area index maps derived from SPOT images for Toolik Lake and Imnavait Creek areas. *Alaska. Polar Record*, 31: 147–54.
Smith, M.-L., M.E. Martin, L. Plourde, and S.V. Ollinger, 2003. Analysis of hyperspectral data for the estimation of temperate forest canopy nitrogen concentration: Comparison between an airborne (AVIRIS) and a spaceborne (Hyperion) sensor. *IEEE Transactions on Geoscience and Remote Sensing*, 41: 1332–1337.
Somers, B., G.P. Asner, R.E. Martin, C.B. Anderson, D.E. Knapp, S.J. Wright, and R. Van De Kerchove, 2015. Mesoscale assessment of changes in tropical tree species richness across a bioclimatic gradient in Panama using airborne imaging spectroscopy. *Remote Sensing of Environment*, 167: 111–120.
Spanner, M.A., L.L. Pierce, D.L. Peterson, and S.W. Running, 1990. Remote sensing of temperate coniferous forest leaf area index: The influence of canopy closure, understory vegetation, and background reflectance. *Remote Sensing of Environment*, 33: 97–112.
Stein, B.R., V.A. Thomas, L.J. Lorentz, and B.D. Strahm, 2014. Predicting macronutrient concentrations from loblolly pine leaf reflectance across local and regional scales. *GIScience & Remote Sensing*, 51(3): 269–287.
Stenberg, P., 1998. Implications of shoot structure on the rate of photosynthesis at different levels in a coniferous canopy using a model incorporating grouping and penumbra. *Functional Ecology*, 12: 82–91.
St-Onge, B.A., and F. Cavayas, 1995. Estimating forest stand structure from high resolution imagery using the directional variogram. *International Journal of Remote Sensing*, 16: 1999–2021.
Syphard, A.D., K.C. Clark, and J. Franklin, 2007a. Simulating fire frequency and urban growth in southern California coastal shrublands, USA. *Landscape Ecology*, 22: 431–445.
Syphard, A.D., V.C. Radeloff, J.E. Keeley, T.J. Hawbaker, M.K. Clayton, S.I. Stewart, and R.B. Hammer, 2007b. Human influence on California Fire Regimes. *Ecological Applications*, 17: 1388–1402.
Terjung, W.H., and S.S.F. Louie, 1972. Potential solar radiation on plant shapes. *International Journal of Biometeorology*, 16: 25–43.
Thenkabail, P.S., E.A. Enclona, M.S. Ashton, C. Legg, and M.J. De Dieu, 2004. Hyperion, IKONOS, ALI, and ETM+ sensors in the study of African rainforests. *Remote Sensing of Environment*, 90: 23–43.
Thenkabail, P.S., M.K. Gumma, P. Teluguntla, and I.A. Mohammed. 2014. Hyperspectral remote sensing of vegetation and agricultural crops. *Photogrammetric Engineering and Remote Sensing*, 80(8): 697–709.
Thomas, V., D.A. Finch, J.H. McCaughey, T. Noland, L. Rich, and P. Treitz, 2006a. Spatial modelling of the fraction of photosynthetically active radiation absorbed by a boreal mixedwood forest using a lidar-hyperspectral approach. *Agricultural and Forest Meteorology*, 140: 287–307.
Thomas, V., P. Treitz, J.H. McCaughey, and I. Morrison, 2006b. Mapping stand-level forest biophysical variables for a mixedwood boreal forest using LiDAR: An examination of scanning density. *Canadian Journal of Forest Research*, 36: 34–47.
Thomas, V., P. Treitz, J.H. McCaughey, T. Noland, and L. Rich, 2008. Canopy chlorophyll concentration estimation using hyperspectral and lidar data for a boreal mixedwood forest in northern Ontario, Canada. *International Journal of Remote Sensing*, 29: 1029–1052.
Thomas, V., J.H. McCaughey, P. Treitz, D.A. Finch, T. Noland, and L. Rich, 2009. Spatial modelling of photosynthesis for a boreal mixedwood forest by integrating micrometeorological, lidar and hyperspectral remote sensing data. *Agricultural and Forest Meteorology*, 149: 639–654.
Thomas, V., T. Noland, P. Treitz, and J.H. McCaughey, 2011. Leaf area and clumping indices for a boreal mixedwood forest: Lidar, hyperspectral, and Landsat models. *International Journal of Remote Sensing*, 32(23): 8271–8297.
Todd, K.W., F. Csillag, and P.M. Atkinson, 2003. Three-dimensional mapping of light transmittance and foliage distribution using lidar. *Canadian Journal of Remote Sensing*, 29(5): 544–555.

Townsend, P.A., J.R. Foster, R.A. Chastain, and W.S. Currie, 2003. Application of imaging spectroscopy to mapping canopy nitrogen in the forests of the central Appalachian Mountains using Hyperion and AVIRIS. *IEEE Transactions on Geoscience and Remote Sensing*, 41: 1347.

Treitz, P.M., and P.J. Howarth, 1999. Hyperspectral remote sensing for estimating biophysical parameters of forest ecosystems. *Progress in Physical Geography*, 23(3): 359–390.

Treitz, P.M., and P.J. Howarth, 2000. High spatial resolution remote sensing data for forest ecosystem classification: An examination of spatial scale. *Remote Sensing of Environment*, 72: 268–289.

Treitz, P.M., 2001. Variogram analysis of high spatial resolution remote sensing data: An examination of boreal forest ecosystems. *International Journal of Remote Sensing*, 22: 3895–3900.

Treitz, P.M., V. Thomas, P.J. Zarco-Tejada, P. Gong, and P.J. Curran, 2010. *Hyperspectral Remote Sensing for Forestry*. ASPRS Monograph Series, 107 pages.

Turner, D.P., W.D. Ritts, W.B Cohen, S.T. Gower, M. Zhao, S.W. Running, S.C. Wofsy, S. Urbanski, A.L. Dunn, and J.W. Munger, 2003a. Scaling Gross Primary Production (GPP) over boreal and deciduous forest landscapes in support of MODIS GPP product validation. *Remote Sensing of Environment*, 88: 256–270.

Turner, D.P., S. Urbanski, D. Bremer, S.C. Wofsy, T. Meyers, S.W. Gower, and M. Gregory, 2003b. A cross-biome comparison of daily light use efficiency for gross primary production. *Global Change Biology*, 9: 383–395.

Tymstra, C., M.D. Flannigan, O.B. Armitage, and K. Logan, 2007. Impact of climate change on area burned in Alberta's boreal forest. *International Journal of Wildland Fire*, 16: 153–160.

Ustin, S.L., and A. Trabucco, 2000. Using hyperspectral data to assess forest structure. *Journal of Forestry*, 98: 47–49.

Ustin, S.L., D.A. Roberts, J.A. Gamon, G.P. Asner, and R.O. Green, 2004. Using imaging spectroscopy to study ecosystem processes and properties. *BioScience*, 54: 523–534.

Valencia, D., P. Martínez, J. Plaza, R.M. Pérez, M.C. Cantero, and R. Paniagua, 2005. Pre-evaluation of wild fires in Monfragüe Regional Park using CHRIS imagery. *Proc. of the 3rd ESA CHRIS/Proba Workshop*, March 21–23, ESRIN, Frascati, Italy (ESA SP-593, June 2005), 4pp.

Varga, T.A., and G.P. Asner. 2008. Hyperspectral and lidar remote sensing of fire fuels in Hawaii Volcanoes National Park. *Ecological Applications*, 18(3): 613–623.

Vose, J.M., P.M. Dougherty, J.N. Long, F.W. Smith, H.L. Gholz, and P.J. Curran, 1994. Factors influencing the amount and distribution of leaf area in pine stands. *Ecological Bulletins (Copenhagen)*, 43: 102–114.

Wang, Q., S. Adiku, J. Tenhunen, and A. Granier, 2005. On the relationship of NDVI with leaf area index in a deciduous forest site. *Remote Sensing of Environment*, 94: 244–255.

Warren, S.D., M. Alt, K.D. Olson, S.D.H. Irl, M.J. Steinbauer, and A. Jentsch, 2014. The relationship between the spectral diversity of satellite imagery, habitat heterogeneity, and plant species richness. *Ecological Informatics*, 24: 160–168.

Wessman, C.A., J.D. Aber, D.L. Peterson, and J. Melillo, 1988. Remote sensing of canopy chemistry and nitrogen cycling in temperate forest ecosystems. *Nature*, 335: 154–156.

White, J. D., S.W. Running, R. Nemani, R.E. Keane, and K.C. Ryan, 1997. Measurement and remote sensing of LAI in rocky mountain montane ecosystems. *Canadian Journal of Forest Research*, 27: 1714–1727.

Woodcock, C.E., and A.H. Strahler, 1987. The factor of scale in remote sensing. *Remote Sensing of Environment*, 21: 333–339.

Woods, M., D. Pitt, M. Penner, K. Lim, D. Nesbitt, D. Etheridge, and P. Treitz. 2011. Operational implementation of a LiDAR inventory in Boreal Ontario. *The Forestry Chronicle*, 87(4): 512–528.

Yuan, X., D. King, and J. Vlcek, 1991. Sugar Maple decline assessment based on spectral and textural analysis of multispectral aerial videography. *Remote Sensing of Environment*, 37: 47–54.

Zarco-Tejada, P.J., J.R. Miller, T.L. Noland, G.H. Mohammed, and P.H. Sampson, 2001. Scaling-up and model inversion methods with narrowband optical indices for chlorophyll content estimation in closed forest canopies with hyperspectral data. *IEEE Transactions on Geoscience and Remote Sensing*, 39: 1491–1507.

Zarco-Tejada, P.J., J.R. Miller, G. Mohammed, T. Noland, and P. Sampson, 2002. Vegetation stress detection through chlorophyll + estimation and fluorescence effects on hyperspectral imagery. *Journal of Environmental Quality*, 31: 1433–41.

Zarco-Tejada, P.J., J.R. Miller, D. Haboudane, N. Tremblay, and S. Apostol, 2004. Detection of chlorophyll fluorescence in vegetation from airborne hyperspectral CASI imagery in the red edge spectral region. *International Geoscience and Remote Sensing Symposium, IGARSS'03, I*, Toulouse, France, pp. 598–600.

Zarco-Tejada, P.J., V. Gonzalez-Dugo, and J.A.J. Bemi, 2012. Fluorescence, temperature and narrow-band indices acquired from a UAV platform for water stress detection using a micro-hyperspectral imager and a thermal camera. *Remote Sensing of Environment*, 117: 322–337.

10 Characterization of Pastures Using Field and Imaging Spectrometers

Izaya Numata

CONTENTS

10.1 INTRODUCTION

Livestock grazing occupies 26% of the noniced terrestrial surface of the planet and is distributed across diverse ecosystems [1]. This land-cover type is of interest in terms of feed availability for livestock production and plays an important role in regional ecosystem functioning. It is critical to be able to measure accurately pasture biophysical and biochemical properties and their changes under human-environment interactions in order to provide information relevant to management practices. Remote sensing has played an important role in monitoring pasture dynamics and providing estimates of pasture properties using multispectral sensors. From the pasture management perspective, one of the ultimate goals for the use of remote sensing would be the quantification of biophysical and biochemical properties of pasture based solely on remotely sensed data, data that are available at low cost and that do not require time-consuming and costly field sampling and subsequent laboratory analysis. Near-infrared spectroscopy (NIRS) has been used as a laboratory method for the rapid evaluation of chemical composition and widely adopted to estimate forage quality parameters [2]. Despite the contributions of previous studies based upon multispectral sensors, the accuracy of broadband remote sensing data for grass estimation is still limited due to their spatial and spectral resolution [3].

In the past decade, hyperspectral remote sensing of pasture has made significant progress in the detection and quantification of pasture biophysical (e.g., biomass) and biochemical (e.g., nutrients and water) variables by identifying critical wavebands and exploiting new absorption features. Despite these efforts, an integrated and comprehensive knowledge of hyperspectral remote sensing of pasture is needed. This chapter describes the state of the art of hyperspectral remote sensing of pasturelands, highlighting recent advances made using field-based and imaging spectrometers.

10.2 FIELD AND IMAGING SPECTROMETERS FOR PASTURE CHARACTERIZATION

Laboratory or in situ spectrometer measurements are necessary steps in identifying basic spectral characteristics of different vegetation species and establish relationships between vegetation attributes and hyperspectral measures [4,5]. A number of studies on pasture/grass characterization have been conducted using field spectrometers for a wide range of focuses such as the estimates of biophysical [biomass and leaf area index (LAI)] and biochemical concentrations (pigments, nutrients, and water content), fractional cover, litter-based estimates to pasture degradation, and more [6–11]. As most grasslands and pastures are composed of vegetation 1 m or less in height, field-based analysis of pasture at the canopy level via a field spectrometer is a practical way to assess relationships between field data and remote sensing measurements. Moreover, sources of errors and their impacts on analyzed relationships can be assessed easily at field levels. Field-based experiments are used for broader spatial scales using imaging spectrometers or used in the inversion of radioactive transfer (RT) models to estimate vegetation properties. Reflectance spectra from distinct grass species or vegetation materials measured in the field can be stored in spectral libraries and used as reference or ideal spectra to calibrate imaging spectrometers and map distinct species at larger spatial scales.

While laboratory analysis is performed under controlled conditions, vegetation spectra measured in situ are governed by many factors, such as canopy vertical and spatial structure, the presence of live and dead materials, and the diversity of background (discussed in the next section). Thus, to have good relationships between grass biophysical and biochemical parameters and field spectra, reliable measurements of both field grass and field spectrometers should be taken.

Imaging spectroscopy provides biophysical and biochemical measurements of landscapes. Compared to field experiments for pasture, a small number of studies still use imaging spectrometers for pasture characterization. Currently, several airborne [e.g., Airborne Visible/Infrared Imaging Spectrometer (AVIRIS), Hydice, HyMap, CASI) and satellite (e.g., Hyperion) imaging spectrometers with different spectral and spatial resolutions are available. One of the goals in using hyperspectral remote sensing for pastures is to extract biophysical and biochemical attributes (biomass, LAI, biochemical concentrations) and detect their spatial distributions through quantitative analytical methods [9,12,13]. Another goal is to improve discrimination of grasslands from other land-cover types that are spectrally ambiguous in the broad spectral band domain. For example, discrimination between dry pasture and bare soil and between green pasture and secondary forest is a real challenge for land-cover mapping in the Amazon using broadband sensors like Landsat [14,15]. Spatial patterns of pasture characteristics at landscape scales provide relevant information for land owners to make decisions on management strategies.

10.3 CONTROLLING FACTORS FOR BIOPHYSICAL AND BIOCHEMICAL CHARACTERISTICS OF PASTURE

Several factors alter grass biophysical and biochemical characteristics, which also directly affect grass spectral reflectance signatures (Table 10.1). Hill [16] lists several grass physical characteristics important for remote sensing, including (a) height and variation in height, (b) proportion of bare

TABLE 10.1
Physical and Chemical Features That Influence the Interaction between Grassland Vegetation and Radiation Sources for Remote Sensing

Physical and chemical features
Grass height and variation in height
Soil coverage
Leaf area
Leaf orientation or leaf angle distribution
Density of reflective and absorptive structures
Proportion of senescent or dead materials
Moisture content
Pigmentation
Spatial arrangement of structures
Variability of all of the preceding due to species diversity and management

Source: Hill, M.J. 2004. In S.L. Ustin (Ed.), *Manual of Remote Sensing Volume 4. Remote Sensing for Natural Resource Management and Environmental Monitoring* (pp. 449–530). Hoboken, NJ: John Wiley & Sons [16].

soil, (c) leaf area, (d) leaf orientation or leaf angle distribution, (e) density of reflective or absorptive structures, (f) proportion of live and dead materials, and (g) spatial arrangement of structures. In addition, natural and human-related factors such as climate, soils, species composition in pastureland, variable grazing pressures, and age of grass all influence grass biophysical and biochemical properties. Some of the issues will be highlighted in what follows.

10.3.1 Structure

Asner [6] and Asner and Heidebrecht [17] provide excellent summaries of vegetation spectral properties and their changes as a function of vegetation structure. The authors emphasized the complexity of these factors by using an inverse radiative model. Grass structure influences interactions between radiation and grass canopy and the responses of radiation (reflected or scattered) to remote sensors. Vertical structure and the spatial arrangement of structures of grass vary according to grass species. Figures 10.1 and 10.2 illustrate structures of two predominant grass species used in Brazilian Amazon pastures.

Brachiaria brizantha, an African grass species widely distributed in the Brazilian Amazon, presents a highly heterogeneous surface. This species has stout erect culms and forms bunched crowns. This creates a tufted structure that does not cover the soil surface evenly, which results in the significant effect of soil background on vegetation reflectance (Figures 10.1a, 10.2a, b). *Brachiaria decumbens* is low growing and more decumbent and forms a dense cover, creating a

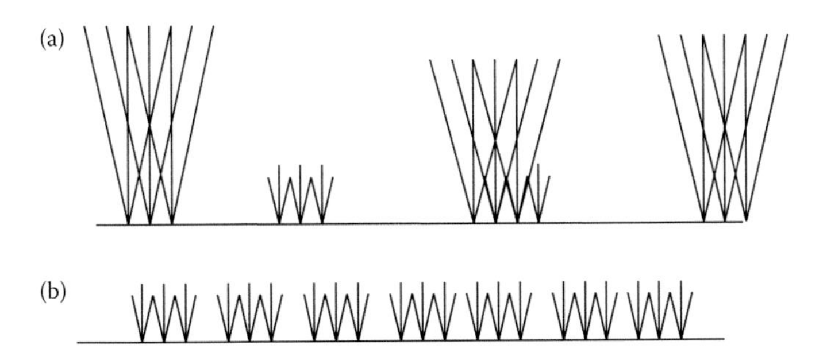

FIGURE 10.1 Grass structure: (a) *Brachiaria brizantha*; (b) *Brachiaria decumbens*.

FIGURE 10.2 (a) Canopy of *B. brizantha*; (b) overview of *B. brizantha*; (c) canopy of *B. decumbens*; (d) overview of *B. decumbens*; (e) averaged reflectance of *B. brizantha* and *B. decumbent*; (f) standard deviation from averaged reflectance of two species (number of sample spectra = 69).

more homogeneous canopy surface (Figures 10.1b, 10.2c,d). In pastures, grass structures are heavily altered by grazing [14]. Average reflectance signatures and the spectral variability of the two species are strongly related to their structural differences especially at the in situ scale (Figure 10.2c,d). Numata et al. [14] observed that the variation in canopy structure within the field of view of a field spectrometer contributes to the spectral variability of canopy reflectance, even for those areas with the same amount of biomass. In the case of these two species, the heterogeneous and complex canopy structure of *B. brizantha* makes biomass estimation more challenging [11].

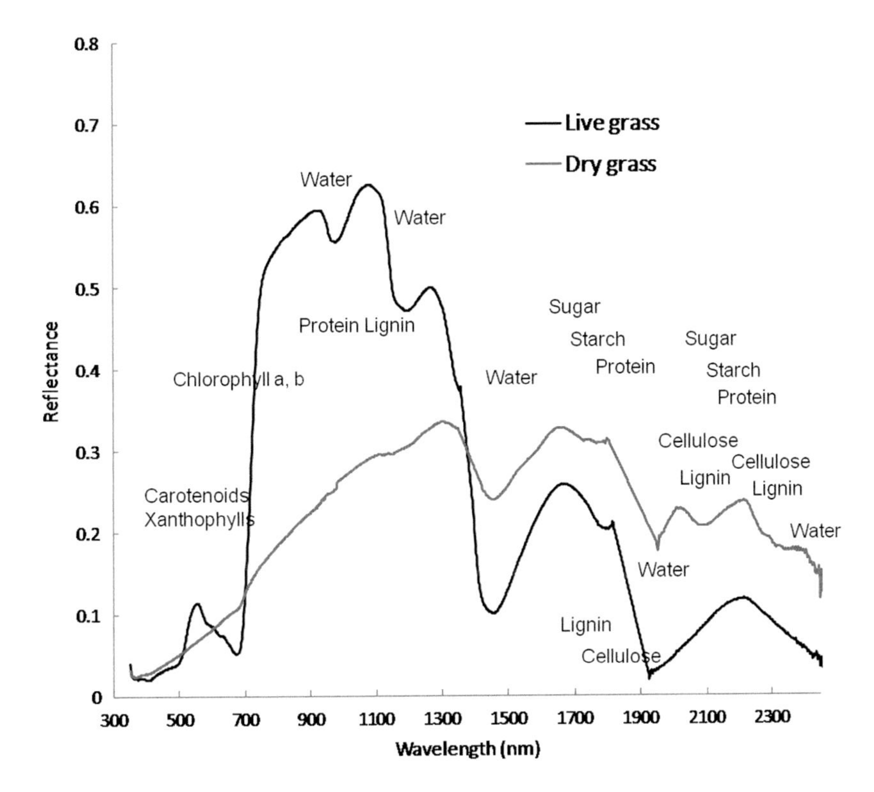

FIGURE 10.3 Reflectance spectra with characteristic absorption features associated with plant chemical constituents for live and dry grass. (Adapted from Hill, M.J. 2004. In S.L. Ustin (Ed.), *Manual of Remote Sensing Volume 4. Remote Sensing for Natural Resource Management and Environmental Monitoring* (pp. 449–530). Hoboken, NJ: John Wiley & Sons [16].)

10.3.2 Foliar Chemical Composition

Chemical constituents such as chlorophyll a and b, protein, lignin, cellulose, water, and others influence vegetation reflectance and are related to pasture quality (Figure 10.3). The absorption features found in the visible spectrum (430–660 nm) such as chlorophyll absorptions are correlated with major nutrients important for animal production, while the shortwave infrared (SWIR) region (1700–2400 nm) is highly characterized by lignin-cellulose absorptions [4]. Green and live leaf reflectance is determined primarily by water, pigment, and carbon content, while dry leaf reflectance shows strong signals of lignin and cellulose content (Figure 10.3). In green leaf, the absorption features of nitrogen (N), phosphorus (P), lignin, cellulose, and other constituents in NIR and SWIR are obscured by leaf water content [18–20]. These chemical constituents determine forage quality and shape characteristic absorption features in the reflectance spectra of forage plants. These spectral characteristics associated with chemical composition indicate grass physiological conditions and vigor and provide information on nutrient status important for animal productivity. Thus, absorption features and their shapes (area and depth) are used to estimate the concentrations of chemical compounds as well as to distinguish grass from different land covers [21] or different grass species [22].

10.3.3 Nonphotosynthetic Vegetation and Background Effects

Most studies of grass/pasture characterization using remote sensing emphasize only green materials to estimate biomass or chemical elements. However, nongreen or nonphotosynthetic vegetation (NPV), referring to senesced or dead grass and litter, is an important component in pasture especially in dry regions or in dry weather conditions and plays a critical role in shaping overall grass spectral reflectance. NPV has its greatest effect in the SWIR region between 2000 and 2400 nm, mainly in connection with the concentration of lignin-cellulose in dry plant residue, as discussed earlier (Figure 10.3). Variation in NPV has a significant impact on vegetation indices (VIs) such as the normalized difference vegetation index (NDVI) and the soil-adjusted vegetation index (SAVI) [23]. Changes in substrate reflectance beneath the grass canopy, including litter on the surface and exposed soil, affect the reflectance of grass canopies with lower vegetation coverage. Litter and soil have very high spectral signatures throughout the range 400–2500 nm, and their signals can dominate the reflectance of grass canopies of low coverage [24,25]. Neglecting the effects of background such as litter and soil fractions in grassland may cause erroneous estimation of biophysical characteristics such as grass biomass through remotely sensed data. Hyperspectral data have the potential to detect vegetation covers, green vegetation NPV, and soil cover, as well as their combinations [21,24,26]. Monitoring of changes in these covers can provide better characterization of grass ecosystem change.

10.4 HYPERSPECTRAL APPROACHES TO PASTURE CHARACTERIZATION

A large number of spectral bands in hyperspectral systems provide an opportunity to develop a range of new measurements or refine conventional approaches to the estimation of grass properties (Table 10.2). In what follows, some typical approaches important for pasture/grass characterization are presented.

10.4.1 Vegetation Indices

VIs, such as simple ratios and the NDVI, with two or more bands are widely used to estimate the biophysical and biochemical properties of vegetation. These VIs explore the contrast between two

TABLE 10.2
Key Spectral Bands Related to Vegetation Properties and Forage Quality

Band (nm)	Vegetation Parameters	References
430	Chlorophyll a, nitrogen	Knox et al. [82]
460	Chlorophyll b, nitrogen	
470	Total plant pigment concentration	Blackburn [83]
530	Chlorophyll-a absorption	Gamon et al. [84]
660	Nitrogen	Carter [85]
695	Crude Protein	Punagalli et al. [64]
700	Total chlorophyll, nitrogen	Carter [85]
720	Total chlorophyll, leaf mass	Horler et al. [86]
775	Crude protein	Kawamura et al. [47]
800	Lignin	Punagalli et al. [64]
820	Leaf mass, leaf area index, Lipid	Carter [85], Punagalli et al. [64]
970	Phosphorus	Knox et al. [82]
990	Crude protein	Punagalli et al. [64]
1540	Cellulose, vegetation water content	Carter [85]
1740	Crude protein	
2060	Protein, nitrogen	Carter [85])
2270	Crude protein	
2280	Cellulose, sugar, starch, leaf mass	Carter [85]
2300	Leaf mass, vegetation water content	Carter [85]
2450	Cellulose, protein, nitrogen	Carter [85]
2470	Cellulose, protein	Kumar et al. [87]

or more spectral bands. In the case of the conventional NDVI, low reflectance in the red, due to chlorophyll absorption, and high reflectance in the NIR, related to multiple scattering effects, are used to estimate vegetation greenness [27]. One advantage of using hyperspectral data is that we can develop a new NDVI with different combinations of two narrowbands from the red and NIR regions that are averaged out over the broadbands of multispectral sensors. Many research studies of hyperspectral remote sensing of pasture employ spectral indices to estimate specific vegetation properties [12,28].

A typical approach to determining the best narrowband VIs is to calculate all possible combinations of two bands and identify a combination that has the highest coefficient of determination (R^2) with a target variable [29,30]. Many two-band combinations derived from hyperspectral data for the estimation of a target variable in a recent study showed much better performance than the traditional red-NIR band combination. For example, Mutanga and Skidmore [29] found that the standard red-NIR-based NDVIs derived from a laboratory-based hyperspectral analysis performed poorly in estimating the dense biomass of tall grass due to the saturation level observed in dense vegetation. However, a modified NDVI with 746 and 755 nm bands had a high R^2 (0.78 compared to 0.25 with the standard NDVI), showing the potential of hyperspectral data to overcome saturation problems with a high-density grass canopy. Darvishzadeh et al. [12] successfully developed narrowband-derived NDVIs and SAVI with two bands selected from two-dimensional correlation plots for the prediction of grass LAI in Majella National Park using HyMap data. Fava et al. [31] analyzed the variability of reflectance and vegetation properties in different pasture growth stages, determined the impact of this variability on VI-based assessment of pasture properties, and evaluated the potential of narrowband NDVI and SR for assessing biomass and LAI as well as canopy N.

10.4.2 Red Edge

Like all green vegetation covers, green grasses have been characterized by a maximum slope in the red edge between 680 and 750 nm. Chlorophyll concentration is strongly correlated with the point of maximum slope between very low reflectance in the red resulting from chlorophyll absorption and very high reflectance in the NIR due to internal cellular scattering in this region [4]. This region of chlorophyll absorption deepens and expands as chlorophyll concentration increases and consequently the red-edge position moves to a longer wavelength [32].

The structure of the chlorophyll red edge is best identified through the first derivative of vegetation reflectance. Due to the strong relationship between chlorophyll concentration and plant productivity, the location of a red-edge point has been used to estimate nutritional status [33,34]. Additionally, LAI and biomass have been found to be well correlated to red-edge parameters in the first derivative reflectance curves or VIs with red-edge wavebands [33,35]. Jago et al. [36] generated red-edge position images by a linear equation for chlorophyll estimation and observed high correlation with grassland canopy chlorophyll concentration ($r = 0.84$).

Cho and Skidmore [35] compared red-edge positions extracted by two methods (the Lagrangian and linear extrapolation) from HyMap images acquired in two different years and found a high correlation with field grass biomass ($R^2 > 0.50$). These results indicate the red-edge-based method may be widely used for grassland monitoring of vigor, nutritional status, and biomass production as imaging spectrometer data become more available [16].

The red edge has also been used to study plant stress due to nutrient deficiency [37,38] and contamination with pollutants such as gas and metals [39,40]. A shift in red-edge position may be used as an indicator of plant stress. One of the typical approaches is to identify nutrient deficiency in grass based upon the shift of the red-edge position. Mutanga and Skidmore [41] related the red-edge position to N supply to *Cenchus ciliaris* grass in a greenhouse. They observed that the red-edge position of grass canopies was shifted from the control at 703 nm to the high N treatment at 725 nm.

Kooistra et al. [39] studied the effects of soil metal concentrations on grass and other vegetation based upon red-edge positions derived from the first derivative calculated from the 690–720 nm absorption feature and other VIs. Some satisfactory relationships were found between the red-edge position and soil metals such as Pb ($R^2 = 0.61$) and Cu ($R^2 = 0.51$). The researchers also observed that the red-edge positions from grass reflectance increased as soil metal concentrations decreased. Smith et al. [40] used ratios of the magnitude of the derivatives from the red-edge region as an index of plant stress responses to soil-oxygen depletion from natural gas leakage. They found that the ratios of the magnitude derivative at 725 nm to that at 702 nm were less in areas where gas was present. The plant stress responses based upon these ratios were identified for long-term leaks in all studied crops but for short-term leaks only in grass. These studies demonstrate the potential of hyperspectral remote sensing for plant stress study.

10.4.3 Spectral Transformations

Across a full spectral range in the optical wavelength (400–2500 nm), vegetation reflectance presents several spectral absorptions associated with biochemical attributes. Several transformation approaches are used to enhance the use of spectral features of vegetation spectra. Common approaches include first and second derivatives, the logarithm of reciprocal reflectance, continuum removal, and other combinations. The depth and the area of these absorptions and indices based upon these features have been increasingly employed for pasture characterization. A study by Kokaly and Clark [19] has been one of the most important references for many research studies on hyperspectral remote sensing of vegetation. Their methodology was developed originally to enhance and standardize known chemical absorption features usually affected by the effects of factors such as water on carbon related absorption features in the SWIR region and exposed soil. The approach uses a continuum removal method [42] that normalizes the spectral curves of the absorption features by establishing a common baseline between the edges of the absorption region (Figure 10.4). In this way, differences in absorption strengths are enhanced. Absorption depth is the normalized depth of an absorption feature from the common baseline. Band depths within absorption features are divided by the band depth at the center of the feature, called normalized band depths. Stepwise multiple linear regression (MLR) is used to analyze normalized band depths for all wavelengths in continuum-removed absorption features and select the most sensitive wavelengths to a target vegetation parameter in each absorption feature [4,20]. Then linear equations are developed between band depth and associated vegetation measurements.

The method has contributed to improving the estimation of chemical constituents such as N, P, crude protein, digestibility, lignin and cellulose [8–10,43–46], biomass [29,31,47], and grass species discrimination [22]. The use of this methodology for imaging spectrometers for the estimation of biochemical concentration in different vegetation types has been successful [48,49], which indicates the applicability of this technique for grassland and pasture to landscape scales.

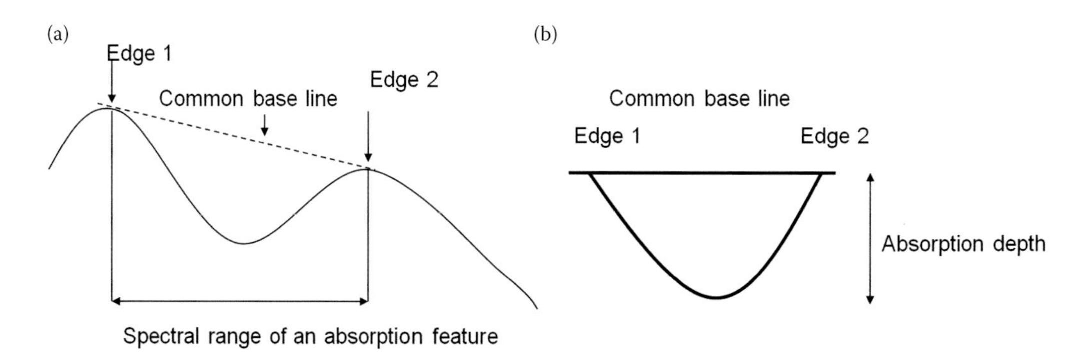

FIGURE 10.4 Illustration of a normalized spectral absorption depth: (a) a spectral absorption feature with established common line and (b) normalized spectral curve by common baseline.

10.4.4 Spectral Mixture Analysis

Spectral mixture analysis (SMA) provides fractional cover measurements related to a land surface within a pixel or the field of view of a sensor [50,51]. Mixture modeling is based upon an assumption that a measured spectral signal is the sum of the signals from the components weighted by their fractions [50]. For an ecosystem like grazing pasture, usually four components are considered, including green vegetation (GV), senesced vegetation or NPV, soil, and shade. SMA has been found to be very useful for estimating fractional covers of the main components of a landscape, including pastures and grasslands [11,15,21,52,53].

SMA involves the following basic steps: (1) selection of endmembers that represent major materials or components existing in the instantaneous field of view or pixel and (2) unmixing or solving the mixing equation, linear or nonlinear, for the fractions of the selected endmembers. This method has been widely used for multispectral data, and the potential of hyperspectral data to accurately estimate vegetation covers using SMA has been evaluated by several researchers. Their results indicate that hyperspectral data provide more accurate estimates of fractional cover measurements compared to multispectral data [11,17]. For example, NPV and soil fractions are not easily separated spectrally in the visible and NIR region in the broadband domain, but these materials can be differentiated based on lignin-cellulose absorption bands in the SWIR [21,24,26]. Using AVIRIS as a high-performance hyperspectral airborne sensor with high signal-to-noise ratios, Asner and Heidebrecht [17] found that the SWIR 2000–2300 nm was a crucial spectral region to estimate accurate fractional covers of PV, NPV, and bare soil for shrub and grassland sites.

Numata et al. [11] compared pasture fractional covers including shade-normalized NPV, GV, and soil estimated by the field reflectance spectra (ASD Inc., Boulder, CO, USA), the Hyperion, and convolved Landsat data from the Hyperion data to field grass covers estimated by charge coupled device (CCD) data (i.e., NPV, GV, and soil) as reference, all measured from the same field transects in Rondônia in the Amazon. NPV and GV fractions derived from field reflectance spectra were the closest values and statistically the same as that of the reference. Between the reference and Hyperion, NPV and GV fractions showed the largest differences and soil fractions had smaller, observable differences in fractions that were not statistically significant. However, Landsat-derived fractions such as NPV and GV showed larger gaps and were statistically different from the same fractions compared to the reference, overestimating NPV and underestimating GV (Table 10.3). The results indicate that hyperspectral data provided more accurate grass fractional covers than Landsat data. This is especially true for pastures in dry regions or degraded pastures where senesced grass and bare soil are present and affect spectral signatures of these pastures [15]. Fraction images derived from EO-1 Hyperion imagery for a pasture area in the Amazon are shown in Figure 10.5.

10.4.5 Statistical Methods

Statistical models using original spectral bands or transformations from hyperspectral data as independent variables have been employed to improve estimation of vegetation parameters or to develop predictive relationships between hyperspectral reflectance and plant constituents of grassland. One of the challenges of hyperspectral data is the extraction of critical spectral information. Multiple regression with hyperspectral data suffers from multicollinearity or spectral overfitting when the number of observations is smaller than the number of wavelengths studied and when input data show a high correlation [4,54]. To avoid this problem, the selection of a few contiguous regions of known absorption features is recommended [10,19]. Stepwise multiple regression has been widely used to select an optimal set of spectral bands for estimating vegetation parameters [19,55]. The advantage of this technique is that the derived features are easily interpretable from a physical point of view [56]. However, this approach still suffers from multi-collinearity and the extensive spectral overlaps of individual properties [47].

TABLE 10.3
Shade-Normalized Fractional Covers[a] of NPV, GV, and Soil for CCD, ASD, Hyperion/EO-1, and Convolved ETM+/Landsat 7 Spectra from Hyperion

Measurements		Mean	Std. Dev.	*p*-Value
Field measurements (CCD)	NPV	0.81	0.06	–
(63 measurements)	GV	0.16	0.05	–
	Soil	0.04	0.07	–
ASD field spectrometer	NPV	0.80	0.17	0.8086
(63 measurements)	GV	0.13	0.13	0.3878
	Soil	0.08	0.12	0.2380
Hyperion/EO-1	NPV	0.85	0.11	0.3589
(12 pixels)	GV	0.10	0.07	0.1241
	Soil	0.05	0.06	0.8419
ETM+/Landsat 7	NPV	0.94*[b]	0.14	0.0078
(12 pixels)	GV	0.07*	0.07	0.0095
	Soil	−0.01	0.09	0.1951

Source: Numata, I. et al. 2008. *Remote Sensing of Environment*, 112, 1569–1583 [11].

[a] The NPV, GV, and soil fractions were generated from full optical range reflectance (400–2500 nm) and were normalized by shade fraction to minimize illumination problems [11]. Fraction values vary from 0 to 1.

[b] Statistically significant mean differences relative to CCD fraction determined using a *t*-test at 0.95 level are marked by *.

Partial least-squares regression (PLSR) [57] is another approach that can be used to reduce a large number of spectral bands with a high degree of collinearity to a smaller number of noncorrelated latent variables. This method operates similarly to principal component analysis, but instead of decomposing the spectra into a set of eigenvectors, scoring and regressing them against response variables, for example vegetation parameters, as a separate step, PLSR uses the response variable information during data decomposition. However, the physical interpretation of the latent variables is difficult, as with principal components [57]. The PLSR method has high predictive abilities and has been used to develop predictive models of feed quality including N, crude protein, lignin, cellulose [25,45,58], and LAI [59,60]. Kawamura et al. [60] coupled PLSR with other approaches, such as genetic algorithms and iterative stepwise elimination, to further improve spectral band selection and removal of redundant spectral bands from hyperspectral data. The successive projections algorithm (SPA) is also a method to reduce variable collinearity. SPA selects important wavelengths by using projection operators in a vector space with the maximum projection values on the orthogonal space of the previous selected wavelengths to avoid collinearity. When coupled with MLR (SPA-MLR), predictive relationships are established between the wavelengths selected by SPA and target constituents of vegetation. Wang et al. [28] observed that SPA-MLR outperformed PLS and SMLR from the perspective of prediction accuracy, model simplicity, and robustness for canopy-level grass nutrient estimation.

10.5 APPLICATIONS OF HYPERSPECTRAL REMOTE SENSING FOR PASTURE ESTIMATION

10.5.1 Pasture Quality

Pasture nutritional quality is an indicator of grass nutrient deficiency and degradation as well as animal grazing distribution patterns [61]. The concentrations of numerous nutrients have been found

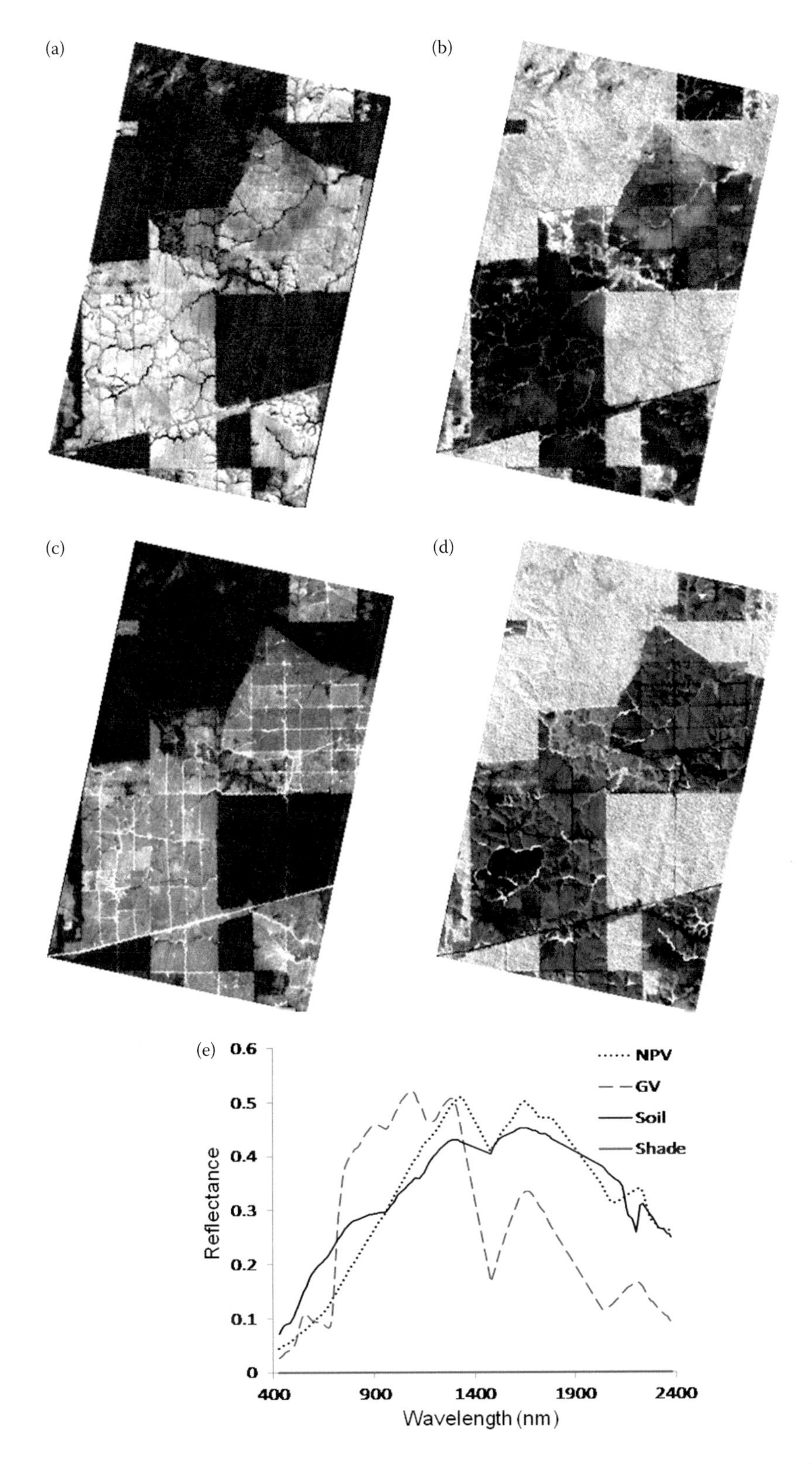

FIGURE 10.5 Fraction images of a pasture area in the southwestern Amazon derived from EO-1 Hyperion imagery. (a) NPV. (b) GV. (c) Soil, (d) Shade, and (e) Endmembers.

to be well correlated with spectral features, reflectance indices, and spectral transforms derived from hyperspectral data [10,19,48,54]. N is one of the most important elements and is strongly related to chlorophyll activity and often associated with protein, which promotes the photosynthetic process. As chlorophyll determines spectral reflectance in the visible region, strong relationships between visible absorption bands and N concentration have been identified [10,54]. P is a fundamental element in tissue composition in addition to being one of the components of nucleic acids and enzymes. Compared to N, P has received less attention but is an important nutrient for pasture management. This has been considered a limiting element in forage production in the tropics like Amazonia and the African savanna [62].

Most narrowbands and absorption features highly correlated with nutrients are concentrated in the visible spectrum, particularly in the red-edge region [8,10,19,63]. Mutanga et al. [63] found that within the red-edge region, those bands selected by stepwise linear regression for the prediction of nutrients were most frequently located around 680 nm, a region of pigment absorption. The green reflectance region (550–580 nm) was another important region for nutrient prediction.

On the other hand, weak correlations between nutrients and hyperspectral data have been observed in the SWIR [10]. For example, in fresh grass canopies, leaf mineral contents such as P, K, and S, are usually difficult to be estimated due to the presence of water that masks the biochemical absorption features, particularly in the SWIR region [19,20]. Additionally, internal scattering and mixing of spectral signatures obscures the absorption signal of nutrients due to differences in the physical structure of the canopies of different species [8]. Continuum-removal methodology by Kokaly and Clark [19] minimizes these effects on biochemical absorption features and enhances absorption strengths.

Mutanga et al. [10] evaluated four absorption variables derived from continuum-removed absorption features to predict canopy N, P, potassium (K), calcium (Ca), and magnesium (Mg) concentrations in five African grass species in the field through MLRs. The continuum removal derivative reflectance (CRDR) variable yielded the highest coefficients of determination of 0.7, 0.8, 0.64, 0.5, and 0.68, with low errors for N, P, K, Ca, and Mg, respectively. In a similar study, Kawamura et al. [8] used PLSR models for the prediction of N, P, K, and sulfur (S) and biomass from grass with absolute reflectance, first derivative reflectance (FDR), and CRDR as input variables for PLSR models. Again, CRDR had the highest R^2 values for all minerals; 0.90, 0.94, 0.81, and 0.94 for N, P, K, and S, respectively.

Besides these major nutrient elements, many efforts have been made to develop predictive models for the estimation of other constituents such as crude protein, fibers, and digestibility using hyperspectral data and statistical methods at different spatial scales [13,45–47,58,64]. Crude protein involves N protein and nonprotein nitrogen. Digestibility is a measure of how much of a forage can be digested, usually measured as dry matter digestibility. Fiber fractions consisting of lignin, cellulose, and cutin are indigestible and negatively correlated to major nutrients including N, crude protein, and digestibility.

Thulin et al. [45] used PLSR and spectral transforms derived from field spectrometers to develop predictive models for crude protein, fiber (lignin and cellulose), and digestibility from temperate pastures in Victoria, Australia. They obtained the best predictive models with continuum removal with spectral bands normalized to the depth of the absorption features for digestibility [adjusted $R^2 = 0.82$, root-mean-square error of prediction (RMSEP) = 3.94), CRDR for crude protein (adjusted $R^2 = 0.62$, RMSEP = 1.87), and cellulose (adjusted $R^2 = 0.73$, RMSEP = 2.37). Other studies also obtained moderately good results for predictive models for crude protein [47] from field spectra in temperate pastures.

These studies recommended specific spectral regions sensitive to different pasture quality parameters. For example, the concentration of crude protein is correlated with wavelengths in the visible and NIR, but also SWIR, including 1721, 1738, 2266–2277, and 1950–2400 nm [47,58,64]. For the prediction of digestibility and fiber contents (i.e., lignin and cellulose) using field and imaging spectrometers, Thulin et al. [45,46] found the spectral regions of the chlorophyll absorption and

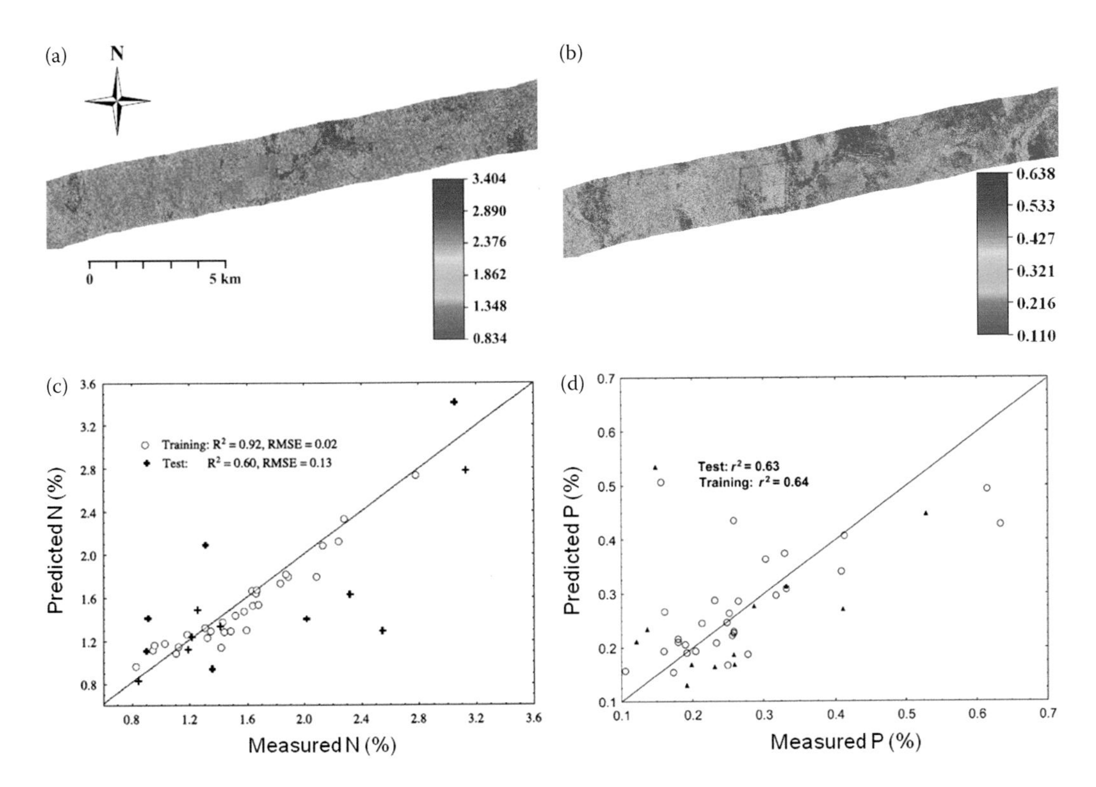

FIGURE 10.6 Maps showing spatial distribution of concentration (%) of nitrogen (a) and phosphorus (b) and scatterplots obtained from the best-trained neural network used for mapping. Scatterplots of nitrogen (%) (c) and phosphorus (%) (d). (From Mutanga, O. and Skidmore, A.K. 2004a. *Remote Sensing of Environment*, 90, 104–115 [9]; Mutanga, O. and Kumar, L. 2007. *International Journal of Remote Sensing*, 28, 4897–4911 [65].)

red edge and SWIR as the main wavelengths. The same spectral regions were found useful for the prediction of N, P, K, Ca, and Mg in African grasses [10].

While most studies on pasture quality assessment using hyperspectral sensors have been done in laboratory and in situ levels, efforts have also been made to map grass chemical constituents using spectrometers. Mutanga and Skidmore [9] integrated continuum removal absorption features from the visible (550–757 nm), SWIR (2015–2199 nm), and red-edge position derived from HyMap imagery and neural networks to map grass N concentration in an African savanna rangeland (Figure 10.6a). While the method used obtained a high coefficient of determination ($R^2 = 0.92$) with an RMSE of 0.02 for the training dataset, the predictive capability with the test data set indicated 60% of the variation in grass nitrogen concentration with an RMSE of 0.13 (Figure 10.6c). Using the same method, Mutanga and Kumar [65] estimated and mapped grass P concentration in the same African rangeland and obtained a coefficient of determination of 0.63 with an RMSE of 0.07 for the test data set (Figure 10.6b,d). They also found that the input of SWIR bands greatly contributed to improving the estimation of grass P concentration, and the prediction errors were drastically reduced when the visible and SWIR bands were used together compared with using the visible input only.

10.5.2 Leaf Area Index

LAI is one of the main drivers of canopy primary productivity and has been a key variable in most ecosystem models. LAI can be directly measured by optical remote sensing of vegetation, including grasslands and pastures. VIs, such as simple band ratio (SR), NDVI, SAVI, and others derived from multispectral sensors, have been widely used to estimate LAI over large regions. Narrow spectral bands derived VIs potentially provide additional improvements over two broadband-based

VIs to estimate LAI. Fava et al. [31] had the best performance for LAI estimation with an SR of 895 nm/730 nm ($R^2 = 0.76$), compared to thea more widely used near-infrared – red ratio, 780 nm/680 nm with ($R^2 = 0.39$). They found that the best combination of VIs for LAI were found in the NIR to SWIR regions, 1105 nm/1229 nm for NDVI ($R^2 = 0.61$) and 1998 nm/1402 nm for SAVI ($R^2 = 0.64$).

Besides simple relationships with these VIs, grass LAI estimation using hyperspectral data can be performed using statistical and physical models. Darvishzadeh et al. [12,59] investigated PLSR and the inversion of the PROSAIL radiative transfer model for LAI estimation in a Mediterranean grassland using both field spectrometer and HyMap airborne images. These results indicate the potential of hyperspectral data to improve LAI estimation by remote sensing.

LAI estimates may be very challenging in mixed grassland, where soil and litter effects significantly influence grass reflectance. He et al. [7] evaluated the performance of 15 different VIs in estimating the LAI of grassland in the semiarid Western region of Canada. Although the relationships between grassland LAI and studied VIs were statistically significant, their predictive capabilities were low ($R^2 = 037–0.44$). A new VI was developed in this study that incorporates the cellulose absorption index (CAI) that varies as a function of the proportion of litter as a litter factor in the adjusted transformed soil-adjusted vegetation index (ATSAVI). This index improved the LAI estimation capability by about 10% ($R^2 = 55\%$). The results indicate the potential contribution of hyperspectral data to improving the LAI estimation by minimizing the effects of litter.

10.5.3 Biomass

Pasture biomass is directly related to pasture and animal productivity (meat and milk), and its quantification is one of the most important, but also challenging applications of remote sensing for pasture research (Table 10.4). Several efforts have been made to estimate biomass in pastures and grassland using multispectral satellite data, and most of them have established generally good relationships between field data and remote sensing derived measures [67,68]. The potential of hyperspectral data for the estimation of grass biomass has been evaluated across different spatial scales, such as controlled laboratory, in situ, and landscape scales [9,28,44,47,56,66,69].

Although reflectance is directly related to LAI, the relationship between reflectance and biomass is indirect. This implies that the same LAI may be representative of different amounts of biomass, depending on the relationships between various canopy structural and density characteristics [16]. Biomass estimation is more problematic particularly for pasture areas with high vegetation density.

TABLE 10.4
Results of Biomass Estimation Using Hyperspectral Derived Measures

	Authors	Sensor	Location	Grass Type	Best Band Combination (nm)	R^2	R^2 (NDVI[a])
NDVI	Mutanga and Skidmore [9]	Field	South Africa	Dense canopy grass	745/755	0.78	0.25
	Cho et al. [71]	HyMap	Italy	Mixed grass	771/740	0.7	0.4
	Cho et al. [71]	HyMap	Italy	Mixed grass	695/786 in 2004 740/786 in 2005	0.56 in 2004 0.64 in 2005	
SR	Fava et al. [31]	Field	Italy	Mixed grass	920/729	0.77	0.35 (standard SR)
NBD[b]	Mutanga and Skidmore [29]	Field	South Africa	Dense canopy	744,689,653,556 (selected by stepwise linear regression)	0.86	0.31–0.32

[a] R^2 for standard NDVI.
[b] NBD = normalized band depth [19].

This is because the conventional NDVI using red and NIR bands reaches a level of saturation above certain LAI levels (2–3) [70]. Thus, the conventional NDVI is not appropriate for biomass estimation in dense grass [3,68]. To overcome this problem, two new narrowband combinations for NDVI have been tested from hyperspectral data (Table 10.2). Mutanga and Skidmore [9] found that, although the standard red-NIR-based NDVIs derived from the laboratory hyperspectral analysis performed poorly in estimating dense biomass of tall grass (*Cenchrus ciliaris*), a modified NDVI with 746 and 755 nm bands had a high R^2 (0.78 compared to 0.25 with the standard NDVI). However, they found that SR yielded higher coefficients of determination ($R^2 = 0.80$ on average) with biomass as compared to NDVI (average $R^2 = 0.77$). Narrowbands in the red edge (680–780 nm) have been found to be more sensitive to canopy biomass compared to red-NIR bands [9,68].

In a similar study, Mutanga and Skidmore [9] found that continuum-removed absorption features, such as band depth ratio, band depth index, and band area, calculated with the bands selected by stepwise linear regression from the red-edge region, had much higher coefficients of determination relationships ($R^2 > 0.80$), with dense grass biomass (*Cenchrus ciliaris*) measured in the laboratory compared to standard red-NIR NDVIs ($R^2 = 0.31$–0.32). Kawamura et al. [8] also found high prediction capabilities of standing biomass ($R^2 > 0.85$) using first-derivative reflectance and continuum-removed derivative reflectance in a PLSR model in New Zealand.

Cho et al. [71] used HyMap, an airborne hyperspectral sensor, to identify the best spectral measures for the prediction of grass biomass in Majella National Park, Italy. Like other laboratory- and field-based studies, those NDVIs were derived from the red-edge region (725–800 nm) and had much higher correlations compared to the traditional NDVI. The researchers also found that PLSR models with the six selected continuum-removed bands produced the highest correlation and lowest standard error for biomass prediction, compared to single variables such as the original reflectance, first derivative, and continuum-removed reflectance.

In the dry season in the Amazon, where a significant amount of dry grass material is found in pasture, Numata et al. [11] evaluated hyperspectral data to estimate live biomass, dead biomass individually, and both combined at the canopy level for two grass species. The results were highly affected by structural differences between two species (Figure 10.2) (Section 10.3). *Brachiaria decumbens* (Figure 10.2b) had better coefficients of determination between biomass and hyperspectral data than *Brachiaria brizantha* in general. Continuum removal of water absorption depth (WAD) and area (WAA) derived from the water absorption region (1100–1250 nm) showed the highest coefficients of determination for aboveground biomass, including total biomass ($R^2 = 0.35$–0.57) and live biomass ($R^2 = 0.31$–0.54), whereas lignin-cellulose absorption depth and area (2105–2230 nm) performed best for senesced biomass ($R^2 = 0.25$–0.64), and NDVI performed very poorly. Beeri et al. [66] estimated photosynthetic vegetation (PV) and NPV biomass based upon an accumulated continuum-removal reflectance between 991 and 1306 nm and broad- and narrowband-based NDVIs derived from HyMap in the Northwestern Glaciated Plains and the Northwestern Great Plains. Again, the continuum-removal-based data performed best for PV and NPV combined biomass data and for PV biomass data. The lowest relative error was found when PV live biomass was measured alone. The researchers also note that the performance of NDVIs was affected by the presence of NPV, which masks spectral responses in the red and NIR.

10.5.4 Pasture Degradation Analysis

About 20% of the world's pastures and rangelands are in some stage of degradation [1], and several factors cause pasture degradation, such as overgrazing, compaction, and erosion caused by livestock action, soil, and climate. Multispectral satellite sensors have been used to assess the effects of land degradation around watering points, grazing intensity, soil biogeochemistry, and climate on grass biophysical changes [14,24,68,72].

Grazing is one of the main driving factors in pasture biomass change and can lead to pasture degradation. Hyperspectral remote sensing has been utilized to assess the impacts of grazing intensity on vegetation changes at pastureland and ecosystem levels. Most studies of this sort have analyzed impacts on pasture structural changes based upon fractional covers of PV, NPV, bare soil, and, in some studies, shade derived from SMA. Elmore and Asner [73] investigated the effects of grazing intensity on soil carbon stocks in Hawaii by estimating plant litter cover based upon NPV derived from AVIRIS. They observed that intensively grazed areas were characterized by higher exposed substrate or soil fraction and lower NPV fraction. As the distance from a grazing center of a pasture area increases, substrate fraction decreased, whereas NPV fraction increased. Furthermore, NPV had the strongest relationship with grazing intensity. The researchers concluded that high levels of NPV could be used to identify areas of lower grazing intensity in their study area (Hawaii). A similar study conducted by Harris and Asner [74] detected a grazing gradient with fractional covers derived from AVIRIS in a rangeland in Utah and demonstrated the potential of airborne hyperspectral sensors' ability to assess rangeland conditions based upon accurately estimated fractional covers sensitive to grazing.

In a field study, Asner et al. [75] characterized the vegetation structures of Amazonian pastures with different planting ages in different soil types based upon NPV and PV derived from photon inverse models applied to grass reflectance measured by a field spectrometer. LAI and NPV area index (NPVAI) estimated from spectral reflectance through photon transport modeling were highly correlated with field LAI and NPVAI, and these fractional covers varied according to grass ages and soil texture. Furthermore, the variation of soil biogeochemical elements P and Ca across the sample pastures were well correlated with canopy LAI+NPVAI inversion calculated from hyperspectral data.

10.5.5 Species Discrimination

Understanding the distribution of different grass species in landscapes is essential for measuring ecological characteristics such as plant functional types. Hyperspectral data have been used in species distinction by developing spectral libraries for spectrally distinct species and creating species maps derived from airborne hyperspectral sensors [76]. In a laboratory-based analysis, Schmidt and Skidmore [22] measured spectral reflectance from eight African grass species to compare the reflectance and the continuum-removed reflectance curves for each of all possible two-species pairs to assess whether these species were spectrally separable. They found that bands that maximized the discrimination between species occurred in the visible region (550–680 nm), indicating that pigment concentrations vary between species.

On the other hand, the normalization of the absorption curves in the NIR region by continuum removal improves our ability to discriminate grass species. Yamano et al. [77] used derivative reflectance in the visible and NIR regions for the distinction analysis of four predominant grass species in Inner Mongolia. They found that fourth-derivative peaks around 670 and 720 nm were an effective discriminator for distinguishing the grass species *Caragana microphylla* from others. However, the discrimination capability of hyperspectral data may largely depend on the season and requires ad hoc calibrations to select a specific model and set of bands for species discrimination [78]. A recent study suggests that hyperspectral data have been a useful tool in the detection of invasive species within pasture and cropland [79–81]. At the landscape scale, species discrimination by hyperspectral data can be further complicated by background diversity [23].

10.6 CONCLUSIONS

The advances made through hyperspectral remote sensing in characterizing pastures have shown the potential to improve our ability to more accurately estimate biophysical and biochemical properties compared to broadband systems, which may help increase livestock productivity through informed management. Some specific conclusions include the following:

Chlorophyll, red-edge, and SWIR regions are recommended for the estimation and prediction of forage nutrients, digestibility, and fibers. In particular, narrowbands in the red edge (680–780 nm) have been found to be more sensitive to pigments, nutrients, and even canopy biomass compared to red-NIR bands.

PLSR combined with absorption-based spectral transformation data enhances the predictability of pasture attributes.

The ability to accurately estimate NPV and bare soil is one of the greatest advantages of hyperspectral remote sensing of pasture over multispectral remote sensing, especially for pastures in dry regions or degraded pastures where senesced grass and bare soil are present and affect spectral signatures of these pastures.

Despite the potential improvement of pasture characterization via hyperspectral data, the results of research studies discussed in this chapter represent a site- and time specific-biophysical condition of pasture, and currently available imaging spectrometers are not adequate for regularly monitoring pastures and grasslands. The ultimate goal of hyperspectral remote sensing of pastureland would be to map and monitor spatial and temporal patterns of pasture quality. This can be addressed in the near future as the data of new hyperspectral sensors such as HyspIRI, EnMap, and PRISM,satellite hyperspectral sensors become available.

REFERENCES

1. FAO 2006. *Livestock's Long Shadow: Environmental Issues and Options*. FAO: Rome, Italy. p. 391.
2. Norris, K.H., Barnes, R.F., Moore, J.E., and Shenk, J.S. 1976. Predicting forage quality by infrared reflectance spectroscopy. *Journal of Animal Science*, 43, 889–897.
3. Gao, J. 2006. Quantification of grassland properties: How it can benefit from geoinformatic technologies? *International Journal of Remote Sensing*, 27, 1351–1365.
4. Curran, P.J. 1989. Remote sensing of foliar chemistry. *Remote Sensing of Environment*, 30, 271–2785.
5. Elvidge, C.D. 1990. Visible and near-infrared reflectance characteristics of dry plant materials. *International Journal of Remote Sensing*, 11, 1775–1795.
6. Asner, G.P. 1998. Biophysical and biochemical sources of variability in canopy reflectance. *Remote Sensing of Environment*, 64, 234–53.
7. He, Y.H., Guo, X.L., Wilmshurst, J., and Si, B.C. 2006. Studying mixed grassland ecosystems II: Optimum pixel size. *Canadian Journal of Remote Sensing*, 32, 108–115.
8. Kawamura, K., Betteridge, K., Sanches, I.D., Tuohy, M.P., Costall, D., and Inoue, Y. 2009. Field radiometer with canopy pasture probe as a potential tool to estimate and map pasture biomass and mineral components: A case study in the Lake Taupo catchment, New Zealand. *New Zealand Journal of Agricultural Research*, 52, 417–434.
9. Mutanga, O. and Skidmore, A.K. 2004. Integrating imaging spectroscopy and neural networks to map grass quality in the Kruger National Park, South Africa. *Remote Sensing of Environment*, 90, 104–115.
10. Mutanga, O., Skidmore, A.K., and Prins, H.H.T. 2004. Predicting *in situ* pasture quality in the Kruger National Park, South Africa, using continuum-removed absorption features. *Remote Sensing of Environment*, 89, 393–408.
11. Numata, I., Roberts, D.A., Chadwick, O.A., Schimel, J.P., Galvao, L.S., and Soares, J.V. 2008. Evaluation of hyperspectral data for pasture estimate in the Brazilian Amazon using field and imaging spectrometers. *Remote Sensing of Environment*, 112, 1569–1583.
12. Darvishzadeh, R., Atzberger, C., Skidmore, A., and Schlerf, M. 2011. Mapping grassland leaf area index with airborne hyperspectral imagery: A comparison study of statistical approaches and inversion of radiative transfer models. *ISPRS Journal of Photogrammetry and Remote Sensing*, 66, 894–906.
13. Knox, N.M., Skidmore, A.K., Prins, H.H.T., Asner, G.P., van der Werff, H.M.A., de Boer, W.F., van der Waal, C. et al. 2011. Dry season mapping of savanna forage quality, using the hyperspectral Carnegie Airborne Observatory sensor. *Remote Sensing of Environment*, 115, 1478–1488.
14. Numata, I., Roberts, D.A., Chadwick, O.A., Schimel, J., Sampaio, F.R., Leonidas, F.C., and Soares, J.V. 2007. Characterization of pasture biophysical properties and the impact of grazing intensity using remotely sensed data. *Remote Sensing of Environment*, 109, 314–327.
15. Roberts, D.A., Numata, I., Holmes, K., Batista, G., Krug, T., Monteiro, A., Powell, B., and Chadwick, O.A. 2002. Large area mapping of land-cover change in Rondonia using multitemporal spectral mixture analysis and decision tree classifiers. *Journal of Geophysical Research-Atmospheres*, 107.

16. Hill, M.J. 2004. Grazing agriculture: Managed pasture, grassland, and rangeland. In S.L. Ustin (Ed.), *Manual of Remote Sensing Volume 4. Remote Sensing for Natural Resource Management and Environmental Monitoring* (pp. 449–530). Hoboken, NJ: John Wiley & Sons.
17. Asner, G.P. and Heidebrecht, K.B. 2002. Spectral unmixing of vegetation, soil and dry carbon cover in arid regions: Comparing multispectral and hyperspectral observations. *International Journal of Remote Sensing*, 23, 3939–3958.
18. Fourty, T., Baret, F., and Verdebout, J. 1996. Leaf optical properties with explicit description of its biochemical composition: Direct and inverse problems. *Remote Sensing of Environment*, 56, 104–116.
19. Kokaly, R.F., and Clark, R.N. 1999. Spectroscopic determination of leaf biochemistry using band-depth analysis of absorption features and stepwise multiple linear regression. *Remote Sensing of Environment*, 67, 267–287.
20. Ramoelo, A., Skidmore, A.K., Schlerf, M., Mathieu, R., and Heitkonig, I.M.A. 2011. Water-removed spectra increase the retrieval accuracy when estimating savanna grass nitrogen and phosphorus concentrations. *ISPRS Journal of Photogrammetry and Remote Sensing*, 66, 408–417.
21. Roberts, D.A., Smith, M.O., and Adams, J.B. 1993. Green vegetation, nonphotosynthetic vegetation, and soils in AVIRIS data. *Remote Sensing of Environment*, 44, 255–269.
22. Schmidt, K.S. and Skidmore, A.K. 2001. Exploring spectral discrimination of grass species in African rangelands. *International Journal of Remote Sensing*, 22, 3421–3434.
23. van Leeuwen, W.J.D. and Huete, A.R. 1996. Effects of stading litter on the biophysical interpretation of plant canopies with spectral indices. *Remote Sensing of Environment*, 55, 123–138.
24. Asner, G.P. and Lobell, D.B. 2000. A biogeophysical approach for automated SWIR unmixing of soils and vegetation. *Remote Sensing of Environment*, 74, 99–112.
25. Okin, G.S., Roberts, D.A., Murray, B., and Okin, W.J. 2001. Practical limits on hyperspectral vegetation discrimination in arid and semiarid environments. *Remote Sensing of Environment*, 77, 212–225.
26. Nagler, P.L., Daughtry, C.S.T., and Goward, S.N. 2000. Plant litter and soil reflectance. *Remote Sensing of Environment*, 1, 207–215.
27. Rouse, J.W., Haas, R.H., Schell, J.A., and Deering, D.W. 1973. Monitoring vegetation systems in the Great Plains with ERTS. In *Proceedings of the third ERTS Symposium*, pp. 309–317.
28. Wang, J.J., Wang, T.J., Skidmore, A.K., Shi, T.Z., and Wu, G.F. 2015. Evaluating different methods for grass nutrient estimation from canopy hyperspectral reflectance. *Remote Sensing*, 7, 5901–5917.
29. Mutanga, O. and Skidmore, A.K. 2004. Narrow band vegetation indices overcome the saturation problem in biomass estimation. *International Journal of Remote Sensing*, 25, 3999–4014.
30. Thenkabail, P.S., Smith, R.B., and De Pauw, E. 2000. Hyperspectral vegetation indices and their relationships with agricultural crop characteristics. *Remote Sensing of Environment*, 71, 158–182.
31. Fava, F., Colombo, R., Bocchi, S., Meroni, M., Sitzia, M., Fois, N., and Zucca, C. 2009. Identification of hyperspectral vegetation indices for Mediterranean pasture characterization. *International Journal of Applied Earth Observation and Geoinformation*, 11, 233–243.
32. Pinar, A. and Curran, P.J. 1996. Grass chlorophyll and the reflectance red-edge. *International Journal of Remote Sensing*, 17, 351–357.
33. Filella, I. and Penuelas, J. 1994. The red-edge position and shape as indicators of plant chlorophyll content, biomass and hydric status. *International Journal of Remote Sensing*, 15, 1459–1470.
34. Lamb, D.W., Steyn-Ross, M., Schaare, P., Hanna, M.M., Silvester, W., and Steyn-Ross, A. 2002. Estimating leaf nitrogen concentration in ryegrass (*Lolium* spp) pasture using the chlorophyll red-edge: Theoretical modelling and experimental observations. *International Journal of Remote Sensing*, 23, 3619–3648.
35. Cho, M.A. and Skidmore, A.K. 2009. Hyperspectral predictors for monitoring biomass production in Mediterranean mountain grasslands: Majella National Park, Italy. *International Journal of Remote Sensing*, 30, 499–515.
36. Jago, R.A., Cutler, M.E.J., and Curran, P.J. 1999. Estimating canopy chlorophyll concentration from field and laboratory spectra. *Remote Sensing of Environment*, 68, 217–224.
37. Schut, A.G.T. and Ketelaars, J. 2003. Imaging spectroscopy for early detection of nitrogen deficiency in grass swards. *Njas-Wageningen Journal of Life Sciences*, 51, 297–317.
38. Schut, A.G.T., van der Heijden, G., Hoving, I., Stienezen, M.W.J., van Evert, F.K., and Meuleman, J. 2006. Imaging spectroscopy for on-farm measurement of grassland yield and quality. *Agronomy Journal*, 98, 1318–1325.
39. Kooistra, L., Salas, E.A.L., Clevers, J., Wehrens, R., Leuven, R., Nienhuis, P.H., and Buydens, L.M.C. 2004. Exploring field vegetation reflectance as an indicator of soil contamination in river floodplains. *Environmental Pollution*, 127, 281–290.

40. Smith, K.L., Steven, M.D., and Colls, J.J. 2004. Use of hyperspectral derivative ratios in the red-edge region to identify plant stress responses to gas leaks. *Remote Sensing of Environment*, 92, 207–217.
41. Mutanga, O. and Skidmore, A.K. 2007. Red edge shift and biochemical content in grass canopies. *ISPRS Journal of Photogrammetry and Remote Sensing*, 62, 34–42.
42. Clark, R.N. and Roush, T.L. 1984. Reflectance spectroscopy: Quantitative analysis techniques for remote sensing applications. *Journal of Geophysical Research*, 89, 6329–6340.
44. Mutanga, O., Skidmore, A.K., and van Wieren, S. 2003. Discriminating tropical grass (*Cenchrus ciliaris*) canopies grown under different nitrogen treatments using spectroradiometry. *ISPRS Journal of Photogrammetry and Remote Sensing*, 57, 263–272.
43. Mutanga, O., Prins, H.H.T., Skidmore, A.K., Wieren, S., Huizing, H., Grant, R., Peel, M., and Biggs, H. 2004. Explaining grass-nutrient patterns in a savanna rangeland of southern Africa. *Journal of Biogeography*, 31, 819–829.
45. Thulin, S., Hill, M.J., Held, A., Jones, S., and Woodgate, P. 2012. Hyperspectral determination of feed quality constituents in temperate pastures: Effect of processing methods on predictive relationships from partial least squares regression. *International Journal of Applied Earth Observation and Geoinformation*, 19, 322–334.
46. Thulin, S., Hill, M.J., Held, A., Jones, S., and Woodgate, P. 2014. Predicting levels of crude protein, digestibility, lignin and cellulose in temperate pastures using hyperspectral image data. *American Journal of Plant Sciences*, 5, 997–1019.
47. Kawamura, K., Watanabe, N., Sakanoue, S., and Inoue, Y. 2008. Estimating forage biomass and quality in a mixed sown pasture based on partial least squares regression with waveband selection. *Grassland Science*, 54, 131–145.
48. Huang, Z., Turner, B.J., Dury, S.J., Wallis, I.R., and Foley, W.J. 2004. Estimating foliage nitrogen concentration from HYMAP data using continuum removal analysis. *Remote Sensing of Environment*, 93, 18–29.
49. Kokaly, R.F., Despain, D.G., Clark, R.N., and Livo, K.E. 2003. Mapping vegetation in Yellowstone National Park using spectral feature analysis of AVIRIS data. *Remote Sensing of Environment*, 84, 437–456.
50. Adams, J.B., Sabol, D.E., Kapos, V., Almeida, R., Roberts, D.A., Smith, M.O., and Gillespie, A.R. 1995. Classification of multispectral images based on fractions of endmembers—Application to land-cover change in the Brazilian Amazon. *Remote Sensing of Environment*, 52, 137–154.
51. Roberts, D.A., Batista, J.L., Pereira, J.L.G., Waller, E., and Nelson, B. 1998. Change identification using multitemporal spectral mixture analysis: Applications in Eastern Amazonia. In C.D. Elvidge, R. Lunetta (Ed.), *Remote Sensing Change Detection: Environmental Monitoring Applications and Methods* (pp. 137–161). Ann Arbor: Ann Arbor Press.
52. Numata, I., Roberts, D.A., Chadwick, O.A., and Hatzel, Y. 2004. Spectral characterization of changes in grassland under climatic and soil gradients in Kohala, Hawaii. In *Proceedings of 2004 AVIRIS Workshop* (pp. 29–31). NASA JPL, Pasadena, CA.
53. Wessman, C.A., Bateson, C.A., and Benning, T.L. 1997. Detecting fire and grazing patterns in tallgrass prairie using spectral mixture analysis. *Ecological Applications*, 7, 493–511.
54. Curran, P.J. 2001. Remote sensing: Using the spatial domain. *Environmental and Ecological Statistics*, 8, 331–344.
55. Zhao D., Starks P.J., Brown M.A., Phillips W.A., and Coleman S.W. 2007. Assessment of forage biomass and quality parameters of bermudagrass using proximal sensing of pasture canopy reflectance. *Grassland Science*, 53, 39–49.
56. Clevers, J.G.P.W., van der Haijden, G.W.A.M., Verzakov, S., and Schaepman, M.E. 2007. Estimating grassland biomass using SVM band shaving of hyperspectral data. *Photogrammetric Engineering and Remote Sensing*, 73, 1141–1148.
57. Geraldi, P. and Kowalski, B.R. 1986. Partial least-squares regression: A tutorial. *Analytical Chimistry Acta*, 185, 1–17.
58. Adjorlolo, C., Mutanga, O., and Cho, M.A. 2015. Predicting C3 and C4 grass nutrient variability using *in situ* canopy reflectance and partial least squares regression. *International Journal of Remote Sensing*, 36, 1743–1761.
59. Darvishzadeh, R., Skidmore, A., Schlerf, M., Atzberger, C., Corsi, F., and Cho, M. 2008. LAI and chlorophyll estimation for a heterogeneous grassland using hyperspectral measurements. *ISPRS Journal of Photogrammetry and Remote Sensing*, 63, 409–426.
60. Kawamura, K., Watanabe, N., Sakanoue, S., Lee, H.J., Lim, J., and Yoshitoshi, R. 2013. Genetic algorithm-based partial least squares regression for estimating legume content in a grass-legume mixture using field hyperspectral measurements. *Grassland Science*, 59, 166–172.

61. McNaughton, S.J. and Banyikwa, F.F. 1995. Plant communities and herbivory. In A.R.E. Sinclair, P. Arcese (Eds.), *Serengeti II—Dynamics, Management, and Conservation of an Ecosystem* (pp. 49–70). Chicago: University of Chicago Press.
62. Dias Filho, M., Davidson, E.A., and de Carvalho, C.J.R. 2000. Linking biogeochemical cyclyes to cattle pasture management and sustainability in the Amazon Basin. In M. McClain, R.L. Victoria, J.E. Ritchey (Eds.), *Biogeochemistry of the Amazon Basin*. New York: Oxford University Press.
63. Mutanga, O., Skidmore, A.K., Kumar, L., and Ferwerda, J. 2005. Estimating tropical pasture quality at canopy level using band depth analysis with continuum removal in the visible domain. *International Journal of Remote Sensing*, 26, 1093–1108.
64. Pullanagari, R.R., Yule, I.J., Tuohy, M.P., Hedley, M.J., Dynes, R.A., and King, W.M. 2012. In-field hyperspectral proximal sensing for estimating quality parameters of mixed pasture. *Precision Agriculture*, 13, 351–369.
65. Mutanga, O. and Kumar, L. 2007. Estimating and mapping grass phosphorus concentration in an African savanna using hyperspectral image data. *International Journal of Remote Sensing*, 28, 4897–4911.
66. Beeri, O., Phillips, R., Hendrickson, J., Frank, A.B., and Kronberg, S. 2007. Estimating forage quantity and quality using aerial hyperspectral imagery for northern mixed-grass prairie. *Remote Sensing of Environment*, 110, 216–225.
67. Field, C.B., Randerson, J.T., and Malmstrom, C.M. 1995. Global net primary production: Combining ecology and remote sensing. *Remote Sensing of Environment*, 51, 74–88.
68. Todd, S., Hoffer, R.M., and Milchunas, D.G. 1998. Biomass estimation on grazed and ungrazed rangelands using spectral indices. *International Journal of Remote Sensing*, 19, 427–438.
69. Cho, M.A. and Skidmore, A.K. 2006. A new technique for extracting the red edge position from hyperspectral data: The linear extrapolation method. *Remote Sensing of Environment*, 101, 181–193.
70. Franklin, J., Prince, S.D., Strahler, A.H., Hanan, N.P., and Simonett, D.S. 1991. Reflectance and transmission properties of West African savanna trees from ground radiometer measurements. *International Journal of Remote Sensing*, 12, 1369–1385.
71. Cho, M.A., Skidmore, A., Corsi, F., van Wieren, S.E., and Sobhan, I. 2007. Estimation of green grass/herb biomass from airborne hyperspectral imagery using spectral indices and partial least squares regression. *International Journal of Applied Earth Observation and Geoinformation*, 9, 414–424.
72. Pickup, G., Bastin, G.N., and Chewings, V.H. 1998. Identifying trends in land degradation in non-equilibrium rangelands. *Journal of Applied Ecology*, 35, 365–377.
73. Elmore, A.J. and Asner, G.P. 2006. Effects of grazing intensity on soil carbon stocks following deforestation of a Hawaiian dry tropical forest. *Global Change Biology*, 12, 1761–1772.
74. Harris, A.T. and Asner, G.P. 2003. Grazing gradient detection with airborne imaging spectroscopy on a semi-arid rangeland. *Journal of Arid Environments*, 55, 391–404.
75. Asner, G.P., Townsend, A.R., and Bustamante, M.M.C. 1999. Spectrometry of pasture condition and biogeochemistry in the Central Amazon. *Geophysical Research Letters*, 26, 2769–2772.
76. Dennison, P.E. and Roberts, D.A. 2003. Endmember selection for multiple endmember spectral mixture analysis using Endmember Average RMSE. *Remote Sensing of Environment*, 87, 295–309.
77. Yamano, H., Chen, J., and Tamura, M. 2003. Hyperspectral identification of grassland vegetation in Xilinhot, Inner Mongolia, China. *International Journal of Remote Sensing*, 24, 3171–3178.
78. Irisarri, J.G.N., Oesterheld, M., Veron, S.R., and Paruelo, J.M. 2009. Grass species differentiation through canopy hyperspectral reflectance. *International Journal of Remote Sensing*, 30, 5959–5975.
79. Lass, L.W., Prather, T.S., Glenn, N.F., Weber, K.T., Mundt, J.T., and Pettingill, J. 2005. A review of remote sensing of invasive weeds and example of the early detection of spotted knapweed (*Centaurea maculosa*) and babysbreath (*Gypsophila paniculata*) with a hyperspectral sensor. *Weed Science*, 53, 242–251.
80. Lopez-Granados, F., Jurado-Exposito, M., Pena-Barragan, J.M., and Garcia-Torres, L. 2006. Using remote sensing for identification of late-season grass weed patches in wheat. *Weed Science*, 54, 346–353.
81. Wang, C.Z., Zhou, B., and Palm, H.L. 2008. Detecting invasive sericea lespedeza (*Lespedeza cuneata*) in Mid-Missouri pastureland using hyperspectral imagery. *Environmental Management*, 41, 853–862.
82. Knox, N.M., Skidmore, A.K., Prins, H.H.T., Heitkonig, I.M.A., Slotow, R., van der Waal, C., & de Boer, W.F. 2012. Remote sensing of forage nutrients: Combining ecological and spectral absorption feature data. *ISPRS Journal of Photogrammetry and Remote Sensing*, 72, 27–35.
83. Blackburn, G.A. 1998. Spectral indices for estimating photosynthetic pigment concentrations: A test using senescent tree leaves. *International Journal of Remote Sensing*, 19, 657–675.
84. Gamon, J.A., Serrano, L., & Surfus, J.S. 1997. The photochemical reflectance index: An optical indicator of photosynthetic radiation use efficiency across species, functional types, and nutrient levels. *Oecologia*, 112, 492–501.

85. Curran, P.J. 1994. Imaging spectrometry. *Progress in Physical Geography*, 18, 247–266.
86. Horler, D.N.H., Dockray, M., & Barber, J. 1983. The red edge of plant leaf reflectance. *International Journal of Remote Sensing*, 4, 273–288.
87. Kumar, L., Schmidt, K.S., Dury, S., & Skidmore, A.K. 2001. Review of hyperspectral remote sensing and vegetation Science. In: Van Der Meer, F.D., & De Jong, S.M. (eds) *Imaging Spectrometry: Basic Principles and Prospective Applications*. Kluwer, Dordrecht, The Netherlands.

11 Hyperspectral Remote Sensing of Wetland Vegetation

Elijah Ramsey III and Amina Rangoonwala

CONTENTS

11.1 INTRODUCTION

Wetlands exert a higher influence on biogeochemical fluxes among land, atmosphere, and hydrologic systems than their 1% worldwide occurrence would suggest (Sahagian and Melack 1996). Despite their importance, wetlands continue to face high detrimental pressures from natural and human-induced forces (Ramsey 1998, Kumar and Sinha 2014). Remote sensing offers the single best source of timely, synoptic wetland status and trend information at a variety of spatial and temporal scales (Wickland 1991, Adam et al. 2010, Guo et al. 2017).

The remote sensing of wetlands does not generally differ in technique or process from remote-sensing-based mapping of other terrestrial features (e.g., Ramsey 2005). Differences exist because of the higher spectral and spatial variability of wetlands due to their occupying a unique interface, or ecotone, between aquatic and upland ecosystems (Mitsch and Gosselink 2000, Adam et al. 2010). Although there are environmental factors that affect all vegetation, such as climate, soils, and geology, the uniqueness of wetland vegetation stems from the biophysical features that define this ecotone. Near the coast, infrequent to near-constant inundation by fresh to saline waters promotes adaptations that set wetland plants apart from all other terrestrial plants. As the spatial and temporal complexities in flushing strength and salinity increase, so do the variety and complexity of wetland species, forms, and associations. Lacustrine and riparian wetlands reveal that this uniqueness of species is primarily in response to seasonal and longer-term cycles of hydrology or changing water inputs into the systems. Coastal wetlands experience complexities common to inland lacustrine and riparian wetlands, but their proximity to the sea adds to their complexity (Mitsch and Gosselink 2000, Federal Geographic Data Committee 2013).

Tidal flushing that carries pulses of elevated saltwater dominates the dynamics of coastal wetlands. Tidal periods and amplitudes vary in time and space; however, all coastal regions experience tidal flooding and most a 28-day cycle exhibiting neap (low) to spring (high) amplitudes. Especially near rivers, coastal wetlands may experience freshwater flooding, sometimes alternating with tidal flooding. Storms can augment the tidal amplitudes because of associated wind and storm surges. Storm surges carrying water with elevated salinity can invade the normally fresher wetland zones impacting less salt-tolerant wetland species. Excessive flood duration, or water logging, can also adversely impact coastal wetlands. Drought, fire, invasive plants, and human development are additional forces that contribute to the spatial complexity of these coastal wetland systems.

The overall results of these spatially and temporally varying forces are highly dynamic and diverse wetland ecotones that transition from the coastal ocean to the upland ecosystem. Effective management for the preservation of these coastal wetlands requires monitoring that can discern local changes on a regional scale. A variety of remote sensing mapping techniques can provide that monitoring discernment, including those applications uniquely available when using airborne and satellite hyperspectral image data.

11.1.1 Benefits of Hyperspectral Data

The benefits of hyperspectral imaging (HSI) narrowband data compared to broadband multispectral (MS) data depend on the spectral resolutions of the HSI and MS sensors. Both sensor types can record in the same reflected electromagnetic (EM) energy range from around 400 to 2500 nm; MS sensors record that EM information in fewer spectral bands (about 3–16 bands) than HSI sensors (often over 100 bands). In addition, to accommodate this high number of spectral bands, HSI sensor bands are narrower (typically around or less than 10 nm) than those used in common MS sensors (Figure 11.1) (Liu et al. 2009). In essence, an HSI sensor records a more continuous spectral record, whereas an MS will record only selected broad areas of the reflected EM range. MS bands normally include the blue, green, and red visible (VIS) bands (400–700 nm), one or more near-infrared (NIR) bands (700–1300 nm), and one or more shortwave infrared (SWIR) bands (1300–2500 nm), which often are

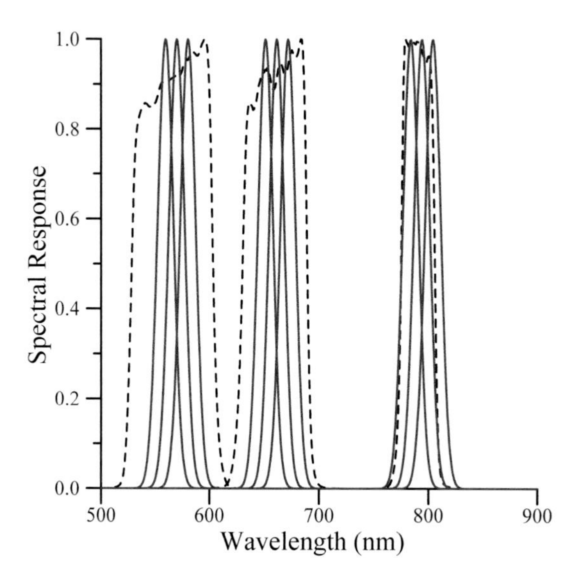

FIGURE 11.1 Spectral response curves of EO-1 Advanced Land Imager (ALI) sensor bands 2 (green reflectance band), 3 (red reflectance band), and 4 (NIR reflectance band) (dashed lines) and similar curves for colocated EO-1 Hyperion sensor (solid lines) approximated as Gaussian distributions (Liu et al. 2009). Three Hyperion bands are plotted for each ALI band.

centered on SWIR transmission peaks (approximately 1500–1750 nm, SWIR1, approximately 2000–2500 nm, SWIR2). Well-known MS satellite sensors are carried by the Landsat, Quickbird, SPOT, and Sentinel satellites. Heavily used airborne HSI sensors include the National Aeronautics and Space Administration's (NASA) Airborne Visible/Infrared Imaging Spectrometer (AVIRIS) and the privately owned Compact Airborne Spectrographic Imager (CASI). The recently decommissioned EO-1 satellite carried both the Hyperion hyperspectral sensor and the Advanced Land Imager (ALI) broadband MS sensor. The Hyperion and ALI sensor combination illustrates the tradeoff in the quantity of spectral information extracted from each ground-resolution element (or pixel) and the imaged area (e.g., Figure 11.1). Both sensors had 30-m pixel resolutions; however, the Hyperion sensor with 224 bands had a nominal 7.6-km-swath width, whereas the ALI with nine MS bands and one panchromatic band collected data over a much larger 37-km-swath width.

By using a high-density, multi-narrowband spectral recording, HSI sensors provide spectral discrimination similar to laboratory spectrophotometers (e.g., Figure 11.1). If the HSI sensor on board an aircraft or satellite were moved to within a few hundred meters of the Earth's surface, the HSI record would simulate a spectrophotometer. The main difference would be the variable illumination and the diffuse nature of the surface target. Because there is a higher number of spectral channels (or bands) with higher spectral resolutions, many mapping applications operationally performed with broadband sensor systems have been enhanced with HSI image data. Some of these classical broadband applications include land-cover-type mapping, change detection, biomass determination, and prototyping sensors for planned systems.

Regarding hyperspectral imagery, specialized processing of high-resolution spectra can map spectral variations within a pixel. This specialized processing can partition each pixel into percent occurrences of various target compositions on the basis of determined spectral differences. By applying linear processing to HSI data, the classical broadband point classifications are transformed into continuous classifications of percent occurrences (i.e., fractional abundances) of land-cover features (i.e., types and conditions), which is particularly useful for characterizing wetlands. Analogously to spectral analyses in a laboratory, HSI sensor data can also be used to detect subtle variation in the reflectance of plant leaves if the canopy structure and background influences can be removed from the recorded reflectance.

Although it is not always necessary for spectral changes in leaves to be linked to abnormal plant canopy change, it is always helpful (Ramsey and Rangoonwala 2005). Defining the biophysical indicator of change at the plant leaf level allows for the construction of a more targeted monitoring strategy. If the vegetation change cannot be defined at the plant leaf level, the ability to document the change at the canopy level via synoptic optical remote sensing is hindered (Ramsey and Rangoonwala 2009). Successful HSI applications account for changes of biochemical properties at the plant leaf level (Ramsey and Rangoonwala 2005), changes in biophysical structure (density and orientation) at the plant canopy level (Ramsey et al. 2015), and atmospheric influences in the surface reflectance (Ramsey and Rangoonwala 2006).

The advantage of HSI spectral alignment over that of MS sensors is realized only if the spectral autocorrelation within the EM response from the terrestrial target does not eliminate the more effective spectral isolation available from HSI sensors (e.g., Warner and Shank 1997). Unlike mineral reflectance spectra from nonvegetated landscapes that often exhibit well-defined spectral features, vegetation reflectance spectra tend to have wide and more subtle features that reflect pigment absorptions. Even though features are apt to be spectrally broad, the use of available image processing techniques can remove or at least diminish spectral autocorrelation in HSI data, enhancing the effectiveness of the classification (e.g., Warner and Shank 1997, Wang and Sousa 2009, Thenkabail et al. 2013).

The near spectrally continuous HSI data can be used to determine single bands and combinations of spectral bands that best address the mapping objective (Thenkabail et al. 2013). If successful mapping can be performed with a small number of spectral bands, then the mapping can be greatly simplified. If these selected bands are replicated on operational multispectral satellite sensors, the mapping process can be further simplified by transfer to that operational remote sensing system. HSI data analysis provides sufficient narrowband coverage for a trained image analyst to define the most appropriate spectral bands (or regions) for any mapping exercise. This unique, advanced spectral analysis function cannot be replicated by spectral analyses that are based on broadband MS data.

11.1.2 Chapter Outline

This chapter focuses on those applications of HSI that illustrate the more common vegetation canopy compositional (characteristic spectra-based mapping) and inferential prediction analysis. Within that focus, the application of hyperspectral remote sensing to mapping and monitoring wetlands is demonstrated in US coastal wetland forests and marshes that share a wide variety of similarities with many coastal and inland wetland vegetation types worldwide (Figure 11.2).

This chapter includes descriptions of mangrove biophysical mapping, invasive vegetation detection and mapping, marsh dieback onset and progression mapping, and other applied research projects that present how hyperspectral information can increase the detail over that available from broadband spectral analyses. The conveyance of HSI applications is facilitated by partitioning these wetlands into three broad components: plant leaf, plant canopy, and the nonplant background. In each project description, the plant leaf and canopy reflectance are coupled. Background and shaded vegetation reflectance contributions to the canopy reflectance are discussed where appropriate. In all cases the canopy structure is considered necessary for interpreting the vegetation canopy reflectance; in two cases, this coupling is demonstrated. These examples incorporate the application of light-interaction models (based on radiative transfer equations), direct measurements, and above the top of canopy (TOC) reflectance.

Even though spectral and spatial resolutions and timing are explicitly stated in all project descriptions, in this discussion, the focus is on the advantages of biophysical HSI mapping. Multiple HSI studies have specifically considered the tradeoffs of spectral and spatial resolutions (e.g., Rosso et al. 2005, Belluco et al. 2006, Adam et al. 2010, Kumar and Sinha 2014) and timing (e.g., Artigas and Yang 2005, Flores-de-Santiago et al. 2013, Tuominen and Lipping 2016). In addition, results

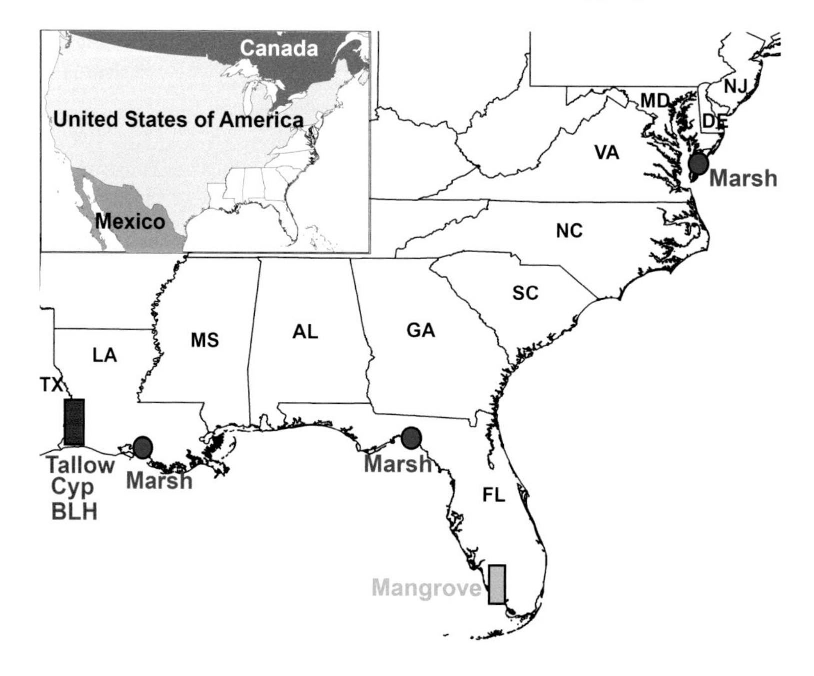

FIGURE 11.2 Locations of USGS projects that used hyperspectral remote sensing techniques in wetland settings. BLH = bottomland hardwoods; Cyp = Cypress.

described in this discussion are most relevant to the environment and conditions of the project. Direct comparability may be diminished in contrasting environmental conditions (Yang et al. 2009) or difference in season promoting a contrast in pigment (Flores-de-Santiago et al. 2013) or water content (Adam et al. 2010).

11.2 HYPERSPECTRAL REMOTE SENSING OF WETLAND FORESTS

Wetland forests occupy various flood and salinity regimes. Of the swamp types, only mangroves are evergreen and tolerant of saline waters. Mangroves occupy 75% of the world's coastlines between 25°N and 25°S latitude (Ramsey and Jensen 1996). Although new world mangroves (red mangrove [*Rhizophora mangle L.*], black mangrove [*Avicennia germinans L.*], white mangrove [*Laguncularia racemosa*]) thrive in the more southern portions of the Gulf of Mexico (hereafter Gulf), black mangroves have tenaciously established footholds as far north as 30° latitude. Situated at the ocean and land interface, mangrove swamps directly experience frequent tidal flooding and the many perturbations of the coastal ocean.

Baldcypress (*Taxodium distichum*) and bottomland-hardwood-dominated forests can make up major parts of deciduous freshwater swamps, and, in contrast to mangrove forests, these swamps extend well into the temperate zones. Baldcypress swamps can occupy more permanently flooded portions of a wetland forest or isolated oxbows that constitute the remnant of former meanders in riparian flood plains (Mitsch and Gosselink 2000). Bottomland hardwoods (BLHs) are found in intermittently flooded floodplain forests that occur along rivers and streams throughout the central and southern United States (King and Keeland 1999). These forests can occur on an elevation gradient between drier upland hardwood forests and more persistently flooded swamps.

11.2.1 Mangrove Forests

HSI mapping of mangroves has been carried out worldwide (Yang et al. 2009), some in regions where mangrove species dominances are completely or predominantly physically separable (Held et al. 2003, Koedsin and Vaiphasa 2013, Zhang and Xie 2013), some in areas where a single mangrove species occurrence is physically separable from nonmangrove land covers (Yang et al. 2009). In areas where mangrove species co-occur as communities within the HSI ground resolution, these inseparable mangrove species are combined into single classes (Kumar et al. 2013).

In a study of mangroves located in the southern Gulf off the southwest coast of Florida (Figure 11.2), ground-based measurements of canopy closure, laboratory leaf spectral measurements, and helicopter-based spectroradiometer measurements were obtained primarily in the month of October in the five mangrove types defined by Lugo and Snedaker (1974) as basin, overwash, fringe, riverine, and dwarf (Ramsey and Jensen 1996). Within the five community types, strands dominantly comprised a mixture of red and black mangroves with sporadic and most often minor inclusion of white mangroves. An initial finding of the study was high correspondence (83% R^2) between the mangrove leaf area index (LAI), or the number of canopy leaf layers, and normalized difference vegetation index (NDVI) obtained from Satellite Pour l'Observation de la Terre (SPOT) red and NIR image data (Jensen et al. 1991). Field and image data used to calculate LAI and NDVI were collected within all community types and species mixtures. The question was why the NDVI-LAI relationship appeared to be nonspecies specific. In response to that question, this study's objectives were to determine (1) the leaf spectral and structural (LAI and LAD [leaf angle distribution or leaf canopy orientation]) changes within and between mangrove species and (2) the relationships between the canopy VIS and NIR (VNIR) spectral variability and canopy species composition and structural variability.

Leaves were collected at 23 nondisturbed mangrove sites throughout the northern Ten Thousand Islands off the southwest coast of Florida (Ramsey and Jensen 1996). Typically, eight to ten leaves taken from small branches near the TOC composed each of the three spectral leaf samples per tree.

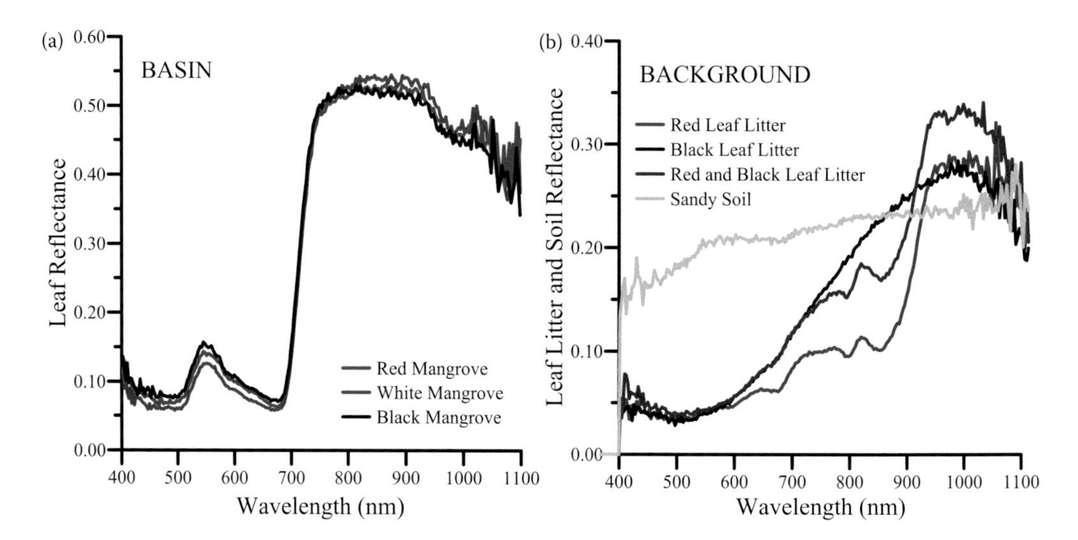

FIGURE 11.3 (a) (Basin) Mangrove top of canopy leaf spectra acquired using a stacked-plate reflectance design. The basin site contained about 40% red, 30% white, and 30% black mangroves. (b) (Background) Figure 11.2 shows the location of the mangrove sites in southwest Florida. (Adapted from Ramsey, E., III and Jensen J. 1996. *Photogramm Eng Rem S*, 62(8), 939–948.)

Limited to the 400–1000 nm portion of the VNIR, spectral reflectance of the 18 red (from 16 sites), 10 black (separate sites), and 7 white (separate sites) leaf samples were obtained using flat-plate methods (Ramsey and Jensen 1996). The reflectance spectra from the multiple samples per tree were averaged. The multiple addition of reflected light from the lower second layer of leaves in the stack was estimated by adaption of a method introduced by Lillesaeter (1982). The estimated second leaf reflectance addition was subtracted from the flat-plate reflectance to obtain the leaf reflectance estimate.

Results of the leaf spectral analyses showed that the average high to low reflectance range was 0.021 and 0.024 in the VIS and 0.04 and 0.05 in the NIR within the black and red species, respectively (Ramsey and Jensen 1996). The white mangrove leaf reflectance ranges were 0.04 in the VIS and 0.05 in the NIR (Ramsey, unpublished data). The red-black, red-white, and black-white average leaf reflectance differences were −0.01, −0.02, −0.01 in the VIS and −0.02, 0.04, 0.06 in the NIR. The average VIS and NIR reflectance ranges within each species relative to the differences between each species' average reflectance suggest that differentiation of the species in the mixed composition stands would be difficult if only the leaf spectral characteristics were considered (an example is given in Figure 11.3a).

Results presented here indicate that between-species mean reflectance differences at best equaled the within-species leaf reflectance variances in the VNIR. Even though possible environmental or seasonal differences may have weakened the comparison with our results, successful classification of red, black, and white mangrove leaves with the addition of SWIR to the VNIR spectral record was reported (Wang and Sousa 2009, Zhang et al. 2014). If that success is carried forward, then the necessary base for species spectral discrimination at the canopy level will be met (Ramsey and Rangoonwala 2009).

Extension of species discrimination at the leaf level to the mangrove canopy level is not always straightforward and can involve adjustment of determined leaf-level relationships to accommodate the varied contributions to the canopy reflectance (Ramsey and Rangoonwala 2006, Wang and Sousa 2009). Where mangrove species dominances are physically separated as monotonic regions, successful species classifications have resulted. Of these, one explicitly showed the importance of canopy structure by coordinated use of HSI and polarimetric synthetic aperture radar (SAR) (Held et al. 2003), while another included texture in the classification

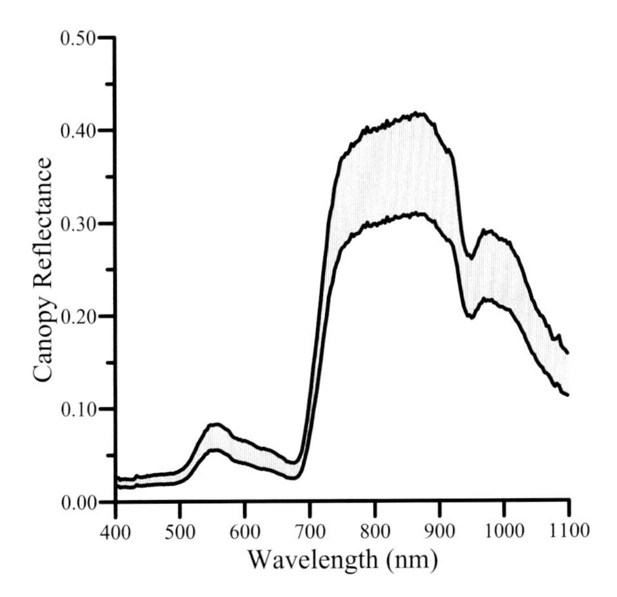

FIGURE 11.4 Range of mangrove canopy reflectance spectra calculated from top of canopy site-specific upwelling (~20-m instantaneous field of view from helicopter platform) and surface downwelling light measurements (mean ± one standard deviation, n = 23). The 23 sites included basin, overwash, fringe, riverine, and dwarf forest sites containing mixtures of black, red, and white mangroves. Location of mangrove sites shown in Figure 11.2. (Adapted from Ramsey, E., III and Jensen J. 1996. *Photogramm Eng Rem S*, 62(8), 939–948.)

by implicitly incorporating canopy structure (Zhang and Xie 2013). In environments of mixed mangrove species compositions, canopy structure is likely to be less discriminating than where mangroves occur in monospecies clusters. In those cases, spectral identity increases its importance in discriminating species.

Another result of the laboratory spectral analyses was related to the leaf-litter spectra (Figure 11.3b) (Ramsey and Jensen 1996). Interestingly, even though spectra of the leaves could not separate mangrove species, the leaf litter collected within stands dominated by each of the single species spectrally differed throughout the VNIR wavelengths. Although no linkage to the length of decomposition in each of the litter samples was considered, these spectral differences illustrate the importance of considering the changing background reflectance as a spectral component of the canopy reflectance (Ramsey and Rangoonwala 2006, Adam et al. 2010).

Even though VNIR leaf spectral properties would not singularly separate the mangrove species, canopy VNIR spectra representing a 20-m instantaneous field of view (IFOV) of the TOC were highly variable (Figure 11.4) (Ramsey and Jensen 1996). The canopy spectra were collected from a helicopter platform with a handheld radiometer at the 23 mangrove sites. The question was what caused the high variability in canopy reflectance.

To understand the cause of the canopy reflectance differences, a radiative transfer (RT) model was implemented by following the construction outlined in Goudriaan (1977). Particulars of the model, assumptions, inputs, and validations are described in Ramsey and Jensen (1995). RT model predictions of the helicopter-based mangrove canopy reflectance spectra were >97%. Even though the analyses showed that canopy reflectance was most sensitive to VNIR leaf reflectance and only moderately sensitive to LAI or LAD, the high spectral variance observed in the canopy reflectance was largely a function of LAI. The LAI dominance implied the lack of consistent leaf reflectance differences, and the modeled stability of LAD as spherical across black, red, and white mangrove canopies limited their influence on the canopy reflectance. Because of that, LAI was highly correlated to canopy reflectance but not to the species of mangrove.

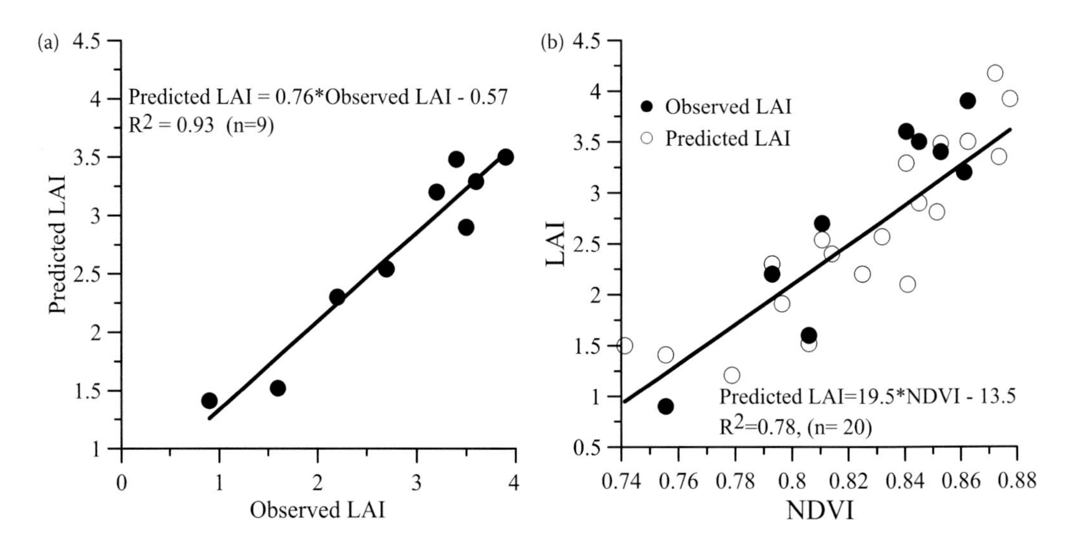

FIGURE 11.5 (a) Mangrove canopy leaf area index (LAI) observed as calculated from site canopy closure measurements and predicted with radiative transfer (RT) canopy model using measured background (e.g., Figure 11.3b Background), leaf (e.g., Figure 11.3a Basin), and canopy reflectance spectra (e.g., Figure 11.4) inputs. (Adapted from Ramsey, E., III and Jensen J. 1996. *Photogramm Eng Rem Sen*, 62(8), 939–948.) (b) NDVI versus predicted LAI (open circles) by RT model and calculated LAI from field measurements (solid circles). (From Ramsey, E., III and Jensen J. 1995. in J. Lyon and J. McCarthy (Eds.), *Wetland and Environmental Applications of GIS*. CRC Press, Boca Raton, FL, 61–81.)

LAI values predicted with the RT model were validated with ground-based measurements at 93% accuracy (Figure 11.5a) (Ramsey and Jensen 1995, 1996). The NDVI extracted from canopy reflectance spectra (e.g., Figure 11.4) within defined narrow bandwidths had a correspondence with the predicted canopy LAIs of 78% (Figure 11.5b) (Ramsey and Jensen 1995). Field-observed LAIs are overlain for comparison (not used in R^2 calculation). The NDVI and LAI correspondence was limited to NDVI > 0.73. The field analyses and RT modeling based on hyperspectral data explained the high correspondence found by Jensen et al. (1991) between field mangrove LAI and broadband NDVI values obtained from SPOT image data.

11.2.2 Baldcypress Forests

The cypress family includes 142 conifer species collectively exhibiting a near-global distribution (Cupressaceae, http://www.conifers.org/cu/Cupressaceae.php [Accessed October 10, 2017]). Baldcypress (*Taxodium distichum* [L.] Rich.) is a species of the cypress family that is widely distributed in the eastern and southeastern United States, especially in river floodplains (Plants Profile, http://plants.usda.gov/java/profile?symbol=TADI2 [Accessed October 10, 2017]). These trees grow in a wide variety of coastal and inland environments; however, in southeastern US wetland forests, baldcypress trees commonly occupy the more frequently flooded areas, where they often occur with water tupelo (*Nyssa aquatica*) trees in swamps along the northern Gulf. Baldcypress are deciduous and typically exhibit more open canopies than observed in mature, minimally disturbed BLH forests. Baldcypress leaves are needlelike, whereas tupelo leaves are somewhat narrow and elliptical.

The spectral detection of baldcypress swamps is enhanced by its unique environment and canopy structures. Basic baldcypress forest mapping is not usually problematic for broadband optical sensors; however, hyperspectral mapping may reveal subtle changes in these swamps that could indicate detrimental and abnormal stresses not obtainable with broadband sensors. Saltwater intrusion related to relative sea-level rise and land subsidence has caused widespread diebacks of coastal baldcypress

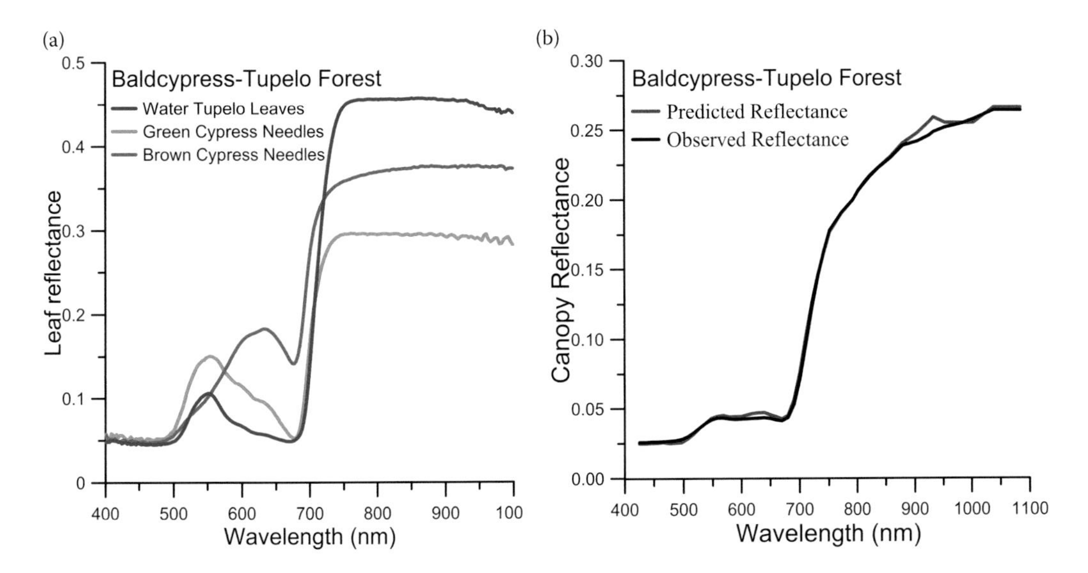

FIGURE 11.6 (a) The complexity of leaf spectra comprising a senescing baldcypress-water tupelo forest. The leaf spectra were obtained following Daughtry et al. (1989). Typically, three leaf spectral samples per tree were collected. Each cypress leaf sample contained multiple needles, while a single leaf composed each tupelo sample. A constant offset (−0.07) was applied in order to align the baldcypress leaf spectra with the tupelo spectrum in the blue wavelength region. (b) Canopy reflectance spectra obtained from radiometer data collected from helicopter platform (observed) and predicted from EO-1 Hyperion image data using methods described in Ramsey and Nelson (2005).

swamps, commonly termed "ghost forests" (Krauss et al. 1998). HSI would be the most appropriate tool to survey broad areas to detect abnormal changes in these forests before irreversible change becomes imminent. Hindrances to consistently detecting subtle changes in the more vulnerable baldcypress swamps are the typically open canopies and the persistent subcanopy flooding. Other obstructions to the detection of abnormal change are the commonly mixed tupelo and baldcypress stands. The combination of these factors can create high variability in the canopy reflectance that complicates the determination of leaf spectral responses to salt stress. For example, leaves composing a mixed baldcypress and tupelo forest stand exhibit significant differences in the green to red-edge wavelength region (Figure 11.6a), a region particularly important in detecting leaf pigment changes (e.g., Ramsey and Rangoonwala 2005). Canopy reflectance based on TOC recordings from a helicopter (observed) and EO-1 Hyperion sensor (predicted) exemplify that complexity (Figure 11.6b). The flat green to red-edge wavelength region within the canopy reflectance spectra reflects the high proportion of baldcypress (approximately 75%), a third of those exhibiting browning, the low water tupelo proportion, and the low canopy closure resulting in a relatively high surface contribution. To detect baldcypress-water tupelo degradation near its onset, the canopy reflectance needs to separate not only the leaf and background components but also the leaf component by species. Though challenging, plant-leaf spectral monitoring based on hyperspectral mapping would be useful for monitoring, conservation, and restoration of these at-risk coastal forests.

11.2.3 Bottomland Hardwood Forests

BLH forests occupy about 2.8 million ha in the Lower Mississippi River Alluvial Valley of the United States and comprise a variety of deciduous species. As in mangrove mapping, the ability to monitor separately the different bottomland communities would greatly enhance the value of the information obtainable from remote sensing data. BLH forests act as carbon sinks, provide habitat for fish and wildlife, offer flood protection, and make many other contributions to the natural and human environments (DeWeese et al. 2007). While preservation of these contributions depends on maintaining BLH composition and species density, introduction of invasive species and human-mediated hydrologic modification change species compositions, and thereby BLH function, altering their ecological contributions (DeWeese et al. 2007). For example, White (1983) classified the BLH forests in southern Louisiana and on the Mississippi border into two community classes. The classification represented two different BLH forests in terms of function and species composition. The cause of the stand composition differences of the two BLH forest classes was related to hydrologic modification resulting in different hydroperiods (i.e., hydrologic regimes) for these two types, each dominating physically separate portions of the wetland landscape (White 1983).

As part of a hurricane impact study, attempts were made to separately classify the two BLH forest classes with prehurricane Landsat and SAR 25-m image data (Ramsey et al. 2009b). These classification attempts were unsuccessful. In a separate study that combined the 25-m damage classes with 3 years of daily Moderate Resolution Imaging Spectroradiometer (MODIS) data, a temporal difference in foliage onset of the two BLH classes was observed in the early spring (Ramsey et al. 2011). Even though the daily temporal resolution of the MODIS product provided discrimination of the two classes, the spatial scale is not conducive to the needed stand-level (around 25 m) monitoring. Further, the 25-m optical sensor systems cannot provide the high-frequency data required.

Remote sensing classifications of hardwood forests have been made at the dominant overstory species and mixed community levels (Hill et al. 2010, Shao et al. 2014). In each of these, multitemporal classification of the broadband image data produced the highest accuracies. The best two-date classification combined spring and fall for discrimination of hardwood species and communities. A more effective mapping strategy could be to collect an HSI image in the spring, the period of maximum spectral contrast as observed with daily MODIS data, and apply spectral analysis to fully discriminate the two BLH forest communities.

11.3 HYPERSPECTRAL REMOTE SENSING OF INVASIVE PLANTS

An applied research study of the mapping of Chinese tallow (*Triadica sebifera*), an invasive tree species, was undertaken in the north-central Gulf coastal region (Figure 11.2) (Ramsey and Nelson 2005, Ramsey et al. 2005a,b). The wetland environments within the study area included baldcypress and BLH forests along with palustrine and estuarine marshes.

The initial mapping of Chinese tallow was conducted with color-infrared (CIR) aerial photography to prove that tallow was separable from all other co-occurring vegetation (Ramsey et al. 2002). The CIR mapping was successful because it occurred during fall senescence when tallow leaves progressively turned from green to red (Figure 11.7), and the 1-m spatial resolution was fine enough to capture small clumps of red leaves.

Although successful, the 1-m CIR mapping was costly and spatially limited. To be cost-effective for resource management, the detection system must offer regional coverage and high repeatability at low cost. A satellite mapping system can provide cost-effective regional coverage; however, the tradeoff is a coarser spatial resolution. The CIR mapping showed that tallow is scattered or occurs in small clumps of a few trees dispersed throughout marsh and wetland forests. In that case, the satellite regional mapping must detect tallow within an image pixel containing mixed and varying compositions of vegetation. A late-fall collection of EO-1 satellite Hyperion sensor hyperspectral data with a 30-m ground spatial resolution was chosen to accomplish that regional monitoring of Chinese tallow.

Mapping success was set at detection of tallow comprising at and above 10% of the TOC occurrences within each 30-m pixel. To be detected consistently at that level, the VIS data must be reproducible to <1% and the NIR to <5%. To obtain that level of reproducibility, the Hyperion data were transformed to TOC reflectance estimates (Ramsey and Nelson 2005).

Transformation entailed the establishment of 34 calibration sites spatially distributed throughout the EO-1 coverage. Helicopter-based upwelling radiance and downwelling irradiance spectra were

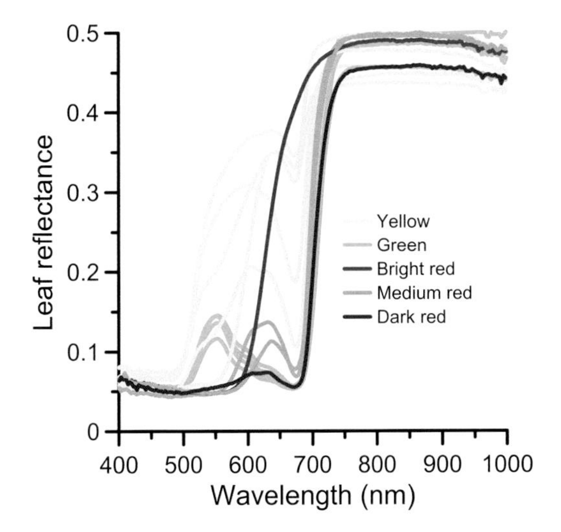

FIGURE 11.7 Leaf reflectance of single tallow leaves collected from a few trees growing in fall senescence period. Yellow, green, and bright, medium, and dark red refer to the mix of different leaf colors exhibited by tallow at one site and at one time. Only red and possibly yellow leaves contrasted with the surrounding vegetation. The location of the tallow sites is shown in Figure 11.2. (Adapted from Ramsey, E., III et al. 2005b. *Int J Remote Sens*, 26, 1611–1636.)

collected the same day as the Hyperion data (Figure 11.8). The helicopter-based data were used to calculated TOC reflectance at each site. Similar to the RT construction used in the mangrove canopy reflectance simulation, programs built on equations derived by Turner and Spencer (1972) were enclosed within an optimization procedure (Himmelblau 1972, Ramsey and Nelson 2005).

The optimization minimized the difference between the helicopter-based and predicted TOC reflectance spectra of each site. The optimized atmospheric variables were used to transform the

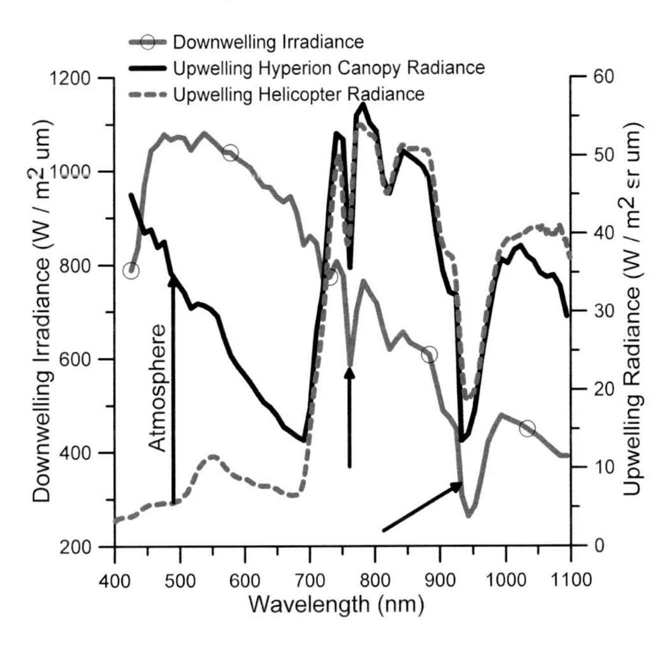

FIGURE 11.8 Near concurrent measurements of downwelling irradiance and upwelling reflected radiance as measured by Hyperion sensor and helicopter-based radiometer over a forest site in southwest Louisiana (a slight bias was added to the helicopter spectra for clarity). The forest site location is shown in Figure 11.2.

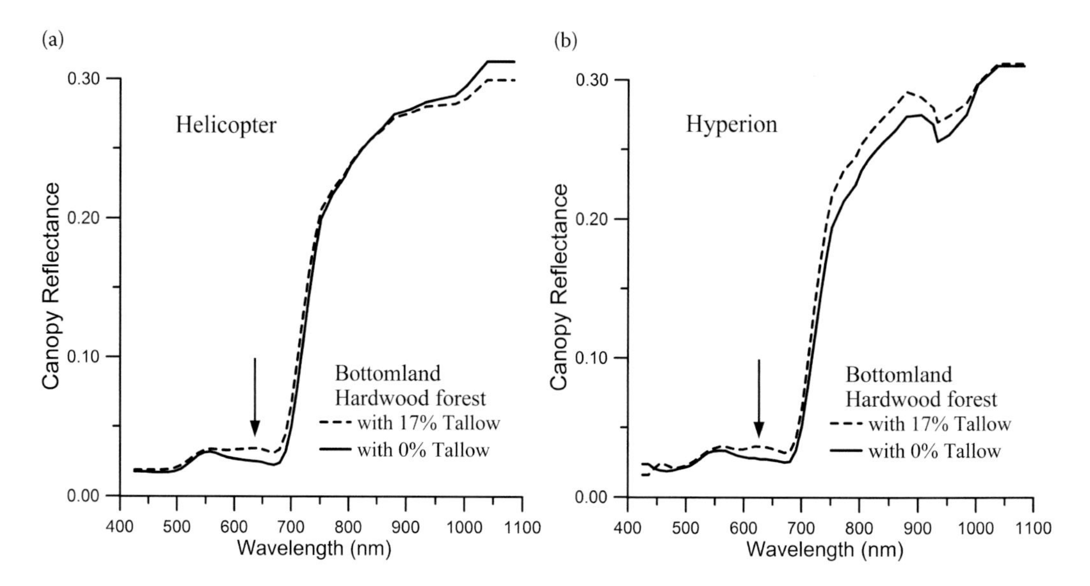

FIGURE 11.9 Canopy reflectance of two bottomland hardwood sites, one with 17% tallow and one without tallow. (a) Helicopter-based canopy reflectance of the two sites. (b) Same comparison of the two sites, however, Hyperion-based canopy reflectance. The arrows indicate the slight differences in reflectance related to the tallow percent occurrence differences at the two sites. (Adapted from Ramsey, E., III and Nelson G. 2005. *Int J Remote Sens*, 26, 1589–1610.)

Hyperion data to TOC reflectance data. Validation confirmed that the TOC maximum reflectance error was <1% in the VIS and <5% in the NIR (Figure 11.9a,b) (Ramsey and Nelson 2005).

The CIR mapping determined that leaf senescence and tallow occurrence patterns were extremely varied and that large (>30 m) monotypic stands of tallow did not exist within the region (Ramsey et al. 2002). These extreme physical heterogeneities and the lack of large tallow stands in the Hyperion scene excluded the more conventional creation of hyperspectral characteristic spectra from homogeneous stands within the image or from leaf spectral libraries.

A more appropriate approach was provided by a multivariate analysis technique, polytopic vector analysis (PVA). PVA extracted characteristic spectra directly from the input canopy reflectance without requiring that spectra form and type be defined a priori as in spectral libraries or as extracted from homogeneous imaged areas (Ehrlich and Crabtree 2000, Ramsey et al. 2005a,b). The PVA-calculated characteristic spectra corresponded to percent occurrences per 30-m Hyperion pixel of red tallow, live vegetation, and senescent vegetation. The senescent vegetation occurred in marshes and most cypress-tupelo forests. Live vegetation included canopy shadows. A fourth characteristic spectrum was created in the PVA that exhibited a form similar to yellow vegetation; however, this spectrum corresponded primarily to noise.

Loadings representing the tendency of each TOC reflectance spectrum to align with the tallow, live, and senescent characteristic spectra were compared to classifications of photography collected at each site at the time of the helicopter-based upwelling radiance recordings (Ramsey et al. 2005a,b). The loadings represented the percent occurrences of tallow, live, and senescent vegetation. Correspondence demonstrated that 78% of the tallow percent occurrences were mapped correctly. Calculated confidence limits for individual predicted values indicated that tallow occurrences made up 10% (<10 × 10 m) of the 30-m pixel were detected 68% of the time and 15% (<12 × 12 m) occurrences were detected 85% of the time. In addition, 92% of live and 82% of senescent vegetation occurrences were mapped accurately.

The application of the characteristic spectra transformed the Hyperion image into three images that represented continuous percent occurrences per 30-m pixel of tallow, live, and senescent

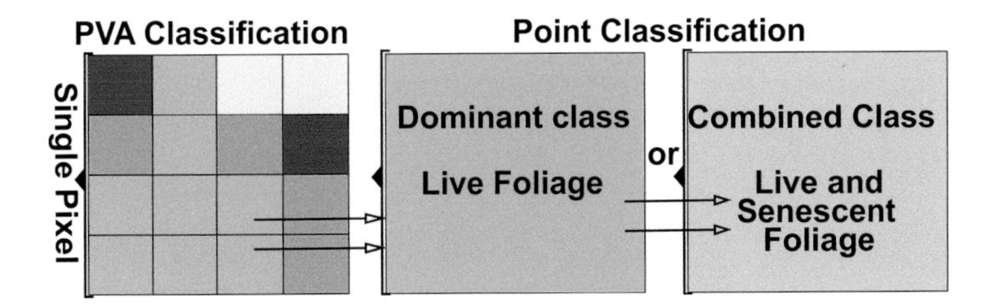

FIGURE 11.10 An illustration of classifications of a pixel based on hyperspectral and broadband sensors. (Left) PVA percent composition classification (green 62.5% live vegetation, red 12.5% red tallow, brown 25% senescing foliage, and yellow 12.5% shadowed foliage) based on hyperspectral data. (Right) A point classification (single or combined class per pixel) based on broadband image data.

vegetation (Ramsey et al. 2005a). In contrast to a point classifier, the continuous classifier ranged from 0% to 100% (Figure 11.10).

The final validation was performed with oblique photography to confirm the predicted occurrences of tallow outside the selected 34 validation sites. The comparison showed high correspondence between the mapped tallow occurrences and the photography spatial distribution. The final step associated the mapped tallow percent occurrences with land-cover types and possible land-based activities that promoted tallow establishment (Figure 11.11). The latter information relied on Landsat land-cover classifications and knowledge of land-cover activities drawn from successive map production every 3 years over a 9-year period (Ramsey et al. 2001). The land-cover mapping followed protocols outlined by the National Oceanic and Atmospheric Administration's Coastal Change Analysis Program (C-CAP) (Klemas et al. 1993).

Land-cover-type association was accomplished by spatially cross tabulating occurrences of tallow with each of the defined C-CAP land covers (Ramsey et al. 2005a). The cross tabulations were based solely on senescent red tallow occurrences and therefore were conservative estimates. Results showed that the highest correspondence of tallow was with wetland forests (BLH and cypress) (Figures 11.11 and 11.12a–c). The highest tallow occurrences were in BLH forests. That tallow was present in the flooded baldcypress forests is a testament to the invasiveness of the Chinese tallow. Cross tabulation suggested that relatively high occurrences of tallow within the coastal marsh were located on the numerous cheniers and scattered topographically higher coastal lands (Figures 11.11 and 11.12e,f). Additionally, even though percent occurrences were low, tallow was associated with the C-CAP water class. These associations were linked to tallow occurring on artificial levees lining the ubiquitous canals within the marsh. These narrow and linear features are mixed with the surrounding water in the Landsat pixel and subsequently hidden within the C-CAP water class. By linking tallow occurrences determined from satellite HSI to land-cover type produced from broadband Landsat data, we were able to take the first steps toward determining the relationship of various wetland land covers and activities to the establishment and location of tallow trees.

11.4 HYPERSPECTRAL REMOTE SENSING OF MARSH WETLANDS

Multiple marsh types can occupy estuarine to lacustrine coastal wetlands; however, dominant types were used to encapsulate hyperspectral methods most relatable to marshes (Figures 11.13a–e). Within saline marshes, smooth cordgrass (*Spartina alterniflora*) and black needlerush (*Juncus romerianus*) produce nearly constant rates of live and dead turnover, showing no clear seasonal trends after reaching maturity (Ramsey et al. 2004). Needlerush and cordgrass exhibit more vertical than horizontal canopy orientations; however, the dominant leaf orientation can change from top to bottom. The brackish salt-hay or saltmeadow cordgrass (*Spartina patens*) and fresh maidencane (*Panicum hemitomon*) marshes occupy the more interior coastal marshes of the Gulf coast (Ramsey et al. 2004). Salt-hay marshes tend to be hummocky with vertical shoots rising above a layer of thick and lodged (nearly horizontal) dead

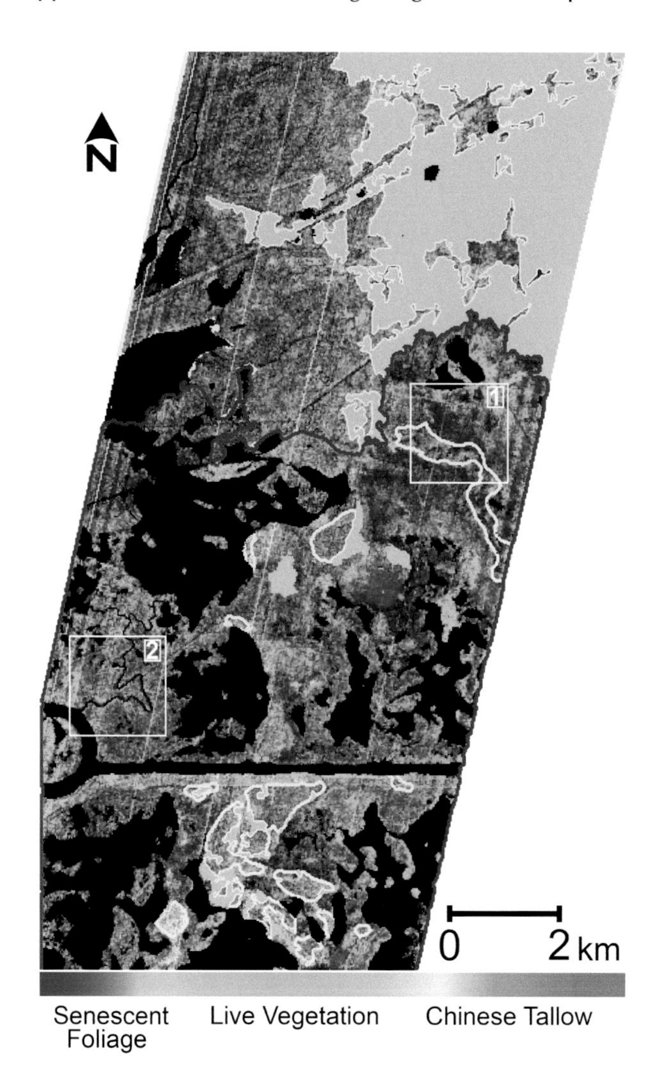

FIGURE 11.11 Chinese tallow in wetland forest and marsh. The continuous color compositions of varying mixtures of the red, blue, and green hues represent the varying percent occurrences of tallow, live vegetation, senescent foliage, and senescing cypress-tupelo forest (see legend and Figure 11.10 for interpretation). The yellow vector outlines BLH and cypress wetland forests and the red vector the palustrine and estuarine marshes. Wetland coverages determined from the Landsat Thematic Mapper C-CAP classifications. (Adapted from Ramsey, E., III et al. 2005a. *Int J Remote Sens*, 26, 1637–1657.) White box outlines reference Figure 11.12. Gray denotes upland mask.

material. As in needlerush and cordgrass marshes, salt-hay marshes appear to have low turnover with little seasonal pattern in live and dead composition. Maidencane marsh canopies exhibit yearly turnover. Beginning with nearly vertical shoots in the early spring, the canopy gains height and increasingly adds mixed orientations and density through the late spring to summer, then begins fall senescence. The maidencane seasonal turnover and its implications in interpreting canopy hyperspectral reflectance spectra demonstrate canopy reflectance and structure coupling.

11.4.1 Canopy Reflectance and Structure

Although spectral and structural variability is high between as well as within the marsh types, observed seasonal and growth patterns suggest a commonality of these variables within each marsh.

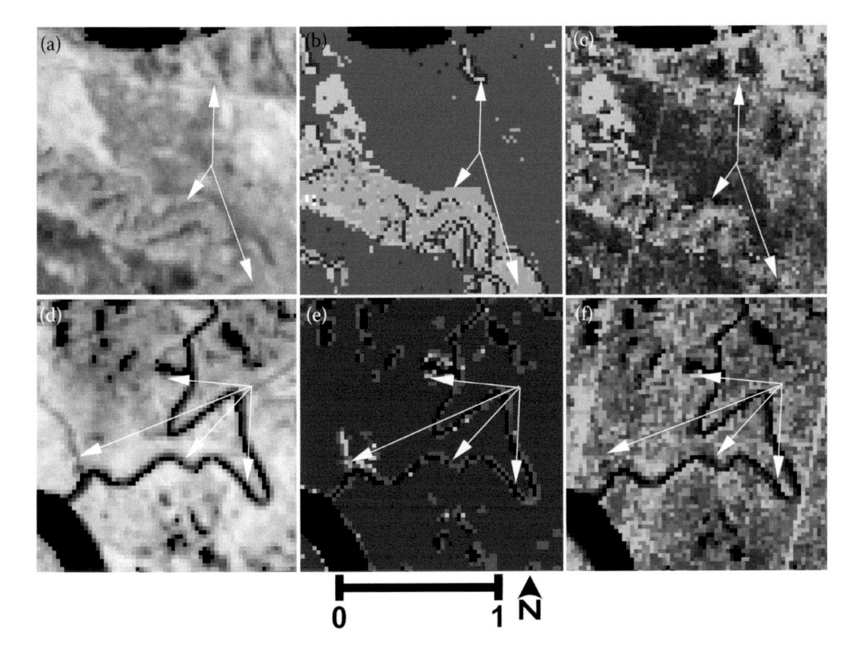

FIGURE 11.12 (Top, Figure 11.11 box 1) Tallow in a wetland forest (cyan) and palustrine marsh (light purple). Arrows highlight tallow occurrences. (a) 2005 TM image (5, 4,1), (b) C-CAP classified TM image, (c) continuous color composition (see legend in Figure 11.11). (Bottom, Figure 11.11 box 2) Tallow in an estuarine marsh (dark purple). (d–f) As previously. Only the highest tallow percent occurrences are highlighted. The overwhelming majority of occurrences are low and not discernable in these depictions. (Adapted from Ramsey, E., III et al. 2005a. *Int J Remote Sens*, 26, 1637–1657.)

To best encapsulate the range and covariance of canopy spectra and structure profiles, seasonal changes of a maidencane fresh marsh structure are described with respect to canopy hyperspectral reflectance differences (Figure 11.14a,b). Outside of external factors that result in dieback and regrowth, particularly burns (Ramsey et al. 2009a), the only marsh in the four introduced that completes a full senescence and regrowth cycle is maidencane (see Figure 11.2 for general locations).

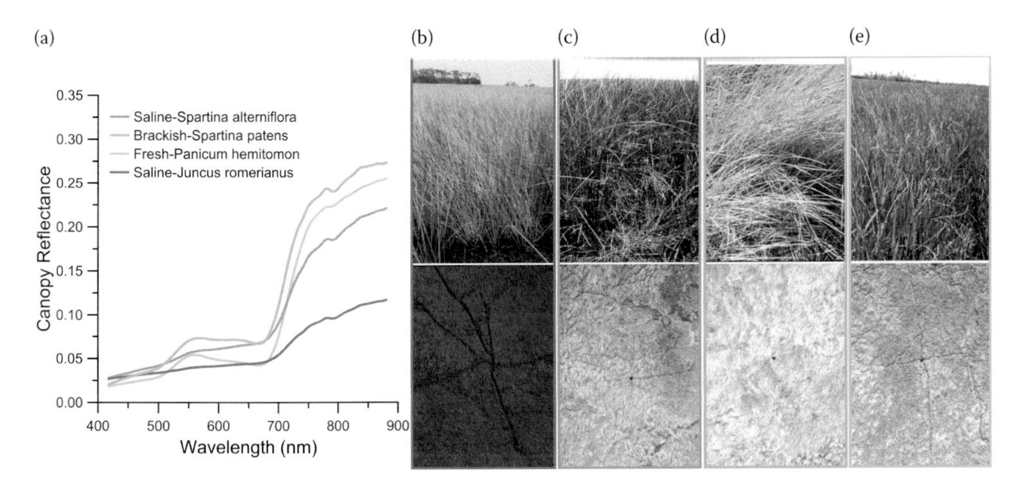

FIGURE 11.13 (a) Marsh canopy reflectance. Reflectance calculated from helicopter upwelling radiance (an average of up to seven per site) and field downwelling irradiance measurements in late spring. Ground and helicopter (red dot locates site center) views of (b) *Juncus romerianus*, (c) *Spartina alterniflora*, (d) *Spartina patens*, and (e) *Panicum hemitomon marshes.*

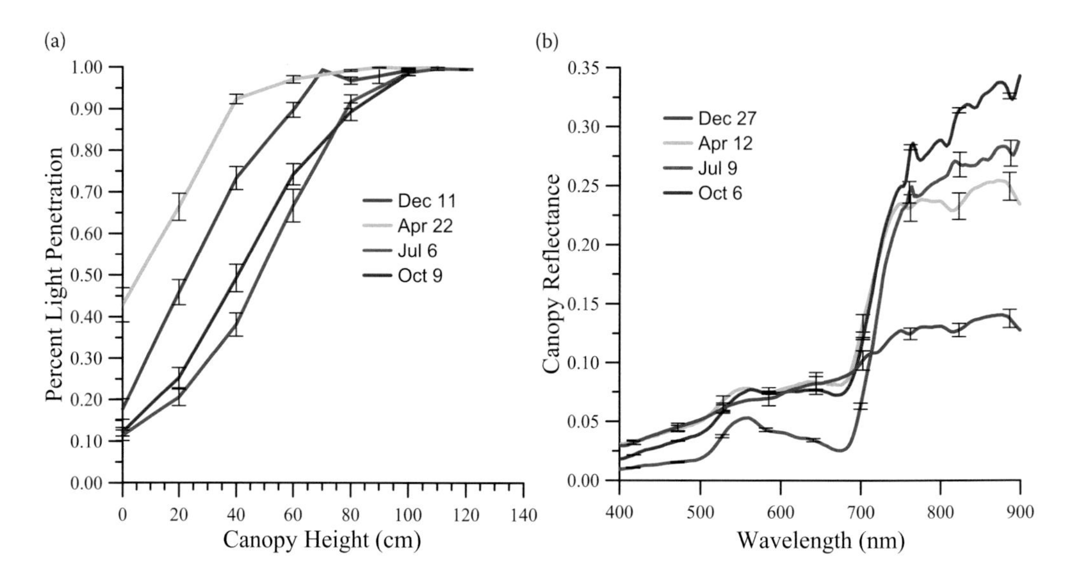

FIGURE 11.14 Light penetration and canopy reflectance of same single 30 × 30 m maidencane marsh (*Panicum hemitomon*) site (location of marsh shown in Figure 11.2). The error bars on the graphs represent variance of replicates. Note that light penetration represents the combined leaf angle distribution (LAD) and leaf area index (LAI) as a canopy structure indicator. (a) The light penetration curves illustrate winter dieback, spring turnover of dead material, summer regrowth, and the beginning of fall senescence. (b) The canopy reflectance spectra show complementary changes in green leaf material. Note the change in VIS reflectance amplitudes and the movement of the red edge (far red, at around 700 nm) from higher to lower wavelengths with the loss and gain of live plant material.

To document canopy structure, up to 22 light attenuation profiles (LAPs) were obtained at each maidencane marsh site in coordination with helicopter-based radiometer upwelling light recordings from the same sites. Collection techniques and analyses of these LAPs have been fully described (Ramsey et al. 2004). The canopy LAPs are represented as a function of canopy LAI and LAD as

$$L_z = L_{sun} \cdot EXP - (LAI \cdot LAD) \cdot Z, \tag{11.1}$$

where L_z is the measured photosynthetically active radiation (PAR) (sunlight from 400 to 700 nm) at a height = Z above the ground surface (0 cm). L_{sun} PAR is the sunlight illuminating the TOC (Figure 11.14a). The L_z/L_{sun} ratio is the fraction of PAR that penetrates to any depth below the TOC.

Canopy reflectance was estimated by dividing the helicopter-based recorded upwelling light with simultaneous downwelling recordings of a separate radiometrically and spectrally aligned radiometer or by using before and after recordings of downwelling sunlight reflected from diffuse cards of constant reflectance (Ramsey and Jensen 1996, Ramsey and Nelson 2005, Ramsey and Rangoonwala 2006) (Figure 11.14b). The upwelling radiometer recordings ranged from about 380 to 1100 nm with nominal band center spacing of 2.6 nm and an estimated spectral bandpass of 10 nm (Markham et al. 1995). No surface flooding was observed during the collection of the LAP and radiance recordings. Canopy LAPs and reflectance spectra of a single maidencane marsh site illustrate the covariance of these two variables. In this example, the coupled response of the marsh canopy LAI and LAD were not separated. This coupling suited the purpose of the example; however, decoupling of these canopy structure variables has been accomplished (Ramsey et al. 2016).

The combined LAP and reflectance information shown in the graphics indicates that the canopy comprised standing dead stalks in winter (December). By early spring (April), the canopy had added new green shoots and lost dead plant material. The summer (July) canopy LAP indicates a denser (or higher LAI) and less vertical (more mixed orientation components or changing LAD) canopy. The July reflectance spectrum indicates the canopy contained a higher percentage of live biomass

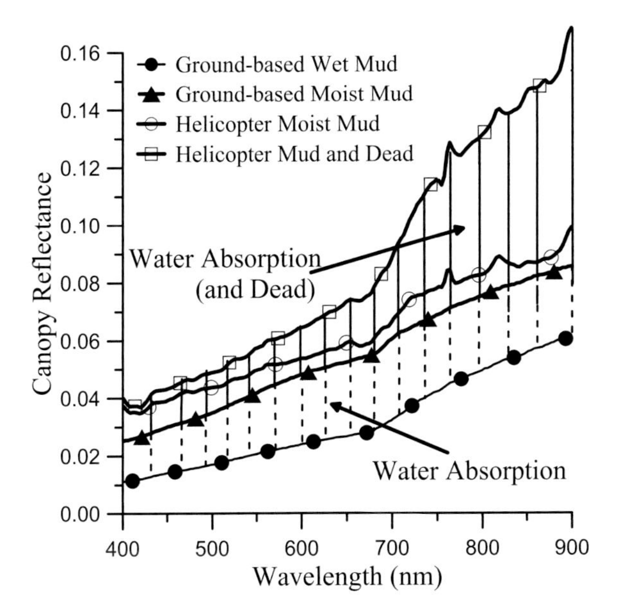

FIGURE 11.15 Reflectance as a function of mud background and water content. Wet and moist refer to high and moderate mud-water contents, respectively, based on visual observations. "Mud and dead" refers to relatively low mud-water content and the presence of dead marsh vegetation. Ground-based refers to measurements obtained about 2 m above the ground level. Note the abrupt reflectance increase around 700 nm in the "mud and dead" spectrum indicates live leaf material was present, even though visually the marsh was brown. (Adapted from Ramsey, E., III, and Rangoonwala A. 2009. in *Remote Sensing of Coastal Environments*, Y. Wang, (Eds.), CRC Press, Remote Sensing Applications Series.)

than either the winter or spring canopies. The dramatic change in the early fall (October) reflectance signified a high decrease in live plant material. The concurrent LAP suggests that the decrease in live material did not represent a large loss of canopy plant material. Instead, the combined information of the LAP and reflectance suggests that the live material senesced in place.

This maidencane example helps to illustrate the need for canopy structure data as well as canopy hyperspectral data to correctly and fully interpret changes in the marsh condition. These changes in the canopy reflectance foretell adverse changes in the plant canopy condition, which is more fully illustrated in the smooth cordgrass example.

11.4.2 Detecting Subtle Changes

In contrast to maidencane marsh, saline smooth cordgrass marsh maintains a more stable canopy reflectance and structure throughout the year (Ramsey et al. 2004). Although reflectance and structure are more stable temporally, they are not uniform spatially. Like many marsh types, cordgrass exists in a variety of forms that have adapted to the differences in inundation and salinities. Canopy structures are short to tall and dense to moderately sparse, and changes in form can occur in response to changes in topography as small as a centimeter (Stout 1984). Interlaced within these spatially changing forms is the varying exposure of the background, which is primarily mud. Further, the varying moisture content of mud mixed with litter affects the spectral nature of the reflectance intensity (Figure 11.15).

11.4.2.1 Tall Form Cordgrass Marsh Dieback

Our first demonstration of applying hyperspectral techniques for mapping abnormal marsh change took place in expansive stands of tall form cordgrass marsh in the north central Gulf (Ramsey and Rangoonwala 2005). Overlain on the tapestry of structural forms, a spatially distributed and seemingly spatially heterogeneous coastal cordgrass marsh dieback occurred (see Figure 11.2 for general location). The reasons for the sudden dieback were unknown. It spread rapidly, but without obvious patterns. Only its predominant association with the cordgrass marsh was somewhat unifying. Reconnaissance of the dieback distribution began soon after the recognition that the phenomenon was occurring at the landscape scale. Visually, the occurrences of dieback seemed to be fairly certain; however, the myriad of dieback severities was difficult to describe. Complex color-coded classifications were used to define the visual progression of dieback, such as yellow-green or brown-green, for example. There was a need to quantify the onset and progression of dieback on a landscape scale.

An in situ remote sensing strategy was developed to determine whether spectral indicators of dieback onset and progression existed at the leaf level, and if so, whether these indicators could be transferred to satellite hyperspectral or broadband sensor systems. The strategy first established a metric for assessing change and the relative change magnitude (Ramsey and Rangoonwala 2005, 2006, 2009). The metric was based on the concept that each isolated dieback occurrence began in a localized area and spread outward. A transect from the most severe portion of the dieback through progressively less-impacted marsh and finally to the local healthy marsh simulated the temporal dieback progression. The inclusion of the local healthy marsh was to account for the naturally occurring cordgrass spatial differences in the observed disparities in dieback progression from site to site. The conceptual model was tested at four dieback sites scattered within a 12 × 12 km region of cordgrass marsh experiencing sudden dieback (Ramsey and Rangoonwala 2005).

Leaf reflectance and transmittance spectra were obtained from two to three plants collected every 5 m along transects spanning the dead to local healthy marsh (Ramsey and Rangoonwala 2005). Three leaf samples each containing the three to five greenest leaves from the plant sample per transect location were analyzed. The reflectance spectra from the most extensive dieback site showed a nearly progressive change in blue (400–500 nm) and red (600–700 nm) magnitudes (Figure 11.16).

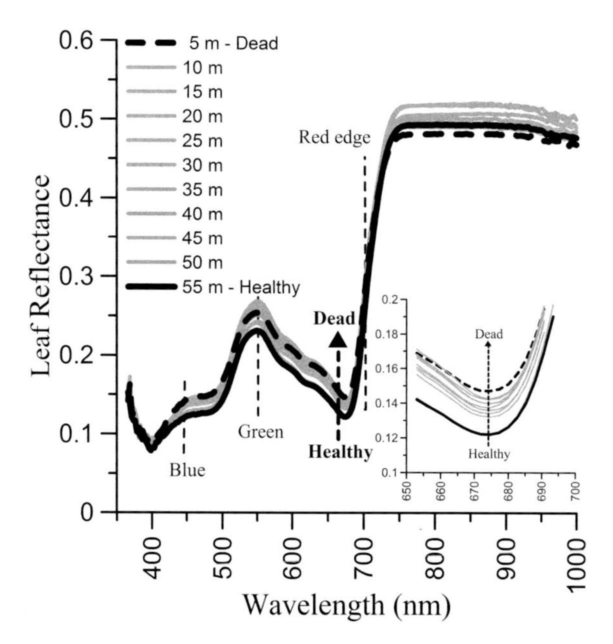

FIGURE 11.16 Leaf reflectance spectra from dead (dashed line) to local healthy (solid line) marsh at a dieback site. The insert exhibits the red band reflectance increase with dieback progression. The same progression was shown in the blue band and similar but less monotonic in the green and red-edge bands (vertical lines and labels locate band centers noted in Figure 11.18). (Adapted from Ramsey, E., III and Rangoonwala, A. 2005. *Photogramm Eng Rem S*, 71, 299–311.)

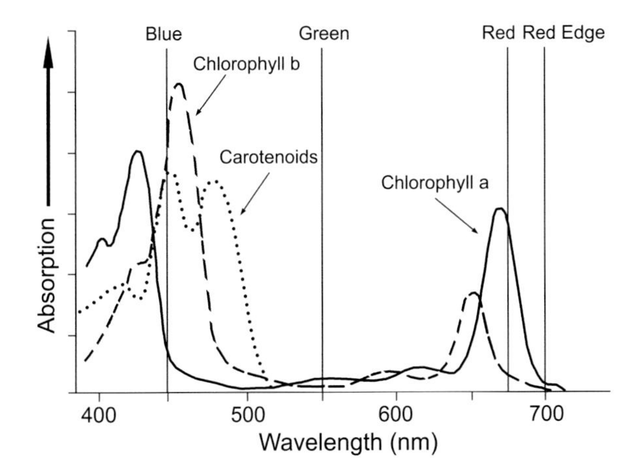

FIGURE 11.17 Absorption spectra of chlorophyll and carotene. Note the blue and red bands are located near peak absorptions, and the green and red-edge bands are located at the tails of these absorptions. (Adapted from Kirk, J. 1994. in *Light & Photosynthesis in Aquatic Ecosystems*, Second Edition, Cambridge University Press, Cambridge, 229–233.)

In both blue and red wavelength regions, plant leaf reflectance decreased from the severest dieback region to the healthiest marsh. Similar but more variable correspondences were exhibited in the green (500–600 nm) and red-edge (around 770 nm) spectral bands. The progressive changes of these VIS reflectance bands with transect distance supported the concept of dieback onset in a localized area as well as its progressive outward spread.

The progressive changes also confirmed that a spectral indicator of the dieback progression was obtainable (Ramsey and Rangoonwala 2005). The direction of changes and their similar but different correspondences of the blue-red and green–red-edge bands to the transect progression also supported the existence of a spectral indicator of dieback onset and progression. Considering only the most common and usually dominant green leaf pigments, carotene (CAR) and chlorophyll a and b (CHL) (Gitelson et al. 2002), a reason for the similarities and differences of leaf reflectance with respect to the dieback was proposed based on the representation of absorption in the reflectance spectrum (Figure 11.17) (Kirk 1994).

CAR and CHL pigments exhibit high absorptions in the blue band, with a maximum between 440 and 460 nm. Secondary CHL absorption peaks are exhibited between about 600–700 nm, with the main CHL-a peak near 670 nm. Green and red-edge bands also experience CHL absorption, although situated in the tails of the pigment absorption spectra. The spectrally varying strengths of the pigment absorptions indicated that small changes in pigment concentrations would be represented as dramatic changes in the green and red-edge bands but by relatively small changes in the blue and red bands. In fact, this difference in spectral sensitivity as suggested by Gitelson and Kaufman (1998) was used to provide greater discernment of the dieback onset and progression.

Changes in leaf pigment absorptions are related to changes in the leaf reflectance and transmittance per wavelength (λ) via Kirchoff's radiation law as

$$\%\text{absorption}(\lambda) = 1 - [\%\text{reflectance}(\lambda) + \%\text{transmittance}(\lambda)]. \tag{11.2}$$

The relationship is also illustrated by combining the three spectral variables in the same graphic (Figure 11.18). Inspection of all leaf spectra representing the four dieback transects found that narrow-wavelength bands located in the blue, green, red, red-edge, and NIR maximized the spectral information obtainable as related to changes in the CHL and CAR pigment concentrations. Even without the need to use the whole spectrum (400–1100 nm), the availability of hyperspectral information provided the ability to closely define the spectral positions and widths best suited for

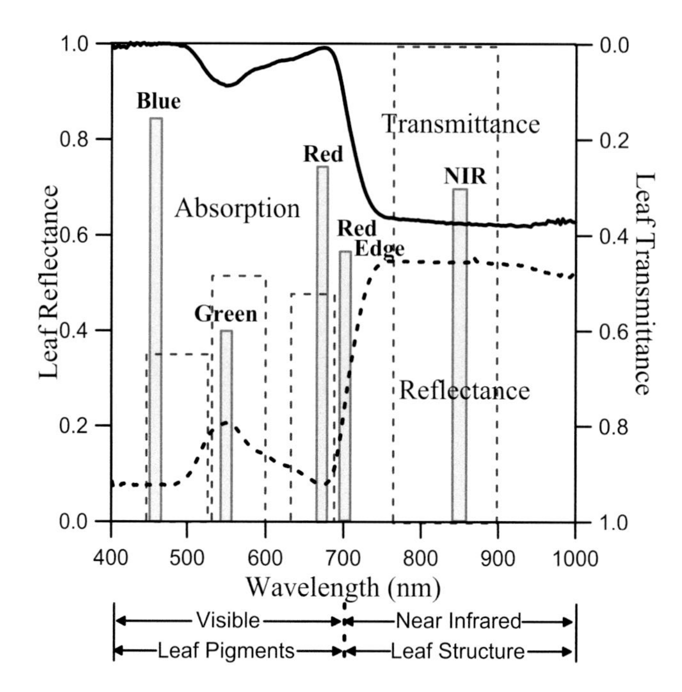

FIGURE 11.18 reflectance and transmittance spectra calculated from direct measurements of smooth cordgrass leaves. Solid bars: bands selected to determine changes in leaf that were indicative of dieback progression; dashed-line bars: VIS and NIR bands available on Landsat sensor. Note the lack of a Landsat red-edge band. The leaf sample was obtained from a dieback site in coastal Louisiana (Figure 11.2). (Adapted from Ramsey, E., III and Rangoonwala, A. 2005. *Photogramm Eng Rem S*, 71, 299–311.)

the dieback onset and progression mapping. Landsat bands are overlain on the same leaf spectra to illustrate their general lack of specificity. The Landsat overlay also points out the lack of a red-edge band, an important spectral region in this study, as in many others. Leaf spectral analyses indicated that the red and blue bands best represented the larger and presumably older diebacks and the green and red-edge bands best represented the presumably younger diebacks. It is also noteworthy that trends in CHL and CAR concentrations estimated from a selection of narrow VIS reflectance bands (Gitelson et al. 2002) indicated that the younger diebacks were in fact in transition (Mendelssohn and McKee 1988). This further supports the older to younger dieback depiction drawn from the blue-red band and green–red-edge band comparisons. HSI provided the bands and bandwidths appropriate for the algorithm.

To provide a single indicator conducive to most operational broadband sensors, an NIR/green simple ratio was used (Gitelson and Kaufman 1998). Overall, the NIR/green ratio performed better than the more common NIR/red simple ratio and provided an indication of the dieback onset at each location (Ramsey and Rangoonwala 2005). Both ratios help desensitize dieback onset and progression mapping to atmospheric influences, variable soil background reflectance, and changing canopy compositions (plant and background) (e.g., Ehrlich et al. 1994). The next step was to transfer this spectral information from the plant leaf level to the plant canopy level.

The method used to transfer the plant leaf results to the plant canopy was helicopter-based radiance collections (Ramsey and Rangoonwala 2006, 2009). This technique allowed targeted and controlled collections with little atmospheric influence. Photographs of each marsh site collected with the radiance recordings were classified into generalized amounts of green to yellow to brown marsh and visible background classes corresponding to the spectral composition classes (Ramsey and Rangoonwala 2006). The photographic classes provided validation of the composition percentages calculated from the canopy reflectance spectra.

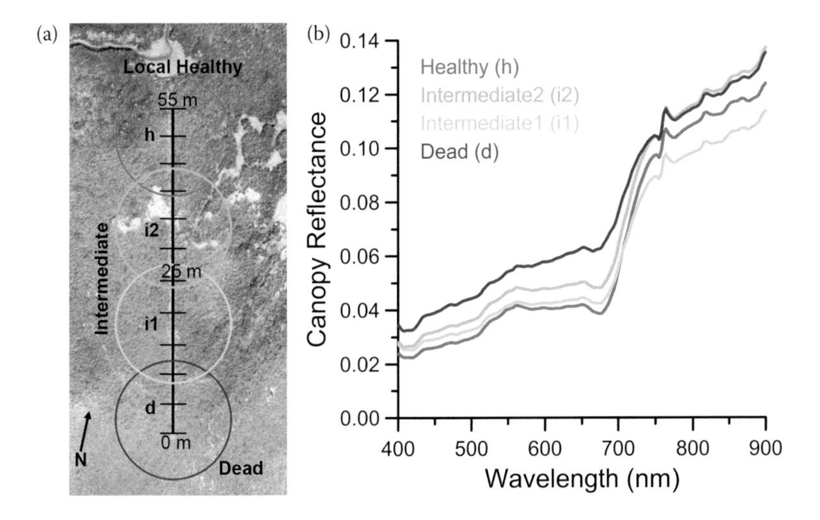

FIGURE 11.19 Marsh canopy sampling strategy and transect canopy reflectance spectra. In the left portion (a), circles along the transect overlain onto photo taken during helicopter overflight represent IFOV of helicopter radiometer, dead (4d); intermediate-1 (4i1); intermediate-2 (4i2); and local healthy (4 h). On the right (b), canopy reflectance associated with each of the four target areas depicted by circles on the transect. Note that the progressively higher VIS canopy reflectance spectra simulate the leaf transect reflectance spectra shown in Figure 11.16. The dieback transect site was located in coastal Louisiana (Figure 11.2). (Adapted from Ramsey, E., III and Rangoonwala A. 2006. *Photogramm Eng Rem S*, 72, 641–652.)

The circles in Figure 11.19a illustrate the general location and spatial extent of the canopy reflectance spectra in Figure 11.19b. As expected from the plant leaf spectral analyses, the canopy VIS reflectance increases as the impact becomes more severe. The canopy reflectance associated with the dead transect portion is dominated by the mud background, whereas the canopy reflectance at the healthy end represents the local healthy marsh reflectance.

As in the invasive mapping with Hyperion data, PVA was used to extract the healthy and dead characteristic reflectance spectra (Ramsey and Rangoonwala 2006). The characteristic spectra defined from canopy reflectance spectra of the transect sites and a more inland reference cordgrass site provided a method to deemphasize the background variability and to isolate those influences related to plant leaf reflectance within the marsh canopy (Ramsey and Rangoonwala 2006). These two PVA characteristic spectra were applied to 13 transect canopy reflectance spectra and 10 additional sites depicting cordgrass marsh at various stages of dieback.

Loadings based on the PVA hyperspectral analyses representing the tendency of the canopy reflectance to align with healthy or dead spectra were compared to marsh classifications of each of the 23 sites (Ramsey and Rangoonwala 2006). In addition, NIR/green and NIR/red ratios calculated for each site from the pertinent canopy reflectance spectrum were compared to the marsh classifications. The PVA loadings better aligned with the plant leaf transect results and the marsh classifications than did NIR/green or NIR/red indicators of marsh dieback progression (Ramsey and Rangoonwala 2006). Of the two ratios, the NIR/green ratio exhibited higher alignment with the classifications than NIR/red and had a higher correspondence with the PVA results (Ramsey and Rangoonwala 2009). PVA analysis based on the whole canopy reflectance spectra (400–1000 nm) produced the best depiction of marsh dieback progression, and the NIR/green band ratio produced lower but acceptable results.

Classification of the PVA results was accomplished with a distance-clustering procedure (SAS® Enterprise). The classification defined three broad categories: healthy, impacted, and severely impacted (including dead) (Ramsey and Rangoonwala 2006, 2009). Overall, based on hyperspectral data analyses, differences in the severity of dieback impact were indicated in the impacted and dead marsh categories. In addition, results based on the PVA whole spectra analyses captured the spectral variability in the nonimpacted or healthy cordgrass marshes, which was expected given the high variability in physical controls. Based on that natural spectral variability, marsh status based on visual color-coded classification did not provide a consistent strategy for monitoring the status of these coastal resources.

11.4.2.2 Short Form Cordgrass Marsh Dieback

The second demonstration of applying hyperspectral techniques for mapping abnormal marsh change took place in a short form cordgrass marsh lying at the apex of a tidal creek connected to the DelMarVA lagoon off the northern coast of Virginia (Figure 11.2) (Marsh et al. 2016). In contrast to the approximately twice daily flooding of the Gulf Coast marsh, flooding by saline lagoon waters is more irregular, occurring at higher tides. Similar to the Gulf Coast event but on a much smaller scale, a dieback occurred in 2004, as evidenced by fairly large areas of dead and completely denuded marsh scattered throughout the cordgrass marsh. Near the dieback region, a large region of cordgrass marsh existed visually unaffected by the dieback. Beginning with the appearance of the dieback and in tandem with ecologists' collecting biophysical information along transects within the marsh dieback, leaf spectra were collected within and outside the dieback region over 4 years from 2004 to 2007.

As in the Gulf study (Ramsey and Rangoonwala 2005), leaf samples were collected along transects in the dieback marsh; however, collections were at sites designated as healthy, intermediate, and dead marsh along transects established by ecological researchers (Marsh et al. 2016). Outside the dieback, leaf samples were collected at three marsh platform sites and at one site at the tidal creek apex at times nearly coincident with collections in the dieback marsh.

The spectral analysis techniques and spectral indicators applied were those substantiated in the Gulf Coast dieback study (Ramsey and Rangoonwala 2005). In this study, the availability of biophysical indicators of marsh condition provided a means of testing the performance of the spectral indicators to determine marsh condition. Of the Gulf Coast indicator suite, the red band, NIR/red vegetation index (VI), and chlorophyll pigment spectral based indices had the highest discriminatory strengths in the marsh condition (Marsh et al. 2016). Based on the collections in 2005, spectral thresholds for healthy, intermediate, and dead marsh were calculated for each of the three indicators (Figure 11.20a).

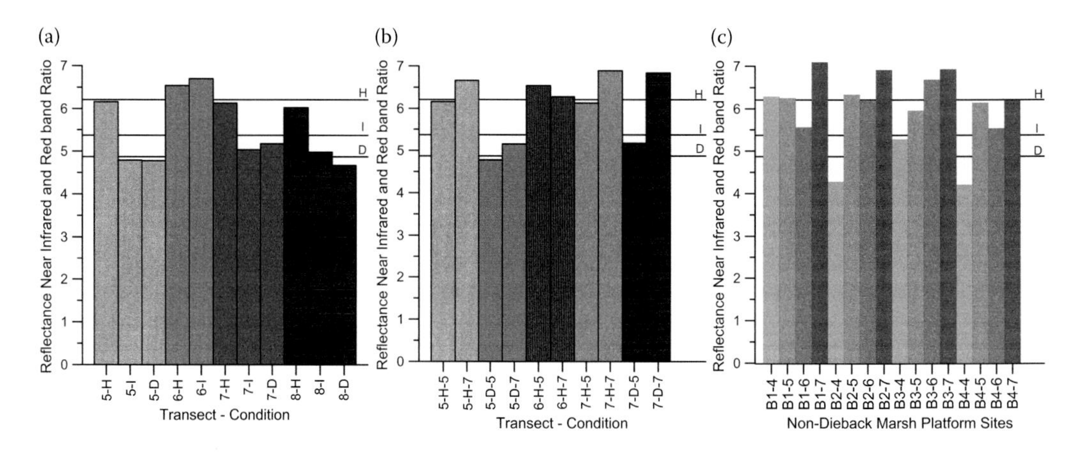

FIGURE 11.20 Only the NIR/red indicator of marsh condition is shown. (a) Spectral indicators of healthy (H), intermediate (I), and dieback (D) marsh at dieback transects T5, T6, T7, and T8 in 2005. The horizontal lines locate the H, I, and D means of each condition class calculated from the 2005 transect samples. X-axis labels represent transect number and condition class (e.g., 5-H). (b) The spectral indicators and spectral-condition thresholds were validated by comparison to samples collected in 2007 and field biophysical measurements (transect number-condition class-year, e.g., 5-H-7). (c) Calibrated and validated thresholds applied to nondieback marsh samples collected from 2004 to 2007 (site-year, e.g., B1–4). (Adapted from Marsh, Q. et al. 2016. *J Coast Conserv*, 20(4), 335–350.)

An evaluation of the consistency of the spectral thresholds was provided by application to spectral indicators based on collections in the final year of the dieback in 2007 compared to 2005 spectral indicators (Figure 11.20b) (Marsh et al. 2016). All spectral indices were highly consistent from 2005 to 2007, except for a dramatic improvement in condition at a site that had been classified as dieback in 2005. Independent biophysical measurements confirmed that the marsh at that site had improved from a dead state in 2005 (7-D-5) to healthy in 2007 (7-D-7), exactly tracking the spectral-condition indicator change.

These spectral-condition indicators and thresholds were applied to the samples collected outside the dieback (Figure 11.20c) (Marsh et al. 2016). In 2004, none of the four sites visually exhibited signs of marsh dieback; however, the spectral indices for these nondieback marsh platform sites indicated stress. This interpretation based on the spectral indices and thresholds was supported by biophysical measurements in the area. Additional evidence corroborating the utility of the spectral indicators and thresholds was the capture of conditions resembling a small-scale dieback in 2006, 2 years after the dieback peak (Figure 11.20c). While the change in spectral indicator magnitudes did not reach dieback levels, they suggested a downturn in marsh condition. This downturn was supported by biophysical measurements. The ability of the spectral indicators to document coordinated changes in marsh condition at multiple sites located in separate marsh areas is a critical result of this work. The alignment of spectral indicator results between dieback and nondieback areas, combined with the proven robustness of the thresholds, demonstrates the effectiveness of the spectral thresholds to quickly assess marsh condition status. These spectral indicators and thresholds are extendable to satellite remote sensing data platforms.

11.5 SUMMARY AND FUTURE DIRECTIONS

The benefits of HSI to remote sensing mapping of wetlands are based on the uniqueness of vegetation (at the species or cover-type level), spectral responses to heterogeneous growing conditions, and phenodynamics of the wetland system. The mixtures of vegetation types, densities, heights and forms, high spatial complexity, and dominant role of flooding in controlling the wetland landscape resulted in the development of unique remote sensing solutions to detect, map, and monitor these dynamic systems. As demonstrated in this chapter, the ability to overcome these complexities relies on targeted applications of HSI that are enhanced when supplemented with pertinent plant leaf spectral and canopy structure and background information.

Five studies were used to demonstrate the advantage of HSI in mapping and monitoring wetland ecosystems. The first coupled hyperspectral image and field data within an RT model. Limited to the VNIR wavelength range, the application of the RT model determined that the non-species-specific relationship between VNIR broadband image data and mangrove canopy LAI resulted from a high conformance in canopy orientation (LAD) and a lack of species specificity in VNIR leaf spectral properties. The RT model-predicted LAIs had a 93% correspondence to field-based LAIs and a 78% correspondence with NDVI values obtained from VNIR hyperspectral canopy reflectance spectra. The RT results also indicated that, although non-species-specific, leaf VNIR reflectance site-specific differences may provide stand-level indicators of mangrove condition and changes. Studies that extended the reflectance from the VNIR to include the SWIR showed successful discrimination of the mangrove species at the leaf level. Classification discrimination at the canopy level was accomplished when species dominances were physically separable at the pixel level, and in two cases, structure information was included in the classification.

The second study that mapped the occurrence of invasive plants in a wetland landscape illustrated how broadband mapping requiring a high spatial resolution (1 m) could be regionalized with moderate-spatial-resolution (30 m) hyperspectral data and image processing. The regional hyperspectral mapping provided not only the invasive tallow detection as obtained in the broadband 1-m mapping but also provided a continuous classification of tallow, live, and senescing vegetation percent occurrences within the 30-m pixel. The continuous tallow percent occurrence classification was used to create an added-value product for resource managers. An integration of land resource information created from broadband data and tallow occurrences created from the hyperspectral data provided new information about the risk of tallow establishment by land-cover type and land activities.

The third study illustrated the coupling between marsh canopy structure and hyperspectral reflectance. The demonstrated coupling showed that the correct interpretation of marsh trends can require concurrent information describing the canopy structure (e.g., LAI and leaf orientation distribution). The study found that increased hyperspectral mapping performance can be realized by utilizing remote sensing data that are more revealing about the marsh canopy structure, such as radar or lidar, in concert with HSI collections.

The fourth and fifth studies demonstrated the ability of hyperspectral data processing to capture subtle leaf and canopy spectral changes related to the onset and progression of dieback in spectrally complex marshes. Although analyses based on the whole VNIR spectra best predicted dieback onset and progression, enhanced mapping performance compared to broadband sensor systems was illustrated when the hyperspectral data were transformed to user-specified narrow-width bands. The bands were placed at leaf reflectance features exhibiting high variability and closely aligned with the major leaf pigment spectral absorption regions in the VIS. The heightened information captured by the hyperspectral reflectance spectra enabled tracking of dieback progression and recovery, revealed subtle changes in marsh condition, and showed that the same marsh experiencing similar stress levels can produce dramatically different outcomes.

Finally, baldcypress and BLH wetland forests were described in terms of where HSI mapping could be advantageously applied. One of those advantages is the use of hyperspectral data and processing to detect the onset of baldcypress stress caused by saltwater intrusion into these freshwater swamps. Early stress detection would promote successful remediation before irreversible change. More effective separation of BLHs by community could also be an advantage of HSI mapping. Mapping could be applied in a period of known maximum spectral contrast to establish the locations and extents of the separate BLH communities. Once established, HSI mapping could monitor the BLH forest communities in order to detect detrimental change. Reestablishment of community extents would occur each year during maximum spectral contrast.

HSI provided increased information about the status or condition of a wetland canopy beyond what was obtainable from broadband sensors in all described comparisons. In each case, the hyperspectral mapping advantage was based on the use of high-quality canopy reflectance spectra largely free from atmospheric distortions and illumination variability. After the high-fidelity reflectance data sets were obtained, specialized tools were required that could transform the reflectance data into information about the wetland condition with enough precision so that subtle changes could be detected. These transformations to reflectance and spectral characteristics are often arduous and detailed. As demonstrated, however, obtaining the mapping objective often depends on the use of HSI. Also introduced was the need at times to incorporate new and more detailed field collection and analysis techniques. Often, particularly in investigative works, mapping success relies on targeted and well-structured field campaigns that provide consistent and pertinent biophysical data that can be related to the hyperspectral reflectance data. Those biophysical and reflectance relationships provide quantitative interpretation of the reflectance data and the ultimate transformation of the reflectance data into meaningful products and information that can only be provided by HSI.

In addition to using hyperspectral data collections in wetland mapping projects, applied research should increase integration of hyperspectral data with broadband optical, radar, and lidar data sources. Broadband and HSI integration can be simple, as demonstrated in the invasive tallow mapping (Ramsey et al. 2005a) or using lidar-generated topography to improve the hyperspectral classification of a marsh (Yang and Artigas 2010). More refined and complete integrations would use hyperspectral data to enhance the information of broadband data and extend that enhanced mapping performance throughout the broadband swath extent, providing wider coverage and thereby higher repeat frequency (Ramsey and Rangoonwala 2009). HSI integration with radar would help account for the marsh structure (Ramsey et al. 2016) and provide an additional source of wetland forest structure (Held et al. 2003, Simard et al. 2010), particularly when the lack of solar illumination or clouds obscure optical collections. Radar data could also be useful for monitoring the hydrologic state of observed wetlands, such as flooding in marshes and swamps. Similar to what was discovered with broadband and HSI integration, current research is pursuing the integration of optical and radar in an effort to provide a more consistent and on-demand land-cover-monitoring system. This integrated mapping should include HSI as well. The full integration of hyperspectral, broadband, and radar image data would advance the relevance of remote sensing for the unique spatial, compositional, and functional setting of dynamic coastal wetlands.

ACKNOWLEDGMENTS

We thank Joseph Spruce (self affiliated) and Beth Middleton (USGS) for their help in preparing this manuscript for publication. Partial funding for the described works was provided by the Louisiana Department of Natural Resources (Agreement 2512-01-11), the National Aeronautics and Space Administration (Grant EO-1-0100-0042), and by USGS Hurricane Sandy Supplemental Funds. Any use of trade, firm, or product name is for descriptive purposes only and does not imply endorsement by the US government.

REFERENCES

Adam, E., Mutanga, O., and Rugege, D. 2010. Multispectral and hyperspectral remote sensing for identification and mapping of wetland vegetation: A review. *Wetlands Ecol Manage*, 18, 281–296.

Artigas, F. J., and Yang, J. S. 2005. Hyperspectral remote sensing of marsh species and plant vigour gradient in the New Jersey Meadowlands. *Int J Remote Sen*, 26(23), 5209–5220.

Belluco, E., Camuffo, M., Ferrari, S., Modenese, L., Silvestri, S., Marani, A., and Marani, M. 2006. Mapping salt-marsh vegetation by multispectral and hyperspectral sensing. *Remote Sens Environ*, 105(1), 54–67.

Daughtry, C., Ranson, K., and Biehl, L. 1989. A new technique to measure the spectral properties of conifer needles. *Remote Sens Environ*, 27, 81–91.

DeWeese, G. G., Grissino-Mayer, H. D., and Lam, N. 2007. Historical land-use/land-cover changes in a bottomland hardwood forest, Bayou Fountain, Louisiana. *Phys Geogr*, 28, 345–359.

Ehrlich, D., Estes, J., and Singh, A. 1994. Applications of NOAA-AVHRR 1 km data for environmental monitoring. *Int J Remote Sens*, 15, 145–161.

Ehrlich, R., and Crabtree, S. 2000. *The PVA Multivariate Unmixing System, Self-Training Classification*, Tramontane, Inc., and C & E Enterprises, Salt Lake City, UT.

Federal Geographic Data Committee. 2013. *Classification of Wetlands and Deepwater Habitats of the United States*. FGDC-STD-004-2013. Second Edition. Wetlands Subcommittee, Federal Geographic Data Committee and U.S. Fish and Wildlife Service, Washington, DC.

Flores-de-Santiago, F., Kovacs, J. M., and Flores-Verdugo, F. 2013. The influence of seasonality in estimating mangrove leaf chlorophyll-a content from hyperspectral data. *Wetlands Ecol Manage*, 21, 193–207.

Gitelson, A., and Kaufman J. 1998. MODIS NDVI optimization to fit the AVHRR data series-spectral considerations. *Remote Sens Environ*, 66, 343–350.

Gitelson, A., Zur Y., Chivkunova, O., and Merzlyak, M. 2002. Assessing Carotenoid content in plant leaves with reflectance spectroscopy. *Photochem Photobiol*, 75, 272–281.

Goudriaan, J. 1977. *Crop Micrometeorology: A Simulation Study*, Netherlands Centre for Agricultural Publishing and Documentation, Wageningen, 249 pp.

Guo, M., Li, J., Sheng, C., Xu, J., and Wu, L. 2017. A review of wetland remote sensing. *Sensors*, 17(4).

Held, A., Ticehurst, C., Lymburner, L., and Williams, N. 2003. High resolution mapping of tropical mangrove ecosystems using hyperspectral and radar remote sensing. *Int J Remote Sens*, 24(13), 2739–2759.

Hill, R. A., Wilson, A. K., George, M., and Hinsley, S. A. 2010. Mapping tree species in temperate deciduous woodland using time-series multi-spectral data. *Appl Veg Sci*, 13, 86–99.

Himmelblau, D. M. 1972. *Applied Nonlinear Programming*, McGraw-Hill, New York.

Jensen, J. R., Lin, H., Yang, X., Ramsey III, E., Davis, B. A., and Thoemke, C. W. 1991. *Geocato International*, 6(2), 13–21.

King, S., and Keeland, B. 1999. Evaluation of reforestation in the Lower Mississippi River Alluvial Valley. *Restor Ecol*, 7, 348–359.

Kirk, J. 1994. In *Light & Photosynthesis in Aquatic Ecosystems*, Second Edition, Cambridge University Press, Cambridge, 229–233.

Klemas, V., Dobson, J., Ferguson, R., and Haddad K. 1993. A coastal land cover classification system for the NOAA coast watch change analysis project. *J Coastal Res*, 9, 862–872.

Koedsin, W., and Vaiphasa, C. 2013. Discrimination of tropical mangroves at the species level with EO-1 Hyperion data. *Remote Sens*, 5, 3562–3582.

Krauss, K., Chambers, J., and Allen, J. 1998. Salinity effects and differential germination of several half-sib families of baldcypress from different seed sources. *New For*, 15(1), 53–68.

Kumar, T., Panigrahy, S., Kumar, P., and Parihar, J. S. 2013. Classification of floristic composition of mangrove forests using hyperspectral data: Case study of Bhitarkanika National Park, India. *J Coast Conserv*, 17, 121–132.

Kumar, L., and Sinha, P. 2014. Mapping salt-marsh land-cover vegetation using high-spatial and hyperspectral satellite data to assist wetland inventory. *GIScience & Remote Sens*, 51(5), 483–497.

Lillesaeter, O. 1982. Spectral reflectance of partly transmitting leaves: Laboratory measurements and mathematical modeling. *Remote Sens Environ*, 1, 247–254.

Liu, B., Zhang, L., Zhang, X., Zhang, B., and Tong, Q. 2009. Simulation of EO-1 Hyperion data from ALI multispectral data based on the spectral reconstruction approach. *Sensors*, 9, 3090–3108.

Lugo, A., and Snedaker, S. 1974. The ecology of mangroves. *Annu Rev Ecol Syst*, 5, 39–64.

Markham, B., Williams, D., Schafer, J., Wood, F., and Kim, M. 1995. Radiometric characterization of diode-array field spectroradiometers. *Remote Sens Environ*, 51, 317–330.

Marsh, Q., Blum, A., Christian, R., Ramsey III, E., and Rangoonwala, A. 2016. Response and resilience of *Spartina alterniflora* to sudden dieback. *J Coast Conserv*, 20(4), 335–350.

Mendelssohn, I., and McKee K. 1988. *Spartina alterniflora* die-back in Louisiana: Time-course investigation of soil waterlogging effects. *J Ecol*, 76, 509–521.

Mitsch, W., and Gosselink, J. 2000. *Wetlands*, Third Edition. John Wiley & Sons, New York.

Ramsey, E., III, and Jensen, J. 1995. Modeling mangrove canopy reflectance using a light interaction model and an optimization technique, in J. Lyon and J. McCarthy (Eds.), *Wetland and Environmental Applications of GIS*. CRC Press, Boca Raton, FL, 61–81.

Ramsey, E., III, and Jensen, J. 1996. Remote sensing of mangroves: Relating canopy spectra to site-specific data. *Photogramm Eng Rem S*, 62(8), 939–948.

Ramsey, E., III. 1998. Radar remote sensing of wetlands, in *Remote Sensing Change Detection: Environmental Monitoring Methods and Applications*, R. Lunetta and C. Elvidge (Eds.), Ann Arbor Press, Ann Arbor, MI, 211–243.

Ramsey, E., III, Nelson, G., and Sapkota, S. 2001. Coastal change analysis program implemented in Louisiana. *J Coastal Res*, 17, 55–71.

Ramsey, E., III, Nelson, G., Sapkota, S., Seeger E., and Martella, K. 2002. Mapping Chinese tallow with color-infrared photography. *Photogramm Eng Rem S*, 68(3), 251–255.

Ramsey, E., III, Nelson, G., Baarnes, F., and Spell, R. 2004. Light attenuation profiling as an indicator of structural changes in coastal marshes, in *Remote Sensing and GIS Accuracy Assessment*, R. Lunetta and J. Lyon (Eds.), CRC Press, Boca Raton, FL, 59–73.

Ramsey, E., III. 2005. Remote sensing of coastal environments, in *Encyclopedia of Coastal Science*, M.L. Schwartz (Ed.), Kluwer Academic Publishers, Amsterdam, The Netherlands, 797–803.

Ramsey, E., III, and Nelson, G. 2005. A whole image approach for transforming EO1 Hyperion hyperspectral data into highly accurate reflectance data with site-specific measurements. *Int J Remote Sens*, 26, 1589–1610.

Ramsey, E., III, and Rangoonwala, A. 2005. Leaf optical property changes associated with the occurrence of *Spartina alterniflora* dieback in coastal Louisiana related to remote sensing mapping. *Photogramm Eng Rem S*, 71, 299–311.

Ramsey, E., III, Rangoonwala A., Nelson, G., and Ehrlich, R. 2005a. Mapping the invasive species, Chinese tallow with EO1 satellite Hyperion hyperspectral image data and relating tallow percent occurrences to a classified Landsat Thematic Mapper landcover map. *Int J Remote Sens*, 26, 1637–1657.

Ramsey, E., III, Rangoonwala, A., Nelson, G., Ehrlich, R., and Martella, K. 2005b. Generation and validation of characteristic spectra from EO1 Hyperion image data for detecting the percent occurrence of invasive species, specifically Chinese tallow. *Int J Remote Sens*, 26, 1611–1636.

Ramsey, E., III, and Rangoonwala, A. 2006. Site-specific canopy reflectance related to marsh dieback onset and progression in coastal Louisiana. *Photogramm Eng Rem S*, 72, 641–652.

Ramsey, E., III, and Rangoonwala, A. 2009. Mapping the onset and progression of marsh dieback, in *Remote Sensing of Coastal Environments*, Y. Wang, (Ed.), CRC Press, Remote Sensing Applications Series.

Ramsey, E., III, Rangoonwala, A., Baarnes, F., and Spell, R. 2009a. Mapping, Mapping fire scars and marsh recovery with remote sensing image data, in *Remote Sensing and GIS for Coastal Ecosystem Assessment and Management*, X. Yang (Ed.), Springer, "Lecture Notes in Geoinformation and Cartography," 415–438.

Ramsey, E., III, Rangoonwala, A., Middleton, B., and Lu, Z. 2009b. Satellite optical and radar image data of forested wetland impact on and short-term recovery from Hurricane Katrina in the lower Pearl River flood plain of Louisiana. USA. *Wetlands*, 29, 66–79.

Ramsey, E., III, Spruce, J., Rangoonwala, A., Suzuoki, Y., Smoot, J., Gasser, J., and Bannister, T. 2011. Monitoring wetland forest recovery along the lower Pearl River with daily MODIS satellite data. *Photogramm Eng Rem S*, 77(11), 1133–1143.

Ramsey, E., III, Rangoonwala, A., Jones, C. E., and Banister, T. 2015. Marsh canopy leaf area and orientation calculated for improved marsh structure mapping. *Photogramm Eng Rem S*, 81(10), 807–816.

Ramsey, E., III, Rangoonwala, A., and Jones, C. E. 2016. Marsh canopy structure changes and the deepwater horizon oil spill. *Remote Sens Environ*, 186, 350–357.

Rosso, P. H., Ustin, S. L., and Hastings, A. 2005. Mapping marshland vegetation of San Francisco Bay, California, using hyperspectral data. *Int J Remote Sen*, 26(23), 5169–5191.

Sahagian, D., and Melack, J. 1996. *Global Wetland Distribution and Functional Characterization: Trace Gases and the Hydrologic Cycle*, Report of the Joint GAIM-DIS-BAHC-IGAC-LUCC Workshop held in Santa Barbara CA, on 16-20 May.

Shao, G., Pauli, B. P., Haulton, G. S., Zollner, P. A., and Shao, G. 2014. Mapping hardwood forests through a two-stage unsupervised classification by integrating Landsat Thematic Mapper and forest inventory data. *J Appl Remote Sens*, 8, 83546-1-14.

Simard, M., Fatoyinbo, L. E., and Pinto, N. 2010. Mangrove canopy 3D structure and ecosystem productivity using active remote sensing. Y. Wang (Ed.), CRC Press, Remote Sensing Applications Series, pp. 61–78.

Stout, J. P. 1984. *The ecology of irregularly flooded salt marshes of the Northeastern Gulf of Mexico: A community profile*, U.S Fish Wildlife Service Biology Report 85.

Thenkabail, P. S., Mariotto, I., Gumma, M. K., Middleton, E. M., Landis, D. R., and Huemmrich, K. F. 2013. Selection of hyperspectral narrowbands (HNBs) and composition of hyperspectral twoband vegetation indices (HVIs) for biophysical characterization and discrimination of crop types using field reflectance and Hyperion/EO-1 data. *IEEE J. Sel Topics Appl Earth Observ in Remote Sens*, 6(2), 427–439.

Tuominen, J., and Lipping, T. 2016. Spectral characteristics of common reed beds: Studies on spatial and temporal variability. *Remote Sens*, 8, 181.

Turner, R., and Spencer, M. 1972. Atmospheric model for correction of spacecraft data, in *Proceedings of Eighth International Symposium of Remote Sensing of Environment, Environmental Research Institute of Michigan*, Ann Arbor, MI, 895–934.

Yang, J., and Artigas, F. J. 2010. Mapping saltmarsh vegetation by integrating hyperspectral and LIDAR remote sensing. Y. Wang (Ed.), CRC Press, Remote Sensing Applications Series, pp. 173–187.

Yang, C., Everitt, J. H., Fletcher, R. S., Jensen, R. R., and Mausel, P. W. 2009. Evaluating AISA + hyperspectral imagery for mapping black mangrove along the south Texas gulf coast. *Photogramm Eng Rem S*, 75(4), 425–435.

Wang, L., and Sousa, W. P. 2009. Distinguishing mangrove species with laboratory measurements of hyperspectral leaf reflectance. *Int J Remote Sens*, 30(5), 1267–1281.

Warner, T., and Shank, M. 1997. Spatial autocorrelation analysis of hyperspectral imagery for feature selection. *Remote Sens Environ*, 60, 58–70.

White, D. 1983. Plant communities of the lower Pearl River basin, Louisiana. *Am Midl Nat*, 110, 381–96.

Wickland, D. 1991. Mission to planet earth: The ecological perspective. *Ecology*, 72, 1923–33.

Zhang, C., and Xie, Z. 2013. Data fusion and classifier ensemble techniques for vegetation mapping in the coastal Everglades. *Geocarto Int*.

Zhang, C., Kovacs, J. M., Yali, L., Flores-Verdugo, F., and Flores-de-Santiago, F. 2014. Separating mangrove species and condition using laboratory hyperspectral data: A case study of degraded mangrove forest of the Mexican Pacific. *Remote Sens*, 6, 11673–11688.

Section IV

Thermal, SWIR, and Visible Remote Sensing

12 Hyperspectral Remote Sensing of Fire
A Review

Sander Veraverbeke, Philip Dennison, Ioannis Gitas, Glynn Hulley, Olga Kalashnikova, Thomas Katagis, Le Kuai, Ran Meng, Dar Roberts, and Natasha Stavros

CONTENTS

12.1 INTRODUCTION

Fire is a ubiquitous disturbance agent in the terrestrial biosphere, and it occurs in ecosystems that range from tropical rainforests to deserts and boreal forests (Bond and Keeley, 2005; Bowman et al., 2009). Fire occurs in a variety of forms, including high-intensity crown fires to long-duration ground fires in organic soils with relatively low intensity (van der Werf et al., 2017). Ecosystems and fire regimes are rapidly changing at historically unprecedented rates (Gillett et al., 2004; Westerling, 2006; Dennison et al., 2014; Stavros et al., 2014). For example, fire activity has significantly increased in boreal forest ecosystems (Gillett et al., 2004; Turetsky et al., 2011; Veraverbeke et al., 2017) and declined in savannas (Andela and van der Werf, 2014; Andela et al., 2017). Agricultural fires are an increasing threat to the sustainable management of tropical forests and peatlands (van Marle et al., 2017). In many regions, recent changes in fire regimes have caused permanent shifts in species composition (Zedler et al., 1983; Tulloch et al., 2016), resulting in changes in landscape heterogeneity (Airey Lauvaux et al., 2016; Boucher et al., 2017). These ecological changes caused by fire have strong implications for biodiversity, land use, carbon and water cycling, and climate. Intensification of the fire regime in many areas contributes to increasing atmospheric carbon dioxide (Keppel-Aleks et al., 2014; Veraverbeke et al., 2015), while vegetation regrowth following a fire can be a multidecadal carbon sink (O'Halloran et al., 2012).

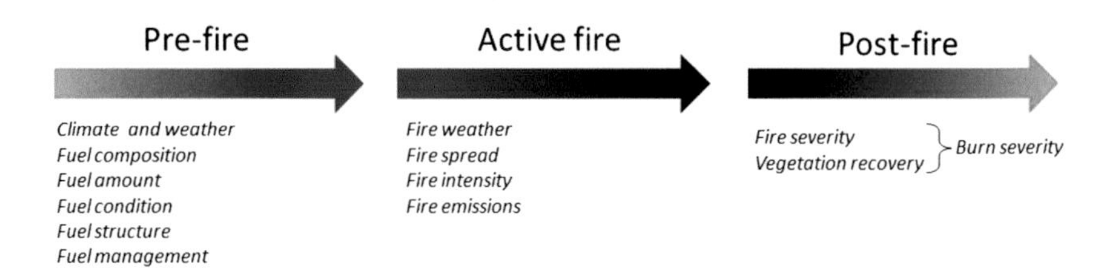

FIGURE 12.1 Temporal phases in fire disturbance continuum.

Fire is a temporally dynamic process in which fuels, climate and weather, and humans interact at timescales ranging from milliseconds to millennia (Bowman et al., 2009). The combustion process itself, which occurs when heat, oxygen, and fuel are combined, is relatively short-lived. In years to weeks before fire, however, droughts and heat waves increase the susceptibility of fuels to ignition and subsequent fire spread (van der Werf et al., 2003; Abatzoglou and Kolden, 2011). Fires spread over landscapes for several hours to months depending on fuel continuity and synoptic weather patterns (Parks, 2014; Veraverbeke et al., 2014a). The first weeks to years after a fire, post-fire landscapes are often susceptible to landslides and floods (Barro and Conard, 1991; Doerr et al., 2006; Diakakis et al., 2017). At decadal timescales, fuel biomass accumulates over decades in synergy with climate and sometimes land management decisions (Hanson and Odion, 2014). At millennial timescales, fires mediate biome boundaries and continental vegetation composition and structure (Lehmann et al., 2014; Rogers et al., 2015; Veraverbeke et al., 2017).

The fire disturbance continuum discriminates between discrete temporal phases during which fire processes occur (Jain et al., 2004). The fire disturbance continuum includes pre-fire, active, and post-fire environments (Figure 12.1). The pre-fire environment refers to the type, structure, condition, and amount of fuels as influenced by climate, weather, and land management. The active fire environment is the phase during which fires spread over the landscape. Topography, fuels, and fire weather influence active fire behavior and intensity. Fire intensity describes the physical combustion process of energy release from organic matter (Keeley, 2009) and is directly related to fire emissions (Wooster et al., 2005). Finally, the post-fire environment is what is left after the fire is extinguished. The post-fire environment is often described interchangeably with the terms fire and burn severity (Boer et al., 2008; Keeley, 2009). Here we define fire severity as the degree of environmental change caused by a fire as evidenced immediately after the fire without recovery effects (Lentile et al., 2006; Veraverbeke et al., 2010; Morgan et al., 2014). Conversely, burn severity gauges both the immediate fire-induced change and vegetation recovery.

Remote sensing has been successfully applied in all stages of the fire disturbance continuum for several decades. Success stories include fuel type mapping (Roberts et al., 2003; Mitri and Gitas, 2006; Peterson et al., 2013; Marino et al., 2016), fire risk assessments (Chuvieco et al., 2004; Meng et al., 2017; Yu et al., 2017), active fire detection (Giglio et al., 2003; Schroeder et al., 2014), burned area mapping (Roy et al., 2005; Gitas et al., 2008; Giglio et al., 2009; Katagis et al., 2014), fire/burn severity assessments (Eidenshink et al., 2007; Veraverbeke et al., 2010), and vegetation recovery mapping (Riaño et al., 2002; van Leeuwen et al., 2010; Veraverbeke et al., 2012a). These applications have primarily capitalized upon broadband multispectral remote sensing data. Broadband multispectral remote sensing is the simultaneous acquisition of calibrated radiance units in a limited number (generally on the order of between 3 and 15) of non-contiguous broad (generally wider than 20 nm) spectral bands. In contrast, narrowband hyperspectral remote sensing is the simultaneous acquisition of calibrated radiance units in many (generally more than 100) narrow (generally 20 nm or smaller) spectrally contiguous bands. Hyperspectral imaging, or imaging spectroscopy, refers to the acquisition of coregistered images over contiguous narrow spectral bands (Schaepman et al., 2009). Hyperspectral remote sensing has proven its utility in a wide range of Earth system science domains, among others those focused on greenhouse gases (Roberts et al., 2010; Dennison et al., 2013), plants (Johnson et al., 1994; Roberts et al., 1998; Asner et al., 2000; Somers et al., 2010; Ustin et al., 2012), minerals (Hook et al., 1991; van der Meer and

Bakker, 1997; Baugh et al., 1998), snow and ice (Painter et al., 2003; Dozier et al., 2009), coastal and inland water (Hoogenboom et al., 1998; Salem et al., 2005; Kudela et al., 2015), urban environments (Roessner et al., 2001; Roberts et al., 2012), and fire (Dennison and Roberts, 2009; Schepers et al., 2014; Veraverbeke et al., 2014b). These studies were conducted based on airborne hyperspectral (AHS) remote sensing, often based on data from the Airborne Visible/Infrared Imaging Spectrometer (AVIRIS) (Green et al., 1998) or the Airborne Prism Experiment (APEX) (Itten et al., 2008). To date, Hyperion on the Earth-Observing One (EO-1) platform, acquiring data between 2000 and 2017, has been the only spaceborne hyperspectral imager that acquired data in the visible to shortwave infrared (VSWIR) spectral range (approximately between 0.4 and 2.5 μm) (Pearlman et al., 2003). In the next few years, several spaceborne hyperspectral sensors may be launched: Environmental Mapping And Analysis Program (EnMAP) (Stuffler et al., 2007), Hyperspectral Imager Suite (HISUI) (Iwasaki et al., 2011), Hyperspectral Infrared Imager (HyspIRI) (Lee et al., 2015), Precursore Iperspettrale Della Missione Applicativa (PRISMA) (Labate et al., 2009), and the Space-borne Hyperspectral Applicative Land and Ocean Mission (SHALOM) (Feingersh and Ben Dor, 2016). These missions will greatly increase the availability and application of hyperspectral data.

With upcoming spaceborne hyperspectral missions and the proven utility of hyperspectral data in fire applications, we provide a review of the current state of the art in hyperspectral remote sensing of fire. We therefore review developments in the pre-fire, active fire, and post-fire stages of the fire disturbance continuum. The primary focus of this review is on applications where hyperspectral data provide a clear improvement over multispectral data or on novel opportunities that arise from hyperspectral data that are not possible based on multispectral data. We also propose avenues for further research.

12.2 HYPERSPECTRAL FIRE APPLICATIONS

12.2.1 Pre-Fire Applications

The pre-fire environment refers to fuel composition, condition, amount, and structure (Chandler et al., 1983) and how these change through time as a function of climate, weather, land management, and land use.

First, fuel composition refers to the type (e.g., grass or shrub) and size of fuels. The composition of fuels affects its chemical composition and, thus, the available energy content that then affects fire intensity, the physical combustion process of energy release from organic matter (Agee, 1993; Keeley, 2009). Fuel composition influences fire management strategies. For example, it may be desirable to protect against high-severity canopy fires by prescribing fire to reduce fuels in many ecosystems. Certain functional environments, such as old growth forests, however, are home to protected or endangered species (Tews et al., 2004), and failure to protect these environments during prescribed burns would result in diversity loss (Kennedy et al., 2008; Rockweit et al., 2017). The size of fuels, defined as the ratio between surface area and volume, will influence fire intensity and moisture loss (Agee, 1993). Second, fuel condition refers to the moisture content and the live or dead fuel status. These parameters influence fuel drying and combustion (Pickett et al., 2010). Moisture content affects the flammability of fuels and, thus, fire behavior such as ignition probability, fire spread rate, and consequent smoke impacts (Anderson, 1970; Forkel et al., 2012). Furthermore, fuel moisture content can influence the vulnerability of plants to disease and mortality and, thus, vulnerability to fire (McDowell et al., 2008). Third, fuel amount refers to the amount of fuels available to burn and is often associated with biomass. Fourth, fuel structure refers to the vertical and horizontal connectivity of fuels, whereby vertical connectivity is defined by the connection of ladder fuels, and horizontal connectivity is defined by the fractional cover and density of vegetation. Fuel structure influences fire spread rates.

Multispectral remote sensing of fuel composition by mapping plant functional types has capitalized upon classification and vegetation index approaches (Hansen and Reed, 2000; Loveland et al., 2000; Friedl et al., 2002; Bartholomé and Belward, 2005; Rollins et al., 2006; Nelson et al., 2013; Ryan and Opperman, 2013). Similarly, retrieving fuel moisture and photosynthetic, that is, live, versus nonphotosynthetic, that is, dead, vegetation from multispectral data is often based on spectral indices (Gao, 1995; Liu and Kogan, 1996; Anderson et al., 2004; Jackson et al., 2004), sometimes augmented with land surface temperature data from thermal bands (Verbesselt et al., 2002; Chuvieco et al., 2003). Retrievals of fuel amount can be based on empirical relationships between field and vegetation index data that are then spatially extrapolated (Dong et al., 2003; Wessels et al., 2006). In a similar way, vegetation indices have also been used as predictors of canopy cover (Larsson, 1993). Robust relationships and useful applications of multispectral pre-fire remote sensing are abundant. However, the fact that reflectance data from a few bands, often combined in a spectral index, results in high correlations with multiple fuel attributes demonstrates that some of these attributes are highly correlated, yet broadband remote sensing may not be able to fully capture subtle differences that may exist within fuel attributes and plant species.

The multiple and narrow spectral bands from hyperspectral remote sensing allow different approaches to determine fuel attributes like fuel composition and condition (Schimel et al., 2013; Thompson et al., 2017). Fuel composition and fractional cover can be determined from spectral mixture analysis (SMA) (Roberts et al., 2006). SMA calculates cover fractions of different ground cover classes and thereby leverages the spectral information over multiple wavebands (Adams et al., 1986; Roberts et al., 1998). Because there is inherent variability within a ground cover class, often many sample spectra, or endmembers, can be used for a single class. Multiple endmember SMA (MESMA) is an extension of SMA that accounts for variability within endmember classes (Roberts et al., 1998; Rogge et al., 2006), which is beneficial for fuel type composition and cover fraction retrievals. For fuel type composition mapping, the ground cover classes can be as specific as ecosystem type and contain multiple slightly variable spectra (Figure 12.2). Ground cover classes can also be defined as green vegetation (GV), nonphotosynthetic vegetation (NPV), and substrate (Roberts et al., 1998). The spatial distribution of GV, NPV, and substrates in a Californian landscape is shown in Figure 12.3. Multitemporal mapping of these ground cover fractions demonstrates drought effects on fuels and their flammability. Figure 12.4, for example, shows the effects of the California drought between 2013 and 2016 on the fraction of alive vegetation, defined as the ratio between the GV fraction and the sum of the GV and NPV fractions. Fuel moisture content can be derived from hyperspectral signals in spectral regions with high water absorption (Yebra et al., 2013). This approach necessitates a clear discrimination between atmospheric and fuel moisture, for example through radiative transfer modeling (Ustin et al., 1998). An alternative approach is to retrieve equivalent water thickness from

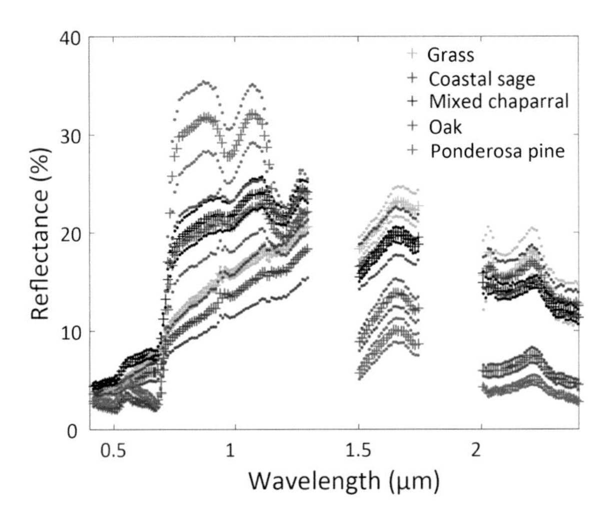

FIGURE 12.2 Example of a vegetation spectral library. The mean (+) and mean plus/minus one standard deviation (·) of five spectra per vegetation type are plotted. Atmospheric water vapor absorption regions were removed.

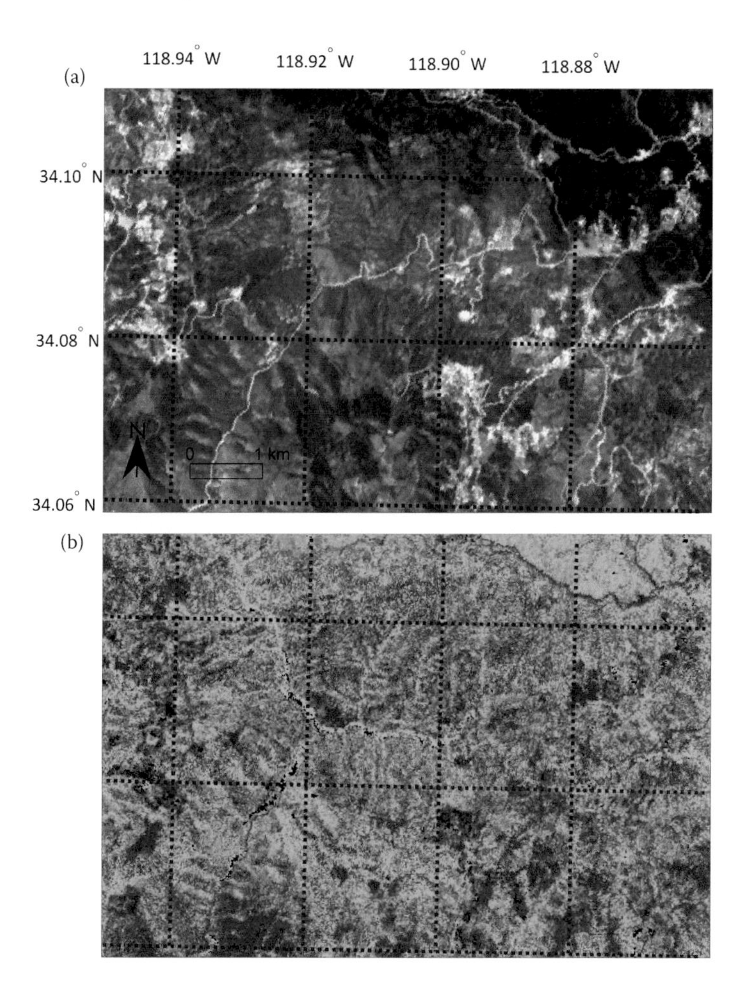

FIGURE 12.3 (a) True color composite from the Airborne Visible/Infrared Imaging Spectrometer over parts of Santa Monica mountains in California, USA. The composite used the bands centered at 0.65 μm (red), 0.55 μm (green), and 0.45 μm (blue). (B) False color composite inputting NPV (red), GV (green), and substrate (blue) covers fractions retrieved from MESMA. Suboptimal retrievals were masked in black.

SWIR bands between approximately 1.2 and 2.5 μm, a spectral region that is highly influenced by water absorption (Gao and Goetz, 1990, 1995). Hyperspectral data can be further useful in deriving variables that relate to fuel accumulation. For example, leaf area index (LAI) as a proxy of stand productivity (Gitelson et al., 2014), and thus fuel accumulation, can be derived from the water content retrieval from hyperspectral data (Roberts et al., 2004). Another indirect way in which hyperspectral data can improve the retrieval of biomass is by providing additional information (e.g., fractional cover of species) that informs the use of allometric equations, for example, in synergy when using other sensor technologies such as light detection and ranging (LiDAR) (Koch, 2010).

12.2.2 Active Fire Applications

Applications developed for active fires fall into two major categories: (1) fire detection, in which the goal is to identify spectra containing active fire (Dennison and Roberts, 2009), and (2) fire temperature retrieval, in which the goal is to model fire temperature from emitted radiance (Dennison et al., 2006). Fire temperature retrieval may provide additional outputs from modeling, such as subpixel fire fractional area.

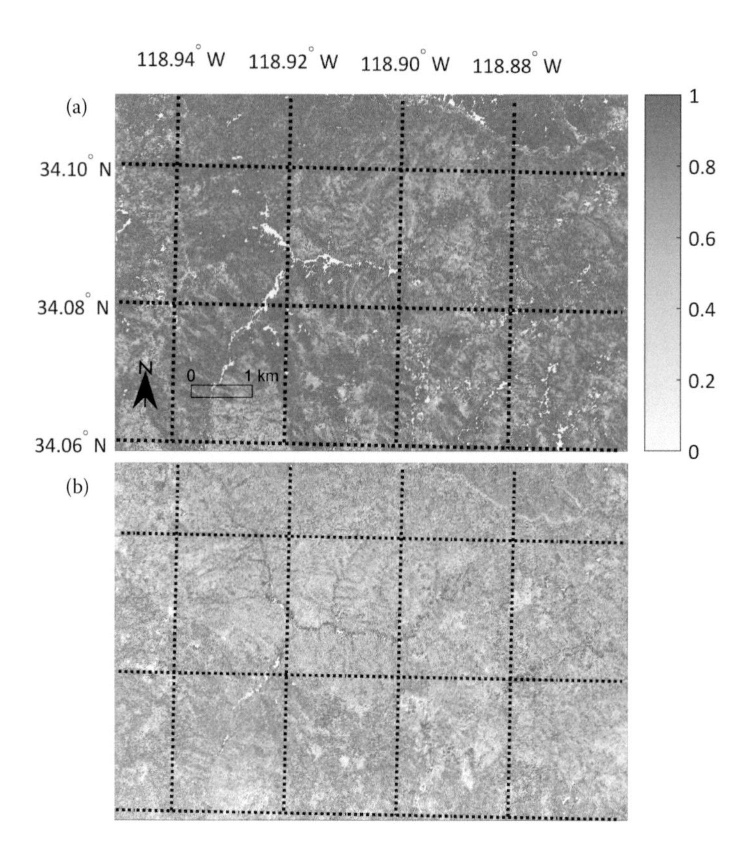

FIGURE 12.4 Decreases in alive vegetation fraction, defined as the ratio between the GV fraction and the sum of the green and non-photosynthetic vegetation fractions, derived from multiple endmember spectral mixture analysis, in California's Santa Monica Mountains between (a) 2013 and (b) 2016 during the California drought. Suboptimal retrievals were masked in white.

Radiance emitted directly from a fire is dependent on the temperature of the burning biomass and flame, the emissivity of the burning biomass and flame, and the depth of the flame. Traditional broadband methods for detecting fire and estimating fire temperature and intensity use brightness temperature, which assumes that fire is a blackbody emitter (Giglio et al., 2003; Zhukov et al., 2006; Roberts and Wooster, 2008; Schroeder et al., 2014). For a blackbody, emitted radiance increases and peak emission shifts to shorter wavelengths as fire temperature increases (Figure 12.5).

Background blackbody surfaces at typical Earth surface temperatures emit most of their radiance in the thermal infrared (TIR) (8–12 μm) and longer wavelengths (Figure 12.5). As temperature increases above 600 K, radiance in the mid-infrared (MIR) (3–5 μm) and SWIR sharply increases. Smoke is a strong scatterer and absorber at wavelengths shorter than 1.2 μm, such that the visible (0.4–0.7 μm) and near-infrared (NIR) (0.7–1.2 μm) spectral regions have limited utility for active fire applications. At longer wavelengths, smoke transmittance is very high, and smoke has a very minor impact on emitted radiance. Thus, the SWIR and MIR spectral regions are most useful for measuring flaming and smoldering combustion (Dennison and Matheson, 2011). The SWIR spectral region demonstrates a very strong difference in emitted spectral radiance across the range of temperatures typical for smoldering and flaming combustion (650 K to higher than 1500 K).

While multiple active fire applications of hyperspectral data have assumed that fire within an instantaneous field of view (i.e., a pixel) is a single temperature blackbody (Dennison et al., 2006; Dennison and Matheson, 2011; Matheson and Dennison, 2012), the actual shape of emitted radiance is a critically important question. Emitted radiance from a single pixel will be a composite of multiple flaming and smoldering elements, each with their own range of temperatures. At-sensor radiance

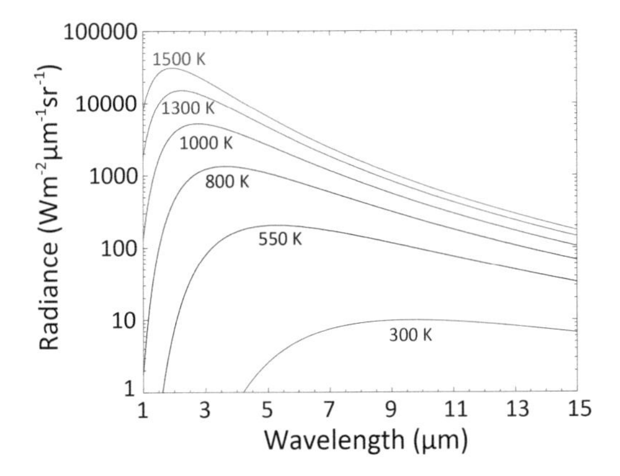

FIGURE 12.5 Blackbody emission curves across a range of temperatures. (After Dennison, P.E., Matheson, D.S., 2011. *Remote Sens. Environ.* 115, 876–886. https://doi.org/10.1016/j.rse.2010.11.015.)

itself will be a combination of burning and nonburning surfaces within the instantaneous field of view. Reflected solar radiance will also be included in at-sensor radiance if a fire is imaged during the day. Flames have lower emissivity than background materials, but the impact of flame emissivity will depend on path length through flame and the soot content of the flame (Giglio and Kendall, 2001; Riggan et al., 2004).

Dennison and Matheson (2011) tested the blackbody assumption by comparing emitted radiance acquired simultaneously by AVIRIS and the MODIS/ASTER airborne simulator (MASTER) (Hook et al., 2001) over a single fire. Fire temperatures modeled using AVIRIS SWIR data were compared to fire temperatures modeled using MASTER multispectral MIR and TIR data. Both models assumed blackbody emission, and temperatures retrieved from the two data sets were found to be poorly correlated below 800 K. Matheson and Dennison (2012) investigated spatial scaling of AVIRIS data over four fires and found decreases in modeled fire temperatures as spectra were aggregated up to coarser spatial resolutions. Based on fire complexity across spatial scales, uncertain emissivity, and results from previous experiments, the blackbody assumption should be regarded with caution (especially for cooler fires).

Regardless of whether blackbody emission approximates actual fire emitted radiance, maximum spectral radiance values produced by fire are an important concern for remote measurement. Sensors designed for measuring land and water surfaces typically do not have a high enough saturation threshold to adequately capture peak emission from wildfires, especially at a finer spatial resolution, where fire can comprise a higher percentage of individual pixels (Realmuto et al., 2015). For example, SWIR bands in AVIRIS data frequently saturate over the hottest parts of wildfires, especially when flown on the lower altitude Twin Otter platform (Dennison et al., 2006; Matheson and Dennison, 2012).

Detection of fire within hyperspectral data presents two challenges. Typically, hyperspectral data are acquired during the day, in which case emitted radiance must be reliably separated from the reflected solar radiance background to accurately detect fire. Also, fire may represent a small percentage of a larger pixel, which effectively dilutes the strength of the emitted radiance signal and makes emitted radiance more difficult to separate from the reflected solar radiance background. Dennison and Roberts (2009) compared all possible normalized difference index combinations of AVIRIS bands in radiance data acquired over the 2003 Simi fire in California, USA. The presence of fire in a pixel increases spectral radiance in longer-wavelength SWIR bands more rapidly than in shorter-wavelength SWIR bands, and indices combining two bands spanning the range of the SWIR spectral region produced the most accurate detection of pixels containing fire. They named the most accurate index the Hyperspectral Fire Detection Index (HFDI):

$$\mathrm{HFDI} = \frac{L_{2.43\mu m} - L_{2.06\mu m}}{L_{2.43\mu m} + L_{2.06\mu m}}, \tag{12.1}$$

where $L_{2.43\mu m}$ is the spectral radiance around 2.43 μm, and $L_{2.06\mu m}$ is the spectral radiance around 2.06 μm.

Dennison and Roberts (2009) further performed a sensitivity analysis on the HFDI, noting the impacts of fire temperature, subpixel fractional area, atmospheric path length and water vapor, and solar zenith angle on the index. HFDI takes advantage of an atmospheric carbon dioxide absorption feature at 2.06 μm. Reflected solar radiance experiences atmospheric carbon dioxide absorption on both the downwelling and upwelling paths, while emitted radiance only experiences absorption on the upwelling path (Dennison, 2006). This difference allows improved separation of fire from the background surface, but it also effectively prohibits remote measurement of carbon dioxide emissions directly over a fire.

Excited potassium has line emission features at 0.767 and 0.770 μm. Potassium emission has been detected over fires using imaging spectrometer data, including AVIRIS (Vodacek et al., 2002) and EO-1 Hyperion (Amici et al., 2011). Finer spectral resolution improves discrimination of the potassium emission feature, but the primary limitation of this technique remains scattering and absorption by smoke within the NIR spectral region (Dennison and Roberts, 2009).

Hyperspectral temperature retrieval methods are based on a spectral mixing model approach (Giglio et al., 2003). The general form of spectral mixing models used for spectral radiance (L_λ) is

$$L_\lambda = \sum_{i=1}^{n} f_i L_{i,\lambda} + \varepsilon_\lambda, \tag{12.2}$$

where $L_{i,\lambda}$ is the radiance of endmember i at wavelength λ, f_i is the fraction of endmember i, n is the number of endmembers, and ε_λ is the residual error. A background endmember can include emitted radiance or reflected solar radiance, depending on the wavelength regions included in the model (Dennison and Matheson, 2011). Most models use a single blackbody emitted radiance endmember for fire (i.e., $n = 2$). In simple two-band models, the temperature of the fire is solved for using the brightness temperature of the measured pixel and the brightness temperature of the background. In models with more than two bands, endmembers spanning a range of temperatures can be compared, and the endmember that produces the lowest model error is used subsequently to assign the fire temperature to a pixel (Dennison et al., 2006). In either case, each endmember is multiplied by a fractional cover. All fractions in the model sum to one, and fractional cover of the fire endmember represents the subpixel fire percentage. This type of mixing model assumes that radiance from fire and background endmembers mix linearly, an assumption that is difficult to test in reality. The mixing of model retrievals of fire temperature has also frequently been applied to multispectral airborne and satellite remotely sensed data (Matson and Holben, 1987; Oertel et al., 2004; Riggan et al., 2004; Zhukov et al., 2006; Eckmann et al., 2008, 2009; Giglio and Schroeder, 2014).

The larger spectral dimensionality provided by hyperspectral data permits the application of more complex mixing models. For example, Dennison et al. (2006) created a three-endmember multiple endmember mixing model that included background endmembers for different vegetation types, soil and ash, a range of emitted radiance endmembers that included modeled atmospheric absorption, and a shade endmember that controlled for the absolute level of radiance. A major advantage of this approach is that it simultaneously produces maps of fire temperature, fire fractional area, and background land-cover type (Figure 12.6). Dennison and Matheson (2011) applied a similar three-endmember model but extended both the background and fire endmembers through the MIR and TIR and expanded the number of potential endmembers to include both smoke- and non-smoke-covered backgrounds. However, the lack of reference data for validation of retrieved fire temperature remains

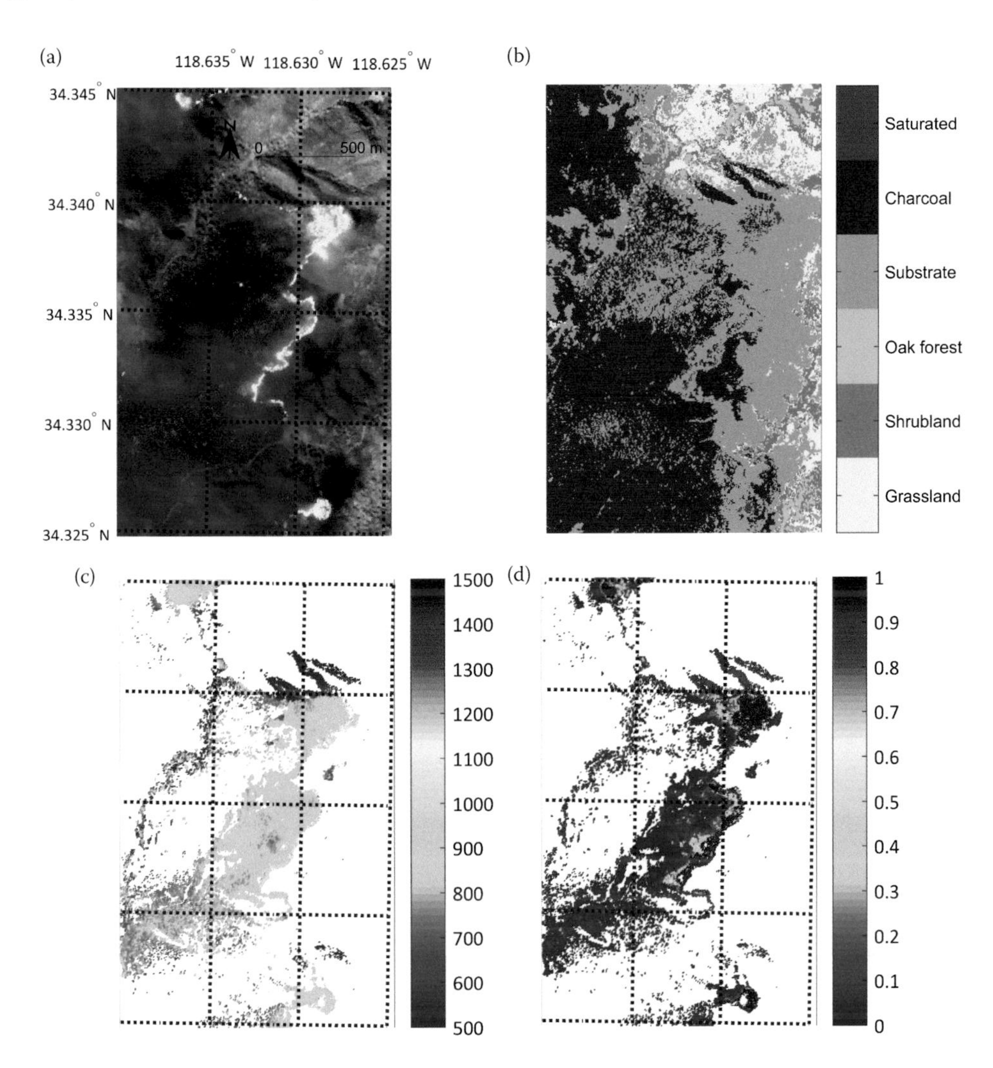

FIGURE 12.6 (a) Color composite from the Airborne Visible/Infrared Imaging Spectrometer over parts of active 2003 Simi fire in California, USA. The composite used the bands centered at 1.70 μm (red), 1.10 μm (green), and 0.66 μm (blue). (b) Land cover, (c) fire temperature (K), and (d) fire fraction derived from multiple endmember spectral mixture analysis. (After Matheson, D.S., Dennison, P.E., 2012. *Remote Sens. Environ.* 124, 780–792. https://doi.org/10.1016/j.rse.2012.06.026.)

a limitation. To date, in situ temperature measurements and hyperspectral imagery have not been collected concurrently over a fire.

12.2.3 Post-fire Applications

12.2.3.1 Fire and Burn Severity

Fire and burn severity are often loosely defined as the degree of environmental change caused by a fire (Key and Benson, 2006). Fire severity refers to the fire-induced change without vegetation recovery effects, while burn severity represents the combined effect of the immediate fire impact and longer-term recovery (Lentile et al., 2006; Veraverbeke et al., 2010; Morgan et al., 2014). Severity often refers to different ecosystem characteristics depending on ecoregion and application. In grasslands, for example, combustion completeness, the ratio of combusted to available biomass, is an important indicator of severity. In temperate ecosystems, severity often refers to tree mortality, while in boreal ecosystems, the depth of burning in organic soils is the main indicator of severity (Turetsky et al., 2011; Rogers et al., 2014; Veraverbeke et al., 2015). Severity data are used in two main applications. First, in the USA, severity maps are operationally used by burned area emergency response (BAER) teams to diagnose risk to infrastructure and safety and prioritize post-fire rehabilitation efforts (Eidenshink et al., 2007). Second, several studies recognize the potential of severity maps to optimize fire emissions estimates (French et al., 2008; De Santis et al., 2010). De Santis et al. (2010) used remotely sensed estimates of severity to optimize combustion values. Similarly, Veraverbeke and Hook (2013) used a remotely sensed indicator of tree mortality in a fire emissions model and demonstrated that emissions estimates with a remotely sensed tree mortality layer were more than 50% lower than without. Rogers et al. (2014) demonstrated the utility of a severity spectral index for estimating fire emissions in boreal ecosystems, and Veraverbeke et al. (2015, 2017) further integrated this spectral index in a statistical model of carbon combustion together with other environmental layers for these ecosystems.

In multispectral remote sensing, the Normalized Burn Ratio (NBR) (López García and Caselles, 1991) has become the most widely used spectral index for assessing fire and burn severity, often applied on Landsat imagery (French et al., 2008). NBR relates to vegetation vigor and moisture by combining NIR and SWIR reflectance. After a fire, there is generally a decrease in NIR reflectance and an increase in SWIR reflectance. The differenced NBR (dNBR) (Key and Benson, 2006) is obtained after bitemporal differencing pre- and post-fire NBR images. The principles of the dNBR index are transferable to hyperspectral remote sensing. The limited availability of spaceborne hyperspectral data and the need for advance planning in airborne campaigns have resulted in very few opportunities to test the performance of a hyperspectral dNBR (Stavros et al., 2016). A rare example of such a pre-/post-fire airborne image acquisition is from van Wagtendonk et al. (2004) for a fire in Yosemite National Park in the USA. They, however, found no increased sensitivity of a hyperspectral dNBR to ground measurements of severity in comparison with the Landsat-derived dNBR. Further opportunities to optimize the hyperspectral dNBR recently arose from acquisitions over two large California wildfires in areas that were part of the HyspIRI preparatory airborne campaign (Figure 12.7). Schepers et al. (2014) tested several hyperspectral indices derived from a post-fire image over a heathland ecosystem in Belgium. They found that the strength and form of the relationships between spectral indices and ground measures of severity varied by vegetation type, necessitating a vegetation stratification to derive optimal results.

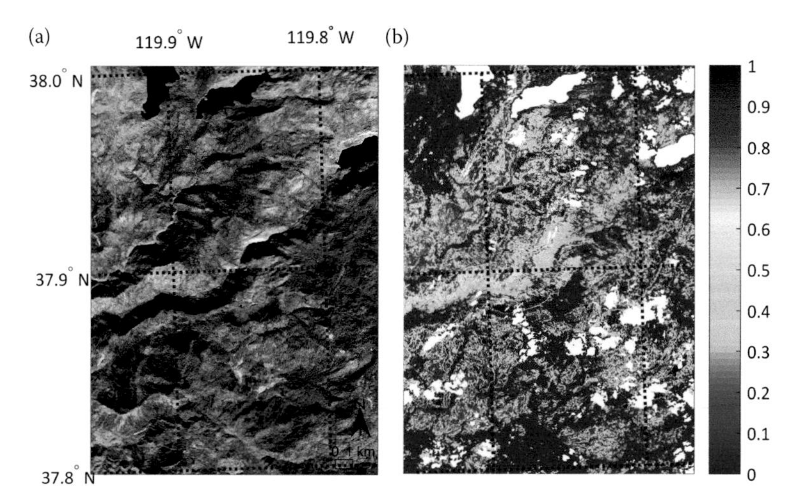

FIGURE 12.7 (a) Post-fire color composite from the Airborne Visible/Infrared Imaging Spectrometer over parts of 2013 Rim fire in California, USA. The composite used the bands centered at 2.10 μm (red), 0.88 μm (green), and 0.69 μm (blue). (b) Hyperspectral differenced Normalized Burn Ratio over the same area. Clouds and water bodies were masked and are depicted in white.

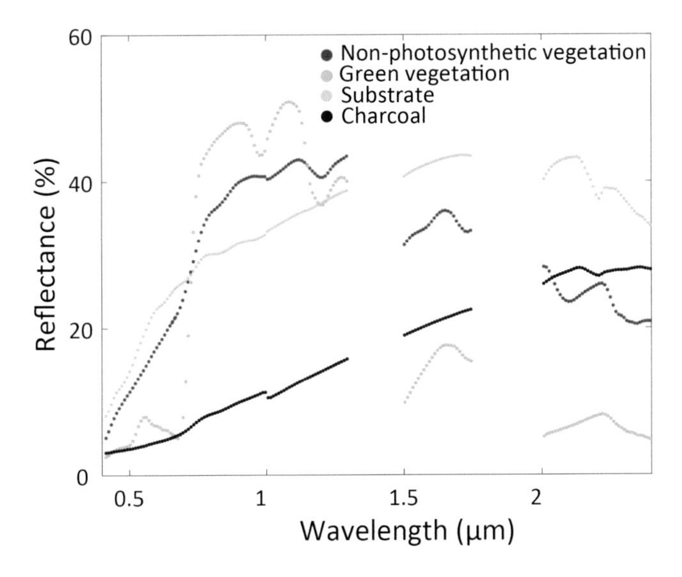

FIGURE 12.8 Spectral signatures of non-photosynthetic vegetation, green vegetation, substrate, and charcoal. Atmospheric water vapor absorption regions were removed.

While spectral indices can be powerful proxies of biophysical properties, they only use spectral information from two or three bands. By doing so, they do not take advantage of the wealth of spectral information available in hyperspectral remote sensing (Figure 12.8). SMA is a powerful analysis tool for severity assessments with the additional advantage that the output of SMA are quantitative abundance estimates of the ground cover classes, without the need for calibration with field data, as with spectral indices (Solans Vila and Barbosa, 2010; Somers et al., 2010) (Figure 12.9). SMA has been applied on multispectral post-fire imagery (e.g., Smith et al., 2007; Fernandez-Manso et al., 2009; Quintano et al., 2013; Veraverbeke and Hook, 2013; Meng et al., 2017); however, a few studies have leveraged the higher spectral resolution from hyperspectral remote sensing (Kokaly et al., 2007; Lewis et al., 2007, 2008, 2011; Robichaud et al., 2007; Veraverbeke et al., 2014b). Kokaly et al. (2007) used AVIRIS data in a hyperspectral classification of ground cover classes. Lewis et al. (2007, 2008, 2011), Robichaud et al. (2007), and Veraverbeke et al. (2014a,b) derived cover fractions of ground classes, including charcoal, ash, GV, scorched vegetation, NPV, soil, and substrates. These estimates, and especially the GV and charcoal fractions, were significantly correlated with ground measurements of severity in a variety of case studies in temperate and boreal ecosystems (Figure 12.9b). Lewis et al. (2008) also found a relationship between ash cover derived from hyperspectral SMA and soil water repellency. Veraverbeke et al. (2014a,b) demonstrated improvements of 7% to 44% in estimating ground cover fractions from hyperspectral data compared to multispectral data. These improvements were the result of the high dimensionality of hyperspectral data that aids discrimination between ground cover classes (Figure 12.10). Discrimination of charcoal hinges on its characteristic low reflectance for NIR wavelengths (Figures 12.8 and 12.10a). GV is spectrally different from other ground cover classes owing to its combined high NIR and low SWIR reflectance (Figures 12.8 and 12.10b). Separability of NPV from substrate is usually more challenging in visible to SWIR spectral regions (Figure 12.10c,d) (Roberts et al., 1993).

12.2.3.2 Vegetation Recovery

Various fire-affected variables can be measured and modeled in post-fire environments. Following the prior definition of burn severity, vegetation recovery can be part of a severity assessment (Lentile et al., 2006; Veraverbeke et al., 2010; Morgan et al., 2014). Post-fire species structure and composition and vegetation succession are crucial variables in understanding ecosystem responses to

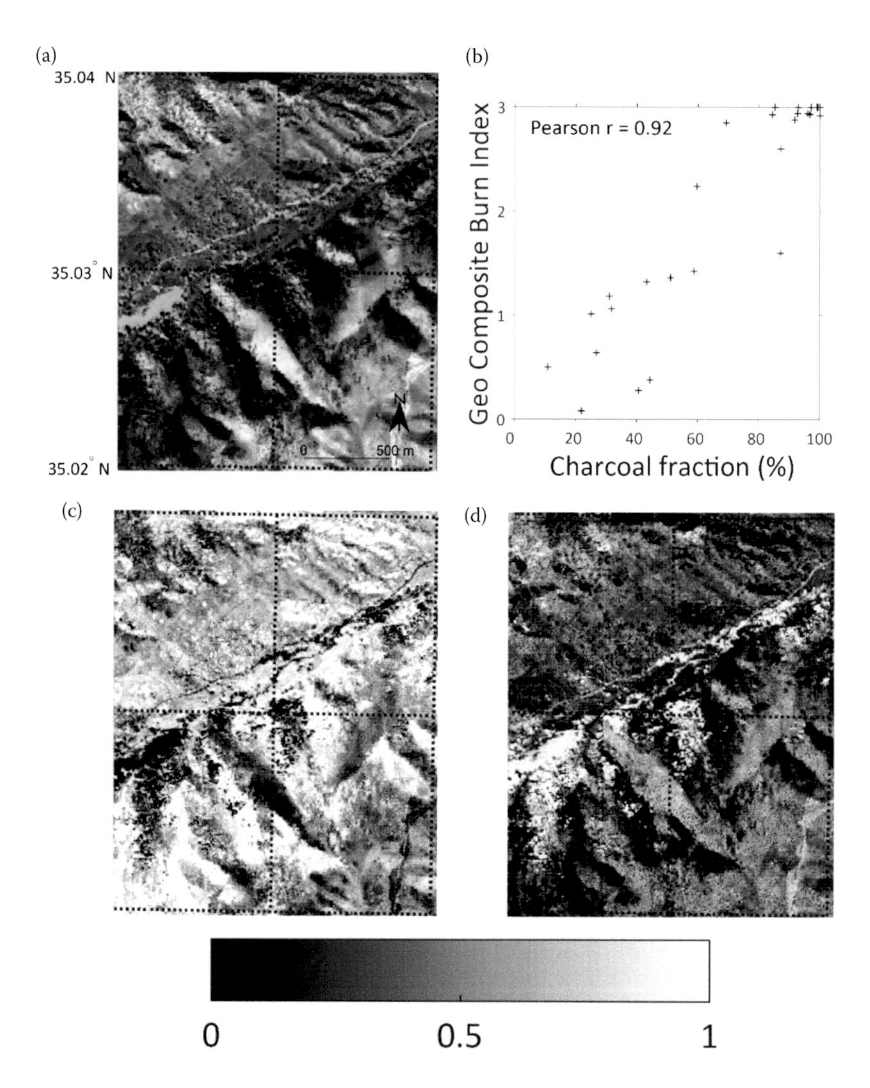

FIGURE 12.9 (a) Post-fire color composite from the Airborne Visible/Infrared Imaging Spectrometer over parts of 2011 Canyon fire in California, USA. The composite used the bands centered at 2.10 μm (red), 0.88 μm (green), and 0.69 μm (blue). (c) Ground cover fractions of charcoal and (d) green vegetation derived from spectral mixture analysis. (b) Correlation between charcoal fraction and Geo Composite Burn Index, a field measurement of fire severity. (De Santis, A., Chuvieco, E., 2009. *Remote Sens. Environ.* 113, 554–562. https://doi.org/10.1016/j.rse.2008.10.011.)

fire disturbance and climate change (Capitanio and Carcaillet, 2008; Chu and Guo, 2013). Especially in forest ecosystems, the spatial distribution of various forest components, such as tree height and sapling density, defines forest structure, while species richness and abundance characterize forest composition and biodiversity (McElhinny et al., 2005).

The terms vegetation recovery, vegetation regrowth, and vegetation regeneration are used, often interchangeably, to describe the various stages of post-fire vegetation succession in fire-affected ecosystems (Johnstone and Kasischke, 2005; Mitri and Gitas, 2012; Veraverbeke et al., 2012b). These terms refer to the recovery process of species or ecosystems to a pre-disturbance state. Fires can also lead to permanent changes in vegetation composition and structure, decreased vegetation cover, biomass loss, and the alteration of landscape patterns (Pérez-Cabello et al., 2009). Consequently, detailed monitoring of post-fire vegetation dynamics helps define the ecological impact of fires on ecosystem functioning and allows implementation of effective restoration measures (Gouveia et al.,

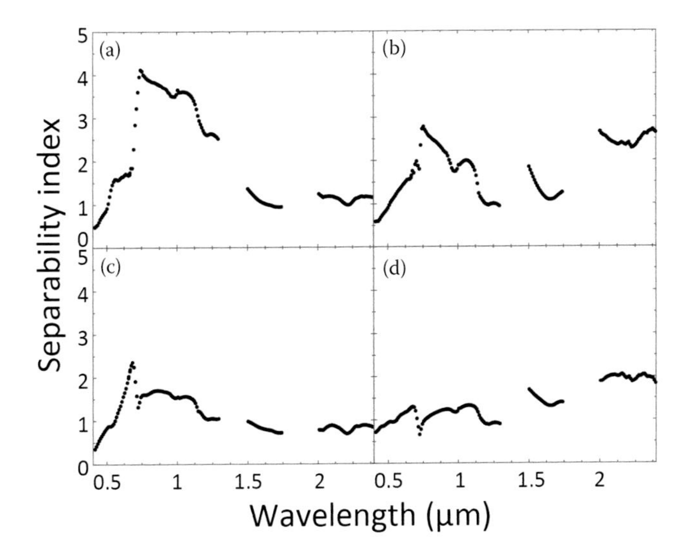

FIGURE 12.10 Spectral separability of (a) charcoal, (b) green vegetation, (c) non-photosynthetic vegetation, and (d) substrate. Spectral separability was calculated from between- and within-class variability of endmember spectra. (Somers, B. et al., 2009. *Int. J. Remote Sens.* 30, 139–147. https://doi.org/10.1080/01431160802304625; Veraverbeke, S. et al., 2014b. *Remote Sens. Environ.* 154, 153–163. https://doi.org/10.1016/j.rse.2014.08.019.)

2010; van Leeuwen et al., 2010; Gitas et al., 2012; Veraverbeke et al., 2012b). Fire severity, post-fire meteorological conditions, fuel type, topography, and soil properties are all factors that influence recovery patterns (Moreira et al., 2009; Veraverbeke et al., 2010; Pausas and Fernández-Muñoz, 2012).

Broadband multispectral remote sensing has been used extensively to assess post-fire vegetation recovery at various temporal and spatial scales (Díaz-Delgado et al., 2003; Goetz et al., 2006; Gouveia et al., 2010; Leon et al., 2012). The existence of long time series of multispectral imagery spanning consecutive decades, combined with free data distribution policies, has substantially increased its use in post-fire recovery applications. Most of these studies rely on vegetation abundance proxies derived from vegetation indices like the normalized difference vegetation index (NDVI) and NBR, often processed using advanced trajectory analysis algorithms (Leon et al., 2012; Katagis et al., 2014; Storey et al., 2016; Zhao et al., 2016).

Few post-fire monitoring applications have used hyperspectral data owing to limited data availability. Hyperspectral data have been useful in vegetation studies unrelated to fires since they provide detailed information about vegetation abundance and composition (Elvidge and Chen, 1995; Thenkabail et al., 2004). For example, hyperspectral data have been successfully used for forest species mapping (Stagakis et al., 2016), land cover classification (Dennison and Roberts, 2003; Stavrakoudis et al., 2012; Pontius et al., 2017), plant stress detection (Hernández-Clemente et al., 2011), and forest photosynthesis monitoring (Hernandez-Clemente et al., 2016).

In a preliminary study on monitoring post-fire succession in California's Santa Monica Mountains, AVIRIS-derived vegetation indices were tested for their ability to detect variations in the photosynthetic activity of chaparral (Qiu et al., 1998). The NDVI was used along with specific narrowband indices, the photochemical reflectance index (PRI) (Gamon et al., 1997), and the water band index (WBI). The findings indicated that including narrowband indices facilitated detection of sensitive changes in photosynthetic activity that were not associated with changes in canopy structure. Multitemporal AVIRIS imagery was used in another post-fire vegetation regeneration in the Santa Monica Mountains (Riaño et al., 2002). Riaño et al. (2002) found that GV fraction performed equally well in both northern mixed chaparral and south coastal sage scrub communities, as opposed to NDVI measurements, which were affected by phenological variations. Also in California, Somers et al. (2016) demonstrated the utility of MESMA-derived cover fraction to

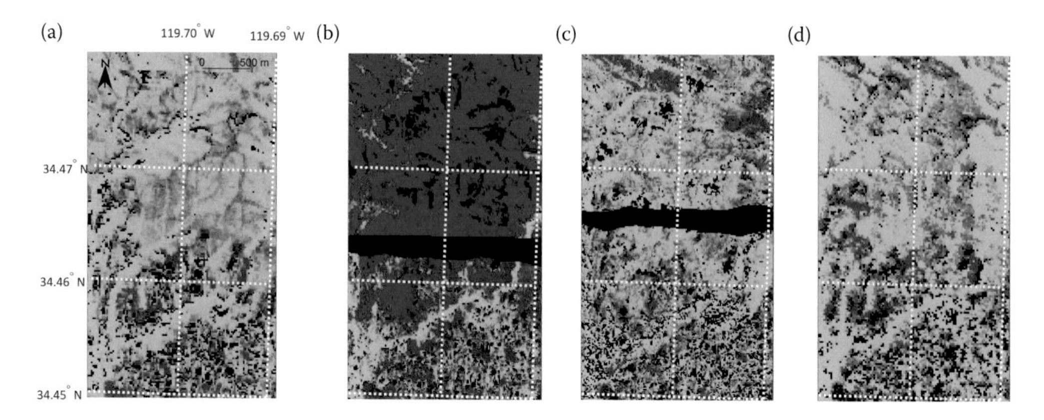

FIGURE 12.11 Time series of surface composition between 2004 and 2013 in area of 2009 Jesusita fire in California, USA, as derived from Airborne Visible/Infrared Imaging Spectrometer (a) on August 6, 2004, (b) just after the fire on August 26, 2009, (c) on April 30, 2010, and (d) on June 6, 2013. The false color composites input non-photosynthetic vegetation (red), green vegetation (green), and substrate and ash (blue) cover fractions retrieved from multiple endmember spectral mixture analysis. Suboptimal retrievals and missing data were masked in black. (After Somers, B. et al., 2016. in: Ruckebush, C. (Ed.), *Data Handling in Science and Technology*. Elsevier, pp. 551–577. https://doi.org/10.1016/B978-0-444-63638-6.00017-6.)

monitor post-fire vegetation recovery (Figure 12.11). In a similar application, multitemporal AHS imagery and SMA were combined to monitor post-fire recovery in Spain (Huesca et al., 2013).

AVIRIS data sets were further used for mapping of sapling density regeneration (Potter et al., 2012) and assessment of coarse woody debris (CWD) distribution (Huang et al., 2009) in Yellowstone National Park, USA. Potter et al. (2012) built statistical relationships between ten different vegetation indices and field estimations of sapling density. The resulting statistical models had high coefficients of determination (R^2) ranging between 0.78 and 0.83. Huang et al. (2009) performed fusion of synthetic aperture radar and AVIRIS data sets for quantifying CWD.

To track ecological changes caused by fire across multiple spatial and temporal scales, Lewis et al. (2017) made combined use of AHS data, spaceborne high, that is, QuickBird, and medium, that is, Landsat, spatial resolution multispectral imagery collected on anniversary dates spanning 10 years after three large wildfires in Montana, USA, 2003. They used MESMA to derive post-fire char, soil, GV, and NPV fraction maps at different recovery times. Retrievals from hyperspectral imagery had stronger correlations with ground measurements one year post-fire, and with vegetation canopy data 10 years post-fire.

Mitri and Gitas (2010, 2012) applied object-based image analysis on single-date EO-1 Hyperion images for classifying forest regeneration and vegetation recovery in fire-affected areas on the Greek island of Thasos. The use of objects offers several advantages over traditional pixel-based methods and allows for a more realistic representation of the real world, especially when applied on high spatial resolution images (Benz et al., 2004). In their initial work, Mitri and Gitas (2010) also showed promising results in mapping regeneration of different pine species and discriminating areas of young and mature pine stands. In their subsequent work, mapping results were improved by combining hyperspectral information from Hyperion with high resolution multispectral images (Mitri and Gitas, 2012). Vegetation indices derived from Hyperion were also useful for estimating post-fire recovery in Mato Grosso in the southern Amazon in Brazil (Numata et al., 2011). In this study, selected narrowband indices outperformed the NDVI in detecting subtle changes in the physiological properties of disturbed forests.

Despite the limited number of applications due to data limitations, hyperspectral data provide opportunities for post-fire monitoring that are not supported by broadband multispectral imagery. The reviewed applications mostly hinged on single-date or a small number of hyperspectral airborne or spaceborne images. Current limitations in deploying airborne campaigns limit further exploration of the possibilities that arise from hyperspectral data in post-fire recovery studies. The upcoming hyperspectral satellite missions in an era of open data policies will enable more systematic exploitation of hyperspectral information in post-fire monitoring studies.

12.3 FUTURE PERSPECTIVES

12.3.1 Automated Image-Based Endmember Extraction for Spectral Mixture Analysis

SMA, and in particular MESMA that accounts for within-class endmember variability, is a popular image analysis technique in all phases of the fire disturbance continuum: before, during, and after a fire. In the studies reviewed here, endmember spectra were most often measured in the field using a field spectroradiometer or derived from imagery or existing spectral libraries (Baldridge et al., 2009). This approach was feasible and effective for local-scale studies, like those based on the limited air- and spaceborne hyperspectral data opportunities reviewed here. The large data sets that will originate from the upcoming spaceborne hyperspectral missions, however, will require automated and image-specific endmember bundle retrievals. The importance of this concept has been discussed, and image-based endmember retrieval techniques have been developed (Bateson et al., 2000; Roth et al., 2012; Somers et al., 2012), yet without large-scale application and thorough validation over space and time. Techniques like MESMA may be applied to generate standardized products for future hyperspectral missions to, for example, quantify GV, NPV, and substrate fractions. However, to achieve this, spectral libraries and revised methods are required to provide fractions that are comparable across large geographic regions, multiple years, and seasons (Dudley et al., 2015). Fire studies focused on pre-fire composition or post-fire recovery will likely be able to leverage information from these products. Specific attention will be required for active fire and severity studies because these applications need additional endmembers that are foreign to vegetation or substrate studies. Spectral signatures of active fires can be modeled from the Planck function for different temperatures (Figure 12.5). Inclusion of the active fire endmember could consist of spectral signatures from multiple temperatures or could capitalize upon iterative optimization techniques. Fire severity studies need the inclusion of a charcoal or ash endmember, which could simply be implemented by extending the three-endmember model of GV, NPV, and substrate to a four-endmember model that adds the charcoal/ash endmember. The implementation of this addition could be restricted to burned areas only, especially when a separate burned area retrieval would be available from the suite of satellite products.

12.3.2 Optimizing Hyperspectral Fire Severity Indices

Hyperspectral data are powerful for fire severity assessments because they allow accurate within-pixel fractional cover estimates of ground cover classes, among others charcoal, that are indicative of severity (Veraverbeke et al., 2014b; Lewis et al., 2017; Meng et al., 2017). A more traditional method of mapping fire and burn severity is the dNBR. The dNBR has the advantage of conceptual simplicity and computational efficiency and may therefore complement more sophisticated retrievals (Veraverbeke and Hook, 2013). The Landsat dNBR is the most often used approach to assessing fire and burn severity (Key and Benson, 2006; French et al., 2008). In multispectral remote sensing, commonly one band combination per sensor allows for the calculation of the dNBR. In hyperspectral remote sensing, however, several band combinations lead to multiple dNBR definitions that are slightly different. Perhaps there exists an optimal combination of NIR and SWIR narrowbands. So far, this exercise has not been undertaken partly because of the limited availability of pre-/post-fire image pairs required for dNBR calculation (Stavros et al., 2016). However, opportunities that arise from two recent California wildfires imaged by AVIRIS as part of the HyspIRI preparatory airborne campaign allow such an investigation. Investigations could focus on relationships with field measurements of severity and spectral index optimality for multiple band combinations (Pinty and Verstraete, 1992; Roy et al., 2006).

12.3.3 Synergy between Hyperspectral and Light Detection and Ranging Data

Hyperspectral and LiDAR data are complementary. Hyperspectral data can discriminate between fuel composition, amount, and condition, yet LiDAR data can provide additional information regarding the three-dimensional (3D) structure of fuels (Riaño et al., 2003). By doing so, the synergy between hyperspectral and LiDAR technologies can effectively be referred to as 3D imaging spectroscopy. This synergy has been explored in a few pre-fire applications (Varga and Asner, 2008; Colgan et al., 2012; Levick et al., 2015). Colgan et al. (2012) used LiDAR to identify individual tree crowns and estimate canopy height, while hyperspectral data guided tree species discrimination. This approach was successful for mapping fuels in Kruger National Park in South Africa. Levick et al. (2015) applied similar methods in the same study area and found that areas with higher fire frequency were associated with reduced tree cover and shifts in canopy height distribution. Varga and Asner (2008) combined hyperspectral and LiDAR data to map the 3D structure of grass fuels in Hawaii Volcanoes National Park, USA. They therefore combined fractional cover estimates of NPV from SMA on hyperspectral data with canopy heights from LiDAR. Their derived fire fuel index is a proxy of flammability and fire spread potential. Combined hyperspectral and LiDAR data have rarely been exploited in post-fire applications. This research gap is likely explained by shortages in synergistic image acquisitions, especially from both before and after fires. Post-fire charcoal fractional cover or dNBR derived from hyperspectral images, combined with changes in canopy height distribution, could significantly refine carbon emission estimates from fires, especially if these post-fire retrievals are supplemented with knowledge on pre-fire fuel composition and amount. Chen (2017) provided some initial insight on complementarities between the dNBR and MESMA fractional covers derived from hyperspectral imagery and canopy height derived from LiDAR in a post-fire environment (Figure 12.12). The MESMA output allows greater separation in riparian areas compared to the dNBR, whereas the canopy heights from LiDAR are indicative of residual standing biomass, even in areas mapped as ash by MESMA. Such synergistic research opportunities are rare within airborne campaigns; however, AHS and LiDAR data from before and after fires cover large parts of two recent California fires and present an ideal case study (Stavros et al., 2016).

12.3.4 Hyperspectral Thermal Applications

Hyperspectral TIR data can provide complementary information to visible and SWIR data in fire applications. These applications include detection of water- and temperature-induced stress in plant species based on spectral changes in TIR emissivity (Ullah et al., 2012; Buitrago et al., 2016; Meerdink et al., 2016) and the detection and quantification of particulate and gaseous emissions from active fires (Hulley et al., 2016; Kuai et al., 2016). Identifying plant species (Ullah et al., 2012; Meerdink et al., 2016) and detecting water- and temperature-induced stress in plant species (Buitrago et al., 2016) using spectral emissivity have so far only been demonstrated with laboratory measurements in controlled environments, but they have the potential to provide information on plant stress and moisture content and, thus, pre- and post-fire fuel condition. For example, Buitrago et al. (2016) found that plants exposed to water and temperature stress showed significant changes in their TIR spectra, which were linked to changes in cuticle thickness and structure. More work is required to apply these methods to hyperspectral TIR data from air- or spaceborne platforms.

Hyperspectral TIR measurements over active fires are presently limited due to detector saturation limits and the lack of suitable air- and spaceborne instrumentation. Wildfires have thus far not been a prime target for most airborne thermal missions, and wildfire occurrence is ephemeral, limiting acquisition windows for airborne campaigns. The Hyperspectral Thermal Emission Spectrometer

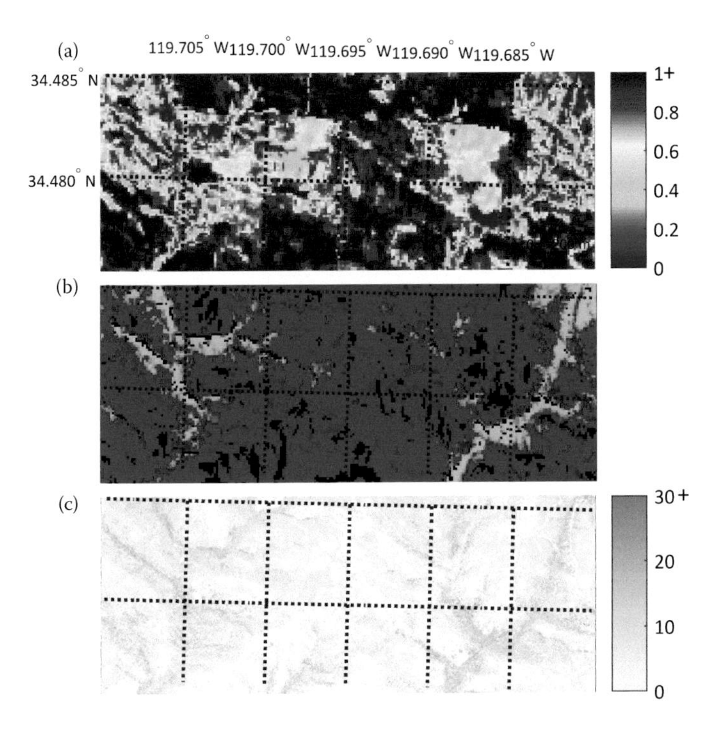

FIGURE 12.12 (a) Hyperspectral differenced NBR and (b) surface composition (as in Figure 12.11b) derived over the 2009 Jesusita burned area in California, USA, as derived from Airborne Visible/Infrared Imaging Spectrometer. (c) Canopy height model (m) derived from an airborne light detection and ranging acquisition in December 2009. (After Chen, M., 2017. *Reconstructing Fire Severity and Post-Fire Recovery in a Southern California Watershed Using Hyperspectral Imagery and LiDAR*. University of California, Santa Barbara.)

(HyTES), an airborne imaging spectrometer with high spectral resolution (256 bands between 7.5 and 12 μm), wide swath (1–2 km), and high spatial resolution (2 m at 1 km altitude flying height), has acquired data over four small active fires in California since deployment in 2013, but these have only been considered targets of opportunity acquired en route to other destinations. Hyperspectral satellite sensors like the Infrared Atmospheric Sounding Interferometer (IASI), (Aires et al., 2002), the Tropospheric Emission Spectrometer (TES), (Beer, 2006), and the Atmospheric Infrared Sounder (AIRS), (Tobin et al., 2006) have the capability to observe large gaseous emissions from fires, but they are limited by their coarse spatial resolutions of 10 km or more and insensitivity to near-surface concentrations due to sensor saturation issues. AHS TIR sensors such as HyTES, on the other hand, have the imaging capability to detect gaseous emission sources at pixel sizes of a few meters and have sufficient spectral information to resolve the spectral absorption signatures of a variety of different trace gases, including methane (CH_4), ammonia (NH_3), hydrogen sulfide (H_2S), sulfur dioxide (SO_2), nitrogen dioxide (NO_2), and nitrous oxide (N_2O) (Hulley et al., 2016; Kuai et al., 2016). The primary gas species emitted from wildfires, CO_2 and CO, do not exhibit spectral absorption features in the TIR region; however, other biomass burning gases, such as CH_4 and NH_3, are detectable with high confidence (Hulley et al., 2016). Biomass burning is a major source of atmospheric NH_3 (Hegg et al., 1988; Whitburn et al., 2015). Examples of the absorption features of CH_4 and NH_3 are shown in Figure 12.13. AHS TIR data have the ability to discriminate these gases within a single plume pixel. A further unique advantage of TIR data for fire applications is that nighttime observations allow easier detection of gas emissions since the collapsed nocturnal planetary boundary layer results in higher near-surface concentrations. In addition, the ability to detect fires is greater at night since during the day active fires can be confused with warm ground surfaces, especially with lower spatial resolution sensors.

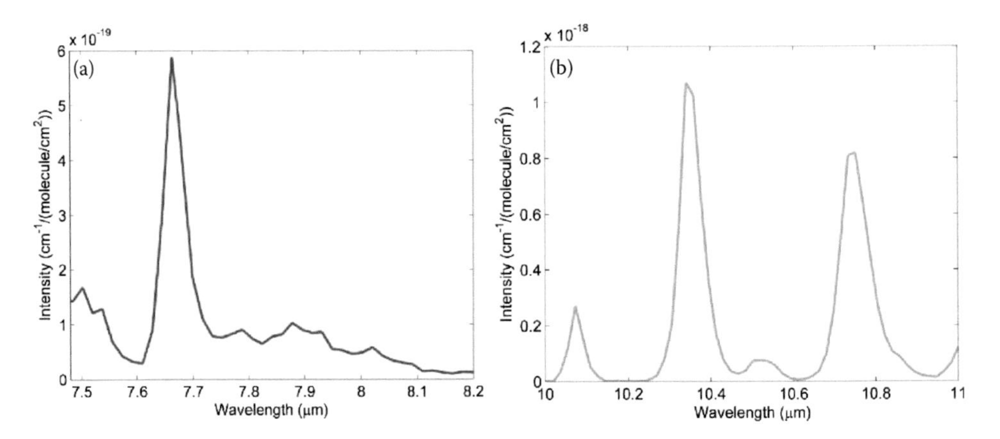

FIGURE 12.13 Absorption features of (a) CH_4 and (b) NH_3 extracted from high-resolution transmission molecular absorption 2012 database (HITRAN) (Rothman et al., 2013) convolved to spectral response functions of Hyperspectral Thermal Emission Spectrometer. Spectral regions with high intensity represent absorption features.

During active fires, hyperspectral TIR data have the ability to quantify surface and near-surface air temperature in the vicinity of the fires and downwind concentrations of NH_3 and CH_4. For the latter, a hybrid clutter matched filter and plume dilation algorithm is first used to identify target plume pixels (Hulley et al., 2016), followed by a more computationally intensive quantitative retrieval (QR) of gas concentrations. The QR algorithm has been successfully applied to CH_4 with an error between 20% and 25% and for NH_3 with an error between 50% and 80% due to uncertainties from instrument noise and spectral interferences from air temperature, surface emissivity, and atmospheric water vapor (Kuai et al., 2016).

HyTES detected an NH_3 plume over the Gulch fire, a small fire in southern Utah, USA, in July 2014 (Figure 12.14). The fire plume exhibited NH_3 mole fraction enhancements of up to 5.5 ppb. This is approximately 10 ppb lower than emissions from the El Segundo natural gas power plant in Los Angeles, USA, observed in prior HyTES campaigns. The magnitude of NH_3 and particulate emissions are primarily determined by combustion type (Yokelson et al., 1997; Reid et al., 2005; Liu et al., 2014). Incomplete combustion products include CO, CH_4, NH_3, C2–C3 hydrocarbons, methanol (CH_3OH), formic and acetic acids, and formaldehyde (CH_2O) (Yokelson et al., 1997; Bertschi et al.,

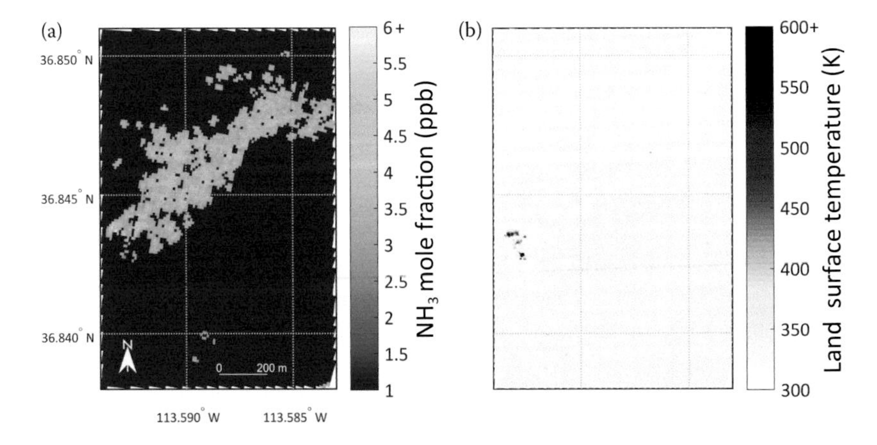

FIGURE 12.14 (a) An ammonia (NH_3) plume from fire emissions and (b) land surface temperature over active fire region in Utah, USA, in July 2014 derived from Hyperspectral Thermal Emission Spectrometer imagery.

2003). The observed NH_3 values are within expectations for the creeping and smoldering conditions and incomplete combustion of the Gulch fire.

Because TIR spectrometers rely on the thermal emission and thermal contrast between the ground and gas for detection, particulate scattering from smoke has little effect on the signal. This suggests potential for combined analysis of both the particulate and gaseous emissions from fires by flying HyTES with a multiangle polarimeter imager such as the Airborne Multi-angle Spectro Polarimetric Imager (AirMSPI) (Diner et al., 2013). AirMSPI is an airborne prototype instrument used for obtaining multiangle polarization imagery. AirMSPI could be used to assess the relative contribution of organic, non-organic, and black carbon particles to total airborne particle emissions, while HyTES could provide information on concentrations of gaseous emissions and temperature. Synergistic use of hyperspectral thermal and multiangle observations would help constrain biomass burning emissions and particulate composition of smoke to help model and predict the impacts of future emissions on air quality and climate change. Another interesting synergy is between hyperspectral VSWIR, MIR, and TIR data. The VSWIR spectral region is more sensitive to high temperatures between approximately 800 and 1500 K and, thus, ideally suited for hot flaming fires. The MIR and TIR spectral regions, in contrast, are more sensitive to lower temperatures between approximately 300 and 800 K, and thus better suited for cooler smoldering fires. The combined use of hyperspectral VSWIR, MIR, and TIR data thus offers opportunities to better characterize the full range of fire temperatures on Earth.

12.4 CONCLUSIONS

Hyperspectral remote sensing has proven utility in all temporal stages of the fire disturbance continuum. In pre-fire applications, hyperspectral data allow for detailed assessment of fuel composition, amount, and condition. Fire temperatures and gaseous emissions can be determined from active fires with hyperspectral data. After a fire, hyperspectral information from charcoal, ash, and vegetation are indicative of fire severity and ecosystem recovery. So far, hyperspectral fire applications have almost exclusively leveraged airborne data in the visible to SWIR. The number of studies is limited because airborne campaign planning generally does not include ephemeral fire occurrence. Despite the limited number of studies, these examples demonstrate the feasibility and maturity of hyperspectral data processing for large-scale applications when such data sets would become available from spaceborne platforms. Scheduled missions like EnMAP, HyspIRI, and PRISMA will provide opportunities to further explore linkages between ecosystem properties and fires at regional to global scales. The maturity of applications based on visible to SWIR regions is contrasted by upcoming innovative developments in the mid- to thermal infrared regions. Recent AHS TIR developments show the potential for significant advances in retrieving fire temperature and gaseous emissions. Further research should focus on preparing the readiness of processing techniques for large-scale hyperspectral applications in the visible to SWIR, increasing airborne acquisition and data exploration of fires with hyperspectral thermal data, and building synergistic capacities between hyperspectral data and structural data from light or radio detection and ranging instruments.

ACKNOWLEDGMENTS

We would like to thank Natalie Queally for providing input for the pre-fire figures, Erin Wetherley for the post-fire recovery figure, and Mingquan Chen for the figure showing the synergy between hyperspectral and light detection and ranging imagery. Parts of this work were carried out at Jet Propulsion Laboratory (JPL), California Institute of Technology, under a contract with the National Aeronautics and Space Administration.

REFERENCES

Abatzoglou, J.T., Kolden, C.A., 2011. Relative importance of weather and climate on wildfire growth in interior Alaska. *Int. J. Wildl. Fire* 20, 479. https://doi.org/10.1071/WF10046

Adams, J.B., Smith, M.O., Johnson, P.E., 1986. Spectral mixture modeling: A new analysis of rock and soil types at the Viking Lander 1 Site. *J. Geophys. Res.* 91, 8098. https://doi.org/10.1029/JB091iB08p08098

Agee, J.K., 1993. *Fire ecology of Pacific Northwest Forests*. Island Press, Washington, DC.

Aires, F., Rossow, W., Scott, N., Chedin, A., 2002. Remote sensing from the infrared atmospheric sounding interferometer instrument 2. Simultaneous retrieval of temperature, water vapor, and ozone atmospheric profiles. *J. Geophys. Res.* 107, 4620. https://doi.org/10.1029/2001JD001591

Airey Lauvaux, C., Skinner, C.N., Taylor, A.H., 2016. High severity fire and mixed conifer forest-chaparral dynamics in the southern Cascade Range, USA. *For. Ecol. Manage.* 363, 74–85. https://doi.org/10.1016/j.foreco.2015.12.016

Amici, S., Wooster, M.J., Piscini, A., 2011. Multi-resolution spectral analysis of wildfire potassium emission signatures using laboratory, airborne and spaceborne remote sensing. *Remote Sens. Environ.* 115, 1811–1823. https://doi.org/10.1016/j.rse.2011.02.022

Andela, N., van der Werf, G.R., 2014. Recent trends in African fires driven by cropland expansion and El Niño to La Niña transition. *Nat. Clim. Chang.* 4, 791–795. https://doi.org/10.1038/nclimate2313

Andela, N., Morton, D.C., Giglio, L., Chen, Y., van der Werf, G.R., Kasibhatla, P.S., DeFries, R.S. et al., 2017. A human-driven decline in global burned area. *Science* 80, 356.

Anderson, H.E., 1970. Forest fuel ignitibility. *Fire. Technol.* 6, 312–319. https://doi.org/10.1007/BF02588932

Anderson, M.C., Neale, C.M.U., Li, F., Norman, J.M., Kustas, W.P., Jayanthi, H., Chavez, J., 2004. Upscaling ground observations of vegetation water content, canopy height, and leaf area index during SMEX02 using aircraft and Landsat imagery. *Remote Sens. Environ.* 92, 447–464. https://doi.org/10.1016/j.rse.2004.03.019

Asner, G.P., Wessman, C.A., Bateson, C.A., Privette, J.L., 2000. Impact of tissue, canopy, and landscape factors on the hyperspectral reflectance variability of arid ecosystems. *Remote Sens. Environ.* 74, 69–84. https://doi.org/10.1016/S0034-4257(00)00124-3

Baldridge, A.M., Hook, S.J., Grove, C.I., Rivera, G., 2009. The ASTER spectral library version 2.0. *Remote Sens. Environ.* 113, 711–715. https://doi.org/10.1016/j.rse.2008.11.007

Barro, S.C., Conard, S.G., 1991. Fire effects on California chaparral systems: An overview. *Environ. Int.* 17, 135–149. https://doi.org/10.1016/0160-4120(91)90096-9

Bartholomé, E., Belward, A.S., 2005. GLC2000: A new approach to global land cover mapping from Earth observation data. *Int. J. Remote Sens.* 26, 1959–1977. https://doi.org/10.1080/01431160412331291297

Bateson, C.A., Asner, G.P., Wessman, C.A., 2000. Endmember bundles: A new approach to incorporating endmember variability into spectral mixture analysis. *IEEE Trans. Geosci. Remote Sens.* 38, 1083–1094. https://doi.org/10.1109/36.841987

Baugh, W.M., Kruse, F.A., Atkinson, W.W., 1998. Quantitative geochemical mapping of ammonium minerals in the southern Cedar mountains, Nevada, using the Airborne Visible/Infrared Imaging Spectrometer (AVIRIS). *Remote Sens. Environ.* 65, 292–308. https://doi.org/10.1016/S0034-4257(98)00039-X

Beer, R., 2006. TES on the aura mission: Scientific objectives, measurements, and analysis overview. *IEEE Trans. Geosci. Remote Sens.* 44, 1102–1105. https://doi.org/10.1109/TGRS.2005.863716

Benz, U.C., Hofmann, P., Willhauck, G., Lingenfelder, I., Heynen, M., 2004. Multi-resolution, object-oriented fuzzy analysis of remote sensing data for GIS-ready information. *ISPRS J. Photogramm. Remote Sens.* 58, 239–258. https://doi.org/10.1016/j.isprsjprs.2003.10.002

Bertschi, I., Yokelson, R.J., Ward, D.E., Babbitt, R.E., Susott, R.A., Goode, J.G., Hao, W.M., 2003. Trace gas and particle emissions from fires in large diameter and belowground biomass fuels. *J. Geophys. Res. Atmos.* 108. https://doi.org/10.1029/2002JD002100

Boer, M., Macfarlane, C., Norris, J., Sadler, R., Wallace, J., Grierson, P., 2008. Mapping burned areas and burn severity patterns in SW Australian eucalypt forest using remotely-sensed changes in leaf area index. *Remote Sens. Environ.* 112, 4358–4369. https://doi.org/10.1016/j.rse.2008.08.005

Bond, W., Keeley, J., 2005. Fire as a global "herbivore": The ecology and evolution of flammable ecosystems. *Trends Ecol. Evol.* 20, 387–394. https://doi.org/10.1016/j.tree.2005.04.025

Boucher, Y., Perrault-Hébert, M., Fournier, R., Drapeau, P., Auger, I., 2017. Cumulative patterns of logging and fire (1940–2009): Consequences on the structure of the eastern Canadian boreal forest. *Landsc. Ecol.* 32, 361–375. https://doi.org/10.1007/s10980-016-0448-9

Bowman, D.M.J.S., Balch, J.K., Artaxo, P., Bond, W.J., Carlson, J.M., Cochrane, M.A., D'Antonio, C.M. et al., 2009. Fire in the earth system. *Science* 324, 481–4. https://doi.org/10.1126/science.1163886

Buitrago, M.F., Groen, T.A., Hecker, C.A., Skidmore, A.K., 2016. Changes in thermal infrared spectra of plants caused by temperature and water stress. *ISPRS J. Photogramm. Remote Sens.* 111, 22–31. https://doi.org/10.1016/j.isprsjprs.2015.11.003

Capitanio, R., Carcaillet, C., 2008. Post-fire Mediterranean vegetation dynamics and diversity: A discussion of succession models. *For. Ecol. Manage.* 255, 431–439. https://doi.org/10.1016/j.foreco.2007.09.010

Chandler, C., Cheney, P., Thomas, P., Trabaud, L., Williams, D., 1983. *Fire in Forestry.* John Wiley & Sons, Inc., New York.

Chen, M., 2017. *Reconstructing fire Severity and Post-Fire Recovery in a Southern California Watershed Using Hyperspectral Imagery and LiDAR.* University of California, Santa Barbara.

Chu, T., Guo, X., 2013. Remote Sensing Techniques in Monitoring Post-Fire Effects and Patterns of Forest Recovery in Boreal Forest Regions: A Review. *Remote Sens.* 6, 470–520. https://doi.org/10.3390/rs6010470

Chuvieco, E., Aguado, I.A., Cocero, D., Rian, D., Riaño, D., 2003. Design of an empirical index to estimate fuel moisture content from NOAA-AVHRR images in forest fire danger studies. *Int. J. Remote Sens.* 24, 1621–1637.

Chuvieco, E., Cocero, D., Riaño, D., Martin, P., Martínez-Vega, J., de la Riva, J., Pérez, F., 2004. Combining NDVI and surface temperature for the estimation of live fuel moisture content in forest fire danger rating. *Remote Sens. Environ.* 92, 322–331. https://doi.org/10.1016/j.rse.2004.01.019

Colgan, M., Baldeck, C., Féret, J.-B., Asner, G., 2012. Mapping savanna tree species at ecosystem scales using support vector machine classification and BRDF correction on airborne hyperspectral and LiDAR data. *Remote Sens.* 4, 3462–3480. https://doi.org/10.3390/rs4113462

Dennison, P.E., Roberts, D.A., 2003. The effects of vegetation phenology on endmember selection and species mapping in southern California chaparral. *Remote Sens. Environ.* 87, 295–309. https://doi.org/10.1016/j.rse.2003.07.001

Dennison, P.E., 2006. Fire detection in imaging spectrometer data using atmospheric carbon dioxide absorption. *Int. J. Remote Sens.* 27, 3049–3055. https://doi.org/10.1080/01431160600660871

Dennison, P., Charoensiri, K., Roberts, D., Peterson, S., Green, R., 2006. Wildfire temperature and land cover modeling using hyperspectral data. *Remote Sens. Environ.* 100, 212–222. https://doi.org/10.1016/j.rse.2005.10.007

Dennison, P.E., Roberts, D.A., 2009. Daytime fire detection using airborne hyperspectral data. *Remote Sens. Environ.* 113, 1646–1657. https://doi.org/10.1016/j.rse.2009.03.010

Dennison, P.E., Matheson, D.S., 2011. Comparison of fire temperature and fractional area modeled from SWIR, MIR, and TIR multispectral and SWIR hyperspectral airborne data. *Remote Sens. Environ.* 115, 876–886. https://doi.org/10.1016/j.rse.2010.11.015

Dennison, P.E., Thorpe, A.K., Pardyjak, E.R., Roberts, D.A., Qi, Y., Green, R.O., Bradley, E.S., Funk, C.C., 2013. High spatial resolution mapping of elevated atmospheric carbon dioxide using airborne imaging spectroscopy: Radiative transfer modeling and power plant plume detection. *Remote Sens. Environ.* 139, 116–129. https://doi.org/10.1016/j.rse.2013.08.001

Dennison, P.E., Brewer, S.C., Arnold, J.D., Moritz, M.A., 2014. Large wildfire trends in the western United States, 1984–2011. *Geophys. Res. Lett.* 41, 2928–2933. https://doi.org/10.1002/2014GL059576

De Santis, A., Chuvieco, E., 2009. GeoCBI: A modified version of the Composite Burn Index for the initial assessment of the short-term burn severity from remotely sensed data. *Remote Sens. Environ.* 113, 554–562. https://doi.org/10.1016/j.rse.2008.10.011

De Santis, A., Asner, G.P., Vaughan, P.J., Knapp, D.E., 2010. Mapping burn severity and burning efficiency in California using simulation models and Landsat imagery. *Remote Sens. Environ.* 114, 1535–1545. https://doi.org/10.1016/j.rse.2010.02.008

Diakakis, M., Nikolopoulos, E.I., Mavroulis, S., Vassilakis, E., Korakaki, E., 2017. Observational evidence on the effects of mega-fires on the frequency of hydrogeomorphic hazards. The case of the Peloponnese fires of 2007 in Greece. *Sci. Total Environ.* 592, 262–276. https://doi.org/10.1016/j.scitotenv.2017.03.070

Díaz-Delgado, R., Lloret, F., Pons, X., 2003. Influence of fire severity on plant regeneration by means of remote sensing imagery. *Int. J. Remote Sens.* 24, 1751–1763. https://doi.org/10.1080/01431160210144732

Diner, D.J., Xu, F., Garay, M.J., Martonchik, J. V., Rheingans, B.E., Geier, S., Davis, A. et al., 2013. The Airborne Multiangle SpectroPolarimetric Imager (AirMSPI): A new tool for aerosol and cloud remote sensing. *Atmos. Meas. Tech* 6, 2007–2025. https://doi.org/10.5194/amt-6-2007-2013

Doerr, S.H., Shakesby, R.A., Blake, W.H., Chafer, C.J., Humphreys, G.S., Wallbrink, P.J., 2006. Effects of differing wildfire severities on soil wettability and implications for hydrological response. *J. Hydrol.* 319, 295–311. https://doi.org/10.1016/j.jhydrol.2005.06.038

Dong, J., Kaufmann, R.K., Myneni, R.B., Tucker, C.J., Kauppi, P.E., Liski, J., Buermann, W., Alexeyev, V., Hughes, M.K., 2003. Remote sensing estimates of boreal and temperate forest woody biomass: Carbon pools, sources, and sinks. *Remote Sens. Environ.* 84, 393–410. https://doi.org/10.1016/S0034-4257(02)00130-X

Dozier, J., Green, R.O., Nolin, A.W., Painter, T.H., 2009. Interpretation of snow properties from imaging spectrometry. *Remote Sens. Environ.* 113, S25–S37. https://doi.org/10.1016/j.rse.2007.07.029

Dudley, K., Dennison, P.E., Roth, K., Roberts, D.A., Coates, A., 2015. A multi-temporal spectral library approach for mapping vegetation species across spatial and temporal phenological gradients. *Remote Sens. Environ.* 167, 121–134. https://doi.org/10.1016/J.RSE.2015.05.004

Eckmann, T., Roberts, D., Still, C., 2008. Using multiple endmember spectral mixture analysis to retrieve subpixel fire properties from MODIS. *Remote Sens. Environ.* 112, 3773–3783. https://doi.org/10.1016/j.rse.2008.05.008

Eckmann, T.C., Roberts, D.A., Still, C.J., 2009. Estimating subpixel fire sizes and temperatures from ASTER using multiple endmember spectral mixture analysis. *Int. J. Remote Sens.* 30, 5851–5864. https://doi.org/10.1080/01431160902748531

Eidenshink, J., Schwind, B., Brewer, K., Zhu, Z., Quayle, B., Howard, S., 2007. A project for monitoring trends in burn severity. *Fire Ecol.* 3, 3–21.

Elvidge, C.D., Chen, Z., 1995. Comparison of broad-band and narrow-band red and near-infrared vegetation indices. *Remote Sens. Environ.* 54, 38–48. https://doi.org/10.1016/0034-4257(95)00132-K

Feingersh, T., Ben Dor, E., 2016. SHALOM—A commercial hyperspectral space mission, in: Qian, S. (Ed.), *Optical Payloads for Space Missions*. John Wiley & Sons, Pondicherry, India, pp. 247–263.

Fernandez-Manso, O., Quintano, C., Frenandez-Manso, A., 2009. Combining spectral mixture analysis and object-based classification for fire severity mapping. *Investig. Agrar. Y Recur. For.* 18, 296–313.

Forkel, M., Thonicke, K., Beer, C., Cramer, W., Bartalev, S., Schmullius, C., 2012. Extreme fire events are related to previous-year surface moisture conditions in permafrost-underlain larch forests of Siberia. *Environ. Res. Lett.* 7, 44021. https://doi.org/10.1088/1748-9326/7/4/044021

French, N.H.F., Kasischke, E.S., Hall, R.J., Murphy, K.A., Verbyla, D.L., Hoy, E.E., Allen, J.L., 2008. Using Landsat data to assess fire and burn severity in the North American boreal forest region: An overview and summary of results. *Int. J. Wildl. Fire* 17, 443. https://doi.org/10.1071/WF08007

Friedl, M., McIver, D., Hodges, J.C., Zhang, X., Muchoney, D., Strahler, A., Woodcock, C. et al., 2002. Global land cover mapping from MODIS: Algorithms and early results. *Remote Sens. Environ.* 83, 287–302. https://doi.org/10.1016/S0034-4257(02)00078-0

Gamon, J.A., Serrano, L., Surfus, J.S., 1997. The photochemical reflectance index: An optical indicator of photosynthetic radiation use efficiency across species, functional types, and nutrient levels. *Oecologia* 112, 492–501. https://doi.org/10.1007/s004420050337

Gao, B.-C., Goetz, A.F.H., 1990. Column atmospheric water vapor and vegetation liquid water retrievals from Airborne Imaging Spectrometer data. *J. Geophys. Res.* 95, 3549. https://doi.org/10.1029/JD095iD04p03549

Gao, B.-C., 1995. A normalized difference water index for remote sensing of vegetation liquid water from space, in: *Proc. SPIE 2480, Imaging Spectrometry, 225* (June 12, 1995). Orlando, Florida, USA, pp. 225–236. https://doi.org/10.1117/12.210877

Gao, B.-C., Goetz, A.F.H., 1995. Retrieval of equivalent water thickness and information related to biochemical components of vegetation canopies from AVIRIS data. *Remote Sens. Environ.* 52, 155–162. https://doi.org/10.1016/0034-4257(95)00039-4

Giglio, L., Kendall, J.D., 2001. Application of the Dozier retrieval to wildfire characterization: A sensitivity analysis. *Remote Sens. Environ.* 77, 34–49. https://doi.org/10.1016/S0034-4257(01)00192-4

Giglio, L., Descloitres, J., Justice, C.O., Kaufman, Y.J., 2003. An Enhanced Contextual Fire Detection Algorithm for MODIS. *Remote Sens. Environ.* 87, 273–282. https://doi.org/10.1016/S0034-4257(03)00184-6

Giglio, L., Loboda, T., Roy, D.P., Quayle, B., Justice, C.O., 2009. An active-fire based burned area mapping algorithm for the MODIS sensor. *Remote Sens. Environ.* 113, 408–420. https://doi.org/10.1016/j.rse.2008.10.006

Giglio, L., Schroeder, W., 2014. A global feasibility assessment of the bi-spectral fire temperature and area retrieval using MODIS data. *Remote Sens. Environ.* 152, 166–173. https://doi.org/10.1016/j.rse.2014.06.010

Gillett, N.P., Weaver, A.J., Zwiers, F.W., Flannigan, M.D., 2004. Detecting the effect of climate change on Canadian forest fires. *Geophys. Res. Lett.* 31, L18211. https://doi.org/10.1029/2004GL020876

Gitas, I.Z., Polychronaki, A., Katagis, T., Mallinis, G., 2008. Contribution of remote sensing to disaster management activities: A case study of the large fires in the Peloponnese, Greece. *Int. J. Remote Sens.* 29, 1847–1853. https://doi.org/10.1080/01431160701874553

Gitas, I., Mitri, G., Veraverbeke, S., Polychronaki, A., 2012. Advances in remote sensing of post-fire vegetation recovery monitoring–a review. *Remote Sens. Biomass–Principles Appl.* 322.

Gitelson, A.A., Peng, Y., Arkebauer, T.J., Schepers, J., 2014. Relationships between gross primary production, green LAI, and canopy chlorophyll content in maize: Implications for remote sensing of primary production. *Remote Sens. Environ.* 144, 65–72. https://doi.org/10.1016/j.rse.2014.01.004

Goetz, S.J., Fiske, G.J., Bunn, A.G., 2006. Using satellite time-series data sets to analyze fire disturbance and forest recovery across Canada. *Remote Sens. Environ.* 101, 352–365. https://doi.org/10.1016/j.rse.2006.01.011

Gouveia, C., DaCamara, C.C., Trigo, R.M., 2010. Post-fire vegetation recovery in Portugal based on spot/ vegetation data. *Nat. Hazards Earth Syst. Sci.* 10, 673–684. https://doi.org/10.5194/nhess-10-673-2010

Green, R.O., Eastwood, M.L., Sarture, C.M., Chrien, T.G., Aronsson, M., Chippendale, B.J., Faust, J.A. et al., 1998. Imaging spectroscopy and the Airborne Visible/Infrared Imaging Spectrometer (AVIRIS). *Remote Sens. Environ.* 65, 227–248. https://doi.org/10.1016/S0034-4257(98)00064-9

Hansen, M.C., Reed, B., 2000. A comparison of the IGBP DISCover and University of Maryland 1 km global land cover products. *Int. J. Remote Sens.* 21, 1365–1373. https://doi.org/10.1080/014311600210218

Hanson, C.T., Odion, D.C., 2014. Is fire severity increasing in the Sierra Nevada, California, USA? *Int. J. Wildl. Fire* 23, 1. https://doi.org/10.1071/WF13016

Hegg, D.A., Radke, L.F., Hobbs, P. V., Riggan, P.J., 1988. Ammonia emissions from biomass burning. *Geophys. Res. Lett.* 15, 335–337. https://doi.org/10.1029/GL015i004p00335

Hernández-Clemente, R., Navarro-Cerrillo, R.M., Suárez, L., Morales, F., Zarco-Tejada, P.J., 2011. Assessing structural effects on PRI for stress detection in conifer forests. *Remote Sens. Environ.* 115, 2360–2375. https://doi.org/10.1016/j.rse.2011.04.036

Hernandez-Clemente, R., Kolari, P., Porcar-Castell, A., Korhonen, L., Mottus, M., 2016. Tracking the Seasonal Dynamics of Boreal Forest Photosynthesis Using EO-1 Hyperion Reflectance: Sensitivity to Structural and Illumination Effects. *IEEE Trans. Geosci. Remote Sens.* 54, 5105–5116. https://doi.org/10.1109/ TGRS.2016.2554466

Hoogenboom, H.J., Dekker, A.G., Althuis, I.A., 1998. Simulation of AVIRIS sensitivity for detecting chlorophyll over coastal and inland waters. *Remote Sens. Environ.* 65, 333–340. https://doi.org/10.1016/ S0034-4257(98)00042-X

Hook, S.J., Elvidge, C.D., Rast, M., Watanabe, H., 1991. An evaluation of short-wave-infrared (SWIR) data from the AVIRIS and GEOSCAN instruments for mineralogical mapping at Cuprite, Nevada. *Geophysics* 56, 1432–1440. https://doi.org/10.1190/1.1443163

Hook, S.J., Myers, J.J., Thome, K.J., Fitzgerald, M., Kahle, A.B., 2001. The MODIS/ASTER airborne simulator (MASTER)—A new instrument for earth science studies. *Remote Sens. Environ.* 76, 93–102. https://doi. org/10.1016/S0034-4257(00)00195-4

Huang, S., Crabtree, R.L., Potter, C., Gross, P., 2009. Estimating the quantity and quality of coarse woody debris in Yellowstone post-fire forest ecosystem from fusion of SAR and optical data. *Remote Sens. Environ.* 113, 1926–1938. https://doi.org/10.1016/j.rse.2009.05.001

Huesca, M., Merino-de-Miguel, S., González-Alonso, F., Martínez, S., Miguel Cuevas, J., Calle, A., 2013. Using AHS hyper-spectral images to study forest vegetation recovery after a fire. *Int. J. Remote Sens.* 34, 4025–4048. https://doi.org/10.1080/01431161.2013.772313

Hulley, G.C., Duren, R.M., Hopkins, F.M., Hook, S.J., Vance, N., Guillevic, P., Johnson, W.R. et al., 2016. High spatial resolution imaging of methane and other trace gases with the airborne Hyperspectral Thermal Emission Spectrometer (HyTES). *Atmos. Meas. Tech.* 9, 2393–2408. https://doi.org/10.5194/ amt-9-2393-2016

Itten, K.I., Dell'Endice, F., Hueni, A., Kneubühler, M., Schläpfer, D., Odermatt, D., Seidel, F. et al., 2008. APEX—The hyperspectral ESA Airborne Prism Experiment. *Sensors* 8, 6235–6259. https://doi. org/10.3390/s8106235

Iwasaki, A., Ohgi, N., Tanii, J., Kawashima, T., Inada, H., 2011. Hyperspectral imager suite (HISUI) -Japanese hyper-multi spectral radiometer, in: *2011 IEEE International Geoscience and Remote Sensing Symposium*. IEEE, pp. 1025–1028. https://doi.org/10.1109/IGARSS.2011.6049308

Jackson, T.J., Chen, D., Cosh, M., Li, F., Anderson, M., Walthall, C., Doriaswamy, P., Hunt, E.R., 2004. Vegetation water content mapping using Landsat data derived normalized difference water index for corn and soybeans. *Remote Sens. Environ.* 92, 475–482. https://doi.org/10.1016/j.rse.2003.10.021

Jain, T., Graham, R., Pilliod, D., 2004. Tongue-tied: Confused meanings for common fire terminology can lead to fuels mismanagement. Wildfire July/August, 22–26.

Johnson, L.F., Hlavka, C.A., Peterson, D.L., 1994. Multivariate analysis of AVIRIS data for canopy biochemical estimation along the oregon transect. *Remote Sens. Environ.* 47, 216–230. https://doi. org/10.1016/0034-4257(94)90157-0

Johnstone, J.F., Kasischke, E.S., 2005. Stand-level effects of soil burn severity on postfire regeneration in a recently burned black spruce forest. *Can. J. For. Res.* 35, 2151–2163. https://doi.org/10.1139/x05-087

Katagis, T., Gitas, I.Z., Toukiloglou, P., Veraverbeke, S., Goossens, R., 2014. Trend analysis of medium- and coarse-resolution time series image data for burned area mapping in a Mediterranean ecosystem. *Int. J. Wildl. Fire* 23, 668–677.

Keeley, J.E., 2009. Fire intensity, fire severity and burn severity: A brief review and suggested usage. *Int. J. Wildl. Fire* 18, 116. https://doi.org/10.1071/WF07049

Kennedy, M.C., Ford, E.D., Singleton, P., Finney, M.A., Agee, J.K., 2008. Informed multi-objective decision-making in environmental management using Pareto optimality. *J. Appl. Ecol.* 45, 181–192. https://doi.org/10.1111/j.1365-2664.2007.01367.x

Keppel-Aleks, G., Wolf, A.S., Mu, M., Doney, S.C., Morton, D.C., Kasibhatla, P.S., Miller, J.B., Dlugokencky, E.J., Randerson, J.T., 2014. Separating the influence of temperature, drought, and fire on interannual variability in atmospheric CO2. *Global Biogeochem. Cycles* 28, 1295–1310. https://doi.org/10.1002/2014GB004890

Key, C.H., Benson, N.C., 2006. Landscape assessment: Ground measure of severity; the Composite Burn Index, and remote sensing of severity, the Normalized Burn Index, in: Lutes, D., Keane, R., Caratti, J., Key, C., Benson, N., Sutherland, S., Grangi, L. (Eds.), *FIREMON: Fire Effects Monitoring and Inventory System.* USDA Forest Service, pp. 1–51.

Koch, B., 2010. Status and future of laser scanning, synthetic aperture radar and hyperspectral remote sensing data for forest biomass assessment. *ISPRS J. Photogramm. Remote Sens.* 65, 581–590. https://doi.org/10.1016/j.isprsjprs.2010.09.001

Kokaly, R.F., Rockwell, B.W., Haire, S.L., King, T.V.V., 2007. Characterization of post-fire surface cover, soils, and burn severity at the Cerro Grande Fire, New Mexico, using hyperspectral and multispectral remote sensing. *Remote Sens. Environ.* 106, 305–325. https://doi.org/10.1016/j.rse.2006.08.006

Kuai, L., Worden, J.R., Li, K.-F., Hulley, G.C., Hopkins, F.M., Miller, C.E., Hook, S.J., Duren, R.M., Aubrey, A.D., 2016. Characterization of anthropogenic methane plumes with the Hyperspectral Thermal Emission Spectrometer (HyTES): A retrieval method and error analysis. *Atmos. Meas. Tech.* 9, 3165–3173. https://doi.org/10.5194/amt-9-3165-2016

Kudela, R.M., Palacios, S.L., Austerberry, D.C., Accorsi, E.K., Guild, L.S., Torres-Perez, J., 2015. Application of hyperspectral remote sensing to cyanobacterial blooms in inland waters. *Remote Sens. Environ.* 167, 196–205. https://doi.org/10.1016/j.rse.2015.01.025

Labate, D., Ceccherini, M., Cisbani, A., De Cosmo, V., Galeazzi, C., Giunti, L., Melozzi, M., Pieraccini, S., Stagi, M., 2009. The PRISMA payload optomechanical design, a high performance instrument for a new hyperspectral mission. *Acta Astronaut.* 65, 1429–1436. https://doi.org/10.1016/j.actaastro.2009.03.077

Larsson, H., 1993. Linear regressions for canopy cover estimation in Acacia woodlands using Landsat-TM, -MSS and SPOT HRV XS data. *Int. J. Remote Sens.* 14, 2129–2136. https://doi.org/10.1080/01431169308954025

Lee, C.M., Cable, M.L., Hook, S.J., Green, R.O., Ustin, S.L., Mandl, D.J., Middleton, E.M., 2015. An introduction to the NASA Hyperspectral InfraRed Imager (HyspIRI) mission and preparatory activities. *Remote Sens. Environ.* 167, 6–19. https://doi.org/10.1016/j.rse.2015.06.012

Lehmann, C.E.R., Anderson, T.M., Sankaran, M., Higgins, S.I., Archibald, S., Hoffmann, W.A., Hanan, N.P. et al., 2014. Savanna vegetation-fire-climate relationships differ among continents. *Science (80).* 343, 548–552. https://doi.org/10.1126/science.1247355

Lentile, L.B., Holden, Z.A., Smith, A.M.S., Falkowski, M.J., Hudak, A.T., Morgan, P., Lewis, S.A., Gessler, P.E., Benson, N.C., 2006. Remote sensing techniques to assess active fire characteristics and post-fire effects. *Int. J. Wildl. Fire* 15, 319. https://doi.org/10.1071/WF05097

Leon, J.R.R., van Leeuwen, W.J.D., Casady, G.M., 2012. Using MODIS-NDVI for the Modeling of Post-Wildfire Vegetation Response as a Function of Environmental Conditions and Pre-Fire Restoration Treatments. *Remote Sens.* 4, 598–621. https://doi.org/10.3390/rs4030598

Levick, S.R., Baldeck, C.A., Asner, G.P., 2015. Demographic legacies of fire history in an African savanna. *Funct. Ecol.* 29, 131–139. https://doi.org/10.1111/1365-2435.12306

Lewis, S.A., Lentile, L.B., Hudak, A.T., Robichaud, P.R., Morgan, P., Bobbitt, M.J., 2007. Mapping ground cover using hyperspectral remote sensing after the 2003 Simi and Old wildfires in Southern California. *Fire Ecol.* 3, 109–128. https://doi.org/10.4996/fireecology.0301109

Lewis, S.A., Robichaud, P.R., Frazier, B.E., Wu, J.Q., Laes, D.Y.M., 2008. Using hyperspectral imagery to predict post-wildfire soil water repellency. *Geomorphology* 95, 192–205. https://doi.org/10.1016/j.geomorph.2007.06.002

Lewis, S.A., Hudak, A.T., Ottmar, R.D., Robichaud, P.R., Lentile, L.B., Hood, S.M., Cronan, J.B., Morgan, P., 2011. Using hyperspectral imagery to estimate forest floor consumption from wildfire in boreal forests of Alaska, USA. *Int. J. Wildl. Fire* 20, 255. https://doi.org/10.1071/WF09081

Lewis, S.A., Hudak, A.T., Robichaud, P.R., Morgan, P., Satterberg, K.L., Strand, E.K., Smith, A.M.S., Zamudio, J.A., Lentile, L.B., 2017. Indicators of burn severity at extended temporal scales: A decade of ecosystem response in mixed-conifer forests of western Montana. *Int. J. Wildl. Fire* 26, 755. https://doi.org/10.1071/WF17019

Liu, W.T., Kogan, F.N., 1996. Monitoring regional drought using the Vegetation Condition Index. *Int. J. Remote Sens.* 17, 2761–2782. https://doi.org/10.1080/01431169608949106

Liu, S., Aiken, A.C., Arata, C., Dubey, M.K., Stockwell, C.E., Yokelson, R.J., Stone, E.A. et al., 2014. Aerosol single scattering albedo dependence on biomass combustion efficiency: Laboratory and field studies. *Geophys. Res. Lett.* 41, 742–748. https://doi.org/10.1002/2013GL058392

López García, M.J., Caselles, V., 1991. Mapping burns and natural reforestation using thematic Mapper data. *Geocarto Int.* 6, 31–37. https://doi.org/10.1080/10106049109354290

Loveland, T.R., Reed, B.C., Brown, J.F., Ohlen, D.O., Zhu, Z., Yang, L., Merchant, J.W., 2000. Development of a global land cover characteristics database and IGBP DISCover from 1 km AVHRR data. *Int. J. Remote Sens.* 21, 1303–1330. https://doi.org/10.1080/014311600210191

Marino, E., Ranz, P., Tomé, J.L., Noriega, M.Á., Esteban, J., Madrigal, J., 2016. Generation of high-resolution fuel model maps from discrete airborne laser scanner and Landsat-8 OLI: A low-cost and highly updated methodology for large areas. *Remote Sens. Environ.* 187, 267–280. https://doi.org/10.1016/j.rse.2016.10.020

Matheson, D.S., Dennison, P.E., 2012. Evaluating the effects of spatial resolution on hyperspectral fire detection and temperature retrieval. *Remote Sens. Environ.* 124, 780–792. https://doi.org/10.1016/j.rse.2012.06.026

Matson, M., Holben, B., 1987. Satellite detection of tropical burning in Brazil. *Int. J. Remote Sens.* 8, 509–516. https://doi.org/10.1080/01431168708948657

McDowell, N., Pockman, W.T., Allen, C.D., Breshears, D.D., Cobb, N., Kolb, T., Plaut, J. et al., 2008. Mechanisms of plant survival and mortality during drought: Why do some plants survive while others succumb to drought? *New Phytol.* 178, 719–739. https://doi.org/10.1111/j.1469-8137.2008.02436.x

McElhinny, C., Gibbons, P., Brack, C., Bauhus, J., 2005. Forest and woodland stand structural complexity: Its definition and measurement. *For. Ecol. Manage.* 218, 1–24. https://doi.org/10.1016/j.foreco.2005.08.034

Meerdink, S.K., Roberts, D.A., King, J.Y., Roth, K.L., Dennison, P.E., Amaral, C.H., Hook, S.J., 2016. Linking seasonal foliar traits to VSWIR-TIR spectroscopy across California ecosystems. *Remote Sens. Environ.* 186, 322–338. https://doi.org/10.1016/j.rse.2016.08.003

Meng, R., Wu, J., Schwager, K., Zhao, F., Dennison, P., Cook, B., Brewster, K., Green, T., Serbin, S., 2017. Using high spatial resolution satellite imagery to map forest burn severity across spatial scales in a Pine Barrens ecosystem. *Remote Sens. Environ.* 191, 95–109.

Mitri, G.H., Gitas, I.Z., 2006. Fire type mapping using object-based classification of Ikonos imagery. *Int. J. Wildl. Fire* 15, 457. https://doi.org/10.1071/WF05085

Mitri, G.H., Gitas, I.Z., 2010. Mapping Postfire Vegetation Recovery Using EO-1 Hyperion Imagery. *Geosci. Remote Sensing, IEEE Trans.* 48, 1613–1618. https://doi.org/10.1109/TGRS.2009.2031557

Mitri, G.H., Gitas, I.Z., 2012. Mapping post-fire forest regeneration and vegetation recovery using a combination of very high spatial resolution and hyperspectral satellite imagery. *Int. J. Appl. Earth Obs. Geoinf.* 20, 60–66. https://doi.org/10.1016/j.jag.2011.09.001

Moreira, F., Catry, F., Duarte, I., Acácio, V., Silva, J.S., 2009. A conceptual model of sprouting responses in relation to fire damage: An example with cork oak (Quercus suber L.) trees in Southern Portugal, in: *Forest Ecology: Recent Advances in Plant Ecology*. pp. 77–85. https://doi.org/10.1007/978-90-481-2795-5_7

Morgan, P., Keane, R.E., Dillon, G.K., Jain, T.B., Hudak, A.T., Karau, E.C., Sikkink, P.G., Holden, Z.A., Strand, E.K., 2014. Challenges of assessing fire and burn severity using field measures, remote sensing and modelling. *Int. J. Wildl. Fire* 23, 1045. https://doi.org/10.1071/WF13058

Nelson, K.J., Connot, J., Peterson, B., Martin, C., 2013. The LANDFIRE refresh Strategy: Updating the National dataset. *Fire Ecol.* 9, 80–101. https://doi.org/10.4996/fireecology.090280

Numata, I., Cochrane, M.A., Galvão, L.S., Numata, I., Cochrane, M.A., Galvão, L.S., 2011. Analyzing the Impacts of Frequency and Severity of Forest Fire on the Recovery of Disturbed Forest using Landsat Time Series and EO-1 Hyperion in the Southern Brazilian Amazon. *Earth Interact.* 15, 1–17. https://doi.org/10.1175/2010EI372.1

Oertel, D., Zhukov, B., Thamm, H.-P., Roehrig, J., Orthmann, B., 2004. Space-borne high resolution fire remote sensing in Benin, West Africa. *Int. J. Remote Sens.* 25, 2209–2216. https://doi.org/10.1080/01431160320001647741

O'Halloran, T.L., Law, B.E., Goulden, M.L., Wang, Z., Barr, J.G., Schaaf, C., Brown, M. et al., 2012. Radiative forcing of natural forest disturbances. *Glob. Chang. Biol.* 18, 555–565. https://doi.org/10.1111/j.1365-2486.2011.02577.x

Painter, T.H., Dozier, J., Roberts, D.A., Davis, R.E., Green, R.O., 2003. Retrieval of subpixel snow-covered area and grain size from imaging spectrometer data. *Remote Sens. Environ.* 85, 64–77. https://doi.org/10.1016/S0034-4257(02)00187-6

Parks, S.A., 2014. Mapping day-of-burning with coarse-resolution satellite fire-detection data. *Int. J. Wildl. Fire* 23, 215. https://doi.org/10.1071/WF13138

Pausas, J., Fernández-Muñoz, S., 2012. Fire regime changes in the Western Mediterranean Basin: From fuel-limited to drought-driven fire regime. *Clim. Change* 110, 215–226. https://doi.org/10.1007/s10584-011-0060-6

Pearlman, J.S., Barry, P.S., Segal, C.C., Shepanski, J., Beiso, D., Carman, S.L., 2003. Hyperion, a space-based imaging spectrometer. *IEEE Trans. Geosci. Remote Sens.* 41, 1160–1173. https://doi.org/10.1109/TGRS.2003.815018

Pérez-Cabello, F., Echeverría, M.T., Ibarra, P., de la Riva, J., 2009. Effects of fire on vegetation, soil and hydrogeomorphological behavior in Mediterranean ecosystems, in: *Earth Observation of Wildland Fires in Mediterranean Ecosystems*. Springer, Berlin Heidelberg, pp. 111–128. https://doi.org/10.1007/978-3-642-01754-4_9

Peterson, S.H., Franklin, J., Roberts, D.A., van Wagtendonk, J.W., 2013. Mapping fuels in Yosemite National Park. *Can. J. For. Res.* 43, 7–17. https://doi.org/10.1139/cjfr-2012-0213

Pickett, B.M., Isackson, C., Wunder, R., Fletcher, T.H., Butler, B.W., Weise, D.R., 2010. Experimental measurements during combustion of moist individual foliage samples. *Int. J. Wildl. Fire* 19, 153. https://doi.org/10.1071/WF07121

Pinty, B., Verstraete, M.M., 1992. GEMI: A non-linear index to monitor global vegetation from satellites. *Vegetatio* 101, 15–20. https://doi.org/10.1007/BF00031911

Pontius, J., Hanavan, R.P., Hallett, R.A., Cook, B.D., Corp, L.A., 2017. High spatial resolution spectral unmixing for mapping ash species across a complex urban environment. *Remote Sens. Environ.* 199, 360–369. https://doi.org/10.1016/j.rse.2017.07.027

Potter, C., Li, S., Huang, S., Crabtree, R.L., 2012. Analysis of sapling density regeneration in Yellowstone National Park with hyperspectral remote sensing data. *Remote Sens. Environ.* 121, 61–68. https://doi.org/10.1016/j.rse.2012.01.019

Qiu, H., Gamon, J.A., Roberts, D.A., Luna, M., 1998. Monitoring postfire succession in the Santa Monica mountains using hyperspectral imagery, in: Green, R.O., Tong, Q. (Eds.), *International Society for Optics and Photonics*, pp. 201–208. https://doi.org/10.1117/12.317812

Quintano, C., Fernández-Manso, A., Roberts, D.A., 2013. Multiple endmember spectral mixture analysis (MESMA) to map burn severity levels from Landsat images in Mediterranean countries. *Remote Sens. Environ.* 136, 76–88. https://doi.org/10.1016/j.rse.2013.04.017

Realmuto, V.J., Dennison, P.E., Foote, M., Ramsey, M.S., Wooster, M.J., Wright, R., 2015. Specifying the saturation temperature for the HyspIRI 4-μm channel. *Remote Sens. Environ.* 167, 40–52. https://doi.org/10.1016/j.rse.2015.04.028

Reid, J.S., Koppmann, R., Eck, T.F., Eleuterio, D.P., 2005. A review of biomass burning emissions part II: Intensive physical properties of biomass burning particles. *Atmos. Chem. Phys* 5, 799–825.

Riaño, D., Chuvieco, E., Ustin, S., Zomer, R., Dennison, P., Roberts, D., Salas, J., 2002. Assessment of vegetation regeneration after fire through multitemporal analysis of AVIRIS images in the Santa Monica Mountains. *Remote Sens. Environ.* 79, 60–71. https://doi.org/10.1016/S0034-4257(01)00239-5

Riaño, D., Meier, E., Allgoewer, B., Chuvieco, E., Ustin, S., 2003. Modeling airborne laser scanning data for the spatial generation of critical forest parameters in fire behavior modeling. *Remote Sens. Environ.* 86, 177–186. https://doi.org/10.1016/S0034-4257(03)00098-1

Riggan, P.J., Tissell, R.G., Lockwood, R.N., Brass, J.A., Pereira, J.A.R., Miranda, H.S., Miranda, A.C., Campos, T., Higgins, R., 2004. Remote measurement of energy and carbon flux from wildfires in Brazil. *Ecol. Appl.* 14, 855–872. https://doi.org/10.1890/02-5162

Roberts, D.A., Smith, M.O., Adams, J.B., 1993. Green vegetation, nonphotosynthetic vegetation, and soils in AVIRIS data. *Remote Sens. Environ.* 44, 255–269. https://doi.org/10.1016/0034-4257(93)90020-X

Roberts, D.A., Gardner, M., Church, R., Ustin, S., Scheer, G., Green, R.O., 1998. Mapping chaparral in the Santa Monica mountains using multiple endmember spectral mixture models. *Remote Sens. Environ.* 65, 267–279. https://doi.org/10.1016/S0034-4257(98)00037-6

Roberts, D.A., Dennison, P.E., Gardner, M.E., Hetzel, Y., Ustin, S.L., Lee, C.T., 2003. Evaluation of the potential of Hyperion for fire danger assessment by comparison to the Airborne Visible/Infrared Imaging Spectrometer. *IEEE Trans. Geosci. Remote Sens.* 41, 1297–1310. https://doi.org/10.1109/TGRS.2003.812904

Roberts, D.A., Ustin, S.L., Ogunjemiyo, S., Greenberg, J., Dobrowski, S.Z., Chen, J., Hinckley, T.M., 2004. Spectral and structural measures of Northwest forest vegetation at leaf to landscape scales. *Ecosystems* 7. https://doi.org/10.1007/s10021-004-0144-5

Roberts, D.A., Dennison, P.E., Peterson, S., Sweeney, S., Rechel, J., 2006. Evaluation of airborne visible/infrared imaging spectrometer (AVIRIS) and moderate resolution imaging spectrometer (MODIS) measures of live fuel moisture and fuel condition in a shrubland ecosystem in southern California. *J. Geophys. Res. Biogeosciences* 111. https://doi.org/10.1029/2005JG000113

Roberts, G.J., Wooster, M.J., 2008. Fire detection and fire characterization over Africa using Meteosat SEVIRI. *IEEE Trans. Geosci. Remote Sens.* 46, 1200–1218. https://doi.org/10.1109/TGRS.2008.915751

Roberts, D.A., Bradley, E.S., Cheung, R., Leifer, I., Dennison, P.E., Margolis, J.S., 2010. Mapping methane emissions from a marine geological seep source using imaging spectrometry. *Remote Sens. Environ.* 114, 592–606. https://doi.org/10.1016/j.rse.2009.10.015

Roberts, D.A., Quattrochi, D.A., Hulley, G.C., Hook, S.J., Green, R.O., 2012. Synergies between VSWIR and TIR data for the urban environment: An evaluation of the potential for the Hyperspectral Infrared Imager (HyspIRI) Decadal Survey mission. *Remote Sens. Environ.* 117, 83–101. https://doi.org/10.1016/j.rse.2011.07.021

Robichaud, P.R., Lewis, S.A., Laes, D.Y.M., Hudak, A.T., Kokaly, R.F., Zamudio, J.A., 2007. Postfire soil burn severity mapping with hyperspectral image unmixing. *Remote Sens. Environ.* 108, 467–480. https://doi.org/10.1016/j.rse.2006.11.027

Rockweit, J.T., Franklin, A.B., Carlson, P.C., 2017. Differential impacts of wildfire on the population dynamics of an old-forest species. *Ecology* 98, 1574–1582. https://doi.org/10.1002/ecy.1805

Roessner, S., Segl, K., Heiden, U., Kaufmann, H., 2001. Automated differentiation of urban surfaces based on airborne hyperspectral imagery. *IEEE Trans. Geosci. Remote Sens.* 39, 1525–1532. https://doi.org/10.1109/36.934082

Rogge, D.M., Rivard, B., Zhang, J., Feng, J., 2006. Iterative Spectral Unmixing for Optimizing Per-Pixel Endmember Sets. *IEEE Trans. Geosci. Remote Sens.* 44, 3725–3736. https://doi.org/10.1109/TGRS.2006.881123

Rogers, B.M., Veraverbeke, S., Azzari, G., Czimczik, C.I., Holden, S.R., Mouteva, G.O., Sedano, F., Treseder, K.K., Randerson, J.T., 2014. Quantifying fire-wide carbon emissions in interior Alaska using field measurements and Landsat imagery. *J. Geophys. Res.* 119, 1608–1629.

Rogers, B.M., Soja, A.J., Goulden, M.L., Randerson, J.T., 2015. Influence of tree species on continental differences in boreal fires and climate feedbacks. *Nat. Geosci.* 8, 228–234. https://doi.org/10.1038/ngeo2352

Rollins, M.G., B.C., Frame, C.K., Keane, R.E., Zhu, Z., Menakis, J.P., B., Caratti, J.F., Holsinger, L. et al., 2006. *The LANDFIRE prototype project: Nationally consistent and locally relevant geospatial data for wildland fire management.* USDA For. Serv.—Gen. Tech. Rep. RMRS-GTR 1–418.

Roth, K.L., Dennison, P.E., Roberts, D.A., 2012. Comparing endmember selection techniques for accurate mapping of plant species and land cover using imaging spectrometer data. *Remote Sens. Environ.* 127, 139–152. https://doi.org/10.1016/j.rse.2012.08.030

Rothman, L.S., Gordon, I.E., Babikov, Y., Barbe, A., Chris Benner, D., Bernath, P.F., Birk, M. et al., 2013. The HITRAN2012 molecular spectroscopic database. *J. Quant. Spectrosc. Radiat. Transf.* 130, 4–50. https://doi.org/10.1016/j.jqsrt.2013.07.002

Roy, D.P., Jin, Y., Lewis, P.E., Justice, C.O., 2005. Prototyping a global algorithm for systematic fire-affected area mapping using MODIS time series data. *Remote Sens. Environ.* 97, 137–162. https://doi.org/10.1016/j.rse.2005.04.007

Roy, D.P., Boschetti, L., Trigg, S.N., 2006. Remote sensing of fire severity: Assessing the performance of the Normalized Burn Ratio. *IEEE Geosci. Remote Sens. Lett.* 3, 112–116. https://doi.org/10.1109/LGRS.2005.858485

Ryan, K.C., Opperman, T.S., 2013. LANDFIRE—A national vegetation/fuels data base for use in fuels treatment, restoration, and suppression planning. *For. Ecol. Manage.* 294, 208–216. https://doi.org/10.1016/j.foreco.2012.11.003

Salem, F., Kafatos, M., El-Ghazawi, T., Gomez, R., Yang, R., 2005. Hyperspectral image assessment of oil-contaminated wetland. *Int. J. Remote Sens.* 26, 811–821. https://doi.org/10.1080/01431160512331316883

Schaepman, M.E., Ustin, S.L., Plaza, A.J., Painter, T.H., Verrelst, J., Liang, S., 2009. Earth system science related imaging spectroscopy—An assessment. *Remote Sens. Environ.* 113, S123–S137. https://doi.org/10.1016/j.rse.2009.03.001

Schepers, L., Haest, B., Veraverbeke, S., Spanhove, T., Vanden Borre, J., Goossens, R., 2014. Burned Area Detection and Burn Severity Assessment of a Heathland Fire in Belgium Using Airborne Imaging Spectroscopy (APEX). *Remote Sens.* 6, 1803–1826. https://doi.org/10.3390/rs6031803

Schimel, D.S., Asner, G.P., Moorcroft, P., 2013. Observing changing ecological diversity in the Anthropocene. *Front. Ecol. Environ.* 11, 129–137. https://doi.org/10.1890/120111

Schroeder, W., Oliva, P., Giglio, L., Csiszar, I.A., 2014. The New VIIRS 375 m active fire detection data product: Algorithm description and initial assessment. *Remote Sens. Environ.* 143, 85–96. https://doi.org/10.1016/j.rse.2013.12.008

Smith, A.M.S., Lentile, L.B., Hudak, A.T., Morgan, P., 2007. Evaluation of linear spectral unmixing and ΔNBR for predicting post-fire recovery in a North American ponderosa pine forest. *Int. J. Remote Sens.* 28, 5159–5166. https://doi.org/10.1080/01431160701395161

Solans Vila, J.P., Barbosa, P., 2010. Post-fire vegetation regrowth detection in the Deiva Marina region (Liguria-Italy) using Landsat TM and ETM+ data. *Ecol. Modell.* 221, 75–84. https://doi.org/10.1016/j.ecolmodel.2009.03.011

Somers, B., Delalieux, S., Stuckens, J., Verstraeten, W.W., Coppin, P., 2009. A weighted linear spectral mixture analysis approach to address endmember variability in agricultural production systems. *Int. J. Remote Sens.* 30, 139–147. https://doi.org/10.1080/01431160802304625

Somers, B., Verbesselt, J., Ampe, E.M., Sims, N., Verstraeten, W.W., Coppin, P., 2010. Spectral mixture analysis to monitor defoliation in mixed-aged Eucalyptus globulus Labill plantations in southern Australia using Landsat 5-TM and EO-1 Hyperion data. *Int. J. Appl. Earth Obs. Geoinf.* 12, 270–277. https://doi.org/10.1016/j.jag.2010.03.005

Somers, B., Zortea, M., Plaza, A., Asner, G.P., 2012. Automated extraction of image-based endmember bundles for improved spectral unmixing. *IEEE J. Sel. Top. Appl. Earth Obs. Remote Sens.* 5, 396–408. https://doi.org/10.1109/JSTARS.2011.2181340

Somers, B., Tits, L., Roberts, D., Wetherley, E., 2016. Endmember library approaches to resolve spectral mixing problems in remotely sensed data: Potential, challenges, and applications, in: Ruckebush, C. (Ed.), *Data Handling in Science and Technology*. Elsevier, pp. 551–577. https://doi.org/10.1016/B978-0-444-63638-6.00017-6

Stagakis, S., Vanikiotis, T., Sykioti, O., 2016. Estimating forest species abundance through linear unmixing of CHRIS/PROBA imagery. *ISPRS J. Photogramm. Remote Sens.* 119, 79–89. https://doi.org/10.1016/j.isprsjprs.2016.05.013

Stavrakoudis, D.G., Galidaki, G.N., Gitas, I.Z., Theocharis, J.B., 2012. A genetic fuzzy-rule-based classifier for land cover classification from hyperspectral imagery. *IEEE Trans. Geosci. Remote Sens.* 50, 130–148. https://doi.org/10.1109/TGRS.2011.2159613

Stavros, E.N., Abatzoglou, J.T., McKenzie, D., Larkin, N.K., 2014. Regional projections of the likelihood of very large wildland fires under a changing climate in the contiguous Western United States. *Clim. Change* 126, 455–468.

Stavros, E.N., Tane, Z., Kane, V.R., Veraverbeke, S., McGaughey, R.J., Lutz, J.A., Ramirez, C., Schimel, D., 2016. Unprecedented remote sensing data over King and Rim megafires in the Sierra Nevada Mountains of California. *Ecology* 97, 3244–3244. https://doi.org/10.1002/ecy.1577

Storey, E.A., Stow, D.A., O'Leary, J.F., 2016. Assessing postfire recovery of chamise chaparral using multi-temporal spectral vegetation index trajectories derived from Landsat imagery. *Remote Sens. Environ.* 183, 53–64. https://doi.org/10.1016/j.rse.2016.05.018

Stuffler, T., Kaufmann, C., Hofer, S., Förster, K.P., Schreier, G., Mueller, A., Eckardt, A. et al., 2007. The EnMAP hyperspectral imager—An advanced optical payload for future applications in Earth observation programmes. *Acta Astronaut.* 61, 115–120. https://doi.org/10.1016/j.actaastro.2007.01.033

Tews, J., Brose, U., Grimm, V., Tielbörger, K., Wichmann, M.C., Schwager, M., Jeltsch, F., 2004. Animal species diversity driven by habitat heterogeneity/diversity: The importance of keystone structures. *J. Biogeogr.* 31, 79–92. https://doi.org/10.1046/j.0305-0270.2003.00994.x

Thenkabail, P.S., Enclona, E.A., Ashton, M.S., Van Der Meer, B., 2004. Accuracy assessments of hyperspectral waveband performance for vegetation analysis applications. *Remote Sens. Environ.* 91, 354–376. https://doi.org/10.1016/j.rse.2004.03.013

Thompson, D.R., Boardman, J.W., Eastwood, M.L., Green, R.O., 2017. A large airborne survey of Earth's visible-infrared spectral dimensionality. *Opt. Express* 25, 9186. https://doi.org/10.1364/OE.25.009186

Tobin, D.C., Revercomb, H.E., Knuteson, R.O., Lesht, B.M., Strow, L.L., Hannon, S.E., Feltz, W.F., Moy, L.A., Fetzer, E.J., Cress, T.S., 2006. Atmospheric Radiation Measurement site atmospheric state best estimates for Atmospheric Infrared Sounder temperature and water vapor retrieval validation. *J. Geophys. Res.* 111, D09S14. https://doi.org/10.1029/2005JD006103

Tulloch, A.I.T., Pichancourt, J.-B., Gosper, C.R., Sanders, A., Chadès, I., 2016. Fire management strategies to maintain species population processes in a fragmented landscape of fire-interval extremes. *Ecol. Appl.* 26, 2175–2189. https://doi.org/10.1002/eap.1362

Turetsky, M.R., Kane, E.S., Harden, J.W., Ottmar, R.D., Manies, K.L., Hoy, E., Kasischke, E.S., 2011. Recent acceleration of biomass burning and carbon losses in Alaskan forests and peatlands. *Nat. Geosci.* 4, 27–31. https://doi.org/10.1038/ngeo1027

Ullah, S., Schlerf, M., Skidmore, A.K., Hecker, C., 2012. Identifying plant species using mid-wave infrared (2.5–6µm) and thermal infrared (8–14µm) emissivity spectra. *Remote Sens. Environ.* 118, 95–102. https://doi.org/10.1016/j.rse.2011.11.008

Ustin, S.L., Roberts, D.A., Pinzón, J., Jacquemoud, S., Gardner, M., Scheer, G., Castañeda, C.M., Palacios-Orueta, A., 1998. Estimating canopy water content of chaparral shrubs using optical methods. *Remote Sens. Environ.* 65, 280–291. https://doi.org/10.1016/S0034-4257(98)00038-8

Ustin, S.L., Riaño, D., Hunt, E.R., 2012. Estimating canopy water content from spectroscopy. *Isr. J. Plant Sci.* 60, 9–23. https://doi.org/10.1560/IJPS.60.1-2.9

van der Meer, F., Bakker, W., 1997. Cross correlogram spectral matching: Application to surface mineralogical mapping by using AVIRIS data from Cuprite, Nevada. *Remote Sens. Environ.* 61, 371–382. https://doi.org/10.1016/S0034-4257(97)00047-3

van der Werf, G., Randerson, J.T., Collatz, G.J., Giglio, L., Kasibhatla, P.S., Arellano, A.F., Olsen, S.C., Kasischke, E.S., 2003. Continental-scale partitioning of fire emissions during the 1997–2001 El Niño/La Niña period. *Science* 300, 1242–3. https://doi.org/10.1126/science.1084460

van der Werf, G.R., Randerson, J.T., Giglio, L., van Leeuwen, T.T., Chen, Y., Rogers, B.M., Mu, M. et al., 2017. Global fire emissions estimates during 1997–2016. *Earth Syst. Sci. Data* 9, 697–720. https://doi.org/10.5194/essd-9-697-2017

van Leeuwen, W.J.D., Casady, G.M., Neary, D.G., Bautista, S., Alloza, J.A., Carmel, Y., Wittenberg, L., Malkinson, D., Orr, B.J., 2010. Monitoring post-wildfire vegetation response with remotely sensed time-series data in Spain, USA and Israel. *Int. J. Wildl. Fire* 19, 75. https://doi.org/10.1071/WF08078

van Marle, M.J.E., Field, R.D., van der Werf, G.R., Estrada de Wagt, I.A., Houghton, R.A., Rizzo, L. V., Artaxo, P., Tsigaridis, K., 2017. Fire and deforestation dynamics in Amazonia (1973-2014). *Global Biogeochem. Cycles* 31, 24–38. https://doi.org/10.1002/2016GB005445

van Wagtendonk, J.W., Root, R.R., Key, C.H., 2004. Comparison of AVIRIS and Landsat ETM + detection capabilities for burn severity. *Remote Sens. Environ.* 92, 397–408. https://doi.org/10.1016/j.rse.2003.12.015

Varga, T.A., Asner, G.P., 2008. Hyperspectral and LiDAR remote sensing of fire fuels in Hawaii Volcanoes National Park. *Ecol. Appl.* 18, 613–623. https://doi.org/10.1890/07-1280.1

Veraverbeke, S., Lhermitte, S., Verstraeten, W.W., Goossens, R., 2010. The temporal dimension of differenced Normalized Burn Ratio (dNBR) fire/burn severity studies: The case of the large 2007 Peloponnese wildfires in Greece. *Remote Sens. Environ.* 114, 2548–2563. https://doi.org/10.1016/j.rse.2010.05.029

Veraverbeke, S., Gitas, I., Katagis, T., Polychronaki, A., Somers, B., Goossens, R., 2012a. Assessing post-fire vegetation recovery using red–near infrared vegetation indices: Accounting for background and vegetation variability. *ISPRS J. Photogramm. Remote Sens.* 68, 28–39. https://doi.org/10.1016/j.isprsjprs.2011.12.007

Veraverbeke, S., Somers, B., Gitas, I., Katagis, T., Polychronaki, A., Goossens, R., 2012b. Spectral mixture analysis to assess post-fire vegetation regeneration using Landsat Thematic Mapper imagery: Accounting for soil brightness variation. *Int. J. Appl. Earth Obs. Geoinf.* 14, 1–11. https://doi.org/10.1016/j.jag.2011.08.004

Veraverbeke, S., Hook, S.J., 2013. Evaluating spectral indices and spectral mixture analysis for assessing fire severity, combustion completeness and carbon emissions. *Int. J. Wildl. Fire* 22, 707. https://doi.org/10.1071/WF12168

Veraverbeke, S., Sedano, F., Hook, S.J., Randerson, J.T., Jin, Y., Rogers, B.M., 2014a. Mapping the daily progression of large wildland fires using MODIS active fire data. *Int. J. Wildl. Fire* 23, 655–667.

Veraverbeke, S., Stavros, E.N., Hook, S.J., 2014b. Assessing fire severity using imaging spectroscopy data from the Airborne Visible/Infrared Imaging Spectrometer (AVIRIS) and comparison with multispectral capabilities. *Remote Sens. Environ.* 154, 153–163. https://doi.org/10.1016/j.rse.2014.08.019

Veraverbeke, S., Rogers, B.M., Randerson, J.T., 2015. Daily burned area and carbon emissions from boreal fires in Alaska. *Biogeosciences* 12, 3579–3601. https://doi.org/10.5194/bg-12-3579-2015

Veraverbeke, S., Rogers, B.M., Goulden, M.L., Jandt, R.R., Miller, C.E., Wiggins, E.B., Randerson, J.T., 2017. Lightning as a major driver of recent large fire years in North American boreal forests. *Nat. Clim. Chang.* 7, 529–534. https://doi.org/10.1038/NCLIMATE3329

Verbesselt, J., Fleck, S., Coppin, P., 2002. Estimation of fuel moisture content towards fire risk assessment: A review. in *Proc. 6th Int. Conf. For. Fire Res.* 1–11.

Vodacek, A., Kremens, R.L., Fordham, A.J., Vangorden, S.C., Luisi, D., Schott, J.R., Latham, D.J., 2002. Remote optical detection of biomass burning using a potassium emission signature. *Int. J. Remote Sens.* 23, 2721–2726. https://doi.org/10.1080/01431160110109633

Wessels, K.J., Prince, S.D., Zambatis, N., MacFadyen, S., Frost, P.E., Van Zyl, D., 2006. Relationship between herbaceous biomass and 1-km^2 advanced very high resolution radiometer (AVHRR) NDVI in Kruger National Park, South Africa. *Int. J. Remote Sens.* 27, 951–973. https://doi.org/10.1080/01431160500169098

Westerling, A.L., 2006. Warming and earlier spring increase western US forest wildfire activity. *Science* 80, 313. https://doi.org/10.1126/science.1128834

Whitburn, S., Van Damme, M., Kaiser, J.W., van der Werf, G.R., Turquety, S., Hurtmans, D., Clarisse, L., Clerbaux, C., Coheur, P.-F., 2015. Ammonia emissions in tropical biomass burning regions: Comparison between satellite-derived emissions and bottom-up fire inventories. *Atmos. Environ.* 121, 42–54. https://doi.org/10.1016/j.atmosenv.2015.03.015

Wooster, M.J., Roberts, G., Perry, G.L.W., Kaufman, Y.J., 2005. Retrieval of biomass combustion rates and totals from fire radiative power observations: FRP derivation and calibration relationships between biomass consumption and fire radiative energy release. *J. Geophys. Res.* 110, D24311. https://doi.org/10.1029/2005JD006318

Yebra, M., Dennison, P.E., Chuvieco, E., Riaño, D., Zylstra, P., Hunt, E.R., Danson, F.M., Qi, Y., Jurdao, S., 2013. A global review of remote sensing of live fuel moisture content for fire danger assessment: Moving towards operational products. *Remote Sens. Environ.* 136, 455–468. https://doi.org/10.1016/j.rse.2013.05.029

Yokelson, R.J., Susott, R., Ward, D.E., Reardon, J., Griffith, D.W.T., 1997. Emissions from smoldering combustion of biomass measured by open-path Fourier transform infrared spectroscopy. *J. Geophys. Res. Atmos.* 102, 18865–18877. https://doi.org/10.1029/97JD00852

Yu, B., Chen, F., Li, B., Wang, L., Wu, M., 2017. Fire risk prediction using remote sensed products: A case of Cambodia. *Photogramm. Eng. Remote Sens.* 83, 19–25. https://doi.org/10.14358/PERS.83.1.19

Zedler, P.H., Gautier, C.R., McMaster, G.S., 1983. Vegetation change in response to extreme events: The effect of a short interval between fires in California chaparral and coastal scrub. *Ecology* 64, 809–818. https://doi.org/10.2307/1937204

Zhao, F., Meng, R., Huang, C., Zhao, M., Zhao, F., Gong, P., Yu, L., Zhu, Z., 2016. Long-term post-disturbance forest recovery in the greater Yellowstone ecosystem analyzed using Landsat time series stack. *Remote Sens.* 8, 898. https://doi.org/10.3390/rs8110898

Zhukov, B., Lorenz, E., Oertel, D., Wooster, M., Roberts, G., 2006. Spaceborne detection and characterization of fires during the bi-spectral infrared detection (BIRD) experimental small satellite mission (2001–2004). *Remote Sens. Environ.* 100, 29–51. https://doi.org/10.1016/j.rse.2005.09.019

Section V

Hyperspectral Data in Global Change Studies

13 Hyperspectral Data in Long-Term, Cross-Sensor Continuity Studies

Tomoaki Miura and Hiroki Yoshioka

CONTENTS

13.1 INTRODUCTION

Numerous satellite optical sensors have been launched and planned for launch for monitoring and characterization of the Earth system and its behaviors. These sensors have been providing and will continue to provide systematic observations of terrestrial vegetation at various spatial, spectral, and temporal resolutions. Spectral vegetation indices (VIs) are among the most widely used satellite data products in monitoring temporal and spatial variations of vegetation photosynthetic activities and biophysical properties. VIs are optical measures of vegetation canopy "greeness," a direct measure of photosynthetic potential resulting from the composite property of total leaf chlorophyll, leaf area, canopy cover, and structure (Huete et al., 2014). Although they are not intrinsic physical quantities, VIs are widely used as proxies in the assessments of many canopy state and biophysical process variables, including leaf area index (LAI), fraction of absorbed photosynthetically active radiation, vegetation fraction, and gross primary production (e.g., Myneni et al., 1997; Sims et al., 2008).

The uses of these observations greatly increase when data sets from multiple sensors are combined, for example, multidecadal land-cover characterization and change detection via multisensor data sources (e.g., Bhattarai et al., 2009; Jepson et al., 2010; Paudel and Andersen, 2010), synergistic applications of multiresolution remote sensing for forest and rangeland inventory (e.g., DeFries et al., 2007; Miettinen and Liew, 2009), and development of multisensor, long-term data records for climate studies (Eidenshink, 2006; Pinzon and Tucker, 2014; Zhang et al., 2014).

Applications of multisensor observations, however, require consideration and account of continuity and compatibility due to differences in sensor/platform characteristics that include band position, spatial resolution, and overpass time (e.g., Batra et al., 2006; Teillet et al., 1997). Multisensor VI continuity becomes a critical and complicated issue because it involves consideration of differences in both sensor/platform characteristics and product generation algorithms, a requirement that needs to be addressed (Swinnen and Veroustraete, 2008). The underlying issue in multisensor VI continuity is that VI values for the same targets will not be directly comparable because input reflectance values differ from sensor to sensor (Teillet et al., 1997; Yoshioka et al., 2003). Swinnen and Veroustraete (2008) provide a comprehensive list of factors to be taken into consideration for extending the Système Pour l'Observation de la Terre (SPOT) VEGETATION normalized difference vegetation index (NDVI) time series back in time with National Oceanic and Atmospheric Administration (NOAA) advanced very-high-resolution radiometer (AVHRR) data.

Hyperspectral remote sensing, in particular imaging sensors, has great potential in addressing several key issues of multisensor VI continuity and providing deeper insights and understanding of the issues. An ultimate advantage of using hyperspectral remote sensing for multisensor continuity studies is that it allows one to analyze the effects of multiple factors simultaneously (Teillet et al., 1997). Specific issues of multisensor VI continuity that can be addressed with hyperspectral remote sensing include the following:

1. Spectral: A large number of narrow spectral bands that continuously cover the visible–near-infrared–shortwave infrared (VIS-NIR-SWIR) wavelength regions can be spectrally convolved to simulate spectral responses of virtually any broadband sensors. The simulated data can be used for multisensor comparisons devoid of misregistration (e.g., Kim et al., 2010). It should be noted that, although the word *simulation* is used here, the resultant, spectrally aggregated values are actual observations.
2. Spatial: Current and future hyperspectral sensors provide or will provide medium-resolution images (3–100 m spatial resolution, with 30 m being typical) with swaths of 3–150 km, with 30 km being typical. These resolutions are fine enough and these swath widths are wide enough to allow for the simulation of various pixel footprint sizes via spatial aggregation. The aggregated data can be used to examine VI compatibility across multiple resolutions (Huete et al., 2005).
3. Algorithmic: The effects of algorithmic differences (e.g., atmospheric correction schemes) can be examined on spectrally or spatially aggregated data from hyperspectral imagery, although limited in types of algorithms that could be tested (e.g., Miura et al., 2013).
4. Angular: Many current and future satellite hyperspectral sensor systems have or will have a cross-track pointing capability. Although limited in the range of possible observation geometry, multiangular hyperspectral observations could be used to address bidirectional reflectance distribution function (BRDF) effects on multisensor VI continuity.

The purpose of this chapter is to discuss the potential uses of hyperspectral remote sensing data in long-term VI continuity for global change studies. We present analysis results obtained from a regional set of Earth Observing-One (EO-1) hyperspectral Hyperion images (Pearlman et al., 2003; Ungar et al., 2003) over the conterminous United States along with literature reviews for this purpose.

13.2 MATERIALS

Five sites within the conterminous United States were selected based upon the availability of nearly cloud-free Hyperion scenes, the availability of in situ atmospheric measurements from the Aerosol Robotic Network (AERONET) (Holben et al., 2001), and a diversity of land-cover types. Level 1R Hyperion scenes were obtained for the five sites for the dates listed in Table 13.1. For each Hyperion scene, Level 2 AERONET data were acquired for a 2 h time period bracketing the image acquisition time (±1 h) (Table 13.1).

Hyperion images were spectrally convolved to spectral bandpasses of various satellite sensors described in Section 13.3. The spectral response curves of these satellite sensors were splined to Hyperion band center wavelengths for each Hyperion pixel (Miura et al., 2013) because each pixel had a slightly different spectral calibration (spectral smile) (Pearlman et al., 2003).

TABLE 13.1
List of Study Sites, Hyperion Image Properties, and *In Situ* Atmospheric Properties

Geographic Location	Latitude/ Longitude (degrees)	Elev.[a] (m)	Biome Type	Date (yyyy/mm/dd)	θ_s/θ_v[b] (degrees)	[c]Ozone (Dobson)	[c]W.V.[d] (cm-atm)	[c]AOT[e] (550 nm)
Harvard Forest, MA	42.532/−72.188	322	Broadleaf Forest	2001/09/05	40.5/3.6	302	0.76	0.03
				2008/05/07	32.2/12.1	362	0.76	0.13
				2008/05/25	28.5/4.8	353	0.76	0.05
				2008/05/30	28.6/12.8	351	1.28	0.17
				2008/06/07	26.2/10.1	347	3.29	0.17
				2008/12/03	67.1/5.2	303	0.64	0.04
Walker Branch, TN	35.958/−84.287	365	Broadleaf Forest	2001/08/14	31.3/2.3	308	2.46	0.27
Maricopa, AZ	33.069/−111.972	360	Broadleaf Cropland/ Open Shrubland	2001/05/24	23.3/5.5	317	1.11	0.06
				2001/07/27	26.4/5.4	295	2.77	0.07
				2001/08/28	32.3/5.0	287	2.76	0.09
				2001/12/02	58.6/4.6	275	0.98	0.05
				2001/12/18	60.5/4.7	280	0.41	0.04
Konza Prairie, KS	39.102/−96.610	341	Prairie Grassland/ Cereal Crop	2002/10/19	52.3/2.6	285	0.98	0.02
				2009/05/08	29.4/1.3	340	2.12	0.10
Sevilleta, NM	34.355/−106.885	1477	Semi-arid Grassland/ Open Shrubland/ Cereal Crop	2001/10/19	48.8/5.0	275	0.46	0.02
				2009/01/16	61.5/16.9	290	0.50	0.02
				2009/09/25	40.8/4.3	278	0.91	0.03
				2009/10/05	45.2/18.6	277	1.11	0.02
				2009/11/05	54.2/12.1	271	0.69	0.02
				2009/12/06	60.8/5.7	276	0.47	0.04

[a] Elevation.
[b] θ_s: solar zenith angle, θ_v: view zenith angle.
[c] The values in these columns were obtained from the Aerosol Robotic Network (AERONET) website (http://aeronet.gsfc.nasa.gov/) (Holben et al., 2001).
[d] Atmospheric water vapor.
[e] Aerosol optical thickness.

The convolved images were first converted to top-of-atmosphere (TOA) reflectances and then corrected for atmosphere with the 6S radiative transfer code (Vermote et al., 2006). The 6S radiative transfer code was constrained by scene-specific geometric conditions extracted from the corresponding image metadata and in situ AERONET atmospheric data (Table 13.1). We performed three types of atmospheric correction: (1) partial correction for molecular scattering and ozone absorption; (2) partial correction for molecular scattering, and ozone and water vapor absorptions; and (3) total correction including aerosol scattering and absorption. The continental aerosol model was assumed for all aerosol corrections based on the aerosol model selection criteria described in Kaufman et al. (1997).

Three VIs were computed and evaluated in this study. The normalized difference vegetation index (NDVI) was computed from the TOA and atmospherically corrected reflectances as (Tucker, 1979)

$$\mathrm{NDVI} = \frac{\rho_{NIR} - \rho_{red}}{\rho_{NIR} + \rho_{red}}, \tag{13.1}$$

where ρ_{red} and ρ_{NIR} are the red and NIR reflectances, respectively. The enhanced vegetation index (EVI), developed as a standard satellite vegetation product for Terra and Aqua MODIS, was computed from the atmospherically corrected reflectances (Huete et al., 2002):

$$EVI = 2.5 \frac{\rho_{NIR} - \rho_{red}}{\rho_{NIR} + 6\rho_{red} - 7.5\rho_{blue} + 1}, \tag{13.2}$$

where ρ_{blue} is the blue reflectance to correct for aerosol influences. Jiang et al. (2008) developed a two-band EVI without a blue band (EVI2), which achieves the best similarity with the EVI and, thus, is applicable to sensors without a blue band such as AVHRR (e.g., Kim et al., 2014):

$$EVI2 = 2.5 \frac{\rho_{NIR} - \rho_{red}}{\rho_{NIR} + 2.4\rho_{red} + 1}. \tag{13.3}$$

EVI2 was computed from the atmospherically corrected reflectances.

13.3 SPECTRAL COMPATIBILITY ANALYSES

One key sensor characteristic that varies widely among sensors is the spectral bandpass filters and many previous studies focused on this "spectral" issue (e.g., Gallo et al., 2005; Gao, 2000; Gitelson and Kaufman, 1998; Gunther and Maier, 2007; Ji et al., 2008). Figure 13.1 shows the normalized spectral response curves of red, NIR, and blue (when available) bands for moderate- to coarse-resolution satellite sensors designed or used for monitoring and biophysical characterization of global vegetation. The bandwidths of AVHRR/2 channels (Figure 13.1a) are the widest, followed by those of the AVHRR/3 sensors (Figure 13.1b), by SPOT-4 and -5 VEGETATION and Advanced Earth Observing Satellite (ADEOS)-II Global Imager (GLI) 250 m bands (Figure 13.1c), and by Terra and Aqua MODIS (Figure 13.1c). The narrowest are the spectral bands of those sensors designed for both oceanic and terrestrial measurements, that is, SeaWiFS, ADEOS-II GLI (1 km), GOSAT CAI, and GOCM-C SGLI (Figure 13.1d). The blue bands of these oceanic/terrestrial sensors, except for GLI, are located at slightly longer wavelengths than those of the terrestrial sensors (i.e., MODIS, VEGETATION, and GLI 250 m). Uniquely positioned are the Visible Infrared Imaging Radiometer Suite (VIIRS) spectral bands. While the VIIRS NIR band is similar to that of MODIS, the VIIRS red band is more similar to the AVHRR/3 counterpart than the MODIS counterpart, and the VIIRS blue band is positioned at slightly longer wavelengths, similar to those of the oceanic/terrestrial sensors (Figure 13.1d).

Spectral convolution of hyperspectral data has been one of the standard methodologies used for assessing and evaluating the effects of these spectral bandpass differences on VI compatibility and continuity (Kim et al., 2010; Miura et al., 2006; Steven et al., 2003; Trishchenko et al., 2002; Trishchenko, 2009; Yoshioka et al., 2006). This approach is advantageous because it makes it possible to examine the continuity/compatibility of pairs of sensors that do not have actual overlapping periods of observations.

Previous studies that used this methodology can be divided into two major categories: (1) empirical studies and (2) theoretical studies. Empirical studies have focused on predicting target sensor reflectance or VI values from those of a source sensor by regression. Polynomials have been assumed as an analytical form that relates reflectance or VI values from two different sensors. Some studies used first-order polynomials and concluded that simple linear relationships would hold for relating the NDVI from two different sensors (Gallo et al., 2005; Steven et al., 2003; van Leeuwen et al., 2006), whereas other studies used second-order polynomials as they found nonlinearity in intersensor NDVI and reflectance relationships (Miura et al., 2006; Trishchenko et al., 2002; Trishchenko, 2009). Kim et al. (2010) showed that the EVI and EVI2 cross-sensor relationships were also modeled satisfactorily well with the first-order polynomial model. Whereas

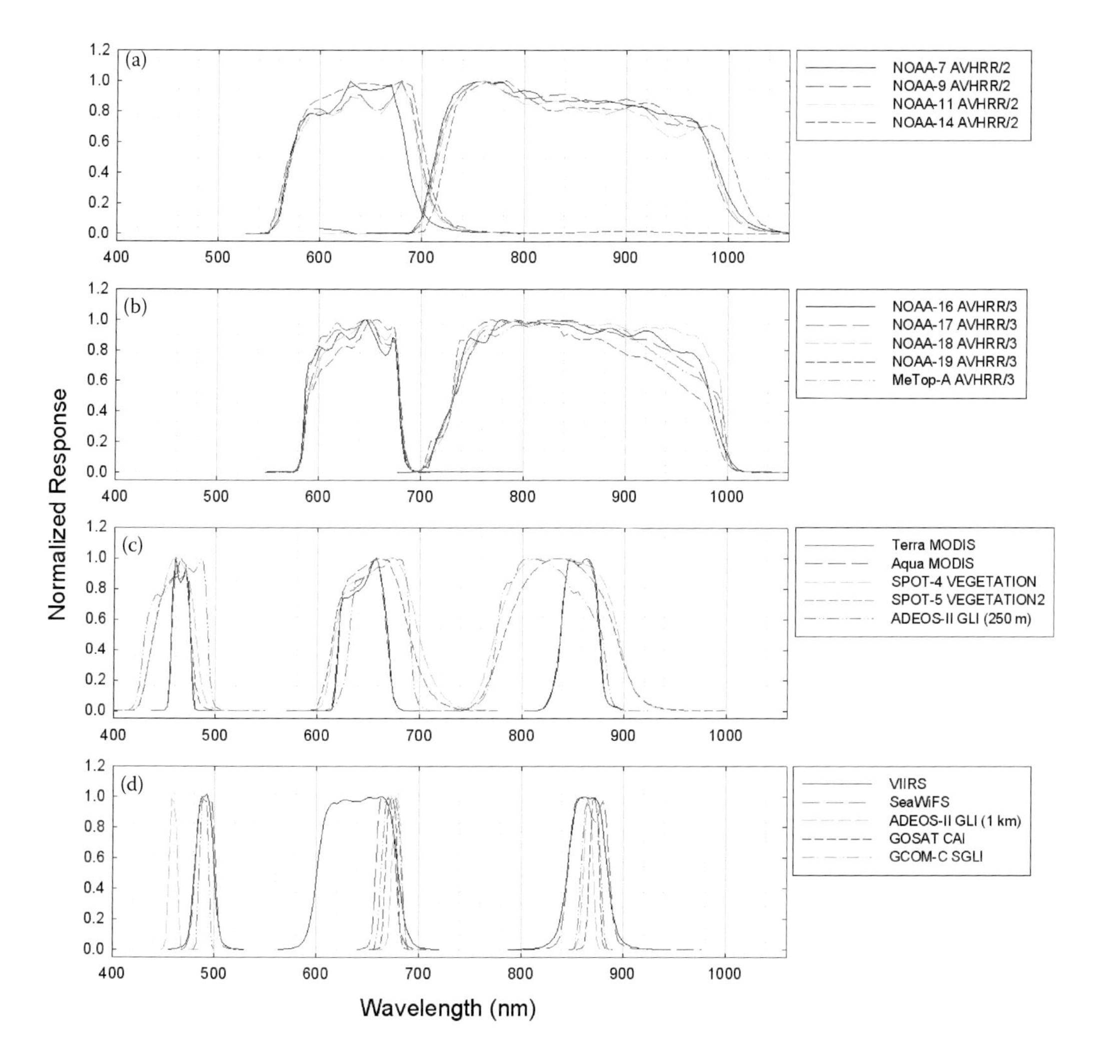

FIGURE 13.1 Normalized spectral response curves of red, near-infrared, and blue bands for select moderate-resolution sensors.

these studies used the ordinary least-squares approach to fit polynomial models, Ji and Gallo (2006) asserted that considering measurement errors of independent variables more accurately characterized intersensor NDVI relationships (i.e., unbiased) and proposed a set of new statistics, "agreement coefficients."

Theoretical studies were motivated to take into account ecosystem parameters (e.g., LAI and soil brightness) in developing a spectral transformation algorithm that theoretically guarantees "exact" translations (Yoshioka et al., 2003, 2012). Based on the physics of atmosphere-vegetation-photon interactions, Yoshioka et al. (2005) theoretically justified the existence of and derived the functional form of interrelating VIs from two sensors. They also showed that this "vegetation isoline" approach to interrelating VIs across sensors resulted in an approximately 50% reduction in variability around the trend in cross-sensor VI relationships. The "exactness" of the translation results with this technique was also demonstrated using a simulated hyperspectral data set (Miura et al., 2008). Noting that the isoline-based translation equation is a ratio of two polynomials, Yoshioka et al. (2006) reduced the isoline-based translation equation into a quadratic polynomial. Although the coefficients of the polynomial could vary with surface and atmospheric conditions, this work theoretically justified the use of a polynomial form for multisensor translations of VIs.

An issue with these previous studies, however, is that each of them was limited in their spatial extent, seasonal coverage, and land-cover type. Therefore, an extension of the results to different land-cover types, geographic areas, or seasons is questionable. Satellite hyperspectral remote sensing has the great potential to be an excellent data source for expanding continuity analyses to global, full-season analyses.

Using the Hyperion data set described in Section 13.2, NDVI, EVI2, and EVI relationships of the sensors in Figure 13.1 to the Terra MODIS sensor are examined (MODIS VI minus source sensor VI plotted against source sensor values) in Figures 13.2 through 13.4. These comparisons assumed a total atmospheric correction scenario at 1 km spatial resolution with the MODIS point spread function (PSF) (see the next section for the MODIS PSF). The figures show how cross-sensor VI relationships (including magnitude and linearity) vary as a function of spectral bandpass. For example, Terra and Aqua MODIS are spectrally perfectly compatible for all three VIs examined in this chapter (Figures 13.2j, 13.3j, and 13.4a), and it can also be seen that cross-sensor relationships to MODIS are generally more linear for EVI2 and EVI than for NDVI (Figures 13.2–13.4).

The simulated data set if expanded to include more scenes can be divided into subsets based on geographic areas, seasons, or land-cover types to examine geographic, seasonal, or land-cover dependencies of cross-sensor VI relationships.

13.4 SPATIAL COMPATIBILITY ANALYSES

Another key sensor characteristic that varies across sensors is spatial resolution (PSF). Although it is critical, the spatial issue of continuity has not received as much attention as the spectral issue. Central to VI spatial compatibility is the scale-invariance properties or scaling uncertainties of VIs with the influence of land surface heterogeneity (Chen, 1999; Friedl et al., 1995; Hall et al., 1992; Hu and Islam, 1997). The VI scaling uncertainties arise when a VI involves a nonlinear transformation of input reflectance data. In other words, an average of fine-resolution VI values is not equal to a VI value computed from coarser-resolution reflectance (Hu and Islam, 1997), and the degree of this difference is expected to vary with surface heterogeneity and VI formula (Friedl et al., 1995; Huete et al., 2005; Obata et al., 2012a). In what follows, we compare the fine-grained VIs and the coarse-scale VIs using the NDVI as an example.

The fine-grained NDVI can be aggregated to a coarser-resolution pixel by (Hu and Islam, 1997; Huete et al., 2005)

$$\begin{aligned}\mathrm{NDVI}_{fine} &= f_1\cdot\mathrm{NDVI}_1 + f_2\cdot\mathrm{NDVI}_2\\ &= \frac{f_1\cdot(\rho_{NIR,1}-\rho_{red,1})}{\rho_{NIR,1}+\rho_{red,1}} + \frac{f_2\cdot(\rho_{NIR,2}-\rho_{red,2})}{\rho_{NIR,2}+\rho_{red,2}},\end{aligned} \tag{13.4}$$

where two surface types with the fractional amounts of f_1 and $f_2 (f_1 + f_2 = 1)$ are assumed. This quantity is not generally equal to the coarser-resolution NDVI computed from the reflectances at the resolution analyzed, which can be expressed using the fine-grained reflectances as

$$\begin{aligned}\mathrm{NDVI}_{coarse} &= \frac{(f_1\cdot\rho_{NIR,1}+f_2\cdot\rho_{NIR,2})-(f_1\cdot\rho_{red,1}+f_2\cdot\rho_{red,2})}{(f_1\cdot\rho_{NIR,1}+f_2\cdot\rho_{NIR,2})+(f_1\cdot\rho_{red,1}+f_2\cdot\rho_{red,2})}\\ &= \frac{f_1\cdot(\rho_{NIR,1}-\rho_{red,1})+f_2\cdot(\rho_{NIR,2}-\rho_{red,2})}{f_1\cdot(\rho_{NIR,1}+\rho_{red,1})+f_2\cdot(\rho_{NIR,2}+\rho_{red,2})},\end{aligned} \tag{13.5}$$

and, thus,

$$\mathrm{NDVI}_{coarse} \neq \mathrm{NDVI}_{fine} \quad \text{or} \quad D \equiv \mathrm{NDVI}_{coarse} - \mathrm{NDVI}_{fine} \neq 0. \tag{13.6}$$

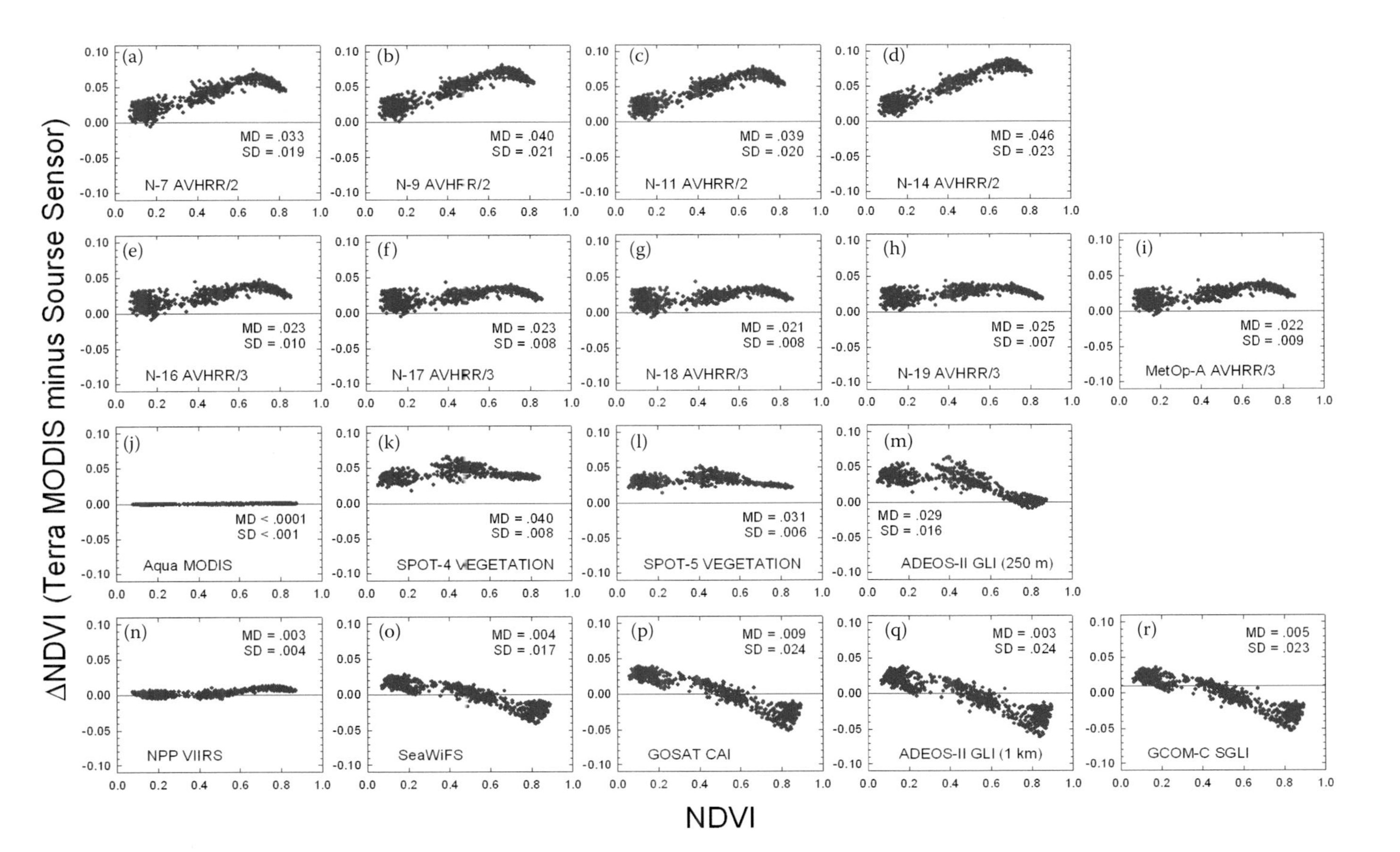

FIGURE 13.2 NDVI differences between Terra MODIS and other moderate-resolution sensors (source sensors) plotted against source sensors. MD and SD are mean differences and standard deviations of the differences, respectively.

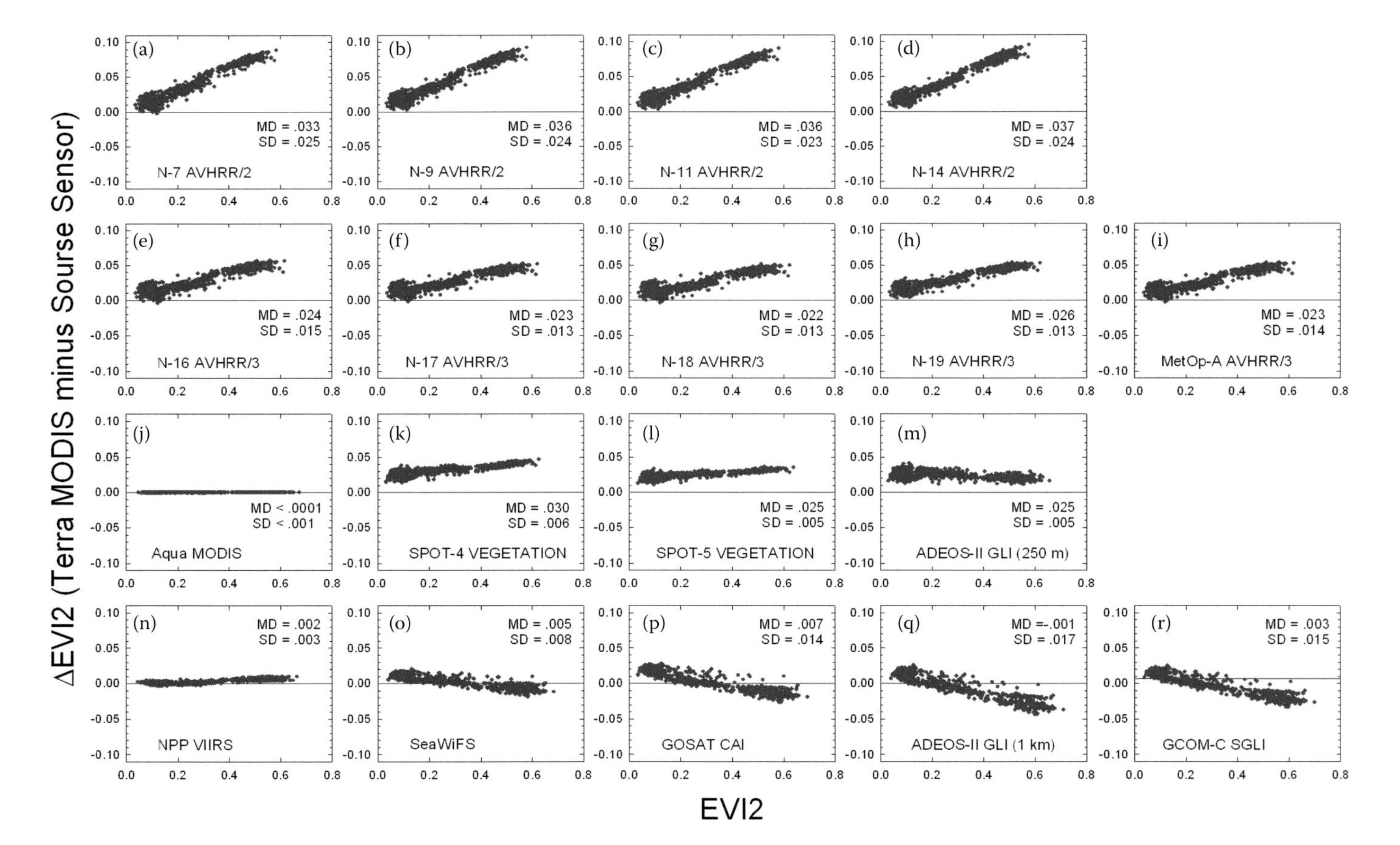

FIGURE 13.3 Same as Figure 13.2, but for the EVI2.

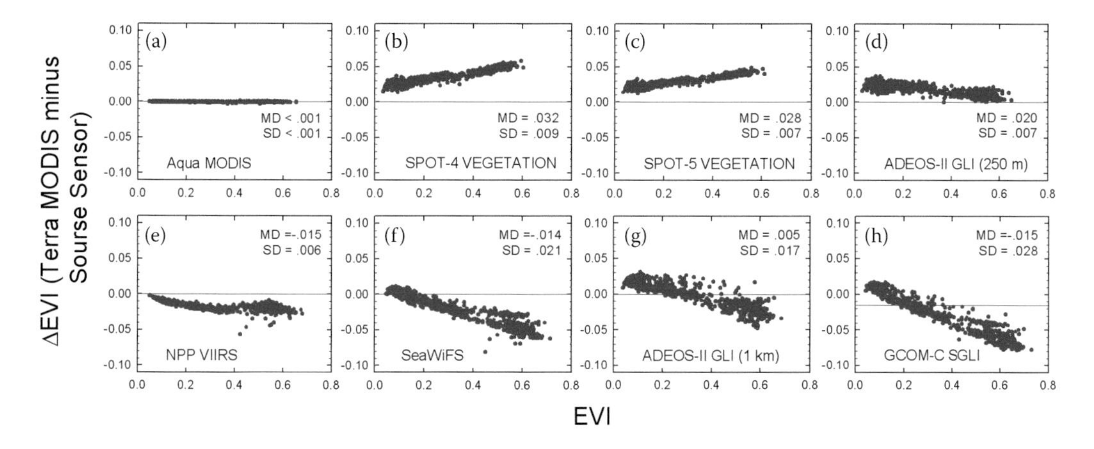

FIGURE 13.4 Same as Figure 13.2, but for the EVI.

These two quantities are equal, or the quantity D is equal to zero (1) when either f_1 or f_2 is equal to zero (i.e., homogeneous case) (Huete et al., 2005) or (2) when the L1 norms of the two endmember spectra are equal (i.e., $\rho_{NIR,1} + \rho_{red,1} = \rho_{NIR,2} + \rho_{red,2}$) (Obata et al., 2012b; Yoshioka et al., 2008). Theoretically, at least, the former applies to the EVI and EVI2 formulas (Huete et al., 2005).

In practice, this implies that VIs derived at a higher resolution could not be used simply as an enhanced resolution of VIs at a lower resolution, that is, their sensitivities to actual surface changes are different. This in turn requires an investigation into whether VIs from multiple sensors with various spatial resolutions and PSFs show compatible or different sensitivities to actual vegetation changes for VI-based change detections. The existing and planned hyperspectral sensors provide images at medium resolutions (30–60 m), which can spatially be aggregated to simulate various resolutions as low as their swath widths. Therefore, with hyperspectral imagery, scaling uncertainties can be analyzed separately (single-factor analysis) and simultaneously with the effect of spectral bandpass differences (two-factor analysis). In what follows, two examples of such hyperspectral data analyses are provided for the demonstration purpose.

In the first example, the Hyperion scenes in Table 13.1 were spatially aggregated to 60, 120, 240, 480, and 960 m spatial resolutions at the VI (fine-grained data) and reflectance (coarse-grained data) levels assuming a square PSF and a total atmospheric correction scenario. The fine-grained VIs were subtracted from the coarser-grained counterparts to assess scale-induced deviations (differences) at the different resolutions (D in Equation 13.6). In Figure 13.5, the derived differences where the MODIS spectral responses were assumed for both the fine- and coarse-grained data (single-factor case) are plotted for the NDVI, EVI2, and EVI for the Maricopa scene of May 24, 2001. The plotted differences were the largest for the NDVI and the smallest for the EVI2, suggesting that the NDVI is subject to larger scaling uncertainties than the EVI2. For all three VIs, in general, mean, maximum, and minimum differences decreased with increasing resolutions, while standard deviations of the difference increased (Figure 13.5). This indicates that the scale-induced deviation is generally larger for higher-resolution differences (i.e., 30 m vs. 960 m), although extremely large deviations are more likely to be encountered for smaller resolutions (i.e., 30 m vs. 60–130 m). The nature of this monotonic change of NDVI scaling errors is discussed in detail for a two-endmember linear mixture model in Obata et al. (2012b) and Yoshioka et al. (2008).

Plotted in Figure 13.6 are the derived differences for the same Maricopa scene in which the MODIS spectral responses were assumed for the coarse-grained data, but the Landsat-5 TM spectral responses for the fine-grained data, simulating synergistic applications of multiresolution remote sensing (two-factor case). Two differences from Figure 13.5 can be observed. First, large systematic differences were introduced due to spectral bandpass differences, which were approximately 0.035 for the NDVI (Figure 13.6a), around 0.03 for the EVI2 (Figure 13.6b), and approximately 0.02

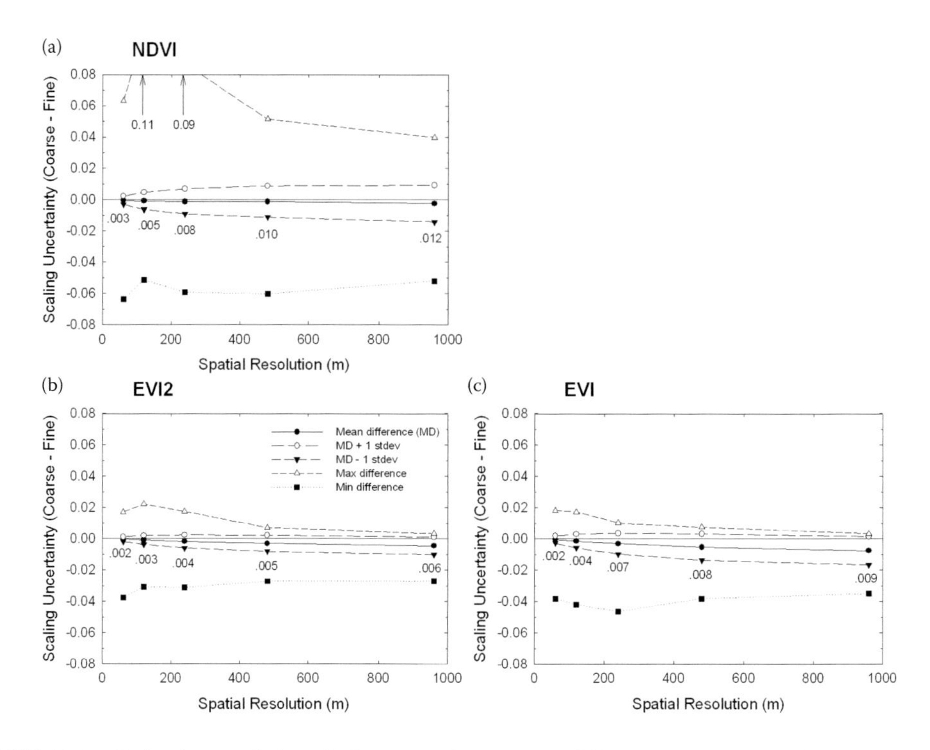

FIGURE 13.5 Univariate statistics of NDVI, EVI2, and EVI differences between their coarse- (MODIS bandpasses) and fine-grained (MODIS bandpasses) averaged values computed from the May 24, 2001, Hyperion image over Maricopa, AZ, USA. The numbers accompanying the filled triangles are standard deviations of the differences: (a) NDVI; (b) EVI2; and (c) EVI.

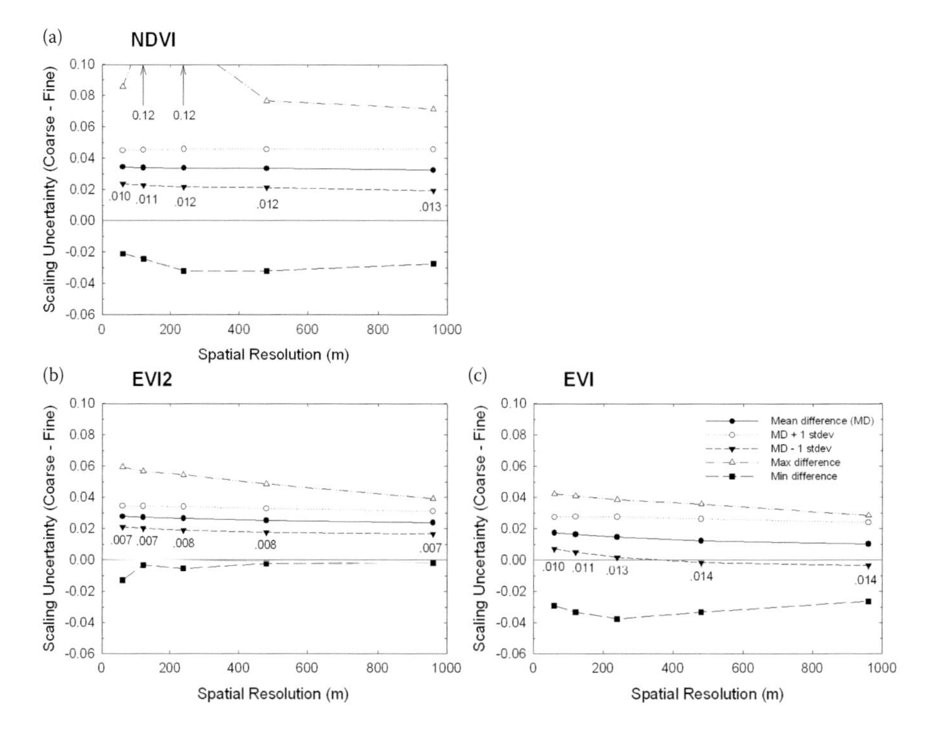

FIGURE 13.6 Same as Figure 13.5, but differences between the coarse- (MODIS bandpasses) and fine-grained (Landsat-5 TM bandpasses) averaged values.

for the EVI (Figure 13.6c). Second, standard deviations of the differences for this two-factor case (Figure 13.6) were larger and more uniform across the resolutions than those for the single-factor case (Figure 13.5). These results simply suggest that scale-induced deviations in VIs can be larger when comparing multiresolution sensor data with different spectral bandpasses than with the same bandpasses. A more thorough analysis is required to understand the mechanism by which scaling uncertainties are affected by spectral bandpass differences.

In the second example, the Hyperion scenes in Table 13.1 were used to simulate the AVHRR Global Area Coverage (GAC) sampling scheme and the MODIS Climate Modeling Grid (CMG) aggregation for assessing the scaling uncertainties between these two products. A GAC pixel value represents the mean of four out of each five consecutive samples along the scan line, and only data from each third scan line are processed and stored, which are performed onboard the sensor in real time (Pinheiro et al., 2006). As a result, the spatial resolution of GAC data near nadir is about 1.1 × 4 km with a 3 km gap between pixels along the track. In contrast, a MODIS CMG pixel is created by aggregating all pixels inside the 0.05° CMG grid (approximately 5 × 5 km) (Didan, K., personal communication).

Spectrally convolved Hyperion scenes were first aggregated to AVHRR 1.1 km and MODIS 500 m resolutions, which were then aggregated to GAC and CMG pixels. We assumed a bell-shaped PSF and a triangular PSF in the scan direction for AVHRR and MODIS, respectively, and a rectangular PSF in the track direction for both AVHRR and MODIS (Schowengerdt, 2006; Wolfe et al., 2002). VIs were computed from the simulated GAC pixels, and five of these pixels were averaged to generate fine-grained VI values approximately equal to the CMG pixel size. These spatial aggregations and averaging of Hyperion pixels were performed carefully and systematically so that the derived fine-grained GAC and coarse-grained CMG pixels were colocated without misregistration.

In Figure 13.7a,b, only the effects of the spectral bandpass difference between Terra MODIS and NOAA-14 AVHRR on the NDVI and EVI2, respectively, were assessed for the two spatial resolutions. MODIS-AVHRR cross-sensor NDVI and EVI2 relationships for the GAC resolution were basically the same as those for the CMG resolution, that is, the trends in the relationships were the same for the two resolutions. In fact, these trends in cross-sensor relationships were also very similar to those observed for 1 km resolution (see Figures 13.2d and 13.3d for the NDVI and EVI2, respectively). In Figure 13.7c,d, only the effects of the spatial resolution difference between the CMG and GAC resolutions (scaling uncertainties) are assessed for the NDVI and EVI2, respectively, by fixing the spectral bandpasses. There were large variations in scale-induced differences for all four cases (MODIS NDVI, AVHRR NDVI, MODIS EVI2, and AVHRR EVI2), ranging from −0.05 to 0.05 at most for MODIS NDVI (Figure 13.7c); however, these scaling uncertainties did not appear to introduce any systematic differences (i.e., mean differences ≈0). In Figure 13.7e,f, the combined effects of the spectral bandpass and spatial resolution differences between MODIS CMG and AVHRR GAC VIs are assessed for the NDVI and EVI2, respectively. For both the NDVI and EVI2, the trends in cross-sensor relationships remained similar to those due only to the spectral bandpass difference; however, the secondary scattering about the mean trends became larger due to the scale-induced variations. These results suggest that MODIS CMG and AVHRR GAC VIs can be combined to generate a long-term data record but would be accompanied by added uncertainties due to scaling differences.

The two preceding examples can be expanded to a larger data set to obtain more reliable estimates of scaling uncertainties. The demonstrated capability of hyperspectral data to analyze multiple factors one at a time and all at once will also make it possible to derive error budgets for those multiple factors. Although an example is not provided here, hyperspectral data can also be used in the same way as described earlier to assess the impact of misregistration on cross-sensor VI continuity.

13.5 ALGORITHM DIFFERENCES

The impacts of algorithm differences posed by sensor characteristic differences on multisensor VI continuity/compatibility are another area that requires careful and thorough investigations.

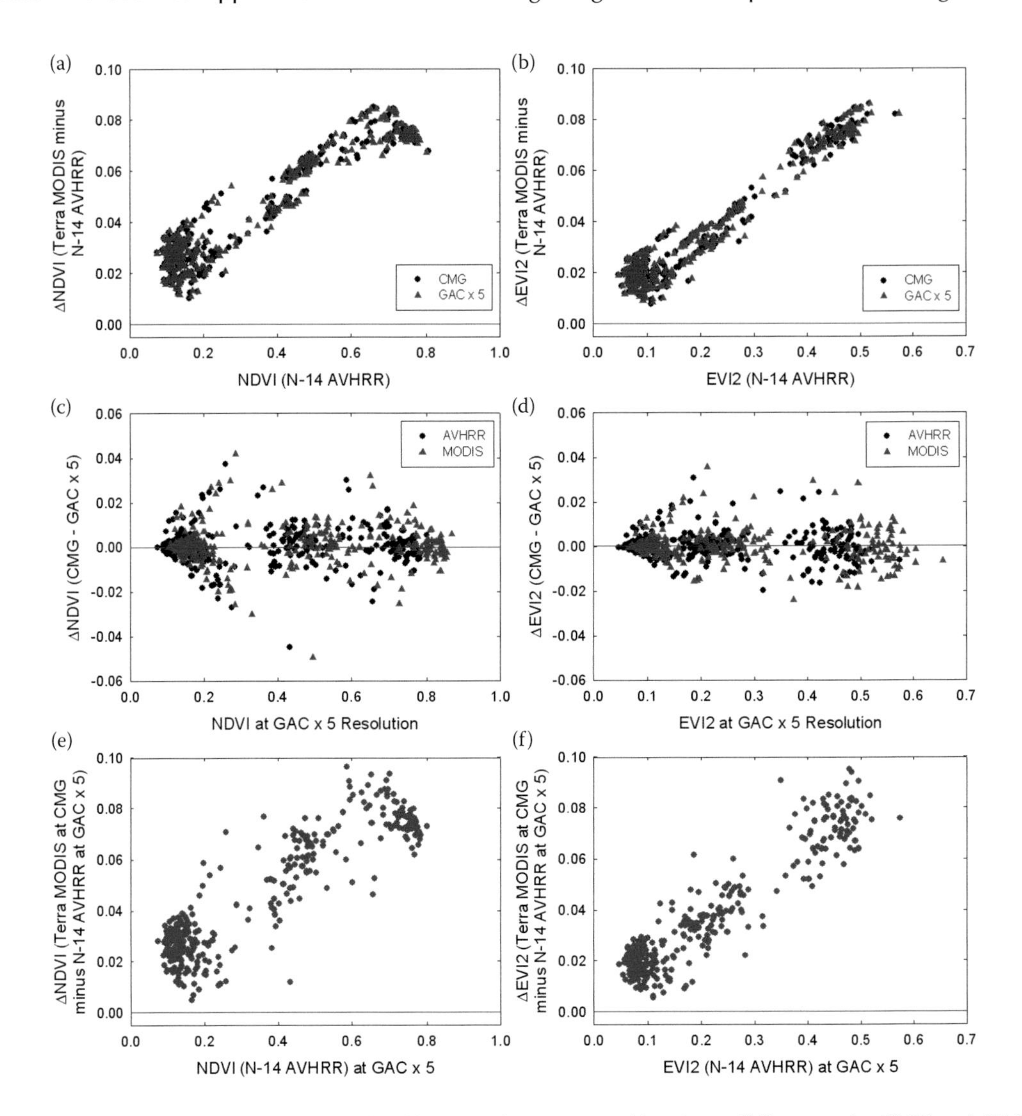

FIGURE 13.7 (a) NDVI and (b) EVI2 differences due to spectral bandpass differences for CMG and GAC resolutions; (c) NDVI and (d) EVI2 differences due to resolution differences for AVHRR and MODIS spectral bandpasses; (e) NDVI and (f) EVI2 differences due to both spectral bandpass (MODIS vs. AVHRR) and spatial resolution (CMG vs. GAC) differences. Here, a GAC pixel is an average of five GAC pixels and, thus, written as "GAC × 5."

Atmospheric correction is one key algorithm difference that exists among sensor products. Hyperspectral remote sensing can be an effective means to address the impacts of various atmospheric correction schemes on cross-sensor VI continuity/compatibility. In what follows, we demonstrate this using the Hyperion scenes in Table 13.1.

Various atmospheric correction schemes for satellite remote sensing have been developed, but the work by Kaufman and Sendra (1988) and Tanré et al. (1992) can be considered to have laid the foundation for operational atmospheric corrections of multispectral data in the solar reflective region over global land surface. The Global Inventory Modeling and Mapping Studies AVHRR NDVI product is corrected only for stratospheric aerosol effects (Pinzon and Tucker, 2014), whereas other AVHRR products, including the Long-Term Data Records NDVI (Franch et al., 2017) and the Conterminous USA and Alaska 1 km AVHRR (Eidenshink, 2006) are corrected for molecular

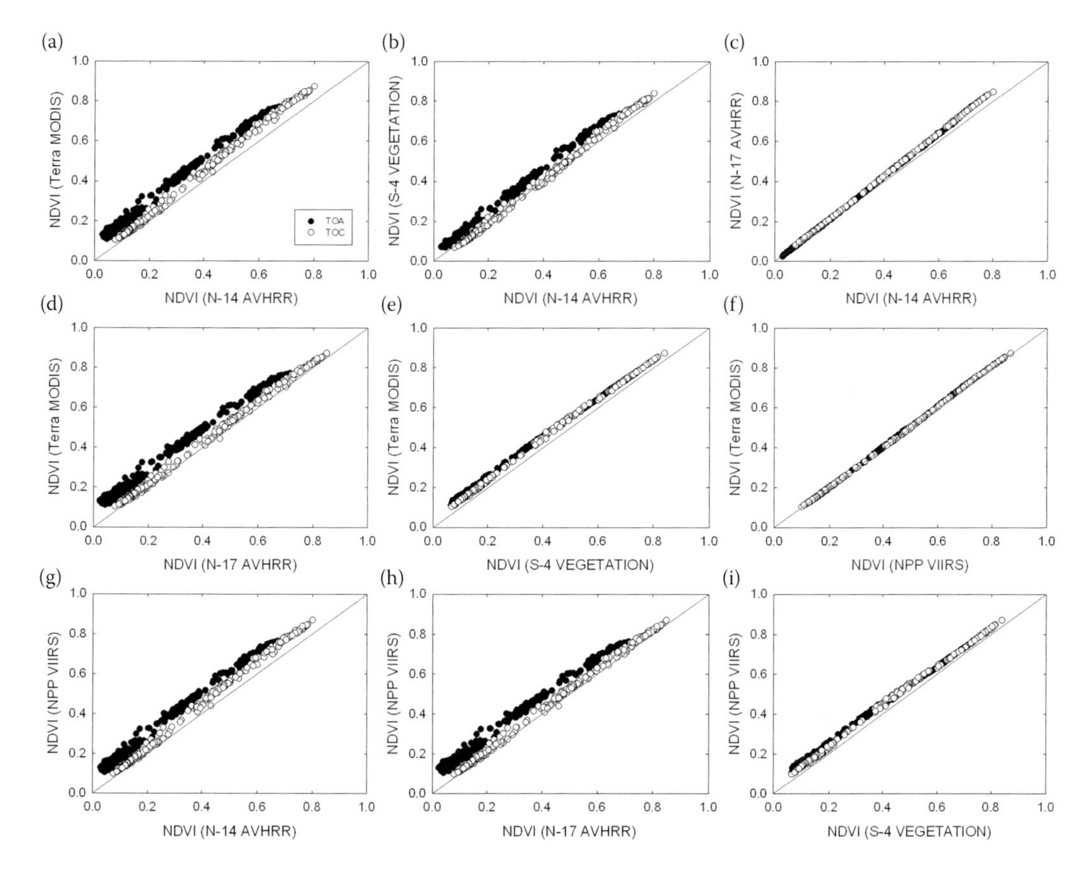

FIGURE 13.8 Cross-sensor NDVI plots for two atmospheric correction schemes: (a) Terra MODIS vs. NOAA-14 AVHRR/2; (b) SPOT-4 VEGETATION vs. NOAA-14 AVHRR/2; (c) NOAA-17 AVHRR/3 vs. NOAA-14 AVHRR/2; (d) Terra MODIS vs. NOAA-17 AVHRR/3; (e) Terra MODIS vs. SPOT-4 VEGETATION; (f) Terra MODIS vs. NPP VIIRS; (g) NPP VIIRS vs. NOAA-14 AVHRR/2; (h) NPP VIIRS vs. NOAA-17 AVHRR/3; (i) NPP VIIRS vs. SPOT-4 VEGETATION.

scattering, and ozone and water vapor absorptions. A total atmospheric correction scheme was implemented for the MODIS VI and VEGETATION NDVI products; however, their algorithms and atmospheric data sources are different (Maisongrande et al., 2004; Vermote and Saleous, 2006). For National Polar-orbiting Partnership (NPP) and Joint Polar Satellite System VIIRS, three VI products are generated: a TOA NDVI without any atmospheric correction and an atmospherically corrected, top-of-canopy (TOC) NDVI and TOC EVI (Vargas et al., 2013).

In Figure 13.8, cross-sensor NDVI relationships are examined for two atmospheric correction schemes: no correction (TOA NDVI) and total correction (TOC NDVI). A GAC sampling scheme was assumed for the simulated AVHRR data, whereas a CMG aggregation scheme was used for all the other simulated sensor data (Section 13.4). Cross-sensor NDVI relationships of both NOAA-14 AVHRR/2 and NOAA-17 AVHRR/3 with other sensors changed with atmospheric correction schemes (Figure 13.8a,b,d,g,h), and thus separate cross-calibrations are required for establishing continuity for the TOA-NDVI and TOC-NDVI. In contrast, NOAA-14 AVHRR and NOAA-17 AVHRR, and Terra MODIS and NPP VIIRS had excellent continuity/compatibility; their cross-sensor NDVI relationships were the same regardless of the atmospheric correction schemes (Figure 13.8c,f). Once a cross-sensor NDVI relationship is established, this can be used to relate the NDVI at either the TOA or TOC level for these sensor pairs. The SPOT-4 VEGETATION sensor also showed relatively robust cross-sensor NDVI relationships with Terra MODIS and NPP VIIRS (Figure 13.8e,i).

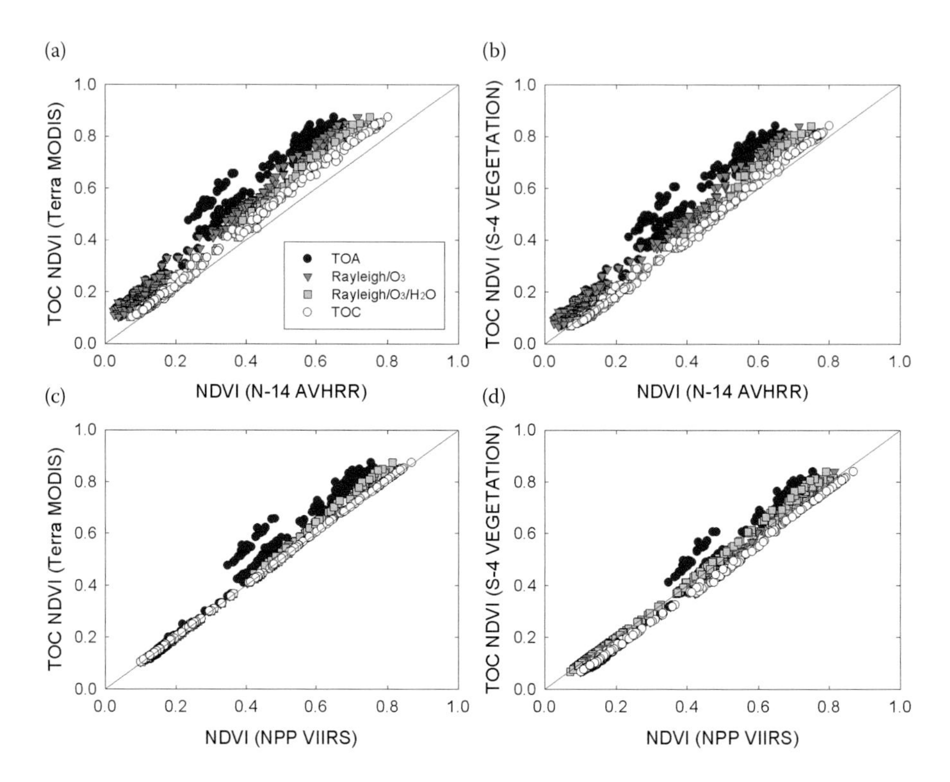

FIGURE 13.9 Cross-sensor NDVI plots for four different atmospheric correction schemes: (a) Terra MODIS vs. NOAA-14 AVHRR/2; (b) SPOT-4 VEGETATION vs. NOAA-14 AVHRR/2; (c) Terra MODIS vs. NPP VIIRS; and (d) SPOT-4 VEGETATION vs. NPP VIIRS. Only NOAA-14 AVHRR and NPP VIIRS values were subjected to various atmospheric correction schemes.

In Figures 13.9 and 13.10, we examine a scenario where total atmosphere-corrected VIs from Terra MODIS and SPOT-4 VEGETATION were compared with partial atmosphere-corrected VIs from NOAA-14 AVHRR and NPP VIIRS. For all the sensor pairs examined, cross-sensor NDVI and EVI2 relationships varied with atmospheric corrections. For both the NDVI and EVI2, larger changes were observed for the relationships involving NOAA-14 AVHRR/2 (Figures 13.9a,b and 13.10a,b). The smallest change was observed for cross-sensor EVI2 relationships involving NPP VIIRS (Figure 13.10c,d).

Overall, these example analyses show important implications for cross-sensor VI continuity/compatibility. It is worthwhile to note that, based on the Hyperion simulation analyses presented here, AVHRR/2 and AVHRR/3, as well as MODIS and VIIRS, maintain excellent continuity for the NDVI regardless of atmospheric corrections.

13.6 ANGULAR EFFECTS

Other sensor and platform characteristics that require consideration are the sun-target-sensor geometry and temporal resolution. While the polar-orbiting wide field of view sensors such as AVHRR, MODIS, VEGETATION, and VIIRS, discussed earlier, provides nearly daily global coverage, they acquire and build up sequential angular views over a period of hours to days (Diner et al., 1999). For time series applications of VIs, this varying observation geometry is considered a source of noise, so various attempts have been made to normalize this noise, including multidate temporal compositing (Holben, 1986; van Leeuwen et al., 1999) or BRDF-based adjustments to nadir-viewing geometry (Schaaf et al., 2002; Vermote et al., 2009). Since the BRDF changes as a function of wavelength and since orbital and scanning characteristics differ from sensor to sensor

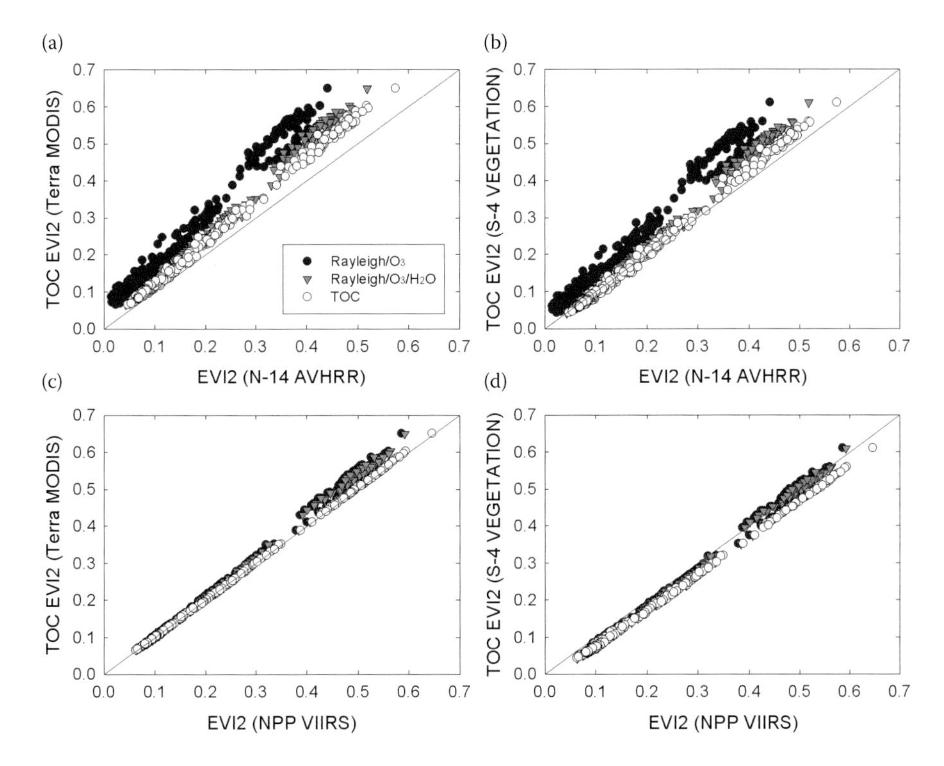

FIGURE 13.10 Same as Figure 13.9, but for EVI2 with three different atmospheric correction schemes.

(including overpass time), the BRDF effect is another issue of multisensor VI continuity that needs to be addressed. Satellite hyperspectral remote sensing has great potential to contribute to this issue. The Hyperion sensor onboard the EO-1 satellite, for example, was capable of imaging within one adjacent World Reference System-2 (WRS-2) path in both the east and west directions, giving the sensor three imaging opportunities with three different view angles per its 16-day orbital repeat cycle. It is, however, unlikely that a large enough number of multiangular hyperspectral images can be obtained over the same target for a short period of time from a single sensor, considering the attainable imaging frequency together with the influence of cloud cover (Galvão et al., 2009). Such observations may only be realized after several hyperspectral sensors are successfully launched in the future.

13.7 DISCUSSIONS

In this chapter, we have discussed the potential uses of satellite hyperspectral remote sensing in multisensor VI continuity/compatibility studies for global long-term monitoring. Some aspects of such uses have been demonstrated using the regional Hyperion data set over the conterminous United States. They included spectral compatibility and spatial compatibility analyses, scaling uncertainty analysis, and the effects of atmospheric correction algorithm differences.

It is important to note that upcoming hyperspectral missions can serve not only as a means of detailed and highly precise characterization of terrestrial vegetation but also as a spaceborne reference for establishing multisensor continuity and compatibility among current, past, and future multispectral sensors. Future hyperspectal missions will provide wider imaging swaths to cover larger land surface areas than the currently achieved coverage. The multiangular hyperspectral observation capability may be one of the next important steps in the field of hyperspectral remote sensing.

ACKNOWLEDGMENTS

The authors would like to thank Alfredo Huete for his continuing discussions about this research topic. The authors would also like to thank Joshua Turner for his assistance in assembling the data set used in this study. We thank the PIs of the five AERONET sites used in this study for their efforts in establishing and maintaining these sites.

REFERENCES

Batra, N., S. Islam, V. Venturini, G. Bisht, and L. Jiang. 2006. Estimation and comparison of evapotranspiration from MODIS and AVHRR sensors for clear sky days over the Southern Great Plains, *Remote Sensing of Environment*, 103(1):1–15. doi:10.1016/j.rse.2006.02.019

Bhattarai, K., D. Conway, and M. Yousef. 2009. Determinants of deforestation in Nepal's central development region, *Journal of Environmental Management*, 91(2):471–488. doi:10.1016/j.jenvman.2009.09.016

Chen, J. M. 1999. Spatial scaling of a remotely sensed surface parameter by contexture, *Remote Sensing of Environment*, 69(1):30–42. doi:10.1016/S0034-4257(99)00006-1

DeFries, R., F. Achard, S. Brown, M. Herold, D. Murdiyarso, B. Schlamadinger, and S. J. de. Carlos. 2007. Earth observations for estimating greenhouse gas emissions from deforestation in developing countries, *Environmental Science & Policy*, 10(4):385–394. doi:10.1016/j.envsci.2007.01.010

Diner, D. J., G. P. Asner, R. Davies, Y. Knyazikhin, J.-P. Muller, A. W. Nolin, B. Pinty, C. B. Schaaf, and J. Stroeve. 1999. New directions in earth observing: Scientific applications of multiangle remote sensing, *Bulletin of the American Meteorological Society*, 80(11):2209–2228. doi:10.1175/1520-0477(1999)080<2209:NDIEOS>2.0.CO;2

Eidenshink, J. 2006. A 16-year time series of 1 km AVHRR satellite data of the conterminous United States and Alaska, *Photogrammetric Engineering and Remote Sensing*, 72(9):1027–1035.

Franch, B., F. E. Vermote, J.-C. Roger, E. Murphy, I. Becker-Reshef, C. Justice, M. Claverie et al. 2017. A 30+ year AVHRR land surface reflectance climate data record and its application to wheat yield monitoring, *Remote Sensing*, 9(3):296. doi:10.3390/rs9030296

Friedl, M. A., F. W. Davis, J. Michaelsen, and M. A. Moritz. 1995. Scaling and uncertainty in the relationship between the NDVI and land surface biophysical variables: An analysis using a scene simulation model and data from FIFE, *Remote Sensing of Environment*, 54(3):233–246.

Gallo, K., L. Ji, B. Reed, J. Eidenshink, and J. Dwyer. 2005. Multi-platform comparisons of MODIS and AVHRR normalized difference vegetation index data, *Remote Sensing of Environment*, 99(3):221–231.

Galvão, L. S., D. A. Roberts, A. R. Formaggio, I. Numata, and F. M. Breunig. 2009. View angle effects on the discrimination of soybean varieties and on the relationships between vegetation indices and yield using off-nadir Hyperion data, *Remote Sensing of Environment*, 113(4):846–856. doi:10.1016/j.rse.2008.12.010

Gao, B. C. 2000. A practical method for simulating AVHRR-consistent NDVI data series using narrow MODIS channels in the 0.5–1.0 μm spectral range, *Remote Sensing of Environment*, 38:1969–1975.

Gitelson, A. A., and Y. J. Kaufman. 1998. MODIS NDVI optimization to fit the AVHRR data series – spectral considerations, *Remote Sensing of Environment*, 66:343–350.

Gunther, K. P., and S. W. Maier. 2007. AVHRR compatible vegetation index derived from MERIS data, *International Journal of Remote Sensing*, 28(3/4):693–708.

Hall, F. G., K. F. Huemmrich, S. J. Goetz, P. J. Sellers, and J. E. Nickeson. 1992. Satellite remote sensing of surface energy balance: Success, failures, and unresolved issues in FIFE, *Journal of Geophysical Research*, 97(D17):19061–19089.

Holben, B. N. 1986. Characteristics of maximum-value composite images from temporal AVHRR data, *International Journal of Remote Sensing*, 7(11):1417–1434.

Holben, B. N., D. Tanré, A. Smirnov, T. F. Eck, I. Slutsker, N. Abuhassan, W. W. Newcomb et al. 2001. An emerging ground-based aerosol climatology: Aerosol Optical Depth from AERONET, *Journal of Geophysical Research*, 106:12,067–12,097.

Hu, Z., and S. Islam. 1997. A framework for analyzing and designing scale invariant remote sensing algorithms, *IEEE Transactions on Geoscience and Remote Sensing*, 35(3):747–755.

Huete, A., K. Didan, T. Miura, E. P. Rodriguez, X. Gao, and L. G. Ferreira. 2002. Overview of the radiometric and biophysical performance of the MODIS vegetation indices, *Remote Sensing of Environment*, 83(1–2):195–213. doi:10.1016/S0034-4257(02)00096-2

Huete, A., K. Ho-Jin, and T. Miura. 2005. Scaling dependencies and uncertainties in vegetation index - biophysical retrievals in heterogeneous environments, *Proceedings of IGARSS '05*, 7:5029–5032.

Huete, A., T. Miura, H. Yoshioka, P. Ratana, and M. Broich. 2014. Indices of vegetation activity, *Biophysical Applications of Satellite Remote Sensing* (Hanes, J. M., editor), Springer, pp. 1–41.

Jepson, W., C. Brannstrom, and A. Filippi. 2010. Access regimes and regional land change in the Brazilian Cerrado, 1972–2002, *Annals of the Association of American Geographers*, 100(1):87–111. doi:10.1080/00045600903378960

Ji, L., and K. Gallo. 2006. An agreement coefficient for image comparison, *Photogrammetric Engineering and Remote Sensing*, 72(7):823–833.

Ji, L., K. Gallo, J. C. Eidenshink, and J. Dwyer. 2008. Agreement evaluation of AVHRR and MODIS 16-day composite NDVI data sets, *International Journal of Remote Sensing*, 29(16):4839–4861.

Jiang, Z., A. R. Huete, K. Didan, and T. Miura. 2008. Development of a two-band enhanced vegetation index without a blue band, *Remote Sensing of Environment*, 112(10):3833–3845. doi:10.1016/j.rse.2008.06.006

Kaufman, Y. J., and C. Sendra. 1988. *Algorithm for automatic atmospheric corrections to visible and near-IR satellite imagery, International Journal of Remote Sensing*, 9(8):1357–1381.

Kaufman, Y. J., D. Tanré, L. A. Remer, E. F. Vermote, A. Chu, and B. N. Holben. 1997. Operational remote sensing of tropospheric aerosol over land from EOS moderate resolution imaging spectroradiometer, *Journal of Geophysical Research*, 102(D14):17051–17067.

Kim, Y., A. R. Huete, T. Miura, and Z. Jiang. 2010. Spectral compatibility of vegetation indices across sensors: A band decomposition analysis with Hyperion data, *Journal of Applied Remote Sensing*, 4:043520. doi:10.1117/1.3400635

Kim, Y., J. S. Kimball, K. Didan, and G. M. Henebry. 2014. Response of vegetation growth and productivity to spring climate indicators in the conterminous United States derived from satellite remote sensing data fusion, *Agricultural and Forest Meteorology*, 194(0):132–143.

Maisongrande, P., B. Duchemin, and G. Dedieu. 2004. VEGETATION/SPOT: An operational mission for the Earth monitoring; presentation of new standard products, *International Journal of Remote Sensing*, 25(1):9–14.

Miettinen, J., and S. C. Liew. 2009. Estimation of biomass distribution in Peninsular Malaysia and in the islands of Sumatra, Java and Borneo based on multi-resolution remote sensing land cover analysis, *Mitigation and Adaptation Strategies for Global Change*, 14(4):357–373. doi:10.1007/s11027-009-9169-6

Miura, T., A. Huete, and H. Yoshioka. 2006. An empirical investigation of cross-sensor relationships of NDVI and red/near-infrared reflectance using EO-1 Hyperion data, *Remote Sensing of Environment*, 100(2):223–236.

Miura, T., H. Yoshioka, and T. Suzuki. 2008. Evaluation of spectral vegetation index translation equations for the development of long-term data records, *IGARSS*, 3:III–712.

Miura, T., J. P. Turner, and A. R. Huete. 2013. Spectral compatibility of the NDVI across VIIRS, MODIS, and AVHRR: An analysis of atmospheric effects using EO-1 Hyperion, *IEEE Transactions on Geoscience and Remote Sensing*, 51(3):1349–1359. doi:10.1109/TGRS.2012.2224118

Myneni, R. B., R. R. Nemani, and S. W. Running. 1997. Estimation of global leaf area index and absorbed par using radiative transfer models, *IEEE Transactions on Geoscince and Remote Sensing*, 35(6): 1380–1393.

Obata, K., T. Miura, and H. Yoshioka. 2012a. Scaling effects in area-averaged values of two-band spectral vegetation indices represented in a general form, *Journal of Applied Remote Sensing*, 6(1):063585–063585. doi:10.1117/1.JRS.6.063585

Obata, K., T. Wada, T. Miura, and H. Yoshioka. 2012b. Scaling effect of area-averaged NDVI: Monotonicity along the spatial resolution, *Remote Sensing*, 4(1):160–179. doi:10.3390/rs4010160

Paudel, K. P., and P. Andersen. 2010. Assessing rangeland degradation using multi temporal satellite images and grazing pressure surface model in Upper Mustang, Trans Himalaya, Nepal, *Remote Sensing of Environment*, 114(8):1845–1855. doi:10.1016/j.rse.2010.03.011

Pearlman, J. S., P. S. Barry, C. C. Segal, J. Shepanski, D. Beiso, and S. L. Carman. 2003. Hyperion, a space-based imaging spectrometer, *IEEE Transactions on Geoscience and Remote Sensing*, 41(6):1160–1173.

Pinheiro, A. C. T., R. Mahoney, J. L. Privette, and C. J. Tucker. 2006. Development of a daily long term record of NOAA-14 AVHRR land surface temperature over Africa, *Remote Sensing of Environment*, 103(2):153–164. doi:10.1016/j.rse.2006.03.009

Pinzon, E. J., and J. C. Tucker. 2014. A non-stationary 1981–2012 AVHRR NDVI3g time series, *Remote Sensing*, 6(8):6929–6960. doi:10.3390/rs6086929

Schaaf, C. B., F. Gao, A. H. Strahler, W. Lucht, X. Li, T. Tsang, N. C. Strugnell et al. 2002. First operational BRDF, albedo nadir reflectance products from MODIS, *Remote Sensing of Environment*, 83(1–2):135–148. doi:10.1016/S0034-4257(02)00091-3

Schowengerdt, R. A. 2006. *Remote Sensing: Models and Methods for Image Processing*, 560p.

Sims, D. A., A. F. Rahman, V. D. Cordova, B. Z. El-Masri, D. D. Baldocchi, P. V. Bolstad, L. B. Flanagan et al. 2008. A new model of gross primary productivity for North American ecosystems based solely on the enhanced vegetation index and land surface temperature from MODIS, *Remote Sensing of Environment*, 112(4):1633–1646. doi:10.1016/j.rse.2007.08.004

Steven, M. D., T. J. Malthus, F. Baret, H. Xu, and M. J. Chopping. 2003. Intercalibration of vegetation indices from different sensor systems, *Remote Sensing of Environment*, 88(4):412–422.

Swinnen, E., and F. Veroustraete. 2008. Extending the SPOT-VEGETATION NDVI time series (1998–2006) back in time with NOAA-AVHRR data (1985–1998) for Southern Africa, *IEEE Transactions on Geoscience and Remote Sensing*, 46(2):558–572. doi:10.1109/TGRS.2007.909948

Tanré, D., B. N. Holben, and Y. J. Kaufman. 1992. Atmospheric correction algorithm for NOAA-AVHRR products: Theory and application, *IEEE Transactions on Geoscience and Remote Sensing*, 30(2):231–248.

Teillet, P. M., K. Staenz, and D. J. Williams. 1997. Effects of spectral, spatial, and radiometric characteristics on remote sensing vegetation indices of forested regions, *Remote Sensing of Environment*, 61(1):139–149.

Trishchenko, A. P., J. Cihlar, and Z. Li. 2002. Effects of spectral response function on surface reflectance and NDVI measured with moderate resolution satellite sensors, *Remote Sensing of Environment*, 81(1):1–18.

Trishchenko, A. P. 2009. Effects of spectral response function on surface reflectance and NDVI measured with moderate resolution satellite sensors: Extension to AVHRR NOAA-17, 18 and METOP-A, *Remote Sensing of Environment*, 113(2):335–341. doi:10.1016/j.rse.2008.10.002

Tucker, C. J. 1979. Red and photographic infrared linear combinations for monitoring vegetation, *Remote Sensing of Environment*, 8(2):127–150. doi:10.1016/0034-4257(79)90013-0

Ungar, S. G., J. S. Pearlman, J. A. Mendenhall, and D. Reuter. 2003. Overview of the Earth Observing One (EO-1) mission, *IEEE Transactions on Geoscience and Remote Sensing*, 41(6):1149–1159.

van Leeuwen, W. J. D., A. R. Huete, and T. W. Laing. 1999. MODIS vegetation index compositing approach: A prototype with AVHRR data, *Remote Sensing of Environment*, 69(3):264–280. doi:10.1016/S0034-4257(99)00022-X

van Leeuwen, W. J. D., B. J. Orr, S. E. Marsh, and S. M. Herrmann. 2006. Multi-sensor NDVI data continuity: Uncertainties and implications for vegetation monitoring applications, *Remote Sensing of Environment*, 100(1):67–81. doi:10.1016/j.rse.2005.10.002

Vargas, M., T. Miura, N. Shabanov, and A. Kato. 2013. An initial assessment of Suomi NPP VIIRS vegetation index EDR, *Journal of Geophysical Research-Atmospheres*, 118(22):12,301–12,316. doi:10.1002/2013JD020439

Vermote, E., D. Tanré, J. L. Deuzé, M. Herman, J. J. Morcrette, and S. Y. Kotchenova. 2006. Second Simulation of a Satellite Signal in the Solar Spectrum - Vector (6SV) User Guide Version 3.

Vermote, E. F., and N. Z. Saleous. 2006. Operational atmospheric correction of MODIS visible to middle infrared land surface data in the case of an infinite Lambertian target, *Earth Science Satellite Remote Sensing, Science and Instruments*, Volume 1 (al, J. J. Q. E., editor), Tsinghua University Press, Beijing, pp. 123–153.

Vermote, E., C. O. Justice, and F. M. Breon. 2009. Towards a generalized approach for correction of the BRDF effect in MODIS directional reflectances, *IEEE Transactions on Geoscience and Remote Sensing*, 47(3):898–908. doi:10.1109/TGRS.2008.2005977

Wolfe, R. E., M. Nishihama, A. J. Fleig, J. A. Kuyper, D. P. Roy, J. C. Storey, and F. S. Patt. 2002. Achieving sub-pixel geolocation accuracy in support of MODIS land science, *Remote Sensing of Environment*, 83(1–2):31–49. doi:10.1016/S0034-4257(02)00085-8

Yoshioka, H., T. Miura, and A. R. Huete. 2003. An isoline-based translation technique of spectral vegetation index using EO-1 Hyperion data, *IEEE Transactions on Geoscience and Remote Sensing*, 41(6):1363–1372. doi:10.1109/TGRS.2003.813212

Yoshioka, H., T. Miura, and H. Yamamoto. 2005. Relationships of spectral vegetation indices for continuity and compatibility of satellite data products, *Multispectral and Hyperspectral Remote Sensing Instruments and Applications II, Proceedings of SPIE* vol. 5655 (Larar, A. M., M. Suzuki and Q. Tong, editors), pp. 233–240. doi:10.1117/12.578630

Yoshioka, H., T. Miura, and H. Yamamoto. 2006. Investigation on functional form in cross-calibration of spectral vegetation index, *Proc. SPIE Remote Sensing and Modeling of Ecosystems for Sustainability III*, 6298(1):629813–629819. doi:10.1117/12.681564

Yoshioka, H., T. Wada, K. Obata, and T. Miura. 2008. Monotonicity of area averaged NDVI as a function of spatial resolution based on a variable endmember linear mixture model, *IGARSS*, 3:III–415.

Yoshioka, H., T. Miura, and K. Obata. 2012. Derivation of relationships between spectral vegetation indices from multiple sensors based on vegetation isolines, *Remote Sensing*, 4(3):583–597. doi:10.3390/rs4030583

Zhang, X., B. Tan, and Y. Yu. 2014. Interannual variations and trends in global land surface phenology derived from enhanced vegetation index during 1982–2010, *Int J Biometeorol*, 58(4):547–564. doi:10.1007/s00484-014-0802-z

Section VI

Hyperspectral Remote Sensing of Other Planets

14 Hyperspectral Analysis of Rocky Surfaces on Earth and Other Planetary Bodies

R. Greg Vaughan, Timothy N. Titus, Jeffrey R. Johnson, Justin J. Hagerty, Laurence A. Soderblom, Paul E. Geissler, David P. Mayer, and Will M. Grundy

CONTENTS

14.1 INTRODUCTION

This book is focused on studies of vegetation on Earth using hyperspectral remote sensing methods. However, it is appropriate to extend the application of these methods to other rocky bodies in our solar system for a variety of reasons. First, minerals, soils, and rocks form the substrate on which vegetation grows on Earth. Compositional analyses of these components with hyperspectral data provide essential background information for distinguishing, identifying, and removing their effects on vegetation spectra. Second, variation in the distribution and chemical and physical properties of soil and rock has been demonstrated to have a significant effect on factors such as moisture retention, dust production, and the presence and distribution of biological species ranging from bacteria, fungi, grasses, shrubs, trees, and small mammals to humans [1]. These factors in turn can have profound influences on human health.

Recent advances in the development and use of hyperspectral data for rocks, soils, and minerals have led to improvements in our understanding of how such data are collected, calibrated and otherwise processed, and applied to understand the geology and biology of materials on Earth and other planetary surfaces. Geologic remote sensing research over the last 40 years has developed or improved upon methods for image processing and analysis, data set validation, remote and in situ data collection, and field calibration and has established nominal wavelengths for the detection of common minerals [2–9]. Early studies used multispectral measurements in the visible, near-infrared, and shortwave infrared (VNIR/SWIR) range (400–2500 nm) to map weathering and alteration minerals [4,10–12]. The advent of hyperspectral VNIR/SWIR measurements from airborne and spaceborne platforms and better calibration enabled the emergence of imaging spectroscopy, and images with high-resolution spectral data for each pixel could be directly compared to the reference spectra of pure minerals and used for remote mineral identification [6,13]. For the past 20 years, much research has been focused on the development of image analysis and processing techniques for hyperspectral VNIR/SWIR data, with particular attention paid to atmospheric correction and unique mineralogical identification [14–17].

Primary rock-forming minerals as well as many secondary weathering and alteration minerals exhibit wavelength-dependent, or spectral, absorption features throughout the visible and infrared wavelengths [18–22]. These features result from the selective absorption of photons with discrete energy levels and are dependent on the elemental composition, crystal structure, and chemical bonding characteristics of a mineral and are, therefore, diagnostic of mineralogy [7,20,23,24]. The identification of a material based on its infrared spectrum requires access to the spectra of well-characterized specimens. Thus, the infrared spectral properties for many natural and artificial materials have been measured in the laboratory and constitute the empirical basis for the surface mapping applications of infrared remote sensing applied to Earth and other planets [18–22,25–27]. Archives of infrared spectral data for rocks, minerals, and vegetation are available for reference from the United States Geological Survey (USGS) spectral library (http://speclab.cr.usgs.gov) [26] and the NASA Jet Propulsion Laboratory (JPL) Advanced Spaceborne Thermal Emission and Reflection Radiometer (ASTER) spectral library (http://speclib.jpl.nasa.gov) [27]. (Note: The "ASTER spectral library" was renamed the "ECOSTRESS spectral library" and version 1.0 released on February 2, 2018. The ECOsystem Spaceborne Thermal Radiometer Experiment on Space Station (ECOSTRESS) is a multispectral thermal infrared instrument that was launched to the International Space Station on June 28, 2018.)

As described here, geologic analyses of remotely acquired multi- and hyperspectral data on planetary bodies have resulted in major scientific discoveries that help to put Earth science into a broader context. For example, correlating orbital remote sensing data for Mars from the Mars Reconnaissance Orbiter, Compact Reconnaissance Imaging Spectrometer for Mars (CRISM) instrument with those of high-resolution views of boulders, craters, and sediment layers by the High-Resolution Imaging Science Experiment (HiRISE) has helped to identify the likely presence of water at the surface of Mars through the discovery of minerals such as gray hematite, clays, and sulfur-rich soils [28,29]. Similar nodules have been observed in southern Utah [30,31], where they appear to have formed by precipitation from fluid that leached out iron-rich minerals and precipitated them at more chemically favorable locations in the host sandstone. This discovery continues to be explored, in part because on Earth it is known that bacteria can make such concretions form more quickly [31,32]. The recent discovery of the presence of water on the surface of the Moon using hyperspectral Moon Mineralogy Mapper (M^3) data [33] indicates that water has been much more common on that body and elsewhere in our solar system than previously thought. These and other discoveries tell us that we still have much to learn about the geology and biology of Earth and other planets in our solar system. Fortunately, remote sensing analyses such as those described in this book provide the tools needed for such future discoveries.

14.1.1 Planetary Bodies and Hyperspectral Instruments

14.1.1.1 Earth

There is currently a host of Earth observing instruments that acquire hyperspectral images in the 400–2500 nm wavelength range [6,34–36]. There has been an increasing number of commercial companies (too numerous to count) designing, building, and flying imaging spectrometers on aircraft as well as uninhabited aerial vehicles. Table 14.1 summarizes the specifications for a selection of these terrestrial imaging spectrometers.

14.1.1.2 Mercury

The MErcury Surface, Space ENvironment, Geochemistry, and Ranging (MESSENGER) spacecraft performed two flybys of Mercury in 2008 and a third in September 2009. MESSENGER became the first spacecraft to enter orbit around Mercury in March 2011. After completing its primary and two extended missions, MESSENGER used its remaining fuel for a planned deorbit and impacted onto Mercury in April 2015. The Mercury Atmospheric and Surface Composition Spectrometer (MASCS) instrument obtained spectra (300–1450 nm) of the surface. The purpose of the MASCS was to help determine the surface mineralogy of Mercury and to help characterize the exosphere (i.e., the outermost layer of Mercury's tenuous atmosphere). The MASCS consisted of two instruments: an Ultraviolet-visible Spectrometer (UVVS) designed primarily for observations of the exosphere,

TABLE 14.1
Summary of Earth and Planetary Hyperspectral Remote Sensing Instruments

		Hyperspectral Instrument	Spectral Range (nm)	Number of Channels	Spectral Bandpass (nm)	Pixel Size(s)	Operational Dates
Earth							
	Airborne	AVIRIS-CL[a]	380–2510	224	10	4–20 m	1989–present
		AVIRIS-NG[b]	380–2510	480	5	4–20 m	2009–present
		ProSpecTIR-VS[c]	400–2450	256	2.3–20	1–10 m	~2000–present
		HyMap[d]	400–2500	128	15	3–10 m	~1997–present
		CASI-1500 h[e]	380–1050	288	2–12	0.5–10 m	~1990–present
	Spaceborne	EO-1—Hyperion[f]	400–2500	220	10	30 m	2001–2017
Mercury		MESSENGER—MASCS[g]	220–1450	768	0.2–0.5	1–650 km	2004–2015
Moon		Chandrayaan-1—Moon Mineralogy Mapper[h]	400–2900	260	10	70–140 m	2008–2009
Mars		Mars Express—OMEGA[i]	350–5100	352	7–20	300 m–4.8 km	2003–present
		Mars Reconnaissance Orbiter—CRISM[j]	362–3920	545	6.55	15.7 m–200 m	2005–present
Ceres and Vesta		Dawn—VIR VNIR[k]	250–1000	432	1.9	35 m–11 km	2007–present
		Dawn—VIR SWIR[k]	950–5000	432	9.5	35 m–11 km	2007–present
Jupiter		Galileo—NIMS[l]	700–5200	1–408	12.5 & 25	50–500 km	1989–2003
Saturn		Cassini—VIMS[m]	300–5100	352	7 & 14	10–20 km	1997–2017
Pluto		New Horizons—Ralph[n]	1200–2500	256	7 & 14	10–20 km	2007–present

[a] Airborne Visible/Infrared Imaging Spectrometer—Classic (http://aviris.jpl.nasa.gov).
[b] Airborne Visible/Infrared Imaging Spectrometer—Next Generation (https://aviris-ng.jpl.nasa.gov).
[c] Spectral Technology and Innovative Research Corporation Hyperspectral Imaging Spectrometer (http://www.spectir.com/technology/hyperspectral-sensor).
[d] HyVista corporation Hyperspectral Mapper, Developed by Integrated Spectronics (http://www.hyvista.com/technology/sensors/hymap).
[e] Compact Airborne Spectrographic Imager (http://www.itres.com/imagers).
[f] Hyperion (https://eo1.usgs.gov/sensors/hyperion).
[g] Mercury Atmospheric and Surface Composition Spectrometer (http://www.messenger-education.org/instruments/mascs.htm).
[h] M^3 (https://solarsystem.nasa.gov/moon/newsdisplay.cfm?Subsite_News_ID=49330&SiteID=6&iSiteID=1).
[i] Observatoire pour la Minéralogie, l'Eau, les Glaces et l'Activité (http://sci.esa.int/science-e/www/object/index.cfm?fobjectid=34826&fbodylongid=1598).
[j] Compact Reconnaissance Imaging Spectrometer for Mars (http://crism.jhuapl.edu).
[k] Visible and Infrared Spectrometer (https://dawn.jpl.nasa.gov/spacecraft/vir.html).
[l] Near-Infrared Mapping Spectrometer (http://www2.jpl.nasa.gov/galileo/instruments/nims.html).
[m] Visual and Infrared Mapping Spectrometer (http://wwwvims.lpl.arizona.edu).
[n] Ralph (https://www.nasa.gov/images/nh-ralph-instrument).

and a Visible-InfraRed Spectrograph (VIRS) intended to provide observations to determine surface mineralogy. http://www.messenger-education.org/instruments/mascs.htm.

14.1.1.3 Moon

The NASA M^3 was a hyperspectral, push-broom imaging spectrometer that was a guest instrument on Chandrayaan-1, the first mission to the Moon from the Indian Space Research Organization (ISRO). Intended for both global mapping and targeted operation, the main objective of M^3 was to allow scientists to examine lunar mineralogy at high spatial and spectral resolution [37]. M^3 targets included such features as outcrops exposed at the walls and central peaks of large craters, complex volcanic terrain, boundaries where different kinds of rocks converge, unusual or rare compositions, and lunar polar regions [37].

14.1.1.4 Mars

Mars has intrigued humankind since before the beginning of recorded history. Until the 1960s, Mars was believed to be a lush planet with canals and agricultural irrigation [38]. This belief became the basis for such science fiction classics as *The War of the Worlds* by H.G. Wells. The first flyby of Mars by Mariner IV in 1965 put an end to such fantasies, revealing the red planet as a cold, arid desert. While the use of remote sensing observations to track vegetation is meaningless for Mars, there are many other applications, such as the monitoring of seasonal changes in surface volatiles or the identification of minerals. The understanding of NIR observations of a planetary surface devoid of vegetation may provide insights into remote sensing of terrestrial surfaces with sparse or nonexistent vegetation, for example, extremely arid deserts and ice-covered regions. Several spacecraft have visited Mars over the last five decades. Since 1997, three rovers, one lander, and four orbiters have successfully arrived at Mars and sent back valuable data to Earth. Most of these spacecraft far exceeded their nominal mission lifetimes (Table 14.1). Two of these spacecraft, Mars Express and Mars Reconnaissance Orbiter, carried VNIR/SWIR imaging spectrometers, which are still functional at the time of this writing. The OMEGA (Observatoire pour la Minéralogie, l'Eau, les Glaces et l'Activité) instrument is a visible and infrared mineral mapper on board the Mars Express orbiter designed to globally map minerals, water (hydrated minerals or ice), and ice. The Compact Reconnaissance Imaging Spectrometer for Mars (CRISM) is a high-spatial-resolution (20–200 m/pixel) imaging spectrometer designed to target regions to detect and map minerals and ice.

14.1.1.5 Main Belt Asteroids: Ceres and Vesta

The Dawn spacecraft has visited the two largest of the main asteroid belt (MAB) objects: asteroid 4 Vesta and 1 Ceres. Ceres is sufficiently large that it is also classified as a dwarf planet, containing 30% of the total mass of the MAB. Dawn carries three types of instruments: two German Framing Cameras (FC1 & FC2), an Italian VNIR/SWIR slit spectrometer (VIR), and an American Gamma Ray and Neutron Detector (GRaND). The primary focus has been comparing and contrasting the surface of these two very different and distinct planetary bodies.

14.1.1.6 Jupiter

The Near-Infrared Mapping Spectrometer (NIMS) instrument on the Galileo spacecraft was the first imaging spectrometer to be sent to the outer solar system. Galileo was launched in 1989 on a grand tour of the solar system that crossed paths with Venus, Earth and the Moon (twice), and two small asteroids before finally arriving at Jupiter in late 1996, just in time to witness the kamikaze comet Shoemaker-Levy 9 crash into the giant planet. For more than 6 years, Galileo orbited Jupiter and made 29 successful close flybys of its major moons and many distant observations of Jupiter's atmosphere and rings and entourage of satellites. Each of the objects visited by Galileo became targets for scientific observations and calibration exercises for the novel NIMS instrument; in Section 14.2.6 we focus on results from the Jupiter system, the primary scientific objective.

14.1.1.7 Saturn

The goals of the Visual and Infrared Mapping Spectrometer (VIMS) aboard the Cassini spacecraft included study of the composition, dynamics, clouds, and thermophysics of the atmospheres of Saturn and Titan and the identification and mapping of surface compositions of Titan, the other icy Saturnian satellites, and Saturn's rings. VIMS was an imaging spectrometer that covered the 350–5170 nm spectral region [39]. The VIMS visible-light channel was a multispectral imager that covered the spectral range from 350 to 1050 nm and used a frame transfer charge-coupled device (CCD) detector on which spatial and spectral information was simultaneously stored. The VIMS infrared channel was a spatial-scanning spot spectrometer that covered the wavelength range from 850 to 5170 nm.

14.1.1.8 Pluto and the Kuiper Belt

NASA's New Horizons spacecraft was launched in 2006. After a 2007 Jupiter flyby and gravity assist, it encountered the Pluto system in 2015 [40]. Following the Pluto flyby, the mission was extended to send the spacecraft to explore a small Kuiper belt object in early 2019. New Horizons maps spectral reflectance at visible and near-infrared wavelengths using an imaging spectrometer system called Ralph [41], which provides four visible-wavelength channels from 400 to 975 nm and 256 near-infrared channels from 1200 to 2500 nm. Ralph collects data without use of moving parts to minimize risk of potential failures during the long cruise to the outer solar system. Spatial and spectral scanning are accomplished by rolling the entire spacecraft. The immobile design saves mass and provides great durability, but it creates challenges for data processing, described in Section 14.2.8. Table 14.1 also summarizes the specifications for imaging spectrometers on planetary missions.

14.1.1.9 Future/Forthcoming Instruments

One future Earth observing hyperspectral sensor mission recommended by the National Research Council, Decadal Survey (Earth Science and Applications from Space: National Imperatives for the Next Decade and Beyond) is the Hyperspectral Infrared Imager (HyspIRI) mission [42]. HyspIRI will have (1) a hyperspectral imaging spectrometer measuring radiance in the 380–2500 nm range with 10 nm spectral resolution, 60 m pixels, and a 19 day equatorial revisit time and (2) a multispectral imager measuring radiance in seven channels in the 7,500–12,000 nm (thermal infrared region) plus one channel at 4000 nm (mid-infrared region) for measuring hot spots due to fires or volcanic activity, with 60 m pixels and a 5 day revisit time (due to a wider swath than the VNIR/SWIR hyperspectral imager). The goals of the HyspIRI mission as related to VNIR/SWIR hyperspectral measurements are to detect responses of ecosystems to human land management and climate change variability, including studying the patterns and spatial distribution of ecosystems and their components; ecosystem function, physiology, seasonal activity, and relation to human health; biogeochemical cycles; and land surface composition [42]. See the HyspIRI website for more information: https://hyspiri.jpl.nasa.gov.

In 2018, the Italian Space Agency (ASI: Agenzia Spaziale Italiana) plans to launch the PRecursore IperSpettrale della Missione Applicativa (PRISMA) spacecraft into sun-synchronous Earth orbit as a preoperational technology demonstration mission. PRISMA will be a medium-resolution (30 m/pixel) hyperspectral imaging spectrometer with 250 channels in the VNIR/SWIR (400–2500 nm) spectral region. The instrument will also acquire coregistered panchromatic imagery at 5 m/pixel [43]. Additionally, the German space agency is planning a 2019 launch of its hyperspectral satellite mission, Environmental Mapping and Analysis Program (EnMAP), which will also feature a VNIR/SWIR hyperspectral imaging spectrometer capable of acquiring 30 m pixel data on a global scale.

The Europa Jupiter System Mission (EJSM) planned for launch in 2020 will include the Jupiter Europa Orbiter and the Jupiter Ganymede Orbiter. The mission will explore the Jupiter system, focusing on potential habitable environments among the planet's icy satellites, with special emphasis on Europa and Ganymede because of the possibility they support internally active oceans. As part of the planned science payloads, hyperspectral imaging instruments are planned (via competitive selection) that will nominally cover the 800–2500 nm spectral region, with a targeted spectral resolution of 4 nm and instantaneous field of view (FOV) smaller than 100 m (https://solarsystem.nasa.gov/missions/ejsm).

NASA's Lucy mission is planned for launch in 2021 to tour the Trojan asteroids orbiting in Jupiter's L4 and L5 Lagrange points. It will carry an updated version of the Ralph instrument called L'Ralph to distinguish it from the New Horizons version. Key changes include use of a much larger format infrared detector array (2048 × 2048 pixels, rather than 256 × 256) that is sensitive out to longer wavelengths (at least 3400 nm, compared with 2500 nm). L'Ralph will also make use of a scan platform and a scan mirror to provide more control over the scan rate.

14.1.1.10 Synergy with Remote Measurements outside the VNIR/SWIR Spectral Range

14.1.1.10.1 Thermal Infrared (TIR) Imaging and Spectroscopy

The thermal infrared (TIR) spectral range (~7,000–14,000 nm) has long been used for terrestrial and planetary geologic studies to map surface materials based on differences in wavelength-dependent spectral features. Numerous studies have demonstrated the use of TIR imaging for applications including mineral and lithologic mapping, geothermal site characterization, determination of mineral and soil properties, estimation of energy fluxes, estimation of evapotranspiration and soil moisture, drought monitoring, detection of fires and volcanic thermal features, and mapping vegetation species [44–54]. Improvements in infrared detector technology over the last 30 years have led to the advent of hyperspectral imaging spectrometers operating in these longer wavelength regions. In addition, recent studies have made advances in the identification of plant species using TIR spectral measurements [54,55]. Currently there are a number of airborne hyperspectral TIR imaging spectrometers in use, including the Spatially Enhanced Broadband Array Spectrograph System (SEBASS) [56], the airborne hyperspectral imager (AHI) [57], the Itres TASI-600 (http://www.itres.com), the Specim AisaOWL (http://www.specim.fi), the Telops Hyper-Cam (http://telops.com/products/hyperspectral-cameras), and the Hyperspectral Thermal Emission Spectrometer (HyTES), built and operated by NASA's Jet Propulsion Laboratory (https://hytes.jpl.nasa.gov) [58].

The VNIR/SWIR and TIR spectral ranges have traditionally been treated separately for a variety of reasons. Differences in the source of radiance (solar reflection vs. thermal emission) require different approaches to the acquisition, calibration, and processing of data. Also, infrared detector technology has played a role in the spectral ranges that are measured by different instruments. The complementary nature of these two wavelength regions has been explored in detail using multi-instrument packages or instruments capable of making multichannel spectral measurements in both wavelength regions [45,59–61]. Some minerals exhibit diagnostic spectral features in one wavelength region, but not the other [62]. Also, some minerals that often occur together in nature have overlapping spectral features, making unique mineral identification ambiguous using one wavelength region alone. Therefore, the synthesis of VNIR/SWIR and TIR provides a way to identify minerals more uniquely and remotely map more minerals and mineral assemblages [62].

14.1.1.10.2 Mars Orbiter Laser Altimeter (MOLA) and Lunar Orbiter Laser Altimeter (LOLA)

Laser altimeters placed in orbit around Mars and the Moon provide not only high-spatial-resolution topographic models but also are important in determining accurate placement of images and geometric registration among multiple imaging and hyperspectral planetary data sets. Sophisticated cartographic techniques allow spatial projection of images onto regional digital elevation models that improve overall accuracy of feature positions and minimize distortions or seams in large image mosaics. This is often essential to enable understanding particular surface features observed in multiple data sets (often acquired at different spatial resolutions). Further, knowledge of local topographic slopes is key in radiometrically calibrating data to compensate for photometric effects induced by millimeter-scale surface properties, light reflected from nearby features (e.g., crater rims), or light reflected by atmospheres (e.g., Mars, Titan).

14.1.1.10.3 Neutron/Gamma Ray

The use of gamma ray and neutron detectors to map surface and near-surface (typically, less than a meter deep) elemental abundances is seeing wide application on Mars [63,64], the Moon [65,66], and asteroids [67]. The detector footprint of the surface is altitude dependent but typically varies from 600 km in diameter for Mars to 5 km for the Moon.

14.1.2 Overview of Hyperspectral Analysis Techniques

14.1.2.1 Radiometric Calibration

Radiometric calibration of hyperspectral imagers varies among instruments, but most employ some type of onboard calibration using integrating spheres illuminated either by sunlight or onboard illumination. Shutter systems are often used for the acquisition of dark measurements to track real-time noise variations with sensor temperature and exposure times. Stray-light corrections can be necessary, often using a combination of preflight measurements or models and in-flight observations of stars or planetary surfaces with minimal spectral or textural variations to serve as working "flat field" observations as supplements to any similar onboard observations. Calibrated scene radiances are often converted to relative reflectance by convolving the solar flux at the planet's solar distance with specific bandpasses.

14.1.2.2 Atmospheric Correction/Compensation

In the VNIR/SWIR wavelength range, radiance measured at the sensor contains information about surface-reflected radiance as well as radiance scattered by molecules and aerosols in the atmosphere and absorbed by atmospheric gases. Of all the gaseous constituents of Earth's atmosphere, seven gases produce strong absorption of radiance in the VNIR/SWIR region: H_2O, CO_2, O_3, CH_4, N_2O, CO, and O_2 [68]. Although H_2O is not the most abundant gas, it is one of the most important in terms of how much it decreases the transmissivity of the atmosphere due to strong absorption. There is also a significant amount of variation in the distribution of these gases, especially H_2O, both spatially and temporally. In addition to the absorbing effects of atmospheric gases, molecules and aerosols in the atmosphere scatter solar radiance. The modeling of atmospheric absorption and scattering processes is known as radiative transfer theory and is reviewed in more detail by Hapke [69–70].

For the atmospheric correction of terrestrial VNIR/SWIR hyperspectral data, a radiative transfer modeling approach is commonly used to correct for the effects of atmospheric conditions, backscattered radiance, albedo, and viewing geometry [71].

14.1.2.3 Spectral Indices

Similar to many terrestrial studies, first-order analyses of planetary data sets often involve the use of ratios or slopes between bands, often in combination with band depths computed at wavelengths diagnostic of particular minerals [72]. Such techniques provide computationally efficient means of identifying overall trends in spectral signatures, covering potentially large regions of planetary surfaces. Such products often provide necessary overviews of regions, from which anomalous spectral features can subsequently be investigated using higher spectral or spatial resolution observations.

In early lunar studies, ratios of narrow band telescopic images acquired at 400 and 560 nm were used to estimate the weight percent of TiO_2 in the mare [73,74]. Early Martian spectroscopy made use of ratios and slopes to estimate dust contamination and movement on local and regional surfaces [75], whereas more recent analyses using CRISM hyperspectral data use a combination of slopes, ratios, and band depths to routinely provide maps of a suite of mineral and surface types [76]. In the outer solar system, relatively lower albedo contaminants on Europa's trailing hemisphere exhibit steep positive spectral slopes (red color) up to 1000 nm. Slight differences in this slope are likely owing to variations in the water abundance and grain size of ice/contaminant mixtures [77].

14.1.2.4 Spectral Mixture Modeling

Natural geologic surfaces are often partially covered with nongeologic materials (e.g., vegetation) or composed of mixtures of minerals with varying grain sizes and differing degrees of compaction/solidification. These factors influence remote spectral measurements and can limit the number of pixels that can be classified and mapped accurately.

Mixing can exist at various scales and affects the measured infrared spectral properties of an area. When minerals in a FOV are physically separated such that there is no scattering between components, the spectral signature of the area represents the sum of the fractions of each component and is thus a linear mixture. Intimate mixing occurs at smaller scales when different minerals are in close contact on a single scattering surface. The presence of rock coatings causes another type of mixing that varies depending on the thickness of the coating and the wavelength of the scattered radiation. Moreover, at the smallest scales, molecular mixing occurs when a liquid such as water is adsorbed onto a mineral surface or vegetation [24].

Even high-spatial-resolution (<2 m pixels) remote sensing measurements can have contributions from multiple subpixel-scale components. A simple semiquantitative approach can be used by calculating linear mixtures of reference spectra from spectral libraries and comparing them to remote and field spectra. Reference spectra for pure minerals can be chosen based on initial spectral analyses or a priori knowledge of the geology and surface minerals expected in a study area. Linear combinations of these pure mineral spectra can be compared to image spectra, and spectral mixtures can be matched to image spectra to identify the dominant mineral phases present and estimate percentages of different mineral constituents. This semiquantitative treatment of linear mixtures for mineral mapping relies primarily on spectral feature shapes and locations, which are sufficient to detect the presence or absence of the dominant mineralogy of mixed pixels and produce mineral maps used to interpret geological and geochemical environments. Using TIR data [78] and both VNIR/SWIR and TIR data [62] showed that a linear unmixing technique that models the percentage of each endmember composition can be used to identify individual surface minerals within a single pixel. More quantitative modeling requires well-calibrated spectral measurements and rigorous solutions to the equations of radiative transfer theory [70].

14.2 HYPERSPECTRAL MISSIONS AND CASE STUDIES

14.2.1 Earth

14.2.1.1 AVIRIS, AVIRIS-NG, and Hyperion

The Airborne Visible Infrared Imaging Spectrometer (AVIRIS) is a hyperspectral imaging spectrometer developed by NASA's JPL and has been in continuous operation since 1989. It measures radiance in 224 continuous channels between 380 and 2500 nm with 10 nm spectral sampling (Table 14.1). It has an instantaneous field of view (IFOV) of 1 mrad (0.057°) and a total field of view (TFOV) of 33° [14]. The signal-to-noise ratio (SNR) at 600 nm is >1000:1 and at 2200 nm is >400:1 [79].

The AVIRIS Next Generation (AVIRIS-NG) instrument is the successor to AVIRIS and has been in operation since 2011 [79]. AVIRIS-NG measures radiance in 427 continuous channels between 380 and 2510 nm with 5 nm spectral sampling. The instrument has IFOV and TFOV characteristics similar to those of its predecessor, but newer detector technology permits a factor of 2 improvement in SNR.

Hyperion was a spaceborne hyperspectral imager that was launched on the NASA EO-1 satellite in 2000 (Table 14.1). It measured radiance in 220 spectral channels from 400 to 2500 nm with 30 m pixels across a 7.5 km swath [36].

14.2.1.2 Calibration and Analysis Techniques

For AVIRIS data, the fast line-of-sight atmospheric analysis of spectral hypercubes (FLAASH) method can be used [71]. The FLAASH method uses a version of the moderate resolution atmospheric radiance and transmission (MODTRAN) model [80,81] to calculate the atmospheric parameters and at-surface reflectance. For MODTRAN calculations the average atmospheric profile for a typical continental location at midlatitudes during the summer, a visibility of 23 km, and a typical aerosol profile model for a rural area are commonly assumed. In addition, a spectral "polishing" routine [71,82] can be used to eliminate spectral artifacts that remain after atmospheric correction and to account for random channel-to-channel noise. The FLAASH method uses a running average across nine adjacent spectral channels to effect this polishing.

Calibrated surface reflectance spectra can be directly compared to reference spectra for pure minerals that are archived in spectral libraries. Both the USGS and JPL have spectral libraries available via the Internet and contain VNIR/SWIR spectra for over 2000 minerals and rocks combined. The most recent version of the USGS digital spectral library can be found at http://speclab.cr.usgs.gov and is described in [26]. The JPL ASTER spectral library (recently renamed to the ECOSTRESS spectral library) is a compilation of spectra of rocks and minerals measured at JPL, the USGS, and Johns Hopkins University and can be found at http://speclib.jpl.nasa.gov and is described in [27].

A comprehensive analysis of multichannel image data incorporates classification techniques based on spectral variability within a scene, the physical laws governing radiative transfer and scattering, its relation to field site measurements, and reliance on reference spectral libraries for matching band position, depth, and shape. The information contained in the spatial and spectral data can be displayed in three different ways: (1) in image space, where the spatial relationships between pixels are shown, (2) in spectral space, where spectral variations within a single pixel are shown, or (3) in feature space, where the spectral variations of each pixel are plotted as points (or vectors) in n-dimensional space, where n is the number of wavelength channels [83]. All three of these display methods can be utilized to maximize the amount of information extracted from multichannel image data.

A series of processing steps has become standard in hyperspectral data analysis that yields reproducible results, although some subjective decisions are required of the user [84,85]. These methods, from [85,86], can be implemented in the Environment for Visualizing Images (ENVI®) software package, currently owned by Harris Geospatial Solutions (http://www.harrisgeospatial.com/SoftwareandTechnology/ENVI.aspx). The methods can be summarized as follows: (1) atmospheric correction and calibration to reflectance, (2) transformation to minimize noise and reduce data dimensionality, (3) location of spectrally "pure" pixels, (4) selection of spectral endmembers, and (5) pixel classification and mapping of spectral endmembers. The purpose of this methodology is to focus only on information that is relevant to characteristic mineralogical features within an image.

14.2.1.3 Case Study

Hyperspectral VNIR/SWIR data have been used to identify and map a wide range of surface weathering and hydrothermal alteration minerals associated with acid mine drainage [87,88], mineral exploration targets, and the surface expression of active and fossil geothermal systems [62,86,89].

In the example shown in Figure 14.1, AVIRIS data were used to map surface minerals associated with the active geothermal system at Steamboat Springs, Nevada. Steamboat Springs is an active geothermal system about 16 km south of Reno. It is characterized by exposures of recent siliceous sinter deposits, hydrothermally altered country rock, and structurally controlled open fissures venting H_2S-rich steam. The Steamboat Springs system has been described as a modern analog to ancient hydrothermal systems associated with epithermal precious metal deposits throughout the Great Basin of the western United States [90–92]. Recent sinter deposits are composed of opaline silica, which transforms to β-cristobalite and chalcedony with increasing depth and age [92]. Also, as a result of the remote mineral mapping from [62], hydrous Na-Al sulfate crusts (tamarugite and alunogen) have been discovered forming around some active fumaroles. In the subsurface, sulfuric acid (H_2SO_4) solutions produced by H_2S reaction with atmospheric O_2 above the water table leaches the surrounding rock leaving opal, residual quartz, or quartz + alunite, adularia, or kaolinite/montmorillonite alteration assemblages. These minerals are characteristic of a steam-heated acid sulfate-type alteration system [93] and are exposed at the surface locally as outcrops of argillized and acid-leached granodiorite and basaltic andesite.

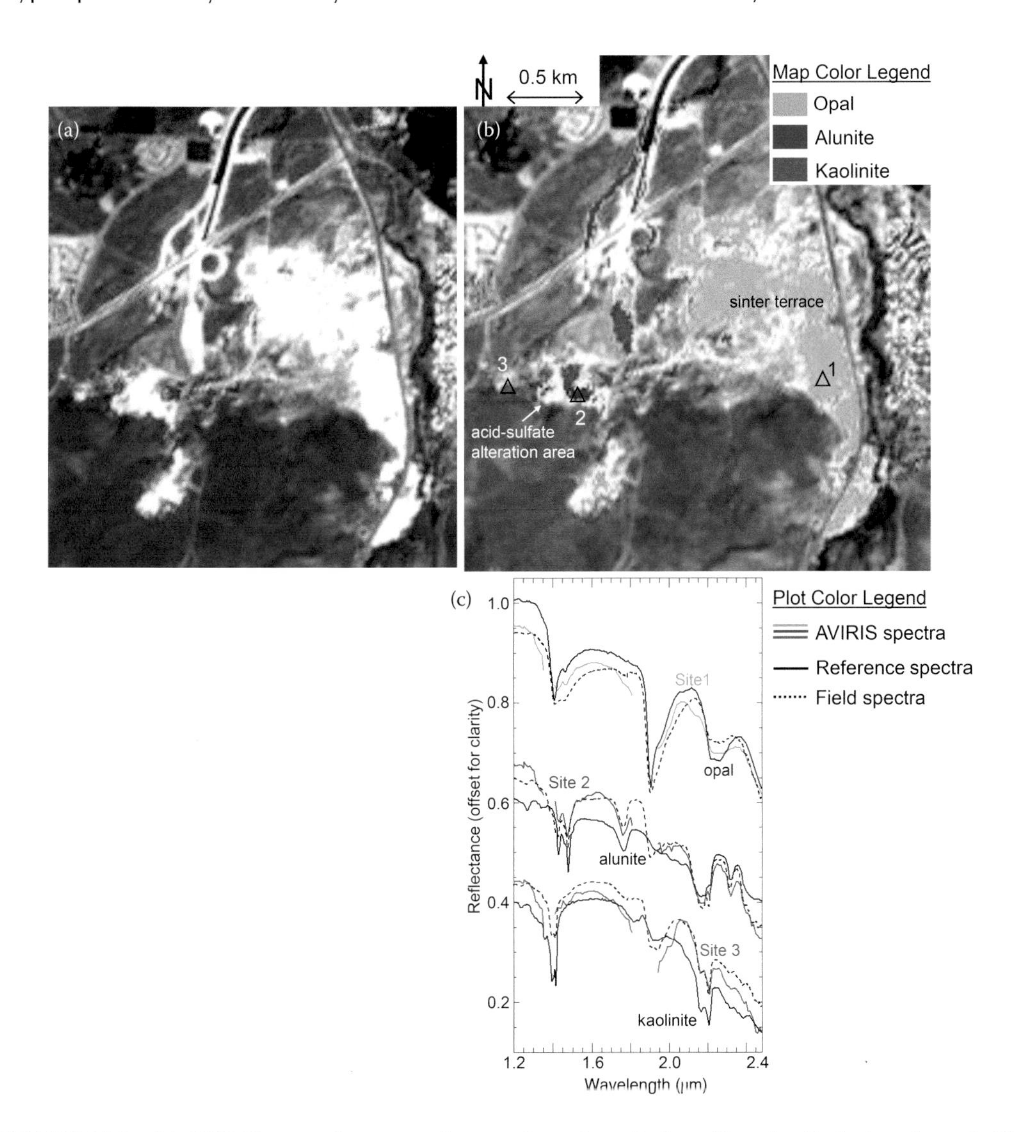

FIGURE 14.1 (a) AVIRIS true color composite over Steamboat Springs, Nevada, displaying channels 28, 18, 8 as RGB, respectively. Roads and urban areas are evident by their spatial patterns and textures. Healthy vegetation is in various shades of green, dry desert vegetation is dark purple to greenish-brown, and bright white indicates areas of exposed rock or soil. The spatial resolution for the AVIRIS image is 18 m. (b) AVIRIS spectral mineral map. The three regions mapped are distinguished by their dominant mineralogy, noted in the legend. (c) AVIRIS spectra (left) from representative field sites (1, 2, and 3—triangles on map) are color coordinated to the mineral map legend. (From Vaughan, R.G. 2004. Surface Mineral Mapping at Virginia City and Steamboat Springs, Nevada with Multi-Wavelength Infrared Remote Sensing Image Data. PhD dissertation, University of Nevada, Reno, 273 pp. [62])

In the Steamboat Springs region, AVIRIS data mapped the distribution of opal, kaolinite, and alunite in generally unvegetated areas (Figure 14.1). The mapped regions are displayed over a grayscale image of AVIRIS reflectance channel 28 (650 nm). The AVIRIS spectrum in orange, from the main sinter terrace, matches the reference spectrum from the USGS spectral library for opal. The broad spectral feature around 2250 nm and the strong features at 1400 and 1900 nm are indicative of opal. The Analytical Spectral Devices (ASD) (Boulder, CO, USA) field spectrometer data from both sites also match the opal spectrum. X-ray diffraction (XRD) analyses indicate that opal is the dominant mineral phase present at this site. The AVIRIS spectrum in blue, from the acid-sulfate alteration area, matches the USGS library reference spectrum for alunite. The broad, asymmetric doublet feature around 2200 nm, and the secondary features at 1760 nm are indicative of alunite. The ASD field spectrum from this site also matches the AVIRIS spectrum, and XRD analyses of samples from this site indicate the presence of alunite, quartz, kaolinite, and minor opal. The AVIRIS spectrum in magenta, also from the acid sulfate alteration area, matches the reference spectrum for kaolinite. The sharp doublet feature around 2200 nm is indicative of kaolinite. The ASD spectrum from this site matches the AVIRIS spectrum, and XRD analyses indicate the presence of kaolinite [62].

14.2.2 Mercury

14.2.2.1 MASCS

The MASCS (Table 14.1) instrument aboard the MESSENGER spacecraft used a compact Cassegrain telescope to focus reflected light from Mercury or its atmosphere for analysis by both the UVVS and the VIRS. The UVVS employed a scanning grating monochromator with three spectral channels (115–190, 160–320, and 250–600 nm) that provided a spectral resolution varying from 0.2 nm at UV wavelengths to 0.5 nm at visible wavelengths [94]. VIRS was a point spectrometer with a 0.023° FOV covering the wavelength range 320–1450 nm at 5 nm spectral resolution. It used two linear diode arrays in the visible (320–950 nm) and near infrared (900–1450 nm) [95,96]. The first two channels of the UVVS instrument were used in combination with the VIRS instrument for surface studies.

14.2.2.2 Calibration and Analysis Techniques

Calibration of VIRS data to radiance involves the use of dark current measurements and corrections for scattered light within the spectrograph. In-flight observations of standard stars validated laboratory data acquired prelaunch to within 10%. Reflectance was calculated as radiance factor (I/F), that is, the ratio of measured surface radiance to that from a perfect Lambertian surface normally illuminated by the Sun. Comparison of MASCS data on the Moon (acquired during its flight to Mercury) to ground-based lunar observations [96] provided confidence in the calibration of the data. However, Domingue et al. [97] pointed out some differences between MASCS and ground-based observations of Mercury in the near infrared.

14.2.2.3 Case Study

The UVVS instrument's observations of the exosphere of Mercury demonstrated in detail spatial variations in sodium, calcium, and magnesium atoms and ions liberated by solar wind and micrometeorite interactions with the surface [98,99]. The unusual differences between the temporal and spatial distributions of these elements were studied while MESSENGER was in orbit around Mercury.

The MASCS VNIR spectra of Mercury did not exhibit any diagnostic spectral absorption features that could be attributed to specific minerals. Holsclaw et al. [96] used analyses of oxygen-metal charge transfer absorptions and subtle spectral slopes in the visible wavelengths (and the lack of identifiable 1000 nm absorptions) to suggest a surface composition with low amounts of Fe^{2+}-bearing silicates and relatively abundant, spectrally neutral opaque minerals. Such low-Fe^{2+} silicates could span the range from plagioclase feldspars to Mg-rich olivine and pyroxenes. Izenberg et al. [100] used MASCS spectra to differentiate four spectral units suggestive of compositional variations. They did not find evidence for sulfide absorption bands in the bright, blue, irregular depressions known as "hollows" where such materials are hypothesized to be abundant [101]. They confirmed evidence for an ultraviolet absorption likely consistent with $<2–3$ wt% FeO in silicates, as well as submiscroscopic metallic iron or opaque minerals. These observations were also modeled by Trang et al. [102] to suggest that both submicroscopic Fe and C were needed, with global averages of 3.5 wt% and 1.9 wt%, respectively. Space weathering effects on Mercury are expected to be similar (if not enhanced) to those on the Moon, where vapor-deposited iron-rich nanophase particles coat many grains, resulting in few spectral features.

14.2.3 Moon

14.2.3.1 Moon Mineralogy Mapper (M^3)

The M^3 imaging spectrometer (Table 14.1) used a HgCdTe detector array for measuring electromagnetic radiation with wavelengths from 430 to 3000 nm (0.43–3 μm) [37]. This wavelength range covers the spectral range where diagnostic absorption features occur for all common lunar rock-forming minerals and hydrous phases. For global mapping, M^3 obtained data in 86 spectral channels. For targeted operation, M^3 divided the approximately 2600 nm range to which it is sensitive into 260 discrete bands, each of which is only 10 nm wide. This is considered very high spectral resolution and was designed to enable M^3 to detect the fine detail required for mineral identification. Unlike previous lunar spectrometers, M^3 included sensitivity at the longer wavelength range from 2500 to 3000 nm, which is sensitive to small amounts of OH and H_2O. This spectral region is dominated by solar reflection, although a small component of emitted thermal radiation was also noted at these longer wavelengths when the lunar surface was warmer than ~250–300 K. M^3 was intended to map the entire lunar surface from an altitude of 100 km at 140 m spatial sampling and 40 nm spectral sampling, with selected targets mapped at 70 m spatial and 10 nm spectral resolution. Although more than 80% coverage of the Moon was obtained by M^3 at low-sun and a spatial resolution of ~140 m/pixel and significant scientific discoveries were made, the Chandrayaan-1 spacecraft suffered from technical difficulties throughout the mission; these difficulties precluded accomplishment of many of the goals of M^3.

Designed for simplicity, reliability, and accuracy, M^3 used a compact system of optics known as an Offner design, which produces little or no distortion, either spatially or spectrally [37]. M^3 used the so-called push-broom method of image acquisition in which the instrument passively sweeps the scene below as it flies, recording 600 pixels of data simultaneously. Each of those 600 pixels simultaneously records images in each of 260 spectral channels. The primary M^3 product was an "image cube" that was 600 pixels wide, infinitely long over time as the instrument flies, and 260 spectral channels deep. The FOV was 24° (or 40 km on the ground at 100 km altitude), allowing contiguous orbit-to-orbit measurements at the equator that minimized variations in lighting conditions.

14.2.3.2 Calibration and Analysis Techniques

Prior to launch, laboratory calibration measurements were made to determine the spectral, radiometric, spatial, and uniformity characteristics of M^3 [37]. The spectral range for the 260 channels was determined to span from 404 to 2993 nm with 9.96 nm sampling. The absolute radiometric calibration was determined with respect to a US National Institute of Science and Technology (NIST) traceable standard at the 5% uncertainty level. The FOV of the M^3 was measured to be 24.3°, and the cross-track sampling was measured as 0.698 mrad. Spectral cross-track uniformity and spectral IFOV uniformity of the M^3 are critical calibration characteristics.

The M^3 ground calibration data files allowed M^3 data to be calibrated to the at-sensor radiance and were further calibrated to the equivalent of reflectance data for scientific analysis. This calibration method involved dividing the radiance data by a solar spectrum and photometrically correcting all data to the same viewing geometry to eliminate variations due to lighting conditions [37]. For warm surfaces, a small thermal component at wavelengths beyond 2000 nm was removed [103]. Lunar samples returned to Earth by the Apollo missions were used as "ground truth" for calibrating the M^3 data. Assuming that lunar samples are representative of specific portions of the lunar surface, the sample properties were used to calibrate the remote sensing data [37]. The Apollo 16 site was well suited for this because it is largely dominated by one type of material (feldspathic breccias), unlike most other landing sites, which contain diverse lithologies [37]. The Apollo 16 region is one of the prime Lunar International Science Calibration/Coordination Targets proposed for cross calibration of lunar data obtained by various missions [37].

14.2.3.3 Case Studies

14.2.3.3.1 Water

"The search for water on the surface of the anhydrous Moon remained an unfulfilled quest for 40 years" [33]. However, M^3 on Chandrayaan-1 detected absorption features near 2.8–3.0 μm on the surface of the Moon during some portions of the day (Figure 14.2). The 3.0 μm absorption feature was identified and the measurement was extended to longer wavelengths by two independent spacecraft: the NASA Cassini mission VIMS and the High-Resolution Instrument Infrared (HRI-IR) spectrometer on the NASA Deep Impact EPOXI mission [104,105]. For silicate bodies, such features are typically attributed to hydroxyl- or water-bearing materials. On the Moon, the feature is seen as a widely distributed absorption that appears strongest at cooler high latitudes and at several fresh feldspathic craters [33]. The general lack of correlation of this feature in sunlit M^3 data with neutron spectrometer hydrogen abundance data suggests that the formation and retention of hydroxyl and water are ongoing surficial processes [33]. The hydration signatures were observed by HRI-IR to be dynamic, with diurnal changes that differed for mare and highland units and returned to a steady state entirely between local morning and evening [105]. The observed hydration variation requires a ready daytime source of water group ions and is considered consistent with a solar wind origin [104,105]. In this scenario, hydrogen ions from the Sun are carried by the solar wind to the Moon, where they interact with oxygen-rich minerals at the top millimeters of the lunar soil to produce the observed H_2O and OH molecules [104,105]. Hydroxyl/water production processes may feed polar cold traps and make the lunar regolith a candidate source of volatiles for human exploration [33].

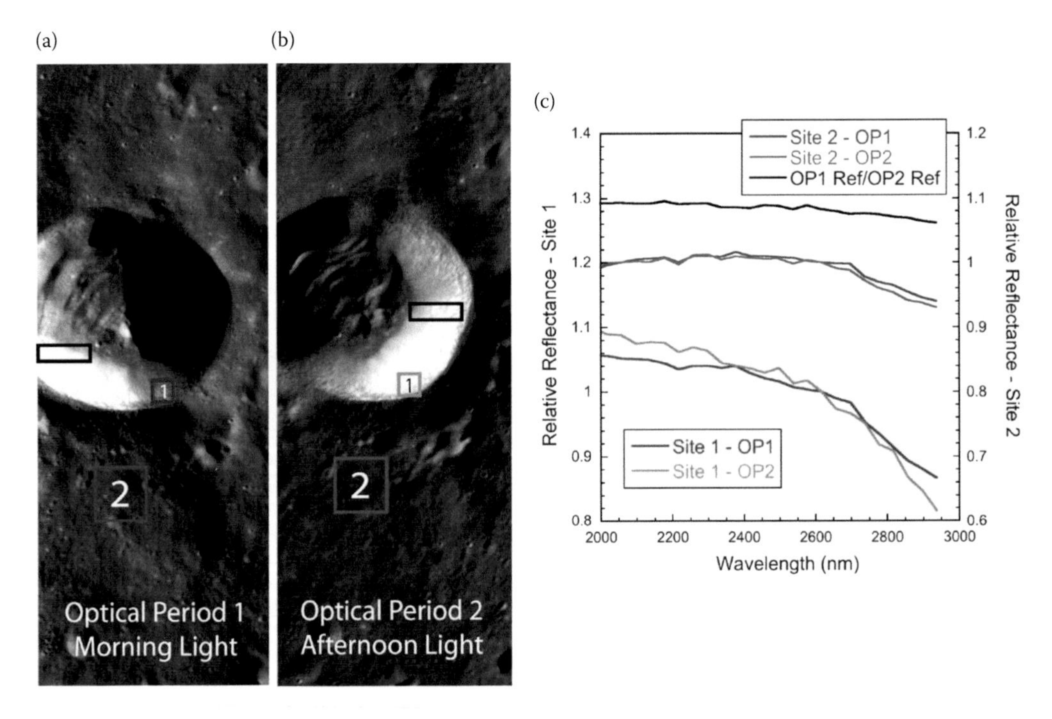

FIGURE 14.2 M^3 data taken 2 months apart during morning (a) [optical period 1 (OP1)] and afternoon (b) [optical period 2 (OP2)] solar illumination. The large Chadwick crater (diameter 30 km) is located on the lunar far side at 258.7°E, 52.7°S. Spectra in panel (c) are scaled reflectance for areas 1 and 2, relative to a local area of strong solar illumination that exhibits a relatively weak 3 μm band in the scene (this reference location varies with geometry). Background soil region 2 (50 × 50 pixels) exhibits a moderately weak and consistent 3 μm band strength. Region 1 within the crater (20 × 20 pixels) exhibits a more prominent apparent band strength, perhaps sensitive to solar illumination. Black boxes in (a) and (b) (50 × 15 pixels) indicate the reference area selected for spectral ratios. (Modified from Pieters, C.M. et al. 2009. *Science*, 326, p. 568, doi: 10.1126/science.1178658. [33])

Such hydration by solar wind particles may occur throughout the solar system on all airless bodies with oxygen-bearing minerals on their surfaces. Although abundances are not definitively known, as much as 1000 water molecules parts per million (0.1%) could be present [104].

14.2.3.3.2 Spinels

M^3 data were also used to search for unusual and, in some cases, new rock types. Recent research has identified two spinel-rich rock types at widely spaced locations on the lunar surface [106]. The first is a pink spinel rich in magnesium and iron, and the second is a black or very dark chromite-rich spinel [106]. Both types of spinel have distinctive spectral absorption features in the M^3 data near 2 μm [106]. As observed in 2 μm band depth maps, the pink spinels occur near Mare Moscoviense on the lunar far side as several small, diffuse deposits that are not obviously dark or associated with any crater or steep slope that has exposed fresh material [106].

Elsewhere in the solar system such spinel-rich surfaces have been observed in a few main-belt asteroids, where the spinel absorption feature near 2 μm is thought to indicate the presence of abundant calcium- and aluminum-rich inclusions (CAIs) such as those found in carbonaceous chondrite meteorites [107]. Current ideas about the origin of these deposits include the preservation of unusual deep-crustal or plutonic materials exposed on the surface or the presence of primitive material that has been deposited onto the lunar surface [108].

Dark, chromite-rich spinels are observed in the Sinus Aestuum region on the lunar near side [109]. These deposits are distinctly different from those on the far side in that they have lower albedo and additional spectral features at visible and 1000 nm wavelengths [109]. They are observed in a region with prominent pyroclastic volcanic deposits; these are believed to have been derived from depths of several hundred kilometers within the Moon and thus provide a link to mantle compositions [109].

14.2.4 Mars

14.2.4.1 Hyperspectral Instruments

14.2.4.1.1 OMEGA

The OMEGA instrument (Table 14.1) is a visible and infrared mineral mapper on board the Mars Express orbiter designed to globally map minerals and water (hydrated minerals or ice). The OMEGA imaging spectrometer uses two bore-sighted telescopes for a spectral range from 500 to 5200 nm. The first telescope uses a silicon CCD (whisk broom) to image the spectral range from 0.5 to ~1 μm. The second telescope uses two InSb detector arrays (push broom) to image the spectral range from ~1000 to 5200 nm [110]. OMEGA has an IFOV of 1.2 mrad, resulting in a spatial resolution of ~300 m close to periapsis (250 km) and ~5 km at an altitude of 4000 km.

14.2.4.1.2 CRISM

The CRISM (Table 14.1) is a high-spatial-resolution (9–200 m/pixel) imaging spectrometer designed to target regions to detect and map minerals and ices. CRISM uses a 10 cm diameter Ritchey-Critien telescope that feeds a pair of Offner convex-grating spectrometers. One spectrometer uses a Silicon detector array (VNIR), and the second spectrometer (IR) uses an array of HgCdTe diodes. A fully gimbaled optical system allows for motion compensation that enables a spatial resolution of 9–19.7 m. CRISM can also operate in mapping mode (no motion compensation) at 100 or 200 m/pix. CRISM's spectral range is from 362 to 3920 nm [9].

14.2.4.2 Calibration and Analysis Techniques

OMEGA Calibration. OMEGA data are processed using standard processes. DN levels are converted to radiances [$W·sr^{-1}·m^{-2}$], and I/F is calculated by dividing by the solar flux [$W·sr^{-1}·m^{-2}$] corrected for the appropriate Sun-Mars distance [110,111]. Unlike most hyperspectral imagers, where the calibrated data are provided, OMEGA provides the raw data and the calibration software to construct both calibrated radiances and geometric coordinates for each spectrum. The advantage of this methodology is that as calibration software is updated, users do not need to download new image cubes, but instead they download the new software and reprocess their raw cubes.

CRISM Calibration. A complete description of CRISM calibration can be found in [9]. CRISM uses a shutter and an integrating sphere as part of the internal calibration systems. The shutter in its closed position allows for the measurement of bias, dark current, and thermal background. A partially opened shutter acts as a mirror, allowing the full view of the integrating sphere.

Spectral Smile. Imaging spectrometers are ideally designed such that one dimension of the CCD is spatial and the other dimension is spectral. Ideally, this means that any row of pixels will correspond to a common wavelength while any column corresponds to a common surface location. Due to minor flaws in optics and alignment, this is almost never the case, causing wavelength positions to shift. If one were to create a contour of a single wavelength across the CCD, it would appear as a smile (or frown).

Atmospheric Corrections. The atmosphere of Mars is composed of 95% CO_2 [112], which produces several absorption features at or near surface spectral features. The presence of these overlapping atmospheric features can introduce error into surface spectral indices by either creating a deeper absorption feature (atmospheric line directly overlaps the surface spectral feature) or introducing a slope into the continuum estimate (atmospheric line is adjacent to the surface feature). In addition, atmospheric scattering can introduce a color slope, causing the surface to appear bluer.

One simple technique to remove most of the atmospheric lines, but not spectral slopes due to scattering, is the volcano scan [113]. The volcano scan technique involves calculating the ratio between the absorption line strengths observed at the top of a volcano (e.g., Olympus Mons) and the base of the volcano. If the surface on top of the volcano has the same photometric properties (e.g., albedo) as the base (which is the case for many volcanos on Mars since these regions are typically covered by thick layers of dust), then the ratio of the line strengths is due to the atmosphere. A few of the absorption lines are at wavelengths that do not overlap with surface spectral features. These line strengths are then used to calculate the ratio among the volcano scan line strengths, which can then be used to correct the atmospheric line strengths that do overlap with surface spectral features. The volcano scan method is useful for a quick analysis of spectral features but is inadequate to remove scattered light in the continuum. One approach is to use a Monte Carlo model designed to simulate multiple scattering of light from dust and ice aerosols in the atmosphere [114,115]. This approach is often combined with a volcano scan method to correct spectra for both scattering and gas absorption. Another method is to use a physics-based radiative transfer model that incorporates the optical properties of the gases and the aerosols (e.g., DISORT) [116]. This method requires a priori knowledge of atmospheric optical depths. This approach, while the most accurate (similar to MODTRAN), is also the most computer intensive.

One of the most common vegetation indices is the normalized difference vegetation index (NDVI), where one compares the reduction of reflectivity in the NIR versus that in the red. This reduction in reflectivity is due to the absorption of water. Surface volatiles and exposed minerals can be identified in a similar manner, using absorption features unique to either a specific mineral or a suite of minerals. Viviano-Beck et al. [117] outlined the most common spectral indices used by CRISM to identify ice and minerals. Many spectral indices measure the depth of absorption bands, while others measure spectral slope. Many of these indices are less sensitive to atmospheric corrections, especially if the spectral feature is not near an atmospheric spectral feature.

14.2.4.3 Case Studies

Ehlman and Edwards [118] compiled a review of the mineral composition of the Martian surface based on both spacecraft and in situ measurements (Figure 14.3). Much of this story was a result of NIR and SWIR spectroscopy identifying and mapping the distributions of phyllosilicates and sulfates.

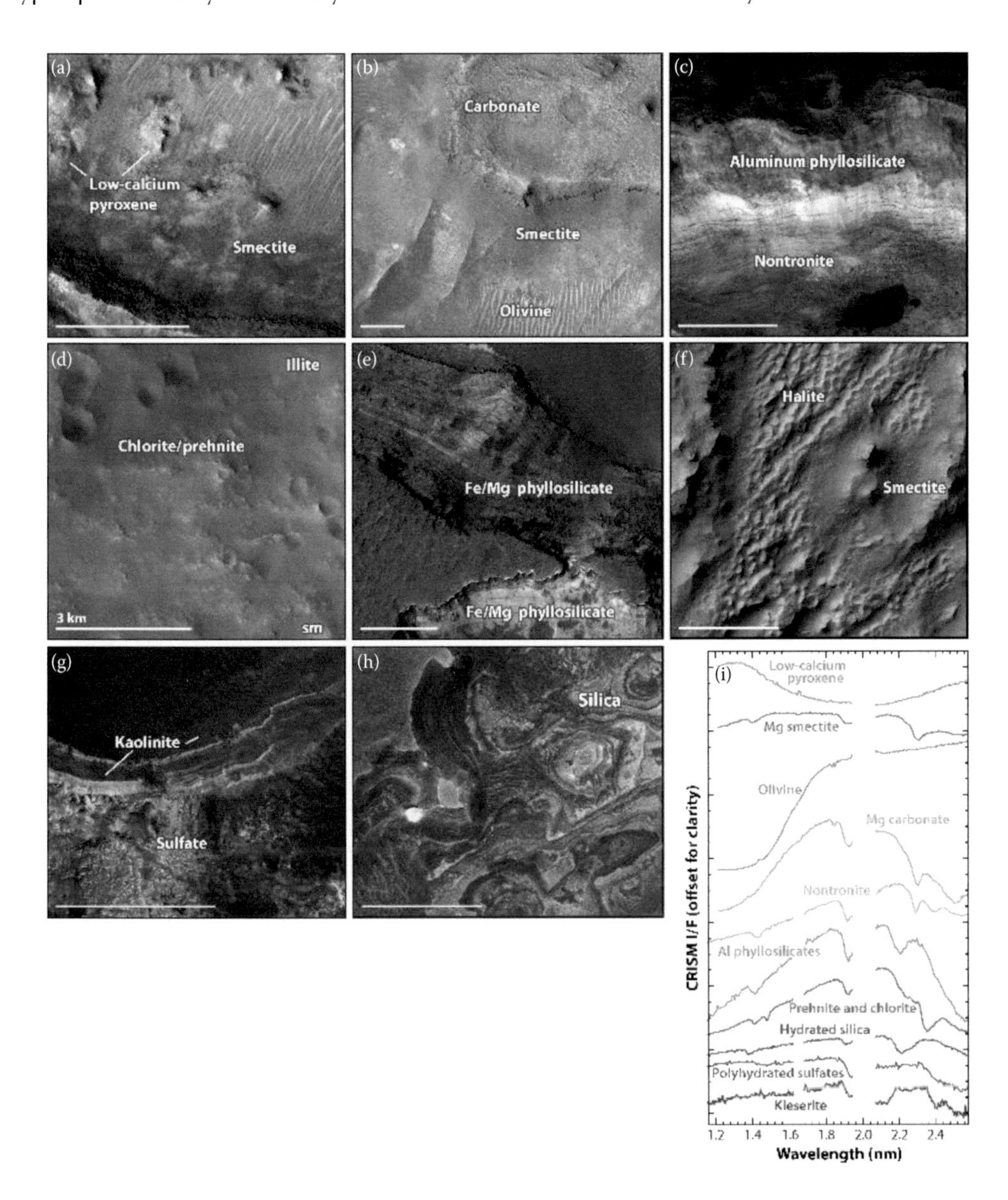

FIGURE 14.3 Mars's aqueous environments as seen by HiRISE and CRISM false color IR images and composites. (a) Rocks disrupted by the Isidis impact and exposed 500 m deep in the Nili Fossae trough contain low-calcium pyroxene breccia blocks (green) partially altered to Fe/Mg smectite and contained within a smectite-bearing matrix. (b) An eroded stratigraphy in Nili Fossae with Mg carbonates (green) formed by alteration of an eroded olivine-rich unit that in turn overlies Mg-smectite clays (blue). Ridges indicate conduits of subsurface fluid flow. Olivine is also enriched in sand dunes. (c) Aluminum phyllosilicates overlie nontronite-bearing sediments at Mawrth Vallis and may have formed by top-down, near-surface leaching. (d) Prehnite, chlorite, and illite are associated with the ejecta of small craters north of Syrtis Major and indicate excavation of hydrothermally altered materials. (e) Sedimentary beds in the Holden crater fan deposit host Fe/Mg phyllosilicates. (f) Chloride salt, possibly halite, overlies older smectite clay–bearing knobs in a shallow depression near 205°E, 33°S. (g) Interbedded sulfate- and kaolinite-bearing sediments in Columbus crater paleo lake deposits. (h) Silica as seen from orbit in layered units around Valles Marineris. Scale bar is 200 m unless otherwise indicated. (i) CRISM visible/SWIR spectra of ancient Martian terrains show mineralogic diversity. Data are not shown at the 1.65 μm CRISM filter boundary or within the 2.0 μm atmospheric CO_2 absorption. (Modified from Ehlman, B.L. and Edwards, C.S. 2014. *Annual Review of Earth and Planetary Sciences*, 42, pp. 291–315, doi: 10.1146/annurev-earth-060313-055024. [118])

14.2.4.3.1 Phyllosilicates, Sulfates, and Alteration Minerals

Prior to the arrival of imaging infrared spectrometers (e.g., OMEGA), Martian history was defined by epochs based on visible imaging—morphology, crater counts, and so forth. With the arrival of the French-led instrument OMEGA on board the European spacecraft Mars Express, a new way of viewing Martian history emerged. This new view was based on mineralogical composition derived from spectral absorption features. Not only were broad classes of minerals identified and mapped (e.g., olivines and pyroxenes), but specific minerals (e.g., nontronite, Fe-rich chlorites, saponite, and montmorillonite) were identified and mapped [119,120]. The new view of composition loosely correlates with the timeline determined from visible imaging but tells a story that surface materials changed from neutral-pH (or slightly alkaline) conditions early in Mars history to acidic alteration at a later epoch [121,122].

In addition to providing the first high-spectral-resolution details of Mars' primary igneous mineralogy (e.g., olivine and pyroxene), OMEGA found evidence for spatially extensive hydration and alteration of the oldest Martian terrains, in the form of hydrated clay minerals [119,120]. These OMEGA-based detections motivated targeted observations by the later CRISM instrument. The higher spatial resolution of CRISM (18 m/pixel in targeted mode) has helped confirm most of OMEGA's clay mineral detections, revealed numerous, less extensive occurrences of clay minerals, and allowed these to be placed in a stratigraphic context.

CRISM data have extended earlier observations by OMEGA that aluminum-bearing clay minerals such as kaolinite and beidellite tend to occur stratigraphically above iron and magnesium-bearing clay minerals [118]. Whereas OMEGA was only able to reliably detect this mineral sequence in large exposures within the Valles Marineris region, the higher spatial resolution of CRISM has enabled detections in smaller valley systems and in the walls of impact craters throughout the oldest Martian terrains [118]. This sequence suggests a near-global transition early in Mars' history from neutral to more acidic alteration conditions, possibly coupled with enhanced leaching of the surface crust [121–123].

14.2.4.3.2 Ices

The polar caps are the most active regions on Mars and the most dynamic processes visible from Earth. The annual cycling of atmospheric CO_2 into the seasonal CO_2 ice caps is a driving force of the Martian climate. The polar layered deposits, with thousands of layers whose thickness is only resolvable with submeter spatial resolution from orbit, may contain a record of past climates. The polar regions contain the majority of known H_2O ice deposits, distributed between the residual caps and near-surface ice in the regolith. CO_2 ice has several spectral features throughout the infrared (e.g., 1435, 2340 nm) Unfortunately, CO_2 ice has many spectral features located at or near the same wavelengths as the gas bands (e.g., 1435 nm). One of the few exceptions is the doublet at 2345 nm. The presence of H_2O ice is easily detected by the presence of broad spectral features at 1500 and 2000 nm.

While the seasonal monitoring and mapping of the seasonal caps date back to Herschel (1784) [124], the ability to correctly identify the seasonal cap composition as CO_2 ice did not occur until the 1960s. See references [125] and [126] for a historical overview. In 2001, Kieffer and Titus [127] recognized that the northern seasonal cap was surrounded by a H_2O-ice annulus during the springtime retreat. This conclusion was based on comparing thermal observations (CO_2 ice is typically at ~145 K) with visible observations. The bright edge of the seasonal cap was too warm to be CO_2 ice and was hypothesized to be H_2O. This hypothesis was later spectrally confirmed by OMEGA in 2005. In addition to spectrally confirming the presence of the H_2O-ice annulus, OMEGA also observed the presence and distribution of H_2O ice intimately mixed with the dominant CO_2 ice through much of the seasonal cap.

14.2.5 Main Belt Asteroids: Ceres and Vesta

The Dawn mission to the MAB was a first in solar system exploration in that it was the first spacecraft to orbit two separate planetary bodies. All other planetary spacecraft missions consist of landers, orbiters, or flybys. Flyby missions occasionally end by entering orbit around the last body visited. Dawn visited and orbited the two largest asteroids within the MAB, 1 Ceres and 4 Vesta. Vesta lies in the dryer inner part of the MAB, while Ceres is in the outer "icier" part. This provided a unique opportunity in comparative planetology.

14.2.5.1 Dawn Mission Visible and InfraRed Spectrometer

The Dawn VIR slit spectrometer is composed of two subsystems: VNIR and SWIR. The VNIR subsystem has a spectral range from 250 to 1000 nm, while the SWIR subsystem measures from 950 to 5000 nm. The spectral frames are 256 pixels (spatial) by 432 pixels (spectral) with a slit width of 64 mrad. The VIR heritage can be traced back to the Cassini VIMS instrument.

14.2.5.2 Calibration and Analysis Techniques

The calibration of VIR spectra consists of a number of standard steps. The first step is to remove the effects of optical misalignment. Defective pixels and data spikes due to high-energy particles are also removed. Saturated pixels are identified and removed. Finally, the calibration matrix is applied. As with many instruments, the calibration coefficients for even and odd detectors are different. VIR makes use of several filters sandwiched into the CCD to ensure only the correct wavelength passes through. While this is necessary to reduce the effects of multiple orders of wavelengths, the boundaries between the filters result in scattered light. Early calibration efforts did not correct for this effect. A complete discussion of calibration can be found in [128]. Once the data have been radiometrically calibrated, each band must be projected into either a common image space or a map. This step is necessary to ensure that each pixel spectrum is from the same location on the surface (i.e., spectral/spatial alignment).

There are three commonly used approaches in the analysis of VIR data: spectral indices, linear spectral fitting, and Bayesian spectral fits. Several spectral indices are used to map minerals on surfaces. The primary indices focus on either band depth or band center. Examples of this are spectral indices for olivine, pyroxenes, howardite–eucrite–diogenite (HED) meteorites, and a range of hydrated minerals. Pyroxenes have broad spectral features at 1000 and 2000 nm. Olivine has a broad feature at 1000 nm. The band centers of the spectral features shift with composition, allowing identification of a suite of minerals within each mineral family. Hydrated minerals provide several narrower features that correspond to hydroxyl and absorbed water bonds.

Spectral fitting is another useful approach, though it is more time intensive. The use of linear combination [129] of a known library of mineral spectra is useful when trying to identify small outcrops of a mineral. This is especially useful when the mineral is spectrally bland within the wavelength region of interest. Dark carbonaceous chondritic material is featureless within VIR's spectral range.

The last commonly used approach is Bayesian nonlinear spectral fitting. This approach has mainly been used to separate the reflection component from the thermal emission component at the longer wavelengths (3000–5000 nm) [130]. Because the Dawn spacecraft does not have a TIR instrument, all surface temperatures are derived from the VIR 4000–5000 nm spectral region. In addition, removal of thermal emission from the 3000 nm region is important in determining the presence and abundances of water ice and hydrated minerals. Since this spectral region is on the Wein side of the Planck function, VIR is not sensitive to surface temperatures below 180 K. This essentially limits surface temperature determinations to those observations at the low to middle latitudes during local midday.

14.2.5.3 Case Studies

The surface of Vesta was expected to be composed mainly of pyroxene minerals based on spectral reflectivity [131]. It was suggested early on that Vesta was the source of HED meteorites, based on similar reflection spectra [132]. Dawn VIR observations confirmed that the surface of Vesta was largely composed of iron-bearing pyroxene-rich minerals comparable to HEDs [133]. In general, the surface can be divided into regions of bright material and dark material. The dark albedo material is composed of carbonaceous chondritelike opaque material [129]. These dark areas also tend to be higher in hydrated minerals, also consistent with a carbonaceous chondritic composition [133]. The bright material could represent pristine areas and are generally composed of howardite-rich eucrites [134]. Additionally, there are outcrops of olivine-rich materials in several craters, including Bellicia and Arruntia craters [135]. The lack of large olivine outcrops, especially in the area surrounding the deep impact crater, Rheasilvia, puts definite constraints on the interior structure of Vesta.

The surface temperatures derived from VIR reveal a surface covered in a fine regolith with a thermal inertia of 30 +/– 10 $m^{-2} s^{-0.5} K^{-1}$ [136]. There is a range of thermal inertia on Vesta, with regions north of the equator having lower thermal inertia, suggesting more weathering or processing, and outcrops of higher thermal inertia around the relatively young Marcia crater [136]. The higher thermal inertia is most likely due to compaction following degassing of near surface volatiles. These results suggest a surface similar to the Moon with few or no rocks on the surface.

Ceres, based on estimated planetary mass and size, was expected to be a very icy world with an extremely young relaxed surface [137,138]. Telescopic observations have purportedly observed plumes of water vapor that are transient but appear to originate from the surface at midlatitudes [139]. Instead, Dawn found a heavily cratered surface that could not have more than 30% ice to rock mixtures [140].

Unlike Vesta, which is quite bright, Ceres has an optical albedo of 0.09 and a bond albedo of 0.03. VIR spectra confirm a surface consistent with carbonaceous chondrites. Spectral features observed by VIR suggest the widespread presence of ammoniated phyllosilicates, which is generally found in the outer Solar System [141]. This suggests either that Ceres formed in the outer part of the Solar System and migrated inward to its current orbit or the ammoniated phyllosilicates are exogenic in origin and were delivered to the surface via impacts.

There are small areas that are bright, the largest ones occurring in Oxo and Occator craters [142,143]. VIR has identified these bright deposits as containing both sodium carbonates and ammonium salts [144]. It has been suggested that these deposits are evaporates, the result of sublimation from an aqueous solution [144–146]. While water is not generally stable on the surface of Ceres [147,148], small deposits of exposed water ice have been discovered in permanently shadowed regions [149,150]. Observations of Ceres' surface temperatures and estimates of thermal inertia are preliminary but generally suggest a surface with low thermal inertia [151].

14.2.6 Jupiter

14.2.6.1 NIMS

A detailed description of the NIMS instrument (Table 14.1) design, calibration, and operation was presented in [152]. NIMS's spectral range, from 700 to 5200 nm, allowed measurements of both reflected sunlight and emitted thermal radiation. The instrument used a 228 mm aperture, f/3.5 Ritchey-Chretien telescope with an equivalent focal length of 800 mm. The incident beam was reflected onto a wobbling secondary mirror that could scan through 20 fixed positions to provide whisk-broom spatial coverage in one direction, while the motion of a scan platform provided spatial coverage in either direction. The beam was directed through a chopper to a wobbling diffraction grating that was rotated in steps to generate spectral coverage at the commanded spectral resolution, up to 408 wavelengths. The chopper ensured that no light reached the detectors while the diffraction grating was in motion. The beam was focused onto an array of 17 detectors: 2 silicon detectors for NIR wavelengths and 15 indium antimonide (InSb) detectors sensitive to longer wavelengths. The focal plane assembly, including the detectors and their preamplifiers, was radiatively cooled to 64 K. The system had an IFOV of 0.5 × 0.5 mrad (leading to spatial resolutions of tens to hundreds of kilometers per "nimsel" for many of the Jovian satellite encounters) and an angular FOV of 0.5 × 10 mrad. A "nimsel" is a NIMS pixel with all the spectral wavelengths spatially colocated.

14.2.6.2 Calibration and Analysis Techniques

NIMS faced many challenges in its mission, including the harsh radiation environment of Jupiter. The instrument was shielded within a 3 mm thick tantalum enclosure, but even so, the SNR degraded sharply during encounters with satellites in the inner reaches of Jupiter's radiation belts. The spacecraft suffered a downlink bottleneck caused by the failure of its high-gain antenna, severely limiting the data volume returned by each instrument during the mission. Contamination was also a concern, and NIMS was equipped with covers, shields, and shades to protect it from thruster firings and thermal influences from the spacecraft. Spectral calibration (subject to change, due to thermal effects or vibration of the diffraction grating during launch) was provided by an onboard InGaAs LED lamp that emitted at a known wavelength. A radiometric calibration target, consisting of an extended, near-field blackbody source heated to a known temperature, was mounted on the spacecraft within the NIMS FOV. A photometric calibration target that diffusely reflected sunlight was also placed in view of the scan platform. Independent verification of many of the spectral detections made by NIMS took place during the millennium passage of the Cassini spacecraft en route to Saturn. Cassini's VIMS confirmed several weak absorption bands in the spectra of the Galilean satellites Io, Europa, Ganymede, and Callisto [153]. The NIMS instrument operated successfully throughout Galileo's 14-year mission despite failures of two of the InSb detectors and the temperature sensor of the radiometric calibration target.

14.2.6.3 Case Study

NIMS is credited with the first detection of ammonia ice clouds in the atmosphere of Jupiter [154] and made important observations of water vapor that demonstrated moist convection in Jovian lightning storms [155,156]. NIMS thermal observations were critical to understanding the nature and distribution of active volcanos on Io [157] and the energetics of their eruptions [158]. NIMS did not identify the mineralogy of the lavas but provided an indirect indication that they must be made up of silicates, since their eruption temperatures were too high for other candidate materials. NIMS produced global maps of Io's surface SO_2 ice abundance and grain size at spatial resolutions from 100 to 350 km/nimsel, using the deep absorption band centered at 4100 nm together with the relative strengths of weaker bands [159,160]. Water ice grain sizes were measured on the icy satellites Europa, Ganymede, and Callisto using a similar approach [161]. On Europa, NIMS found that the water ice bands were distorted, leading to suggestions of the presence of hydrated phases of salts [162] or sulfuric acid [163]. NIMS did not specifically identify the visibly red endogenic material that is associated with tectonic ridges and other young geological features on Europa but found associations between these features and the degree of distortion of the water ice bands. Several compounds attributed to radiolysis were identified on the surfaces of the icy moons, including H_2O_2 (through a band at 3500 nm) [164], SO_2 (4050 nm), and CO_2 (4260 nm). CO_2 signatures were particularly strong on Callisto and the older terrain on Ganymede [165,166], perhaps due to radiolysis of meteoritic carbonaceous material. A weak feature attributed to CN (4570 nm) was also seen on Callisto.

NIMS's mission ended in September 2003, when the Galileo spacecraft plunged deliberately into Jupiter to ensure that the risk of biological contamination of Europa was eliminated.

14.2.7 Saturn

14.2.7.1 VIMS

Cassini's VIMS (Table 14.1) was a primary orbital instrument to study the composition, dynamics, clouds, and thermophysics of the atmospheres of Saturn and Titan and the identification and mapping of surface compositions of Titan, the other icy Saturnian satellites and rings. VIMS consisted of two instrumental subsystems covering different spectral ranges [39]. The VIMS-VIS (visible) channel was a multispectral imager that covered the spectral range from 350 to 1050 nm with 96 spectral channels with a spectral resolution of 7.3 nm and used a frame transfer CCD detector on which spatial and spectral information was simultaneously stored. The VIMS-IR (infrared) channel covered a wavelength range from 850 to 5170 nm using 256 spectral channels with a spectral resolution of 16.6 nm. The two spectrometers had many modes, but they typically employed effective IFOVs of 0.5 × 0.5 mrad. The largest spectral cubes made were 64 × 64 (lines and samples), although much smaller spatial views were often used to save on data volume. During Cassini's approach to Titan, VIMS observational sequences commenced several hours before closest approach, providing hemispheric mapping with full spectral cubes with spatial resolutions $\leq$50 km/pixel. About 4 h before closest approach, 2 × 2 mosaics provided coverage of the disk. Resolutions of ~10 km/pixel were achieved ~0.5 h before closest approach when regional mosaics could be acquired. Isolated cubes, rarely acquired during close flybys, could yield resolutions as high as 250 m/pixel.

14.2.7.2 Calibration and Analysis Techniques

The two imaging spectrometers that composed VIMS were calibrated separately [39]. The VIMS-VIS channel was calibrated in Italy prior to integration at JPL with the VIMS-IR channel that was also calibrated by JPL. Final tests of the integrated instrument included only filter transmission and mineral target measurements. Spectral calibration characterized the bandpass for each spectral channel over the FOV and as a function of temperature. Central wavelengths of the VIMS-IR channels varied by <1 nm; because the sampling interval was about 16 nm, this was a small shift. Likewise, the VIMS-VIS wavelength variations were <0.3 nm compared to the 7 nm sampling interval. Preflight calibration included radiometric/flat field response, geometric, polarimetric, spectral, and solar port response. Cooling the thermal vacuum chamber walls with liquid N_2 simulated the flightlike thermal environment. The simulated thermal environment was quite accurate as the in-flight temperatures of optics and focal planes were within a few Kelvin of those in the test chamber. In the case of analysis of VIMS data by the authors of this chapter, ISIS-3 (the USGS publicly available Integrated Software for Imagers and Spectrometers) has been used for radiometric and geometric corrections and for subsequent spectral analysis and correlations with other Cassini data sets. For VIMS cubes, ISIS-3 provides a set of end-to-end tools (https://isis.astrogeology.usgs.gov/Installation/index.html) that start with raw data publicly available from the NASA PDS and generate high-level products that have been radiometrically corrected and transformed to a wide variety of map projections. ISIS-3 also provides a suite of interactive tools for spectral analysis, generation of control nets, and mosaicking to cartographic standards.

14.2.7.3 Case Study

Titan affords a rich solar system laboratory to study active organic synthesis on a global scale—it may exhibit chemical pathways holding clues to the primordial prebiotic organic chemistry that led to the emergence of life on Earth. Methane makes up a few percent of Titan's thick cold nitrogen atmosphere. Moving in a global cycle, the methane forms clouds, rain, rivers, lakes, and seas akin to Earth's hydrological cycle. In the upper atmosphere, methane and nitrogen are energetically broken down and recombine to form a vast spectrum of organics ranging from simple gases to large complex molecules. These form mixtures of organic liquids and aerosol solids that rain onto the surface. Spectral evidence for the composition of the lakes derived from VIMS observations is the subject here.

Able to penetrate the thickly absorbing, hazy atmosphere through a series of atmospheric windows, VIMS is the primary Cassini instrument used to study Titan's surface composition. Sunlight penetrates to the surface through narrow transmission windows that are separated by deep methane absorption bands. Within atmospheric windows with wavelengths <1000 nm, surface signals are swamped by multiple scattering from aerosols. Aerosol scattering becomes decreasingly effective at longer wavelengths. As a result the surface is visible only through atmospheric windows centered at 940, 1080, 1280, 1600, 2000, 2700, 2800, and 5000 nm; all of these are in the spectral range of the VIMS-IR channel.

Long before Cassini-Huygens arrived at Saturn in 2004 there was the expectation of finding hydrocarbon lakes. In mid-2005 the Cassini Imaging Science Subsystem photographed a large dark feature suggestive of a lake near the south pole (later named Ontario Lacus) [167]. In 2006 Cassini synthetic aperture radar images of the north polar region revealed a vast array of lakelike features north of 75° [168]. But the evidence for their being liquid was based morphological pattern and extremely low radar cross section. VIMS provided additional evidence for the composition of the lakes.

It had been predicted that the lakes would consist largely of methane and nitrogen with several tens of percent ethane [169]. Methane would be hard to detect owing to the abundant methane in the atmosphere, and nitrogen exhibits no features in the atmospheric windows, but ethane was a possibility. During a close flyby of Titan in late 2007, VIMS observed Ontario Lacus and collected a sequence of four spectral cubes [170]. Spectra from inside and outside the lake are quite similar, dominated by the absorption bands in the atmosphere. Spectral differences do exist, however, in particular in the 2000 and 5000 nm windows. Ratios of spectra from the interior of the lake and from a nearby area outside cancel effects of the strong atmospheric absorptions and reveal spectral features in these two bands. Spectra with nearly identical path lengths were used to minimize residuals from strong atmospheric absorptions in the ratios. Ratios are shown for the dark lake interior and for a narrow annulus resembling a beach just inside the bright shoreline. The ratio spectra are nearly flat across the rest of the windows, showing that the strong atmospheric absorptions were mostly canceled out.

Brown et al. [170] identified two well-developed spectral features in the ratios: a narrow absorption at 2018 nm and a broad absorption in the 5000 nm window that shows a steep drop at ~4800 nm, continuing downward to the end of the VIMS spectral range. Clark et al. [171] provided derived optical constants for liquid ethane measured in the laboratory. Brown et al. [170] derived model ethane spectra from these optical constants (red line in Figure 7.3)—most of ethane's narrow absorption bands fall within the deep atmospheric window, with the narrow exception of a feature near the 2000 nm window that matches what is seen in the ratio spectra of the lake. The steep drop in reflectance beyond 4800 nm is a strong indicator of lake composition. This is characteristic of alkanes, including ethane, propane, and butane [171,172]. The presence of propane, butane, and higher-order alkanes could explain the continued drop beyond the ethane absorption feature modeled in the 5000 nm window by Brown et al. [170]. Although the VIMS spectra cannot detect liquid methane, theoretical work by Mitri et al. [173] on Titan's hydrocarbon lakes shows that if ethane is present, methane is most probably a major component as well.

14.2.8 Pluto

14.2.8.1 New Horizons Ralph

Ralph has a 75 mm aperture reflecting telescope with a 650 mm focal length. A dichroic beamsplitter divides the beam between two separate focal planes, the Multi-spectral Visible Imaging Camera (MVIC) and the Linear Etalon Imaging Spectral Array (LEISA), both of them passively cooled. Unlike most other instruments described in this chapter, Ralph obtains data without use of any internal moving parts. The entire spacecraft is rolled to provide spatial and spectral coverage. The MVIC focal plane has seven CCD arrays, most of them operated in time-delay integration (TDI) mode. MVIC's FOV is scanned across a scene while reading out the CCD arrays at a matched rate. Four of the CCDs, operated in concert, are fitted with color filters to cover wavelengths from 400 to 975 nm. LEISA, Ralph's infrared component, uses a 256 × 256 HgCdTe detector array. Linear variable interference filters are affixed to the array, making each row of pixels sensitive to a specific wavelength of infrared light between 1.2 and 2.5 μm. As with MVIC, the FOV of LEISA is swept across the scene by rolling the spacecraft. Frames are recorded at a rate so that the scene has moved by about a pixel between each successive frame. LEISA's primary filter provides a spectral resolution ($\lambda/\Delta\lambda$) of 240 over the full wavelength range. An additional filter provides a higher spectral resolution of 560 over a narrow wavelength range from 2.1 to 2.25 μm, where diagnostic absorption bands for ices of N_2, NH_3, and CH_4 are located.

14.2.8.2 Calibration and Analysis Techniques

Ralph radiometric calibration is based on a combination of preflight ground measurements and in-flight observations. MVIC is sufficiently sensitive that observations of stellar clusters can provide numerous simultaneous observations of stars with known brightnesses. Repeated observations of clusters NGC 6405, 6475, and 3532 were used to monitor instrument calibration [174]. LEISA is less sensitive, requiring relatively bright stars. Its calibration is based primarily on Vega and Arcturus observations. Flat fields for both MVIC and LEISA obtained prior to launch were found to be less satisfactory than hoped for, and with no mechanism to insert a flat field target into Ralph's FOV, the only way to get in-flight flat fields was to use solar system targets during flybys. Flat field observations were done during the Jupiter flyby by slewing the Ralph field rapidly across Jupiter perpendicular to the usual scan orientation, in order to blur out spatial information. This worked reasonably well, but not for all wavelengths. Jupiter reflects too little light at the wavelengths of its strong CH_4 absorption bands. Thus, the procedure was repeated in the Pluto system, using Pluto itself as the flat field source.

A particular challenge with New Horizons Ralph is that the spacecraft must be rolled to sweep the instrument's FOV and wavelength channels across a scene. This motion is driven by monopropellant thrusters rather than reaction wheels and can only be controlled to within a deadband. The tighter the deadband, the more frequently thruster firings are required. Whenever a thruster fires, the rate and direction of motion change, although the readout rate remains constant. The spacecraft records its orientation at a 1 Hz rate, and this information can be used by the ISIS-3 software to map-project each pixel of each frame to the appropriate location on the target body's surface. One additional complication is that New Horizons flies past its targets at roughly 14 km/s. The finite duration of Ralph scans, especially for the LEISA observations of infrared wavelengths, means that the spacecraft-target geometry changes appreciably over the course of the scan. One way to handle such a data set would be to compile a long list of spatial plus spectral samples and test models against that list. But it is far more convenient to work with image cubes in which two axes are spatial and one is spectral. To create image cubes that minimize the distortion from spacecraft motion, ISIS-3 software is used to project the data to an orthographic or point-perspective view of the target body from the spacecraft's location at the midtime of the scan. This procedure merges together pixels from different phase and emission angles, leading to modest geometric errors. For Pluto, the last LEISA scan before closest approach was the most affected, with the image cube combining footprints differing in phase and emission angle from midscan values by as much as $\pm 1.6°$.

14.2.8.3 Case Study

LEISA was designed to be sensitive to characteristic vibrational absorption bands of ices of CH_4, CO, and N_2, all known from preencounter telescopic observations to exist on Pluto and to be heterogeneously distributed in longitude [175,176]. During the hours before closest approach, New Horizons scanned LEISA across the encounter hemisphere of Pluto at spatial scales ranging from 3 to 10 km/pixel. These observations provided detailed maps of the geographic distribution of the three volatile ices, providing insights into the processes governing their distribution. Each of the ices has distinct distributions in latitude, longitude, and altitude [177], pointing to the complex influence of seasonally varying patterns of insolation [178,179] driving sublimation and condensation at distinct rates for the different ices, with N_2 being the most volatile and CH_4 being the least volatile of the three.

Pluto's equatorial latitudes were observed to be mostly devoid of volatile ices [180], featuring a belt of dark, reddish maculae, typified by Cthulhu Macula on the encounter hemisphere. But low latitudes are not entirely free of volatile ices. Notable exceptions include the bladed terrain of Tartarus Dorsa, strikingly rugged high-altitude ridges rich in CH_4 ice, possibly analogous to terrestrial penitentes [181]. Another site of equatorial volatile ices is the deep basin of Sputnik Planitia, where N_2 appears to be the dominant species, but CH_4 and CO are also present [182]. These ices form a sufficiently deep deposit to be actively convecting, a style of glaciation not previously seen anywhere in the Solar System [183–185].

At midlatitudes, CH_4 ice preferentially appears on mountain tops and crater rims, and further north it occurs across broad plains. Retreating scarps cut into those plains, as in Piri Planitia [186], exposing a substrate featuring absorptions by H_2O ice, which is inert at Pluto surface temperatures and is likely representative of the underlying "bedrock." Such scarps could be related to the million year Milankovich-like "mega-seasonal" cycles that arise from Pluto's polar precession and the evolution of its eccentric heliocentric orbit [178,179,187]. N_2 and CO ices also appear at these latitudes, but mostly in topographic lows such as Burney crater, where the higher pressure helps stabilize them from sublimation.

Further north, the polar region of Lowell Regio is currently experiencing continuous summer sunlight. CH_4 ice is prevalent there, forming deposits on the order of a kilometer in thickness. These deposits are punctuated by a variety of deep pits including clusters and aligned rows of pits that are clearly not impact craters. They are more likely related to some eruptive process or to subsurface melting and undercutting of the CH_4 ice mantle [188]. There is little evidence of the more volatile N_2 and CO ices at polar latitudes, suggesting that they have already sublimated away from the northern pole in the current summer season [180,182].

New Horizons Ralph was also able to map nonvolatile materials on Pluto's surface. Chief among these was the previously mentioned broad expanses of dark, reddish material that are especially prevalent at equatorial latitudes. This material is thought to be carbon-rich, macromolecular tholins. It likely originates as photochemical haze in Pluto's upper atmosphere, although radiolytic production from Pluto's surface ices is also possible. H_2O ice also has prominent absorptions at LEISA wavelengths, allowing it to be mapped. It is associated with the dark red material in some places, especially toward the peripheries of the maculae. H_2O ice also appears in rugged mountain ranges along the western margin of Sputnik Planitia, such as Al-Idrisi and Baré Montes. Curiously, an area especially rich in H_2O ice is seen around a few of the pits in the CH_4-ice-rich midlatitudes at Supay Facula. The H_2O ice there does not appear to be exposed "bedrock" as in Piri Planitia, as there is no morphological contrast between it and the surrounding CH_4-dominated terrain. It is a puzzle how inert H_2O ice could come to be superposed on seasonally mobile CH_4 ice unless some sort of eruptive process had recently emplaced it. It has to be a relatively young deposit, since any ancient surface would eventually accumulate a veneer of photochemical haze settling out of the atmosphere, as in Cthulhu and the other dark equatorial maculae. (Note: Pluto feature names mentioned in this chapter are a work in progress. They include a mix of formally approved and informal names.)

14.3 CONCLUSIONS AND FUTURE CHALLENGES

Hyperspectral analyses of planetary surfaces in the VNIR and SWIR wavelengths are used to constrain their geology and mineralogy and as a means of understanding their history and evolution. The methods used to acquire and calibrate planetary data sets share many similarities with those used for terrestrial observations of vegetated and rocky surfaces on the Earth. In fact, many of the techniques developed for the study of planetary hyperspectral data sets germinated from studies using Earth-based observations. Even though other planetary bodies in our solar system almost certainly do not have vegetation like the Earth (the focus of this volume), there are several reasons for scientific interactions between those studying imaging spectroscopy (or hyperspectral remote sensing) applications for the Earth and those studying other planetary bodies using similar sensors. First, most of the design of hyperspectral sensors that gather data in the 400–2500 nm range is the same whether they are deployed to study the Earth or other planetary bodies. Second, the methods and techniques of hyperspectral data processing, interpretation, and analysis are similar irrespective of which planetary bodies are being studied. For example, hyperspectral data analysis methods discussed in various chapters of this volume, such as subpixel analysis, band ratios, principal component analysis, spectral matching techniques, linear and nonlinear unmixing, and various classification techniques, can be applied across planetary hyperspectral data sets. Third, preprocessing algorithms, such as radiometric, geometric, and atmospheric corrections, are similar or can be used with slight modifications for various planetary bodies. Fourth, as we have learned from previous chapters, the background influence from factors such as soils on spectral reflectivity is significant. Thus, the lessons learned in soil spectral reflectivity studies, considering their chemical-physical-moisture properties, on one planet should be applicable to other planetary bodies as well. For example, certain narrow bands or their combinations best predict soil organic matter or other soil properties. Fifth, hyperspectral narrow bands provide an opportunity to compute hundreds or thousands of material identification indices, such as indices that provide the best results in modeling and mapping soil characteristics. This understanding, too, can be used across planetary bodies. Sixth, the spectral libraries available for soils, rocks, vegetation, and soils with varying moisture content measured from terrestrial materials are also useful in spectral matching techniques for other planetary surfaces to identify and label targets by matching ideal spectra with target spectra. For example, an absorption feature found around 2800–2900 nm in data gathered by the Moon Mineralogy Mapper hyperspectral sensor onboard Chandrayaan-1 led to the detection of hydroxyl/water (about 0.1% of total volume of material) on the Moon.

Planetary bodies range from those without atmospheres to those with atmospheres much thicker than the Earth's. Despite the challenges associated with specific data sets, valuable information regarding the composition and mineralogy of planetary surfaces has been derived. More importantly, many of the data sets acquired now provide well-calibrated observations that will enable future researchers to explore these surfaces in more detail using analytical techniques yet to be developed. Such future work will include improved methods of eliminating interference from atmospheric contamination, particularly on bodies that experience atmospheric variability from seasonal changes. Ongoing work regarding unmixing of complicated rock and mineral mixtures using laboratory analyses and field studies will continue to improve the precision and accuracy of determining compositional variability and mineral abundances using remotely sensed hyperspectral data sets.

The instruments and case studies described here together document a highly active field of research that has important implications for remote sensing studies of the Earth. Not only do the results of such planetary remote sensing studies help us to understand the nature and distribution of materials such as rocks and soils, but also they have been invaluable in detecting evidence of water in places in our solar system we had long believed to be dry. There is a clear link between water and life in our solar system [189] (http://mepag.jpl.nasa.gov/reports/MEPAG_Goals_Document_2010_v17.pdf). Scientists are now in the process of unraveling and extending links between observed geological and biological materials such that we will soon be able to characterize habitable environments both on and off the Earth. Finally, as scientists continue to discover numerous exoplanets, some of which may be habitable, the knowledge of hyperspectral signature becomes pivotal and will play a crucial role in gathering knowledge on the exoplanets.

REFERENCES

1. Nielsen, U.N., Osler, G.H.R., Campbell, C.D., Burslem, D.F.R.P., and van der Wal, R. 2010. The Influence of Vegetation Type, Soil Properties and Precipitation on the Composition of Soil Mite and Microbial Communities at the Landscape Scale. *Journal of Biogeography*, 37, pp. 1317–1328.
2. Abrams, M.J., Ashley, R.P., Rowan, L.C., Goetz, A.F.H., and Kahle, A.B. 1977. Mapping of Hydrothermal Alteration Minerals in the Cuprite Mining District, Nevada, Using Aircraft Scanner Images for the Spectral Region 0.46–2.36 μm. *Geology*, 5, pp. 713–718.
3. Goetz, A.F.H., Rock, B.N., and Rowan, L.C. 1983. Remote Sensing for Exploration: An Overview. *Economic Geology*, 78 (4), pp. 573–590.
4. Marsh, S.E. and McKeon, J.B. 1983. Integrated Analysis of High-Resolution Field and Airborne Spectroradiometer Data for Alteration Mapping. *Economic Geology*, 78 (4), pp. 618–632.
5. Taranik, J.V. 1988. Application of Aerospace Remote Sensing Technology to Exploration for Precious Metal Deposits in the Western United States. In: *Bulk Mineable Precious Metal Deposits of the Western United States*, R.W. Schafer editor, GSN Symposium Proceedings, Geological Society of Nevada, Reno, NV.
6. Kruse, F.A. 1999. *Visible-Infrared Sensors and* Case Studies. Ch 11 in: *Remote Sensing for the Earth Sciences: Manual of Remote Sensing*, A.N. Rencz editor, 3rd ed. vol. 3, John Wiley & Sons, New York.

7. Burns, R.G. 1993. *Mineralogical Applications of Crystal Field Theory*. 2nd ed., Cambridge University Press.
8. Bibring, J.-P., Langevin, Y., Mustard, J.F., Poulet, F., Arvidson, R., Gendrin, A., Gondet, B., Mangold, N., Pinet, P., and Forget, F. 2006. Global Mineralogical and Aqueous Mars History Derived from OMEGA/ Mars Express Data. *Science*, 312 (5772), pp. 400–404.
9. Murchie, S., Arvidson, R., Bedini, P., Beisser, K., Bibring, J.-P., Bishop, J., Boldt, J., Cavender, P., Choo, T., Clancy, R.T., and 40 coauthors. 2007. Compact Reconnaissance Imaging Spectrometer for Mars (CRISM) on Mars Reconnaissance Orbiter (MRO). *Journal of Geophysical Research*, 112 (E5), CiteID E05S03.
10. Abrams, M.J., Brown, D., Lepley, L., and Sadowski, R. 1983. Remote Sensing for Porphyry Copper Deposits in Southern Arizona. *Economic Geology*, 78 (4), pp. 591–604.
11. Podwysocki, M.H., Segal, D.B., and Abrams, M.J. 1983. Use of Multispectral Scanner Images for Assessment of Hydrothermal Alteration in the Marysvale, Utah, Mining Area. *Economic Geology*, 78 (4), pp. 573–590.
12. Hutsinpiller, A. and Taranik, J.V. 1988. Spectral Signatures of Hydrothermal Alteration at Virginia City, Nevada. In: *Bulk Mineable Precious Metal Deposits of the Western United States*, R.W. Schafer editor, Geological Society of Nevada Symposium Proceedings, pp. 505–530.
13. Swayze, G.A., Clark, R.N., Smith, K.S., Hageman, P.L., Sutley, S.J., Pearson, R.M., Rust, R.S. et al. 1998. Using Imaging Spectroscopy to Cost-Effectively Locate Acid-Generating Minerals at Mine Sites: An Example from the California Gulch Superfund Site in Leadville, Colorado. In: *Proceedings of the 7th Airborne Earth Science Workshop*, JPL Publication 97-21 (1), pp. 385–389.
14. Vane, G., Green, R.O., Chrien, T.G., Enmark, H.T., Hansen, E.G., and Porter, W.M. 1993. The Airborne Visible/Infrared Imaging Spectrometer (AVIRIS). *Remote Sensing of Environment*, 44, pp. 127–143.
15. Gao, B., Heidebrecht, K.B., and Goetz, A.F.H. 1993. Derivation of Scaled Surface Reflectances from AVIRIS Data. *Remote Sensing of Environment*, 44, pp. 165–178.
16. King, T.V.V. and Clark, R.N. 2000. Verification of Remotely Sensed Data. Ch 5 in: *Remote Sensing for Site Characterization*, F. Kuehn, T. King, B. Hoerig, and D. Peters editors, Springer-Verlag, Berlin, Germany.
17. Jacobsen, A., Heidebrecht, K.B., and Goetz, A.F.H. 2000. Assessing the Quality of the Radiometric and Spectral Calibration of Casi Data and Retrieval of Surface Reflectance Factors. *Photogrammetric Engineering and Remote Sensing*, 66 (9), pp. 1083–1091.
18. Lyon, R.J.P. 1965. Analysis of Rocks by Spectral Infrared Emission (8–25 microns). *Economic Geology*, 60, pp. 715–736.
19. Farmer, V.C. (Ed.) 1974. *The Infrared Spectra of Minerals*. Mineralogical Society Monograph 4, Mineralogical Society, London.
20. Hunt, G.R. 1980. Electromagnetic Radiation: The Communication Link in Remote Sensing. Ch 2 in: *Remote Sensing in Geology*, B.S. Siegal and A.R. Gillespie editors, John Wiley, New York, pp. 5–45.
21. Clark, R.N., King, T.V.V., Klejwa, M., and Swayze, G.A. 1990. High Spectral Resolution Reflectance Spectroscopy of Minerals. *Journal of Geophysical Research*, 95 (B8), pp. 12653–12680.
22. Salisbury, J.W., Walter, L.S., Vergo, N., and D'Aria, D.M. 1991. *Infrared (2.1–25 μm) Spectra of Minerals*. Johns Hopkins University Press, Baltimore, MD.
23. Burns, R.G. 1993. Origin of Electronic Spectra of Minerals in the Visible to Near-Infrared Region. Ch 1 in: *Topics in Remote Sensing 4—Remote Geochemical Analysis: Elemental and Mineralogical Composition*, C.M. Pieters and P.A.J. Englert editors, Cambridge University Press.
24. Clark, R.N. 1999. Spectroscopy of Rocks and Minerals and Principles of Spectroscopy. Ch 1 in: *Remote Sensing for the Earth Sciences: Manual of Remote Sensing*, A.N. Rencz editor, 3rd edition, vol. 3, John Wiley & Sons, New York.
25. Clark, R.N., Swayze, G.A., Wise, R.A., Live, K.E., Hoefen, T.M., Kokaly, R.F., and Sutley, S.J. 2007. USGS Digital Spectral Library splib06a: U.S. Geological Survey Data Series 231.
26. Kokaly, R.F., Clark, R.N., Swayze, G.A., Livo, K.E., Hoefen, T.M., Pearson, N.C., Wise, R.A. et al. 2017. USGS Spectral Library Version 7: U.S. Geological Survey Data Series 1035, 61 p., https://doi.org/10.3133/ds1035.
27. Baldridge, A.M., Hook, S.J., Grove, C.I., and Rivera, G. 2009. The ASTER Spectral Library Version 2.0. *Remote Sensing of Environment*, 113 (4), pp. 711–715.
28. Roach, L.H., Mustard, J.F., Swayze, G., Milliken, R.E., Bishop, J.L., Murchie, S.L., and Lichtenberg, K. 2010. Hydrated Mineral Stratigraphy of Ius Chasma, Valles Marineris. *Icarus*, 206, pp. 253–268.
29. Milliken, R.E., Grotzinger, J.P., and Thomson, B.J. 2010. Paleoclimate of Mars as Captured by the Stratigraphic Record in Gale Crater. *Geophysical Research Letters*, 37, L04201, doi:10.1029/2009GL041870.
30. Chan, M.A. and Parry, W.T. 2002. *Rainbow of the Rocks: Mysteries of Sandstone Colors and Concretions in Colorado Plateau Canyon Country*, Public Information Series 77, Utah Geological Survey, 17 pp.

31. Chan, M.A., Johnson, C.M., Beard, B.L., Bowman, J.R., and Parry, W.T. 2006. Iron Isotopes Constrain the Pathways and Formation Mechanisms of Terrestrial Oxide Concretions: A Tool for Tracing Iron Cycling on Mars? *Geosphere*, 2 (7), pp. 324–332, doi: 10.1130/GES00051.1.
32. Douka, C.E. 1977. Study of Bacteria from Manganese Concretions. Precipitation of Manganese by Whole Cells and Cell-free Extracts of Isolated Bacteria. *Soil Biology and Biochemisty*, 9, pp. 89–97.
33. Pieters, C.M., Goswami, J.N., Clark, R.N., Annadurai, M., Boardman, J., Buratti, B., Combe, J-P. et al. 2009. Character and Spatial Distribution of OH/H_2O on the Surface of the Moon Seen by M^3 on Chandrayaan-1. *Science*, 326, p. 568, doi: 10.1126/science.1178658.
34. Cocks, T., Jenssen, R., Stewart, A., Wilson, I., and Shields, T. 1998. The HyMap Airborne Hyperspectral Sensor: The System, Calibration and Performance. In: *Proceedings of 1st EARSeL Workshop on Imaging Spectroscopy*, M. Schaepman, D. Schläpfer, and K.I. Itten editors, pp. 37–43.
35. Watts, L.A., Davis, R.O., Granneman, R.D., LaVeigne, J.D., Chandos, R.A., Russell, E.E., and Cairns, B. 2001. Unique VISNIR-SWIR Hyperspectral and Polarimeter Measurements. In: *Proceedings of the 5th Airborne Remote Sensing Conference*, San Francisco, CA.
36. Pearlman, J.S., Barry, P.S., Segal, C.C., Shepanski, J., Beiso, D. and Carman, S.L. 2003. Hyperion, a Space-Based Imaging Spectrometer. *IEEE Transactions on Geoscience and Remote Sensing*, 41, pp. 1160–1172.
37. Pieters, C.M., Boardman, J., Buratti, B., Chatterjee, A., Clark, R., Glavich, T., Green, R. et al. 2009. The Moon Mineralogy Mapper (M^3) on Chandrayaan-1, Current. *Science*, 96 (4), pp. 500–505.
38. Antoniadi, E.M. 1930. *The Planet Mars, Trans.* Keith Reid Ltd., Patrick Moore, Devon, UK, 1975.
39. Brown, R.H., Baines, K.H., Bellucci, G., Bibring, J.-P., Buratti, B.J., Capaccioni, F., Cerroni, P. et al. 2004. The Cassini Visual and Infrared Mapping Spectrometer (VIMS) Investigation. *Space Science Reviews*, 115, pp. 111–168.
40. Stern, S.A., Bagenal, F., Ennico, K., Gladstone, G.R., Grundy, W.M., McKinnon, W.B., Moore, J.M. et al. 2015. The Pluto System: Initial Results from Its Exploration by New Horizons. *Science*, 350, p. 292.
41. Reuter, D.C., Stern, S.A., Scherrer, J., Jennings, D.E., Baer, J.W., Hanley, J., Hardaway, L. et al. 2008. Ralph: A Visible/Infrared Imager for the New Horizons Pluto/Kuiper Belt Mission. *Space Science Reviews*, 140, pp. 129–154.
42. National Research Council (NRC) Decadal Survey. 2007. Earth Science and Applications from Space: National Imperatives for the Next Decade and Beyond, Committee on Earth Science and Applications from Space: A Community Assessment and Strategy for the Future, National Research Council, ISBN: 978-0-309-10387-9, 456 pp.
43. Loizzo, R., Ananasso, C., Guarini, R., Lopinto, E., Candela, L., and Pisani, A.R. 2016. The PRISMA Hyperspectral Mission. In: *Proc. "Living Planet Symposium 2016"*, Prague, Czech Republic, 9–13 May 2016 (ESA SP-740, August 2016).
44. Cudahy, T.J., Okada, K., Yamato, Y., Maekawa, M., Hackwell, J.A., and Huntington, J.F. 2000. Mapping skarn and porphyry alteration mineralogy at Yerington, Nevada, using airborne hyperspectral TIR SEBASS data. CSIRO Exploration and Mining Report 734R, CSIRO Exploration and Mining, Floreat Park, WA, Australia.
45. Rowan, L.C. and Mars, J.C. 2003. Lithologic Mapping in the Mountain Pass, California Area Using Advanced Spaceborne Thermal Emission and Reflection Radiometer (ASTER) Data. *Remote Sensing of Environment*, 84, pp. 350–366.
46. Vaughan, R.G., Calvin, W.M., and Taranik, J.V. 2003. SEBASS Hyperspectral Thermal Infrared Data: Surface Emissivity Measurement and Mineral Mapping. *Remote Sensing of Environment*, 85, pp. 48–63.
47. Vaughan, R.G., Hook, S.J., Calvin, W.M., and Taranik, J.V. 2005. Surface Mineral Mapping at Steamboat Springs, Nevada with Multi-Wavelength Thermal Infrared Images. *Remote Sensing of Environment*, ASTER Special Issue, 99 (1–2), pp. 140–158.
48. Coolbaugh, M.F., Kratt, C., Fallacaro, A., Calvin, W.M., and Taranik, J.V. 2007. Detection of Geothermal Anomalies Using Advanced Spaceborne Thermal Emission and Reflection Radiometer (ASTER) Thermal Infrared Images at Bradys Hot Springs, Nevada, USA. *Remote Sensing of Environment*, 106 (3), pp. 350–359.
49. Anderson, M. and Kustas, W. 2008. Thermal Remote Sensing of Drought and Evapotranspiration. *Eos, Transactions American Geophysical Union*, 89 (26), pp. 233–234.
50. Anderson, M.C., Hain, C., Wardlow, B., Pimstein, A., Mecikalski, J.R., and Kustas, W.P. 2011. Evaluation of Drought Indices Based on Thermal Remote Sensing of Evapotranspiration Over the Continental United States. *Journal of Climate*, 24 (8), pp. 2025–2044.
51. Schlerf, M., Rock, G., Lagueux, P., Ronellenfitsch, F., Gerhards, M., Hoffmann, L., and Udelhoven, T. 2012. A Hyperspectral Thermal Infrared Imaging Instrument for Natural Resources Applications. *Remote Sensing*, 4, pp. 3995–4009, doi:10.3390/rs4123995.

52. Eisele, A., Lau, I., Hewson, R., Carter, D., Wheaton, B., Ong, C., Cudahy, T.J., Chabrillat, S., and Kaufmann, H. 2012. Applicability of the Thermal Infrared Spectral Region for the Prediction of Soil Properties Across Semi-Arid Agricultural Landscapes. *Remote Sensing*, 4, pp. 3265–3286.
53. Ramsey, M.S. and Harris, A.J. 2013. Volcanology 2020: How will Thermal Remote Sensing of Volcanic Surface Activity Evolve Over the Next Decade? *Journal of Volcanology and Geothermal Research*, 249, pp. 217–233.
54. Meerdink, S.K., Roberts, D.A., King, J.Y., Roth, K.L., Dennison, P.E., Amaral, C.H., and Hook, S.J. 2016. Linking Seasonal Foliar Traits to VSWIR-TIR Spectroscopy Across California Ecosystems. *Remote Sensing of Environment*, 186, pp. 322–338.
55. da Luz, B.R. and Crowley, J.K. 2010. Identification of Plant Species by Using High Spatial and Spectral Resolution Thermal Infrared (8.0–13.5 μm) Imagery. *Remote Sensing of Environment*, 114, pp. 404–413.
56. Hackwell, J.A., Warren, D.W., Bongiovi, R.P., Hansel, S.J., Hayhurst, T.L., Mabry, D.J., Sivjee, M.G., and Skinner, J.W. 1996. LWIR/MWIR Imaging Hyperspectral Sensor for Airborne and Ground-Based Remote Sensing. *SPIE*, 2819, pp. 102–107.
57. Lucey, P.G., Williams, T.J., Hinrichs, J.L., Winter, M.E., Steutel, D., and Winter, E.M. 2001. Three Years of Operation of AHI: The University of Hawaii's Airborne Hyperspectral Imager. In: B.F. Andresen, G.F. Fulop, and M. Strojnik editors, *Proceedings of SPIE infrared technology and applications XXVII*, vol. 4369 (pp. 112–120). International Society for Optical Engineering, Bellingham, WA.
58. Hook, S.J., Johnson, W.R., and Abrams, M.J. 2013. NASA's Hyperspectral Thermal Emission Spectrometer (HyTES). In: *Thermal Infrared Remote Sensing: Sensors, Methods, Applications*, C. Kuenzer and S. Dech editors, Springer, pp. 93–115.
59. Kruse, F.A. 2002. Combined SWIR and LWIR Mineral Mapping Using MASTER/ASTER. In: *Proceedings, IGARSS 2002, CD ROM*, Toronto, Canada, pp. 2267–2269.
60. Cudahy, T.J., Wilson, J., Hewson, R., Linton, P., Harris, P., Sears, M., Okada, K., and Hackwell, J.A. 2001. Mapping Porphyry-Skarn Alteration at Yerington, Nevada, Using Airborne Hyperspectral VNIR-SWIR-TIR Imaging Data. In *Geoscience and Remote Sensing Symposium, 2001. IGARSS'01. IEEE 2001 International* (Vol. 2, pp. 631–633). IEEE.
61. Vaughan, R.G. and Calvin, W.M. 2004. Synthesis of High-Spatial Resolution Hyperspectral VNIR/SWIR and TIR Image Data for Mapping Weathering and Alteration Minerals in Virginia City, Nevada. In: *Geoscience and Remote Sensing Symposium, 2004. IGARSS'04. Proceedings. 2004 IEEE International* (Vol. 2, pp. 1296–1299). IEEE.
62. Vaughan, R.G. 2004. Surface Mineral Mapping at Virginia City and Steamboat Springs, Nevada with Multi-Wavelength Infrared Remote Sensing Image Data. *PhD dissertation*, University of Nevada, Reno, 273 pp.
63. Boynton, W.V., Feldman, W.C., Squyres, S.W., Prettyman, T.H., Brückner, J., Evans, L.G., Reedy, R.C., Starr, R., Arnold, J.R., Drake, D.M., and 15 coauthors. 2002. Distribution of Hydrogen in the Near Surface of Mars: Evidence for Subsurface Ice Deposits. *Science*, 297 (5578), pp. 81–85.
64. Boynton, W.V., Taylor, G.J., Evans, L.G., Reedy, R.C., Starr, R., Janes, D.M., Kerry, K.E., Drake, D.M., Kim, K.J., Williams, R.M.S., and 18 coauthors. 2007. Concentration of H, Si, Cl, K, Fe, and Th in the Low- and Mid-Latitude Regions of Mars. *Journal of Geophysical Research*, 112 (E12), CiteID E12S99.
65. Hasebe, N., Shibamura, E., Miyachi, T., Takashima, T., Kobayashi, M., Okudaira, O., Yamashita, N., Kobayashi, S., Ishizaki, T., Sakurai, K., and 13 coauthors. 2008. Gamma-Ray Spectrometer (GRS) for Lunar Polar Orbiter SELENE. *Earth, Planets and Space*, 60, pp. 299–312.
66. Mitrofanov, I.G., Bartels, A., Bobrovnitsky, Y.I., Boynton, W., Chin, G., Enos, H., Evans, L., Floyd, S., Garvin, J., Golovin, D.V., and 26 coauthors. 2010. Lunar Exploration Neutron Detector for the NASA Lunar Reconnaissance Orbiter. *Space Science Reviews*, 150 (1-4), pp. 183–207.
67. Prettyman, T.H., Feldman, W.C., Barraclough, B.L., Capria, M.T., Coradini, A., Enemark, D.C., Fuller, K.R., Lawrence, D.J., Patrick, D.E., Raymond, C.A., and 2 coauthors. 2004. Mapping the Elemental Composition of Ceres and Vesta: Dawn's Gamma Ray and Neutron Detector. In: *Instruments, Science, and Methods for Geospace and Planetary Remote Sensing*, C.A. Nardell, P.G. Lucey, J.-H. Yee, and J. B. Garvin, editors, Proceedings of the SPIE, Vol. 5660, pp. 107–116.
68. Schott, J.R. 1997. *Remote Sensing: The Image Chain Approach.* Oxford University Press, New York.
69. Hapke, B. 1993. Combined Theory of Reflectance and Emittance Spectroscopy. Ch 2 in: *Topics in Remote Sensing 4—Remote Geochemical Analysis: Elemental and Mineralogical Composition*, C.M. Pieters and P.A.J. Englert editors, Cambridge University Press.
70. Hapke, B. 1993. Theory of Reflectance and Emittance Spectroscopy. In: *Topics in Remote Sensing 3*, R.E. Arvidson and M.J. Rycroft editors, Cambridge University Press.

71. Adler-Golden, S., Berk, A., Bernstein, L.S., Richtsmeier, S.C., Acharya, P.K., and Matthew, M.W. 1998. FLAASH, A MODTRAN 4 Atmospheric Correction Package for Hyperspectral Data Retrievals and Simulation. *Proceedings of the 7th JPL Airborne Earth Science Workshop, JPL.*
72. Clark, R.N. and Roush, T.L. 1984. Reflectance Spectroscopy: Quantitative Analysis Techniques for Remote Sensing Applications. *Journal of Geophysical Research*, 89 (B7), pp. 6329–6340, doi:10.1029/JB089iB07p06329.
73. Charette, M.P., McCord, T.B., Pieters, C., and Adams, J. 1974. Application of Remote Spectral Reflectance Measurements to Lunar Geology Classification and Determination of Titanium Content of Lunar Soils. *Journal of Geophysical Research*, 79, pp. 1605–1613.
74. Johnson, J.R., Larson, S.M., and Singer, R.B. 1991. Remote Sensing of Potential Lunar Resources: 1. Near-Side Compositional Properties. *Journal of Geophysical Research*, 96, pp. 18861–18882.
75. Geissler, P.E., Singer, R.B., Komatsu, G., Murchie, S., and Mustard, J. 1993. An Unusual Spectral Unit in West Candor Chasma—Evidence for Aqueous or Hydrothermal Alteration in the Martian Canyons. *Icarus*, 106 (2), pp. 380–391.
76. Pelkey, S.M., Mustard, J.F., Murchie, S., Clancy, R.T., Wolff, M., Smith, M., Milliken, R. et al. 2007. CRISM Multispectral Summary Products: Parameterizing Mineral Diversity on Mars from Reflectance. *Journal of Geophysical Research*, 112, p. E08S14, doi:10.1029/2006JE002831.
77. Geissler, P.E., Greenberg, R., Hoppa, G., McEwen, A., Tufts, R., Phillips, C., Clark, B. et al. 1998. Evolution of Lineaments on Europa: Clues from Galileo Multispectral Imaging Observations. *Icarus*, 135 (1), pp. 107–126, ISSN 0019-1035, doi: 10.1006/icar.
78. Ramsey, M.S. and Christensen, P.R. 1998. Mineral Abundance Determination: Quantitative Deconvolution of Thermal Emission Spectra. *Journal of Geophysical Research*, 103 (B1), pp. 577–596.
79. Hamlin, L., Green, R.O., Mouroulis, P., Eastwood, M., Wilson, D., Dudik, M., and Paine, C. 2011. Imaging Spectrometer Science Measurements for Terrestrial Ecology: AVIRIS and New Developments, paper presented at *Aerospace Conference. IEEE*, 2011, 1–7, http://dx.doi.org/10.1109/AERO.2011.5747395.
80. Berk, A., Bernstein, L.S., and Robertson, D.C. 1989. *MODTRAN: A Moderate Resolution Model for LOWTRAN7.* Rep. GL-TR-89-0122, Air Force Geophysics Lab, Bedford, MA.
81. Berk, A., Anderson, G.P., Bernstein, L.S., Acharya, P.K., Dothe, H., Matthew, M.W., Adler-Golden, S. et al. 1999. MODTRAN4 Radiative Transfer Modeling for Atmospheric Correction. In: *Proceedings of the 8th JPL Airborne Earth Science Workshop*, JPL Publication 99-17.
82. Boardman, J.W. 1998. Post-ATREM Polishing of AVIRIS Apparent Reflectance Data Using EFFORT: A Lesson in Accuracy versus Precision. *Summaries of the 7th JPL Airborne Earth Science Workshop*, JPL Publication 97-21.
83. Landgrebe, D. 2000. Information Extraction Principles and Methods for Multispectral and Hyperspectral Image Data. In: *Information Processing for Remote Sensing*, C.H. Chen, editor, World Scientific Publishing Co., Inc., River Edge, NJ.
84. Kruse, F.A. and Huntington, J.F. 1996. The 1995 AVIRIS Geology Group Shoot. In: *Proceedings of the Sixth JPL Airborne Earth Science Workshop*, JPL Publication 96-4 (1), Pasadena, California, pp. 155–164.
85. Kruse, F.A., Boardman, J.W., and Huntington, J.F. 2003. Comparison of Airborne Hyperspectral Data and EO-1 Hyperion for Mineral Mapping. In: Special Issue, *IEEE Transactions on Geoscience and Remote Sensing*, 41 (6), pp. 1388–1400.
86. Kruse, F.A., Boardman, J.W., and Huntington, J.F. 1999. Fifteen Years of Hyperspectral Data: Northern Grapevine Mountains, Nevada. In: *Proceedings of the 8th JPL Airborne Earth Science Workshop*, JPL publication 99-17, pp. 247–258.
87. Swayze, G.A., Smith, K.S., Clark, R.N., Sutley, S.J., Pearson, R.M., Vance, J.S., Hugeman, P.L. et al. 2000. Using Imaging Spectroscopy to Map Acidic Mine Waste. *Environmental Science & Technology*, 34, pp. 47–54.
88. Montero, I.C., Brimhall, G.H., Alpers, C.N., and Swayze, G.A. 2004. Characterization of Waste Rock Associated with Acid Drainage at the Penn Mine, California, by Ground-Based Visible to Short-Wave Infrared Reflectance Spectroscopy Assisted by Digital Mapping. *Chemical Geology*, 215 (1–4), pp. 453–472.
89. Vaughan, R.G. and Calvin, W.M. 2005. Mapping Weathering and Alteration Minerals in the Comstock and Geiger Grade Areas Using Visible to Thermal Infrared Airborne Remote Sensing Data, in Rhoden, H.N., Steininger, R.C., and Vikre, P.G., eds., *Geological Society of Nevada Symposium 2005: Window to the World*, Reno, Nevada, May 2005.
90. White, D.E. 1981. Active Geothermal Systems and Hydrothermal Ore Deposits. In: *Economic Geology: 75th Anniversary Volume*, B.J. Skinner editor, Economic Geology Publishing Co., pp. 392–423.

91. White, D.E., Thompson, G.A., and Sandberg, C.H. 1964. Rocks, Structure, and Geologic History of Steamboat Springs Thermal Area, Washoe County, Nevada. *USGS Professional Paper*, 458-B, 61 pp.
92. Hudson, D.M. 1987. Steamboat Springs Geothermal Area, Washoe County, Nevada. In: *Bulk Mineable Precious Metal Deposits of the Western United States*, J.L. Johnson editor, Geological Society of Nevada field trip guidebook, pp. 408–412.
93. Rye, R.O., Bethke, P.M., and Wasserman, M.D. 1992. The Stable Isotope Geochemistry of Acid Sulfate Alteration. *Economic Geology*, 87 (2), pp. 225–262.
94. Solomon, S.C., McNutt, R.L., Gold, R.E., and Domingue, D.L. 2007. MESSNEGER Mission Overview, 131, pp. 3–39.
95. McClintock, W.E., Izenberg, N.R., Holsclaw, G.M., Blewett, D.T., Domingue, D.L., Head III, J.W., Helbert, J. et al. 2008. Flyby Reflectance during MESSENGER's First Mercury Spectroscopic Observations of Mercury's Surface. *Science*, 321, pp. 62–65, doi: 10.1126/science.1159933.
96. Holsclaw, G.M., W.M., McClintock, D.L. Domingue, N.R. Izenberg, D.T. Blewett, and A.L. Sprague. 2010. A Comparison of the Ultraviolet to Near-Infrared Spectral Properties of Mercury and the Moon as Observed by MESSENGER. *Icarus*, 209, pp. 179–194, doi: 10.1016/j.icarus.2010.05.001.
97. Domingue, D.L., Vilas, F., Holsclaw, G.M., Warell, J., Izenberg, N.R., Murchie, S.L., Denevi, B.W. et al. 2010. Whole-Disk Spectrophotometric Properties of Mercury: Synthesis of MESSENGER and Ground-Based Observations. *Icarus*, 209, pp. 101–124, doi:10.1016/j.icarus.2010.02.022.
98. McClintock, W.E., Bradley, E.T., Vervack, Jr., R.J., Killen,R.M., Sprague, A.L., Izenberg, N.R., and Solomon, S.C. 2008. Mercury's Exosphere: Observations during MESSENGER's First Mercury Flyby. *Science*, 321, pp. 92–94, doi:10.1126/science.1159467.
99. Vervack, R.J., Jr., McClintock, W.E., Killen, R.M., Sprague, A.L., Anderson, B.J., Burger, M.H., Bradley, E.T. et al. 2010. Mercury's Complex Exosphere: Results from MESSENGER's Third Flyby. *Science*, doi:10.1126/science.1188572.
100. Izenberg, N.R., Klima, R.L., Murchie, S.L., Blewett, D.T., Holsclaw, G.M., McClintock, W.E., Malaret, E. et al. 2014. The Low-Iron, Reduced Surface of Mercury as Seen in Spectral Reflectance by MESSENGER. *Icarus*, 228, pp. 364–374, http://dx.doi.org/10.1016/j.icarus.2013.10.023.
101. Blewett, D.T., Chabot, N.L., Denevi, B.W., Ernst, C.M., Head, J.W., Izenberg, N.R., Murchie, S.L. et al. 2011. Hollows on Mercury: MESSENGER Evidence for Geologically Recent Volatile-Related Activity. *Science*, 333 (6051), pp. 1856–1859.
102. Trang, D., Lucey, P.G., and Izenberg, N.R. 2017. Radiative Transfer Modeling of MESSENGER VIRS Spectra: Detection and Mapping of Submicroscopic Iron and Carbon. *Icarus*, 293, pp. 206–217, http://dx.doi.org/10.1016/j.icarus.2017.04.026.
103. Clark, R., Pieters, C.M., and Green, R.O. 2009. March. Thermal Removal from Moon Mineralogy Mapper (M^3) Data. In *Lunar and Planetary Science Conference* (Vol. 40).
104. Clark, R. 2009. Detection of Adsorbed Water and Hydroxyl on the Moon. *Science*, 326 (#5952), pp. 562–564, doi: 10.1126/science.1178105.
105. Sunshine, J.M., Farnham, T.L., Feaga, L.M., Groussin, O., Merlin, F., Milliken, R.E. and A'Hearn, M.F. 2009. Temporal and Spatial Variability of Lunar Hydration as Observed by the Deep Impact Spacecraft. *Science*, 326 (#5952), pp. 565–568, doi:10.1126/science.1179788.
106. Taylor, L.A., Head, J.W., Pieters, C.M., Sunshine, J.M., Staid, M., Isaacson, P. and Petro, N.E. 2009. The Role of Spinel Minerals in Lunar Magma Evolution, *AGU Fall Meeting 2009*, Abstract #P34A-07.
107. Pieters, C.M., Boardman, J.W., Burratti, B., Clark, R.N., Combe, J., Green, R.O., Head, J.W. et al. 2009. New Mg-Spinel Rock-Type on the Lunar Farside and Implications for Lunar Crustal Evolution, *AGU Fall Meeting 2009*, Abstract #P34A-05.
108. Pieters, C.M., Boardman, J.W., Burratti, B., Cheek, L., Clark, R.N., Combe, J., Green, R.O. et al. 2009. Lunar Magma Ocean Bedrock Anorthosites Detected at Orientale Basin by M^3, *AGU Fall Meeting 2009*, Abstract #P34A-08.
109. Sunshine, J.M., Pieters, C.M., Besse, S., Boardman, J.W., Buratti, B.J., Clark, R.N., Combe, J. et al. 2009. Hidden in Plain Sight: Spinel-Rich Deposits on the Central Nearside of the Moon, *AGU Fall Meeting 2009*, Abstract #P34A-06.
110. Bibring, J.P., Langevin, Y., Gendrin, A., Gondet, B., Poulet, F., Berthé, M., Soufflot, A. et al. 2005. Mars Surface Diversity as Revealed by the OMEGA/Mars Express Observations. *Science*, 307, pp. 1576–1581.
111. Bonello, G., Pierre Bibring, J., Soufflot, A., Langevin, Y., Gondet, B., Berthé, M., and Carabetian, C. 2005. The Ground Calibration Setup of OMEGA and VIRTIS Experiments: Description and Performances. *Planetary and Space Science*, 53 (7), pp. 711–728.

112. Owen, T., Biemann, K., Biller, J.E., Lafleur, A.L., Rushneck, D.R., and Howarth, D.W. 1977. The Composition of the Atmosphere at the Surface of Mars. *Journal of Geophysical Research*, 82, pp. 4635–4639.
113. McGuire, P.C., Bishop, J.L., Brown, A.J., Fraeman, A.A., Marzo, G.A., Frank Morgan, M., Murchie, S.L. et al. 2009. An Improvement to the Volcano-Scan Algorithm for Atmospheric Correction of CRISM and OMEGA Spectral Data. *Planetary and Space Science*, 57 (7), pp. 809–815.
114. Vincendon, M., and Langevin, Y. 2010. A Spherical Monte-Carlo Model of Aerosols: Validation and First Applications to Mars and Titan. *Icarus*, 207 (2), pp. 923–931.
115. Vincendon, M., Langevin, Y., Poulet, F., Bibring, J.-P., and Gondet, B. 2007. Recovery of Surface Reflectance Spectra and Evaluation of the Optical Depth of Aerosols in the Near-IR Using a Monte Carlo Approach: Application to the OMEGA Observations of High-Latitude Regions of Mars. *Journal of Geophysical Research*, 112 (E8), CiteID E08S13.
116. McGuire, P.C., Wolff, M.J., Smith, M.D., Arvidson, R.E., Murchie, S.L., Clancy, R.T., Roush, T.L. et al. 2008. MRO/CRISM Retrieval of Surface Lambert Albedos for Multispectral Mapping of Mars with DISORT-Based Radiative Transfer Modeling: Phase 1—Using Historical Climatology for Temperatures, Aerosol Optical Depths, and Atmospheric Pressures. *IEEE Transactions on Geoscience and Remote Sensing*, 46 (12), pp. 4020–4040.
117. Viviano-Beck, C.E., Seelos, F.P., Murchie, S.L., Kahn, E.G., Seelos, K.D., Taylor, H.W., Taylor, K. et al. 2014. Revised CRISM Spectral Parameters and Summary Products Based on the Currently Detected Mineral Diversity on Mars. *J. Geophys. Res. Planets*, 119, pp. 1403–1431, doi:10.1002/2014JE004627.
118. Ehlman, B.L. and Edwards, C.S. 2014. Mineralogy of the Martian Surface. *Annual Review of Earth and Planetary Sciences*, 42, pp. 291–315, doi: 10.1146/annurev-earth-060313-055024.
119. Mustard, J.F., Murchie, S.L., Pelkey, S.M., Ehlmann, B.L., Milliken, R.E., Grant, J.A., Bibring, J.-P. et al. 2008. Hydrated Silicate Minerals on Mars Observed by the Mars Reconnaissance Orbiter CRISM Instrument. *Nature*, 454, pp. 305–309, doi:10.1038/nature07097.
120. McKeown, N.K., Bishop, J.L., Noe Dobrea, E.Z., Ehlmann, B.L., Parente, M., Mustard, J.F., Murchie, S.L., Swayze, G.A., Bibring, J., and Silver, E.A. 2009. Characterization of Phyllosilicates Observed in the Central Mawrth Vallis Region, Mars, Their Potential Formational Processes and Implications for Past Climate. *Journal of Geophysical Research*, 114, E00D10, doi: 10.1029/2008JE003301.
121. Soderblom, L.A. and Bell, J.F., III. 2008. Exploration of the Martian Surface: 1992–2007, The Martian Surface—Composition. In: *Mineralogy, and Physical Properties*. J. Bell, III editor, Cambridge University Press, 340 line figures, 40 halftones, 76 plates, 652 pages. 9780521866989, p. 3.
122. Bibring, J.-P. and Langevin, Y. 2008. Mineralogy of the Martian Surface from Mars Express OMEGA Observations, *The Martian Surface—Composition, Mineralogy, and Physical Properties*. J. Bell, III editor. Cambridge University Press, 340 line figures, 40 halftones, 76 plates, 652 pages. 9780521866989, p. 153.
123. Ehlmann, B.L., Berger, G., Mangold, N., Michalski, J.R., Catling, D.C., Ruff, S.W., Chassefière, E., Niles, P.B., Chevrier, V. and Poulet, F. 2013. Geochemical Consequences of Widespread Clay Mineral Formation in Mars' Ancient Crust. *Space Science Reviews*, 174, pp. 329–64.
124. Herschel, W. 1784. On the Remarkable Appearances at the Polar Regions of the Planet Mars, the Inclination of Its Axis, the Position of Its Poles, and Its Spheroidical Figure; With a Few Hints Relating to Its Real Diameter and Atmosphere. *Philosophical Transactions of the Royal Society of London*, 74, pp. 233–273.
125. James, P.B., Kieffer, H.H., and Paige, D.A. 1992. The Seasonal Cycle of Carbon Dioxide on Mars, In: *Mars (A93-27852 09-91)*, pp. 934–968.
126. Titus, T.N., Calvin, W.M., Kieffer, H.H., Langevin, Y., and Prettyman, T.H. 2008. *Martian Polar Processes, the Martian Surface—Composition, Mineralogy, and Physical Properties*. J. Bell, III, editor, Cambridge University Press, 340 line figures, 40 halftones, 76 plates, 652 pages. 9780521866989, p. 578.
127. Kieffer, H.H. and Titus, T.N. 2001. TES Mapping of Mars' North Seasonal Cap. *Icarus*, 154 (1), pp. 162–180.
128. Carrozzo, F.G., Raponi, A., De Sanctis, M.C., Ammannito, E., Giardino, M., D'Aversa, F., Fonte, S., and Tosi, F. 2016. Artifacts Reduction in VIR/Dawn Data. *Review of Scientific Instruments*, 87 (12), p. 124501.
129. Zambon, F., Tosi, F., Carli, C., De Sanctis, M.C., Blewett, D.T., Palomba, E., Longobardo, A. et al. 2016. Lithologic Variation Within Bright Material on Vesta Revealed by Linear Spectral Unmixing. *Icarus*, 272, pp. 16–31.
130. Tosi, F., Capria, M.T., De Sanctis, M.C., Combe, J.-Ph., Zambon, F., Nathues, A., Schröder, S.E. et al. 2014. Thermal Measurements of Dark and Bright Surface Features on Vesta as Derived from Dawn/VIR. *Icarus*, 240, pp. 36–57.

131. McCord, T.B., Adams, J.B., and Johnson, T.V. 1970. Asteroid Vesta: Spectral Reflectivity and Compositional Implications. *Science*, 168 (3938), pp. 1445–1447.
132. Consolmagno, G.J. and Drake, M.J. 1977. Composition and Evolution of the Eucrite Parent Body—Evidence from Rare Earth Elements. *Geochimica et Cosmochimica Acta*, 41, pp. 1271–1282.
133. De Sanctis, M.C., Ammannito, E., Capria, M.T., Tosi, F., Capaccioni, F., Zambon, F., Carraro, F. et al. 2012. Spectroscopic Characterization of Mineralogy and Its Diversity Across Vesta. *Science*, 336 (6082), pp. 697–700.
134. Zambon, F., De Sanctis, M.C., Schröder, S., Tosi, F., Longobardo, A., Ammannito, E., Blewett, D.T. et al. 2014. Spectral Analysis of the Bright Materials on the Asteroid Vesta. *Icarus*, 240, pp. 73–85.
135. Ammannito, E., De Sanctis, M.C., Palomba, E., Longobardo, A., Mittlefehldt, D.W., McSween, H.Y., Marchi, S. et al. 2013. Olivine in an Unexpected Location on Vesta's Surface. *Nature*, 504 (7478), pp. 122–125.
136. Capria, M.T., Tosi, F., De Sanctis, M.C., Capaccioni, F., Ammannito, E., Frigeri, A., Zambon, F. et al. 2014. Vesta Surface Thermal Properties Map. *Geophysical Research Letters*, 41 (5), pp. 1438–1443.
137. McCord, T.B., Castillo-Rogez, J., and Rivkin, A. 2011. Ceres: Its Origin, Evolution and Structure and Dawn's Potential Contribution. *Space Science Reviews*, 163 (1–4), pp. 63–76.
138. Bland, M.T. 2013. Predicted Crater Morphologies on Ceres: Probing Internal Structure and Evolution. *Icarus*, 226 (1), pp. 510–521.
139. Küppers, M., O'Rourke, L., Bockelée-Morvan, D., Zakharov, V., Lee, S., von Allmen, P., Carry, B. et al. 2014. Localized Sources of Water Vapour on the Dwarf Planet (1) Ceres. *Nature*, 505 (7484), p. 525.
140. Bland, M.T., Raymond, C.A., Schenk, P.M., Fu, R.R., Kneissl, T., Pasckert, J.H., Hiesinger, H. et al. 2016. Composition and Structure of the Shallow Subsurface of Ceres Revealed by Crater Morphology. *Nature Geoscience*, 9 (7), p. 538.
141. De Sanctis, M.C., Ammannito, E., Raponi, A., Marchi, S., McCord, T.B., McSween, H.Y., Capaccioni, F. et al. 2015. Ammoniated Phyllosilicates with a Likely Outer Solar System Origin on (1) Ceres. *Nature*, 528 (7581), pp. 241–244.
142. Nathues, A., Hoffmann, M., Schaefer, M., Le Corre, L., Reddy, V., Platz, T., Cloutis, E.A. et al. 2015. Sublimation in Bright Spots on (1) Ceres. *Nature*, 528 (7581), pp. 237–240.
143. Russell, C.T., Raymond, C.A., Ammannito, E., Buczkowski, D.L., De Sanctis, M.C., Hiesinger, H., Jaumann, R. et al. 2016. Dawn Arrives at Ceres: Exploration of a Small, Volatile-Rich World. *Science*, 353 (6303), pp. 1008–1010.
144. De Sanctis, M.C., Raponi, A., Ammannito, E., Ciarniello, M., Toplis, M.J., McSween, H.Y., Castillo-Rogez, J.C. et al. 2016. Bright Carbonate Deposits as Evidence of Aqueous Alteration on (1) Ceres. *Nature*, 536 (7614), pp. 54–57.
145. Nathues, A., Platz, T., Hoffmann, M., Thangjam, G., Cloutis, E.A., Applin, D.M., Le Corre, L. et al. 2017. Oxo Crater on (1) Ceres: Geological History and the Role of Water-Ice. *The Astronomical Journal*, 154 (3), article id. 84, 13 pp.
146. Nathues, A., Platz, T., Thangjam, G., Hoffmann, M., Mengel, K., Cloutis, E.A., Le Corre, L., Reddy, V., Kallisch, J., and Crown, D.A. 2017. Evolution of Occator Crater on (1) Ceres. *The Astronomical Journal*, 153 (3), article id. 112, 12 pp.
147. Titus, T.N. 2015. Ceres: Predictions for Near—Surface Water Ice Stability and Implications for Plume Generating Processes. *Geophysical Research Letters*, 42 (7), pp. 2130–2136.
148. Hayne, P.O. and Aharonson, O. 2015. Thermal Stability of Ice on Ceres with Rough Topography. *Journal of Geophysical Research: Planets*, 120 (9), pp. 1567–1584.
149. Combe, J.P., McCord, T.B., Tosi, F., Ammannito, E., Carrozzo, F.G., De Sanctis, M.C., Raponi, A. et al. 2016. Detection of Local H_2O Exposed at the Surface of Ceres. *Science*, 353 (6303), id.aaf3010.
150. Platz, T., Nathues, A., Schorghofer, N., Preusker, F., Mazarico, E., Schröder, S.E., Byrne, S. et al. 2016. Surface Water-Ice Deposits in the Northern Shadowed Regions of Ceres. *Nature Astronomy*, 1, id. 0007.
151. Rognini, E., Capria, M.T., Tosi, F., Frigeri, A., De Sanctis, M.C., Palomba, E., Longobardo, A. et al. 2017. Ceres Surface Temperatures: Comparison between Observation and Theory, *European Planetary Science Congress 2017*, held September 17–22, 2017 in Riga Latvia, id. EPSC2017-643.
152. Carlson, R.W., Weissman, P.R., Smythe, W.D., and Mahoney, J.C. 1992. Near-Infrared Mapping Spectrometer Experiment on Galileo. *Space Science Reviews*, 60 (1–4), pp. 457–502.
153. McCord, T.B., Brown, R.H., Baines, K., Bellucci, G., Bibring, J., Buratti, B., Capaccioni, F. et al. 2001. Galilean Satellite Surface Composition: New Cassini VIMS Observations and Comparison with Galileo NIMS Measurements. *American Geophysical Union, Fall Meeting 2001*, abstract #P12B-0504.

154. Baines, K.H., Carlson, R.W., and Kamp, L.W. 2002. Fresh Ammonia Ice Clouds in Jupiter I. Spectroscopic Identification, Spatial Distribution, and Dynamical Implications. *Icarus*, 159 (1), pp. 74–94.
155. Gierasch, P.J., Ingersoll, A.P., Banfield, D., Ewald, S.P., Helfenstein, P., Simon-Miller, A., Vasavada, A., Breneman, H.H., and Senske, D.A., Galileo Imaging Team. 2000. Observation of Moist Convection in Jupiter's Atmosphere. *Nature*, 403 (6770), pp. 628–630.
156. Ingersoll, A.P., Gierasch, P.J., Banfield, D., and Vasavada, A.R., Galileo Imaging Team. 2000. Moist Convection as an Energy Source for the Large-Scale Motions in Jupiter's Atmosphere. *Nature*, 403 (6770), pp. 630–632.
157. Lopes-Gautier, R., Douté, S., Smythe, W.D., Kamp, L.W., Carlson, R.W., Davies, A.G., Leader, F.E. et al. 2000. A Close-Up Look at Io from Galileo's Near-Infrared Mapping Spectrometer. *Science*, 288 (5469), pp. 1201–1204.
158. Davies, A.G. 2003. Volcanism on Io: Estimation of Eruption Parameters from Galileo NIMS Data. *Journal of Geophysical Research*, 108 (E9), pp. 10–1, doi:10.1029/2001JE001509.
159. Carlson, R.W., Smythe, W.D., Lopes-Gautier, R.M.C., Davies, A.G., Kamp, L.W., Mosher, J.A., Soderblom, L.A. et al. 1997. Distribution of Sulfur Dioxide and Other Infrared Absorbers on the Surface of Io. *Geophysical Research Letters*, 24, p. 2479.
160. Douté, S., Schmitt, B., Lopes-Gautier, R., Carlson, R., Soderblom, L., and Shirley, J. 2001. Mapping SO2 Frost on Io by the Modeling of NIMS Hyperspectral Images. *Icarus*, 149 (1), pp. 107–132.
161. Stephan, K., Jaumann, R., Hibbitts, C.A., and Hansen, G.B. 2005. Band Depths Ratios of Water Ice Absorptions as an Indicator of Variations in Particle Size of Water Ice on the Surface of Ganymede. American Astronomical Society, DPS meeting #37, #58.16. *Bulletin of the American Astronomical Society*, 37, p. 754.
162. McCord, T.B., Hansen, G.B., Fanale, F.P., Carlson, R.W., Matson, D.L., Johnson, T.V., Smythe, W.D. et al. 1998. Salts on Europa's Surface Detected by Galileo's Near Infrared Mapping Spectrometer. *Science*, 280 (5367), p. 1242.
163. Carlson, R.W., Johnson, R.E., and Anderson, M.S. 1999. Sulfuric Acid on Europa and the Radiolytic Sulfur Cycle. *Science*, 286 (5437), pp. 97–99.
164. Carlson, R.W., Anderson, M.S., Johnson, R.E., Smythe, W.D., Hendrix, A.R., Barth, C.A., Soderblom, L.A. et al. 1999. Hydrogen Peroxide on the Surface of Europa. *Science*, 283 (5410), p. 2062.
165. McCord, T.B., Carlson, R., Smythe, W., Hansen, G., Clark, R., Hibbitts, C., Fanale, F. et al. 1997. Organics and Other Molecules in the Surfaces of Callisto and Ganymede. *Science*, 278, pp. 271–275.
166. Hibbitts, C.A., Pappalardo, R.T., Hansen, G.B., and McCord, T.B. 2003. Carbon Dioxide on Ganymede. *Journal of Geophysical Research (Planets)*, 108 (E5), pp. 2–1.
167. Turtle, E.P., Perry, J.E., McEwen, A.S., DelGenio, A.D., Barbara, J., West, R.A., Dawson, D.D., and Porco, C.C. 2009. Cassini Imaging of Titan's High-Latitude Lakes, Clouds, and South-Polar Surface Changes. *Geophysical Research Letters*, 36, p L02204.
168. Stofan, E.R., Elachi, C., Lunine, J.I., Lorenz, R.D., Stiles, B., Mitchell, K.L., Ostro, S. et al. 2007. The Lakes of Titan. *Nature*, 445, pp. 61–64.
169. Lunine, J.I., Stevenson, D.J. and Yung, Y.L. 1983. Ethane Ocean on Titan. *Science*, 222, pp. 1229–1230.
170. Brown, R.H., Soderblom, L.A., Soderblom, J.M., Clark, R.N., Jaumann, R., Barnes, J.W., Sotin, C., Buratti, B., Baines, K.H., and Nicholson, P.D. 2008. The Identification of Liquid Ethane in Titan's Ontario Lacus. *Nature*, 434, pp. 607–610.
171. Clark, R.N., Curchin, J.M., Hoefen, T.M., and Swayze, G.A. 2009. Reflectance Spectroscopy of Organic Compounds I: Alkanes. *Journal of Geophysical Research*, 114, p E03001.
172. Grundy, W.M., Schmitt, B., and Quirico, E. 2002. The Temperature-Dependent Spectrum of Methane Ice I between 0.7 and 5 mm and Opportunities for Near-Infrared Remote Thermometry. *Icarus*, 155, pp. 486–496.
173. Mitri, G., Showman, A.P., Lunine, J.I., and Lorenz, R.D. 2007. Hydrocarbon Lakes on Titan. *Icarus*, 186, pp. 385–394.
174. Howett, C.J.A., Parker, A.H., Olkin, C.B., Reuter, D.C., Ennico, K., Grundy, W.M., Graps, A.M. et al. 2017. Inflight Radiometric Calibration of New Horizons' Multispectral Visible Imaging Camera (MVIC). *Icarus*, 287, pp. 140–151.
175. Owen, T.C., Roush, T.L., Cruikshank, D.P., Elliot, J.L., Young, L.A., de Bergh, C., Schmitt, B. et al. 1993. Surface Ices and Atmospheric Composition of Pluto. *Science*, 261, pp. 745–748.
176. Grundy, W.M. and Buie, M.W. 2001. Distribution and Evolution of CH4, N2, and CO Ices on Pluto's Surface: 1995–1998. *Icarus*, 153, pp. 248–263.
177. Grundy, W.M., Binzel, R.P., Buratti, B.J., Cook, J.C., Cruikshank, D.P., Dalle Ore, C.M., Earle, A.M. et al. 2016. Surface Compositions Across Pluto and Charon. *Science*, 351, p. 1283.

178. Earle, A.M. and Binzel, R.P. 2015. Pluto's Insolation History: Latitudinal Variations and Effects on Atmospheric Pressure. *Icarus*, 250, pp. 405–412.
179. Binzel, R.P., Earle, A.M., Buie, M.W., Young, L.A., Stern, S.A., Olkin, C.B., Ennico, K. et al. 2017. Climate Zones on Pluto and Charon. *Icarus*, 287, pp. 30–36.
180. Schmitt, B., Philippe, S., Grundy, W.M., Reuter, D.C., Côte, R., Quirico, E., Protopapa, S. et al. 2017. Physical State and Distribution of Materials at the Surface of Pluto from New Horizons LEISA Imaging Spectrometer. *Icarus*, 287, pp. 229–260.
181. Moore, J.M., Howard, A.D., Umurhan, O.M., White, O.L., Schenk, P.M., Beyer, R.A., McKinnon, W.B. et al. 2018. Bladed terrain on Pluto: Possible origins and evolution. *Icarus*, 300, pp. 129–144.
182. Protopapa, S., Grundy, W.M., Reuter, D.C., Hamilton, D.C., Dalle Ore, C.M., Cook, J.C., Cruikshank, D.P. et al. 2017. Pluto's Global Surface Composition through Pixel-by-Pixel Hapke Modeling of New Horizons Ralph/LEISA Data. *Icarus*, 287, pp. 218–228.
183. McKinnon, W.B. and 14 co-authors 2016. Convection in a volatile nitrogen-ice-rich layer drives Pluto's geological vigour. *Nature*, 534, pp. 82–85.
184. Trowbridge, A.J., Melosh, H.J., Steckloff, J.K., and Freed, A.M. 2016. Vigorous convection as the explanation for Pluto's polygonal terrain. *Nature*, 534, pp. 79–81.
185. Vilella, K. and Deschamps, F. 2017. Thermal convection as a possible mechanism for the origin of polygonal structures on Pluto's surface. *Journal of Geophysical Research*, 122, pp. 1056–1076.
186. Moore, J.M., Howard, A.D., Umurhan, O.M., White, O.L., Schenk, P.M., Beyer, R.A., McKinnon, W.B. et al. 2017. Sublimation as a landform-shaping process on Pluto. *Icarus*, 287, pp. 320–333.
187. Dobrovolskis, A.R., Peale, S.J., and Harris, A.W. 1997. Dynamics of the Pluto-Charon Binary. In: S.A. Stern and D.J. Tholen editors, *Pluto and Charon*, University of Arizona Press, Tucson, 159–190.
188. Howard, A.D., Moore, J.M., White, O.L., Umurhan, O.M., Schenk, P.M., Grundy, W.M., Schmitt, B. et al. 2017. Pluto: Pits and mantles on uplands north and east of Sputnik Planitia. *Icarus*, 293, pp. 218–230.
189. Hoehler, T.M. and Westall, F. 2010. Mars Exploration Program Analysis Group Goal One: Determine If Life Ever Arose on Mars. *Astrobiology*, 10 (9), pp. 859–867. doi:10.1089/ast.2010.0527.

Section VII

Conclusions

15 Fifty Years of Advances in Hyperspectral Remote Sensing of Agriculture and Vegetation—Summary, Insights, and Highlights of Volume IV

Advanced Applications in Remote Sensing of Agricultural Crops and Natural Vegetation

Prasad S. Thenkabail, John G. Lyon, and Alfredo Huete

CONTENTS

Goals here are to provide readers with an overview of chapters and contents. This they can read in the very beginning, before moving on to the individual chapters. Or they can read it at the very end to refresh their thoughts and to summarize the contents. The chapter also provides editors' perspectives, bringing their collective experiences and expertise to guide the reader. We have kept the summaries brief and illustrative, so that the reader can quickly gather essential knowledge and guidance. The in-depth detail can be found in the individual chapters.

Hyperspectral remote sensing, also referred to as imaging spectroscopy, is a mechanism for gathering data in narrowbands (10 nm or less) across the electromagnetic). Hyperspectral data are typically gathered in 1 to 10 nm bandwidths per band, leading to hundreds or thousands of narrowbands over a wavelength range. Hyperspectral narrowband (HNB) data are gathered throughout the electromagmetic spectrum in the visible (400–700 nm), red-edge (701–760 nm), near-infrared (NIR) (761–920 nm), water-absorption NIR (921–980 nm), far NIR (981–1300 nm), shortwave infrared (SWIR) (1301–3000 nm), middle-infrared (MIR) (3001–5000 nm), thermal-infrared (TIR) (8000–15,000 nm), and so on. However, even a few (e.g., 20) HNBs selected continually along a wavelength range (400–2500 nm) are also called hyperspectral data. It is not so much the number of bands that defines hyperspectral data, but rather an adequate number (e.g., 20 or 30) of targeted nonredundant HNBs (Thenkabail et al., 2013, 2014a,b; Thenkabail 2015) at specific sections of the spectrum along given wavelengths. Nevertheless, in most cases, hyperspectral data are acquired continually in hundreds or thousands of HNBs leading to spectral signatures of various earth features.

Here authors explore the challenges of processing a large amounts of data through the use of various data dimensionality reduction techniques to eliminate redundancies and optimize remaining data for a given application, thereby unlocking rich spectral signatures for analysis. Opportunities to advance remote sensing science in various vegetation and crop studies are presented throughout. The computation of hyperspectral vegetation indices (HVIs) and numerous other methods and techniques are central to most; with recent advances in cloud computing, machine learning, and artificial intelligence have enabled a "paradigm shift" in how remote sensing science is approached, and this greatly benefits hyperspectral remote sensing data analysis and applications.

15.1 USING HYPERSPECTRAL DATA IN PRECISION FARMING APPLICATIONS

Precision farming implies farming with precise knowledge of the within-field variability where every portion of the farm is studied and understood in great detail. The causes of within-field variability are wide ranging and include factors such as soils, management (e.g., tillage versus no tillage, drainage versus no drainage), inputs (e.g., nitrogen, potassium, phosphorous applications), genomics (e.g., cultivars), watering methods (irrigation versus rainfed), composting (e.g., organic or inorganic), mechanization (e.g., tractors or combine versus animal plowing), and a host of other factors (e.g., pests, diseases). Any one or a combination of these factors causes within-field variability. Precision farming requires us to understand and manage these within-field variabilities to ensure optimal productivity of the entire farm. Integrated use of spatial technologies such as remote sensing, geographic information systems (GISs), and the Global Positioning Systems (GPS), along with modeling and decision support systems (DSSs), has now been widely applied to obtain information required for precision farming. Mulla (2013) presents a detailed assessment of advances made over the last 25 years in the use of remote sensing in precision farming, its current level of maturity, and knowledge gaps requiring inputs for further advances. Remote sensing allows for characterization, modeling, and mapping of almost any crop variable; this is due to current advances in sensor data from multiple platforms acquiring data in visible, NIR, SWIR, TIR, Light Detection and Ranging (LiDAR), radar, and microwave parts of the spectrum with adequate spatial and temporal resolution. Spectral resolution data are available both in broadbands, such as in Landsat, and narrowbands (10 nm or less) from imaging spectrometers on various platforms acquiring data throughout the electromagnetic spectrum. As a result, crop characteristics for precision farming are widely studied using both broadband remote sensing (e.g., Figures 15.1 and 15.2) and narrowband hyperspectral remote sensing (e.g., Figures 15.3 and 15.4). Crop characteristics such as biomass and the leaf area index (LAI) of crops are modeled using the Landsat-5 Thematic Mapper (TM) broadband remote sensing indices and mapped as illustrated in Figures 15.1 and 15.2. The geoprecision regarding where these variabilities occur specifically on a farm and accuracies of modeling and mapping them

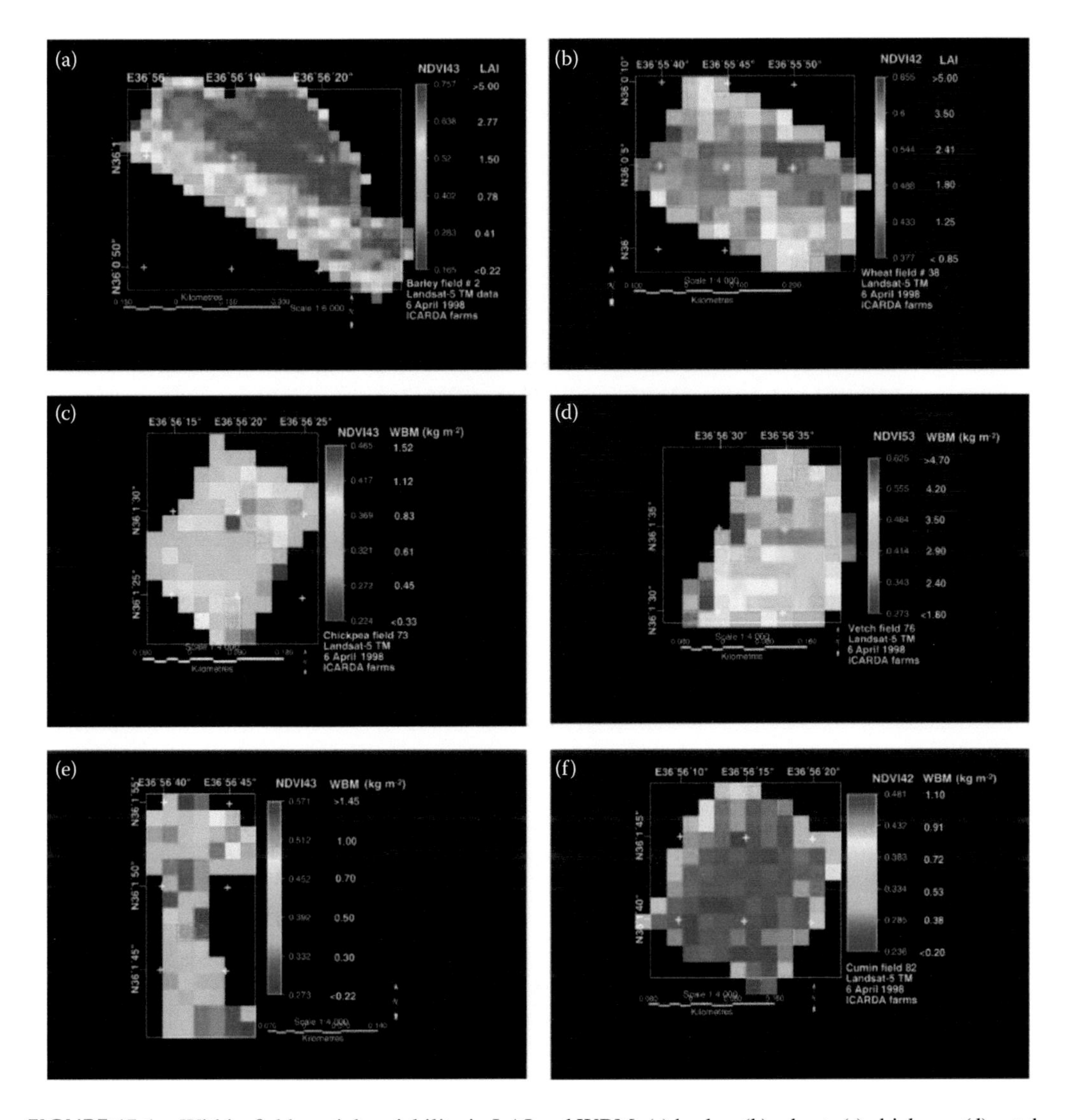

FIGURE 15.1 Within-field spatial variability in LAI and WBM: (a) barley, (b) wheat, (c) chickpea, (d) vetch, (e) lentil, and (f) cumin. Variability of individual farms related to VIs involving any two combinations of Landsat-5 TM bands at 2, 3, 4, and 5 for the April 6, 1998 image. (From Thenkabail, P.S. 2003. *International Journal of Remote Sensing*, Volume 24, Issue 14, Pages 2879–2904.)

have seen significant improvements using HNBs and HVIs for all sorts of crop variables (e.g., Figures 15.3 and 15.4) in a wide range of crops and study sites from across the world (Thenkabail et al., 2000, 2002a,b, 2004a,b, 2012, 2013, 2014a,b; Thenkabail 2015; Thenkabail et al., 2004; Marshall and Thenkabail, 2014; Marshall et al., 2015a,b; Aneece and Thenkabail, 2018). Specific HNBs and HVIs were used in those studies to evaluate a wide array of crop variables required for precision farming. Li et al. (2018) used field spectroscopy data and determined that leaf nitrogen content (LNC) was best studied with seven HNBs centered at 445, 556, 657, 764, 985, 1082, and 1194 nm, and leaf phosphorous content (LPC) was best studied with six HNBs centered at 755, 832, 891, 999, 1196, and 1267 nm. HNB data were also increasingly acquired on a real-time basis through the use of drones or unmanned aerial vehicles (UAVs) that fly miniature sensors of all kinds. This ensures data availability for any place anytime, as long as one can collect, operate, and process data from drones.

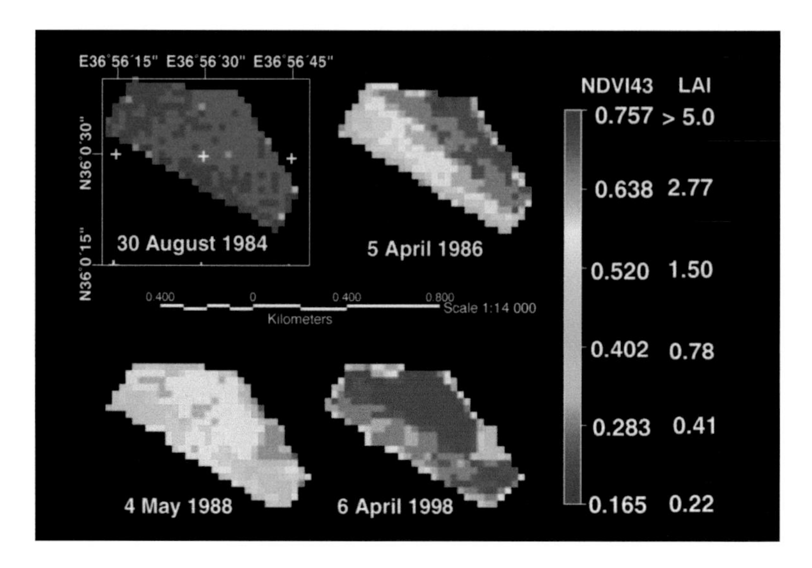

FIGURE 15.2 Temporal variations in LAI and NDVI for 1986, 1988, and 1998 images compared with base 1984 image, which had no crop. (From Thenkabail, P.S. 2003. *International Journal of Remote Sensing*, Volume 24, Issue 14, Pages 2879–2904. DOI: 10.1080/01431160710155974.)

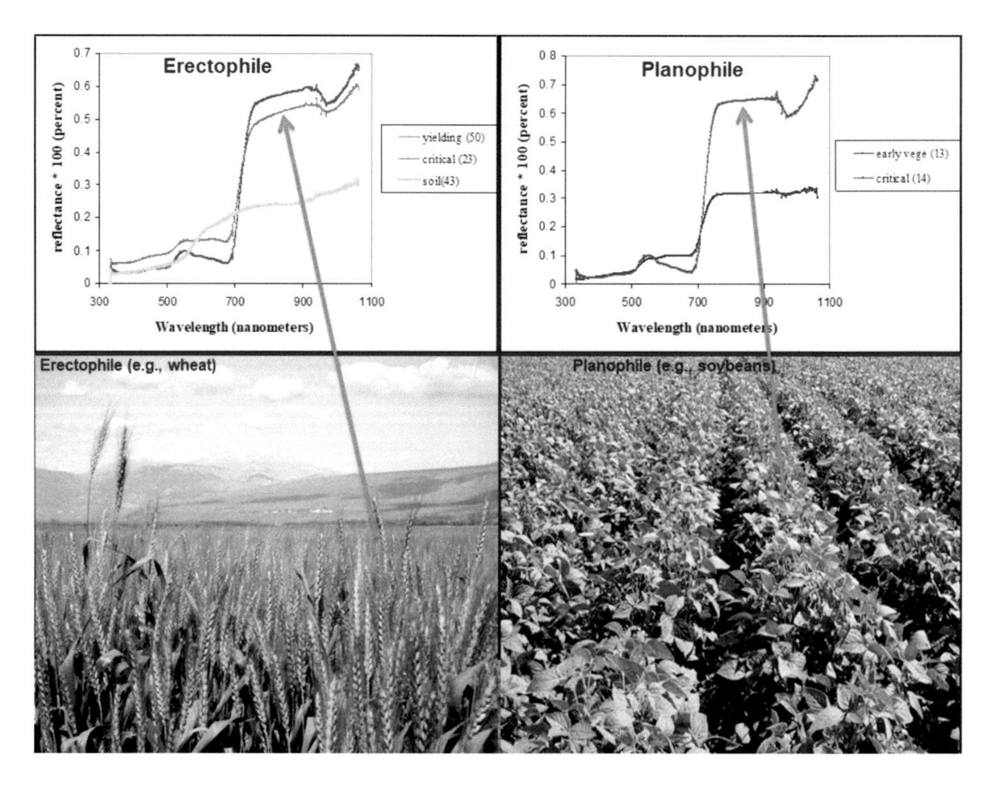

FIGURE 15.3 Hyperspectral data acquired from handheld spectroradiometer for wheat (erectophile) and soybean (planophile) crops at different growth stages. Study sites: Syria. (Modified from Thenkabail, P.S. et al. 2000. *Remote Sensing of Environment*, Volume 71, Issue 2, Pages 158–182.)

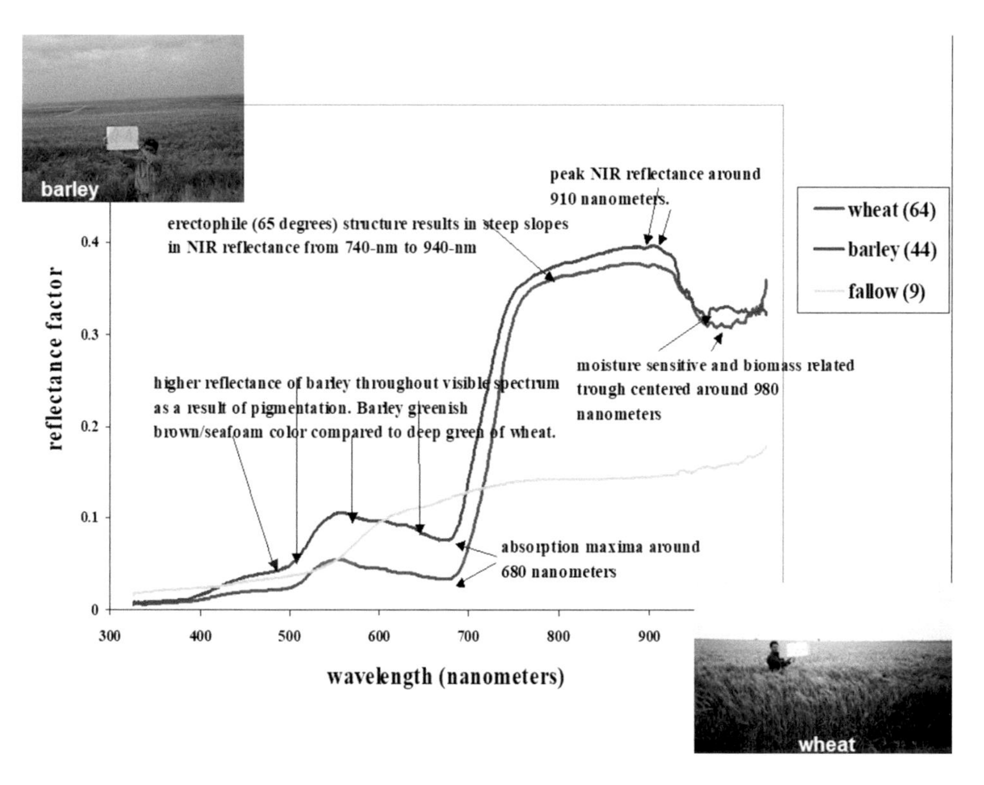

FIGURE 15.4 Wheat and barley crops at critical growth phases. Data acquired using handheld spectroradiometers in Syria. (Modified from Thenkabail, P.S. et al. 2000. *Remote Sensing of Environment*, Volume 71, Issue 2, Pages 158–182.)

In Chapter 1, Dr. Haibo Yao et al. focus on precision farming applications using hyperspectral data. They highlight the need to have high spatial, high spectral, and frequent coverage (temporal) resolutions of imagery for precision farming applications. Preferably, 5 m or less spatial resolution, 10 nm wide or less spectral resolution covering all parts of the spectrum (visible, NIR, SWIR, TIR, LiDAR, radar, microwave), and frequent temporal coverage (e.g., weekly) would be ideal for a wide range of precision farming applications. They also highlight the importance of various unique HVIs to study different crop characteristics required in precision farming. The chapter presents and discusses specific parameters modeled and mapped for precision farming applications and include: (1) soil properties such as organic matter, electrical conductivity, potassium, manganese, pH, soil moisture, and soil salinity; (2) weed studies such as weed species and invasive species; (3) herbicide damage like paraquat injury in corn, glyphosate injury in soybeans, and dicambo injury in cotton; (4) N variability in various crops; (5) yield variability of various crops; and (6) pest infestations like aphids on wheat, bacterial leaf blight, and fungal infections. They also showed how yields of corn crops are mapped during various phenological growth stages using temporal imagery. The proliferation of UAVs or drone-based imaging, where hyperspectral (typically, 400–2500 nm range imaging spectrometers with 1 to 10 nm bandwidths), thermal, and LiDAR sensors were flown to acquire data for precision farming applications, was addressed. Recent advances by CubeSats and SmallSats that have global coverage potential and acquire data from unique sensors, including hyperspectral and hyperspatial examples, are also likely to become popular. When data from these satellites reach a certain maturity level with good calibration and frequency of acquisition, they will become the ideal global platform for acquiring data required for precision farming.

15.2 HYPERSPECTRAL NARROWBANDS AND THEIR INDICES ON ASSESSING NITROGEN CONTENTS OF COTTON CROPS

Plant nitrogen (N) is crucial for crop productivity. The study of N variability in farms and addressing it are essential components of any precision farming effort. At the same time, maintaining optimal N application, without overapplication that leads to polluting the ground and surface water resources, is important in sustainable and environmentally friendly agriculture. Inoue et al. (2012) illustrate how spectral reflectivity changes with N application for rice crops (Figure 15.5). Zhou et al. (2016) showed strong relationships between canopy N content (CNC) and hyperspectral indices of winter wheat that varied with growth stages but was not influenced by cultivars, growing conditions, or nutritional status. They also suggest that growth stages before and after heading growth stages were the best to assess CNC, whereas periods around heading provided poorer results; N needed to be monitored throughout crop growth stages to ensure corrective measures when N was deficient or in excess and was best predicted for each crop at specific growth stages. Zhao et al. (2018) suggested the use of HNBs centered at 710 and 512 nm and a normalized difference spectral index computed using these wavebands to best predict N in corn crops. The best spectral indices for estimating leaf N accumulation (LNA) in wheat were found to be two-band HVIs involving HNBs: 860 and 720 nm, 990 and 720 nm, 736 and 526 nm, and 725 and 516 nm (Yao et al., 2010). Wang et al. (2012) suggested a three-band HVI, $(R_{924} - R_{703} + 2 \times R_{423})/(R_{924} + R_{703} - 2 \times R_{423})$, as the best to study LNC in rice and wheat crops. Optimal bandwidths were 36 nm for 924 nm, 15 nm for 703 nm, and 21 nm for 423 nm. Wang et al. (2018) presented the same for both LNC and CNC of forest canopies. CNC was estimated more accurately than LNC. In contrast, Corti et al. (2017) suggested that N content was best estimated at the leaf level and not at the plant level in studying spinach canopies. Overall, both LNC and CNC of plants were successfully studied by numerous researchers over the years using HNBs or HVIs.

Chapter 2, by Dr. Jianlong Li et al. focuses on determining the N content of cotton crops and study cotton in its various distinct growth stages using hyperspectral measurements from a handheld spectroradiometer. Wet biomass and dry biomass, yield, and various agronomic variables (e.g., chlorophyll a, chlorophyll b, LAI) are studied for various N applications. The dry weight of aboveground biomass of cotton improved significantly with increased N application. The authors establish that canopy spectral reflectance to N application was more sensitive in the NIR than in the visible. Their results suggest that the difference in spectral reflectance between 550–750 nm was improved using a continuum-removal approach (Gomez et al., 2008). They infer that the most sensitive reflectance to N rate application was located at two sites (e.g., 690 and 710 nm) of chlorophyll maximal absorption centered at 680 nm. Overall, they suggest that the best HNBs to study N were around the 620–640 nm and 690–710 nm regions.

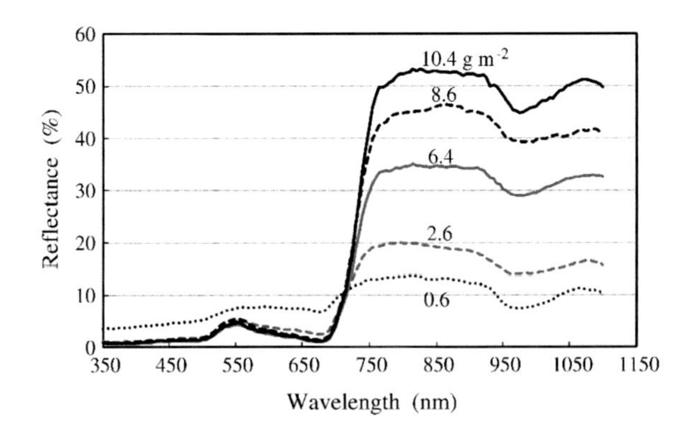

FIGURE 15.5 Some typical reflectance spectra of rice canopies in visible to near-infrared wavelength regions. These spectra are from the ground-based data set in Japan. Numbers indicate the canopy nitrogen content, or CNC (g m^{-2}) values. (From Inoue, Y. et al. 2012. *Remote Sensing of Environment*, Volume 126, Pages 210–221.)

15.3 HYPERSPECTRAL ANALYSIS OF THE EFFECTS OF HEAVY METALS ON VEGETATION REFLECTANCE

The presence of heavy metals, such as mercury (Hg), iron (Fe), lead (Pb), copper (Cu), zinc (Zn), vanadium (V), arsenic (As), manganese (Mn), tallium (TI), chromium (Cr), molybdenum (Mo), strontium (Sr), and cadmium (Cd), is detrimental to the health of soils, vegetation, water, animals, and humans. Various HNBs are sensitive to specific heavy metals in subtle and distinct ways (e.g., Figure 15.6). Researchers (Ren et al., 2009; Li et al., 2010; Hu, 2011; Wang et al. 2018) have shown that (Figure 15.6) (Wang et al., 2018): (1) Pb, Zn, Cu, and As concentrations in paddy leaves lie around 460, 560, 660, and 1100 nm (Ren et al., 2010); (2) Pb in vegetation is best monitored using 450, 550, 670, 760, and 1240 nm (Li et al., 2010); (3) heavy-metal-contaminated plants are best studied using 554, 631, and 557 nm (Hu, 2011); (4) 1240 nm reflectance is negatively but linearly correlated with their metal content (Rosso et al., 2005); and (5) 782 nm was the optimal waveband for detecting Cd in *Brassica rapa* chinesis leaves. Detection of heavy metals using hyperspectral data is the preferred approach compared to broadband remote sensing. This is because HNB data suppress noise and enhance the signal of the targeted metals through spectral derivatives from HNBs, as well as from HVIs (Wang et al., 2018). Thus, the detection of heavy metals is largely dependent on the covariation with the spectrally responsive metals or organic matter in the soil (Wang et al., 2018). Heavy metals are negatively correlated with plant chlorophyll concentrations (Liu et al., 2010). Shi et al. (2016) showed (R716−R568)/(R552−R568) good correlation in estimating soil arsenic in rice fields. Overall, visible and NIR bands often correlated best. However, a number of studies showed the great importance of red-edge (700–760 nm) bands in studying stress due to heavy metals.

In Chapter 3, Dr. Terry Slonecker et al. studied the effects of heavy metals on vegetation. They began by exploring existing methods and techniques for determining heavy metals through vegetation studies of various kinds (Slonecker et al., 2009). Healthy vegetation absorbs greatly in the red portion (630–690 nm) and reflects greatly in the NIR (>760–920 nm). Between the red and the NIR is the red edge (>690 to 760 nm), where there occurs a swift change from high absorption of the red to high reflectance of the NIR. The intermediate region, between the red and the NIR, of the electromagnetic spectrum where this change occurs is called the red edge. Healthy plants experience a so-called red shift in the red-edge portion, whereas stressed plants experience a so-called blue shift in the red-edge portion. The greater the stress (or when the plant is senescing/dry), the greater the blue shift of the spectral signature in the red-edge portion of the spectrum. Naturally, vegetation affected by heavy metals

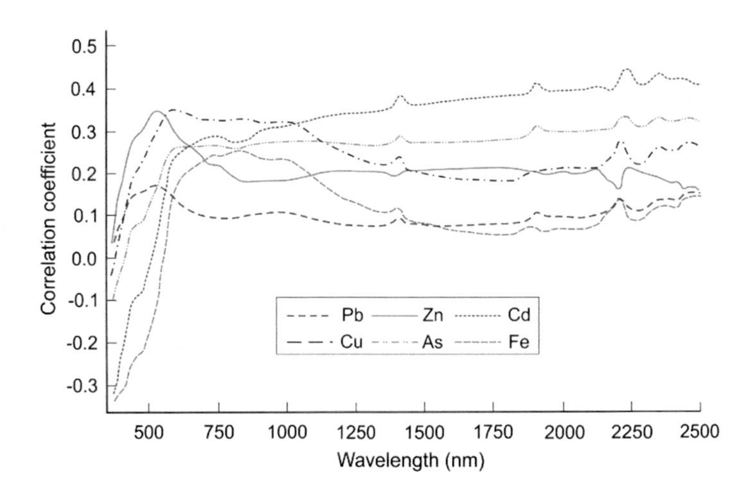

FIGURE 15.6 Correlation coefficient between spectral response and six levels of metal concentration in vegetation over wavelength range 400–2500 nm. (From Ren et al. 2010, Wang, Z. et al. 2018. *Agricultural and Forest Meteorology*, Volumes 253–254, Pages 247–260.)

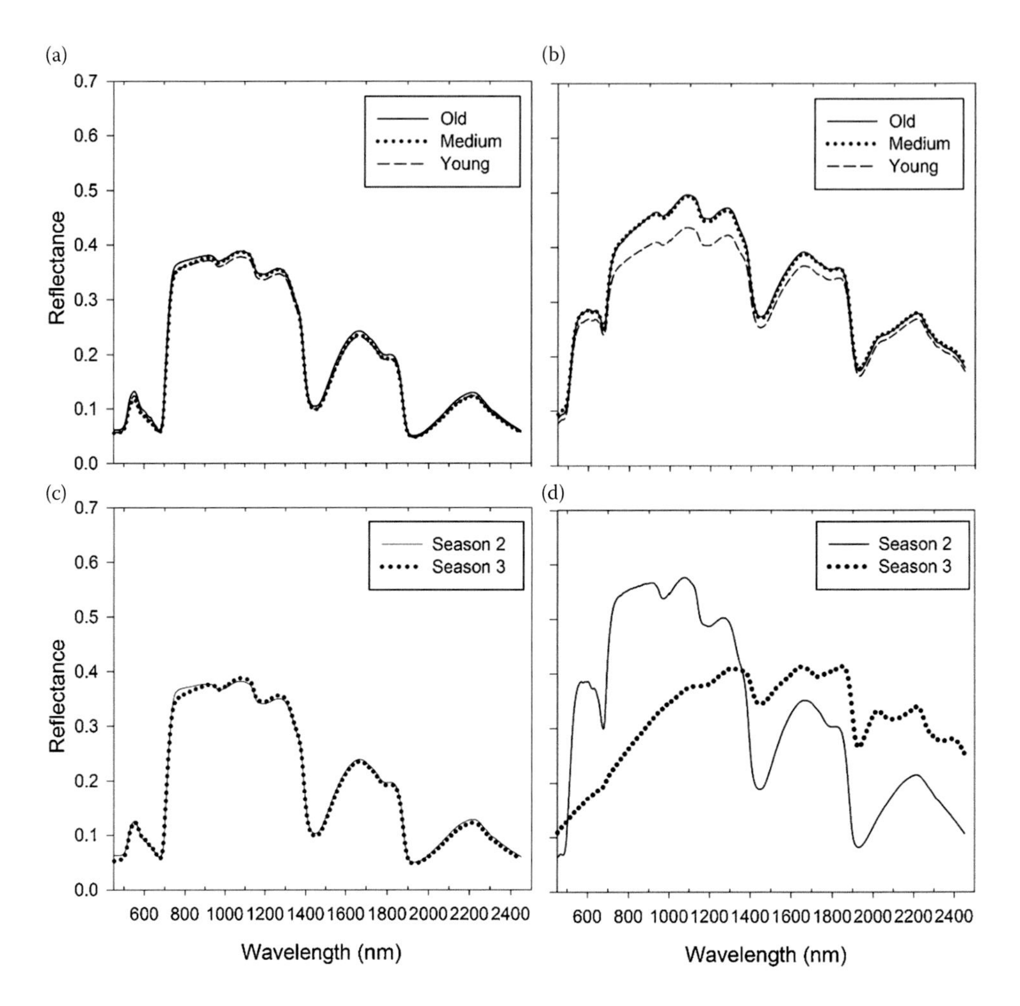

FIGURE 15.7 Mean spectral reflectance of: (a) succession class in leaves, (b) succession class in flower heads, (c) seasons in leaves, and (d) seasons in flowers heads. (From Carvalho et al. 2013. *International Journal of Applied Earth Observation and Geoinformation*, Volume 24, Pages 32–41. ISSN 0303-2434, https://doi.org/10.1016/j.jag.2013.01.005.)

experiences a blue shift. This concept is well explained as the authors discuss the stress-sensitive indices that use the red-edge and other sensitive portions of the spectrum, such as the Red-Edge Position (REP), Red-Edge Vegetation Stress Index (RESI), Photochemical Reflectance Index (PRI), and Normalized Pigment Chlorophyll Index (NPCI), and compare these indices with the performance of a number of other indices. They show that the indices most sensitive to plant stress from heavy metals in the soil were NPCI, PRI, REP, and the continuum-removed band depth at 1730 nm (CR1730).

15.4 MAPPING THE DISTRIBUTION AND ABUNDANCE OF FLOWERING PLANTS USING HYPERSPECTRAL SENSING

Flowers constitute an important commercial enterprise both in natural and controlled (e.g., greenhouse) environments. In many parts of the world, the beginning of flowering is the first indication of the dawn of spring. Flowers have an important function in pollination and are a source of food and energy for bees to sustain and generate honey. Many other creatures, such as butterflies, depend heavily on flowering plants for sustenance and survival. They are widely used in daily life for offerings and to greet loved ones and wish them well. Flowering status, including flowering date and degree of flowering, could reflect ecological processes in assessing plant phenological response to global warming (Chen et al., 2009). Remote sensing is a powerful tool to map flowers. Since flowers have distinct spectral signatures compared to plants, it is feasible to track when plants flower and in mapping and monitoring flower farms (Figure 15.7). The presence of flowers consistently provides better species discrimination using hyperspectral data, with 500–600 nm being the most crucial wavebands when data are acquired in the 475–900 nm range, as shown by Gross and Heumann (2014). Chen et al. (2009) demonstrated high accuracies in mapping the flowering of *Halerpestes tricuspis (Ranunculaceae)* on the Tibetan Plateau. They proposed a hyperspectral flower index (HFI) = 100[(R600 – Rv600)/(1 – L* R550 – R670)], where reflectance data are obtained from 600, 550, and 670 nm waveband centers, and v600 is the reflectance of pure green vegetation at 600 nm (Chen et al., 2009). Campbell and Fearns (2018) determined that to attain high flower presence accuracies with *Apis mellifera* (the European honey bee), having an average forage radius of less than 1 km from their hive, two factors are crucial: (1) image pixel size and (2) vegetation background. When background errors are minimized and pixel resolution is high, the percentage of flowers can be calculated with as little as 2% flower cover (Campbell and Fearns, 2018). Flower head pyrrolizidine alkaloids, which are involved in plant defense against herbivores, can be detected through hyperspectral reflectance (Carvalho et al., 2013) (Figure 15.7).

In Chapter 4, Dr. Tobias Landmann discusses the importance of flower mapping using the distinctive spectral characteristics of flowers, and studies the challenges and limitations of hyperspectral remote sensing in flower mapping. The importance of visible light in the 450–680 nm region for detecting and characterizing flowers is discussed. However, it is better to characterize flower signatures in the entire spectral range (e.g., 400–2500 nm). Flowers, depending on color and maturity, reflect quite distinctly relative to the rest of the plant. Once these characteristics are understood, it becomes easier to categorize and map different flower types from different plants and species. However, hyperspectral data need to be acquired with adequate spatial resolution, which will depend on the field and plant size. For example, farms with such flowers as sunflowers, poppies, tulips, and seasonal grasses, when spread across large fields, can be studied with good accuracy using hyperspectral data acquired at a pixel resolution of 5 to 30 m. In contrast, flower characterization and mapping when shrubs, bushes, and other plants are spread across the landscape intermittently require hyperspectral imaging spectrometer data acquired from imagery with a very fine spatial resolution of less than 5 m. Chapter 4 provides a fine synopsis of the existing state of knowledge in characterizing and mapping flowers using hyperspectral data in different environments using different sensors from various platforms.

15.5 CROP WATER PRODUCTIVITY OF WORLD CROPS

Crop water productivity (CWP) (kg/m^3) is defined as crop productivity (CP) (kg/m^2) divided by crop water use (CWU) (m^3/m^2). In simpler terms it is referred to as "crop per drop." CWP is often confused with water use efficiency (WUE), but they are distinct concepts. WUE is simply the efficiency in water use without any reference to productivity. One definition of WUE is the ratio of the amount of water delivered to a field to the amount of water released at a reservoir. Water is transported from the reservoir to the field through various mechanisms (e.g., open channel flow, closed pipes), during which many losses (e.g., evaporation, percolation) occur. In contrast, CWP is the water used to produce biomass, or grain yield. CWP is also defined in terms of economic value. Thus, the CWP is defined as

$$\text{WP}(\text{kg/m}^3) = \frac{\text{Yield}(\text{kg/m}^3 \text{or kg/pixel}) \text{or economic value (\$)}}{\text{Water use or ET}_{\text{actual}}\,(\text{m}^2/\text{m}^2 \text{ or m}^3/\text{pixel})}$$

Remote sensing is widely used in studies of CWP (Biradar et al., 2008; Platonov et al., 2008; Cai et al., 2009; Teixeira et al., 2015). An overview of the CWP modeling and mapping methods are illustrated in Figures 15.8 and 15.9 and involve: (1) mapping croplands (e.g., irrigated and rainfed), major crop types, and cropping intensities by, for example, developing automated machine learning algorithms

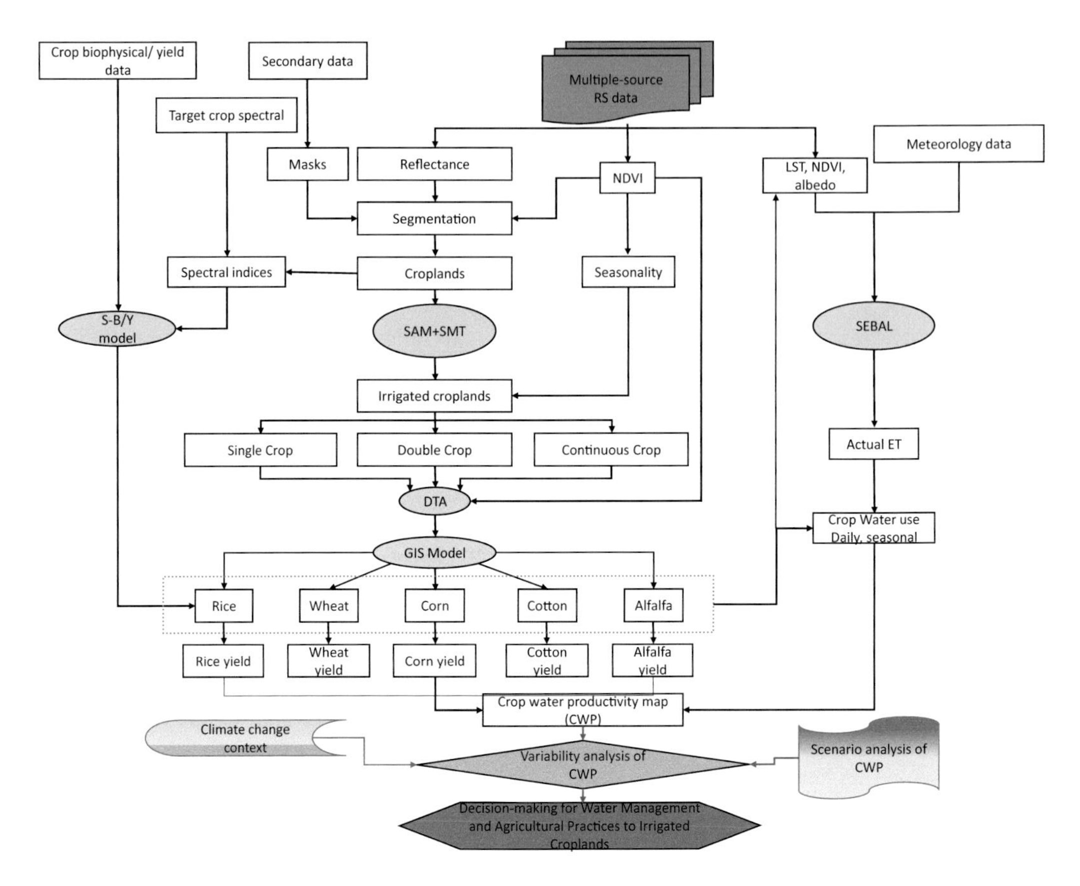

FIGURE 15.8 CWP mapping protocol for selected crops. The logical steps involved in producing the CWP maps of five leading world crops (rice, wheat, corn, cotton, alfalfa) are illustrated. This will "pinpoint" areas of high and low water productivity, which in turn will lead to establishing areas under various levels of low water productivity. Then the scenario analysis will help determine various quanta of water saved if the low water productivity levels are improved to various higher levels of water productivity. The water thus saved ("new water") will then be diverted to alternative water uses (e.g., urban, environments) or simply held as "water banks" for utilization during lean years.

(AMLAs) on the Google Earth Engine (GEE) cloud computing platform (Thenkabail et al., 2012; Thenkabail and Wu, 2012; Teluguntla et al., 2017, 2018; Xiong et al., 2017a); (2) crop productivity (productivity per unit area, kg/m^2) mapping through multisensor remote sensing spectrobiophysical modeling and extrapolating to larger areas using remote sensing (Biradar et al., 2008; Platonov et al., 2008); (3) water use (actual evapotranspiration, or ET_A (m^3/m^2) assessments of irrigated (blue water) and rainfed (green water) crops through SEBAL or surface energy balance modeling (Mariotto et al., 2011; Biggs et al., 2015, 2016); and (4) CWP, productivity per unit of water or "crop per drop" (kg/m^3), modeling and mapping (Biradar et al., 2008; Platonov et al., 2008; Cai et al., 2009; Teixeira et al., 2015). All of these can be modeled and mapped at various spatial resolutions (e.g., 250, 30, and 5 m) using multisensor remote sensing data that "pinpoint" areas of high and low CWP, thereby helping us to: (a) focus on areas of low CWP, (b) assess causes of low CWP, and (c) take remedial measures to raise areas of low CWP to higher levels, leading to a blue revolution. This in turn will lead to informed application of management practices and associated water savings exactly where they are needed. The "new water" can then be diverted to environmental and urban uses or simply held as "water banks" (both above- and below ground) for lean years.

However, recent research has clearly demonstrated the potential for significant advances in modeling and mapping CWP using hyperspectral data (Mariotto et al., 2013; Thenkabail et al.,

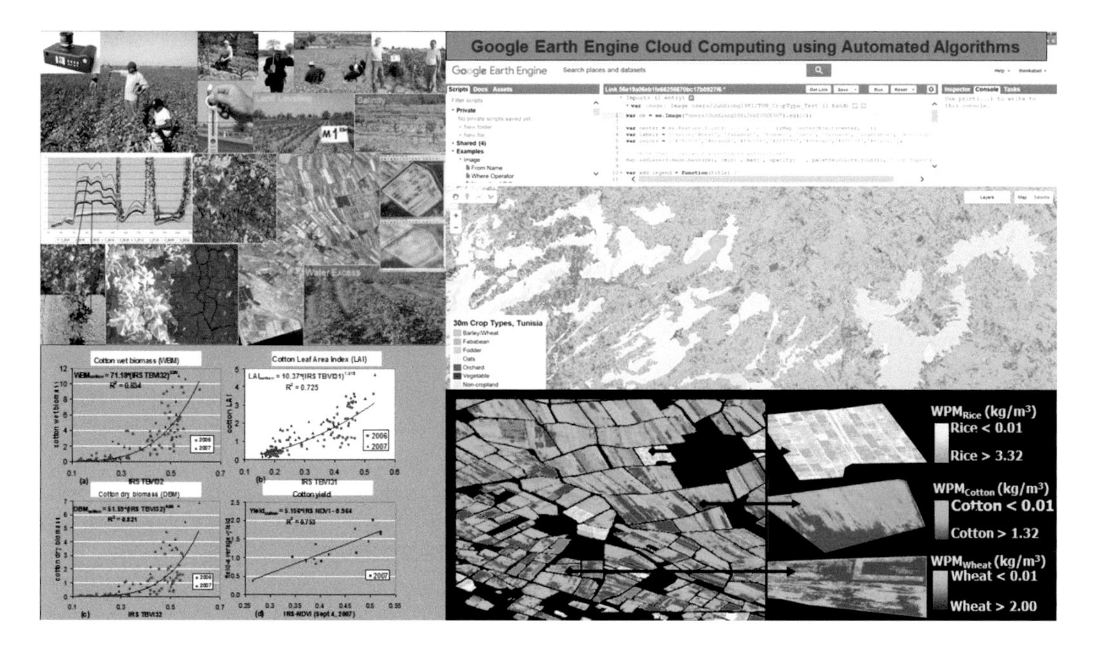

FIGURE 15.9 CWP modeling and mapping using automated algorithms on Google Earth Engine (GEE) cloud computing are illustrated here. The process involved the use of extensive in situ data to model and map: (a) crop types, (b) crop productivity, and (c) water productivity. The GEE cloud computing will allow regional to global CWP studies over very large areas using multisensor remote sensing.

2013; Marshall and Thenkabail, 2014, 2015a,b; Marshall et al., 2016). These studies indicate unique HVIs that clearly increase accuracies of CP and CWP modeling and mapping relative to broadband remote sensing studies reported in the previous paragraph. The computation of unique HVIs is feasible thanks to discrete HNBs in which data are acquired, providing possibilities to gather data from specific portions of the spectrum (e.g., Figure 15.10).

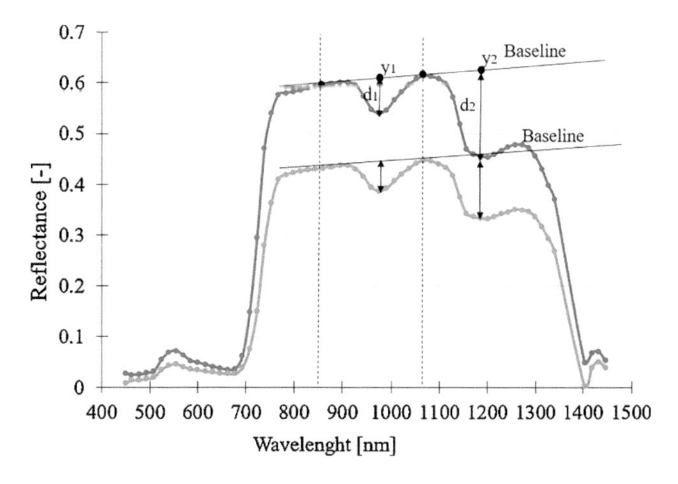

FIGURE 15.10 Graphic representation of DWI, which is the sum of the depths at 970 and at 1200 nm of the top of canopy (TOC) reflectance, with respect to the baseline formed between the peaks at 850 and 1080 nm. The two spectra correspond to different CWC values (blue spectrum 459 g/m^2, orange spectrum 359 g/m^2). (For interpretation of the references to color in this figure legend, the reader is referred to the web version of this article.) (From Pasqualotto, N. et al. 2018. *International Journal of Applied Earth Observation and Geoinformation*, Volume 67, 2018, Pages 69–78.)

In Chapter 5, Dr. Michael Marshall et al. provide a clear methodology for modeling and mapping CWP using remote sensing. The methods and approaches presented in the chapter can be applied to multispectral broadband (MBB) remote sensing or HNB remote sensing. The chapter extensively discusses the hyperspectral approach but illustrates the results using Landsat-7 Enhanced Thematic Mapper Plus (ETM+) data because of the lack of availability of hyperspectral images from space. The methods are illustrated using rice, wheat, corn, and soybean crops. First, these crop types were mapped using remote sensing. Second, CP (kg/m^2) models were developed through relationships between crop yield (Y) (kg/m^2) with various HVIs. In Chapter 5, Marshall et al. showed that the aboveground wet biomass (AWB) and the aboveground dry biomass (ADB) were consistently modeled with higher R^2 values using HNBs relative to MBBs. Improvements were 5%–31% for two-band HVIs and 3%–33% for three-band HVIs. In this regard, it is worthwhile referring to the Marshall and Thenkabail (2015b) study, which showed that the HNB-based AWB (kg/m^2) models ($R^2 = 0.71$–0.98) explained significantly higher variability relative to MBB-based AWB models derived from Landsat ($R^2 = 0.32$–0.82), IKONOS ($R^2 = 0.50$–0.94), GeoEye-1 ($R^2 = 0.55$–0.95), and WorldView-2 ($R^2 = 0.36$–0.87) for rice, wheat, corn, cotton, and alfalfa crops. Similar results were also obtained by Mariotto et al. (2013) and Thenkabail et al. (2013). Third, the most highly correlated HVIs to determine ET_A or CWU were the red-edge band centered at 672 nm and a visible blue band in the 428–478 nm spectral region, which provided an R^2 value of 0.51. CWU (ET_A) was typically obtained using various surface energy balance methods such as the Surface Energy Balance Algorithm (SEBAL) (Bastiaanssen et al., 1998a,b), its modified version, the Mapping Evapotranspiration at high Resolution with Internalized Calibration (METRIC) (Allen et al., 2007), or other methods such as the Simplified Surface Energy Balance (SSEB) (Anderson, 2011; Senay, 2013) and the Atmosphere-Land Exchange Inverse (ALEXI) (Anderson et al., 2011) model. CWP can then be derived by simply dividing CP by CWU. In Chapter 5, to verify and validate CWP, they have demonstrated the use of eddy covariance flux tower data which measured crop biomass and latent heat flux (LE) over a growing season. An interesting outcome presented in the chapter was the CWP values determined for various crops: (1) Wheat crop used the least water, but produced low yields resulting in lowest CWP (1.53 ± 0.15 kg m^3); (2) rice crop used the most water, but also produced high yields resulting in moderate CWP (1.60 kg m^3); (3) Soybean crop used low water quantities, but produced high yields resulting in high CWP (1.80 ± 0.48 kg m^3); (4) Corn crop used high water, but also produced very high yields resulting in highest CWP (3.00 ± 0.89 kg m^3) among the four crops.

15.6 HYPERSPECTRAL REMOTE SENSING TOOLS FOR QUANTIFYING PLANT LITTER AND INVASIVE SPECIES IN ARID ECOSYSTEMS

Plant litter is important for enhancing soil organic matter, naturally enriching soils, stabilizing soils from erosion, practicing organic agriculture, and improving crop productivity. Plant litter can also be used as fuel, fertilizer, and biomass energy. In contrast, when plant litter is burned, as happens quite often in many agricultural fields around the world, carbon is released to the atmosphere. Thus, proper management of crop residues is crucial for sustainable agriculture, improved food production, and avoiding carbon pollution. Conservation tillage accounts for more than 30% of crop residue cover (CRC) in comparison to nonconservation tillage, which accounts for less than 30% of the CRC (Zheng et al., 2012). Over the years, tillage has been mapped using Landsat images, beginning with the pioneering work of Van Deventer et al. (1997). Zheng et al. (2012) used Landsat TM and Landsat ETM+ data to classify CRC into three categories (CRC $<$ 30%, 30% $<$ CRC $<$ 70%, CRC $>$ 70%) and obtained high levels of accuracy with overall accuracies exceeding 90%, producer accuracies of 83%–100%, and user accuracies of 75%–100%. They further showed strong (R^2 of 0.89) linear relationships between CRC and minimum normalized difference tillage index (minNDTI), where $NDTI = (B5 - B7)/(B5 + B7)$, with B5 and B7 being the Landsat TM and ETM+ bands 5 and 7. Various tillage-based indices are shown in Table 15.1. Daughtry et al. (2006) showed that Landsat

TABLE 15.1
Satellite-Based Tillage Indices

Sensor	Tillage Index	Formula	Description	Reference
Landsat	CRIM	SM/SR	SM: distance from point M to soil line; SR: distance between soil and residue lines at point M	Biard and Baret (1997)
	Simple tillage index (STI)	B5/B7	B2: Landsat TM/ETM+ band 2; B4: TM/	Van Deventer et al. (1997)
	NDTI	(B5 − B7)/(B5 + B7)	ETM+ band 4; B5:	Sullivan et al. (2006)
	Modified CRC	(B5 − B2)/(B5 + B2)	TM/ETM+ band 5; B7:	McNairn and Protz (1993)
	NDI5; NDI7	(B4 − B5)/(B4 + B5); (B4 − B7)/(B4 + B7)	TM/ETM+ band 7;	
Hyperion	CAI	$0.5(R_{20} + R_{22}) - R_{21}$	R_{20} and R_{22}; the reflectance on the shoulders at 2021 nm and 2213 nm	Daughtry et al. (2006)
ASTER	LCA	100(2 × B6 − B5 − B8)	B5, B6, B7, B8: ASTER	Daughtry et al. (2005)
	SINDRI	(B6 − B7)/(B6 + B7)	SWIR bands 5, 6, 7, and 8	Serbin et al. (2009)

Source: Zheng, B. et al. *Remote Sensing of Environment*, Volume 117, Pages 177–183.

TM data had poor correlations with corn and soybean crop residues. In contrast, their study showed CRC of corn and soybean fields were linearly related to the cellulose absorption index (CAI) derived using Hyperion data. CAI measures the relative intensity of cellulose and lignin absorption features near 2100 nm (Daughtry et al., 2006). CAI = (R2023 + R2215 nm) − R2100 nm.

Table 15.1 describes the average depth of the cellulose absorption feature, with positive values of CAI indicating the presence of cellulose (Nagler et al., 2003). Nagler et al. (2003) studied litter from four crops (corn, soybean, rice, and wheat) and two tree species (coniferous and deciduous) and showed that the mean CAI of the soils was −2.0, whereas the mean CAI of the plant litter was 5.2. (For an overview of invasive species see Chapter 9.) The big advantage of using hyperspectral data in invasive species studies is the distinct endmember spectra (e.g., 26) one can develop for each species (e.g., Figure 15.11). This in turn helps separate and classify various species (invasive and native) with greater degrees of accuracy compared to broadband remote sensing data (Chapter 9).

Chapter 6 by Dr. Pamela Nagler et al. has distinct and important components on studying plant litter and separating and classifying invasive species. The literature reviewed focuses importantly on separating plant litter from soils as it is difficult yet vital to do so in the 400–1100 nm region since both materials have very similar spectra. Second, fluorescence is better suited than the 400–1100 nm spectra in separating litter from soils, but these spectra suffer from a relatively small fluorescence signal. Third, SWIR (1400–3000 nm) provides specific wavebands where litter is best classified. Chapter 6 demonstrated the strengths of CAI (Table 15.1) in modeling and mapping plant litter and recommends the use of HNB data from sensors such as the Airborne Advanced Visible/Infrared Imaging Spectrometer and spaceborne Hyperion that acquire data in very narrow spectral wavebands (10 nm or finer) or in HNBs to derive CAI. Chapter 6 also provides a detailed study of two invasive species affecting the western United States, including Salt Cedar or tamarisk (*Tamarisk ramosissima*) along riparian vegetation zones and buffelgrass in the Sonoran Desert. Both species expand with detrimental effect to native vegetation, causing ecological and economic harm. Most remote sensing studies have not been very successful at separating Salt Cedar from other plants owing to an absence of time-series images and adequate spatial resolution to map these fragmented species spread across the landscape and not necessarily as contiguous distributions. In particular,

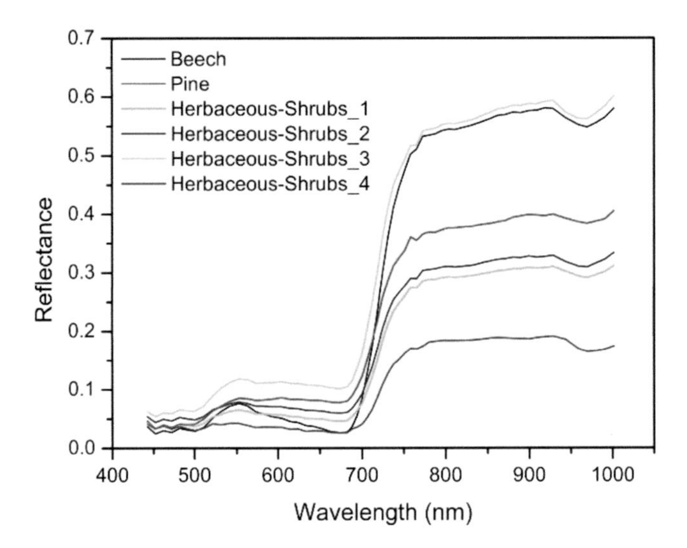

FIGURE 15.11 Spectral signatures of the six endmembers used in BI-ICE algorithm for unmixing CHRIS/PROBA imagery. All spectra are derived from the purest CHRIS pixels for each class. (From Stagakis, S. et al. 2016. *ISPRS Journal of Photogrammetry and Remote Sensing*, Volume 119, 2016, Pages 79–89.)

images acquired during the flowering phenological phase of the tamarisk were found helpful in achieving greater accuracies. Dr. Nagler et al. showed that reflectance in the NIR region of 800–1300 nm were higher for Salt Cedar and for three crops (alfalfa, cotton, and melons) compared to other vegetation in the landscape. Further, reflectances of all crops were higher in the 1300–2500 nm region relative to tamarisk. They recommend specific vegetation indices (e.g., NDVI) (R1,5 and R1,7; where R is reflectance and 1, 5, and 7 are Landsat bands), and color composites help separate tamarisk-infested areas. The second invasive species studied was buffelgrass. It is spread across the Sonoran Desert and often mixed with several other native species and in different proportions. Phenology is important in separating buffelgrass from other species as absorption in 675, 950, 1150, 1450 nm wavelengths for buffelgrass was exemplary during the month of August and was noticeable in September, as reported in Chapter 6. The authors also found that the cellulose/lignin absorption feature at 2050 nm was most the characteristic feature for the buffelgrass on six of the seven dates of the images acquired in different months. Nevertheless, use of HNB data and HVIs along with use of imagery with good temporal coverage and spatial resolution will likely provide the best solution to accurately classifying invasive species.

15.7 HYPERSPECTRAL PHENOLOGY APPLICATIONS

Vegetation phenology is an integrative environmental indicator of climate change, and the long-term observations of these changes help us to understand climate change trends over space and time (Peng et al., 2018; Workie and Debella, 2018). For example, certain plants flower at the start of the spring season. Tracking this flowering event tells us whether the spring seasons started on time, early, or late. Also, when seasons begin and end in different parts of the world, there are inferences one can draw on climate from that. In agriculture, understanding and monitoring crop phenology is crucial to determining their growth, health, seasonality, and productivity. Tracking crop phenology at specific farms helps to determine whether a particular crop is grown or changed across seasons or years. For example, a field growing wheat will have a shorter growing season compared to cotton or sugarcane. Phenology also helps us study whether there was delayed, failed, or normal irrigation or rainfall for crops during a growing season. In recent years, field-based networks of phenocams have been used to gather data on phenology in many parts of the world, such as by the National Phenology Network

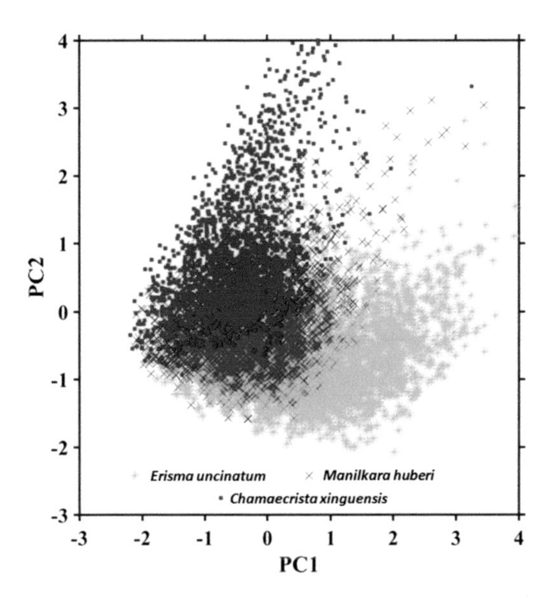

FIGURE 15.12 First two scores from PCA applied to reflectance of 89 bands (437–898 nm) of hyperspectral camera, showing canopy spectral variations between three species ($n = 7650$ pixels from 17 dates). In the observation period (July 29–September 25), brightness increased from the left to the right side of PC1 with gains in leaves. NPV increased from the bottom to the top of PC2 with losses in leaves. (From de Moura, Y.M. et al. 2017. *ISPRS Journal of Photogrammetry and Remote Sensing*, Volume 131, Pages 52–64.)

(NPN) in the United States (Denny et al., 2014). However, time-series remote sensing in various spatial, spectral, and temporal resolutions is widely recognized as the gold standard for monitoring phenology of vegetation of all sorts including agricultural crops. Land surface phenology (LSP) studies using remote sensing have addressed habitat quality (Weber et al., 2018), agricultural crops (Lausch et al., 2015), mangrove forests (Pastor-Guzman et al., 2018), forests (White et al., 2014), and multiple types of other ecosystems (Peng et al., 2018). Obtaining remote sensing data at various phenological growth stages significantly increases classification accuracies of crops and vegetation as well as quantifying their biophysical and biochemical characteristics. Few studies exist on the use of hyperspectral data in phenology studies. However, hyperspectral data hold the key to advancing phenology studies (e.g., Christian et al., 2015; Lausch et al., 2015) using remote sensing because they can be gathered for specific species or crop or vegetation types in HNBs, making it possible to quantify specific biophysical and biochemical quantities as well as offering the potential to study specific phenological measures like flowering, greenup, and maturity. For example, as shown in Figure 15.12, in a 3-species study using 89 hyperspectral bands (de Moura et al., 2017) during a specific observation period (July 29–September 25), brightness increased from the left to the right side of PC1 with gains in leaves (Figure 15.12). Nonphotosynthetic vegetation (NPV) increased from the bottom to the top of PC2 with losses in leaves (Figure 15.12) (de Moura et al., 2017).

In Chapter 7, Dr. Alfredo Huete et al. begin by defining phenology in terms of conventional in situ approaches that study individual plant biological life cycle events, such as bud break, flowering, pollination, and fruiting, whereas landscape phenology (LSP) defines aggregate seasonal vegetation patterns sensed by satellites. The chapter identifies the key phenology matrices studied under LSP using remote sensing data considering the: (1) "timing" in the start of growing season, end of growing season, length of growing season, peak of growing season, minimum greenness value, rate of greenup, and rate of drying or curing and (2) "magnitude" involving the seasonal amplitude of greenness values, peak greenness value, minimum greenness value, and integrals over the growing season. This is followed by a study of phenological matrices using coarse- and moderate-resolution remote sensing. Quantitative parameters like LAI and biomass are widely used in LSP studies using remote sensing. Hyperspectral data are especially useful in species-level studies of phenology at the leaf scale and plant scale and are helpful in advancing phenology studies using remote sensing to the next level by enabling study of such phenological parameters as leaf coloration changes, bud break, onset of new leaves, and flowering.

15.8 LAND COVER, FORESTS, WETLAND, AND URBAN APPLICATIONS USING HYPERSPECTRAL DATA

Determination of land cover and land use (LCLU) and LCLU change (LCLUC) is among the most important, common, and widely used applications of remote sensing. Remote sensing is now widely accepted as the only source of data for mapping LULCC over very large areas. Indeed, LULCC studies are conducted using remote sensing data of a wide range of resolutions (e.g., 10 km, 1 km, 500 m, 250 m, 30 m, 10 m, 1 m) acquired from spaceborne, airborne, drone-mounted, and ground-based sensors. These sensors allow us to study diverse forms of LULC that include croplands, rangelands, forests, wetlands, urban, deserts, water bodies, and combinations of these. Depending on the sensor spectral, spatial, radiometric, and temporal resolutions, details of each LULC, such as crop type, forest type, wetland class, and others, are obtained. Nevertheless, imaging spectroscopy data or hyperspectral data add a new dimension to LULC classification by enabling new capabilities or advancing capabilities or supplementary/complementary capabilities relative to other sensors. These advances in HNB data relative to MBB data include the ability to: (1) map additional LULC classes, (2) discern classes such as species types (e.g., in forests or wetlands or rangelands or croplands), (3) achieve greater accuracies, (4) accurately quantify, model, and map a wide range of biophysical and biochemical characteristics of vegetation, and (5) develop specific HVIs for specific vegetation quantities. For example, in a complex and fragmented ecosystem, Pignatti et al. (2009a,b) showed the potential of Hyperion data in mapping land cover and vegetation diversity up to the fourth level of the COoRdinate INformation on the Environment (CORINE) legend. They compared 30 m, 242 band (acquired in 400–2500 nm spectral range) Hyperion data–derived land cover and vegetation classes with that of Multispectral Infrared Visible Imaging Spectrometer (MIVIS) airborne hyperspectral imagery, which has a 6–7 m spatial resolution with four spectrometers in the visible and near-infrared (VNIR), SWIR, and TIR ranges to demonstrate the capability of Hyperion in mapping complex classes (e.g., Figure 15.13). Jafari and Lewis (2012) derived endmembers with EO-1 Hyperion hyperspectral data in an arid land study area. They showed that one endmember was highly correlated ($R^2 = 0.89$) with cottonbush (*Maireanaaphylla*) vegetation cover that was distributed as patches throughout the study area. Another endmember was highly correlated ($R^2 = 0.68$) with the total vegetation cover of green and gray-green perennial shrubs (e.g., Mulga, *Acacia aneura*). Chen et al. (2017) used high-spatial-resolution airborne hyperspectral data to classify and map land cover in the city of Pavia in northern Italy using a spectral mixture analysis technique and achieved an overall accuracy of 97.24% that involved mapping vegetation, impervious surface, soil, water, and shadow. Urban study capabilities (e.g., Figure 15.14) of hyperspectral data were illustrated by Behling et al. (2015) using airborne HyMap data that had 126 spectral bands with data acquired at a 3–6 m spatial resolution.

Chapter 8 by Dr. Pandey et al. begins by providing an overview of the strengths of hyperspectral data as opposed to multispectral data and lists some of the past, present, and upcoming hyperspectral sensors. This is followed by a brief overview of the existing LULC studies using hyperspectral data and the key pixel-based and object-based classification methods that can be used for hyperspectral or multispectral data analysis. The rest of the chapter overwhelmingly focuses on tools, methods, and approaches of acquiring hyperspectral data from field-based spectroradiometers or through UAVs. They highlight the importance of having a thorough knowledge of spectral libraries and a good understanding of spectral characteristics at the species or vegetation-type level in order to properly study, model, or map LULC in diverse ecosystems of the world.

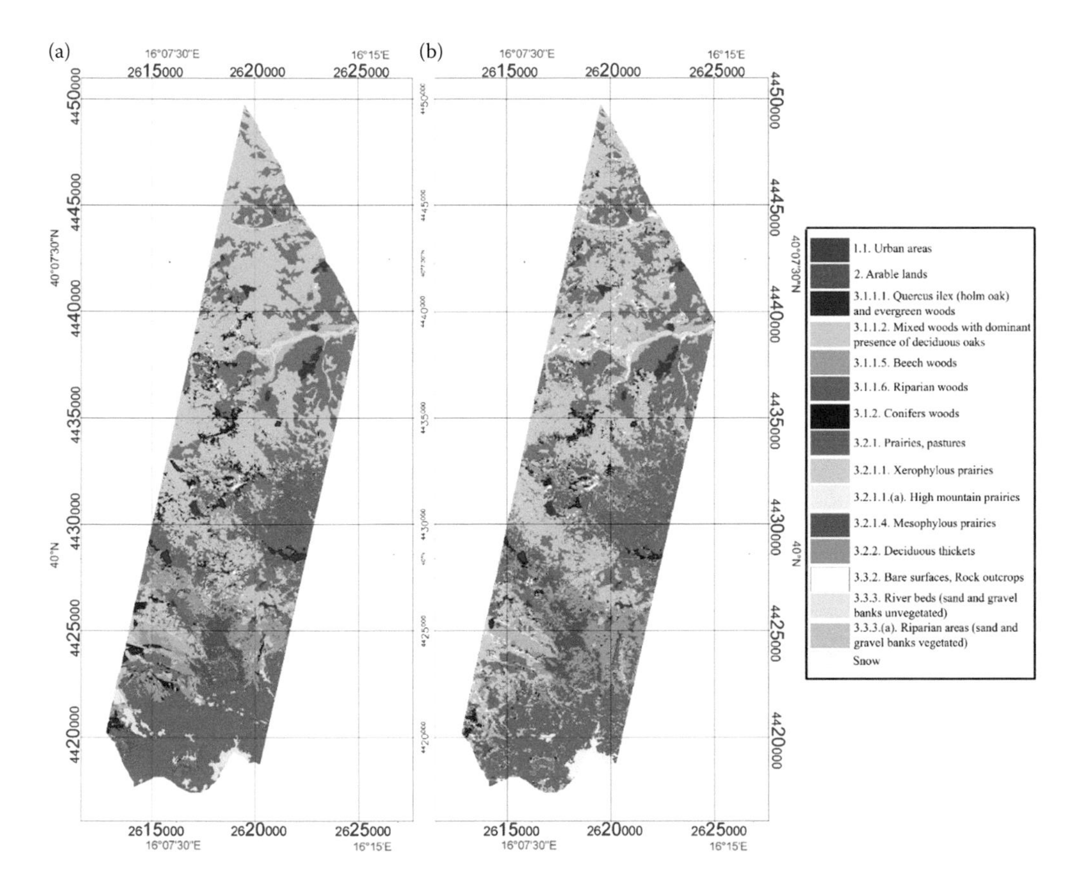

FIGURE 15.13 (a) MIVIS and (b) Hyperion classification maps obtained by applying the MD algorithm with 13 CORINE classes (up to fourth level). The MIVIS thematic map was spatially resampled to the Hyperion spatial resolution. (From Pignatti et al. 2009. *Remote Sensing of Environment*, Volume 113, Issue 3, Pages 622–634.)

15.9 HYPERSPECTRAL REMOTE SENSING FOR FOREST MANAGEMENT

The importance of remote sensing in forest management studies cannot be overemphasized. Remote sensing provides information about forest classification, carbon sequestration or release, selective logging, timber harvesting, sustainable forest management, forest health and disease, species identification, and other characteristics like biomass and LAI estimations. Forests are complex ecosystems, diverse in nature, and are often inaccessible. They are home to the planet's rich flora, fauna, and biodiversity. Forest management is currently conducted using a wide array of remote sensing data such as submeter to 5 m IKONOS (Kayitakire et al., 2006), 5–10 m Satellites Pour l'Observation de la Terre (SPOT) (Wolter et al., 2009), 30 m Landsat (Pasquarella et al., 2018), and 250 m to 1000 m Moderate Resolution Imaging Spectroradiometer (MODIS) (Wheeler et al., 2018). Kayitakire et al. (2006) determined forest structural variables like top height, circumference, stand density, and age variables were estimated using 1 m data of IKONOS-2 with R^2 values of the best models ranging from 0.76 to 0.82. The researchers suggest that the structural variables determined using Landsat TM/ETM and SPOT High Resolution Visible(HRV) sensors have limitations for operational applications. Wolter et al. (2009) studied forests in northern Minnesota, USA, using SPOT-5 5 to 10 m data and established R^2 values of 0.82–0.93 for canopy diameter, 0.82–0.90 for diameter at breast height, 0.69–0.92 for tree height, 0.52–0.68 for vertical length of live crowns, and 0.71–0.74 for basal areas. Landsat has enabled the world's first 30 m forest cover change data set (Hansen et al., 2013). Further, spectral-temporal features derived from Landsat time series of all

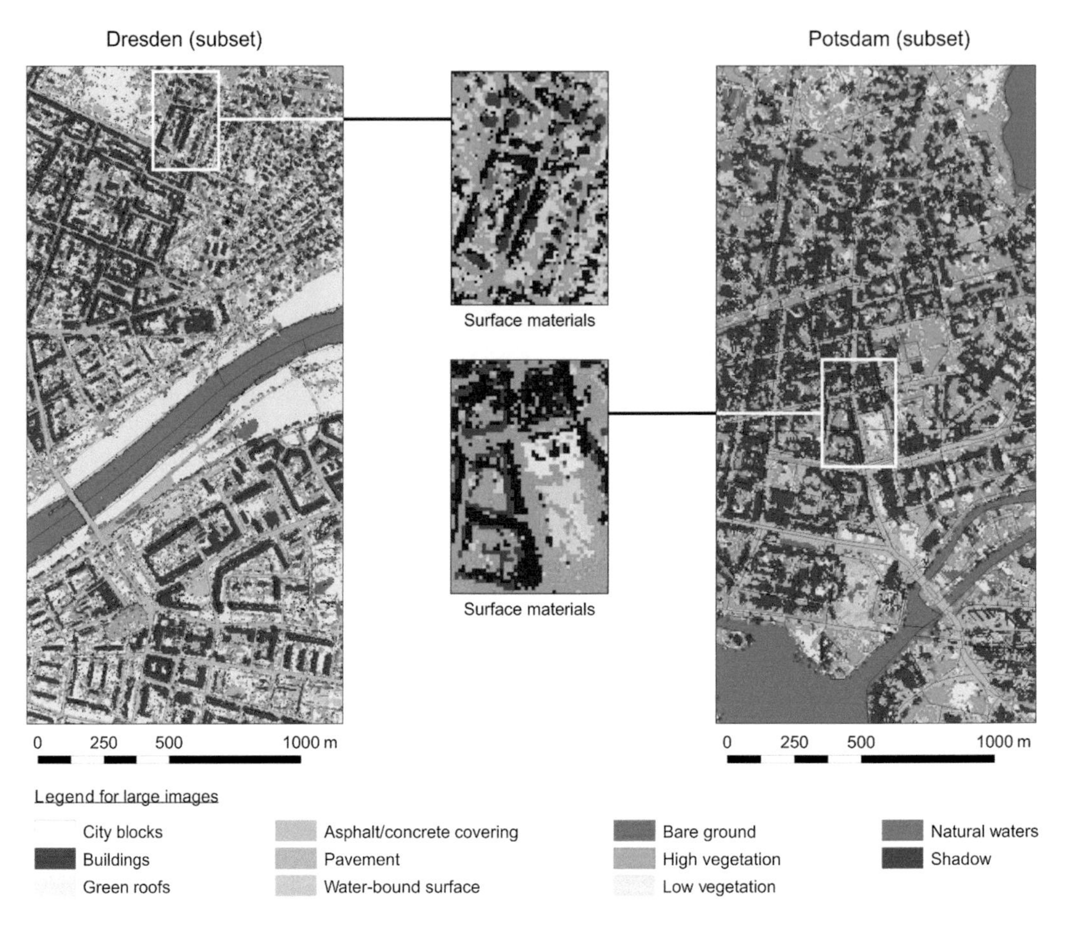

FIGURE 15.14 Linking categories derived from hyperspectral urban surface material classification based on 2004 HyMap data of Dresden (left) and 1999 HyMap data of Potsdam (right). The insets show the original variety of urban surface materials. (From Behling, R. et al. 2015. *Ecological Indicators*, Volume 48, Pages 218–234.)

available observations (1985–2015) relative to more conventional single-date and multidate inputs helped significantly improve forest type mapping (Pasquarella et al., 2018). More frequently available MODIS data at 250 or 500 m can be used to rapidly detect tree cover loss pixels, which can then be field verified (Wheeler et al., 2018).

Forests remove about 30% of greenhouse gas (GHG) emissions from the atmosphere, but also cause about 10% of emissions due to swift forest degradation, and the capacity of remaining forests to act as sinks is dwindling (Union of Concerned Scientists, 2013; Le Quéré et al., 2016; Houghton et al., 2017; Rights and Resources Initiative, 2018). However, forest carbon estimates vary widely (Asner et al., 2004; Houghton, 2008) as a result of knowledge gaps (Bates et al., 2008), data and methods used (Brown et al., 2001), and rapid changes in tropical land use (Nepstad et al., 2008), resulting in "missing" carbon in the global carbon budget (DeFries et al., 2004; Bates et al., 2008). Given forest diversity, complexity, and vast stretches of inaccessibility, remote sensing is the best source of data to study forests. Nevertheless, the issue is complicated by differing spatial, spectral, radiometric, and temporal resolutions of imagery (Malhi and Wright, 2004; Lu, 2006), which are inadequately addressed in studies of various forest characteristics. Future forest carbon cycling trends attributable to losses and regrowth associated with global climate and land-use change are uncertain and have often alternated between sources and sinks in Asia (Dixon et al., 1994; Phillips, 2009). The long-term net flux of carbon between terrestrial ecosystems and the atmosphere has been dominated by two factors: (1) changes in the total area of forests and (2) per-hectare changes in forest biomass resulting from management and regrowth (IPCC, 2007; Houghton, 2008). Apart from regional-level uncertainties (Nepstad et al., 2008), the carbon flux of tropical forests is greatly influenced by uncertainty in the regenerative capacity of forests and in harvest and management policies (Nepstad et al., 2008). The need to remove uncertainties and errors in carbon storage and change calculations from rainforests is more urgent than ever before. Under the United Nations Framework Convention on Climate Change (UNFCCC), countries must report regularly the state of their forest resources and emerging mechanisms, such as Reducing Emissions from Deforestation in Developing Countries (REDD), which are likely to require temporally and spatially fine-grained assessments of carbon stocks (UNFCCC, 2008). Whereas remote sensing is considered the best option for rapid and consistent reporting of spatially explicit carbon flux dynamics in rainforests (Brown et al., 2001; Asner et al., 2004; DeFries et al., 2004, Malhi and Wright, 2004), research on uncertainties and errors (Thenkabail et al., 2004b) are inadequate for an operational system. For example, there are no regionwide data on the transition between primary and secondary forests across the uplands of Asia. Nevertheless, tropical secondary forests and agroforests and plantations are important sinks in the global carbon cycle, affecting global climate, and have an important role within the Clean Development Mechanism (CDM) of the Kyoto Protocol (Brown et al., 2001). Other significant uncertainties exist. Recent research in neotropical forests has shown that during drought years, the net flux of several billion tons of carbon dioxide (CO2) is released into the atmosphere due to reduced tree growth and mortality, more than the combined annual emissions of Japan and Europe (Phillips, 2009).

Nevertheless, Hyperspatial data (<10 m), such as IKONOS or SPOT, or data frequently available with daily coverage of MODIS have significant limitations in modeling and mapping forest biophysical (e.g., biomass, LAI), and biochemical (e.g., chlorophyll, pigments) quantities. These advances are achievable through hyperspectral data (Thenkabail et al., 2004a,b) gathered over a continuous spectrum (Liu and Wu 2018). For example, Näsi et al. (2018) used hyperspectral cameras based on a tunable Fabry-Pérot interferometer to assess outbreaks of destructive bark beetle species that posed a serious threat to urban boreal forests in North America and Fennoscandia. Individual spruces were classified as healthy, infested, or dead with an overall accuracy of 79%, and when there were two groups (healthy versus dead), the overall accuracy increase to 93%. Forest fuel availability was studied to establish forest canopy fuel metrics—such as fuel load, live fuel moisture content, and live-dead ratio—by integrating airborne laser scanning (ALS) and high-resolution airborne hyperspectral data (AHS) (Ramirez et al., 2018). Opportunities to significantly advance carbon storage and flux estimates through improved land-use class (LUC) estimates and modeling exist with the evolution in spaceborne hyperspectral, hyperspatial, and advanced multispectral sensors (Melesse et al., 2007), as a result of improvements in the spatial, spectral, radiometric, and temporal properties as well as in optics and signal-to-noise ratio of data. High spatial resolution allows location while high spectral resolution allows identification of features. Hyperspectral remote sensing sensors also facilitate direct measurement of canopy chemical content (e.g., chlorophyll, nitrogen), forest species, chemistry distribution, timber volumes, and water (Asner and Martin, 2008) and improved biophysical and yield characteristics (Thenkabail et al., 2003, 2004a,b; Asner et al., 2004). Thenkabail et al. (2004b) demonstrated an increased accuracy of about 30% in LUC and biomass when 30 hyperspectral wavebands were used relative to 6 nonthermal Landsat TM bands. Hyperspatial data have demonstrated the capability to extract individual tree crowns from 1 m panchromatic data. Agroforest successional stages have been mapped and their varying carbon sink strengths assessed using IKONOS (Thenkabail et al., 2004a). In contrast, forest structure variables (e.g., biomass, LAI) were poorly predicted by older-generation sensors (Pignatti et al., 2009b). Hyperspectral data, such as from the 242-band Earth Observing-1 (EO-1) Hyperion, offer advanced capabilities to classify complex forests such as tropical forests (Ferreira et al., 2007). A number of machine learning algorithms have been applied on cloud computing platforms to rapidly classify forests. Ferreira et al. (2007), for example, showed that support vector machines and random forests provided the best match with reference data (Figure 15.15) when using four distinct machine learning algorithms (Figure 15.15).

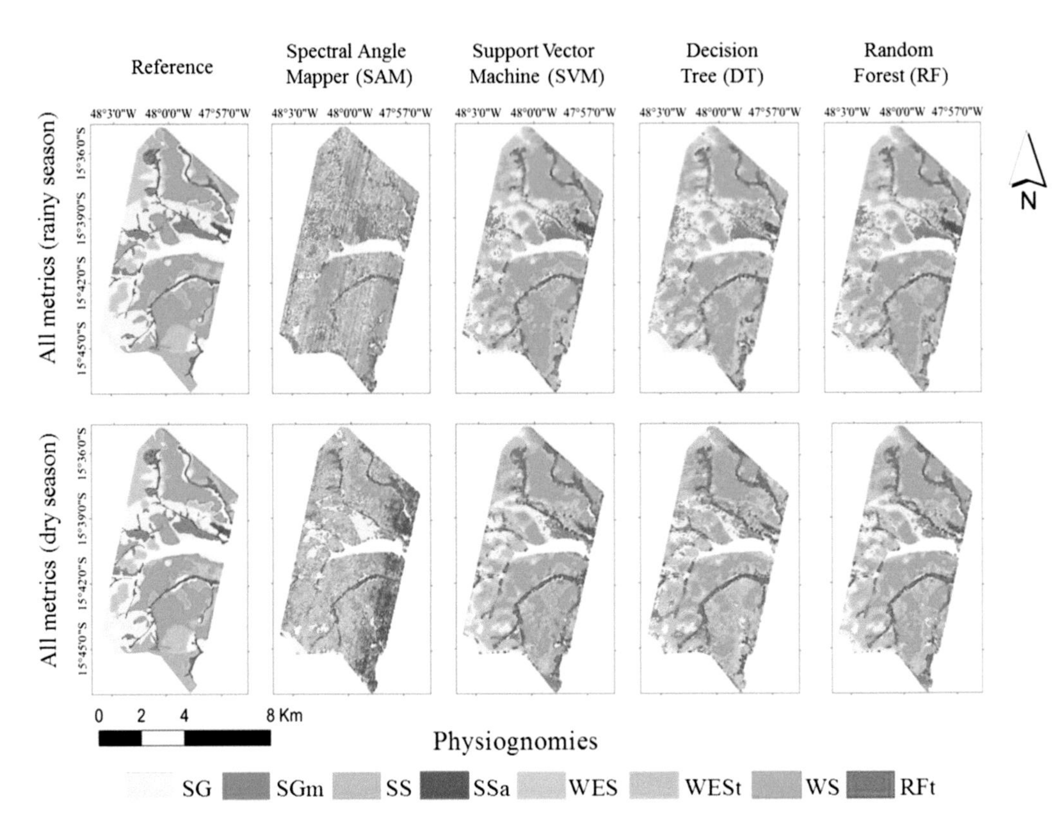

FIGURE 15.15 Classification images in the rainy and dry seasons generated by different techniques using all metrics. (The reference map was published by Ferreira et al. 2007. *International Journal of Remote Sensing*, Volume 28, Pages 413–429.) Note: vegetation physiognomies sensed by Hyperion over Brasilia National Park (BNP): (1) Campo Limpo (Savanna Grassland—SG); (2) Campo Limpo with Murundus (SGm); (3) Campo Sujo (Shrub Savanna—SS); (4) Campo Sujo with *Lychnophora ericoides* (popular name = arnica) and *Vellozia squamata* (canela de ema) (SSa); (5) Campo Cerrado (Wooded Savanna—WES); (6) Campo Cerrado with *Trembleya parviflora* (WESt) (quaresmeira); (7) Cerrado sensu stricto (Woodland Savanna—WS); and (8) Mata de Galeria (Riparian Forest—RFt). (From Toniol, A.C. et al. 2017. *Remote Sensing Applications: Society and Environment*, Volume 8, Pages 20–29.)

In Chapter 9, Dr. Valerie Thomas provides an overview of various forest-type ecological services and their rich and complex systems. Dr. Thomas describes how remote sensing is used in various forest applications such as forest inventories, species identification, and modeling and mapping forest biophysical and biochemical quantities. Forest quantities measured using multispectral and hyperspectral remote sensing included tree height, depth at breast height, aboveground biomass, crown area, basal area, LAI, and stem density. The chapter provides illustrations of a number of these applications like measuring dominant height using derivative chlorophyll index and mean dominant height. This is followed by a discussion of the literature on carbon studies and forest fuel assessments using various hyperspectral remote sensing studies. Dr. Thomas also points to future advances where LiDAR data will be fused with hyperspectral data to advance accuracies of metrics on forest characteristics such as biomass and carbon assessments. Tree height data from LiDAR, when combined with biophysical and biochemical measurement capabilities from Hyperspectral data, will lead to these advances. Further, repeated coverage from the forthcoming hyperspectral sensors like HyspIRI, EnMAP, and PRISMA, as well as repeated acquisitions from UAVs from multiple sensors (e.g., LiDAR, Hyperspectral), are harbingers of a new era in remote sensing. Dr. Thomas concluded that hyperspectral data have many known, proven capabilities to advance forest studies, but utilization of the data for those purposes is very limited at present.

15.10 CHARACTERIZATION OF PASTURES USING FIELD AND IMAGING SPECTROMETERS

Pasture lands are both natural and managed. They contain important and unique flora and fauna with rich biodiversity. Grazing periods of pasture lands last from a few weeks in well-managed rotation systems to years on extensively managed pastures and are superimposed on climate-driven phenology (Jakimow et al., 2018). A good understanding of pasture productivity, phenology, and nutritional quality will help assess carrying capacity and animal health. Worldwide, about 25% of the terrestrial area is pastureland, mainly for animal grazing and for natural habitats of various kinds. In many countries such as Brazil, pasturelands for cattle grazing are in deforested land areas. Dry pasture also easily catches fire that spreads swiftly over large stretches of land. These factors lead to substantial release of GHGs (Davidson et al., 2012). Pastures are mapped with a great degree of accuracy using remote sensing data of various kinds and are one or more of the classes in many remote sensing products (Biradar et al., 2009; Thenkabail et al., 2009a,b; Salmon et al., 2015; Jakimow et al., 2018). However, difficulties are found in mapping pastoral nomadic livestock systems, which are dependent on variable seasonal rains (Hopping et al., 2018). Also, various pasture biophysical and biochemical quantities are best characterized using HNB data that need to be gathered on pasture spectral characteristics, as well as other vegetation (e.g., Figure 15.16). This will help clearly separate distinct spectral characteristics of various vegetation categories. For example, a meta-analysis of 77 grassland and shrubland studies determined that the performance of hyperspectral data depended on the traits being studied, with an R^2 value of 0.79 for LAI, 0.77 for chlorophyll, 0.80 for carotenoids, 0.75 for phosphorous, 0.74 for nitrogen, 0.69 for plant water, and 0.64 for lignin (Cleemput et al., 2018). Fava et al. (2009) showed a simple ratio involving HNBs in the NIR (770–930 nm) and in the red edge (720–740 nm) yielding the best performance for green

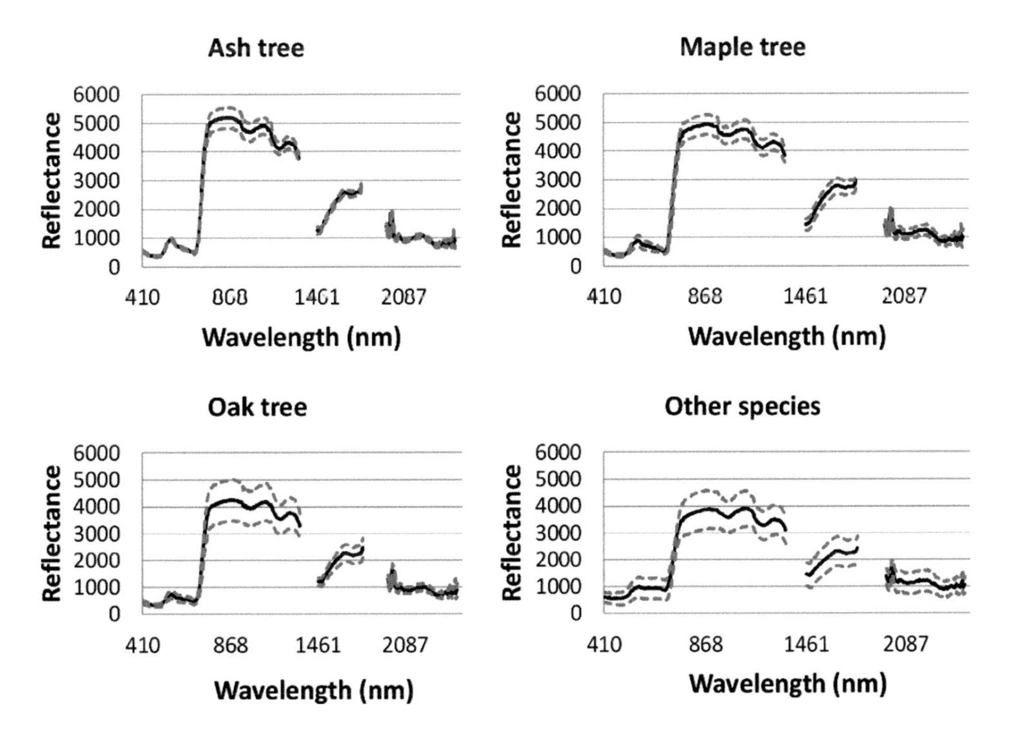

FIGURE 15.16 Pixel-weighted spectra (bold line) and standard deviation (±SD, dashed line) of reflectance by species. Observe the subtle as well as very distinct differences in spectral reflectivity in specific wavelength range (e.g., in 400–700 nm) between various tree types. (From Liu, H., Wu, C. 2018. *International Journal of Applied Earth Observation and Geoinformation*, Volume 68, Pages 298–307. ISSN 0303-2434, https://doi.org/10.1016/j.jag.2017.12.001.)

biomass with an R^2 value of 0.73, LAI with an R^2 value of 0.73, and nitrogen with an R^2 value of 0.73. Pullanagari et al. (2016) used 1 m spatial resolution airborne data in the range 380–2500 nm with 448 HNBs to predict pasture nutrient concentrations such as nitrogen (N), phosphorus (P), potassium (K), sulfur (S), zinc (Zn), sodium (Na), manganese (Mn), copper (Cu), and magnesium (Mg). The majority of nutrients (N, P, K, Zn, Na, Cu, and Mg) were best predicted by random forest regression algorithms followed by S and Mn, which were best predicted by support vector regression (SVR) by Pullanagari

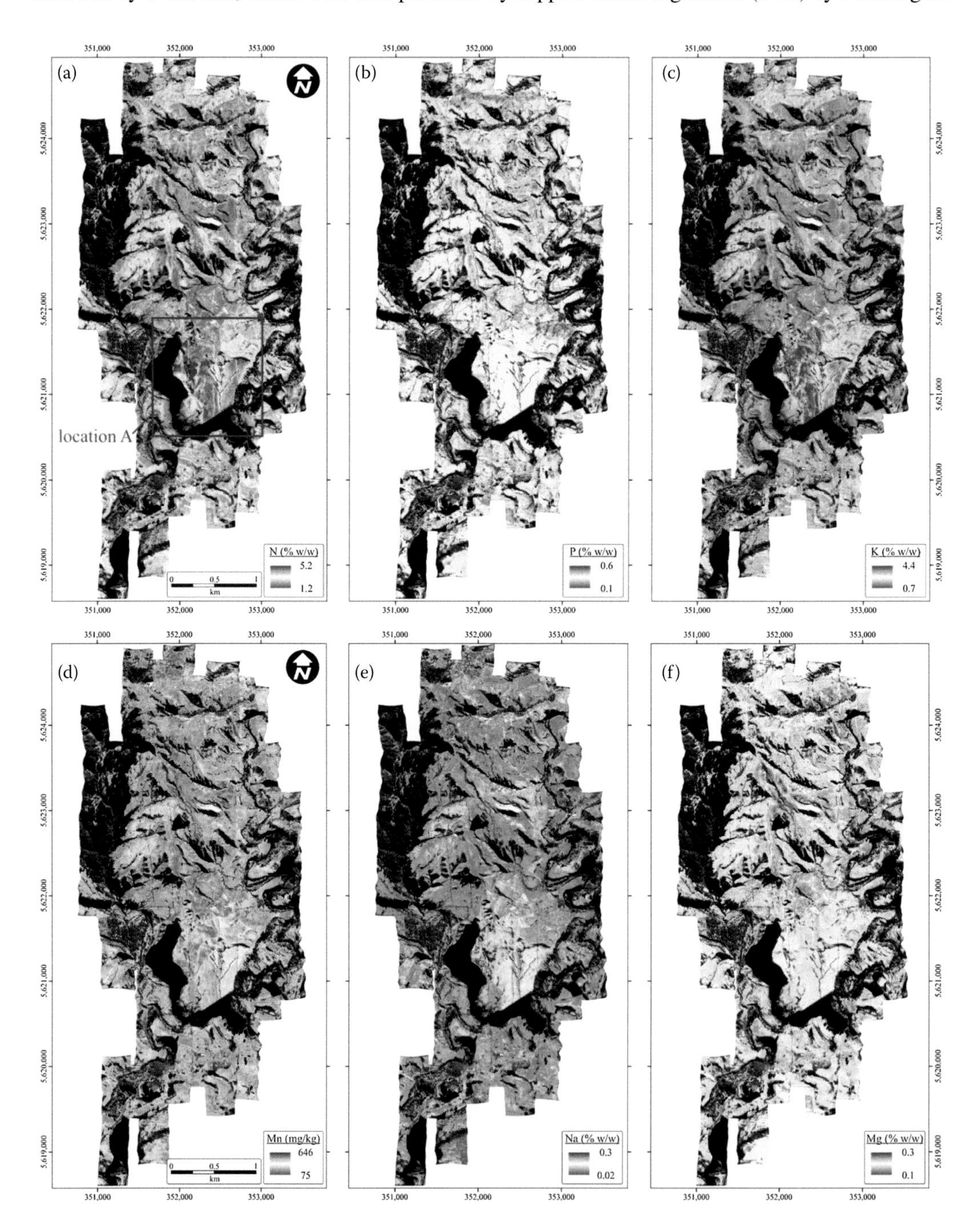

FIGURE 15.17 Spatially registered nutrient concentration maps of (a) nitrogen, (b) phosphorus, (c) potassium, (d) manganese, (e) sodium, (f) magnesium, *(Continued)*

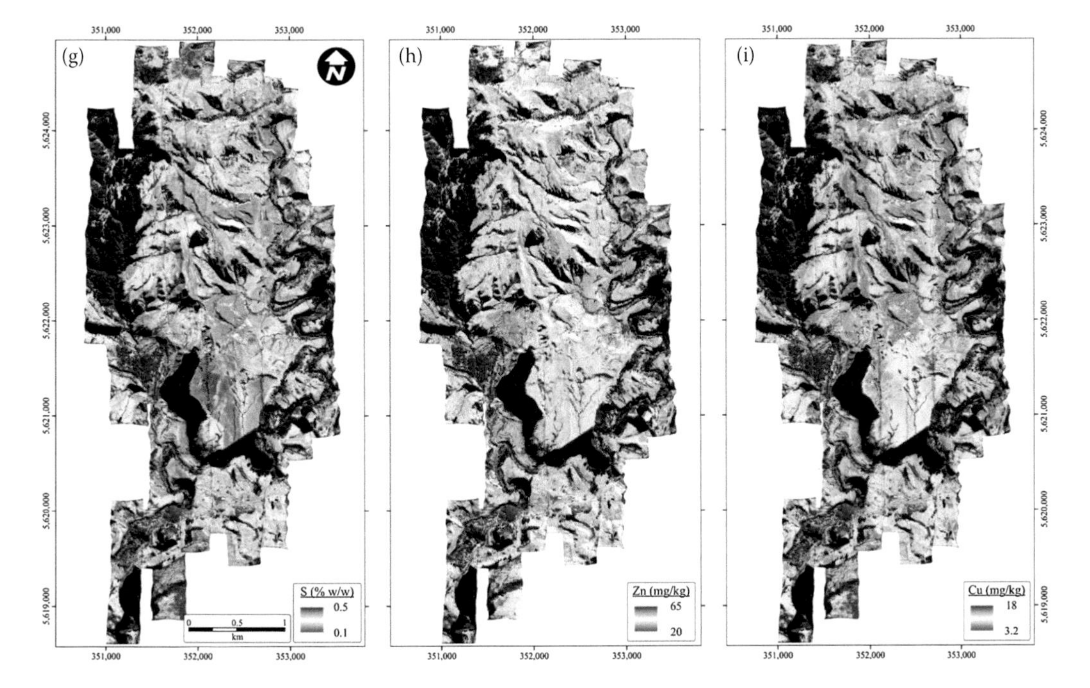

FIGURE 15.17 (*Continued*) Spatially registered nutrient concentration maps of (g) sulfur, (h) zinc, and (i) copper of mixed pastures derived from canopy reflectance acquired by AisaFENIX airborne imaging spectrometer. (From Pullanagari, R.R. et al. 2016. *ISPRS Journal of Photogrammetry and Remote Sensing*, Volume 117, Pages 1–10.)

et al. (2016). Such understandings will help establish nutrient concentration maps (e.g., Figure 15.17) that are so crucial to determining the quality of pasture available as feed.

The focus of Chapter 10 by Dr. Izaya Numata et al. was the study of pasture using field-based and imaging spectroradiometer-based instruments. For this they reviewed and identified biophysical and biochemical variables of pastureland or grasslands. The biophysical and biochemical variables included grass height, soil coverage, leaf area, leaf orientation, moisture content, pigmentation, and several other parameters considered specific to pasture species. They presented studies of species such as *B. brizantha* and *B. decumben*. The chapter outlines detailed methods and approaches used in pasture characterization that are discussed under the following sections: (1) vegetation indices (VIs), (2) red-edge indices, (3) spectral transformations, (4) spectral mixture analysis, and (5) statistical methods. The specific applications of hyperspectral data in pasture studies were: (1) pasture quality, (2) LAI, (3) biomass, (4) pasture degradation analysis, and (5) species discrimination. Pasture quality was assessed in terms of the levels of N, P, K, Ca, and Mg using both field-based spectroscopy and imaging spectroscopy. The researchers highlighted the importance of 680 nm and 550–580 nm wavebands in quantifying N, P, K, Ca, and Mg. The CAI provided the best relationships with LAI. A review of the studies relating pasture biomass to hyperspectral data yielded R^2 values between 0.56–0.86. The current state of pasture degradation studies and species discrimination studies using hyperspectral data are discussed.

15.11 HYPERSPECTRAL REMOTE SENSING OF WETLAND VEGETATION

The Ramsar Convention defined wetlands as "areas of marsh, fen, peatland or water, whether natural or artificial, permanent or temporary, with water that is static or flowing, fresh, brackish or salt, including areas of marine water the depth of which at low tide does not exceed six metres" (Ramsar, 2013). Wetlands occupy 4%–6% of the Earth's land area and globally store about 771 billion tons of

GHGs (20% of all carbon on Earth) (Hansen et al., 2002; Pelley, 2008; Tiner, 2009; Lopez et al., 2013). This is about the same amount of carbon as is now in the atmosphere. However, they also release methane, a GHG (Pelley, 2008) that is 22 times more potent than CO_2, on a per-unit-mass basis, in absorbing long-wave radiation on a 100-year time horizon (Zhuang et al., 2009). Nearly 60% of the planet's wetlands have been destroyed in the past 100 years, mostly due to agriculture. That is because they are widely accepted as rich agroecosystems with unique flora, fauna, biodiversity, rich soils, and abundant water relative to uplands. In Asia, wetlands are extensively used for agriculture, especially rice cultivation, which overwhelmingly takes place in lowlands. In contrast, the wetlands of Africa are increasingly considered "hotspots" for agricultural development and for expediting Africa's Green and Blue Revolutions. In Africa, currently, these wetlands (especially inland valley wetlands) are unutilized or highly underutilized in spite of their rich soils and abundant water availability as a result of: (a) limited road and market access and (b) prevailing diseases such as malaria, trypanosomiasis (sleeping sickness), and onchocerciasis (river blindness). However, the utilization of inland valley wetlands for agriculture is becoming unavoidable in West and Central African countries due to increasing pressure for food from a ballooning human population and as a result of the difficulty in finding arable uplands. Nevertheless, studies are currently under way to prioritize the use of wetlands for conservation versus agricultural development (Thenkabail et al., 2009b). In Africa, since most wetlands are still intact, there is immense pressure to develop them to ensure food security on the continent. Indeed, many consider wetlands as the best hope for Africa's Green and Blue Revolutions (WARDA, 2006) and a far better option for food security than the alternative of building large dams, which would result in greater destruction of pristine rainforests (FAO, 2008).

Several studies discuss methods of wetland mapping using remote sensing (Lunetta et al., 1999; Thenkabail et al., 2000; Harvey and Hill, 2001; Lyon, 2001; Ozesmi and Bauer, 2002; Hirano et al., 2003; May et al., 2003; Töyrä and Pietroniro, 2005; Wagner et al., 2007; Wright and Gallant, 2007; Jones et al., 2009; Lyon and Lyon, 2011; Lopez et al., 2013). High levels of accuracy in identifying and mapping wetlands are feasible when multidate, multisensor, very-high-spatial-resolution images are used (e.g., Lan and Zhang, 2006; Becker et al., 2007; Gilmore et al., 2008). Ramsey et al. (1998) found that an integrated synthetic aperture radar (SAR)-optical (TM and CIR) fusion product improved the accuracy of wetland classes by up to 20%. SAR data are sensitive to relative elevation differences and soil moisture and are ideal for characterizing lowlands (with higher moisture) and uplands (with far lower moisture) (Wagner et al., 2007). Recent research (Thenkabail et al., 2000; Kulawardhana et al., 2007; Islam et al., 2008; Jones et al., 2009; Lyon and Lyon, 2011) has demonstrated the ability to attain high levels of accuracy in mapping wetlands using multiple data. These data included: (1) Global Land Survey 2005 (GLS2005) Landsat 30 m, (2) Japanese Earth Resources Satellite (JERS) SAR 100 m, (3) MODIS Terra/Aqua 250–500 m, (4) Space Shuttle Topographic Mission (SRTM) 90 m, and (5) a suite of secondary data sets (e.g., soils). Automated methods of wetland recognition involved (Lan and Zhang, 2006; Islam et al., 2008; Jones et al., 2009): (a) algorithms to rapidly identify wetland streams using SRTM Digital Elevation Model or DEM data, (b) thresholds of SRTM-derived slopes, (c) thresholds of spectral indices and wavebands, and (d) automated classification techniques. First, wetlands are topographical lowlands, so the DEM data offer a significant opportunity to identify lowlands from uplands. Automated methods involving the SRTM-derived wetland boundaries have four known limitations (Islam et al., 2008): (i) turning out nonexistent or spurious wetlands, (ii) providing non-smooth alignment, (iii) resulting in spatial dislocation of streams, and (iv) absence of stream width. Second, the SRTM DEM data were used to derive local slope maps in degree using the Slope function of ArcInfo Workstation GIS. A threshold-of-degree slope provides areas of wetlands or low-lying areas and nonwetlands. Third, the wetlands in the images can be highlighted by enhancing images (Lyon and McCarthy, 1995; Lunetta et al., 1999; Lyon and Lyon, 2011; Lopez et al., 2013). The thresholds of indices and wavebands will automatically separate wetlands from nonwetlands (Kulawardhana et al., 2007; Schowengerdt, 2007) with adequate accuracies. Numerous researchers have also attempted wetland separation through automated classification techniques on various remotely sensed data (Jensen

FIGURE 15.18 Wetlands of Ghana mapped using Landsat 30 m ETM+ data. In the entire country of Ghana, there is 11.4% (2,714,946 ha) of total geographic area (23,853,300 ha). Only 5% (130,000 ha) of these wetlands are cultivated. Color: red: wetlands with dense natural vegetation, cyan and yellow: wetlands with some cultivation, blue: water body.

et al., 2002; Fuller et al., 2006; Lan and Zhang, 2006) without first identifying and separating wetland areas from other land units. However, as Ozesmi and Bauer (2002) point out, this leads to difficulties of wetland categorization because of spectral confusion (Lan and Zhang, 2006). This is because the automated classification techniques are applied on entire image areas that include wetlands and other land units that often have significantly similar spectral properties. Classification accuracies improve when multitemporal data are used along with ancillary data such as soils and topography (Ozesmi and Bauer, 2002) in a GIS modeling framework (Sader et al., 1995; Lyon, 2001; Fuller et al., 2006; Lopez et al., 2013). Automated methods are rapid, but need to be supplemented by semiautomated methods to increase accuracies and decrease errors of omission and commission. The aforementioned methods were used to map wetlands of Ghana (Figure 15.18) using Landsat ETM+ data and SRTM data. However, to study wetland characteristics like what species of plants exist in the wetlands, one needs to use hyperspectral data (e.g., 2). In Figure 15.19, Rebelo et al. (2018) illustrated hyperspectral data of six wetland species gathered using a 350–2349 nm range handheld spectroradiometer. They demonstrated the capability of the hyperspectral data in establishing plant functional traits (PFTs) of four morphological and three biochemical traits. The morphological traits studied were leaf area, specific leaf area, leaf mass, and leaf length/width ratio. The biochemical traits studied were lignin and cellulose content. The researchers also studied three other traits: total biomass, leaf C/N ratio, and cellulose concentration. HNB data used in the study explained 38%–98% of the variability of these traits. Similarly, hyperspectral data from various platforms can be used to advance our knowledge of wetland characterization and mapping (Lopez et al., 2013).

In Chapter 11, Dr. Elijah Ramsey highlighted the importance of hyperspectral data in wetland vegetation characterization and the study of their traits. For this, they first examined species-level studies pertaining to wetland forest species with a specific focus on mangrove forests, baldcypress forests, and bottom hardwood forests. One needs to consider several issues in these studies that include: (1) platforms from which the data are acquired, (2) resolutions (spatial, spectral, temporal), (3) whether spectra are gathered for leaves or plants, (4) background influences, (5) species type, and (6) other factors (e.g., date on which data are gathered that is linked to plant phenology). These issues are discussed in detail by studying the forest types mentioned. Quantities such as LAI modeled

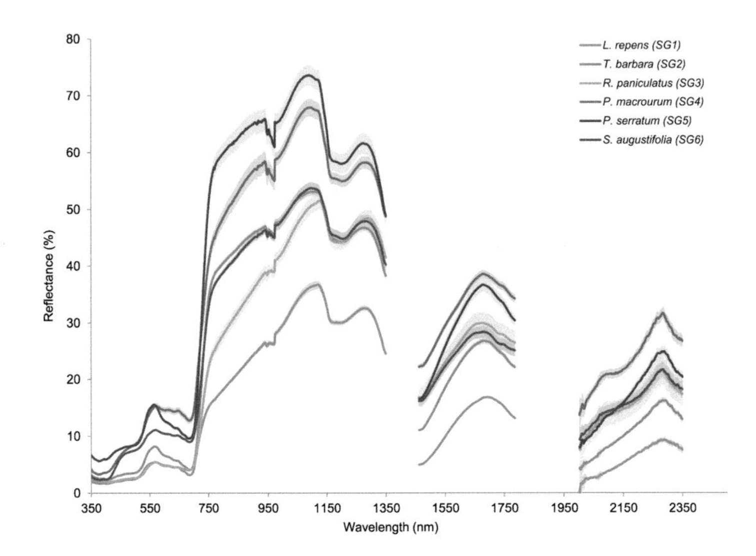

FIGURE 15.19 Illustrative figure showing spectral population of six example species (one for each spectral group (SG1-6)). Solid line: mean spectrum; shaded zone: variation (SD). (From Rebelo et al. 2018. *Remote Sensing of Environment*, Volume 210, Pages 25–34. ISSN 0034-4257, https://doi.org/10.1016/j.rse.2018.02.031.)

using HNBs and HVIs are presented and discussed. Radiative transfer (RT) models predict LAI by explaining 78%–93% of the variability. Second, they show the strength of studying invasive species like Chinese tallow in wetland forest and marsh. Through continuous color compositions of varying mixtures of the red, blue, and green hues, the models showed how occurrences of tallow, live vegetation, and senescent foliage were distinguished. This is followed by the study of marsh wetland species such as *Juncus romerianus*, *Spartina alterniflora*, *Spartina patens*, and *Panicum hemitomon*. Methodological issues pertaining to distinguishing subtle changes in spectra at the leaf and canopy levels for considering the six issues listed earlier were discussed using various wetland plant species as examples. Overall, HNBs and HVIs clearly showed significant advantages in classifying and quantifying various wetland characteristics relative to multispectral broadband data.

15.12 HYPERSPECTRAL REMOTE SENSING OF FIRE: A REVIEW

Detecting and studying fire, both natural (e.g., wildfires) and human-caused (e.g., agricultural stubble burning), are important in terms of understanding and quantifying GHGs such as carbon dioxide (CO_2), carbon monoxide (CO), methane (CH_4), and nitrous oxide (N_2O) released to the atmosphere. Today, an estimated 5.6 gigatons of carbon, globally, are released into the atmosphere each year due to fossil fuel burning and an additional 2.4 gigatons of carbon per year from tropical forest fires (Source: NASA Earth Observatory). Fire activity is determined by a wide range of factors, including long- and short-term climatic conditions, climate seasonality, wind speed and direction, topography, and fuel biomass (Alvarado et al., 2017) determining that drought during ignition season was the single biggest cause for natural fires in the savannas. Over the years, Landsat and similar high-resolution data have been widely used in studying fire occurrence and spread, scars left by fires, and regrowth. Remote sensing is used to study both prefire assessment of potential fire areas as well as postfire assessments of vegetation regrowth. Remote sensing has also been used to study a wide range of fires such as those from coal mines (Wang et al., 2015a,b), agricultural stubble burns (Smith et al., 2007; Vadrevu et al., 2011), volcanoes (Trifonov et al., 2017), and natural fires in forests and savannas. Remote sensing helps assess the damage caused to homes and other property through prefire and postfire. Broadly, remote-sensing-based methods have been investigated for fire danger management activities and are categorized into two major groups: fire danger monitoring

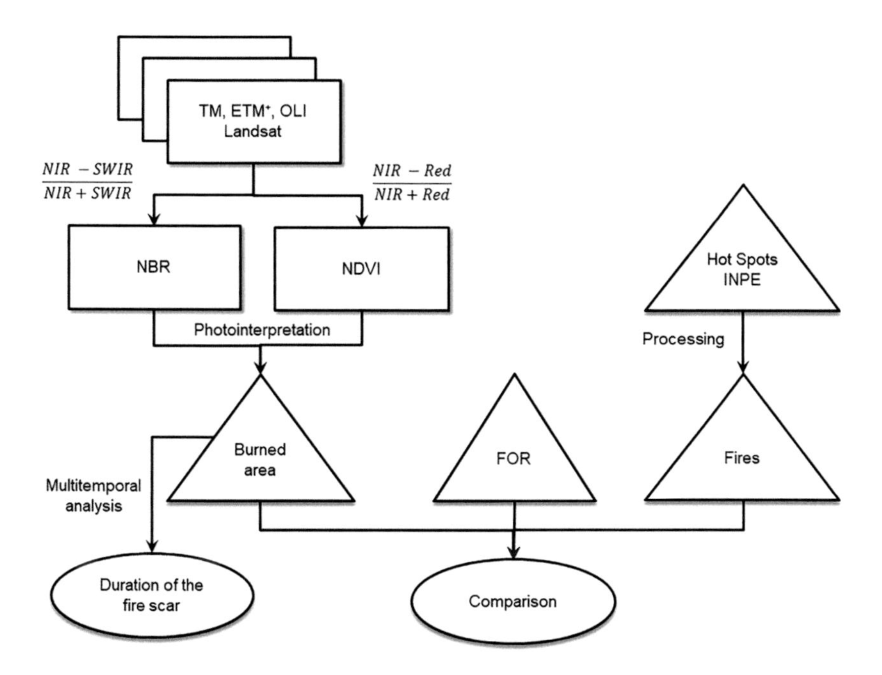

FIGURE 15.20 Fire study methods overview using Landsat and similar data. (From Costa dos Santos, J.F. et al. 2018. *Science of The Total Environment*, Volume 616–617, Pages 1347–1355.)

and fire danger forecasting systems (Chowdhury and Hassan, 2015). Multitemporal images are used to compute normalized difference vegetation index (NDVI) and the normalized burn ratio (NBR) index to study fire occurrence and fire's impacts prior to and following fire occurrence (Cost dos Santos et al., 2018). The two widely used remote sensing indices for pre- and postfire assessments are as follows (Cost dos Santos et al., 2018; Figure 15.20):

$$\text{NDVI} = (\text{NIR}-\text{red})/(\text{NIR} + \text{red});$$

$$\text{NBR} = (\text{NIR-SWIR})/(\text{NIR} + \text{SWIR}).$$

where NIR = near-infrared band, SWIR = shortwave infrared band, and red = red band.

Cooler-temperature objects (i.e., temperatures up to 500 K) emit most of their radiance in the TIR (8–12 μm) and MIR (3–5 μm) regions of the electromagnetic spectrum, whereas hotter-temperature objects (e.g., smoldering and flaming combustion above temperatures of 500 K) emit more of their radiance in the SWIR (1.4–2.5 μm) (Matheson and Dennison, 2012). Hyperspectral data covering a wavelength range of 1.2–2.5 μm can be used to detect fires and model fire temperature and background land cover (Matheson and Dennison, 2012). Dennison and Matheson (2011) improved upon the fire temperature algorithm presented in Dennison et al. (2006) using a hyperspectral fire detection index (HFDI) for fire detection, as well as separate spectral libraries of background endmembers for smoke, nonsmoke, and fire pixels (Matheson and Dennison, 2012):

$$\text{HFDI} = (L_{2429\text{nm}} - L_{2061\text{nm}})/(L_{2429\text{nm}} - 15.\ L_{2061\text{nm}});$$

where L stands for radiance at the specified wavelength (Matheson and Dennison, 2012).

In Chapter 12, Dr. Sander Veraverbeke et al. approached the fire mapping task in three stages: prefire, active fire, and postfire. The researchers reviewed the relevant studies and discussed such topics as fuel types, fire risk assessment, active fires, burned area mapping, and vegetation recovery postfire. Prefire assessments were conducted by mapping PFTs through a VI approach. In contrast,

HNB data fuel composition and condition were determined from spectral mixture analysis (SMA) (Roberts et al., 2006) taking ground cover classes green vegetation, NPV, and substrate. These topics are discussed and then authors highlight the two challenges of detecting active fires using hyperspectral data. First, since hyperspectral data are acquired during the day, the emitted radiance from the objects must be separated from the reflected solar radiance. The second challenge is the need to detect fire from a fraction of a pixel. The HFDI (Matheson and Dennison, 2012), the most accurate HNB index to detect fire, is based on the fact that fire detection capability increases at longer wavelengths versus shorter wavelengths. Postfire studies are conducted using NBR and difference NBR (dNBR). The dNBR detects prefire and postfire differences in NBR. However, SMA will allow for use of data from multiple narrowbands, unlike the two-band VIs. Thus, narrowbands further advance our knowledge of fire studies. Postfire vegetation recovery studies are important to see how ecosystems recover after fire and to establish which species are most resilient. These multispectral broadband studies involve the use of NDVI and NBR in vegetation recovery. However, HNBs offer many more opportunities in terms of numerous additional HNBs, HVIs that help study species types, and vegetation biophysical and biochemical quantities. Finally, the chapter presents and discusses the hyperspectral thermal data that complement the data in prefire, active-fire, and postfire studies.

15.13 HYPERSPECTRAL DATA IN LONG-TERM CROSS-SENSOR VEGETATION INDEX CONTINUITY FOR GLOBAL CHANGE STUDIES

Ever-increasing numbers of satellites offers a huge advantage to the myriad of Earth Resources Applications as the frequency of data acquisition increases. However, utilization of data from multiple sensors becomes feasible only when there is accurate intersensor calibration and normalization. All Earth Observation (EO) satellite data suffer from drift and biases relative to their prelaunch calibration (Gorroño et al., 2017). This calls for the development of a systematic sensor calibration, normalization, and within- and across-sensor calibration relationships. For this, each sensor needs to be continually calibrated and progressive sensor degradation accounted for by applying calibration equations, leading to the normalization of data over time and space. There are many excellent examples of within- and across-sensor calibration studies. For example, Chander et al. (2009) provided calibration coefficients for Landsat multispectral scanner (MSS), TM, ETM+, and EO-1 Advanced Land Imager (ALI). Goward et al. (2012) showed the intersensor relationships involving ResourceSat-1 Advanced Wide Field Sensor (AWiFS) and Landsat TM/ETM+ sensors. Multisensor NDVI applications will benefit most if atmospheric corrections are adequately addressed and translation equations applied (e.g., Figure 15.21) (Van Leeuwen et al., 2006). Some multisensor discrepancies might be more complex, but they can be overcome using error reducing analysis techniques like data smoothing and normalization (z-score) (Van Leeuwen et al., 2006). HNB data can be used to simulate multiple sensors and develop intersensor relationships, as illustrated in Figure 15.21 between high-spatial-resolution IKONOS and multispectral broadband Landsat ETM+. Once such relationships are developed, it becomes feasible to utilize data from multiple satellite sensors for the myriad of Earth Resources Applications.

In Chapter 13, Dr. Tomoaki Miura uses hyperspectral data for cross-sensor studies using VIs from multiple sensors. Hyperion hyperspectral data from five sites in the USA as well as in situ atmospheric measurements from the Aerosol Robotic Network (AERONET) stations were gathered for the study. Three VIs were used including NDVI and two enhanced vegetation indices (EVIs). The VI interrelationships were developed between a number of satellite sensors, such as Terra MODIS, NOAA-14 AVHRR/2, NOAA-17 AVHRR/3, NPP VIIRS, and SPOT-4 VEGETATION. Specific issues of multiple sensors addressed in developing intersensor VI relationships were spectral, spatial, algorithmic, and angular. They highlight that the value of hyperspectral data is not only in improving the characterization and modeling of Earth resource studies but also in facilitating intersensor calibration by developing relationships between the band reflectivity of different sensors or VIs derived from different sensors.

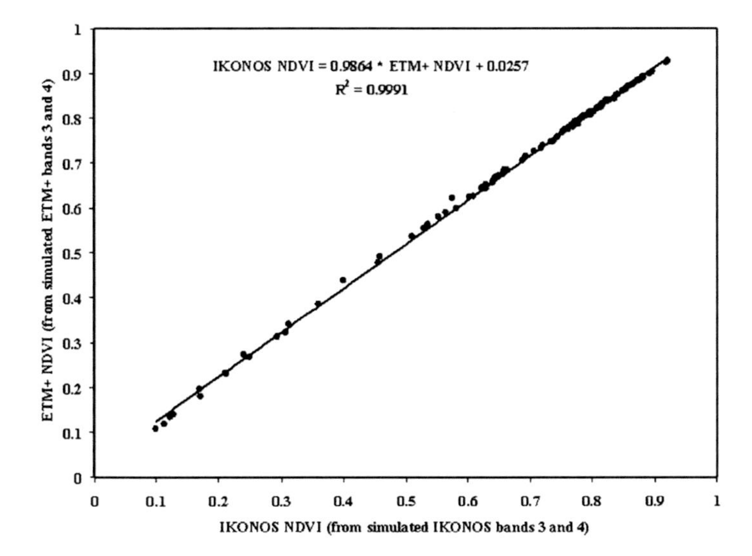

FIGURE 15.21 IKONOS NDVI versus ETM+ NDVI based on simulated data for these bands from African savannas. Hyperspectral spectroradiometer data were used to simulate ETMz and IKONOS bands. Land-cover types used include shrubs, grasses, weeds, agriculture, and soils. (From Thenkabail, P.S. 2004. *International Journal for Remote Sensing*, Volume 25, Issue 2, Pages 389–408.)

15.14 HYPERSPECTRAL ANALYSIS OF ROCKY SURFACES ON EARTH AND OTHER PLANETARY SYSTEMS

Many planets in our Solar System have numerous Earth-like features. For example, Mercury, Venus, Earth's Moon, and Jupiter's moon Io all have channels formed by lava, and Mars has had an extensive fluvial history involving dissection by water (Baker et al., 2015). Sites associated with serpentinization processes, both on Earth and throughout the Solar System, are becoming increasingly compelling for the study of habitability and astrobiology (Amador et al., 2018). Data from sensors like Compact Reconnaissance Imaging Spectrometer for Mars (CRISM) can be used to study serpentinization processes (Amador et al., 2018). Hyperspectral studies in the 400–3000 nm region acquired using three different imaging spectrometers (Goswami and Annadurai, 2008) during the Chandrayaan-1 mission to the Moon were used to detect hydroxyl lunar water. Rocks and minerals found on Earth are also found on other planets, so the same hyperspectral imaging spectroscopy tools and techniques used on Earth can also be used on those planets (e.g., Figure 15.22). There are many compelling synergies in the use of imaging spectroscopy or hyperspectral data of Earth and other planets. The first synergy occurs in the imaging spectroscopy instrumentation itself. The design and development of hyperspectral instrumentation to study various planets have overwhelming similarities. Second, the processes of gathering, storing, and preprocessing of data have many similarities. Third, the methods for data mining, data reduction, and overcoming data redundancies are complementary/supplementary. Fourth, preprocessing methods and techniques, such as atmospheric correction and conversion of radiance to reflectance, are similar. Fifth, application studies of different planets can share image analysis knowledge for analyzing rocks, minerals, soils, ice, hydroxyl water, gases, and others (e.g., extinct volcanoes). Sixth, full-range, solar-reflected hyperspectral microscopy to support Earth remote sensing research is being increasingly discussed (Slonecker et al., 2018). This becomes apparent as we consider the various chapters of this multivolume book and compare those data, methods, and approaches to the study of other planets presented and discussed by Dr. Greg Vaughan coauthors.

Chapter 14, by Dr. Greg Vaughan et al. provides a compelling presentation and discussion of hyperspectral imaging spectroscopy studies of many planets and show us where and how opportunities exist for collaboration with those studying Earth. Each application can inform the other regardless

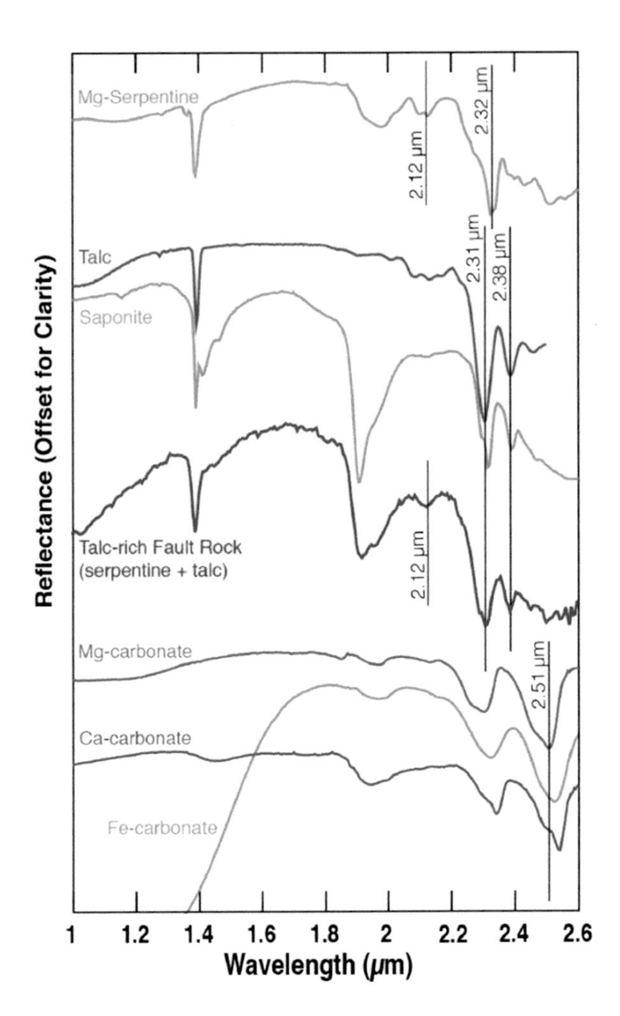

FIGURE 15.22 Near-infrared reflectance library spectra used for target transformation. Black lines and associated wavelength centers indicate important spectral absorptions used to distinguish between spectral types. The ~2.51 µm Mg-carbonate absorption is marked to show how the absorption center shifts to longer wavelengths depending on the carbonate cation. (From Amador, E.S. et al. 2018. *Icarus*, Volume 311, Pages 113–134.)

of planet or moon. Hyperspectral studies such as those of EO-1 Hyperion and Hyperspectral Infrared Imager (HyspIRI), are discussed here, while hyperspectral imagers used for other planetary applications are discussed by Dr. Greg Vaughan et al. including: (1) the MErcury Surface, Space ENvironment, Geochemistry, and Ranging (MESSENGER) for Mercury, (2) Observatoire pour la Minéralogie, l'Eau, les Glaces et l'Activité (OMEGA) for Mars, (3) Near-Infrared Mapping Spectrometer (NIMS) for Jupiter, (4) Visual and Infrared Mapping Spectrometer (VIMS) for Saturn, and (5) New Horizons for Pluto. These and several other forthcoming hyperspectral sensors used in the study of various planets are discussed in the chapter. This is followed by hyperspectral data analysis, such as radiometric and atmospheric corrections. For example, atmospheric correction algorithms, such as the fast line-of-sight atmospheric analysis of spectral hypercubes (FLAASH) or moderate-resolution atmospheric radiance and transmission (MODTRAN), are commonly applied to data acquired from any imaging sensors on the planets. Methods of data analysis, such as spectral indices and spectral mixture analysis, are discussed. Finally, the chapter explores several case studies using imaging spectroscopy data, such as the study of atmospheric compositions, aerosol depth/thickness, minerals, soils, rocks, ice, and hydroxyl water on various planets.

REFERENCES

Allen, R.G., Tasumi, M., Trezza, R. 2007. Satellite-based energy balance for mapping evapotranspiration with internalized calibration (METRIC)—Model. *Journal of Irrigation and Drainage Engineering*, Volume 133, Issue 4, Pages 380–394.

Alvarado, S.T., Fornazari, T., Cóstola, A., Morellato, L.P.C., Silva, T.S.F. 2017. Drivers of fire occurrence in a mountainous Brazilian cerrado savanna: Tracking long-term fire regimes using remote sensing. *Ecological Indicators*, Volume 78, Pages 270–281. ISSN 1470-160X, https://doi.org/10.1016/j.ecolind.2017.02.037.

Amador, E.S., Bandfield, J.L., Thomas, N.H. 2018. A search for minerals associated with serpentinization across Mars using CRISM spectral data. *Icarus*, Volume 311, Pages 113–134. ISSN 0019-1035, https://doi.org/10.1016/j.icarus.2018.03.021.

Aneece, I., and Thenkabail, P.S. 2018. Spaceborne Hyperspectral EO-1 Hyperion data pre-processing: Methods, approaches, and algorithms. Book Chapter 9 in Volume I: Introduction, Sensor Systems, Spectral Libraries, and Data Mining". In Book: "Hyperspectral Remote Sensing of Vegetation (Second Edition, 4 Volume Set). CRC Press- Taylor and Francis group, Boca Raton, London, New York. (Editors: Thenkabail, P.S., Lyon, J.G., and Huete, A.). (Editors: Thenkabail, P.S., Lyon, J.G., and Huete, A.). IP-091722

Anderson, M.C., Kustas, W.P., Norman, J.M., Hain, C.R., Mecikalski, J.R., Schultz, L., Gonzalez-Dugo, M.P. et al. 2011. Mapping daily evapotranspiration at field to continental scales using geostationary and polar orbiting satellite imagery. *Hyrol. Earth Systems Science*, Volume 15, Pages 223–239.

Asner, G., Nepstad, D., Cardinot, G., Ray, D. 2004. Drought stress and carbon uptake in an Amazon forest measured with spaceborne imaging spectroscopy. *Proceedings of the National Academy of Sciences of the United States of America*, Volume 101, Issue 16, Pages 6039–6044.

Asner, G.P., and Martin, R.E. 2008. Spectral and chemical analysis of tropical forests: Scaling from leaf to canopy levels. *Remote Sensing of Environment*. 112 (10): 3958–3970.

Baker, V.R., Hamilton, C.W., Burr, D.M., Gulick, V.C., Komatsu, G., Luo, W., Rice, J.W., Rodriguez, J.A.P. 2015. Fluvial geomorphology on Earth-like planetary surfaces: A review. *Geomorphology*, Volume 245, Pages 149–182. ISSN 0169-555X, https://doi.org/10.1016/j.geomorph.2015.05.002.

Bastiaanssen, W., Menenti, M., Feddes, R., Holtslag, A. 1998a. A remote sensing surface energy balance algorithm for land (SEBAL). 1. Formulation. *Journal of Hydrology*, Volume 212, Pages 198–212.

Bastiaanssen, W., Pelgrum, H., Wang, J., Ma, Y., Moreno, J., Roerink, G., Van der Wal, T., 1998b. A remote sensing surface energy balance algorithm for land (SEBAL): Part 2: Validation. *Journal of Hydrology*, Volume 212, Pages 213–229.

Bates, B.C., Kundzewicz, Z.W., Wu, S., Palutikof, J.P. (Eds.). 2008. *Climate Change and Water. Technical Paper of the Intergovernmental Panel on Climate Change*. Geneva: IPCC Secretariat, 210 pp.

Becker, B.L., Lusch, D.P., Qi, J. 2007. A classification-based assessment of the optimal spectral and spatial resolutions for Great Lakes coastal wetland imagery. *Remote Sensing of Environment*, Volume 108, Issue 1, Pages 111–120.

Behling, R., Bochow, M., Foerster, S., Roessner, S., Kaufmann, H. 2015. Automated GIS-based derivation of urban ecological indicators using hyperspectral remote sensing and height information. *Ecological Indicators*, Volume 48, Pages 218–234. ISSN 1470-160X, https://doi.org/10.1016/j.ecolind.2014.08.003.

Biard, F., Baret, F. 1997. Crop residue estimation using multiband reflectance. *Remote Sensing of Environment*, Volume 59, Pages 530–536.

Biggs, T.W., Petropoulos, G.P., Velpuri, N.M., Marshall, M., Glenn, E.P., Nagler, P., Messina, A. 2015. *Remote Sensing of Actual Evapotranspiration from Croplands. Remote Sensing of Water Resources, Disasters, and Urban Studies*. CRC Press, pp. Pages 59–99.

Biggs, T.W., Marshall, M., Messina, A. 2016. Mapping daily and seasonal evapotranspiration from irrigated crops using global climate grids and satellite imagery: Automation and methods comparison. *Water Resources Research*, Volume 52, Issue 9. http://doi.org/10.1002/2016WR019107.

Biradar, C.M., Thenkabail, P.S., Platonov, A., Xiangming, X., Geerken, R., Vithanage, J., Turral, H., Noojipady, P. 2008. Water productivity mapping methods using remote sensing. *Journal of Applied Remote Sensing*, Volume 2, Pages 023544.

Biradar, C.M., Thenkabail, P.S., Noojipady, P., Yuanjie, L., Dheeravath, V., Velpuri, M., Turral, H., Gumma, M.K. et al. 2009. A global map of rainfed cropland areas (GMRCA) at the end of last millennium using remote sensing. *International Journal of Applied Earth Observation and Geoinformation*, Volume 11, Issue 2, Pages 114–129. Doi: 10.1016/j.jag.2008.11.002.

Brown, S., Iverson, L.R., Prasad, A. 2001. *Geographical Distribution of Biomass Carbon in Tropical Southeast Asian Forests: A Database*. ORNL/CDIAC-119, NDP-068. Oak Ridge, Tennessee, U.S.A.: Carbon Dioxide Information Analysis Center, U.S. Department of Energy, Oak Ridge National Laboratory.

Cai, X.L., Thenkabail, P.S., Biradar, C., Platonov, A., Gumma, M., Dheeravath, V., Cohen, Y. et al. 2009. Water productivity mapping methods and protocols using remote sensing data of various resolutions to support "more crop per drop". *Journal of Applied Remote Sensing*, Volume 3, Issue 1, Pages 033557. Doi: 10.1117/1.3257643. Published October 12, 2009.

Campbell, T., Fearns, P. 2018. Simple remote sensing detection of Corymbia calophylla flowers using common 3–band imaging sensors. *Remote Sensing Applications: Society and Environment*. ISSN 2352-9385, https://doi.org/10.1016/j.rsase.2018.04.009.

Carvalho, S., Schlerf, M., van der Putten, W.H., Skidmore, A.K. 2013. Hyperspectral reflectance of leaves and flowers of an outbreak species discriminates season and successional stage of vegetation. *International Journal of Applied Earth Observation and Geoinformation*, Volume 24, Pages 32–41. ISSN 0303-2434, https://doi.org/10.1016/j.jag.2013.01.005.

Chander, G., Markham, B.L., Helder, D.L. 2009. Summary of current radiometric calibration coefficients for Landsat MSS, TM, ETM+, and EO-1 ALI sensors. *Remote Sensing of Environment*, Volume 113, Issue 5, Pages 893–903. ISSN 0034-4257, https://doi.org/10.1016/j.rse.2009.01.007.

Chen, J., Shen, M., Zhu, X., Tang, Y. 2009. Indicator of flower status derived from in situ hyperspectral measurement in an alpine meadow on the Tibetan Plateau. *Ecological Indicators*, Volume 9, Issue 4, Pages 818–823. ISSN 1470-160X, https://doi.org/10.1016/j.ecolind.2008.09.009.

Chen, F., Wang, K., de Voorde, T.V., Tang, T.F. 2017. Mapping urban land cover from high spatial resolution hyperspectral data: An approach based on simultaneously unmixing similar pixels with jointly sparse spectral mixture analysis. *Remote Sensing of Environment*, Volume 196, Pages 324–342. ISSN 0034-4257, https://doi.org/10.1016/j.rse.2017.05.014.

Chowdhury, E.H., Hassan, Q.K. 2015. Operational perspective of remote sensing-based forest fire danger forecasting systems. *ISPRS Journal of Photogrammetry and Remote Sensing*, Volume 104, Pages 224–236. ISSN 0924-2716, https://doi.org/10.1016/j.isprsjprs.2014.03.011.

Christian, B., Joshi, N., Saini, M., Mehta, N., Goroshi, S., Nidamanuri, R.R., Thenkabail, P., Desai, A.R., Krishnayya, N.S.R. 2015. Seasonal variations in phenology and productivity of a tropical dry deciduous forest from MODIS and Hyperion. *Agricultural and Forest Meteorology*, Volume 214–215, Pages 91–105. ISSN 0168-1923, https://doi.org/10.1016/j.agrformet.2015.08.246.

Cleemput, E.V., Vanierschot, L., Fernández-Castilla, B., Honnay, O., Somers, B. 2018. The functional characterization of grass- and shrubland ecosystems using hyperspectral remote sensing: Trends, accuracy and moderating variables. *Remote Sensing of Environment*, Volume 209, Pages 747–763. ISSN 0034-4257, https://doi.org/10.1016/j.rse.2018.02.030.

Corti, M., Gallina, P.M., Cavalli, D., Cabassi, G. 2017. Hyperspectral imaging of spinach canopy under combined water and nitrogen stress to estimate biomass, water, and nitrogen content. *Biosystems Engineering*, Volume 158, Pages 38–50. ISSN 1537-5110, https://doi.org/10.1016/j.biosystemseng.2017.03.006.

Costa dos Santos, J.F., Nunes Romeiro, J.M., de Assis, J.B., Pereira Torres, F.T., Gleriani, J.M. 2018. Potentials and limitations of remote fire monitoring in protected areas. *Science of The Total Environment*, Volume 616–617, Pages 1347–1355. ISSN 0048-9697, https://doi.org/10.1016/j.scitotenv.2017.10.182.

Daughtry, C.S.T., Hunt Jr., E.R., Doraiswamy, P.C., McMurtrey III, J.E. 2005. Remote sensing the spatial distribution of crop residues. *Agronomy Journal*, Volume 97, Pages 864–871.

Daughtry, C.S.T., Doraiswamy, P.C., Hunt, E.R., Stern, A.J., McMurtrey, J.E., Prueger, J.H. 2006. Remote sensing of crop residue cover and soil tillage intensity. *Soil and Tillage Research*, Volume 91, Issue 1–2, Pages 101–108. ISSN 0167-1987, https://doi.org/10.1016/j.still.2005.11.013.

Davidson, E.A., de Araujo, A.C., Artaxo, P., Balch, J.K., Brown, I.F., Bustamante, M.M.C., Coe, M.T. et al. 2012. The Amazon basin in transition. *Nature*, Volume 481, Pages 321–328.

Denny, E.G., Gerst, K.L., Miller-Rushing, A.J., Tierney, G.L., Crimmins, T.M., Enquist, C.A. et al. 2014. Standardized phenology monitoring methods to track plant and animal activity for science and resource management applications. *Volume International Journal of Biometeorology*, Volume 58, Isuse 4, Pages 591–601.

DeFries, R.S., Houghton, R.A., Hansen, M.C., Field, C.B., Skole, D., Townshend, J. 2004. Carbon emissions from tropical deforestation and regrowth based on satellite observations for the 1980s and 1990s. *Proceedings of the National Academy of Sciences U S A*. 2002 October 29, Volume 99, Issue 22, Pages 14256–14261.

de Moura, Y.M., Galvão, L.S., Hilker, T., Wu, J., Saleska, S., do Amaral, C.H., Nelson, B.W. et al. 2017. Spectral analysis of amazon canopy phenology during the dry season using a tower hyperspectral camera and modis observations. *ISPRS Journal of Photogrammetry and Remote Sensing*, Volume 131, Pages 52–64. ISSN 0924-2716, https://doi.org/10.1016/j.isprsjprs.2017.07.006.

Dennison, P.E., Charoensiri, K., Roberts, D.A., Peterson, S.H., Green, R.O. 2006. Wildfire temperature and land cover modeling using hyperspectral data. *Remote Sensing of Environment*, Volume 100, Pages 212–222.

Dennison, P.E., Matheson, D.S. 2011. Comparison of fire temperature and fractional area modeled from SWIR, MIR, and TIR multispectral and SWIR hyperspectral airborne data. *Remote Sensing of Environment*, Volume 115, Pages 876–886.

Dixon, R.K., Solomon, A.M., Brown, S., Houghton, R.A., Trexier, M.C., Wisniewski, J. 1994. Carbon pools and flux of global forest ecosystems. *Science*, Volume 263, Issue 5144, Pages 185–190.

FAO. 2008. Food and Agriculture Organization (Content source); Jim Kundell (Topic Editor). 2008. *Water profile of Ghana. In: Encyclopedia of Earth*. Cleveland, C.J. (Ed.), Washington, D.C.: Environmental Information Coalition, National Council for Science and the Environment.

Fava, F., Colombo, R., Bocchi, S., Meroni, M., Sitzia, M., Fois, N., Zucca, C. 2009. Identification of hyperspectral vegetation indices for Mediterranean pasture characterization. *International Journal of Applied Earth Observation and Geoinformation*, Volume 11, Issue 4, Pages 233–243. ISSN 0303-2434, https://doi.org/10.1016/j.jag.2009.02.003.

Ferreira, M.E., Ferreira, L.G., Sano, E.E., Shimabukuro, Y.E. 2007. Spectral linear mixture modelling approaches for land cover mapping of tropical savanna areas in Brazil. *International Journal of Remote Sensing*, Volume 28, Pages 413–429.

Fuller, L.M., Morgan, T.R., Aichele, S.S. 2006. Wetland delineation with IKONOS high-resolution satellite imagery, Fort Custer Training Center, Battle Creek, Michigan, 2005. (accessed in 03 Mar 2006). Scientific Investigations Report. United States Geological Survey, p. 20.

Gilmore, M.S., Wilson E.H., Barrett, N., Civco, D.L., Prisloe, S., Hurd, J.D., Chadwick, C. 2008. Integrating multi-temporal spectral and structural information to map wetland vegetation in a lower Connecticut River tidal marsh. *Remote Sensing of Environment*, Volume 112, Issue 11, Pages 4048–4060.

Gomez, C., Lagacherie, P., Coulouma, G. 2008. Continuum removal versus PLSR method for clay and calcium carbonate content estimation from laboratory and airborne hyperspectral measurements. *Geoderma*, Volume 148, Issue 2, Pages 141–148. ISSN 0016-7061, https://doi.org/10.1016/j.geoderma.2008.09.016.

Gorroño, J., Banks, A.C., Fox, N.P., Underwood, C. 2017. Radiometric inter-sensor cross-calibration uncertainty using a traceable high accuracy reference hyperspectral imager. *ISPRS Journal of Photogrammetry and Remote Sensing*, Volume 130, Pages 393–417. ISSN 0924-2716, https://doi.org/10.1016/j.isprsjprs.2017.07.002.

Goswami, J.N., Annadurai, M. 2008. Chandrayaan-1 mission to the Moon. *Acta Astronautica*, Volume 63, Issue 11–12, Pages 1215–1220. ISSN 0094-5765.

Goward, S.N., Chander, G., Pagnutti, M., Marx, A., Ryan, R., Thomas, N., Tetrault, R. 2012. Complementarity of ResourceSat-1 AWiFS and Landsat TM/ETM+ sensors. *Remote Sensing of Environment*, Volume 123, Pages 41–56, http://digitalcommons.unl.edu/usgsstaffpub/523/ or https://pubs.er.usgs.gov/publication/70043219.

Gross, J., Heumann, B. 2014. Can flowers provide better spectral discrimination between herbaceous wetland species than leaves?. *Remote Sensing Letters*, Volume 5, Issue 10, Pages 892–901. Doi: 10.1080/2150704X.2014.973077.

Hansen, D., Uphoff, N., Lal, R. 2002. *Food Security and Environmental Quality in the Developing World*. Boca Raton, FL: CRC Press, p. 480.

Hansen, M.C., Potapov, P. V., Moore, R., Hancher, M., Turubanova, S.A., Tyukavina, A., Thau, D. et al. 2013. High-resolution global maps of 21st-century forest cover change. *Science*, Volume 342, Pages 850–853. Data available on-line from: http://earthenginepartners.appspot.com/science-2013-global-forest.

Harvey, K.R., Hill, G.J.E. 2001. Vegetation mapping of a tropical freshwater swamp in the Northern Territory, Australia: A comparison of aerial photography, Landsat TM and SPOT satellite imagery. *International Journal of Remote Sensing*, Taylor & Francis, Volume 22, Pages 2911–2925.

Hirano, A., Madden, M., Welch, R. 2003. Hyperspectral image data for mapping wetland vegetation. *BioOne Journal*, Volume 23, Issue 2.

Hopping, K.A., Yeh, E.T., Gaerrang, Harris, R.B. 2018. Linking people, pixels, and pastures: A multi-method, interdisciplinary investigation of how rangeland management affects vegetation on the Tibetan Plateau. *Applied Geography*, Volume 94, Pages 147–162. ISSN 0143-6228, https://doi.org/10.1016/j.apgeog.2018.03.013.

Houghton, R.A. 2008. Carbon flux to the atmosphere from land-use changes: 1850–2005. In: *TRENDS: A Compendium of Data on Global Change*. Oak Ridge, Tenn., U.S.A.: Carbon Dioxide Information Analysis Center, Oak Ridge National Laboratory, U.S. Department of Energy.

Houghton, R.A. 2017. Tropical forests are a net carbon source based on aboveground measurements of gain and loss. *Science*, Volume 358.

Houghton, R.A., Birdsey, R.A., Nassikas, A., McGlinchey, D. 2017. Forests and land use: Undervalued assets for global climate stabilization. In: *Policy Brief.* Baccini, A., Walker, W., Carlvalho, L., Farina, M., Sulla-Menashe, D., Houghton, R.A. (Eds.), Falmouth: Woods Hole Research Center.

Hu, Y. 2011. Vegetation stress level monitoring in mine area based on hj-1 hyperspectral data. *Master's thesis*, Qingdao: Shandong University of Science and Technology, p. 57.

Inoue, Y., Sakaiya, E., Zhu, Y., Takahashi, W. 2012. Diagnostic mapping of canopy nitrogen content in rice based on hyperspectral measurements. *Remote Sensing of Environment*, Volume 126, Pages 210–221. ISSN 0034-4257, https://doi.org/10.1016/j.rse.2012.08.026.

IPCC. 2007. Climate Change 2007: Synthesis report. In: *Contribution of Working Groups I, II and III to the Fourth Assessment Report of the Intergovernmental Panel on Climate Change.* Core Writing Team, Pachauri, R.K., Reisinger, A. (Eds.). Geneva, Switzerland: IPCC, 104 pp.

Islam, M.A., Thenkabail, P.S., Kulawardhana, R.W., Alankara, R., Gunasinghe, S., Edussriya, C., Gunawardana, A. 2008. Semi-automated methods for mapping wetlands using Landsat ETM+ and SRTM data. *International Journal of Remote Sensing*, Volume 29, Issue 24, Pages 7077–7106.

Jafari, R., Lewis, M.M. 2012. Arid land characterisation with EO-1 Hyperion hyperspectral data. *International Journal of Applied Earth Observation and Geoinformation*, Volume 19, Pages 298–307. ISSN 0303-2434, https://doi.org/10.1016/j.jag.2012.06.001.

Jensen, J.R., Rutchey, K., Koch, M.S., Narumalani, S. 2002. Inland wetland change detection in the Everglades Water Conservation area 2A using a time series of normalized remotely sensed data. *Phtotogrammetric Engineering and Remote Sensing*, Volume 61, Issue 2, Pages 199–209.

Jakimow, B., Griffiths, P., van der Linden, S., Hostert, P. 2018. Mapping pasture management in the Brazilian Amazon from dense Landsat time series. *Remote Sensing of Environment*, Volume 205, Pages 453–468. ISSN 0034-4257, https://doi.org/10.1016/j.rse.2017.10.009.

Jones, K., Lanthier, Y., Voet, P.V.D., Valkengoed, E.V., Taylor, D., Fernández-Prieto, D. 2009. Monitoring and assessment of wetlands using Earth Observation: The GlobWetland project. *Journal of Environmental Management.*, Volume 90, Issue 7, Pages 2154–2169, Doi: 10.1016/j.jenvman.2007.07.037.

Kayitakire, F., Hamel, C., Defourny, P. 2006. Retrieving forest structure variables based on image texture analysis and IKONOS-2 imagery. *Remote Sensing of Environment*, Volume 102, Issues 3–4, Pages 390–401. ISSN 0034-4257, https://doi.org/10.1016/j.rse.2006.02.022.

Kulawardhana, R.W., Thenkabail, P.S., Vithanage, J., Biradar, C., Islam, Md.A., Gunasinghe, S., Alankara, R. 2007. Evaluation of the Wetland Mapping Methods using Landsat ETM+ and SRTM Data. *Journal of Spatial Hydrology*, Volume 7, Issue 2, Pages 62–96. ISSN 1530-4736.

Lan, Z., Zhang, D., 2006. Study on optimization-based layered classification for separation of wetlands. *International journal of remote sensing* 27, 1511–1520.

Lausch, A., Salbach, C., Schmidt, A., Doktor, D., Merbach, I., Pause, M. 2015. Deriving phenology of barley with imaging hyperspectral remote sensing. *Ecological Modelling*, Volume 295, Pages 123–135. ISSN 0304-3800, https://doi.org/10.1016/j.ecolmodel.2014.10.001.

Le Quéré, C., Andrew, R.M., Canadell, J.G., Sitch, S., Korsbakken, J.I., Peters, G.P., Hauck, V.H. et al. 2016. Global carbon budget. *Earth Systems Science Data*, Volume 8, Pages 605–649.

Li, M., Liu, X., Liu, M. 2010. Fuzzy neural network model for predicting stress levels in rice fields polluted with heavy metals using hyperspectral data. *Acta Scientifica Circum.*, Volume 30, Issue 10, Pages 2108–2115.

Li, N., Lue, J., Altermann, W. 2010. Applications of spectral analysis to monitoring of heavy metal-induced contamination in vegetation. *Spectrosc. Spect. Anal.*, Volume 30, Pages 2508–2511.

Li, L., Wang, S., Ren, T., Wei, Q., Ming, J., Li, J., Li, X., Cong, R., Lu, J. 2018. Ability of models with effective wavelengths to monitor nitrogen and phosphorus status of winter oilseed rape leaves using in situ canopy spectroscopy. *Field Crops Research*, Volume 215, Pages 173–186. ISSN 0378-4290, https://doi.org/10.1016/j.fcr.2017.10.018.

Liu, Y., Chen, H., Wu, G., Wu, X. 2010. Feasibility of estimating heavy metal concentrations in Phragmites australis using laboratory-based hyperspectral data—A case study along Le"an River, China. *International Journal of Applied Earth Observation and Geoinformation*, Volume 12, Supplement 2, Pages S166–S170. ISSN 0303-2434, https://doi.org/10.1016/j.jag.2010.01.003.

Liu, H., Wu, C. 2018. Crown-level tree species classification from AISA hyperspectral imagery using an innovative pixel-weighting approach. *International Journal of Applied Earth Observation and Geoinformation*, Volume 68, Pages 298–307. ISSN 0303-2434, https://doi.org/10.1016/j.jag.2017.12.001.

Lopez, R., Lyon, J., Lyon, L., Lopez, D. 2013. *Wetland Landscape Characterization: Practical Tools, Methods, and Approaches for Landscape Ecology, Second Edition*, Boca Raton, FL: CRC Press, p. 308.

Lu, D. 2006. The potential and challenge of remote sensing based biomass estimation. *International Journal of Remote Sensing*, Volume 27, Pages 1297–1328. [your library's links] [your library's links] [informaworld].
Lunetta, R.S., Balogh, M.E., Merchant, J.W. 1999. Application of multi-temporal Landsat 5 TM imagery for wetland identification. *Photogrammetric Engineering and Remote Sensing*, Volume 65, Pages 1303–1310.
Lyon, J. 2001. *Wetland Landscape Characterization: Techniques and Applications for GIS Mapping. Remote Sensing, and Image Analysis.* CRC Press, p. 135.
Lyon, J., Lyon, L. 2011. *Practical handbook for wetland identification and delineation.* CRC Press, p. 208.
Lyon, J., McCarthy, J. 1995. *Wetland and Environmental Applications of GIS.* CRC Press, p. 400.
Malhi, Y. Wright, J. 2004. Spatial patterns and recent trends in the climate of tropical rainforest regions. *Philosophical Transactions of the Royal Society of London B Biological Sciences*, Volume 359, Issue 1443, Pages 311–329.
Marshall, M.T., Thenkabail, P.S. 2014. Biomass modeling of four leading World crops using hyperspectral narrowbands in support of HyspIRI mission. *Photogrammetric Engineering and Remote Sensing*, Volume 80, Issue 4, Pages 757–772. IP-052043.
Marshall M.T., Thenkabail, P. 2015a. Developing in situ non-destructive estimates of crop biomass to address issues of scale in remote sensing. *Remote Sensing*, Volume 7, Issue 1, Pages 808–835. Doi: 10.3390/rs70100808. IP-060652.
Marshall, M.T., Thenkabail, P.S. 2015b. Advantage of hyperspectral EO-1 Hyperion over multispectral IKONOS, GeoEye-1, WorldView-2, Landsat ETM+, and MODIS vegetation indices in crop biomass estimation. *International Society of Photogrammetry and Remote Sensing (ISPRS) Journal of Photogrammetry and Remote Sensing (ISPRS P&RS)*, Volume 108, Pages 205–218. http://dx.doi.org/10.1016/j.isprsjprs.2015.08.001. IP-060745.
Marshall, M.T., Thenkabail, P.S., Biggs, T., Post, K. 2016. Hyperspectral narrowband and multispectral broadband indices for remote sensing of crop evapotranspiration and its components (transpiration and soil evaporation). *Agricultural and Forest Meteorology*, Volume 218–219, Pages 122–134. IP-065032.
Mariotto I., Gutschick V., Clason D.L. 2011. Mapping evapotranspiration from ASTER data through GIS spatial integration of vegetation and terrain features. *Photogrammetric Engineering & Remote Sensing (PE&RS)*, Volume 77, Issue 5, Pages 483–493.
Marshall M., Thenkabail P. 2014. Biomass modeling of four leading world crops using hyperspectral narrowbands in support of HyspIRI Mission. *Photogrammetric Eng. Remote Sens.*, Volume 80, pages 757–772.
Marshall M., Thenkabail P. 2015. Advantage of hyperspectral EO-1 Hyperion over multispectral IKONOS, GeoEye-1, WorldView-2, Landsat ETM+, and MODIS vegetation indices in crop biomass estimation. *ISPRS J. Photogramm. Remote Sens.*, Volume 108, pages 205–218.
Mariotto, I., Thenkabail, P.S., Huete, H., Slonecker, T., Platonov, A. 2013. Hyperspectral versus multispectral crop- biophysical modeling and type discrimination for the HyspIRI mission. *Remote Sensing of Environment*, Volume 139, Pages 291–305. IP-049224.
Matheson, D.S., Dennison, P.E. 2012. Evaluating the effects of spatial resolution on hyperspectral fire detection and temperature retrieval. *Remote Sensing of Environment*, Volume 124, Pages 780–792. ISSN 0034-4257, https://doi.org/10.1016/j.rse.2012.06.026.
May, D., Wang, J., Kovacs, J., Muter, M. 2003. Mapping wetland extent using IKONOS satellite imagery of the O'donnell point region, Georgian Bay, *Ontario.Proc. 25th Canadian Symp.* Remote Sensing, Canadian Aeronautics and Space Institute: CD-9pp.
McNairn, H., Protz, R. 1993. Mapping corn residue cover on agricultural fields in Oxford County, Ontario, using Thematic Mapper. *Canadian Journal of Remote Sensing*, Volume 19, Pages 152–159.
Melesse, A.M., Weng, Q., Thenkabail, P., Senay, G. 2007. Remote sensing sensors and applications in environmental resources mapping and modelling. Special issue of remote sensing of natural resources and the environment. *Sensors Journal*, Volume 7, Pages 3209–3241. http://www.mdpi.org/sensors/papers/s7123209.pdf.
Mulla, D.J. 2013. Twenty five years of remote sensing in precision agriculture: Key advances and remaining knowledge gaps. *Biosystems Engineering*, Volume 114, Issue 4, Pages 358–371. ISSN 1537-5110, https://doi.org/10.1016/j.biosystemseng.2012.08.009.
Nagler, P.L., Inoue, Y., Glenn, E.P., Russ, A.L., Daughtry, C.S.T. 2003. Cellulose absorption index (CAI) to quantify mixed soil–plant litter scenes. *Remote Sensing of Environment*, Volume 87, Issues 2–3, Pages 310–325. ISSN 0034-4257, https://doi.org/10.1016/j.rse.2003.06.001.
Näsi, R., Honkavaara, E., Blomqvist, M., Lyytikäinen-Saarenmaa, P., Hakala, T., Viljanen, N., Kantola, T., Holopainen, M. 2018. Remote sensing of bark beetle damage in urban forests at individual tree level using a novel hyperspectral camera from UAV and aircraft, Urban Forestry & Urban Greening, Volume 30, Pages 72–83. ISSN 1618-8667, https://doi.org/10.1016/j.ufug.2018.01.010.

Nepstad, D.C., Stickler, C.M., Soares-Filho, B., Merry, F. 2008. Interactions among Amazon land use, forests, and climate: Prospects for a near-term forest tipping point. *Philosophical Transactions Royal Society*, Volume 363, Pages 1737–1746.

Ozesmi, S.L., Bauer, M.E. 2002. Satellite remote sensing of wetlands. *Wetlands Ecology and Management*, Volume 10, Pages 381–402. (stacy did her master's in SENR/OSU and took classes from Lyon).

Pasqualotto, N., Delegido, J., Wittenberghe, S.V., Verrelst, J., Rivera, J.P., Moreno, J. 2018. Retrieval of canopy water content of different crop types with two new hyperspectral indices: Water Absorption Area Index and Depth Water Index. *International Journal of Applied Earth Observation and Geoinformation*, Volume 67, Pages 69–78. ISSN 0303-2434, https://doi.org/10.1016/j.jag.2018.01.002.

Pasquarella, V.J., Holden, C.E., Woodcock, C.E. 2018. Improved mapping of forest type using spectral-temporal Landsat features. *Remote Sensing of Environment*, Volume 210, Pages 193–207. ISSN 0034-4257, https://doi.org/10.1016/j.rse.2018.02.064.

Pastor-Guzman, J., Dash, J., Atkinson, P.M. 2018. Remote sensing of mangrove forest phenology and its environmental drivers. *Remote Sensing of Environment*, Volume 205, Pages 71–84. ISSN 0034-4257, https://doi.org/10.1016/j.rse.2017.11.009.

Pelley, J. 2008. Can wetland restoration cool the planet? *Environmental Science and Technology*, Volume 42, Issue 24, Pages 8994.

Peng, D., Wu, C., Zhang, X., Yu, L., Huete, A.R., Wang, F., Luo, S., Liu, X., Zhang, H. 2018. Scaling up spring phenology derived from remote sensing images. *Agricultural and Forest Meteorology*, Volumes 256–257, Pages 207–219. ISSN 0168-1923.

Phillips, O.L., Malhi, Y., Higuchi, N., Laurance, W.F., Nunez, P.V., Vasquez, R.M., Laurance, S.G. et al. 1998. Changes in the carbon balance of tropical forests: Evidence from long-term plots. *Science*, Volume 282, Pages 439–442.

Phillips, O.L. 2009. Drought Sensitivity of the Amazon Rainforest. *SCIENCE* 323: 1344–1347

Pignatti, S., Cavalli, R.M., Cuomo, V., Fusilli, L., Pascucci, S., Poscolieri, M., Santini, F. 2009. Evaluating Hyperion capability for land cover mapping in a fragmented ecosystem: Pollino National Park, Italy. *Remote Sensing of Environment*, Volume 113, Issue 3, Pages 622–634.

Platonov, A., Thenkabail, P.S., Biradar, C., Cai, X., Gumma, M., Dheeravath, V., Cohen, Y. et al. 2008. Water productivity mapping (WPM) using landsat ETM+ data for the irrigated croplands of the syrdarya river basin in Central Asia. *Sensors Journal*, Volume 8, Issue 12, Pages 8156–8180. Doi: 10.3390/s8128156. http://www.mdpi.com/1424-8220/8/12/8156/pdf.

Pullanagari, R.R., Kereszturi, G., Yule, I.J. 2016. Mapping of macro and micro nutrients of mixed pastures using airborne AisaFENIX hyperspectral imagery. *ISPRS Journal of Photogrammetry and Remote Sensing*, Volume 117, Pages 1–10. ISSN 0924-2716, https://doi.org/10.1016/j.isprsjprs.2016.03.010.

Ramirez, F.J.R., Rafael, M., Navarro-Cerrillo, M., Varo-Martínez, A., Quero, J.L., Doerr, S., Hernández-Clemente, R. 2018. Determination of forest fuels characteristics in mortality-affected Pinus forests using integrated hyperspectral and ALS data. *International Journal of Applied Earth Observation and Geoinformation*, Volume 68, Pages 157–167. ISSN 0303-2434, https://doi.org/10.1016/j.jag.2018.01.003.

Ramsar, 2013. Prof. G. V. T. Matthews, The Ramsar Convention on Wetlands: its History and Development (Ramsar, 1993), PDF version, re-issued 2013.

Ramsey III, E., Nelson, G., Sapkota, S. 1998. Classifying coastal resources by integrating optical and radar imagery and color infrared photography. *Mangroves and Salt Marshes*, Volume 2, Issue 2, Pages 109–119.

Rebelo, L.-M., Finlayson, C.M., Nagabhatla, N. 2009. Remote sensing and GIS for wetland inventory, mapping and change analysis. *Journal of Environmental Management* 90, 2144–2153.

Rebelo, A.J., Ben Somers, Karen J. Esler, Patrick Meire, 2018. Can wetland plant functional groups be spectrally discriminated? *Remote Sensing of Environment,* Volume 210, Pages 25–34, ISSN 0034-4257, https://doi.org/10.1016/j.rse.2018.02.031.

Ren, H.-Y., Zhuang, D., Singh, A.N., Pan, J., Qiu, D., Shi, R. 2009. Estimation of As and Cu contamination in agricultural soils around a mining area by reflectance spectroscopy; a case study. *Pedosphere*, Volume 19, Issue 6, Pages 719–726.

Ren, J., Juan Chen, Lei Han, Mei Wang, Bin Yang, Ping Du, Fasheng Li, 2018. Spatial distribution of heavy metals, salinity and alkalinity in soils around bauxite residue disposal area. *Science of The Total Environment*, Volumes 628–629, Pages 1200–1208, ISSN 0048-9697, https://doi.org/10.1016/j.scitotenv.2018.02.149.

Rights and Resources Initiative. 2018. Uncertainty and opportunity: The status of forest carbon rights and governance frameworks in over half of the world's tropical forests. https://rightsandresources.org/wp-content/uploads/2018/03/EN_Status-of-Forest-Carbon-Rights_RRI_Mar-2018.pdf.

Rosso, P.H., Pushnik, J.C., Lay, M., Ustin, S.L. 2005. Reflectance properties and physiological responses of *Salicornia virginica* to heavy metal and petroleum contamination. *Environmental Pollutyion*, Volume 137, Issue 2, Pages 241–252.

Sader, S.A., Ahl, D., Liou, W.-S. 1995. Accuracy of Landsat-TM and GIS rule-based methods for forest wetland classification in Maine. *Remote Sensing of Environment* 53, 133–144.

Salmon, J.M., Friedl, M.A., Frolking, S., Wisser, D., Douglas, E.M. 2015. Global rain-fed, irrigated, and paddy croplands: A new high resolution map derived from remote sensing, crop inventories and climate data. *International Journal of Applied Earth Observation and Geoinformation*, Volume 38, Pages 321–334. ISSN 0303-2434, https://doi.org/10.1016/j.jag.2015.01.014.

Schowengerdt, R.A. 2007. *Remote Sensing: Models and Methods for Image Processing*. San Diego, California, USA: Academic press (Elsevier), p. 509.

Senay, G.B., Bohma, S., Singh, R.K., Gowda, P.H., Velpuri, N.M., Alemu, H., Verdin, J.P. 2013. Operational evapotranspiration mapping using remote sensing and weather datasets: A new parameterization for the SSEB approach. *JAWRA Journal of the American Water Resources Association*, Volume 49, Issue 3: Paper No. JAWRA-12-0097-P.

Serbin, G., Hunt, E.R., Daughtry, C.S.T., McCarty, G., Doraiswamy, P. 2009. An improved ASTER index for remote sensing of crop residue. *Remote Sensing*, Volume 1, Pages 971–991.

Shi, T., Liu, H., Chen, Y., Wang, J., Wu, G. 2016. Estimation of arsenic in agricultural soils using hyperspectral vegetation indices of rice. *Journal of Hazardous Materials*, Volume 308, Pages 243–252. ISSN 0304-3894, https://doi.org/10.1016/j.jhazmat.2016.01.022.

Slonecker, T., Haack, B. Price, S. 2009. Spectroscopic analysis of arsenic uptake in Pteris ferns. *Remote Sensing*, Volume 1, Issue 4, Pages 644–675.

Slonecker, E.T., Allen, D.W., Resmini, R.G., Rand, R.S., Paine, E. 2018. Full-range, solar-reflected hyperspectral microscopy to support earth remote sensing research. *Journal of Applied Sensing*, Volume 12, Issue 2, Pages 026024. Doi: 10.1117/1.JRS.12.026024.

Smith, R., Adams, M., Maier, S., Craig, R., Kristina, A., Maling, I. 2007. Estimating the area of stubble burning from the number of active fires detected by satellite. *Remote Sensing of Environment*, Volume 109, Issue 1, Pages 95–106. ISSN 0034-4257, https://doi.org/10.1016/j.rse.2006.12.011.

Stagakis, S., Vanikiotis, T., Sykioti, O. 2016. Estimating forest species abundance through linear unmixing of CHRIS/PROBA imagery. *ISPRS Journal of Photogrammetry and Remote Sensing*, Volume 119, Pages 79–89. ISSN 0924-2716, https://doi.org/10.1016/j.isprsjprs.2016.05.013.

Sullivan, D.G., Truman, C.C., Schomberg, H.H., Endale, D.M., Strickland, T.C. 2006. Evaluating techniques for determining tillage regime in the Southeastern Coastal Plain and Piedmont. *Agronomy Journal*, Volume 98, Pages 1236–1246.

Teixeira, A.de.C., Hernandez, F.B.T., Scherer-Warren, M., Andrade, R.G., Leivas, J.F., Victoria, D.C., Bolfe, E.L., Thenkabail, P.S., Franco, R.A.M. 2015. Water productivity studies from earth observation data: Characterization, modeling and mapping water use and water productivity, chapter 4. In: *Remote Sensing Handbook" (Volume III): Remote Sensing of Water Resources, Disasters, and Urban Studies*. Thenkabail, P.S. (Editor-in-Chief), Boca Raton, London, New York: Taylor and Francis Inc./CRC Press, pp. 101–128. ISBN 9781482217919—CAT# K22128. IP-058357.

Teluguntla, P., Thenkabail, P.S., Xiong, J., Gumma, M.K., Congalton, R.G., Oliphant, A., Poehnelt, J., Yadav, K., Rao, M., Massey, R. 2017. Spectral matching techniques (SMTs) and automated cropland classification algorithms (ACCAs) for mapping croplands of Australia using MODIS 250-m time-series (2000–2015) data. *International Journal of Digital Earth*. Doi: 10.1080/17538947.2016.1267269. IP-074181, http://dx.doi.org/10.1080/17538947.2016.1267269.

Teluguntla, P., Thenkabail, P.S., Oliphant, A., Xiong, J., Gumma, M.K. 2018. A 30-m Landsat-derived cropland extent product of australia and china using random forest machine learning algorithm on google earth engine cloud computing platform. *International Journal of Photogrammetry and Remote Sensing (ISPRS) Journal of Photogrammetry and Remote Sensing*, Volume 144, Pages 325–340. https://doi.org/10.1016/j.isprsjprs.2018.07.017 In press.

Thenkabail, P.S., Smith, R.B., Pauw, E.D. 2000. Hyperspectral vegetation indices and their relationships with agricultural crop characteristics. *Remote Sensing of Environment*, Volume 71, Issue 2, Pages 158–182. ISSN 0034-4257, https://doi.org/10.1016/S0034-4257(99)00067-X.

Thenkabail P.S., Smith, R.B., De-Pauw, E. 2000b. Hyperspectral vegetation indices for determining agricultural crop characteristics. *Remote Sensing of Environment*, Volume 71, Pages 158–182.

Thenkabail P.S., Smith, R.B., De-Pauw, E. 2002a. Evaluation of narrowband and broadband vegetation indices for determining optimal hyperspectral wavebands for agricultural crop characterization. *Photogrammetric Engineering and Remote Sensing*, Volume 68, Issue 6, Pages 607–621.

Thenkabail, P.S. 2003. Biophysical and yield information for precision farming from near-real-time and historical Landsat TM images. *International Journal of Remote Sensing*, Volume 24, Issue 14, Pages 2879–2904. Doi: 10.1080/01431160710155974.

Thenkabail, P.S. 2004. Inter-sensor relationships between IKONOS and Landsat-7 ETM+ NDVI data in three ecoregions of Africa. *International Journal for Remote Sensing*, Volume 25, Issue 2, Pages 389–408.

Thenkabail, P., Enclona, A., Ashton, M., Legg, C., De Dieu, M. J. 2004. Hyperion, IKONOS, ALI, and ETM+ sensors in the study of African rainforests. *Remote Sensing of Environment*, Volume 90, Issue 1, Pages 23–43. https://doi.org/10.1016/j.rse.2003.11.018

Thenkabail, P.S., Enclona, E.A., Ashton, M.S., Legg, C., Jean De Dieu, M. 2004a. Hyperion, IKONOS, ALI, and ETM+ sensors in the study of African rainforests. *Remote Sensing of Environment*, Volume 90, Pages 23–43.

Thenkabail, P.S., Enclona, E.A., Ashton, M.S., Van Der Meer, V. 2004b. Accuracy assessments of hyperspectral waveband performance for vegetation analysis applications. *Remote Sensing of Environment*, Volume 91, Issue 2–3, Pages 354–376.

Thenkabail, P.S., Biradar, C.M., Noojipady, P., Dheeravath, V., Li, Y.J., Velpuri, M., Gumma, M. et al. 2009a. Global irrigated area map (GIAM), derived from remote sensing, for the end of the last millennium. *International Journal of Remote Sensing*, Volume 30, Issue 14, Pages 3679–3733.

Thenkabail P.S., Lyon, J., Turral, H., Biradar, C.M. 2009b. *Book entitled: Remote Sensing of Global Croplands for Food Security*. Boca Raton, London, New York: CRC Press- Taylor and Francis Group, p. 556. (48 pages in color).

Thenkabail P.S., Wu Z. 2012. An automated cropland classification algorithm (ACCA) for Tajikistan by combining landsat, MODIS, and secondary data. *Remote Sensing*, Volume 4, Issue 10, Pages 2890–2918. (65%). Download the paper @ this link: http://www.mdpi.com/2072-4292/4/10/2890. ACCA algorithm at this link: http://www.sciencebase.gov/catalog/folder/4f79f1b7e4b0009bd827f548. IP-035313.

Thenkabail P.S., Knox J.W., Ozdogan, M., Gumma, M.K., Congalton, R.G., Wu, Z. et al. 2012. Assessing future risks to agricultural productivity, water resources and food security: How can remote sensing help?. *Photogrammetric Engineering and Remote Sensing*, August 2012 Special Issue on Global Croplands: Highlight Article, Volume 78, Issue 8, Pages 773–782. IP-035587.

Thenkabail, P.S., Mariotto, I., Gumma, M.K., Middleton, E.M., Landis, D.R., Huemmrich, F.K. 2013. Selection of hyperspectral narrowbands (HNBs) and composition of hyperspectral twoband vegetation indices (HVIs) for biophysical characterization and discrimination of crop types using field reflectance and Hyperion/EO-1 data. *IEEE Journal of Selected Topics in Applied Earth Observations and Remote Sensing*, Volume 6, Issue 2, Pages 427–439. Doi: 10.1109/JSTARS.2013.2252601. (80%). IP-037139.

Thenkabail, P.S., Enclona, E.A., Ashton, M.S., Legg, C., Dieu, M.J.D. 2014a. Hyperion, IKONOS, ALI, and ETM+ sensors in the study of African rainforests. *Remote Sensing of Environment*, Volume 90, Issue 1, Pages 23–43. ISSN 0034-4257, https://doi.org/10.1016/j.rse.2003.11.018.

Thenkabail, P.S., Gumma, M.K., Teluguntla, P., Mohammed, I.A. 2014b. Hyperspectral remote sensing of vegetation and agricultural crops. Highlight article. *Photogrammetric Engineering and Remote Sensing*, Volume 80, Isse 4, Pages 697–709. IP-052042.

Thenkabail, P.S. 2015. Hyperspectral remote sensing for terrestrial applications, chapter 9. In: *"Remote Sensing Handbook" (Volume II): Land Resources Monitoring, Modeling, and Mapping with Remote Sensing*. Thenkabail, P.S., (Editor-in-Chief). Boca Raton, London, New York: Taylor and Francis Inc./CRC Press, pp. 201–236. ISBN 9781482217957—CAT# K22130. IP-0606312.

Tiner, R.W. 2009. *Global istribution of Wetlands*. Encyclopedia of Inland Waters, pp. 526–530.

Toniol, A.C., Galvão, L.S., Ponzoni, F.J., Sano, E.E., de Jesus Amore, D. 2017. Potential of hyperspectral metrics and classifiers for mapping Brazilian savannas in the rainy and dry seasons. *Remote Sensing Applications: Society and Environment*, Volume 8, Pages 20–29. ISSN 2352-9385, https://doi.org/10.1016/j.rsase.2017.07.004.

Töyrä, J., Pietroniro, A. 2005. Towards operational monitoring of a northern wetland using geomatics-based techniques. *Remote Sensing of Environment* 97, 174–191.

Trifonov, G.M., Zhizhin, M.N., Melnikov, D.V., Poyda, A.A. 2017. VIIRS nightfire remote sensing volcanoes. *Procedia Computer Science*, Volume 119, Pages 307–314. ISSN 1877-0509, https://doi.org/10.1016/j.procs.2017.11.189.

UNFCCC. 2008. *UNFCCC (United Nations Framework Convention on Climate Change), 2008*. Report of the Conference of the Parties on its thirteenth session, held in Bali from 3 to 15 December 2007. Addendum, Part 2. Document FCCC/CP/2007/6/Add.1. Bonn, Germany: UNFCCC.

Union of Concerned Scientists, 2013. Access at: www.ucsusa.org/palmoilfacts

Vadrevu, K.P., Ellicott, E., Badarinath, K.V.S., Vermote, E. 2011. MODIS derived fire characteristics and aerosol optical depth variations during the agricultural residue burning season, north India. *Environmental Pollution*, Volume 159, Issue 6, Pages 1560–1569. ISSN 0269-7491, https://doi.org/10.1016/j.envpol.2011.03.001.

Van Deventer, A., Ward, D., Gowda, Lyon, J. 1997. Using Thematic Mapper data to identify contrasting soil plains and tillage practices. *Photogrammetric Engineering and Remote Sensing*, Volume 63, Pages 87–93.

Van Leeuwen, W.J.D., Barron J. Orr, Stuart E. Marsh, Stefanie M. Herrmann, Multi-sensor NDVI data continuity: Uncertainties and implications for vegetation monitoring applications. *Remote Sensing of Environment*, Volume 100, Issue 1, 2006, Pages 67–81, ISSN 0034-4257, https://doi.org/10.1016/j.rse.2005.10.002.

Wagner, W., Blöschl, G., Pampaloni, P., Calvet, J.-C., Bizzarri, B., Wigneron, J.-P., Kerr, Y. 2007. Operational readiness of microwave remote sensing of soil moisture for hydrologic applications. *Nordic Hydrology*, Volume 38, Issue 1, Pages 1–20.

Wang, W., Yao, X., Yao, S.F., Tian, Y.C., Liu, X.J., Ni, J., Cao, W.X., Zhu, Y. 2012. Estimating leaf nitrogen concentration with three-band vegetation indices in rice and wheat. *Field Crops Research*, Volume 129, Pages 90–98. ISSN 0378-4290, https://doi.org/10.1016/j.fcr.2012.01.014.

Wang, Y., Tian, F., Huang, Y., Wang, J., Wei, C. 2015a. Monitoring coal fires in Datong coalfield using multi-source remote sensing data. *Transactions of Nonferrous Metals Society of China*, Volume 25, Issue 10, Pages 3421–3428. ISSN 1003-6326, https://doi.org/10.1016/S1003-6326(15)63977-2.

Wang, Z., Skidmore, A.K., Wang, T., Darvishzadeh, R., Hearne, J. 2015b. Applicability of the PROSPECT model for estimating protein and cellulose+lignin in fresh leaves. *Remote Sensing of Environment*, Volume 168, Pages 205–218. ISSN 0034-4257, https://doi.org/10.1016/j.rse.2015.07.007.

Wang, Z., Skidmore, A.K., Darvishzadeh, R., Wang, T. 2018. Mapping forest canopy nitrogen content by inversion of coupled leaf-canopy radiative transfer models from airborne hyperspectral imagery. *Agricultural and Forest Meteorology*, Volumes 253–254, Pages 247–260. ISSN 0168-1923.

WARDA. 2006. *Medium Term Plan 2007–2009. Charting the Future of Rice in Africa. Africa.* Cotonou, Republic of Benin: Rice Center (WARDA).

Weber, D., Schaepman-Strub, G., Ecker, K. 2018. Predicting habitat quality of protected dry grasslands using Landsat NDVI phenology. *Ecological Indicators*, Volume 91, Pages 447–460. ISSN 1470-160X, https://doi.org/10.1016/j.ecolind.2018.03.081.

Wheeler, D., Guzder-Williams, B., Petersen, R., Thau, D. 2018. Rapid MODIS-based detection of tree cover loss. *International Journal of Applied Earth Observation and Geoinformation*, Volume 69, Pages 78–87. ISSN 0303-2434, https://doi.org/10.1016/j.jag.2018.02.007.

White, K., Pontius, J., Schaberg, P. 2014. Remote sensing of spring phenology in northeastern forests: A comparison of methods, field metrics and sources of uncertainty. *Remote Sensing of Environment*, Volume 148, Pages 97–107. ISSN 0034-4257, https://doi.org/10.1016/j.rse.2014.03.017.

Wolter, P.T., Townsend, P.A., Sturtevant, B.R. 2009. Estimation of forest structural parameters using 5 and 10-meter SPOT-5 satellite data. *Remote Sensing of Environment*, Volume 113, Issue 9, Pages 2019–2036. ISSN 0034-4257, https://doi.org/10.1016/j.rse.2009.05.009.

Workie, T.G., Debella, H.J. 2018. Climate change and its effects on vegetation phenology across ecoregions of Ethiopia, global ecology and conservation, Volume 13, Article e00366. ISSN 2351-9894, https://doi.org/10.1016/j.gecco.2017.e00366.

Wright, C., Gallant, A. 2007. Improved wetland remote sensing in Yellowstone National Park using classification trees to combine TM imagery and ancillary environmental data. *Remote Sensing of Environment* 107, 582–605.

Xiong, J., Thenkabail, P.S., Tilton, J.C., Gumma, M.K., Teluguntla, T., Oliphant, A., Congalton, R.G., Yadav, K., Gorelick, N. 2017a. Nominal 30-m Cropland Extent Map of Continental Africa by Integrating Pixel-Based and Object-Based Algorithms Using Sentinel-2 and Landsat-8 Data on Google Earth Engine. *Remote Sensing*, Volume 9, Issue 10, Pages 1065. Doi: 10.3390/rs9101065, http://www.mdpi.com/2072-4292/9/10/1065.

Xiong, J., Thenkabail, P.S., Gumma, M.K., Teluguntla, P., Poehnelt, J., Congalton, R.G., Yadav, K., Thau, D. 2017b. Automated cropland mapping of continental Africa using Google Earth Engine cloud computing. *ISPRS Journal of Photogrammetry and Remote Sensing*, Volume 126, Pages 225–244. http://www.sciencedirect.com/science/article/pii/S0924271616301575.

Yao, X., Zhu, Y., Tian, Y.C., Feng, W., Cao, W.X. 2010. Exploring hyperspectral bands and estimation indices for leaf nitrogen accumulation in wheat. *International Journal of Applied Earth Observation and Geoinformation*, Volume 12, Issue 2, Pages 89–100. ISSN 0303-2434, https://doi.org/10.1016/j.jag.2009.11.008.

Zhao, B., Duan, A., Ata-Ul-Karim, S.T., Liu, Z., Chen, Z., Gong, Z., Zhang, J. et al. 2018. Exploring new spectral bands and vegetation indices for estimating nitrogen nutrition index of summer maize, European Journal of Agronomy, Volume 93, Pages 113–125. ISSN 1161-0301, https://doi.org/10.1016/j.eja.2017.12.006.

Zheng, B., Campbell, J.B., de Beurs, K.M. 2012. Remote sensing of crop residue cover using multi-temporal Landsat imagery. *Remote Sensing of Environment*, Volume 117, Pages 177–183. ISSN 0034-4257, https://doi.org/10.1016/j.rse.2011.09.016.

Zhou, X., Huang, W., Kong, W., Ye, H., Luo, J., Chen, P. 2016. Remote estimation of canopy nitrogen content in winter wheat using airborne hyperspectral reflectance measurements. *Advances in Space Research*, Volume 58, Issue 9, Pages 1627–1637. ISSN 0273-1177, https://doi.org/10.1016/j.asr.2016.06.034.

Zhuang, Q., Melack, J.M., Zimov, S., Sakha, C., Walter, K.M., Butenhoff, C.L., Khalil, A.K. 2009. Global Methane Emissions from Wetlands, Rice Paddies, and Lakes. *Eos*, Volume 90, Issue 5, Pages 37–44.

Index

D

E

F

G

H

I

J

K

L

W

Y